Communications
in Computer and Information Science 169

Archana Mantri   Suman Nandi
Gaurav Kumar   Sandeep Kumar (Eds.)

# High Performance Architecture and Grid Computing

International Conference, HPAGC 2011
Chandigarh, India, July 19-20, 2011
Proceedings

Springer

Volume Editors

Archana Mantri
Suman Nandi
Gaurav Kumar
Sandeep Kumar

Chitkara University
Chandigarh 160 009, India

E-mail:
archana.mantri@chitkara.edu.in
suman.nandi@chitkara.edu.in
gaurav.kumar@chitkara.edu.in
sandeep.mca@chitkara.edu.in

ISSN 1865-0929
e-ISSN 1865-0937
ISBN 978-3-642-22576-5
ISBN 978-3-642-22577-2 (eBook)
DOI 10.1007/978-3-642-22577-2
Springer Heidelberg Dordrecht London New York

Library of Congress Control Number: 2011931730

CR Subject Classification (1998): C.2, H.4, I.2, H.3, D.2, J.1, H.5



*Typesetting:* Camera-ready by author, data conversion by Scientific Publishing Services, Chennai, India

Printed on acid-free paper

Springer is part of Springer Science+Business Media (www.springer.com)

# Message from the General Chair

It is indeed a matter of pride that the Department of Computer Applications, Chitkara University, Punjab, in association with the University of Applied Sciences Osnabrück, Germany, has taken a pioneering initiative in organizing and international conference on the highly significant topic of High-Performance Architecture and Grid Computing, HPAGC-2011, along with a one-day workshop on the latest topics in data mining and cloud computing. In the quest for knowledge, we take immense pride in providing a platform for presenting and exchanging current research themes and experiences and fostering a relationship among universities, research institutes, industry and policy makers to take stock of the current developments and have a look into the future trends in this area.

HPAGC 2011 brought together academic scientists, leading engineers, industry researchers and students to exchange and share their experiences and research results in all aspects of high-performance computing, and to discuss the practical challenges encountered and the solutions adopted.

I want to express my sincere gratitude to Springer, for publishing the proceedings of the conference.

It was a privilege to welcome delegates from India and abroad and I would like to thank the organizers for planning this conference in a highly professional manner.

Madhu Chitkara

# Message from the Volume Editors

It gives us immense pleasure to present the proceedings of the International Conference on High-Performance Architecture and Grid Computing (HPAGC 2011). Chitkara University is indebted to the University of Applied Sciences Osnabrück, Germany, and Springer for their involvement.

The International Conference on High-Performance Architecture and Grid Computing is structured with the aim of presenting and exchanging current research themes and experiences, and it fosters a relationship among universities, research institutes and industry and policy makers to take stock of the current developments and have a look into the future trends in this area. We received 240 papers from researchers from around the world, and 87 manuscripts were selected after a rigorous review process for publication in the conference proceedings.

We express our appreciation and thanks to the Organizing Committee for making HPAGC 2011 a big success and an achievement for Chitkara University.

Archana Mantri
Suman Nandi
Gaurav Kumar
Sandeep Kumar

# Organization

## General Chairs

| | |
|---|---|
| Ashok Chitkara | Chitkara University, India |
| Madhu Chitkara | Chitkara University, India |

## Chief Guest

| | |
|---|---|
| Jaya Panwalkar | Director, nVIDIA, India |

## Conference Chair

| | |
|---|---|
| Archana Mantri | Chitkara University, India |

## Program Chairs

| | |
|---|---|
| Bhanu Kapoor | President, Mimasic, USA and Chitkara University, India |
| Suman Kumar Nandi | Chitkara University, India |

## Technical Chair

| | |
|---|---|
| Gaurav Kumar | Chitkara University, India |

## Publicity Chair

| | |
|---|---|
| Vandana Bajaj | Chitkara University, India |

## Workshop Chair

| | |
|---|---|
| Rajni Duggal | Chitkara University, India |

## Finance Chair

| | |
|---|---|
| Rashmi Aggarwal | Chitkara University, India |

## Advisory and Technical Review Committee

| | |
|---|---|
| Manuel Frutos-Perez | University of the West of England, UK |
| Sigurd Meldal | San Jose State University, USA |
| Thierry PRIOL | EIT ICT Labs, France |
| Michael Uelschen | University of Applied Sciences Osnabrück, Germany |
| Heinz-Josef Eikerling | University of Applied Sciences Osnabrück, Germany |
| A.K. Saxena | I.I.T. Roorkee, India |
| Srikanta Tirthapura | Iowa State University, USA |
| Sumeet Dua | Louisiana State University Health Sciences Center, New Orleans, USA |
| T. Meyyappan | Alagappa University, India |
| C. Lakshmi | SRM University, India |
| Kumar Padmanabh | Infosys Technologies Ltd., India |
| Rohit Gupta | Infosys Technologies Ltd., India |
| K.V. Arya | ABV-IITM, India |
| Leszek T. Lilien | Western Michigan University, USA |
| Veena Goswami | KIIT University, India |
| Dana Petcu | West University of Timisoara, Romania |
| Louise Perkins | Universiti Sains Malaysia, Malaysia |
| Sriram Chellappan | Missouri University, USA |
| Sanjay Madria | Missouri University, USA |
| Seema Bawa | Thapar University, India |
| R.K. Bawa | Punjabi University, India |
| Ashwani Kush | Kurukshetra University, India |
| Bharat Bhargava | Purdue University, USA |
| N. Jaisankar | VIT University, India |
| Amlan Chakrabarti | University of Calcutta, India |
| Natarajan Meghanathan | Jackson State University, USA |
| Krishna Kant | George Mason University, USA |
| Jiannong Cao | Hong Kong Polytechnic University, China |
| David Peleg | Weizmann Institute of Science, Israel |
| Maurice Herlihy | Brown University, USA |
| Elizabeth Buchanan | University of Wisconsin-Stout, USA |
| Maninder Singh | Thapar University, India |
| Kawaljeet Singh Chowdhary | Punjabi University, India |
| Vishal Goyal | Punjabi University, India |
| Himanshu Aggarwal | Punjabi University, India |

## Organizing Committee (Chitkara University, Punjab)

Vikram Mangla
Deepika Chaudhary
Nishu Bali
Preetinder Brar
Jaiteg Singh
Vikas Rattan
Vinay Kukreja
Maninderjit Singh Khanna
Nidhi Arora
Sheilini Jindal
Ravita Chahar

# Table of Contents

## Theme - 1: Grid and Cloud Computing

## Theme - 2: High Performance Architecture

## Theme - 3: Information Management and Network Security

# Era of Cloud Computing

Pramod Kumar Joshi[1] and Sadhana Rana[2]

Asst. Professor-Dept. of Computer Science
Amrapali Institute-AIMCA
pramodjoshijee@gmail.com
Asst. Professor-Dept. of Information Technology
Amrapali Institute-AITS, Haldwani, Nainital, India
sadhanarana@gmail.com

**Abstract.** Cloud Computing offers an entirely new way of looking at IT infrastructure. From a hardware point of view, cloud computing offers seemingly never-ending computing resources available on demand, thereby eliminating the need to budget for hardware that may only be used in high peak timeframes. Cloud computing eliminates an up-front commitment by users, thereby allowing agencies to start small and increase hardware resources only when there is an increase in their needs. Moreover, cloud computing provides the ability to pay for use of computing resources on a short-term basis as needed and release them as needed .In this paper we focus on area , issues and future of Cloud Computing.

**Keywords:** Cloud Computing, CCT, ACC.

## 1 Introduction

Cloud computing is Internet-based computing, using shared resources, software, and information are provided to computers and other devices on demand, like the electricity grid. Cloud computing describes a new supplement, consumption, and delivery model for IT services based on the Internet, and it typically involves over-the-Internet provision of dynamically scalable and often virtualized resources.

This frequently takes the form of web-based tools or applications that users can access and use through a web browser as if it a program installed locally on their own computer. Typical cloud computing providers deliver common business application online that are accessed from another Web service or software like a web browser, while the software and data are stored on servers. A key element of cloud computing is customization and the creation of a user-defined experience. Cloud computing is a term used to describe both a platform and type of application. A Cloud computing platform dynamically provisions, configures and reconfigures servers as needed. Servers in the cloud can be physical machines or virtual machines. Advanced clouds typically include other computing resources such as storage area networks (SANs), network equipment, firewall and other security devices. Cloud computing also describes applications that are extended to be accessible through the Internet. These cloud applications use large data centers and powerful servers that host Web applications and Web

A. Mantri et al. (Eds.): HPAGC 2011, CCIS 169, pp. 1–8, 2011.

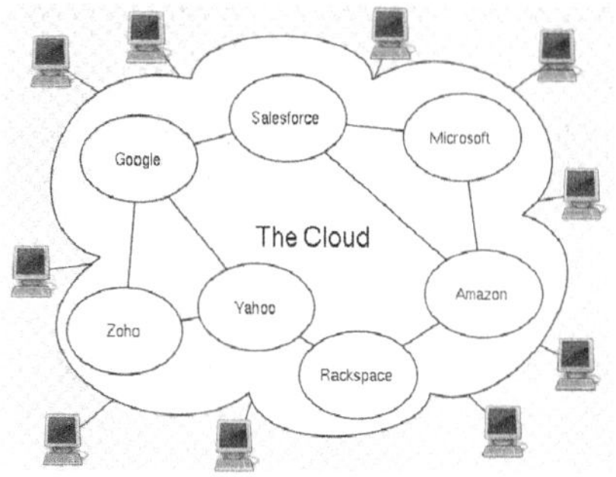


**Fig. 1.** Cloud computing conceptual diagram

services. Anyone with a suitable Internet connection and a standard browser can access a cloud application.

The National Institute of Standards and Technology's (NI ST) Information Technology Laboratory recognizes that cloud computing is an "evolving paradigm." As such, its definition attributes, and characteristics are still being debated by the public and private sectors, and are certain to continue to evolve in the near future. Nevertheless, initial steps have been taken toward constructing a universally accepted explanation of cloud computing's key characteristics, as well as definitions for the various deployment and service models. These definitions have been widely reported but are worth repeating, particularly in a field that is still rapidly developing.

**According to NIST**

Cloud computing is a model for enabling convenient, on-demand network access to a shared pool of configurable computing resources (e.g., networks, servers, storage, applications, and services) that can be rapidly provisioned and released with minimal management effort or service provider interaction. This cloud model promotes availability and is composed of five essential characteristics, three service models, and four deployment models.

## 2 The Five Essential Characteristics

### a) On-demand Self Service

A consumer can unilaterally provision computing capabilities, such as server time and network storage, as needed automatically without requiring human interaction with each service's provider.

### b) Broad Network Access

Capabilities are available over the network and accessed through standard mechanisms that promote use by heterogeneous thin or thick client platforms (e.g., mobile phones, laptops, and PDAs).

### c) Resource Pooling

The provider's computing resources are pooled to serve multiple consumers using a multi-tenant model with different physical and virtual resources dynamically assigned and reassigned according to consumer demand. There is a sense of location independence in that the customer generally has no control or knowledge over the exact location of the provided resources but may be able to specify location at a higher level of abstraction (e.g., country, state, or data center). Examples of resources include storage, processing, memory, network bandwidth, and virtual machines.

### d) Rapid Elasticity

Capabilities can be rapidly and elastically provisioned, in some cases automatically, to quickly scale out, and rapidly released to quickly scale in. To the consumer, the capabilities available for provisioning often appear to be unlimited and can be purchased in any quantity at any time.

### e) Measured Service

Cloud systems automatically control and optimize resource use by leveraging a metering capability at some level of abstraction appropriate to the type of service (e.g., storage, processing, bandwidth, and active user accounts). Resource usage can be monitored, controlled, and reported providing transparency for both the provider and consumer of the utilized service.

## 3 The Three Services Models

### 3.1 Cloud Infrastructure as a Service (IaaS)

The capability provided to the consumer is to provision processing, storage, networks, and other fundamental computing resources where the consumer is able to deploy and run arbitrary software, which can include operating systems and applications. The

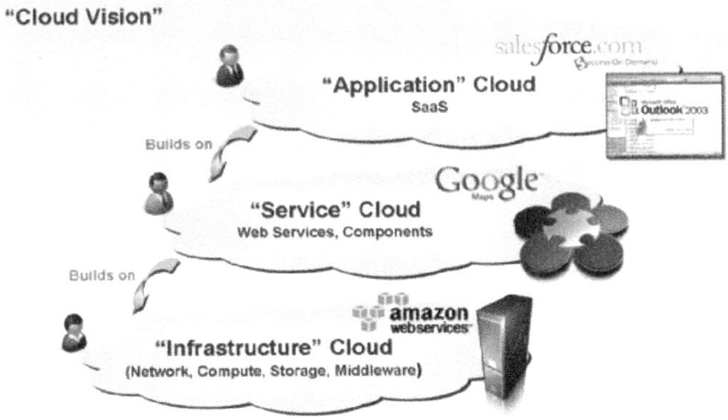


**Fig. 2.** for Services Models

consumer does not manage or control the underlying cloud infrastructure but has control over operating systems, storage, deployed applications, and possibly limited control of select networking components (e.g., host firewalls).

### 3.2 Cloud Software as a Service (SaaS)

The capability provided to the consumer is to use the provider's applications running on a cloud infrastructure. The applications are accessible from various client devices through a thin client interface such as a web browser (e.g., web based email). The consumer does not manage or control the underlying cloud infrastructure including network, servers, operating systems, storage, or even individual application capabilities, with the possible exception of limited userspecific application configuration settings.

### 3.3 Cloud Platform as a Service (PaaS)

The capability provided to the consumer is to deploy onto the cloud infrastructure consumer-created or acquired applications created using programming languages and tools supported by the provider. The consumer does not manage or control the underlying cloud infrastructure including network, servers, operating systems, or storage, but has control over the deployed applications and possibly application hosting environment configurations

## 4 The Four Deployment Models

### a) Private Cloud

The cloud infrastructure is operated solely for an organization. It may be managed by the organization or a third party and may exist on premise

### b) Community Cloud

The cloud infrastructure is shared by several organizations and supports a specific community that has shared concerns (e.g., mission, security requirements, policy, and compliance considerations). It may be managed by the organizations or a third party and may exist on premise or off premise.

### c) Public Cloud

The cloud infrastructure is made available to the general public or a large industry group and is owned by an organization selling cloud services.

### d) Hybrid Cloud

The cloud infrastructure is a composition of two or more clouds (private, community, or public) that remain unique entities but are bound together by standardized or proprietary technology that enables data and application portability (e.g., cloud bursting for load-balancing between clouds).

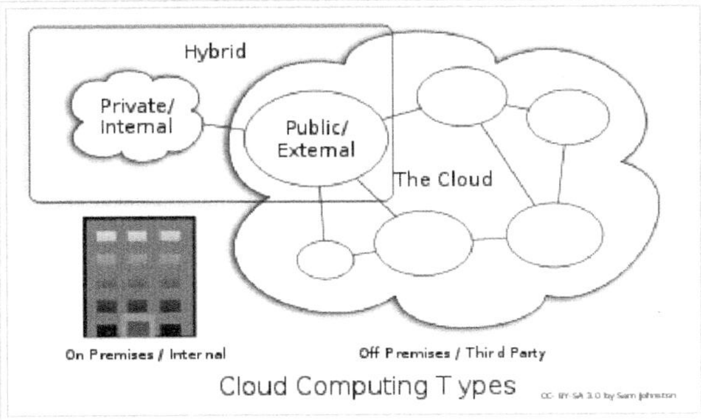


**Fig. 3.** For Deployment models

## 5 The Benefits of Cloud Computing

As cloud computing begins to take hold, several major benefits have become evident:

### a) Costs

The cloud promises to reduce the cost of acquiring, delivering, and maintaining computing power, a benefit of particular importance in times of fiscal uncertainty. By enabling agencies to purchase only the computing services needed, instead of investing in complex and expensive IT infrastructures, agencies can drive down the costs of developing, testing, and maintaining new and existing systems.

### b) Access

The cloud promises universal access to high-powered computing and storage resources for anyone with a network

access device. By providing such capabilities, cloud computing helps to facilitate telework initiatives, as well as bolster an agency's continuity of operations (COOP) demands.

### c) Scalability and Capacity

The cloud is an always-on computing resource that enables users to tailor consumption to their specific needs. Infinitely scalable, cloud computing allows IT infrastructures to be expanded efficiently and expediently without the necessity of making major capital investments. Capacity can be added as resources are needed and completed in a very short period of time. Thus, agencies can avoid the latency, expense, and risk of purchasing hardware and software that takes up data center space -- and can reduce the traditional time required to scale up an application in support of the mission. Cloud computing allows agencies to easily move in the other direction as well, removing capacity, and thus expenses, as needed.

### d) Resource Maximization

Cloud computing eases the burden on IT resources already stretched thin, particularly important for agencies facing shortages of qualified IT professionals.

### e) Collaboration

The cloud presents an environment where users can develop

Software- based services that enhances collaboration and fosters greater information sharing, not only within the agency, but also among other government and private entities.

## 6 Issues And Risks

- One of the key issues in cloud computing is the move towards a multi-sourced IT environment, where some services are provided in house, some from other government entities, and some from a range of infrastructure, application, and process suppliers in the form of private, public, community, or hybrid clouds.
- better suited for providers to deliver, and which lend themselves to the pay-per-use cloud approach? These considerations should be made in conjunction with the imperative to consolidate, simplify, and optimize an agency's IT environment, to reduce operational costs and free up investment for other mission-focused initiatives.
- Implementing a cloud computing IaaS model incurs different risks than managing a dedicated agency data center. Risks associated with the implementation of such a new service delivery model include policy changes, implementation of dynamic applications, and securing the dynamic environment. Most often, the mitigation plan for these risks depends on assessing the IT services needed to support end users and how they will be delivered, establishing proactive program management, and implementing industry best practices and government policies in the management of that program.
- For cloud computing to be widely adopted, assurances must be made that data is not only always accessible, but also totally secure. Agencies will undoubtedly need to actively put in place security measures that will allow dynamic application use and information-sharing to be implemented with the highest degree of security. Indeed, any significant data breach will exacerbate already existing fears about whether data is indeed safe in the cloud.
- To enable the cloud and fully realize its potential, certain fundamental elements must be addressed. To begin with, the cloud must function at levels equal to or better than the current IT systems – and must deliver tangible savings and benefits, including raising energy efficiency and reducing environmental impact. Users must be assured of near-ubiquitous and open access via the Internet, and be able to move among the cloud platforms as needed – with the users' rights to the data clearly defined and protected. Above all, as previously stated, user data must be secure at all times.

## 7 Applications

The applications of cloud computing are practically limitless. With the right middleware, a cloud computing system could execute all the programs a normal computer could run. Potentially, everything from generic word processing software to customized computer programs designed for a specific company could work on a cloud computing system. Why would anyone want to rely on another computer system to run programs and store data? Here are just a few reasons:

- Clients would be able to access their applications and data from anywhere at any time. They could access the cloud computing system using any computer linked to the Internet. Data wouldn't be confined to a hard drive on one user's computer or even a corporation's internal network.
- It could bring hardware costs down. Cloud computing systems would reduce the need for advanced hardware on the client side. You wouldn't need to buy the fastest computer with the most memory, because the cloud system would take care of those needs for you. Instead, you could buy an inexpensive computer terminal. The terminal could include a monitor, input devices like a keyboard and mouse and just enough processing power to run the middleware necessary to connect to the cloud system. You wouldn't need a large hard drive because you'd store all your information on a remote computer.
- Corporations that rely on computers have to make sure they have the right software in place to achieve goals. Cloud computing systems give these organizations company-wide access to computer applications. The companies don't have to buy a set of software or software licenses for every employee. Instead, the company could pay a metered fee to a cloud computing company.
- Servers and digital storage devices take up space. Some companies rent physical space to store servers and databases because they don't have it available on site. Cloud computing gives these companies the option of storing data on someone else's hardware, removing the need for physical space on the front end.
- Corporations might save money on IT support. Streamlined hardware would, in theory, have fewer problems than a network of heterogeneous machines and operating systems.
- If the cloud computing system's back end is a grid computing system, then the client could take advantage of the entire network's processing power. Often, scientists and researchers work with calculations so complex that it would take years for individual computers to complete them. On a grid computing system, the client could send the calculation to the cloud for processing. The cloud system would tap into the processing power of all available computers on the back end, significantly speeding up the calculation.

## Conclusion

We find that Cloud computing can be rapidly provisioned and released with minimal management effort or service provider interaction. Ultimately, with its offering of scalable, real-time, internet-based information technology services and resources, the cloud can satisfy the computing needs of a universe of users, without the users incurring the costs of maintaining the infrastructure.

## References


- Danielson, Krissi (2008-03-26). Distinguishing Cloud Computing from Utility Computing. Ebizq.net. (retrieved August 22, 2010)
- Gruman, Galen (2008-04-07). What cloud computing really means. Journal, InfoWorld (retrieved June 2, 2009)
- Michael, O.:The Rise of Cloud Computing.". April 2010. Windows ITpro.com. 2010-04-26 (retrieved August 22, 2010)
- Lai, Eric (2009-08-27). Google, Amazon, Microsoft beef up cloud services. Infoworld.com (retrieved August 22, 2010)
- Poeter, Damon (2010-07-12). Channel Web: "HP, Microsoft Promise Windows Azure Cloud Platform By Year's End." July 2010. Poeter. Crn.com (retrieved August 22, 2010)
- Goldman, Alex (2010-09-01). Channel Web: "IBM Outlines an Aggressive Cloud Computing Strategy" September 2010. Goldman. internet.com (retrieved September 22, 2010)

# An Overview on Soft Computing Techniques

K. Koteswara Rao and G. SVP Raju

CSE Dept, GMRIT, Rajam, CS&ST Dept, Andhra University
koteswara2003@yahoo.co.in, gsvprajudr9@yahoo.co.in

**Abstract.** Soft computing is a term applied to a field within computer science which is characterized by the use of inexact solutions to computationally-hard tasks such as the solution of NP-complete problems, for which an exact solution cannot be derived in polynomial time. This paper explains about the soft computing and its components briefly, also explains the need use and efficiency of its components. Soft computing differs from conventional (hard) computing in that, unlike hard computing, it is tolerant of imprecision, uncertainty, partial truth, and approximation. In effect, the role model for soft computing is the human mind. The guiding principle of soft computing is: Exploit the tolerance for imprecision, uncertainty, partial truth, and approximation to achieve tractability, robustness and low solution cost.

**Keywords:** soft computing, conventional, imprecision, uncertainty.

## 1 Introduction

Soft Computing became a formal Computer Science area of study in the early 1990's.Earlier computational approaches could model and precisely analyze only relatively simple systems. More complex systems arising in biology, medicine, the humanities, management sciences, and similar fields often remained intractable to conventional mathematical and analytical methods. That said, it should be pointed out that simplicity and complexity of systems are relative, and many conventional mathematical models have been both challenging and very productive.

The basic ideas underlying soft computing in its current incarnation have links to many earlier influences, among them Zadeh's 1965 paper on fuzzy sets; the 1973 paper on the analysis of complex systems and decision processes; and the 1979 report (1981 paper) on possibility theory and soft data analysis. The inclusion of neural computing and genetic computing in soft computing came at a later point.

At this juncture, the principal constituents of Soft Computing (SC) are Fuzzy Logic (FL), Neural Computing (NC), Evolutionary Computation (EC) Machine Learning (ML) and Probabilistic Reasoning (PR), with the latter subsuming belief networks, chaos theory and parts of learning theory. What is important to note is that soft computing is not a mélange (combination). Rather, it is a partnership in which each of the partners contributes a distinct methodology for addressing problems in its domain. In this perspective, the principal constituent methodologies in Soft Computing are complementary rather than competitive. Furthermore, soft computing may be viewed as a foundation component for the emerging field of conceptual intelligence.

A. Mantri et al. (Eds.): HPAGC 2011, CCIS 169, pp. 9–23, 2011.

Soft computing deals with imprecision, uncertainty, partial truth, and approximation to achieve tractability, robustness and low solution cost. Components of soft computing include:

Fuzzy Logic(FL)
Neural networks (NN)
Evolutionary computation (EC), including:
Evolutionary algorithms
Harmony search
Swarm intelligence
Ideas about probability including Bayesian network
Machine Learning

**Importance of Soft Computing**

The complementarities of FL, NC, and PR have an important consequence: in many cases a problem can be solved most effectively by using FL, NC, and PR in combination rather than exclusively. A striking example of a particularly effective combination is what has come to be known as "neurofuzzy systems." Such systems are becoming increasingly visible as consumer products ranging from air conditioners and washing machines to photocopiers and camcorders. Less visible but perhaps even more important are neurofuzzy systems in industrial applications. What is particularly significant is that in both consumer products and industrial systems, the employment of soft computing techniques leads to systems which have high MIQ (Machine Intelligence Quotient). In large measure, it is the high MIQ of SC-based systems that accounts for the rapid growth in the number and variety of applications of soft computing. In many ways, soft computing represents a significant paradigm shift in the aims of computing - a shift which reflects the fact that the human mind, unlike present day computers, possesses a remarkable ability to store and process information which is pervasively imprecise, uncertain and lacking in categoricity.

## 2 Fuzzy Logic

The concept of Fuzzy Logic (FL) was conceived by Lotfi Zadeh, a professor at the University of California at Berkley, and presented not as a control methodology, but as a way of processing data by allowing partial set membership rather than crisp set membership or non-membership. This approach to set theory was not applied to control systems until the 70's due to insufficient small-computer capability prior to that time. Professor Zadeh reasoned that people do not require precise, numerical information input, and yet they are capable of highly adaptive control. If feedback controllers could be programmed to accept noisy, imprecise input, they would be much more effective and perhaps easier to implement.

Fuzzy logic is a superset of conventional (Boolean) logic that has been extended to handle the concept of partial truth -- truth values between "completely true" and "completely false".

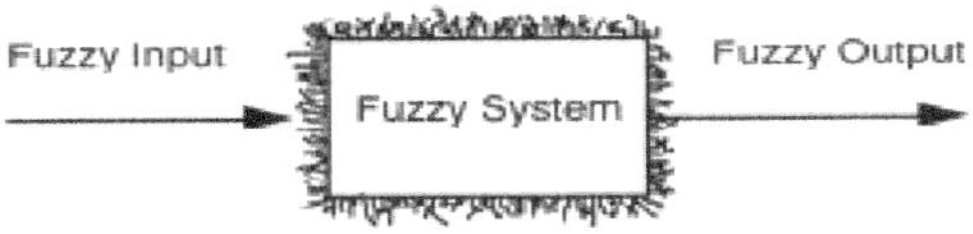


**Fig. 1.** Simple Block Diagram of Fuzzy systems

Fuzzy Logic is a problem-solving control system methodology that lends itself to implementation in systems ranging from simple, small, embedded micro-controllers to large, networked, multi-channel PC or workstation-based data acquisition and control systems. It can be implemented in hardware, software, or a combination of both. FL provides a simple way to arrive at a definite conclusion based upon vague, ambiguous, imprecise, noisy, or missing input information. FL's approach to control problems mimics how a person would make decisions, only much faster.

**Fig. 2.** Simple Fuzzy crisp inputs to get the fuzzy inputs

FL is a logic used to represent the Fuzzy sets to respective logical values of any hedge. A Fuzzy set a can be defined by its membership Ma (x).

To represent FL are use a list of pairs each pair represents value and Fuzzy membership..

Ex : a = { (x1, Ma (x1)... (xn, Ma (xm)}

Ex : representing the heights of 3 member, Joseph, John, James by using Fuzzy logic. A concept such as height which can have values from a range of Fuzzy values including "tall" "medium", "short".

Joseph is of 7 feet, John is of 4 feet, James is of 5.10 feet.

The definition of height of James falls under category of "Tall" for some and "Medium" for some people. By using Fuzzy James can be put in the list of "tall" people by giving rate of "tall" So the Fuzzy values can be defined as{1,0,,0.5} for Joseph, John, James.

**Why to choose fuzzy logic**

1. Because of rule base operation any reasonable number of inputs can be processed and numerous outputs can be generated.

2. FL is a inherently robust and don't require Precise.

3. FL can control non linear systems that would be difficult are impossible to model mathematical.

4. FL can give accurate result.

5. Since FL control processes user defined rules governing the target control system. It may be modified easily to improve performance.

### 2.1 Fuzzy Logic v/s Conventional Control Methods

Fuzzy Logic incorporates a simple, rule-based IF X AND Y THEN Z approach to a solving control problem rather than attempting to model a system mathematically. The FL model is empirically-based, relying on an operator's experience rather than their technical understanding of the system.

For example, rather than dealing with temperature control in terms such as "SP =500F", "T <1000F", or "210C <TEMP <220C", terms like "IF (process is too cool) AND (process is getting colder) THEN (add heat to the process)" or "IF (process is too hot) AND (process is heating rapidly) THEN (cool the process quickly)" are used. These terms are imprecise and yet very descriptive of what must actually happen. Consider what you do in the shower if the temperature is too cold: you will make the water comfortable very quickly with little trouble. FL is capable of mimicking this type of behavior but at very high rate.

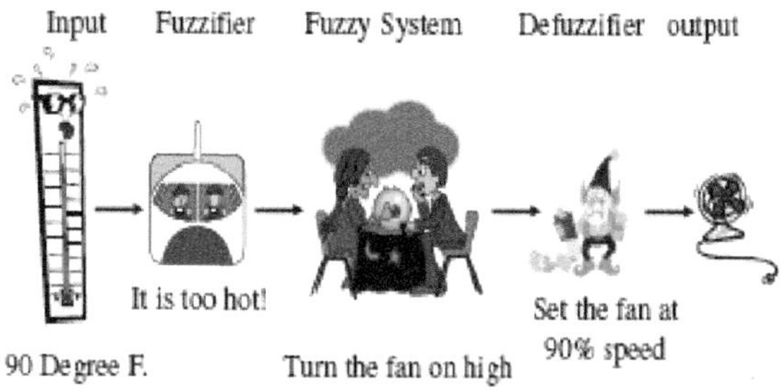


**Fig. 3.** Example Fuzzy Systems

### 2.2 How Does Fuzzy Logic Works

Fuzzy Logic requires some numerical parameters in order to operate such as what is considered significant error and significant rate-of-change-of-error, but exact values of these numbers are usually not critical unless very responsive performance is required in which case empirical tuning would determine them. For example, a simple temperature control system could use a single temperature feedback sensor whose data is subtracted from the command signal to compute "error" and then time-differentiated to yield the error slope or rate-of-change-of-error, hereafter called "error-dot". Error might have units of degs F and a small error considered to be 2F while a large error is 5F. The "error-dot" might then have units of degs/min with a small error-dot being 5F/min and a large one being 15F/min. These values don't have to be symmetrical and can be "tweaked" once the system is operating in order to optimize performance. Generally, FL is so forgiving that the system will probably work the first time without any tweaking.

#### 2.2.1 A Basic Fuzzy Set Operations

The basic operations are equality, union, Intersection, Complement, union

**Fuzzy logic control**

It explains how fuzzy logic control works and its essential components

**Fuzzification**

It is the process of mapping input to the FLC into fuzzy membership values in the various input universe discourse.

**Fuzzy rule base:** It consists of a set of antecedent-consequent linguistic rules of the form

If e is PS AND ce is NS THEN μ is PS

**Fuzzyinference**

It consists to translate all expressions to fuzzy language notation.

**Defuzzification**

It is the procedure for mapping from a set of inferred fuzzy control signals contained within a fuzzy output window to a non-fuzzy control signal.

***Advantages***

It lets the designer describe the desired system behavior with simple 'if-then' relations. In many applications, this gets a simpler solutions in less design time. In addition, the designer can use all available engineering know-how to optimize the system performance directly. Fuzzy logic solutions are easy to verify and optimize.

Combining conventional techniques and artificial neural networks with fuzzy logic, more powerful solutions may be offered. Fuzzy logic provides an approach to approximate reasoning in which the rules of inference are approximate rather than exact.

Fuzzy logic is useful in manipulating information that is incomplete, imprecise or unreliable. The primary motivation of fuzzy set theory is to desire to build a formal quantitative structure capable of capturing the imprecision of human knowledge, that is, the manner in which knowledge is expressed in natural language.
limited responsibilities.

### 2.2.2 How Fuzzy Logic Used

1) Define the control objectives and criteria: What am I trying to control? What do I have to do to control the system? What kind of response do I need? What are the possible (probable) system failure modes?
2) Determine the input and output relationships and choose a minimum number of variables for input to the FL engine (typically error and rate-of-change-of-error).
3) Using the rule-based structure of FL, break the control problem down into a series of IF X AND Y THEN Z rules that define the desired system output response for given system input conditions. The number and complexity of rules depends on the number of input parameters that are to be processed and the number fuzzy variables associated with each parameter. If possible, use at least one variable and its time derivative. Although it is possible to use a single, instantaneous error parameter without knowing its rate of change, this cripples the system's ability to minimize overshoot for a step inputs.

4) Create FL membership functions that define the meaning (values) of Input/Output terms used in the rules.
5) Create the necessary pre- and post-processing FL routines if implementing in software, otherwise program the rules into the FL hardware engine.
6) Test the system, evaluate the results, tune the rules and membership functions, and retest until satisfactory results are obtained.

**Linguistic Variables**

In 1973, Professor Lotfi Zadeh proposed the concept of linguistic or "fuzzy" variables. Think of them as linguistic objects or words, rather than numbers. The sensor input is a noun, e.g. "temperature", "displacement", "velocity", "flow", "pressure", etc. Since error is just the difference, it can be thought of the same way. The fuzzy variables themselves are adjectives that modify the variable (e.g. "large positive" error, "small positive" error ,"zero" error, "small negative" error, and "large negative" error). As a minimum, one could simply have "positive", "zero", and "negative" variables for each of the parameters. Additional ranges such as "very large" and "very small" could also be added to extend the responsiveness to exceptional or very nonlinear conditions, but aren't necessary in a basic system.

## 3 Neural Networks

A neural network is a massively parallel distributed processor that has a natural propensity for storing experiential knowledge and making it available for use. The term **neural network** was traditionally used to refer to a network or circuit of biological neurons. The modern usage of the term often refers to artificial neural networks, which are composed of artificial neurons or nodes. Thus the term has two distinct usages:

1.Biological neural networks are made up of real biological neurons that are connected or functionally related in the peripheral nervous system or the central nervous system. In the field of neuroscience, they are often identified as groups of neurons that perform a specific physiological function in laboratory analysis.

2.Artificial neural networks are composed of interconnecting artificial neurons (programming constructs that mimic the properties of biological neurons). Artificial neural networks may either be used to gain an understanding of biological neural networks, or for solving artificial intelligence problems without necessarily creating a model of a real biological system. The real, biological nervous system is highly complex and includes some features that may seem superfluous based on an understanding of artificial networks. A *neural network* (NN), in the case of artificial neurons called *artificial neural network* (ANN) or *simulated neural network* (SNN), is an interconnected group of natural or artificial neurons that uses a mathematical or computational model for information processing based on a connectionist approach to computation.

In more practical terms neural networks are non-linear statistical data modeling or decision making tools. They can be used to model complex relationships between inputs and outputs or to find patterns in data.

### 3.1 Applications of Natural and of Artificial Neural Networks

The utility of artificial neural network models lies in the fact that they can be used to infer a function from observations and also to use it. This is particularly useful in applications where the complexity of the data or task makes the design of such a function by hand impractical.

*Real life applications*

The tasks to which artificial neural networks are applied tend to fall within the following broad categories:

Function approximation, or regression analysis, including time series prediction and modeling.
Classification, including pattern and sequence recognition, novelty detection and sequential decision making.
Data processing, including filtering, clustering, blind signal separation and compression.

Application areas of ANNs include system identification and control (vehicle control, process control), game-playing and decision making (backgammon, chess, racing), pattern recognition (radar systems, face identification, object recognition, etc.), sequence recognition (gesture, speech, handwritten text recognition), medical diagnosis, financial applications, data mining (or knowledge discovery in databases, "KDD"), visualization and e-mail spam filtering.

### 3.2 Disadvantage of Neural Networks

The cause for a particular behavior cannot be interpreted , nor a neural network can be modified manually to change to a certain desired behavior. Also the selection appropriate net model and setting the parameters of learning algorithm are till 'black art' and require much experience. A common criticism of neural networks, particularly in robotics, is that they require a large diversity of training for real-world operation.

## 4 Evolutionary Computation

In computer science, evolutionary computation is a subfield of artificial intelligence (more particularly computational intelligence) that involves combinatorial optimization problems.

Evolutionary computation uses iterative progress, such as growth or development in a population. This population is then selected in a guided random search using parallel processing to achieve the desired end. Such processes are often inspired by biological mechanisms of evolution.

The use of Darwinian principles for automated problem solving originated in the fifties. It was not until the sixties that three distinct interpretations of this idea started to be developed in three different places.

Evolutionary programming was introduced by Lawrence J. Fogel in the USA, while John Henry Holland called his method a genetic algorithm. In Germany Ingo

Rechenberg and Hans-Paul Schwefel introduced evolution strategies. These areas developed separately for about 15 years. From the early nineties on they are unified as different representatives ("dialects") of one technology, called evolutionary computing. Also in the early nineties, a fourth stream following the general ideas had emerged – genetic programming.

These terminologies denote the field of evolutionary computing and consider evolutionary programming, evolution strategies, genetic algorithms, and genetic programming as sub-areas.

## 4.1 Techniques

Evolutionary techniques mostly involve metaheuristic optimization algorithms such as:

- Evolutionary algorithms (comprising genetic algorithms, evolutionary programming, evolution strategy and Genetic programming)
- Swarm intelligence (comprising ant colony optimization and particle swarm optimization).

## 4.2 Evolutionary Algorithms

Evolutionary algorithms form a subset of evolutionary computation in that they generally only involve techniques implementing mechanisms inspired by biological evolution such as reproduction, mutation, recombination, natural selection and survival of the fittest. Candidate solutions to the optimization problem play the role of individuals in a population, and the cost function determines the environment within which the solutions "live" (see also fitness function). Evolution of the population then takes place after the repeated application of the above operators.

## 4.3 Evolutionary Algorithm Techniques

***Genetic algorithm*** - This is the most popular type of EA. One seeks the solution of a problem in the form of strings of numbers (traditionally binary, although the best representations are usually those that reflect something about the problem being solved), by applying operators such as recombination and mutation (sometimes one, sometimes both). This type of EA is often used in optimization problems;

***Genetic programming*** - Here the solutions are in the form of computer programs, and their fitness is determined by their ability to solve a computational problem.

***Evolutionary programming*** - Similar to genetic programming, but the structure of the program is fixed and its numerical parameters are allowed to evolve;

***Evolution strategy*** - Works with vectors of real numbers as representations of solutions, and typically uses self-adaptive mutation rates;

***Neuroevolution*** - Similar to genetic programming but the genomes represent artificial neural networks by describing structure and connection weights. The genome encoding can be direct or indirect.

### 4.3.1 Genetic Algorithms

*4.3.1.1 Foundations of Science*

In the mid 1800s, 1859 to be exact, a British naturalist named Charles Darwin published a book that would change the way humans view the world. In this book, The Origin of Species, Darwin proposed that humans, and in fact all creatures, were not put on this planet by God and made unchanging, but rather that they evolved from other creatures. At the time, the idea sounded preposterous, but time and time again have we discovered that he may be correct. Advances in technology have made it possible for us to read our DNA and that of other creatures, and what it has shown us is that we aren't as different from other creatures as we think. Over time, creatures change to adapt to their environment to survive and thrive.

One of the most striking examples of this is the Galapagos Islands. Located in the Pacific Ocean, off the coast of Ecuador, this series of islands is one of the most prominent examples of evolution and adaptation. The island contains many species not found anywhere else on the planet, including several species of birds that share many characteristics; too many for it to be a coincidence. It is believed that many birds were blown to the islands by winds and were unable to get back. Over time, the birds spread throughout the islands and began to change to better survive in the differing environments of the islands. Some birds developed large, strong beaks suited to cracking nuts, others long, narrow beaks more suitable for digging bugs out of wood.

The birds that had these characteristics when blown to the island survived longer than other birds. This allowed them to reproduce more and therefore have more offspring that also had this unique characteristic. Those without the characteristic gradually died out from starvation. Eventually all of the birds had a type of beak that helped it survive on its island. This is the process of natural selection and evolution. The individuals themselves do not change, but those that survive better, or have a higher fitness, will survive longer and produce more offspring. This continues to happen, with the individuals becoming more suited to their environment every generation. It was this continuous improvement that inspired computer scientists, one of the most prominent being John Holland, to create genetic algorithms.

*4.3.1.2 Introduction to Genetic algorithms*

Many human inventions were inspired by nature. An artificial neural network is one example. Another example is Genetic Algorithms (GA). GAs search is by simulating evolution, starting from an initial set of solutions or hypotheses, and generating successive "generations" of solutions. This particular branch of AI was inspired by the way living things evolved into more successful organisms in nature. The main idea is survival of the fittest, a.k.a. natural selection.

This is an introduction to genetic algorithm methods for optimization. Genetic algorithms were formally introduced in the United States in the 1970s by John Holland at University of Michigan. The continuing price/performance improvement of computational systems has made them attractive for some types of optimization. In particular, genetic algorithms work very well on mixed (continuous *and* discrete), combinatorial problems. They are less susceptible to getting 'stuck' at local optima than gradient search methods. But they tend to be computationally expensive.

To use a genetic algorithm, you must represent a solution to your problem as a *genome* (or *chromosome*). The genetic algorithm then creates a population of solutions and applies genetic operators such as mutation and crossover to evolve the solutions in order to find the best one(s).

This presentation outlines some of the basics of genetic algorithms. The three most important aspects of using genetic algorithms are:

(1) Definition of the objective function
(2) Definition and implementation of the genetic representation.
(3) Definition and implementation of the genetic operators.

Once these three have been defined, the generic genetic algorithm should work fairly well. Beyond that you can try many different variations to improve performance, find multiple optima (species - if they exist), or parallelize the algorithms.

Genetic algorithms are one of the best ways to solve a problem for which little is known. They are a very general algorithm and so will work well in any search space. All you need to know is what you need the solution to be able to do well, and a genetic algorithm will be able to create a high quality solution. Genetic algorithms use the principles of selection and evolution to produce several solutions to a given problem.

Genetic algorithms tend to thrive in an environment in which there is a very large set of candidate solutions and in which the search space is uneven and has many hills and valleys. True, genetic algorithms will do well in any environment, but they will be greatly outclassed by more situation specific algorithms in the simpler search spaces. Therefore you must keep in mind that genetic algorithms are not always the best choice. Sometimes they can take quite a while to run and are therefore not always feasible for real time use. They are, however, one of the most powerful methods with which to (relatively) quickly create high quality solutions to a problem.

*4.3.1.3 Basics of Genetic Algorithms*

The most common type of genetic algorithm works like this: a population is created with a group of individuals created randomly. The individuals in the population are then evaluated. The evaluation function is provided by the programmer and gives the individuals a score based on how well they perform at the given task. Two individuals are then selected based on their fitness, the higher the fitness, the higher and the chance of being selected. These individuals then "reproduce" to create one or more offspring, after which the offspring are mutated randomly. This continues until a suitable solution has been found or a certain number of generations have passed, depending on the needs of the programmer.

**Selection**

A key idea of the GA is that parents should be selected based on their fitness. While there are many different types of selection, here we will cover the most common type - roulette wheel selection. In roulette wheel selection, individuals are given a probability of being selected that is directly proportionate to their fitness.

Each chromosome i in the population P is assigned a probability of selection pi based on its fitness fi as a proportion of total population fitness:

$$p_i = \frac{f_i}{\sum_{j \in P} f_j}$$

Two individuals are then chosen randomly based on these probabilities and produce offspring. Pseudo-code for a roulette wheel selection algorithm is shown below.

```
    for all members of population
        sum += fitness of this individual
end for
for all members of population
        probability = sum of probabilities + (fitness / sum)
        sum of probabilities += probability
end for
 loop until new population is full
        do this twice
          number = Random between 0 and 1
         for all members of population
           if number > probability but less than next probability
              then you have been selected
         end for
        end
        create offspring
      end loop
```

While this code is very general and will obviously not compile, it illustrates the basic structure of a selection algorithm. Besides, you should write the code yourself, you learn better that way.

Having selected a parent we then reproduce that parent according to our operators, parameters, etc.

According to our parameters we may apply crossover, and/or mutation, and/or any other operators. Often crossover rate (probability of applying crossover during reproduction) is referred to as x Similarly mutation rate (per gene probability of mutation during reproduction) by μ.

**Crossover**

So now you have selected your individuals, and you know that you are supposed to somehow produce offspring with them, but how should you go about doing it? The most common solution is something called crossover, and while there are many different kinds of crossover, the most common type is single point crossover. In single point crossover, you choose a locus at which you swap the remaining alleles from on parent to the other.

This is complex and is best understood visually. As you can see, the children take one section of the chromosome from each parent. The point at which the chromosome is broken depends on the randomly selected crossover point. This particular method is called single point crossover because only one crossover point exists. Sometimes only child 1 or child 2 is created, but oftentimes both offspring are created and put into the new population. Crossover does not always occur, however. Sometimes, based on a set probability, no crossover occurs and the parents are copied directly to the new population. The probability of crossover occurring is usually 60% to 70%.

**Mutation**

Before: 1101101001101110
After: 1101100001101110

After selection and crossover, you now have a new population full of individuals. Some are directly copied, and others are produced by crossover. In order to ensure that the individuals are not all exactly the same, you allow for a small chance of mutation. You loop through all the alleles of all the individuals, and if that allele is selected for mutation, you can either change it by a small amount or replace it with a new value. The probability of mutation is usually between 1 and 2 tenths of a percent. A visual for mutation is shown below.

As you can easily see, mutation is fairly simple. You just change the selected alleles based on what you feel is necessary and move on. Mutation is, however, vital to ensuring genetic diversity within the population.

**Applications**

Genetic algorithms are a very effective way of quickly finding a reasonable solution to a complex problem. Granted they aren't instantaneous, or even close, but they do an excellent job of searching through a large and complex search space. Genetic algorithms are most effective in a search space for which little is known. You may know exactly what you want a solution to do but have no idea how you want it to go about doing it. This is where genetic algorithms thrive. They produce solutions that solve the problem in ways you may never have even considered. Then again, they can also produce solutions that only work within the test environment and flounder once you try to use them in the real world. Put simply: use genetic algorithms for everything you cannot easily do with another algorithm.

## 4.4 Swarm Intelligence

*Ant colony optimization:* Based on the ideas of ant foraging by pheromone communication to form paths. Primarily suited for combinatorial optimization problems.

*Particle swarm optimization:* Based on the ideas of animal flocking behavior. Also primarily suited for numerical optimization problems.

*Harmony search*

In computer science and operations research, **harmony search** (HS) is a phenomenon-mimicking algorithm (also known as metaheuristic algorithm, soft computing algorithm or evolutionary algorithm) inspired by the improvisation process of musicians. In the HS algorithm, each musician (= decision variable) plays (= generates) a note (= a value) for finding a best harmony (= global optimum) all together. The Harmony Search algorithm has the following merits:

- HS does not require differential gradients, thus it can consider discontinuous functions as well as continuous functions.
- HS can handle discrete variables as well as continuous variables.
- HS does not require initial value setting for the variables.
- HS is free from divergence.
- HS may escape local optima.

## 5 Probability

Probability is a way of expressing knowledge or belief that an event will occur or has occurred. The concept has been given an exact mathematical meaning in probability theory, which is used extensively in such areas of study as mathematics, statistics, finance, gambling, science, Artificial intelligence/Machine learning and philosophy to draw conclusions about the likelihood of potential events and the underlying mechanics of complex systems.

Probabilitic logic is a natural extension of traditional logic truth tables:the results they define are derived through probabilistic expressions instead.

The aim of a probabilistic logic is to combine the capacity of probability theory to handle uncertainty with the capacity of deductive logic to exploit structure.

The result is a richer and more expressive formalism with a broad range of possible application areas.

The term probabilistic logic was first used in a paper by Nils Nilsson published in 1986, where the truth values of sentences are probabilities.

*Application areas:*

- Artificial Intelligence
- Bio informatics
- Game theory
- Psychology
- Statistics
- Argumentation theory

**Drawback of probability theory**

It only allows the modeling of stochastic uncertainty, which allows or which deals with the uncertainty of whether a certain event will take place or not.

*Bayesian network*

A Bayesian network, belief network or directed acyclic graphical model is a probabilistic graphical model that represents a set of random variables and their conditional dependencies via a directed acyclic graph (DAG). For example, a Bayesian network could represent the probabilistic relationships between diseases and symptoms. Given symptoms, the network can be used to compute the probabilities of the presence of various diseases.

Formally, Bayesian networks are directed acyclic graphs whose nodes represent random variables in the Bayesian sense: they may be observable quantities, latent variables, unknown parameters or hypotheses. Edges represent conditional dependencies; nodes which are not connected represent variables which are conditionally independent of each other. Each node is associated with a probability function that takes as input a particular set of values for the node's parent variables and gives the probability of the variable represented by the node. For example, if the parents are $m$ Boolean variables then the probability function could be represented by a table of $2^m$ entries, one entry for each of the $2^m$ possible combinations of its parents being true or false.

Efficient algorithms exist that perform inference and learning in Bayesian networks. Bayesian networks that model sequences of variables (*e.g.* speech signals or protein sequences) are called dynamic Bayesian networks. Generalizations of Bayesian networks that can represent and solve decision problems under uncertainty are called influence diagrams.

***Applications***

Bayesian networks are used for modeling knowledge in computational biology and bioinformatics (gene regulatory networks, protein structure, gene expression analysis, medicine , document classification, information retrieval, image processing, data fusion, decision support systems,engineering, gaming and law.

## 6 Machine Learning

It is a branch of artificial intelligence. It is a scientic discipline that is concerned with the design and development of algorithms that allow computers to evolve behaviors based on empirical data such as from sensor data or databases.

A major focus of machine learning research is to automatically learn to recognize complex patterns and make intelligent decisions based on data ,the difficulty lies in the fact that the set of all possible behaviors given all possible inputs is too large to be covered by the set of observed examples .Machine learning require cross-disciplinary proficiency in several areas such as probability theory, statistics etc.,

*Applications:* Machine perception

- Computer vision
- Natural language processing
- Brain-machine interfaces
- Search engines.

## 7 Conclusion

The complementarities of FL, NC, and PR has an important consequence: in many cases a problem can be solved most effectively by using FL, NC, and PR in combination rather than exclusively. A striking example of a particularly effective combination is what has come to be known as "neurofuzzy systems." Such systems are becoming increasingly visible as consumer products ranging from air conditioners and washing machines to photocopiers and camcorders. Less visible but perhaps even more important are neurofuzzy systems in industrial applications. What is particularly significant is that in both consumer products and industrial systems, the employment of soft computing techniques leads to systems which have high MIQ (Machine Intelligence Quotient). In large measure, it is the high MIQ of SC-based systems that accounts for the rapid growth in the number and variety of applications of soft computing. In many ways, soft computing represents a significant paradigm shift in the aims of computing - a shift which reflects the fact that the human mind, unlike present day computers, possesses a remarkable ability to store and process information which is pervasively imprecise, and uncertain.

## References

1. Simha, N.K., Gupta, M.M., Konar: Soft computing & Intelligent systems, Theory & principles, Techniques applications. Acadamic press series. Springer publications, Heidelberg (1999); ISBN-81-312-0847-8
2. Nilsson, N.J.: Neural networks and Fuzzy systems: dynamic Intelligence, approach to machine Intelligence, Bast Kosko. Prentice Hall of India, Englewood Cliffs (2002); ISBN-81-203-0868-9
3. Kartalopoulos, S.V.: Understanding Neural Networks and Fuzzy logic: Basic concepts and applications. Prentice hall of India publications, Englewood Cliffs (2003); ISBN-81-203-1680-0
4. Patterson, D.W.: Artifical Intelligence and expert systems. Prentice hall India, Englewood Cliffs (1999); ISBN-81-203- 0777-1
5. Du, Swamy: Neural Networks in a soft computing framework. Springer International edn. (2008); ISBN-9788181289537
6. Nair, S.: Artifical Intelligence',Elaine Rich,Kevin knight, 3rd edn. Tata Mc Graw Hill (2009); ISBN-13-978-0-07-008770-5
7. Amit: Computational Intelligence and Applications (2007); ISBN-978-81-8128-653-6

# A Novel Approach for Task Processing through NEST Network in a Grid

Tarun Gupta and Vipin Tyagi

Jaypee University of Information Technology,
Solan, Himachal Pradesh,
India - 173215
{tarun7434,dr.vipin.tyagi}@gmail.com

**Abstract.** With the increase in the complexity of task, complex architectures such as grid systems and cluster computing are employed to process huge amount of data. The major problem issues of such task processing systems include heterogeneity, load balancing, synchronization etc. The network employed to perform complex computations are hybrid form of peer network that utilizes the power of peer nodes to perform computations. The proposed architecture is an attempt to process task provided by set of users with a load balancing mechanism and node prioritization for task allocation through the Nest network. The Nest Network proposed for Grid is a peer network that processes the complex task provided by user and returns the processed output.

**Keywords:** Load Balancing, NEST Network, Node prioritization, Task processing.

## 1 Introduction

NEST Network [1] comprises of nests which are systems connected to one another in peer-to-peer network with resources that can perform computations on the given task. These nest are accompanied by the ants which are agent programs generated to monitor the network by their random movement in the network.

The resources offered by such network can be put to solve complex tasks. The peer network [2] proves to be great source of hardware resources which can thus be employed in a managed way to produce efficient results. The heterogeneous grid network [3] [4] [5] is an efficient way to solve complex task processing [6] with efficient Load Balancing [7] [8] [9] mechanism employed to distribute the load to the heterogeneous nodes available for processing. The distributed environment [10] [11] is key for complex task processing as it is independent of the location. The architecture facilitates the distributed environment for task processing.

The architecture was generated by monitoring the ant movement in real life. An ant move randomly and passes the information to the other ants present in the neighborhood. Similarly the ants are used to monitor the nest state and pass the information to the different network actors so that the required actions can be performed.

A. Mantri et al. (Eds.): HPAGC 2011, CCIS 169, pp. 24–29, 2011.

## 2 Proposed Architecture

The base of the architecture is the peer based Ant network [1] that works on the principle of communication between the peer nodes and process the task allocated. The registered user set can interact with the job queue to get their task functioned. The users submit their task request to the job queue and get the result as response from the job queue. The System Manager is the main actor of the system that manages the complete system with the help of various other actors that are having set of described role to play. The other actor present includes Task Scheduler, Nest Manager, Load Distributor, Result Collector and lastly the Nest Network. The communications within the various actors of the system is a well defined process.

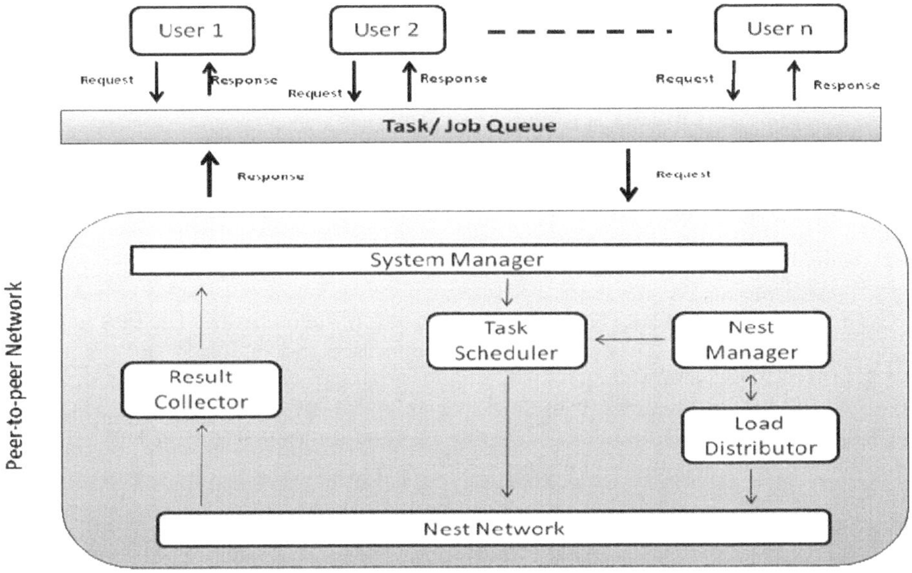


**Fig. 1.** Block Diagram of Proposed Architecture

### 2.1 Work Flow

The environment in Figure 1 facilitates task processing and the load balancing [6] [12], which can help to solve complex task in much less time duration with good efficiency. The work flow of the above architecture can be formulated with the help of below steps:

1. Various users who want to accomplish some task or job passes their request to the Job Queue with the userID and password, these tasks request are entered into the queue. Every job is assigned a JobID and the UserID is bounded with the respective JobID.

2. The Job Queue further passes on the request to the System manager with respective JobID. System Manager performs an authentication check on the user by verifying the userID and password.
3. After successful authentication the job with JobID is passed to Task Scheduler. Task Scheduler selects a Nest that is potential of doing the job with the information stored in the Nest Manager such as Load Status, computation capacity etc. After the selection of the Nest the Job is passed for execution to that Nest and the NestID is stored.
4. A Nest starts processing the job.
5. The Load Distributor monitors the network with the help of various Ants (Agent program or process) present in the network. If any Ant found a Nest in overloaded condition it informs the Load Distributor. Load distributor with the help of information from Nest Manager, distributes the Load of that nest to the other neighboring nest with the help of Ant. Ant takes the extra load from the nest and with the information provided by Load Distributor, distributed it to other Nest in the Network on random basis just as Ant do in reality.
6. After the completion of the Job the result is passed on to the Result Collector with the specific JobID. The Result Collector passes the information about the completion of task to System Manager which in turn passes it to the Job queue and further to Requested User.

## 2.2 Role of Various Actors of the System

1. **System Manager:** The supreme Manager of the network. The task is to take the jobs from the job queue and assign each task a JobID and associate it with a userID. The job is then passed to the network for the completion. System manager passes the task to the Task Scheduler. Another important role of system manager is to collect the result from the Result collector and reply it back to the Job queue from where it is further transferred back to requested user. The complete monitoring of the network is also done by the System Manager which monitors any failure in the network. It manages failure through the various other agents present in the system.
2. **Task Scheduler:** Task scheduler [6] is responsible for scheduling task in the network. Task Scheduler selects a Nest that is potential of doing the job with the help of information stored in the Nest Manager such as Load Status, computation capacity etc. After the selection of the Nest the Job is passed for execution to that Nest and the NestID is stored. It selects the most efficient nest by the process of heuristic knowledge available about that nest.
3. **Nest Manager:** It keeps the information about every nest in the network and maintains their identity. The information about nests is contained by the nest manager. The information such as computational capacity, present Load status, and previous throughput is stored by the nest manager. The Nest manager is updated with this information by the Load Distributor which monitors the network with the help of Ants.

4. **Load Distributor:** Load Distributor [13] plays an important role in the network. It monitors the network with the help of ant agents that move randomly in the network and acknowledge the Nest Manager about the status of each node in the network. It a node is found with extra load its load is redistributed to some other capable node in the network.
5. **Result Collector:** The job of the result collector is to take the competed job back from the network and validate the job completion. The completed job is then returned back to the System Manager which further returns it back to the user which submitted the task.
6. **Job Queue:** A Queue that manages the various task requests. The queue takes the various jobs coming for the completion arrange them in the order of their visit and forward sequentially to the System Manager for further processing.
7. **Nest Network:** A network that comprises of Nest i.e. nodes that do the processing and the ant agents that are used to monitor the network. The collection of Nest and the ants constitute the Nest Network. It is a random network in which nodes are linked to each other in Mesh topology.

## 3 Node Prioritization for Task Allocation

The Task Scheduler is responsible for the allocation of the task in the network. The System Manager provides the task to be executed to the Task Scheduler. Task Scheduler communicates with the Nest Manager to allocate the task to the nest. The Nest Manager provides the information needed by the Task Scheduler to allocate the task.

The parameters according to which the Task Scheduler decides the allocation of task:

1. **Computational Capacity (CC):** The nest computational power is stored in the field. The computational capacity of the node is rated on the scale of 1-10. The rating is relative to the other nodes. The node with the highest computational capacity is assigned the value 10 and lowest with 1.
2. **Load Status (LS):** The Load status contains the status of the Load for the node. The Load status can be:
   0- if the nest is free and do not have any job for processing.
   1- if the nest is occupied but not over-loaded.
   2- if the nest is overloaded.
3. **Performance (P):** The field stores the heuristic performance of the nest. It is a credit based field, every time a nest successfully complete the task a credit is increased or the value of P increases for that node.
4. **Node Distance (ND):** The field stores the distance of each node from the Task Scheduler. The distance is stored in terms of number of nest need to be traversed to reach that particular nest in the network. The distance number is specified in terms of numbers.
5. **Average working time per day (WT):** The field contains the average working time of the node per day. The time is monitored for each node and an average time in hours is computed. The value of the field varies from 0-24.

### 3.1 Allocation Factor (AF)

The allocation factor is computed on the basis of above described factors. The collective monitoring of all the above factors result in the AF.

**Algorithm:**

```
Initialization : number of nest =n;
        for node i to n
        do
              if( LS(i) =0 or LS(i)=1) then
                        AF(i)=CC(i) + P(i) + WT(i)-ND(i);
        enddo
Select Max(AF(i)) and allocate task to i;
```

Allocation Factor (AF) is computed by the above mentioned algorithm and the nest with the maximum Allocation Factor is selected for the task allocation by Task Scheduler.

## 4 User Management

The architecture supports numerous users to process their task simultaneously. As the architecture is distributed it can be accessed by any user present in the network. Any unknown user can also interact with the architecture and may tries to disturb its functioning. So a proper user management in the architecture is required.

This user management is done by the System Manager. The user that needs to process its job must be registered with the System Manager. System Manager assigns each user with a unique userID and password after the registration process that need to be mentioned every time making a task processing request. The System Manager on receipt of the JobID with the integrated userID and password initially authenticate the user by verifying the userID and password. After the successful registration the request is forwarded to the next stage else the user is first asked to register.

## 5 Failure Recovery

The primary failures that can hamper the functioning of the mentioned Architecture include nest failure, actor failure, network overloading. The problems are under consideration for minimization or removal.

## 6 Conclusion

The architecture presented here provides a mechanism to perform complex computations. The computation power of peer based NEST network is utilized to process complex task with efficient load balancing and task distribution mechanism. The strategy can be employed to various set of applications where numerous users want to process their request. The focus is to test the architecture in various real time applications and monitor the behavior during the process and improve it. Further the task division mechanism can be employed to divide the task in sub parts and process each part at different peer nodes after which the result can be aggregated to provide the result. The mechanism will improve the proficiency and the speed of the proposed architecture to a great extent.

## References


1. Montresor, A., Meling, H., Babaoglu, O.: Messor: Load-Balancing through a Swarm of Autonomous Agents. Technical Report UBLCS-02-08, Department of Computer Science, University of Bologna (May 2002)
2. Peer-to-Peer Architecture Case Study: Gnutella Network, Matei Ripeanu
3. Foster, I., Kesselman, C., Tuecke, S.: The anatomy of the Grid: Enabling scalable virtual organization. The International Journal of High Performance Computing Applications 15(3), 200–222 (2001)
4. Venugopal, S., Buyya, R., Winton, L.: A grid service broker for scheduling e-science applications on global data grids. Concurrency and Computation: Practice and Experience (CCPE) 18(6), 685–699 (2006)
5. Foster, I., Kesselman, C.: The Grid: Blueprint for a new computing infrastructure. Morgan Kaufmann Publishers Inc., San Francisco (1999)
6. Ucar, B., Aykanat, C., Kaya, K., Ikinci, M.: Task assignment in heterogeneous computing system. Journal of parallel and Distributed Computing 66, 32–46 (2006)
7. Li, J., Kameda, H.: Load balancing problems for multiclass jobs in distributed/parallel computer systems. IEEE Transactions on Computers 47(3), 322–332 (1998)
8. Yagoubi, B., Lilia, H.T., Moussa Department, H.S.: of Computer Science,University of Oran,Algeria load Balancing in Grid Computing. Asin Journal of Information Technology 5(10), 1095–1103 (2006)
9. Dandamudi, S.P.: Sensitivity evaluation of dynamic load sharing in distributed systems. IEEE Concurrency 6(3), 62–72 (1998)
10. Wu, J.: Distributed System Design. CRC Press, Boca Raton (1999)
11. Distributed processing in support of an imaging project, Brian Culver, Department of Computer Science, Sam Houston State University, Huntsville, Texas 77341-2090
12. Attiya, G., Hamam, Y.: Two phasealgorithm for load balancing in heterogeneous distributed systems. In: Proc. 12th IEEE EUROMICRO conference on Parallel, Distributed and Network-based Processing, Coruna, Spain, pp. 434–439 (2004)
13. Beltran, M., Guzman, A., Bosque, J.L.: Dealing with heterogeneity in load balancing algorithm. In: Proc. 5th IEEE International Symposium on Parallel and Distributed Computing, Timisoara, Romania, pp. 123–132 (2006)

# TCP/IP Security Protocol Suite for Grid Computing Architecture

Vikas Kamra and Amit Chugh

Lecturer (CSE), Lingaya's University, Faridabad
kamra1984@gmail.com, amitchugh_9@rediffmail.com

**Abstract.** Grid computing is a term referring to the combination of computer resources from multiple administrative domains to attain a common goal. The grid can be thought of as a distributed system with non-interactive workloads that involve a large number of files. In this paper, we propose a solution for various security issues found in High Performance Grid Computing Architecture. We analyze different network layers available in Grid Protocol Architecture and identify various security disciplines at its different layers. We also analyze various Security Suites available for TCP/IP Internet Protocol Architecture. The paper intends to achieve two major tasks. First, it defines the various Security Disciplines on different layers of Grid Protocol Architecture. Second, it proposes different Security Suites applicable for different levels of Security Disciplines available in different layers of TCP/IP Security Protocol Suite.

**Keywords:** Grid Computing, TCP/IP Protocol Architecture, TCP/IP Security Suite.

## 1 Grid Computing Architecture: An Introduction

Grid computing combines computers from multiple administrative domains to reach a common goal to solve a single task, and may then disappear just as quickly. One of the main strategies of grid computing is to use middleware to divide and apportion pieces of a program among several computers, sometimes up to many thousands. Grid computing involves computation in a distributed fashion, which may also involve the aggregation of large-scale cluster computing-based systems.

Grids are a form of distributed computing whereby a "super virtual computer" is composed of many networked loosely coupled computers acting together to perform very large tasks. This technology has been applied to computationally intensive scientific, mathematical, and academic problems through volunteer computing, and it is used in commercial enterprises for such diverse applications as drug discovery, economic forecasting, seismic analysis, and back office data processing in support for e-commerce and Web services.

Grid computing appears to be a promising trend for three reasons: (1) its ability to make more cost-effective use of a given amount of computer resources, (2) as a way to solve problems that can't be approached without an enormous amount of computing power, and (3) because it suggests that the resources of many computers can be

A. Mantri et al. (Eds.): HPAGC 2011, CCIS 169, pp. 30–35, 2011.

cooperatively and perhaps synergistically harnessed and managed as a collaboration toward a common objective. In some grid computing systems, the computers may collaborate rather than being directed by one managing computer. One likely area for the use of grid computing will be pervasive computing applications - those in which computers pervade our environment without our necessary awareness.

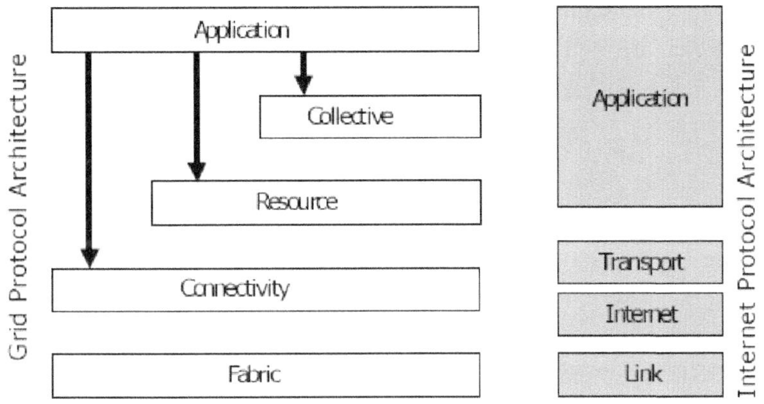


**Fig. 1.** High Performance Grid Computing Architecture

## 2 Security Requirements in Grid Computing Architecture

Security requirements within the Grid environment are driven by the need to support scalable, dynamic, distributed virtual organizations (VOs)—collections of diverse and distributed individuals that seek to share and use diverse resources in a coordinated fashion. From a security perspective, a key attribute of VOs is that participants and resources are governed by the rules and policies of the classical organizations of which they are members. The combination of dynamic policy overlays and dynamically created entities drives the need for three key functions in a Grid security model.

**Multiple security mechanisms.** Organizations participating in a VO often have significant investment in existing security mechanisms and infrastructure. Grid security must interoperate with, rather than replace, those mechanisms.

**Dynamic creation of services.** Users must be able to create new services (e.g., "resources") dynamically without administrator intervention. These services must be coordinated and must interact securely with other services. Thus, we must be able to name the service with an assertable identity and to grant rights to that identity without contradicting the governing local policy.

**Dynamic establishment of trust domains.** In order to coordinate resources, VOs need to establish trust among not only users and resources in the VO but also among the VO's resources, so that they can be coordinated. These trust domains can span multiple organizations and must adapt dynamically as participants join, are created, or leave the VO. Traditional means of security administration that involve manual editing of policy databases or issuance of credentials cannot meet the demands of these dynamic scenarios. We require a user-driven security model that allows users to create entities and policy domains in order to create and coordinate resources within VOs.

The various security disciplines at different levels of grid computing architecture are listed as follows: Authentication, Authorization, Confidentiality, Privacy, Message Integrity, Policy Exchange, Firewall Traversal, Delegation, Single Sign On, Credential life span and renewal, Secure Logging, Assurance, and Manageability.

## 3 TCP/IP Security Protocol Suite

**Kerberos** is an Authentication service designed for use in a Distributed Environment. Kerberos makes use of a trusted third part Authentication service that enables client and server to establish authenticated communication.

**Secure / Multipurpose Internet Mail Extension (S/MIME)** is a security environment to the MIME internet E-Mail format standard, based on technology from Rivest Shannon Algorithm data security.

**Pretty Good Policy (PGP)** provides Confidentiality and Authentication service that can be used for electronic mail and file storage application. PGP provide five services: Authentication, Confidentiality, Compression, E-Mail Compatibility, and Segmentation.

**Secure Electronic Transaction (SET)** is an open Encryption and security specification designed to protect credit card transaction on the internet.

**Secure Socket Layer (SSL)** makes use of TCP to provide reliable end to end secure services. SSL is combination of four Protocols: SSL Record Protocol, SSL Handshake Protocol, SSL Change Cipher Specification Protocol, and SSL Alert Protocol. Application layer security is achieved by all these facilities. The standard accepted version is TLS (Transport Layer Security).

**IPSec** provide security to IP layer. It provides the capability to secure communication across a LAN and across the internet. IPSec encompasses three functional areas Authentication, Confidentiality, Key Management. Authentication Header (AH) protocol, Encryption Security Payload (ESP), Internet Security Association and Key Management Protocol (ISAKMP) are the working protocols of IPSec.

**CheckSum** is a tool which is used to confirm Authentication of Sending Data. It is used to provide authentication services for various server and client terminals over the computer network.

## 4 Proposal for Security at Different Layers of Grid Computing Architecture

There are many security suites that are following different security disciplines at different levels of TCP/IP Protocol Layers. Table 1 is defining the relationship between various security disciplines and security suites which can be applied on different layers of grid protocol architecture.

**Table 1.** Proposal for security at different layers of grid computing architecture

| Grid Architecture Layers | Security Disciplines | Security Suites |
|---|---|---|
| Application Layer & Collective Layer & Resource Layer | Authentication<br>Authorization<br>Confidentiality | Kerberos<br>S/MIME<br>PGP<br>SET |
| Resource Layer & Connectivity Layer | Piracy<br>Message Integrity<br>Policy Exchange<br>Firewall Traversal<br>Delegation<br>Single Sign On<br>Credential Life Span<br>Renewal | SSL |
| Connectivity Layer | Secure Logging<br>Assurance<br>Manageability | IPSec |
| Fabric Layer | Network Authentication | CheckSum |

Different security suites provide security for various security disciplines required at different layers of Grid Computing Architecture. We can represent the proposal for security requirements at different layers of Grid Computing Architecture by defining its relationship with different security suites available in TCP/IP Security Protocol Suite, as shown in the following diagram:

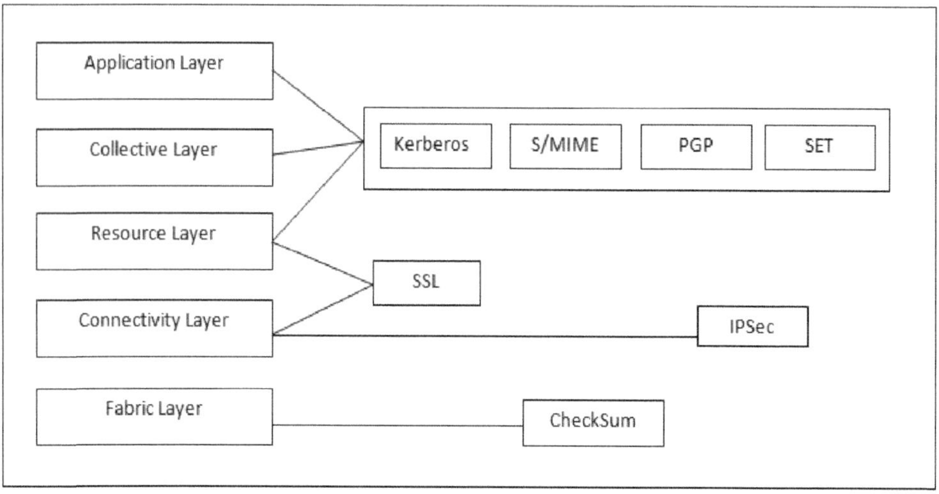


**Fig. 2.** TCP/IP Security Protocol Suite for Grid Computing Architecture

## 5 Conclusion

Requirement of security in grid computing systems is not ignorable. Security is best defined by its discipline. TCP/IP model is well known model for communication. Layer by layer security is applied in TCP/IP model. This paper presents the security level applied on each layer of grid computing architectures. There is a proposal for of applying various security disciplines from TCP/IP Security Suite on grid computing architecture. Every security suite is having their own constraints like, number of users in communication, intrusion detection and policy management.

## References

1. Foster, I., Kesselman, C., Tsudik, G., Tuecke, S.: A Security Architecture for Computational Grids. In: Computational Grids: The Future of High Performance Distributed Computing. Morgan Kaufmann, San Francisco (1998)
2. Joseph, J., Ernest, M., Fellenstein, C., et al.: Evolution of grid computing architecture and grid adoption models. From Open Grid Services Infrastructure to WS-Resource Framework: Refactoring & Evolution (March 2004)
3. Buyya, R., Abramson, D., Giddyet, J., et al.: A Case for Economy Grid Architecture for Service Oriented Grid Computing. Monash University (October 2002)

4. Bellovin, S.M., et al.: Probable Plaintext Cryptanalysis of the IP Security Protocols. In: Koblitz, N. (ed.) CRYPTO 1996. LNCS, vol. 1109, Springer, Heidelberg (1996), http://www.springerlink.com/content/
5. Foster, I., Kesselman, C.: The Grid: Blueprint for a New Computing Infrastructure. Morgan Kaufmann, San Francisco (1999), http://www.nature.com/nature/webmatters/grid/
6. Buyya, R.K.: High performance Grid Computing Architecture, http://www.buyya.com/ecogrid/
7. Fang, X., Yang, S., Guo, L., ZhangResearch, L.: on Security Architecture and Protocols of Grid Computing System
8. Kacsuk, P. (ed.): Journal of Grid Computing Main. Springer, Netherlands ISSN: 1570-7873 (print version) Journal no. 10723

# Security Issues in Cloud Computing

Pardeep Sharma[1], Sandeep K. Sood[2], and Sumeet Kaur[1]

[1] Computer Science & Engineering Guru Kashi Campus,
Punjabi University, Talwandi Sabo, India
shah4ash@yahoo.co.in, purbasumeet@yahoo.co.in
[2] Department of Computer Science & Engineering,
Guru Nanak Dev University, Regional Campus, Gurdaspur, India
san1198@gmail.com

**Abstract.** The cloud is next generation platform that provides dynamic resource pooling, virtualization and high resource availability. It is one of today's most enticing technology areas due to its advantages like cost efficiency and flexibility. There are significant or persistent concerns about the cloud computing those are impeding momentum and will compromise the vision of cloud computing as a new information technology procurement model. A general understanding of cloud computing refers to the concept of grid computing, utility computing, software as a service, storage in cloud and virtualization. It enables the virtual organization to share geographically distributed resources as they pursue common goals, assuming the absence of central location, omniscience and an existing trust relationship. This paper is a survey more specific to the different security issues that has emanated due to the nature of the service delivery models of a cloud computing system.

**Keywords:** Cloud Computing, Hypervisor, Privacy, Security and Virtualization.

## 1 Introduction

Cloud computing is a dream of computing as a utility. It makes software more attractive as a service and shaping the way as information technology hardware is designed and purchased. Cloud computing is defined as applications delivered as services over the Internet, hardware and system software in the datacenters that provides services. These services are called software as a service (SAAS). The datacenter hardware and software is known as a cloud .The foundation of cloud concept is based on the lease manner. The idea of cloud computing was very popular in the late 1960s when researchers thought about the utility computing. But in the mid-1970s this idea was attenuated when it became clear that companies of the day were unable to sustain such a futuristic computing model. However, with the increasing demand of computation resources, the concept has been revitalized. With the growth in Internet technology concepts such as search engines, the term cloud computing began to emerge in technology circles [1- 3]. The concept of cloud computing becomes more understandable when enterprises begins to think about what modern information technology environments always require. Modern information technology environments always like to increase capacity or add capabilities to their infrastructure dynamically, without

A. Mantri et al. (Eds.): HPAGC 2011, CCIS 169, pp. 36–45, 2011.

investing in new infrastructure. Given a solution to the aforementionedneeds, cloud computing models encompass a subscription-based or pay-per-use paradigm [4].It provides a service that can be used over the Internet and extends an information technology shop's existing capabilities. This approach provides a return on investment that companies are aiming for since decade. The tremendous growth of the Web over the last decade has given rise to a new class of web scaling problems and challenges such as supporting thousands of concurrent e-commerce transactions or millions of searchqueries in a minute. It has become a large and growing market because of its value propositions of low costs, increased flexibility, and shorter time to market.Security issues in cloud computing are hampering the interest of perspective organizations. There have been a lot of proven security attacks on different cloud computing providers such as Google (Gmail, App Engine), Amazon Web Services (Amazon S3), Salesforce.com (Salesforce.com) **etc.** Security is one of the main concerns in cloud computing environment [5].

This paper is organized as follows. In Section 2, we introduce the literature review which is classified as different types of effort done in security aspects with their advantages and disadvantages. In Section 3, we describe the challenges of the cloud. In Section 4, we describe the different service models. We describe Deployment models in Section 5. In Section 6, we show the comparison of deployment models in terms of security requirements. In Section 7, we propose future research directions and Section 8 concludes the paper.

## 2 Literature Review

In 2000, Yamaguchi and Hashiyama [6] proposed concept of Reconfigurable Computing technique for encryption processing. Reconfigurable Computing (RC) is capable of accelerating the information processing using dynamic reconfiguration of Field Programmable Gate Arrays (FPGAs). Dividing the target problems into hardware and software processing appropriately, the computation time will be less. It is one of the aims of researchers to have fast and flexible encryption technique in the Internet. Encryption technique generally consumes more computational power and needs specific hardware for feasible implementation. Moreover, these techniques are computational intensive. They implemented RC system onto FPGA board. In this technique, they developed application specific IC (ASIC) but this process has a problem of scaling. It is suitable for real time problems only.

In 2009, Yuefa et al. [7] suggested the concept of hadoop distributed file system (HDFC) architecture for the data security requirement of cloud computing. They use the same file system as the Google uses named as Google file system (GFS). This model works only in open system.

In 2010, Tribhuwan et al. [8] proposed a method to enhance the security of data stored in the cloud by utilizing the concept of homomorphic tokens and distributed verification of erasure coded data. This method attains the integration of storage correctness insurance and data error locations. They introduces a new two way handshake scheme which is based on the token management method but this method does not work properly for maintaining the integrity and confidentiality of data.

In 2010, Brandic and Dustdar [9] proposed a novel approach for compliance management in clouds, termed as Compliant Cloud Computing (C3). They used novel languages for specifying compliance requirements which concernedabout security issues, privacy and trust, by leveraging domain specific languages and compliance level agreements. They proposed C3 middleware architecture. In this, the middleware is responsible for the deployment of certifiable and auditable applications, for provider selection with the user requirements.

In 2010, Ramgovind et **al.** **[10]** purposed overall security perspective for cloud computing with the aim to highlight the security concerns. They tried to address the cloud computing concerns and also remained successful to some extend to realize the full potential of Cloud computing. Some of the most important issues in Cloud computing are like data storage and data localization in the cloud. They also addressed problems like how the organization will deal with new and current cloud compliance risks. It helps for cloud computing implementation. It deals with the potential impact of cloud computing on the business concerning governance and legislation. They discussed how cloud computing may affect the organization in terms of its business intelligence and intellectual property by potentially influencing its market differentiation.

In 2010, Almulla and Yeun [11] **discussed** challenges regarding information security concerns as confidentiality, integrity and availability. Most of the organizations are very much concerned about the security issues and the ownership of the data. However, they have not addressed security challenges for cloud computing including Identity and Access Management (IAM). They presented the current state authentication, authorization and auditing of users by accessing the cloud along with the emerging IAM protocols and standards.

In 2010, Somani et al. [12] suggested the cloud storage and data security in the cloud by implementation of digital signature with RSA algorithm. In Digital Signature, software will crunch down the data, document into just a few lines by a using hashing algorithm. They also suggested cloud challenges and responsibilities. They proposed algorithms for implementing digital signature with RSA algorithm. This technique crunch the document using hash functions, encrypt the message digest with private key then uses RSA algorithm.

In 2010 Sato et al. [13] suggested that one of security concern for cloud can be summarized as social insecurity. It is classified into the multiple stakeholder problems, the open space security problem and the mission critical data handling problem. As a solution of those problems, they proposed a new cloud trust model. They consider both internal trusts model and contracted trust that controls cloud service providers. They present a model named as "Security Aware Cloud." In a security aware cloud, internal trust must be established as the firm base of trust. By implementing security such as identity management and key management on internal trust, they obtain a firm trust model.

## 3 Challenges of the Cloud

**Ultra large-scale:** Larger is the cloud, faster is the cloud. The cloud providers have a large network of servers, which are to give services to users or consumers. The cloud of Google has owned more than one million servers. Even in Amazon, IBM, Microsoft, Yahoo, they have more than hundreds of thousands servers. There are hundreds of servers in an enterprise [**10, 12**].

**Virtualization:** Cloud computing makes user to get service anywhere, through any kind of terminal. It is applied to memory, networks, storage, hardware and operating system .You can do all you want through net service using a notebook computer or a mobile phone [10]. Users can attain or share it safely through an easy way, anytime, anywhere. Virtualization has characteristics like Partitioning (many applications and operating systems are supported in a single physical system by partitioning or separating the available resources) and Isolation (each virtual machine is isolated from its host physical system and other virtualized machine. **therefore if one virtual machine crashes, it doesn't affect the other virtual machines**) [12]. In addition, data is not shared between one virtual container and another. It also provides Encapsulation (a virtual machine can be represented as a single file, so you can identify it easily based on the service it provides. In essence, the encapsulated process could be a business service. The encapsulated virtual machines can be presented to an application as a complete entity. Therefore, it can protect each application so that it does not interfere with another application) [**14, 15**].

**High reliability:** Cloud uses data multi transcript fault tolerant. It replicates the same data at different location or at different machines that ensure high reliability. Chances of data crash become less. It supports the integrity and transaction constraints as well [14].

**Versatility:** Cloud computing can produce various applications supported by cloud and one cloud can support different applications running on it, at the same time. It may be for same problem or for different problems [16].

**High extendibility:** The scale of cloud can extend dynamically to meet the increasingly requirement. This application brings up hundreds of virtual servers on-demand, runs a parallel computation on them. By using an open source distributed processing framework called Hadoop, then shuts down all the virtual servers. Releasing all bound resources back to the cloud with low programming effort and at a very reasonable cost for the caller [**16, 17**].

**On demand service: Cloud** is a large resource pool that you can buy according to your need. Cloud is just like running water, Electric, and gas that can be charged by the amount that you used. It works like pay-as-you go manner, simply as in homes we pay for electricity bills as how much we used. Similar in cloud, we pay as we use the resources of cloud provider. This is also known as utility computing.

**Extremely inexpensive: The** centered management of cloud make the enterprise need not undertake the management cost of data center that increase very fast. The versatility can increase the utilization rate of the available resources compared with traditional system, so users can fully enjoy the low cost advantage [17].

## 4 Service Models

Cloud computing provides three service models that provides different levels of control and security are described ahead.

**Software as a service (SAAS):** The services provided over the Internet is referred as software as a service. It includes the capabilities which are provided to the consumers to use the provider's applications, that running on a cloud infrastructure. Applications accessible from various client devices through a thin client interface such as a Web browser (e.g., web-based email) [4]. The consumer does not manage or control the underlying cloud infrastructure, network, servers, operating systems, storage, or even individual application capabilities, with the possible exception of limited user-specific application configuration settings. The traditional model of software distribution is the software purchased and installed on personal computers, is sometimes referred to as Software-as-a-Product [18].It is a software distribution model in which applications are hosted by a vendor or service provider and made available to customers over a network, typically the Internet. Software as a service is becoming an increasingly dominant delivery model as underlying technologies that support web services and service-oriented architecture (SOA) [1] [19].

**Platform as a service (PAAS):** Platform as a service provides capabilities to the consumers to deploy onto the cloud infrastructure. Various consumers created applications use programming languages and tools supported by the provider. The consumer does not manage or control the underlying cloud infrastructure, network, servers, operating systems and storage. The consumer has control over the deployed applications and possibly application hosting environment configurations. Cloud computing has acquired to includes the platform for building and running custom web-based applications, this concept known as Platform-as-a-Service. PAAS is an outgrowth of the SAAS applications delivery model. The PAAS model has aim to support the complete life cycle of building, delivering web applications and services entirely available from the Internet. Adverse to IAAS model, where developers may create a specific operating system instance with homegrown applications, PAAS developers are concerned only with web based development and generally do not care which operating system is used[4]. Its services allow users to focus on innovation rather than complex infrastructure. Organizations can redirect a significant portion of their budgets for creating applications that provide real business values instead of worrying about all the infrastructure issues. The PAAS model is thus driving a new era of mass innovation. Now, developers around the world can access unlimited computing power. Anyone with an Internet connection can build powerful applications and easily deploy them to users globally [1] [17] [19].

**Infrastructure as a service (IAAS):** Infrastructure as a service provides control over the storage and resources. It provides the consumer to rent processing, storage, networks, and other fundamental computing resources. Where the consumer is able to deploy and run arbitrary software, or control the underlying cloud infrastructure. The consumer has less control over operating systems, storage, deployed applications, and possibly selects networking components (e.g., firewalls, load balancers). IAAS is the delivery of computer infrastructure (typically a platform virtualization environment) as a service. It leverages significant technology, services, and data center investments to deliver information technology (IT) as a service to customers [19]. IAAS is established on a model of service delivery that provisions a predefined and standardized infrastructure that is specifically optimized for the customer's applications. An IAAS provider handles the transition and hosting of selected applications on their infrastructure. Customers maintain ownership and management of their applications while off-loading hosting operations and infrastructure management to the IAAS providers [1].

## 5 Deployment Models

**Private Cloud:** In private clouds the physical infrastructure may be owned by or physically located in the organization's datacenters. These are managed by third party or own personnel with on or off-premise. They provide a single-tenant (dedicated) operating environment with all the benefits and functionality of elasticity, accountability and utility model of Cloud. The consumers of the service are called trusted. Trusted consumers of service are those who are considered part of an organization's legal/contractual umbrella including employees, contractors and business partners. It is easier to align with security, compliance and regulatory requirements and provide more enterprise control over development and use. All cloud resources and applications managed by the organization itself. Utilization on the private cloud can be much more secure than that of the public cloud because of its internal control or exposure. Only the organization and stake holders may have access to operate on a private cloud [15] [19].

**Public Cloud:** Public Clouds are provided by designated service providers. Theyoffer either a single-tenant (dedicated) or multi-tenant (shared) operating environment with the benefits and functionality of elasticity, accountability and utility model of Cloud. In public cloud the physical infrastructure is generally owned by and managed by the designated service provider which is located within the provider's datacenters (off-premise.) Consumers of Public Cloud services are called untrusted. Untrusted consumers are those that may be authorized to consume some or all services but are not logical extensions of the organization. These types of cloud are stand alone or proprietary, run by third party companies such as Google, Amazon etc.

**Managed Cloud:** These types of clouds are established where various organizations have same requirements. They offer both single-tenants (dedicated) or multi-tenant (shared) operating environment with all the benefits and functionality of elasticity, accountability and utility model of Cloud [20]. The physical infrastructure is owned by and/or physically located in the organization's datacenters with an extension of management and security control planes controlled by the designated service provider. Consumers of Managed Clouds may be trusted or untrusted.

**Hybrid Cloud:** Hybrid Clouds are a combination of public and private cloud. It offers transitive information exchange and possibly application compatibility and portability. Cloud service offerings and provides utilizing standard or proprietary methodologies regardless of ownership or location. This model provides an extension of management and security control planes. Consumers of hybrid Clouds are trusted or untrusted [20, 21].

**Table 1.** Cloud Computing Services Models and Their Providers [11]

| | SERVICES | PROVIDERS |
|---|---|---|
| SAAS | • Support running multiple instances on it.<br>• Develop software that run in cloud | • Google Docs<br>• Mobile Me<br>• Zoho |
| PAAS | • Platform which allow developers to create programs that run in the cloud.<br>• It Include several applications services which allow easy development | • Microsoft Azure<br>• Force.com<br>• Google App Engine |
| IAAS | • Consist of database servers and storage<br>• Highly scaled and shared computing infrastructure | • Amazon S3<br>• Sun's Cloud Service |

## 6 Comparison Analysis

We have already described different delivery models of cloud by which different types of services are delivered to the end user. The three delivery models are the SAAS, PAAS and IAAS which provide infrastructure resources, application platform and software as services to the consumer. These service models also place a different level of security requirement in the cloud environment. IAAS is the foundation of all cloud services, with PAAS built upon it and SAAS in turn built upon it. Just as capabilities are inherited, so are the information security issues and risks. There are significant trade-offs to each model in terms of integrated features, complexity, extensibility and security. If the cloud service provider takes care of the security at the lower part of the security architecture of cloud, then the consumers become more responsible for implementing and managing the security capabilities. This paper presents information of the services and providers of services. Comparative analysis on different cloud delivery and deployment models is presented along with security concern [22]. Parameters in security are identification, authorization, confidentiality, integrity, non-repudiation and availability in terms of deployment models. Figure.1 shows that security is required at higher extent in public cloud. This is one the main area of researchers to improve security in public cloud. Especially authorization and integrity in public cloud require great attention of researchers to fulfill the dream of implementation of cloud [17].

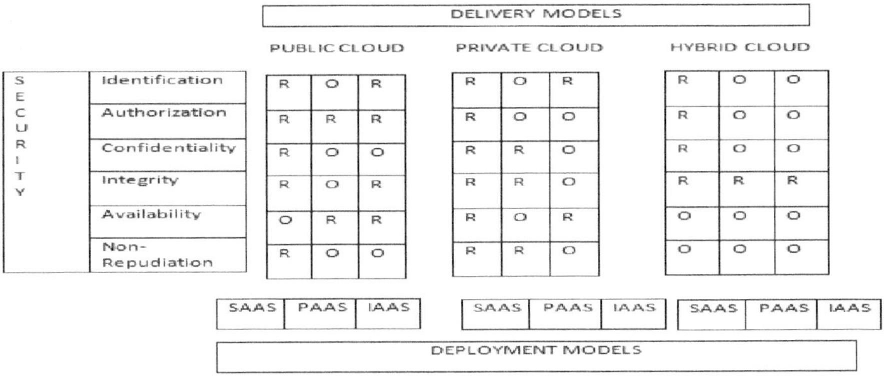


R = Required, O = Optional

**Fig. 1.** Comparison of Security Parameters in Different Delivery Models

## 7 Future Direction

One of the thrust and major area of research is to find technical solutions for the interoperability among the cloud. Cloud enterprises want to assure that there will be exit or a migration strategy across multiple clouds thereby avoiding the perils of vendor lock-in. Second is the enabler ecosystem. There are various complex domains within a cloud data center infrastructure. Some examples of these domains are computing, network, storage, security, software applications and service management. In those domains, there are several areas of complexity including integration, interoperability, operation, scalability, and compliance. Because of this enterprises start adopting private clouds, they would need a healthy ecosystem of cloud solution providers. It would ease the burden of the above mentioned complexitiesof cloud computing. The main area for research in cloud computing is its security. It is great obstacle in implementation of cloud. Different solutions to security have been suggested. These solutions include reconfigurable computing, cryptography, identity access management and various cloud computing models as well. Still, efficient solutions are required for different domains of clouds.

## 8 Conclusion

Although Cloud computing can be seen as a new phenomenon which is set to revolutionize the way we use the Internet, there is much to be cautious about. There are many new technologies emerging at a rapid rate, each with technological advancements and with the potential of making human's lives easier. However one must be very careful to understand the limitations and security risks posed in utilizing these technologies. Cloud computing is no exception. In this paper challenges, deployment models and key security issues which are currently faced by cloud computing are highlighted. We mentioned the requirement of security at different service models which easily give the view where we require more security and concentrate our focus to under developed areas. By following this paper, the insecurities of cloud may be easily expelled, saving business owner time and investment. This service can be easily integrated by different organizations such as banking, search engines and enterprise applications.

## References


1. Julisch, K., Hall, M.: Security and Control in the Cloud. Information Security Journal: A Global Perspective 19(6), 299–309 (2010)
2. Balachandra, R.K., Ramakrishna, P.V., Rakshit, K.: Cloud Security Issues. In: IEEE International Conference on Services Computing, pp. 517–520 (2009)
3. Cheng, G., Jin, H., Zou, D., Zhang, X.: "Building Dynamic and Transparent Integrity Measurement and Protection for Virtualized Platform in Cloud Computing. Concurrency and Computation: Practice and Experience 22, 1893–1910 (2010)
4. Armbrust, M., Fox, A., Griffith, R., Joseph, A.D., Katz, R., Konwinski, L., Lee, G.: Above the clouds: A Berkeley View of Cloud Computing. University of California, Berkeley, Tech. Rep. USB-EECS-2009, vol. 28, pp. 23-29 (2009)
5. Zissis, D., Lekkas, D.: Addressing Cloud Computing Security Issues. Future Generation Computer System (2010) Article in Press, `http://dx.doi.org/10.1016/j.future.2010.12.006`
6. Yamaguchi, T., Hashiyama, T., Okuma, S.: A Study on Reconfigurable Computing System Cryptography. In: IEEE International Conference on Cloud Computing, vol. 4, pp. 2965–2968 (2000)
7. Yuefa, D., Bo, W., Yaqiang, G., Quan, Z.: Data Security Model for Cloud Computing. In: International Workshop on Information Security and Applications, pp. 141–144 (2009)
8. Tribhuwan, M.R., Bhuyar, V.A., Pirzade, S.: Ensuring Data Storage Security in Cloud Computing Through Two Way Handshake Based on Token Management. In: International Conference on Advances in Recent Technology in Communication and Computing, pp. 386–389 (2010)
9. Brandic, I., Dustdar, S., Anstett, T., Schumm, D., Leymann, F.: Compliant Cloud Computing (C3): Architecture and Language Support for User-Driven Compliance Management in Clouds. In: IEEE International Conference on Cloud Computing, pp. 244–251 (2010)
10. Ramgovind, S., Eloff, M., Smith, E.: The Management of Security in Cloud Computing. In: IEEE International Conference on Service Computing, pp. 126–130 (2010)
11. Almulla, S.A., Yeun, C.Y.: Cloud Computing Security Management. In: IEEE International Conference on Service Computing, pp. 121–126 (2010)
12. Somani, U., Lakhani, K., Mundra, M.: Implementing Digital Signature with RSA Encryption Algorithm to Enhance Data Security of Cloud in Cloud Computing. In: IEEE International Conference on Parallel, Distributed and Grid Computing, pp. 85–94 (2010)
13. Sato, H., Kanai, A., Tanimoto, S.: A Cloud Trust Model in a Security Aware Cloud. In: IEEE International symposium on Applications and the Internet, pp. 121–124 (2010)
14. Kaufman, L.M.: Data Security in the World of Cloud Computing. IEEE Security and Privacy 7(4) (2009)
15. Amazon Web Services (AWS), `http://aws.amazon.com`
16. Shen, Y., Li, K., Yang, L.T.: Advanced Topics in Cloud Computing. Journal of Network and Computer Applications 12, 301–310 (2010)

17. Subashini, S., Kavitha, V.: A Survey on Security Issues in Service Delivery Models of Cloud Computing. Journal of Network and Computer Applications 34, 1–11 (2010)
18. Chow, R., Golle, P., Jakobsson, M., Shi, E., Staddon, J.: Controlling Data in the Cloud Outsourcing Computation without Outsourcing Control. In: CCSW, Chicago, Illinois, USA (2009)
19. Lombardi, F., Pietro, R.D.: Secure Virtualization for Cloud Computing. Journal of Network and Computer 12, 407–412 (2010)
20. Google App Engine, `http://code.google.com/appengine`
21. Caslo, V., Rak, M., Vilano, U.: Identity Federation in Cloud Computing. In: IEEE International Conference on Information Assurance and Security, pp. 253–259 (2010)
22. Casola, V., Mazzeo, A., Mazzocca, N., Victoriana, V.: A Security Metric for Public key Infrastructures. Journal of Computer Security 15(2), 78–85 (2007)

# Classification of Software Quality Attributes for Service Oriented Architecture

Satish Kumar, Neeta Singh, and Anuj Kumar

School of Information and Communication Technology
Gautam Buddha University, Greater Noida, U.P. -201308, India
satishkumar.serg@gmail.com, neeta@gbu.ac.in,
anujchandila@gmail.com

**Abstract.** In last few years, the emergence of Service-Oriented Architecture (SOA) is an extensive field in research due to the popularity of supporting wide range of quality attributes. SOA is becoming a popular architectural pattern for developing distributed system with prominent quality attributes. Due to the emergence of Web Service that is implemented by SOA have several quality issues such as performance, security, reliability and degree of interoperability or reusability. This paper presents a comprehensive study about positive or negative effect of software quality attributes (SQA) in developing distributed system. This paper also describes issues related to each quality attribute in developing distributed system. Finally, a classification framework of SQA shows the relationship between SOA and SQA.

**Index Keywords:** Service-Oriented Architecture, Software Quality Attributes, Web Services.

## 1 Introduction

In recent years, the demand of distributed system is increasing day by day, due to the popularity of web services and SOA architectural style that has recently gained popularity because of its potential to maximize reuse, interoperability, scalability and performance [1]. SOA is enabled with several characteristic such as loose-coupling, location transparency, dynamic binding, self-contained and modular that give scalability and cross platform for web service development with high quality attributes such as interoperability, scalability and performance. The importance of SOA is that it provides an easy scalable paradigm for organizing large network of web services that require interoperability to understand the value inherent in the individual component [2][4]. This is a main motivation for adopting SOA for web services but web services have some limitation that cannot implement the full characteristic of SOA. Web service doesn't support the notion of contract lease and no particular official specification provides Quality of Service level for services. All these limitation are come under the organization policy for making better business.

Software architecture is playing an important role as a bridge between business requirement and software system. It is very difficult to selecting and designing of an architecture that fulfils the functional and software quality requirement (performance,

A. Mantri et al. (Eds.): HPAGC 2011, CCIS 169, pp. 46–51, 2011.

reliability and security etc.) is a key to success of the system. Today the first choice of software architect is SOA for developing distributed software due to achieving wide-spread interoperability, performance and scalability which are seamless demand of enterprise in business development. The next section of this paper focuses on current issues of software quality attributes in web services development.

## 2 Interoperability

In general, Interoperability means the ability of component or application to work with other component or application without making any special effort from end user. Interoperability is an essential factor in the success of solutions that are based on web Services and SOA, along with other key factors such as contracts, loose coupling, and reuse [3]. Interoperability always makes a positive impact in web service design and it must be achieved in designing of its architecture. The aim of web services interoperability is to provide a seamless connection between applications on network. There are various technology (XML, WSDL, SOAP, UDDI) and platforms (.NET, J2EE) for achieving the high interoperability in the web application. WSDL is core to maintain the interoperability in web application. Developers define the contract information in WSDL that describes interface of new web application and existing application. WSDL provides the facility to multiple clients to access the service without knowing about the underlying implementation. But in web service development, interoperability cannot be guaranteed due to various reasons like differences in the versions of web service standards and specifications supported, differences in error handling mechanisms, differences in protocol support etc[10]. In distributed environment, there are some interoperability issues such as,

**Data access:** Today, data interoperability is provided by XML and web services between different applications in heterogeneous distributed environment. Toolkit interoperability is a big issue in consuming web services data from one toolkit to another toolkit due to many differences in programming languages, because these languages doesn't exist XML standard schema for representing database [5].

**Encoding style:** A SOAP message is also a famous interoperability issue due to complex message structure and encoding. SOAP messages are structured in two ways: RPC and Document style. The design of RPC message is based according to web service implementation while Document style is more loosely coupled with web service implementation. SOAP message encoding might be concerned to some problem in using object serialization due to different encoding style such as Literal and SOAP.

**Conversion problem:** Information is lost when converting from a XML Schema Definition (XSD) type to a native type or vice versa. This is a typical problem when using different vendor implementation for the web service and client side . This problem is caused because one-to-one mapping of data types are lacking [17]. For example: use of unsigned values (uint, ulong) in .NET that cannot be interpreted in Java, which lack support for unsigned values.

## 3 Reliability

The WS-Reliability and WS-Reliable Messaging Standards are defined by OASIS and the former has borrowed from the ebXML Message Service Specification 2.0 technology. WS-Reliability is a SOAP-based (SOAP 1.1 and SOAP 1.2 Part 1) specification that fulfils reliable messaging requirements critical to some applications of web services [5]. Web services reliability deals with the reliable web service messaging which includes guaranteed delivery of messages, message ordering and elimination of duplicate messages and also provides reliable communication between Web services.

**Service reliability:** It means services are operate correctly and reports any kind of failure during communication to the service user [15]. Services are rely on HTTP which are stateless and follow best effort delivery mechanism but now days these problems are solved by using new reliable protocol such as REST,HTTPR and the use of asynchronous messages queues.

**Message reliability:** It means guaranteed delivery of message to intended user. It provides ordered delivery, duplicate elimination and message state disposition wherever messages are sent between web services [6]. Web application require to reliable messaging in order to fulfil their organization requirement effectively and successfully but lost, unordered and duplicate messages can have negative impact on successful business due to some issues such as messages may get lost during transmission, receiving or sending host may become unavailable, unordered and delayed messages can lead to problem for online transactions [13].

**Load balancing and address messages**: It affects the overall performance of when two web services communicate for information sharing. For example, some time distributors and suppliers make an agreement to exchange information at a particular time. At that time the issue is how to handle the load of several distributors. Web-Addressing is a mechanism that provides the facility of endpoint reference to addressing messages that makes better communication channel without load factor because it affects only network load balancing.

## 4 Security

The security of a web services based system depends not only on the security of the services themselves, but also on the confidentiality and integrity of the XML based SOAP messages used for communication [6]. Web services security describes enhancement to the existing SOAP messaging to provide quality of protection through the application of message integrity, message confidentiality and single message authentication to SOAP messages [16]. Security is a major concern for SOA and web services because some time it makes negative impact on other quality attributes (performance, interoperability and modifiability).For example: If a service consumer cannot provide a security token required by the service provider, interoperability will clearly suffer. Most home-grown service consumers use username and password for authentication. If a service provider starts requiring the use of digital certificate for authentication, that provider's services cannot be accessed by service consumer who does not have the ability to use digital certificates. Security is also a challenge for architect because he should pay attention to some characteristic of SOA that directly impacts security. For instance SOA based distributed system includes services provided by third party. Third party service provider should be authenticated but it is not enough suppose system sends important data to the third party service provider, this data must be protected not only in transmission but also when it is stored. Another issue in SOA based system is implemented with public directory where the services are deployed. It is major security issue for identify the valid publisher to maintain and add new services in the directory.

## 5 Performances

Web service performance measures the capability of the services to provide appropriate processing and response time and throughput when performing its function under stated conditions [8]. It is major issue for both service provider and service consumer because services in SOA are distributed in heterogeneous environment that degrade performance when information is exchanged between web services. In web based distributed application, more services are vulnerable to changing and unpredictable load then the effect of unpredictable load makes degrade performance in composite web services [9][14]. In implementation of Composite Web Services (CWS) have several issues such as sometime CWS deals with the difficulties to identify the specific source of a bottleneck where source code is not available or heavy load on network. There are some other performance issues in web service development.

- Parsing of SOAP messages is expensive operation regarding to time factor. Web services messages are medium for communication and information sharing. Due to load of SOAP messages take long time for parsing which makes the impacts on service performance.
- XML digital signature and XML encryption are major issue for processing because security between endpoints is not free and it can take long time for processing due to some credentials.

## 6 Scalability

Scalability is the ability of SOA based application to continue to function well when the system is changed in size or in volume in order to meet users needs [16]. There is no technology for implement the scalability in SOA application but it can be achieved by using some approaches such as using right communication protocol, using caching, and standard coding style. The goal is to allow the application to handle as much load. Before design an SOA architecture, Software Architect has some consideration for improving both performance and scalability of SOA application by using two approaches which are in practice today[11].

- **Vertical Scalability***:* Vertical Scalability increases the performance of application by adding resources in a single logical unit. For example, to add multiple CPU to an existing server or expanding storage by adding memory.
- **Horizontal Scalability:** Adding multiple logical units of resources and making them work as a single unit.

Designing of SOA based distributed system provides widespread scalability but few issues are arising before achieve the efficient scalability in web application. Some issues are associated with how to develop the SOA based application in effective coding style, but the essential bottlenecks associated with how to store and access data from the application [13]. The architecture of SOA based application is inherently scalable due to supporting wide range of characteristics which provide easy scalability in the development of distributed system. It can grow up onto multiple web servers and heterogeneous datacenters. For access data from one application to another application in heterogeneous environment is a bottleneck of scalability. Bottleneck typically associated with relational database in heterogeneous environment due to access session data and storage of data that data is also another potential scalability bottleneck.

## 7 Classification Framework of SQA

After discussion about some important quality attribute in web service point of views and go through a brief survey of quality attributes that show issues and impact (negative and positive) for SOA implementation. To produce a classified framework of software quality attributes that shows the relationship with SOA.

**Table 1.** (Software Quality Attributes for SOA)

| Impact | Software Quality Attributes |
|---|---|
| Positive impact | Interoperability |
| Neutral impact | Reliability, Scalability, |
| Negative impact | Security, Performance, |

## 8 Conclusion

This paper described Service-Oriented Architecture is a particular architectural style that provides roadmap for web service development with wide range of software quality attributes. So, first choice of software architect is SOA for making successful solution. This paper also gives brief discussion about the impact of quality attributes and issues in web service development. XML security standards are major impacts on the web service performance, it a open research area how to balance the security and performance in web service development.

## References


1. Gomaa, H., Street, J.: Software Architecture Reuse Issues in Service-Oriented Architecture. In: Proceedings of the 41st Hawaii International Conference on System Science (2008)
2. Amirzafari, B., Valipour, M.H.: A Brief Survey of Software Architecture Concepts and Service-Oriented Architecture. In: IEEE International Conference, pp. 34–38 (2009)
3. Shetty, S.D., Vadivel, S.: Interoperability issues seen in Web Services. IJCSNS International Journal of Computer Science and Network Security (2009)
4. Erl, T.: Service-Oriented Architecture: Concepts, Technology, and Design. Prentice Hall PTR, Englewood Cliffs (2005)
5. Blaga, L., Ratiu, I.: Interoperability issues in accessing database through Web Services. Recent Advances in Neural Networks, Fuzzy System & Evaluationary Computing (2008)
6. World Wide Web Consortium: Messaging Framework, 2 edn. version 1.2 part 1, http://www.w3.org/TR/soap12-part1/
7. Atkinson, B., Al, E.: Specification: Web Services Security, version 1.0, http://www.ibm.com/developerworks/webservices/library/specification/ws-secure
8. Her, J.S., Choi, S.W., Kim, S.D.: Modelling QoS Attributes and Metrics for Evaluating Services in SOA Considering Consumers' Perspective as the First Class Requirement. In: IEEE Asia-pacific Services Computing Conference (2007)
9. Smith, D.B., Simanta, S., Balasubramaniam, S.: Challenges for Assuring Quality of Service in a Service-Oriented Environment. In: ICSE Workshop, Canada, pp. 103–106 (2009)
10. Kumar, A., Kumari, G.P., Kuppuraju, S.: Case Study to Verify the Interoperability of a Service Oriented Architecture Stack. In: IEEE International Conference on Services Computing (2007)
11. Writing Scalable Web Applications, http://www.xibl.com/performance/writing-scalable-web-applications/
12. Rossi, G., Buckley, I., Sadjadi, M.: Web Services Reliability Patterns, http://www.cis.fiu.edu/~sadjadi/Publications/WSReliability-SEKE-2009.pdf
13. Khan, I.: Address Scalability Bottlenecks with Distributed Caching. MSDN Magazine, http://msdn.microsoft.com/enus/magazine/ff714590.aspx
14. Ma, K.J., Bartos, R.: Performance Impact of Web Service Migration in Embedded Environments. In: IEEE International Conference on Web Services (2005)
15. Merson, P., Bass, L.: Quality Attributes for Service-Oriented Architectures. In: IEEE International Workshop on System Development in SOA Environments (2007)
16. Nordbotten, N.A.: XML and Web Services Security Standards. IEEE Communications surveys & Tutorials 11 (2009)
17. Wangming, Y.: Web services programming tips and tricks: Improve interoperability between J2EE technology and NET, http://www128.ibm.com/developerworks/xml/library/ws-tip-j2eenet1

# Energy Efficiency for Software and Services on the Cloud

Priyanka Bhati[1], Prerna Sharma[2], Avinash Sharma[3],
Jatin Sutaria[4], and M. Hanumanthapa[5]

[1] Lecturer, Computer Science Department,
Management And Commerce Institute of Global Synergy,
Ajmer, Rajasthan, India
priyanka.goodluck@gmail.com
[2] PG Student, Information Technology Department,
Rajasthan College of Engineering for Women,
Jaipur, Rajasthan, India
prerna.it10@gmail.com
[3] Associate Professor & Research Scholar,
SGVU, Jaipur, Rajasthan, India
sh_avinash@yahoo.com
[4] Senior Software Engineer, SABA Inc. USA
jatinsut@yahoo.com
[5] Associate Professor, Bangalore University,
Bangalore, India

**Abstract.** The market for cloud computing services has continued to expand despite a general decline in economic activity in most of the world. Cloud computing is computation, software, data access, and storage services that do not require end-user knowledge of the physical location and configuration of the system that delivers the services.

This Paper provides an in-depth analysis of the energy efficiency benefits of cloud computing, including an assessment of the software as a service (SaaS), platform as a service (PaaS), and infrastructure as a service (IaaS) markets. It also highlights the key demand drivers and technical developments related to cloud computing, in addition to detailed quantification of energy savings and GHG reduction opportunities under a cloud computing adoption scenario, with a forecast period extending through 2020.

**Keywords:** Cloud Computing, Energy Reduction, Market Forecast, Technology issues.

## 1 Introduction

Cloud computing has recently received considerable attention, as a promising approach for delivering ICT services by improving the utilization of data centre resources. In principle, cloud computing can be an inherently energy-efficient technology for ICT provided that its potential for significant energy savings that have so far focused on hardware aspects, can be fully explored with respect to system operation and networking aspects.

A. Mantri et al. (Eds.): HPAGC 2011, CCIS 169, pp. 52–55, 2011.

Growth in cloud computing has some important consequences for both greenhouse gas (GHG) emissions and sustainability. Thanks to massive investments in new data center technologies, computing clouds in general and public clouds in particular are able to achieve industry-leading rates of efficiency. Simply put, clouds are better utilized and less expensive to operate than traditional data centers.

## 2 Technolgy Issues

Technology that can improve the utilization of resources, and thus reduce the power consumption is virtualization of computer resources. ***Virtualization*** technology allows one to create several Virtual Machines (VMs) on a physical server and, therefore, reduce the amount of hardware in use and improve the utilization of resources. Among the benefits of virtualization are improved fault and performance isolation between applications sharing the same computer node. Terminal servers have also been used in Green IT practices. Cloud computing paradigm leverages virtualization technology and provides the ability to provision resources on-demand on a pay-as-you-go basis. Organizations can outsource their computation needs to the Cloud, thereby eliminating the necessity to maintain own computing infrastructure.

***Virtualization Technology Vendors***

Three most popular virtualization technology solutions: the **Xen hypervisor6**, **VMware solutions7** and **KVM8**. Both of these systems support power management, however, neither allows coordination of VMs' specific calls for power state changes. Other important capabilities supported by the virtualization solutions are offline and live migrations of VMs. They enable transferring VMs from one physical host to another, and thus have facilitated the development of different techniques for virtual machines consolidation and load balancing.

## 3 Power Advantage of Cloud Computing

There has been some debate here over the energy efficiency profile of the cloud, environmental sciences. There are four primary reasons why cloud computing should be a more power-efficient approach than an in-house data center.

a) **Workload diversity:** Because we will have many different sorts of users making use of the cloud resources – different applications, different feature set preferences and different usage volumes – this will improve hardware utilization and therefore make better use of power that we're using any way to keep a server up and running.

b) **Economies of economies of scale:** There are certain fixed costs associated with setting up any physical data center. Implementing technical and organization changes is cheaper per computation for larger organizations than for IT.

c) **Power-management flexibility:** It's easier to manage virtual servers than physical servers from a power perspective. If hardware fails, the load can automatically be deployed elsewhere. Likewise, in theory, we could move all virtual loads to certain servers when loads are light and power-down or idle those that aren't being used.

**d) We can pick the most efficient site possible:** So, for example, if we are a business based in a state that uses primarily coal-powered electricity, do we really want to site our data center there? "If we have a data center in a place that is all coal-powered, this is a big business risk,". In a future where there might actually be a tax on the carbon our company produces, that would certainly be a risk indeed.

## 4 Conclusion

In recent years, energy efficiency has emerged as one of the most important design requirements for modern computing systems, such as data centers and Clouds, as they continue to consume enormous amounts of electrical power. Apart from high operating costs incurred by computing resources, this leads to significant emissions of carbon dioxide into the environment. Efficient power management in computing systems is a well-known and extensively studied in the past problem. Intelligent management of resources may lead to significant reduction of the energy consumption by a system, while meeting the performance requirements. The virtualization technology has advanced the area by introduction of a very effective power saving technique.

## 5 Future Scope

For the future research work we propose the investigation of the following directions. It is crucial to develop intelligent techniques to manage network resources efficiently.

**a)** One of the ways to achieve this for virtualized data centers is to continuously optimize network topologies established between VMs, and thus reduce network communication overhead and load of network devices.
**b)** Another direction for future work, which deals with low-level system design, is improvement of the power supplies efficiency, as well as development of hardware components that support performance scaling proportionally to power consumption. Reduction of the transition overhead caused by switching between different power states and VM migration overhead can greatly advance energy-efficient resource management and has to be also addressed by future research.

## References

[1] Pamlin, D.: The Potential Global CO2 Reductions from ICT Use: Identifying and Assessing the Opportunities to Reduce the First Billion Tonnes of CO2, WWF, Sweden (May 2008)
[2] Accenture, Data Centre Energy Forecast Report. Final Report, Silicon Valley Leadership Group (July 2008)
[3] Hewitt, C.: ORGs for scalable, robust, privacy-friendly client cloud computing. IEEE Internet Comput., 96–99 (September 2008)

[4] Changjiu, X., Yung-Hsiang, L., Zhiyuan, L.: Energy- Aware Scheduling for Real-Time Multiprocessor Systems with Uncertain Task Execution Time. In: Proc. 44th Annual Conf. Design Automation, pp. 664–669. ACM, San Diego (2007)
[5] Merkel, A., Bellosa, F.: Memory-Aware Scheduling for Energy Efficiency on Multicore Processors. In: Proc. Workshop on Power Aware Computing and Systems (HotPower 2008), San Diego, CA, USA, pp. 123–130 (December 2008) USENIX online
[6] Chunlin, L., Layuan, L.: Utility-based scheduling for grid computing under constraints of energy budget and deadline. Comput. Stand. Interfaces (2008)
[7] Koomey, J.: Server Energy Measurement Protocol. In: Version 1.0. Following Energy Efficiency Server Benchmark Technical Workshop, Santa Clara, CA (2006), `http://www.energystar.gov/ia/products/downloads/Finalserverenergyprotocol-v1.pdf` (last accessed August 12, 2009)
[8] Berl, A., de Meer, H.: An Energy-Efficient Distributed Office Environment. In: Proc. European Conf. Universal Multiservice Networks (ECUMN 2009). IEEE Computer Society Press, Sliema (2009)
[9] Berl, A., Weidlich, R., Schrank, M., Hlavacs, H., de Meer, H.: Network virtualization in future home environments. In: Bartolini, C., Gaspary, L.P. (eds.) DSOM 2009. LNCS, vol. 5841, pp. 177–190. Springer, Heidelberg (2009)

# Evaluation of Grid Middleware Frameworks for Execution of MPI Applications

Abhishek Jain[1] and Sathish S. Vadhiyar[2]

[1] Department of Physics
Birla Institute of Technology & Science (BITS), Pilani, India
abhishekjain1988@gmail.com, f2006637@bits-pilani.ac.in
[2] Supercomputer Education and Research Centre
Indian Institute of Science, Bangalore, India
vss@serc.iisc.ernet.in

**Abstract.** Execution of large-scale parallel applications that span multiple distributed sites is important to realize the potential of computational grids. There are various problems being faced by the developers while running the applications on multiple clusters. In the last few years many groups have developed middleware frameworks that enable execution of MPI applications on multiple clusters where the slave nodes of a cluster have private or hidden IP address. This paper evaluates and compares such middleware frameworks for execution of MPI applications and discusses the merits of the solutions.

**Keywords:** Grid middleware frameworks; GRIDMPI; MC-MPI; MPI applications; PACX-MPI.

## 1 Introduction

Parallel architectures are motivated by the applications which often require computational capabilities which will take a long time by a single massively parallel processing system (MPP) or a parallel vector processor (PVP). So, the researchers come up with possible solution to couple different computational resources distributed all over the world. The coupling of MPPs and/or PVPs requires a reliable and, if possible, dedicated network connection between the machines. This result in the formation of clusters distributed geologically all around the world. However, only the master node of a cluster has global IP address while the slave nodes have hidden or private IP addresses and hence cannot be directly contacted by a node of another cluster. This presents a challenge in executing a single MPI parallel application across multiple clusters since a process executing in a slave node of a cluster will not be able to send a message to another process executing in the slave node of another cluster. To overcome the challenges related to hidden IPs, various grid middleware frameworks have been proposed for the execution of MPI parallel applications across multiple clusters.

In this paper, we discuss the proposed solutions, experimentally evaluate and compare the different solutions. We have presented existing Grid middleware frameworks implementing MPI libraries in section 2. It is essential as it provides the different

A. Mantri et al. (Eds.): HPAGC 2011, CCIS 169, pp. 56–61, 2011.

measurement techniques and their implementation details as per proposed by the different middleware frameworks. In Section 3, we have discussed the proposed solutions, evaluated and performance measurements on different clusters. Finally, in Section 4, the conclusion of the work is presented to support the viability of the analysis.

## 2 Grid Middleware Frameworks for MPI Applications

In this section, we describe the various middleware frameworks we considered for execution of MPI applications.

### A. PACX-MPI 1,2,9

PACX-MPI (Parallel Computer eXtension) is designed as a library interfacing between the user application and the local intra-machine MPI implementation. When the application calls a MPI function, the call is intercepted by PACX-MPI and decision is made regarding if there is a need to contact another MPP (or cluster in our case) during the call execution. If not, the library sends the message using the matching MPI call from vendor's MPI implementations are used for all intra-machine communications.

When the MPI call involves another cluster or MPP the communication is forwarded via network by using TCP/IP sockets, but during this process, MPI processes do not exchange messages directly. Instead, on each parallel system two special nodes are reserved, one for every (incoming and outgoing) communication direction. On each of these nodes, a daemon MPI process is executed. The daemon process takes care of communications with the local nodes, compression and decompression of data for remote communication and communication with the peer daemons of other parallel machines. This daemon approach bundles the communication and eliminates the need to open connections between each process on every system, which saves resources and permits to handle security issues centrally.

A drawback of this design is that the use of wide-area, heavy-weight protocol stacks like TCP/IP for inter-cluster communication introduces significant latency.

### B. GRID-MPI 3,4

GridMPI is a new MPI programming environment primarily designed to efficiently run MPI applications in the Grid. GridMPI introduces a latency-aware collective's layer which optimizes the communication performance over the links with non-uniform latency and bandwidth and hides the details of the lower-level communication libraries. GridMPI is developed using YAMPI 5 for intra-cluster communication and supports the Interoperable MPI (IMPI) protocol 6 for inter-cluster communication over TCP/IP. It uses a message relay mechanism to support private IP address clusters, which is transparent for MPI processes that communicate using the IMPI protocol. It uses a user-level proxy implementation called the IMPI Relay.

### C. MC-MPI

MC-MPI 7 is a Grid-enabled implementation of MPI, developed at the University of Tokyo. Its main features include message traversal across firewalls and NATs, and locality aware connection management.MC-MPI constructs an overlay network, allowing nodes behind firewalls and nodes without global IP addresses to participate in computations. MC-MPI automatically probes connectivity, selects which connections to establish, and performs routing. Establishing too many connections, especially wide-area connections, results in many problems, including but not limited to the following: exhaustion of system resources (e.g., file descriptors, memory), high message reception overhead, and congestion between clusters during all-to-all communication. Therefore, MC-MPI limits the number of connections that are established. MC-MPI uses a lazy connects strategy, as a result, fewer connections are established for applications in which few process pairs communicate.

## 3 Experiments and Results

In this section, we describe the platforms and benchmark used for our experiments and the results.

### A. Experimental Setup

We used two clusters located in Grid Applications Research Lab, Supercomputer Education and Research Centre, Indian Institute of Science. The first cluster, *cluster-1*, consists of 8 single-core AMD Opteron based servers with CentOS release 4.3, 1 GB RAM, 2x80 GB hard drive and connected by Gigabit Ethernet. The second cluster, *cluster-2*, consists of 8 dualcore AMD Opteron 1214 based 2.21 GHz Sun Fire servers, with CentOS release 4.3, 2 GB RAM, 250 GB hard drive and connected by Gigabit Ethernet. The master nodes of the two clusters are connected by a 100 Mbps Ethernet switch. Direct communications are possible only between the master nodes. MPIBench 8, a benchmark to evaluate MPI implementations, was used to evaluate the performance of different middleware frameworks. MPIBench uses globally synchronized clock that is based on CPU cycle counters. This allows accurate measurement of individual MPI communications. MPIBench is thus able to provide distributions (histograms) of communication times, rather than just average values. The histograms can provide additional insight into communications performance. We evaluated the performance of PACX-MPI, GridMPI, and MC-MPI using MPIBench .We executed the standard MPIBench programs across 2 clusters, using a total of 16 processes with 8 processes in each cluster. All the observations are in milliseconds. The results shown are average values for 1000 experiments for each message size.

### B. Results

We have found that when inter-cluster communication is evaluated using the standard MPI benchmark, different MPI Communication libraries give significantly different results for various MPI routines. This is primarily due to the different communication techniques used by the different libraries.

We have divided our results into two categories. In the first category, we have evaluated the point-to-point communication for all the three MPI libraries while in the second part we compare the results for different MPI routines used as for collective communications.

### 1) Point-to-Point Communication

MPI_Send/MPI_Recv is used by different MPI benchmark programs to evaluate the MPI middle ware frameworks. It has been observed from the figure 1,2,3,4 that MC-MPI outmatches PACX-MPI and GRIDMPI. With keen observation it is seen, in figure 1, where MPI_ISend routine is evaluated, that initially GridMPI follows the same latency as that of PACX-MPI but, when the message size increases to the level 524288 bytes, there is a sharp rise in latency. There is a massive difference in the results obtained for MPI_Sendrecv routine if we run on two different clusters with increasing number of processes. In figure 3, which describes the result as of running MPI_Isendlocal, it is observed that the large delay in PACX-MPI in comparison to the other two middleware frameworks.

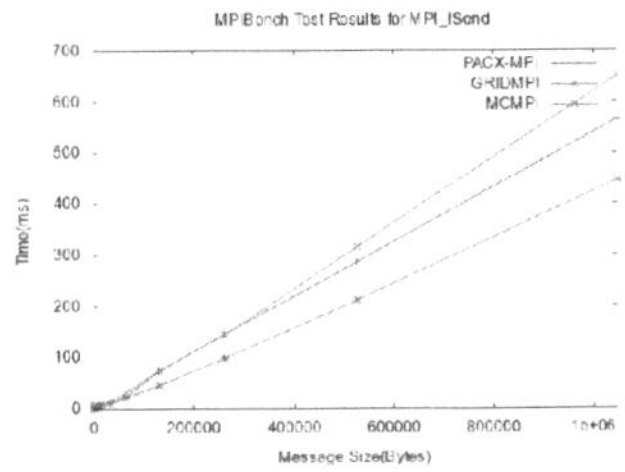


Fig. 1. MPI_ISend on 16 CPUs

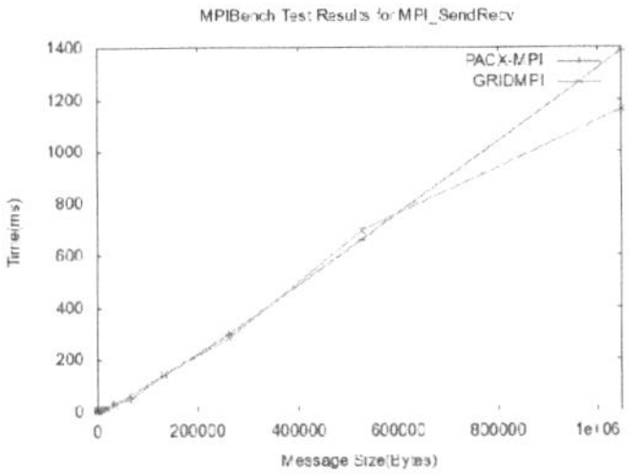


Fig. 2. MPI_SendRecv on 16 CPUs

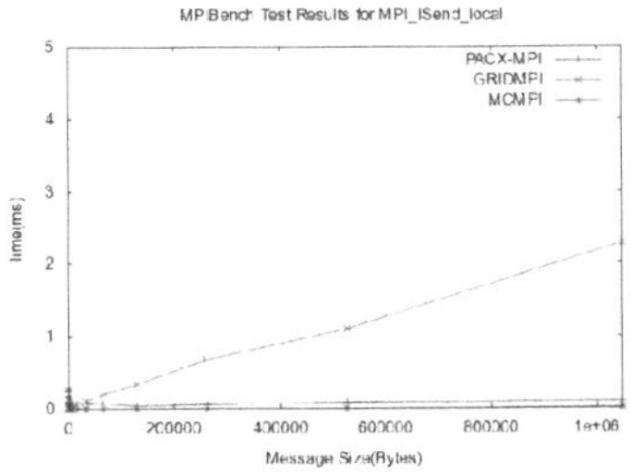


Fig. 3. MPI_ISendlocal on 16 CPUs

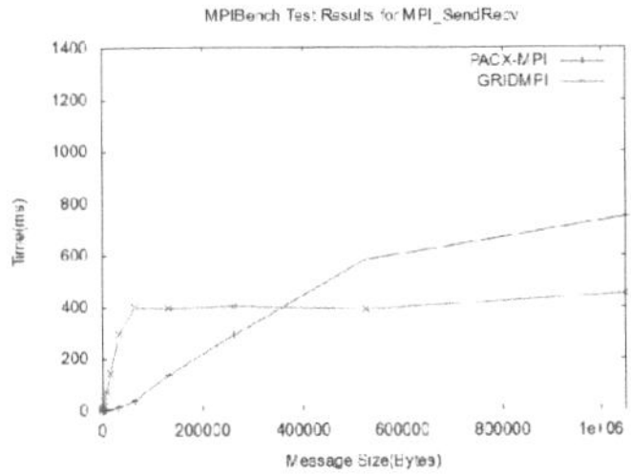


Fig. 4. MPI_SendRecv on 8 CPUs

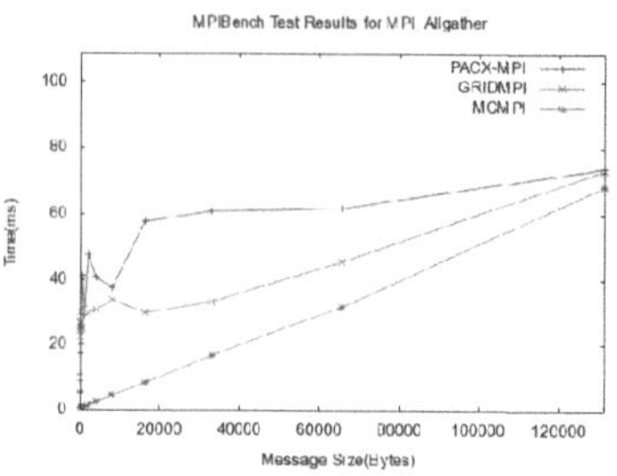


**Fig. 5.** MPI_Allgather on 16 CPUs

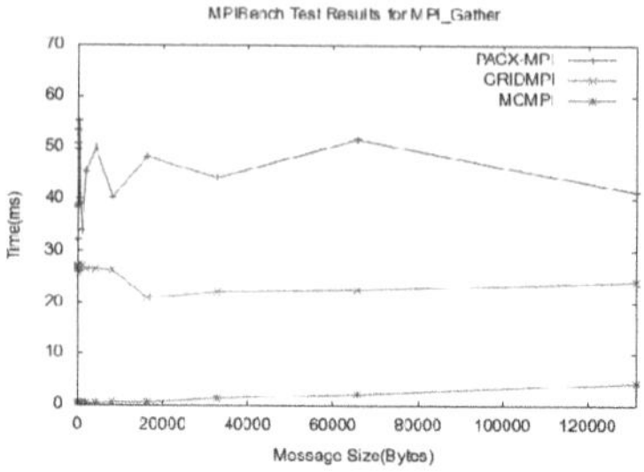


**Fig. 6.** MPI_Gather on 16 CPUs

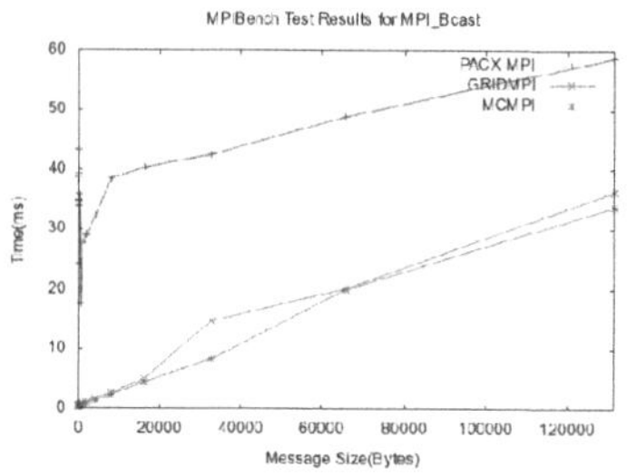


**Fig. 7.** MPI_Bcast on 8 CPUs

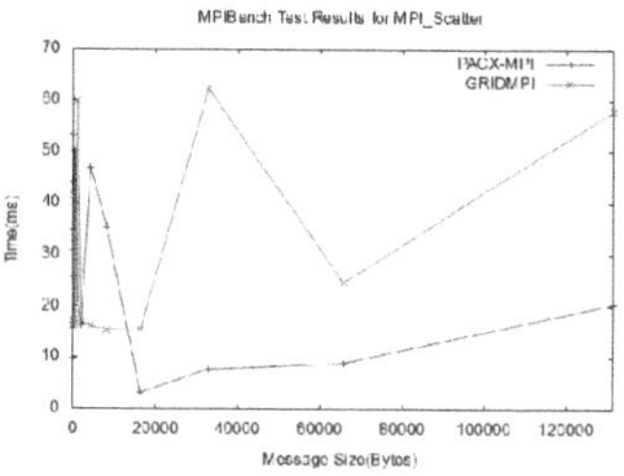


**Fig. 8.** MPI_Scatter on 16 CPUs

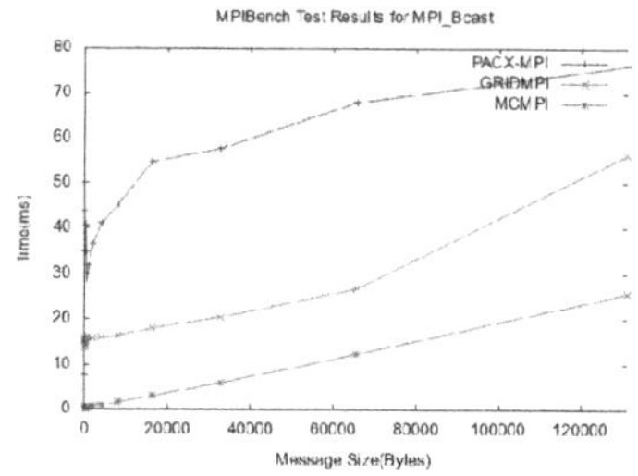


**Fig. 9.** MPI_Bcast on 16 CPUs

## 2) Collective Communications

The detailed analysis in collective communication shows that, MC-MPI outperforms GridMPI and PACX-MPI in most of the MPI routines of MPI-Benchmark. MC-MPI not only takes the least time, but it also has the advantage of being configurable at the user level. GridMPI performed better in some MPI routines. These results are consistent with the theoretical consideration. GridMPI uses its trunking method which is used to improve available inter-cluster bandwidth. Trunking is a connection aggregation technique using multiple pairs of the IMPI Relay. In case of collective communication, it can be observed that, when the message size is less than 1024 bytes, there is a sharp rise in the time taken for GridMPI. This is due to the fact that for these message sizes, GridMPI does not use trunking. After 1024 bytes, we can observe that there is a steady rise in the time with the message size. This is due to the fact that GridMPI enables trunking after the message size of 1024 bytes. Since PACX-MPI, uses wide-area, heavy-weight protocol stacks like TCP/IP for inter-cluster communication, it introduces significant latency overheads. MC-MPI performs locality-aware communication optimizations for only point-to- point operations. Hence less time is taken for communications with MCMPI. Though MC-MPI performs better, it has some drawbacks. Its current version does not support many basic MPI routines.

## 4 Conclusion and Outlook

In this paper, we compared three different grid middleware frameworks, namely, PACX-MPI, GridMPI and MC-MPI, by conducting experiments using MPIBench benchmarks on two gigabit clusters connected by a 100 Mbps network. Our results show that in general, MC-MPI gives the best performance in most cases and has the advantage of being configurable at the user level. Future studies will involve the performance analysis using NAS Parallel Benchmarks and including MPICH-GX and MPICH-Madeleine in the comparison list.

## References


1. Gabriel, E., Resch, M., Beisel, T., Keller, R.: Distributed Computing in a Heterogenous Computing Environment. In: Euro PVMMPI 1998 (1998)
2. Muller, M., Hess, M., Gabriel, E.: Grid enabled MPI Solutions for Clusters. In: CCGRID 2003: Proceedings of the 3rd International Symposium on Cluster Computing and the Grid, p. 18 (2003)
3. Grid MPI, `http://www.gridmpi.org`
4. Takano, R., Matsuda, M., Kudoh, T., Kodama, Y., Okazaki, F., Ishikawa, Y., Yoshizawa, Y.: High Performance Relay Mechanism for MPI Communication Libraries Run on Multiple Private IP Address Clusters. In: CCGRID 2008: Proceedings of the 2008 Eighth IEEE International Symposium on Cluster Computing and the Grid, pp. 401–408 (2008)
5. YAMPI, `http://www.il.is.s.u-tokyo.ac.jp/yampi`
6. George, W., Hagedorn, J., Devaney, J.: IMPI: Making MPI Interoperable. Journal of Research of the National Institute of Standards and Technology 105(3) (2000)
7. Saito, H., Taura, K.: Locality-aware Connection Management and Rank Assignment for Wide-area MPI. In: PPoPP 2007: Proceedings of the 12th ACM SIGPLAN Symposium on Principles and Practice of Parallel Programming, pp. 150–151 (2007)
8. MPIBench, `http://www.dhpc.adelaide.edu.au/projects/mpibench`
9. Balkanski, D., Trams, M., Rehm, W.: Communication middleware systems for heterogeneous clusters: a comparative study. In: Proceedings of 2003 IEEE International Conference on Cluster Computing, pp. 504–507 (2003)

# Virtualization as an Engine to Drive Cloud Computing Security

Snehi Jyoti[1], Snehi Manish[2], and Gill Rupali[1]

[1] Chitkara University, Rajpura, Punjab, India
(09988701479), (09988092468)
jyoti.snehiverma@gmail.com,
rupali.Gill@chitkara.edu.in
[2] Infosys Technologies Ltd., Chandigarh, India
(9569446393)
snehi.manish@gmail.com

**Abstract.** In this paper we have proposed virtualization as an engine to drive cloud based security. Cloud computing is an approach for the delivery of services while virtualization is one possible service that could be delivered. Virtualization enables better security and enables a single user to access multiple physical devices. Large corporations with little downtime tolerance and airtight security requirements may find that virtualization fits them best. Thin clients and software as a service will free users from being tied to their computers, and allow them to access their information anywhere they can find an internet connection. Virtualization is a computing technology that enables a single user to access multiple physical devices. It is a single computer controlling multiple machines, or one operating system utilizing multiple computers to analyze a database. Virtualization may also be used for running multiple applications on each server rather than just one. It enables us to consolidate our servers and do more with less hardware. With growing pressure to move in this direction, we're suggesting Virtualization for cloud based security.

**Keywords:** Cloud, Infrastructure-as-a-Service (IaaS), Platform-as-a-Service (PaaS) and Software-as-a-Service (SaaS), Virtualization.

## 1 Introduction to Cloud

Widespread availability of economical computing power in business and in homes has created the next advance in information technology. Cloud computing is a style of computing in which dynamically scalable and virtualized resources are provided as a service over the Internet. The name cloud computing was inspired by the cloud symbol that is used to represent the Internet in flow charts and diagrams. A server or database can be physically located in a highly-secure, remote location while the data is accessed from a client's computer, using the database's server to retrieve, sort, and analyze the data. A cloud computing provider owns the hardware while providing hosted, managed services to its clients on a usage basis. Cloud computing generally utilizes virtualized IT resources such as networks, servers, and computing devices.2

A. Mantri et al. (Eds.): HPAGC 2011, CCIS 169, pp. 62–66, 2011.

Security is a top concern organizations have about moving critical business applications to the cloud. Cloud computing delivers hosted services over the Internet. These services are broadly divided into three categories: Infrastructure-as-a-Service (IaaS),Platform-as-a-Service (PaaS) and Software-as-a-Service (SaaS).18910.

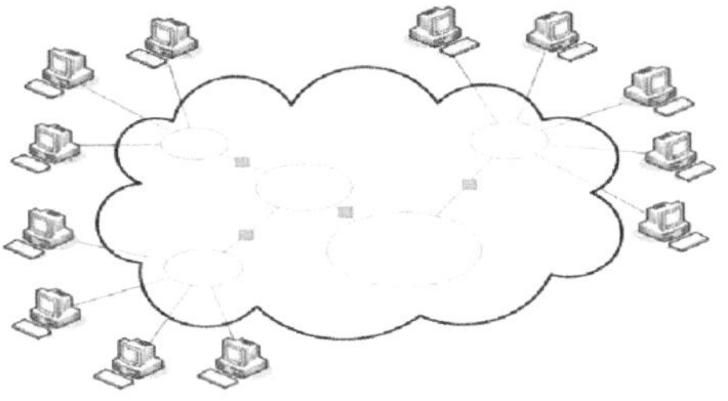

**Fig. 1.** Cloud Structure

**Infrastructure-as-a-Service (IaaS)** is the delivery of computer infrastructure as a service. Instead of purchasing servers, software, data center space or network equipment, clients instead buy those resources as a fully outsourced service. The service is typically billed on a utility computing basis and amount of resources consumed will typically reflect the level of activity.1

**Platform-as-a-service (PaaS)** in the cloud is defined as a set of software and product development tools hosted on the provider's infrastructure. Developers create applications on the provider's platform over the Internet. PaaS providers may use APIs, or gateway software installed on the customer's computer.7

**Software-as-a-service (SaaS)** in this cloud model, the vendor supplies the hardware infrastructure, the software product and interacts with the user through a front-end portal. SaaS is a very broad market. Services can be anything from Web-based email to inventory control and database processing. Because the service provider hosts both the application and the data, the end user is free to use the service from anywhere.

A cloud can be private or public. A public cloud sells services to anyone on the Internet. Eg. Amazon Web Services is the largest public cloud provider. A private cloud is a proprietary network or a data center that supplies hosted services to a limited number of people. When a service provider uses public cloud resources to create their private cloud, the result is called a virtual private cloud.2

## 2 Virtualization

Virtualization is the creation of a virtual version of an operating system, a server, a storage device or network resources. Virtualization is a computing technology that enables a single user to access multiple physical devices. Virtualization provides flexibility that is a great match for cloud computing. Moreover, cloud computing can be defined based on the virtual machine containers created with virtualization. Virtualization provides more servers on the same hardware and cloud computing provides measured resources while paying for what you use.3

Virtualization comes in many types focusing on control and usage schemes that emphasize efficiency. A single terminal runs multiple machines, or a single task running over multiple computers via idle computing power. Virtualization is also seen in a central computer hosting an application to multiple users, preventing the need for that software to be repeatedly installed on each terminal. Data from different hard drives, USB drives, and databases can be coalesced into a central location, both increasing accessibility and security through replication. Physical computer networks can be split into multiple virtual networks, allowing a company's central IT resources to service every department with individual local area networks.311

Virtualization involves deploying programs and operating systems from central machine to several virtual machines through the organization. The immediate benefits of a system similar to this consist of reduced hardware and servicing costs as well as the power to lower downtime and increase the development cycles. Virtualization can be Hardware virtualization, Software Virtualization and server virtualization. There are several approaches to virtualizing servers, including GRID approaches OS –level virtualization sometimes called containers where multiple instances of an application can run in isolation from one another on a single OS instance, and hypervisor-based virtualization.41213

## 3 Security in Cloud Computing

In a cloud, traditional security methodologies do not work as the service providers cannot allow information owners, or clients, to manipulate the security settings of the fabric. If this were allowed, it would be possible for one client to change security settings illicitly in their favor, or change security settings of other clients maliciously. This situation is unacceptable since the information owner cannot manage the security posture of their computing environment. Therefore, a security model is needed that allows for an information owner to protect their data while not interfering with the privacy of other information owners within the cloud. The cloud requires a new model for handling security, one that is shared between operators and clients. Operators need to give clients visibility into the security posture of the fabric while maintaining control. The clients need to have assurance that they can control the privacy and confidentiality of their information at all times and have assurances that if needed, they can remove, destroy, or lock down their data at any time. Security is an integral and separately configurable part of the private cloud fabric, designed for set of on-demand, elastic and programmable services.5 There are a series of points that lead small companies toward cloud-based security solutions, which includes Time to value, Ease of implementation, Zero-effort upgrades, Programmable Infrastructure, Logical Security Policies, On-Demand Elastic Services, Adaptive Trust Zones and Configurable Security Policy Management.91213

## 4 Our Proposal

It is generally considered that Virtualization adds overhead in amount of code and performance. Virtualization software heavily depends on hardware reliability and assumes that hardware does not fail. But in this paper we are proposing Virtualization as an engine to drive cloud security. In this approach we can take a large set of low-cost commodity systems and tie them together into one large supercomputer we can strip down servers to the bare essentials according to need. Certain types of security solutions can be deployed inside the network, such as intrusion detection and prevention systems, application firewalls, and data encryption systems. There are limits to how much a cloud-based solution can be customized to meet a particular company's needs. The degree to which customers' data is protected when it is stored in the cloud and about the identity management mechanisms used to validate users in business applications. As new workloads are introduced into the trust zone, the VM will adapt and cater to the new workload, as it will when individual machines move. Private cloud infrastructure will require security services that are designed to provide separation of workloads of different trust levels as a core capability. The virtualization/cloud stack provider should provide a rich tapestry of robust security capabilities "baked in" to the platform itself, or the virtualization/cloud stack provider should provide security-enabling hooks to enable an ecosystem of security vendors to provide the bulk of security to be "bolted on," or the virtualization stack provider should maximize the security of the underlying virtualization/cloud platform and focus on API security, isolation and availability of service only while pushing the bulk of security up into the higher-level application layers. Key advantage of Virtualization and Cloud Computing is a significant improvement in security, availability, and data protection. A decentralized IT infrastructure managed by an IT service provider that is wholly dedicated to its resilience and availability is immune to physical or data disasters. Replication over multiple systems ensures data backups. A dedicated data center service provider is better able to keep up with the latest security methods and technology upgrades. Through the provision of managed IT services, all of these benefits are embedded in the cloud computing model.

## 5 Conclusion

Virtualization is an approach to consolidating technology resources for improved efficiency and the elimination of redundancy by leveraging every opportunity to utilize idle resources and find places where multiple processes can be run at one time. Virtualization and Cloud Computing is a significant improvement in security, availability, and data protection.

## References

1. Leach, J.: The Rise of Service Oriented IT and the Birth of Infrastructure as a Service, March 20, 2008; 2 J. Heiser and M. Nicolett, Accessing the Security Risks of Cloud Computing, Gartner, Inc., Stamford, CT (2008)
2. Armbrust, M., Fox, A., Griffith, R., et al.: Above the Clouds: A Berkeley View of Cloud Computing. Unversity of California, Berkeley (2009)
3. Berger, S., Cáceres, R., Goldman, K.A., et al.: vTPM: Virtualizing the Trusted Platform Module. In: Proceedings of the 15th USENIX Security Symposium, Vancouver, B.C (2006)

4. Scarlata, V., Rozas, C., Wiseman, M., et al.: TPM Virtualization: Building a General Framework. In: Pohlmann, N., Reimer, H. (eds.) Trusted Computing, pp. 43–56. Vieweg+Teubner, Wiesbaden (2008)
5. Krautheim, F.J., Phatak, D.S.: LoBot: Locator Bot for Securing Cloud Compupting Environments. Submitted 2009 ACM Cloud Computing Security Workshop, Chicago, IL (2009)
6. Zissis, D., Lekkas, D.: Addressing Cloud Computing Security Issues. Future Generation Computer System (2010) Article in Press, `http://dx.doi.org/10.1016/j.future.2010.12.006`
7. Yamaguchi, T., Hashiyama, T., Okuma, S.: A Study On Reconfigurable Computing System Cryptography. In: IEEE International Conference on Cloud Computing, vol. 4, pp. 2965–2968 (2000)
8. Yuefa, D., Bo, W., Yaqiang, G., Quan, Z.: Data Security Model for Cloud Computing. In: International Workshop on Information Security and Applications, pp. 141–144 (2009)
9. Sato, H., Kanai, A., Tanimoto, S.: A Cloud Trust Model in a Security Aware Cloud. In: IEEE International symposium on Applications and the Internet, pp. 121–124 (2010)
10. Kaufman, L.M.: Data Security in the World of Cloud Computing. IEEE Security and Privacy 7(4) (2009)
11. Amazon Web Services (AWS), `http://aws.amazon.com`
12. Shen, Y., Li, K., Yang, L.T.: Advanced Topics in Cloud Computing. Journal of Network and Computer Applications 12, 301–310 (2010)
13. Subashini, S., Kavitha, V.: A Survey on Security Issues in Service Delivery Models of Cloud Computing. Journal of Network and Computer Applications 34, 1–11 (2010)

# Multi-dimensional Grid Quorum Consensus for High Capacity and Availability in a Replica Control Protocol

Vinit Kumar[1] and Ajay Agarwal[2]

[1] Associate Professor with Krishna Engineering Collage, Ghaziabad, India
vinitbaghel@gmail.com
[2] Professor with Krishna Institute of Engineering & Technology, Ghaziabad, India
ajay.aagar@gmail.com

**Abstract.** In distributed systems it is often necessary to provide coordination among the multiple concurrent processes to tolerate the contention, periods of asynchrony and a number of failures. Quorum systems provide a decentralized approach for such coordination. In this paper, we propose a replica control protocol by using a Multi-dimensional-grid-quorum-consensus, which is the generalization of a read-one-write-all (ROWA) protocol, Grid quorum consensus protocol and D-Space quorum consensus protocol. Provides very high read availability and read capacity while maintaining the reconfigurable levels of write availability and fault tolerance.

**Keywords:** Data Replication, Distributed systems, Quorum consensus, Quorum systems.

## 1 Introduction

Quorum systems are the basic tools that provide coordination among multiple concurrent processes in various distributed applications. Quorum systems are attractive because they provide a decentralized approach that tolerates failures. Quorum systems are also interesting for large-scale systems, because it is possible to make the size of quorums increase much slower than the system size. Therefore, it is possible to provide very high availability with a reasonable communication cost.

In general, the commonly accepted observations are that read operations to data occurs more frequently than write operations. Therefore, we require a system that has very high read availability, very high read capacity and improved access efficiency under acceptable write availability conditions. In order to achieve all that we are required to replicate the data at multiple sites. However, the cost of maintaining data consistency counterbalances the benefits of data replication.

Read-one-write-all (ROWA) 1 is the simplest protocol for data replication in which read operations read any copy, while write operations write all the copies. This protocol is very attractive for very high read availability, and very high read capacity. However, this is at the cost of very low write availability. All copies need to be

A. Mantri et al. (Eds.): HPAGC 2011, CCIS 169, pp. 67–78, 2011.

operational for a write to proceed. As the number of failures increases, this protocol is not the right choice to handle the data replication.

To handle the number of failures of the replica sites efficiently, we use the various Quorum consensus protocols 2-10. In quorum consensus, for read operations, we require to read all the copies of the read-quorum and select the latest version among them. While for the write operations, we require to write all copies in the write-quorum with version number. A write-quorum intersects each other write-quorum and each read-quorum, therefore at any time only one write operation is possible even in concurrent scenario. These protocols increases the write availability and fault tolerance but at the cost of degradation of the read availability and read capacity.

In this paper, we present a multi-dimensional-grid-quorum-consensus protocol, for managing replicated data. This protocol provides the very high read availability and very high capacity under the acceptable write availability and fault tolerance. If the reliability of replica sites varies, then we can reconfigure the quorum structure of this protocol very smoothly to achieve the desired write availability.

Organization of the paper is as follows, Next section describes the related work. Section 3 presents the multi dimensional Grid-quorum-systems (MDQS). We describe the system model of the replicated database in Section 4. Section five presents the replica control protocol and Section 6 analyzes the performance of MDQS. The final section is a conclusion.

## 2 Related Work

R.H. Thomas 2 first introduced quorum system in 1979 as majority quorum systems. The quorum size in majority quorum system is $\lceil (n+1)/2 \rceil$ that means majority of processes give the permission for the work to do. D.K. Gifford 3 has introduced the weighed voting in the same year that was the more generalization of the majority quorum systems. We can construct Majority quorum systems by assigning the equal weight to each process. If we assign more votes to one process means we are moving toward centralize system rather than distributed system. H. Garcia Molina has contributed in weighted voting systems in 1985 by explaining how to assign votes in distributed systems 11.

Based on finite projective planes, Mamoru Maekawa has introduced the FPP quorum systems 12 in 1985, which cause the reduction in quorum size up to O ($\sqrt{n}$). This class of quorum systems was fully distributed. The size of quorum in Maekawa's FPP quorum systems is least among fully distributed quorums but the availability is very low in these quorum systems. In the same paper, Maekawa has presented the Grid-quorum-system. S. Y. Cheung, M. Ahamad, and M. H. Ammar further explained this in detail in 1990 for replicated data 6. In such cases, the quorum size is $O\ (2\sqrt{n})$.

D. Agrawal and A. El-Abbadi presented the tree based quorum systems 4 in 1991, which organizes the system nodes into a binary tree structure. Each quorum consists of all the nodes of a path from the root to a leaf node, which uses $\lceil \log(n+1) \rceil$ nodes in a quorum if all such nodes are up. The performance of such quorum system gracefully degrades when failure nodes reaches to a worst case of $\lceil (n+1)/2 \rceil$ nodes. The quorum construction is not fully distributed and load of the quorum is very high.

Akhil Kumar presented the hierarchical quorum systems 5 in 1991 that constructs an n-array tree, where elements are the leaves. A quorum is formed recursively from the root node, obtaining a quorum in a majority of sub-trees. His system has minimum quorum size of $n^{0.63}$. He discusses informally and briefly a restricted definition of cost of failures. His system has a constant cost of failures.

Y. Marcus and D. Peleg presented the wheel Quorum Systems 13 in 1992, wherein the smallest quorum size is two, and the largest quorum size is n-1.

Wu and Belford presented Triangular lattice Quorum systems 14 in 1992, in which the smallest quorum size is of size $O(\sqrt{(2n)})$.

Naor and Wool 15 studied the load of many quorum systems. They presented four quorum systems in 1994 as path quorum systems, B-Grid-quorum-systems, SC-Grid-quorum-systems, And/OR quorum systems with high load and high availability. The smallest quorum size of path quorum system is $O(\sqrt{(2n)})$ and for others it is $O(\sqrt{n})$.

Chang and Chang presented triangular mesh Quorum systems 16 in 1995. The quorum size is $O(\sqrt{(2n)})$.

Cho and Wang presented Triangular Grid-quorum-systems 17 in 1996 having smallest quorum is of $O(2\sqrt{n})$ size.

R.A. Bazzi proposed the planar quorum systems 18 in 1996 based on the connected regions in planar graphs. A particular case is the triangle lattice, consists a large triangle into smaller triangular faces. The analysis shows that both the load and system availability are better than that of the Paths system proposed in 19.

Divyakant Agrawal, Omer Egecioglu, and Amr El Abbadi have presented the billiard quorum systems 20 in 1997. Such systems use the modified grid, and path that resemble billiard ball path instead of horizontal and vertical line segments of rows and columns as the grid scheme. The size of these quorums is $\sqrt{2}\sqrt{n}$.

Peleg and Wool proposed the Crumbling Walls quorum systems 21 in 1997, in which the nodes are in rows of varying widths, and a quorum consists of one full row plus one representative from every row below the full row. CWlog system has a small quorum size and a relatively high availability.

S. D. Lang and L. J. Mao have presented the torus quorum systems 22 in 1998; quorum size is $O(\sqrt{(2n)})$.

A.Fu, T.Lau, G.Ng, and M.H. Wong has presented the Hypercube Quorum systems 8 in 1998, with quorum size $O(n^{0.63})$.

Fu, Wong, and Wong have presented the Diamond Quorum systems [9] in 1999. The smallest read-quorum size is two, and the smallest write-quorum size is $O(\sqrt{(2n)})$.

B. Silaghi, P. Keleher, and B. Bhattacharjee have presented the D-Space quorum consensus protocol 10 in 2004. These researchers have shown that implementing a read-few write-many replica protocol using d-space quorums yields both superior operational availability, as well as message complexity, to the hierarchical quorum consensus method.

Quorum systems that have an asymmetry in their read and write-quorum sizes are more suitable for data replication. Therefore, majority quorum systems 2, tree quorum systems 4, hierarchical quorum systems 5, Grid-quorum-system 6, hypercube quorum systems 8, diamond quorum systems 8, D-space-quorum-systems10 are quite suitable for data replication because these quorum systems have an asymmetry in their read and write-quorum sizes.

## 3 Multi Dimensional Grid-Quorum-Systems

In this section, we have presented the multi dimensional Grid-quorum-system as follows:

**Definition 3.1.** A Multi Dimensional Grid G is a discrete space of k dimensions, and each dimension $d_j$ (where $1 \le j \le k$) in G is indexed from 1 to m, i.e. $N = m^k$ (where N are the total number of elements). Let the index set of each dimension $d_j$ be $D_j = \{1,\ldots, m\}$, (where $1 \le j \le k$). Each element of G can be represented by a vector $\{x_{1,\ldots\ldots}, x_k\}$, where $x_j \in D_j$, $1 \le j \le k$.

**Definition 3.2.** A Multi Dimensional Grid-quorum-system Q is a collection of read-quorums and write-quorums such that $Q = R \cup W$ where R is a set of read-quorums. In addition, W is set of write-quorums. $rq_i \cap wq_j \neq \emptyset$ for every i & j, $rq_i \in R$ & $wq_j \in W$ as well as $wq_i \cap wq_j \neq \emptyset$ for every i & j, $wq_i \in W$ and $wq_j \in W$ where $rq_j$, $wq_j$ for every j can be defined as the collection of elements of G such that

Firstly, we choose s dimensions out of total k dimensions and let S be the set of these chosen dimensions. In addition, let T be the set of remaining k-s dimensions.

We define sub space $A_i$ to be:

$$A_i=\{(x_{1,\ldots,}x_k) | (x_{1,\ldots,}x_k) \in G \wedge (d_j \in S \wedge \forall x_j \in D_j \wedge \forall j \in \{1,..,k\}) \wedge ( d_j \in T \wedge \exists x_j \in D_j \wedge \forall j \in \{1,.., k\}) \} \text{ where } 1 \le i \le m^{k-s} . \tag{1}$$

We call this sub space $A_i$ as read-hyper-plane. In addition, there could be $m^{k-s}$ such read-hyper-planes. Moreover, each read-hyper-plane comprises $m^s$ elements. $A_i \cap A_j = \emptyset$ for any $i \neq j$. Let set of these read-hyper-planes be the Pr.

We define sub space $B_i$ to be:

$$B_i= \{(x_{1,\ldots,}x_k) | (x_{1,\ldots,}x_k) \in G \wedge (d_j \in T \wedge \forall x_j \in D_j \wedge \forall j \in \{1,..k\}) \wedge (d_j \in S \wedge \exists x_j \in Dj \wedge \forall j \in \{1,.., k\}) \} \text{ where } 1 \le i \le m^s . \tag{2}$$

We call this space B as write-hype-planes. In addition, there could be $m^s$ such write-hype-planes. Moreover, each write-hyper-plane comprises $m^{k-s}$ elements. $B_i \cap B_j = \emptyset$ for any $i \neq j$. Let set of these write-hype-planes be the Pw.

Now we choose t such that $1 \le t \le m^s$

Any read-quorum $rq \in R$ is as follows

$$rq= \{A | \exists A \in Pr\} - \{a | a \subseteq A \wedge |a|=t-1\} . \tag{3}$$

Any write-quorum $wq \in W$ is as follows

$$wq= \{A | \exists A \in Pr\} \cup \{B | B \subseteq Pw \wedge |B|=t\} . \tag{4}$$

**Lemma 1:** Each element of read-quorum exists at different write-hype-planes.

Proof: Let there are two elements, on the read-hyper-plane then we now required checking that they may be at same write-hyper-plane or not.

From the definition of read-hyper-plane $A_i$, Refer to "(1)," it is clear that for each dimension $d_j$: $d_j \in T$ we have only one index value that is $\exists x_j$: $\in D_j$ for one read-hyper-plane. Only index values may change on the rest of the dimensions dj: dj ∈ S. So, for the two elements on the read-hyper-plane may have the different index values at dj: dj ∈ S dimensions.

Let these two elements are also on one write-hyper-plane. However, from the definition of write-hyper-plane $B_i$, Refer to "(2)," for the single index values in the dimension $d_j$: $d_j \in T$, we cannot have different index values in the dimension of dj: dj ∈ S. that is a contradiction. Therefore, each element of read-hyper-plane exists at different write-hype-planes.

Since each of the element of read-quorum lies on read-hyper-plane, therefore each element of read-quorum exists at different write-hype-planes ■

**Theorem 1:** In Multidimensional-grid-quorum-system Q, any read-quorum has non-empty intersection with any write-quorum.

Proof: here we required to prove that rq ∩ wq $\neq \emptyset$ where rq ∈ R & wq ∈ W. Refer to "(4)," we have used total t write-hype-planes for write-quorum. Refer to "(3)," Since $|rq| = |A|-|a|$, $|A|= m^s$, and $|a|=t-1$ Therefore $|rq|=m^s-t+1$. As shown in lemma 1 above, each element of read-quorum exists at different write-hype-planes; therefore elements in read-quorum are at total $m^s - t+1$ write-hype-planes.

Total write-hype-planes common in both read and write-quorums= [(total write-hype-planes)-((total write-hype-planes used by write-quorum) + (total write-hype-planes used by read-quorum))] . (5)

Total write-hype-planes common in both read and write-quorums= $[(m^s)-((t) + (m^s - t+1))] = 1$ . (6)

Refer to "(4)," If write-quorum is having a write-hyper-plane then all the elements of write-hyper-plane will be in write-quorum.

Since, we have observed that read-quorum and write-quorums have one write-hyper-plane common. Write-quorum has all the elements of it and as by lemma 1; read-quorum has one element of it. Therefore, read and write-quorums have one element common. That is rq ∩ wq $\neq \emptyset$. Hence, In Multidimensional-grid-quorum-system Q, any read-quorum has non-empty intersection with any write-quorum. ■

**Theorem 2:** In Multidimensional-grid-quorum-system Q any two write-quorums have non-empty intersection.

Proof: here we required to prove that $wq_i \cap wq_j \neq \emptyset$ where $wq_i, wq_j \in W$. Refer to "(4)," it is clear that there exists a read-hyper-plane, which one is the subset of write-quorum. That is $\{A|\exists\ A \in Pr\} \subseteq wq$. Refer to "(3)," there exists a read-quorum, which one is the subset of a given read-hyper-plane. That is $\{rq|\exists\ rq \in A\} \subseteq \{A|\exists\ A \in Pr\}$. From this observation, we can say that $\{rq|\exists\ rq \in A\} \subseteq \{A|\exists\ A \in Pr\} \subseteq wq$. This shows that every write-quorum have some read-quorums.

For $wq_i \in W$, $\{rq|\exists\ rq \in A\} \subseteq \{A|\exists\ A \in Pr\} \subseteq wq_i$. Since, by theorem 1, we can say, $rq \cap wq_j \neq \emptyset$.

$rq \subseteq wq_i, rq \cap wq_j \neq \emptyset \Rightarrow wq_i \cap wq_j \neq \emptyset$.

Hence, In Multidimensional-grid-quorum-system Q any two write-quorums have non-empty intersection. ■

**Theorem 3:** Read-one-write-all (ROWA) structure is a special case of the Multidimensional-grid-quorum-system.

Proof: In multidimensional grid quorum if we choose t= $m^s$, then, Refer to "(4)," write-quorum includes all the write-hype-planes. Refer to "(2)," That means write-quorum wq have all the elements of the G. |wq|=N.

Refer to "(2)," each read-hyper-plane comprises $m^s$ elements. Again Refer to "(3)," $|rq|=|A|-t+1= m^s - m^s +1=1$.

Hence, read-one-write-all (ROWA) structure is a special case of the Multidimensional-grid-quorum-system. ■

**Theorem 4:** Grid-quorum-system is a special case of the Multidimensional-grid-quorum-system, where read-quorum has all the elements of any column and write-quorum has all the elements of any row and any column.

Proof: In multidimensional-grid-quorum-system if we choose k=2, and t=1, then G turn into a two dimensional grid. Refer to "(3), and (4)," quorums generated by this construct a Grid-quorum-systems.

Hence, Grid-quorum-system is a special case of the Multidimensional-grid-quorum-system, where read-quorum has all the elements of any column and write-quorum has all the elements of any row and any column. ■

**Theorem 5:** D-space-quorum-system is a special case of the Multidimensional-grid-quorum-system, where read-quorum has all the elements of any one read-hyper plane and write-quorum has all the elements of any one read-hyper plane and any one write-hyper plane.

Proof: In multidimensional grid quorum if we choose t=1, then. Refer to "(3), and (4)," quorums generated are same as D-space-quorum-systems.

Hence, D-space-quorum-system is a special case of the Multidimensional-grid-quorum-system, where read-quorum has all the elements of any one read-hyper plane and write-quorum has all the elements of any one read-hyper plane and any one write-hyper plane. ■

## 4 System Model

A replicated database consists of a group of n sites, which communicate by exchanging messages. Sites are fail-stop, and site failures can be detected. Sites are connected to each other via a reliable network. We consider a crash-recovery model where sites can recover and re-join the system after synchronizing their state with the one of the running replicas.

The database is fully replicated, and thus, each site contains a copy of the database. We assume that all sites are homogeneous and each site is able to execute t operations per second. Clients interact with the database by issuing either read or write operations. For replicated databases, the correctness criterion is one-copy-serializability. In this criterion, each copy must appear as a single logical copy and the execution of concurrent write operations must be equivalent to a serial execution over all the physical copies. While read operations can be executed concurrently.

A client submits read or write operation to one of the sites in the system, and this site coordinates its actions with the rest of the system. An operation is called *local* at the site it is submitted to, and *remote* at the other sites. We assume that all sites receive the same amount of local operations.

## 5 Replica Control Protocol

This protocol uses the version numbers that remain associated with each copy of data to identify the latest update, and we use locking to ensure mutual exclusion.

If a client issues read or write, operation to a replica site than this local site picks a read-quorum or write-quorum as per the operation. In addition, try to lock all the sites among the respective quorum. If we fail to lock all the sites, we select another respective quorum randomly or deterministically. Again, try to lock all sites among the quorum. We repeat this process until we get all the locks.

After a very long time even if we fail to lock any quorum means the system is unavailable due to massive sites failure. A reconfiguration of quorum structure is required. After that, again we repeat process of getting locks. If we get all locks, than following steps are required to follow.

A read operation will read all the copies in a read-quorum along with its version numbers. Now we require compare the version numbers, and a copy with the highest version number is the result of a read operation.

A write operation will read the version number of all copies in a write-quorum. Now we compare the version numbers, and select the highest among them. After that, we get new highest version number by incrementing the selected highest version number. Finally, we write all the copies in a write-quorum along with the new highest version number.

## 6 Performance Analysis of the Protocol

We would like to analyze the protocol, under various parameters; as quorum size, fault tolerance, read availability, read capacity, and write availability.

### 6.1 Quorum Size

In order to perform a read or a write operation, the number of messages required depends upon the quorum size. Access efficiency of a read or a write operation may be improved by having low quorum size.

#### 6.1.1 Read-Quorum Size

Refer to "(3)," $|rq| = |A|-|a|$, $|A|= m^s$, and $|a|=t-1$, Therefore, $|rq|=m^s-t+1$. Where $1 \le t \le m^s$, Minimum read-quorum size is one, when $t= m^s$, and maximum read-quorum size is $m^s$, when $t=1$.

### 6.1.2 Write-Quorum Size

Refer to "(1), (2), and (4)," |wq| = |A|+t*|B|-t, |A|= $m^s$, and |B|= $m^{k-s}$, Therefore, |wq|=$m^s$+t* $m^{k-s}$ -t. Where $1 \leq t \leq m^s$, minimum write-quorum size is $m^s$+$m^{k-s}$ -1 when t= 1 and maximum write-quorum size is N, when t= $m^s$.

## 6.2 Fault Tolerance

Maximum number of faults that can be tolerated by the system is called fault tolerance.

If the elements of one write-quorum are alive and others are faulty, even then write and read operations may be possible. Since write-quorum also has elements of one read-hyper-plane, therefore, if write-quorum is alive then read-quorum will also be alive.

Fault tolerance= (total number of elements - elements of one write-quorum).

Therefore, fault tolerance= N- ($m^s$+t* $m^{k-s}$ –t).

## 6.3 Read Capacity Analysis

We define Read-capacity as the maximum number of concurrent read operations or maximum number of disjoint read-quorums.

Refer to "(1)," $A_i \cap A_j = \emptyset$ for any $i \neq j$. In addition, refer to "(3)," any read quorum is the subset of some read-hyper-plane $A_i$. By all this, we can conclude that maximum number of disjoint read-quorum= (Total number of read-hyper-planes) * (Total number of disjoint read-quorums in a read-hyper-plane). Total number of read-hyper-planes= $m^{k-s}$.

Maximum number of disjoint read-quorums in a read-hyper-plane $= \frac{\text{Toal number of elements in a read-hyper-plane}}{\text{Toal number of elements in a read-quorum}} = \left\lfloor \frac{m^s}{(m^s)-t+1} \right\rfloor$

$$\text{Maximum number of disjoint read} - \text{quorum} = m^{k-s} * \left( \left\lfloor \frac{m^s}{(m^s) - t + 1} \right\rfloor \right)$$

Read capacity= Maximum number of disjoint read-quorum

$$Read\ capacity = m^{k-s} * \left( \left\lfloor \frac{m^s}{(m^s) - t + 1} \right\rfloor \right)$$

## 6.4 Read Availability Analysis

We define the read availability as the probability that the system is in a state that would allow the read operation to succeed.

Let $p$ be the availability of a node. I.e. $p$ is the probability that a node is in a ready state to perform an operation. We represent the probability of X as $\mathbb{p}$(X). Moreover, $\alpha_G^R$ be the representation of Read availability.

$$\alpha_G^R = \mathbb{p}(at\ least\ one\ operational\ read\ quorum\ exists)$$

There are total $m^{k-s}$ read-hyper-planes in G, which are all disjoints. All the elements of one read-quorum, lies on one read-hyper-plane.

$$\mathbb{p}(at\ least\ one\ operational\ read\ quorum\ exists\ in\ a\ A_i) = \sum_{l=m^s-t+1} \binom{m^s}{l} p^l\ (1-p)^{(m^s-l)}$$

$$\mathbb{p}(no\ operational\ read\ quorum\ exists\ in\ a\ A_i\ ) = \left(1 - \sum_{l=m^s-t+1} \binom{m^s}{l} p^l\ (1-p)^{(m^s-l)}\right)$$

$$\mathbb{p}(no\ operational\ read\ quorum\ exists\ in\ any\ A_i) = \left(\left(1 - \textstyle\sum_{l=m^s-t+1}\binom{m^s}{l} p^l\ (1-p)^{(m^s-l)}\right)\right)^{m^{k-s}}$$

$$\mathbb{p}(at\ least\ one\ operational\ read\ quorum\ exists) = 1 - \mathbb{p}(no\ operational\ read\ quorum\ exists\ in\ any\ A_i)$$

So, $\alpha_G^R = \left(1 - \left(\left(1 - \sum_{l=m^s-t+1}\binom{m^s}{l} p^l\ (1-\ \ p)^{(m^s-l)}\right)\right)^{m^{k-s}}\right)$

## 6.5 Write Availability Analysis

We define the write availability as the probability that the system is in a state that would allow the write operation to succeed. Let $\alpha_G^W$ be the representation of Write availability.

$$\alpha_G^W = \mathbb{p}(at\ least\ one\ operational\ write\ quorum\ exists)$$

There are total $m^s$ write-hype-planes in G, which are all disjoint. Also have the $m^{k-s}$ disjoint read-hyper planes in G. for construction of a write-quorum, we require at least t operational write-hyper planes and at least one operational read-hyper plane.

Write operation is successful if we can access all the sites of one read-hyper plane and $m^{k-s} - t$ sites on the t write hyper-plane. Since, t sites are common among a read-hyper plane and t write-hyper planes.

More generally, write operation is successful if we can access all the sites on the $x$ read hyper-planes, where $1 \leq x \leq m^{k-s}$ and $y * (m^{k-s} - x)$ sites on the $y$ write hyper-plane, where $t \leq y \leq m^s - 1$.

Let $Y = y * (m^{k-s} - x)$ be the sites on the $y$ write hyper-plane that we can access, where $t \leq y \leq m^s - 1$.

Let $Y_1 = y * (m^{k-s} - x)$ be the sites on the $y$ write hyper-plane that we can access, where$0 \leq y \leq t - 1$.

Let $Y_2 = y * (m^{k-s} - x)$ be the sites on the $y$ write hyper-plane that we can be access, where$y = m^s$.

$$\mathbb{p}(\,Y) = 1 - (\mathbb{p}(Y_1) + \mathbb{p}(Y_2))$$

$$\mathbb{p}(Y_1) = \sum_{y=0}^{t-1} \binom{m^s}{y} p^{(y*(m^{k-s}-x))}(1-p)^{((m^k-m^s)-(y*(m^{k-s}-x)))}$$

$$\mathbb{p}(Y_2) = p^{m^k}$$

$$\mathbb{p}(Y) = 1 - \left( \sum_{y=0}^{t-1} \binom{m^s}{y} p^{\left(y*(m^{k-s}-x)\right)} (1-p)^{\left((m^k - m^s) - \left(y*(m^{k-s}-x)\right)\right)} \right) - p^{m^k}$$

Let $X$ =all $(x * m^s)$ sites on the on the x read hyper-plane can be accessed.

$$\mathbb{p}(X) = p^{(x*m^s)}$$

$$\alpha_G^W = \sum_{x=1}^{m^{k-s}} \mathbb{p}(X) * \mathbb{p}(Y)$$

$$\alpha_G^W = \sum_{x=1}^{m^{k-s}} p^{(x*m^s)} * \left( 1 - \left( \sum_{y=0}^{t-1} \binom{m^s}{y} p^{\left(y*(m^{k-s}-x)\right)} (1 - p)^{\left((m^k - m^s) - \left(y*(m^{k-s}-x)\right)\right)} \right) - p^{m^k} \right)$$

## 7 Conclusion

In this paper, we present a multi-dimensional-grid-quorum-consensus protocol. This is a generalization of the read-one-write-all (ROWA), Grid quorum consensus and D-space quorum consensus protocol. All these three given consensus protocols are the special cases of the Multi-dimensional-grid-quorum-consensus protocol, so, this protocol have all those merits which they have.

For the given fault-tolerance level of write operations, the read-quorum size is significantly low, therefore this protocol provides good read capacity, read availability, and read efficiency.

In addition to this, protocol is highly reconfigurable as compared to the above-mentioned protocols. This benefits as, we can smoothly increase the write availability and fault tolerance by the reconfiguration in the quorum structure. This is very much desirable because the increase in write availability and fault tolerance comes at the cost of degradation of the read availability and read capacity. For the read-few and write-many access approaches, read availability and read capacity are very important parameters. These parameters should remain high as much as possible. Therefore, in this sense our protocol is optimal. This approach in designing distributed systems is desirable since it provides fault-tolerance without much imposing unnecessary costs on the failure-free mode of operations.

## References


1. Ahamad, M., Ammar, M., Cheung, S.: Replicated data management in distributed systems. In: Casavant, T.L., Singhal, M. (eds.) Readings in Distributed Computing Systems, pp. 572–591. IEEE Computer Society Press, Los Alamitos (1994)
2. Thomas, R.H.: A majority consensus approach to concurrency control for multiple copy database. ACM Trans. Database Systems 4(2), 180–209 (1979)
3. Gifford, D.K.: Weighted voting for replicated data. In: Proc. 7th ACM Symp. on Operating Systems Principles, pp. 150–162 (December 1979)
4. Agrawal, D., El-Abbadi, A.: An efficient and fault-tolerant solution for distributed mutual exclusion. ACM Trans. Computer Systems 9(1), 1–20 (1991)
5. Kumar, A.: Hierarchical quorum consensus, A new algorithm for managing replicated data. IEEE Trans. Computers 40(9), 996–1004 (1991)
6. Cheung, S.Y., Ahamad, M., Ammar, M.H.: The grid protocol: a high performance scheme for maintaining replicated data. In: Proceedings of the 6th International Conference on Data Engineering (1990)
7. Kumar, A., Cheung, S.Y.: A high availability √n hierarchical grid algorithm for replicated data. Inform. Process. Lett. 40, 311–316 (1991)
8. Fu, A., Lau, T., Ng, G., Wong, M.H.: Hypercube quorum consensus for mutual exclusion and replicated data management. An International Journal Computers and Mathematics with Applications 36(5), 45–59 (1998)
9. Fu, A.W.-C., Wong, Y.S., Wong, M.H.: Diamond Quorum Consensus for High Capacity and Efficient in a Replicated Data System. Distributed and Parallel Databases, pp. 1–25 (1999)
10. Silaghi, B., Keleher, P., Bhattacharjee, B.: Multi-Dimensional Quorum Sets for Read-Few Write-Many Replica Control Protocols. In: Proceedings of the 4th International Workshop on Global and Peer-to-Peer Computing (2004)
11. Garcia-Molina, H., Barbara, D.: How to assign votes in a distributed system. J. ACM 32(4) (1985)
12. Maekawa, M.: A √n algorithm for mutual exclusion in decentralized systems. ACM Trans. Computer Systems 3(2), 145–159 (1985)
13. Marcus, Y., Peleg, D.: Construction Methods for Quorum Systems, Tech. Report CS92 {33, The Weizmann Institute of Science, Rehovot, Israel (1992)
14. Wu, C., Belford, G.: The triangular lattice protocol: a high fault tolerant protocol for replicated data. In: Proc. 11th IEEE Symposi. Reliable and Distributed Systems, pp. 66–73 (1992)
15. Naor, M., Wool, A.: The load, capacity and availability of quorum systems. In: Proc. 35th IEEE Symposium on Foundations of Computer Science, pp. 214–225 (1994)
16. Chang, Y., Chang, Y.: A fault-tolerant triangular mesh protocol for distributed mutual exclusion. In: Proc. 7th IEEE Symp. on Parallel and Distributed Processing, pp. 694–701 (October 1995)
17. Cho, C.H., Wang, J.T.: Triangular grid protocol, an efficient scheme for replica control with uniform access quorums. In: Proc. 2nd Internat. Euro-Par Conf., Lyon, France (August 1996)
18. Bazzi, R.A.: Planar quorums, in Proceedings of the 10th International Workshop on Distributed Algorithms, Bologna, Italy, 1996, Lecture Notes in Comput. In: Babaoğlu, Ö., Marzullo, K. (eds.) WDAG 1996. LNCS, vol. 1151, pp. 251–268. Springer, Heidelberg (1996)

19. Naor, M., Wool, A.: The load, capacity, and availability of quorum systems. SIAM Journal on Computing 27(2), 423–447 (1998)
20. Agrawal, D., Egecioglu, O., Abbadi, A.E.: Billiard quorums on the grid. Information Processing Letters 64(1), 9–16 (1997)
21. Peleg, D., Wool, A.: Crumbling walls: a class of practical and efficient quorum systems. Distrib. Comput. 10(2), 87–97 (1997)
22. Lang, S.D., Mao, L.J.: A Torus Quorum Protocol for Distributed Mutual Exclusion. In: Proc. of the 10th Int'l Conf. on Parallel and Distributed Computing and Systems, pp. 635–638 (1998)
23. Neilsen, M.L.: Quorum structures in distributed systems, Ph.D. Thesis, Department of Computer and Information Sciences, Kansas State University (1992)
24. Ibaraki, T., Kameda, T.: A theory of coteries: mutual exclusion in distributed systems. IEEE Trans. Parallel Distrib. Systems 4(7), 749–779 (1993)
25. Kumar, A., Rabinovich, M., Sinha, R.: A performance study of general grid structures for replicated data. In: Proc. Internat. Conf. Distributed Computing Systems, pp. 178–185 (May 1993)
26. Peleg, D., Wool, A.: The availability of quorum systems. Inform. and Comput. 123(2), 210–223 (1995)
27. Ng, W.K., Ravi Shankar, C.V.: Coterie templates: A new quorum construction method. In: Proceedings of the 15th International Conference on Distributed Computing Systems, Vancouver, Canada, pp. 92–99 (May 1995)
28. Peleg, D., Wool, A.: How to be an efficient snoop, or the probe complexity of quorum systems. In: Proc. 15th ACM Symp. Principles of Distributed Computing, pp. 290–299 (1996)
29. Holzman, R., Marcus, Y., Peleg, D.: Load balancing in quorum systems. SIAM J. Discrete Math. 10, 223–245 (1997)
30. Luk, W.-S., Wong, T.-T.: Two new quorum based algorithms for distributed mutual exclusion. In: Proc. 17th International Conference on Distributed Computing Systems, pp. 100–106. IEEE, Los Alamitos (1997)
31. Wool, A.: Quorum Systems in Replicated Databases: Science or Fiction? IEEE Technical Committee on Data Eng. 21(4), 3–11 (1998)
32. Amir, Y., Wool, A.: Optimal availability quorum systems: Theory and practice. Information Processing Letters 65(5), 223–228 (1998)
33. Naor, M., Wieder, U.: Scalable and dynamic quorum systems. In: Proceedings of the ACM Symposium on Principles of Distributed Computing (2003)

# Efficient Task Scheduling Algorithms for Cloud Computing Environment

S. Sindhu[1] and Saswati Mukherjee[2]

[1] Research Scholar,
Department of Information Science and Technology
sindhu.nss@gmail.com
[2] Professor
College of Engineering,
Guindy, Anna university, Chennai
msaswati@yahoo.com

**Abstract.** Cloud Computing refers to the use of computing, platform, software, as a service. It's a form of utility computing where the customer need not own the necessary infrastructure and pay for only what they use. Computing resources are delivered as virtual machines. In such a scenario, task scheduling algorithms play an important role where the aim is to schedule the tasks effectively so as to reduce the turnaround time and improve resource utilization. This paper presents two scheduling algorithms for scheduling tasks taking into consideration their computational complexity and computing capacity of processing elements. CloudSim toolkit is used for experimentation. Experimental results show that proposed algorithms exhibit good performance under heavy loads.

**Keywords:** Cloud computing, IaaS, Private Cloud, Task Scheduling.

## 1 Introduction

Cloud computing is the latest buzzword in the IT industry. It is an emerging computing paradigm with foundations of grid computing, utility computing, service oriented architecture, virtualization and web 2.0. The user can access all required hardware, software, platform, applications, infrastructure and storage with the ownership of just an internet connection. "A Cloud is a type of parallel and distributed system consisting of a collection of inter-connected and virtualized computers that are dynamically provisioned and presented as one or more unified computing resources based on service-level agreements established through negotiation between the service provider and consumers" [1]. Some of the applications of cloud computing are on-line gaming, social networking, scientific applications. One of the key issues in public clouds are that of security and privacy [2]. In public clouds data centers hold end-users data which otherwise would have been stored on their own computers. Hence there is a growing demand for private clouds. A private cloud is one which is owned and operated within the firewalls of an organization. It allows an organization to manage its internal IT infrastructure effectively and provide services to its local users [3]. A

A. Mantri et al. (Eds.): HPAGC 2011, CCIS 169, pp. 79–83, 2011.

private cloud should support efficient resource allocation policies that adhere to the specific requirements of an organization (high availability, reliability, QoS). Cloud computing relies heavily on virtualization. Clouds are virtual clusters. Hence efficient scheduling of tasks and virtual machines across the various heterogenous physical machines is a crucial task, especially in a private cloud environment where the resources are limited. Very limited research have been done so far in scheduling, in a private cloud environment, except for some generic algorithms being adopted in various tools like Eucalyptus and OpenNebula, [4][5], that are being used as infrastructure to build private and hybrid clouds. We focus on the task scheduling problem in a private cloud environment in this paper. Some performance metrics like high throughput, low response time, minimum makespan and flowtime are the conventional metrics used for task scheduling. Here we present two scheduling algorithms for scheduling tasks in a private cloud environment, where the main aim is to obtain a minimum makespan. We have used the CloudSim simulator to implement our proposed algorithms since it provides the necessary environment to test the scheduling algorithms in a repeated and controlled environment.

The rest of the paper is organized as follows: Section 2 presents the related work. Section 3 describes the proposed scheduling algorithms. Section 4 briefs the experimental setup, results and discussion. Finally, Section 5 gives the conclusion and future work.

## 2 Related Work

Scheduling policies in a cloud environment vary depending on the deployment model of the cloud. This section provides a brief review of some related work done in scheduling in a cloud. Ye Hu, Johnny Wong, Gabriel Iszlai, Marin Litoiu [6] have proposed a new probability dependent priority algorithm to determine the minimum number of servers required to execute the jobs, considering jobs in two different classes, such that the SLAs of both job classes are met. In [7] an optimized algorithm for task scheduling based on Activity Based Costing, is presented, that selects a set of resources, to schedule the tasks, such that the profit is maximized. A heuristic method to schedule bag-of-tasks (tasks with short execution time and no dependencies) in a cloud is presented in [8] so that the number of virtual machines to execute all the tasks, within the budget, is minimum and at the same time speedup is maximum. Hybrid cloud is a model which combines a private cloud and public cloud. That is, during peak load when there are not sufficient resources to execute a task in a private cloud, outsource the same to a public cloud provider and get it done. An optimal scheduling policy based on linear programming, to outsource deadline constrained workloads in a hybrid cloud scenario is proposed in [9]. The scheduling policy used in Eucalyptus [4] are First Fit and Round Robin. In OpenNebula, Haizea can be used as the scheduler backend, which supports advance reservations in the form of a lease, [5]. But none of the existing algorithms have considered the computational complexity of tasks for scheduling. The heuristic algorithms for scheduling jobs on computational grids [10] provides a framework for our investigation. Our proposed work focuses on scheduling tasks in a private cloud environment.

## 3 Scheduling Algorithms

A good scheduling algorithm should lead to better resource utilization and better system throughput. To formulate the problem, let us consider $C_n$={ $C_1$, $C_2$, …,$C_n$} be 'n' cloudlets, $V_m$={ $V_1$, $V_2$, …,$V_m$} be 'm' virtual machines and $PE_p$={ $PE_1$, $PE_2$,…, $PE_p$} be the processing elements across all the hosts in a datacenter. Makespan is defined as the finishing time of the last job in a set of jobs. Let $CT_c$ be the completion time when the last cloudlet 'c' finishes processing. Our objective is to minimize $CT_c$.

### 3.1 Longest Cloudlet Fastest Processing Element (LCFP)

In this algorithm the computational complexity of the cloudlets are considered while making scheduling decisions. The lengthier cloudlets are mapped to Processing Elements (PEs) having high computational power so as to minimize the makespan. In this algorithm the longer jobs finishes quickly when compared with the FCFS where processing requirement of jobs are not considered while making scheduling decisions.

**Algorithm**

1. Sort the cloudlets in descending order of length.
2. Sort the PEs across all the hosts in descending order of processing power.
3. Create virtual machines in the sorted list of PEs by packing as many VMs as possible in the fastest PE.
4. Map the cloudlets from the sorted list to the created VMs.

### 3.2 Shortest Cloudlet Fastest Processing Element (SCFP)

In this algorithm the shorter cloudlets are mapped to PEs having high computational power so as to reduce flowtime (sum of completion time of a set of jobs) while at the same time taking into consideration that longer jobs are not starved.

## 4 Implementation

### 4.1 Cloudsim Simulation Environment

Cloudsim [11] is a generalized, extensible simulation framework that enables modeling, simulation, and experimentation of Cloud computing infrastructures and application services. In cloudsim, Datacenter component is the main hardware infrastructure that provide services for servicing user requests. A Datacenter is composed of a set of hosts, which are responsible for managing VMs during their life cycles. Host is a component that represents a physical computing node in a Cloud. It is assigned a pre-configured processing capability (expressed in million of instructions per second – MIPS), memory, storage. Virtual Machine Provisioner component is responsible for allocation of application-specific VMs to Hosts in a Cloud-based data center. The default policy implemented by the VM Provisioner is a straightforward policy that allocates a VM to the Host in First-Come-First-Serve (FCFS) basis. In Cloudsim user jobs are called as cloudlets. Each cloudlet is assigned an id, length. Its assumed that, larger the length of cloudlets, the higher is the complexity. Cloudlets can be bound to a Virtual Machine explicitly as specified by user or dynamically at run-time. Cloudlet-Scheduler is the component responsible for mapping the cloudlets to the VMs. The default scheduling policy used by CloudletScheduler is First Come First Served (FCFS) i.e., each cloudlet in the queue is mapped to the list of created virtual machines on a FCFS basis. It does not consider the processing requirement of a job while making the scheduling decision. Also FCFS does not consider the processing requirement of user jobs. It suffers from starvation as lengthy jobs ahead in the queue delay shorter jobs with high response time and also results in poor resource utilization.

### 4.2 Experimentation Results

The algorithms for simulation are implemented on an Intel Dual Core machine with 320GB HDD and 2GB RAM on Cent OS 5.5. The experiments are conducted on a simulated Cloud environment provided by CloudSim. The speed of each processing element is expressed in MIPS (Million Instructions Per Second) and the length of each cloudlet is expressed as the number of instructions to be executed . The simulation environment consists of one Data Center with two hosts having three and two Processing Elements respectively. Each Processing Element is assigned varying computing power (varying MIPS). The algorithms are tested by varying the number of cloudlets from 10 to 50 and also randomly varying the length of cloudlets. Also, the number of VMs used to execute the cloudlets, are varied accordingly. The overall makespan to execute the cloudlets is used as the metric to evaluate the performance of the proposed algorithms. It has been observed that, for smaller number of tasks, all the three algorithms exhibit more or less similar performance since the length of the queued cloudlets are less. But as shown in Fig. 1, as the number of tasks increase, LCFP exhibits better performance when compared to SCFP and FCFS since longer tasks complete faster thereby reducing the makespan.

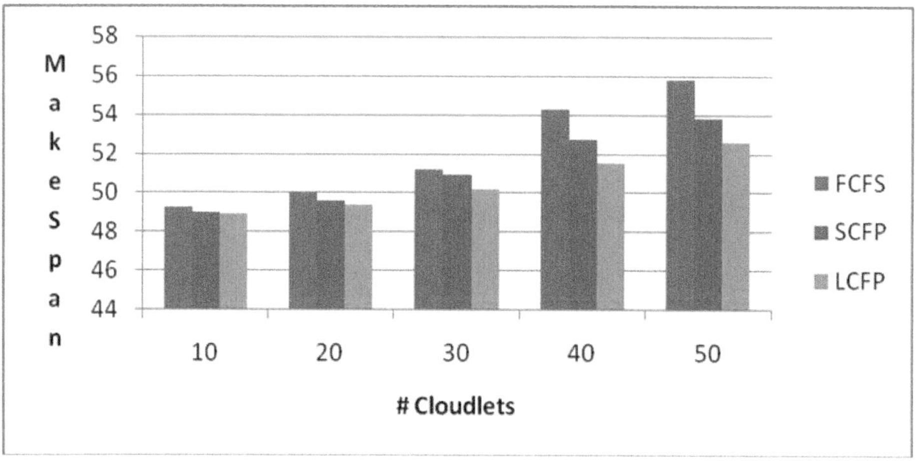


**Fig. 1.** Graph between number of cloudlets submitted and the makespan for FCFS, LCFP and SCFP

## 5 Conclusion

Cloud Computing is an upcoming research area. Scheduling of tasks, so as to improve resource utilization, while at the same time, considering the QoS of tasks, is an important problem, in a private cloud environment, since in a private cloud, the resources are limited. This paper explores the use of two sscheduling algorithms for scheduling tasks which considers the processing requirement of a task and the computational capacity of a resource while making scheduling decisions. In future we would like to experiment with more algorithms that use heuristic methods for scheduling and also consider the priority of tasks.

## References


1. Buyya, R., Yeo, C.S., Venugopal, S., Broberg, J., Brandic, I.: Cloud computing and Emerging IT Platforms: Vision, Hype, and Reality for Delivering Computing as the 5th Utility. In: Future Generation Computer Systems, vol. 25(6), pp. 599–616. Elsevier Science, Amsterdam (2009)
2. Dikaiakos, M.D.: Cloud Computing: Distributed Internet Computing for IT and Scientific Research. IEEE Transactions on Internet Computing 13(5), 10–13 (2009)
3. Sotomayor, B., Montero, R.S., Llorente, I.M., Foster, I.: Virtual Infrastructure Management in Private and Hybrid clouds. IEEE Transactions on Internet Computing 13(5), 14–22 (2009)
4. Nurmi, D., Wolski, R., Grzegorczyk, C., Obertelli, G., So-man, S., Youseff, L., Zagorodnov, D.: The Eucalyptus Open-source Cloud-computing system. In: IEEE International Symposium on Cluster Computing and the Grid, CCGrid 2009 (2009)
5. Open Nebula, `http://www.opennebula.org`
6. Hu, Y., Wong, J., Iszlai, G., Litoiu, M.: Resource provisioning for cloud computing. In: Conference of the Centre for Advanced Studies on Collaborative Research, CASCON 2009, New York (2009)
7. Cao, Q., Wei, Z.-B., Gong, W.-M.: An Optimized Algorithm for Task Scheduling Based On Activity Based Costing In Cloud computing. In: 3rd International Conference on Bioinformatics and Biomedical Engineering, Beijing (2009)
8. Silva, J.N., Veiga, L., Ferreira, P.: Heuristics for Resource Allocation on Utility Computing Infrastructures. In: 6th International Workshop on Middleware for Grid Computing, New York (2008)
9. Van den Bossche, R., Vanmechelen, K., Broeckhove, J.: Cost Optimal Scheduling in Hybrid IaaS Clouds for Deadline Constrained Workloads. In: 3rd IEEE International Conference on Cloud Computing, Miami (July 2010)
10. Abraham, A., Buyya, R., Nath, B.: Nature's Heuristics for Scheduling Jobs on Computational Grids. In: 8th IEEE International Conference on Advanced Computing and Communications, ADCOM 2000, India (December 2000)
11. Calheiros, R.N., Ranjan, R., De Rose, C.A.F., Buyya, R.: CloudSim: A Novel Framework for Modeling and Simulation of Cloud Computing Infrastructures and Services (2009)

# "Cloud Computing: Towards Risk Assessment"

Bharat Chhabra[1] and Bhawna Taneja[2]

[1] Department of Computer Science,
Govt. College Safidon (Jind)
bharat.pnp@gmail.com
[2] Student (M.Tech.), Deptt. Of Computer Sc.,
Kurukshetra Univ. Kurukshetra
tanejabhawna@gmail.com

**Abstract.** Cloud Computing is a revolutionary trend that not only minimizes the processing cost but also enhances the Return on Investment (ROI) despite that several risks are still challenging the paradigm. To ensure the confidentiality, integrity and availability of crucial data in the Cloud, policies and processes must be created to address this expanded reliance on these extended models. Although there is presence of SLA (Service Level Agreement) and NDA (Non-disclosure agreement), "It's not enough for everybody. Some people do want to go deeper." There are lots of questions which are still un-answered. For instance regulatory compliance, location of data centers, its physical and network security etc. Although Cloud Computing is an excellent outsourcing idea, many believe that it also presents a long list of legal and other security concerns. In this paper we are focusing on assessing the various risks present at different layers of Cloud Architecture and their potential consequences and their plausible remedial actions too.

**Keywords:** Cloud Architecture, Non Disclosure Agreement, Service Level Agreements.

## 1 Introduction

Cloud computing [1][2] is the vital turn of the IT industry. It shows the diligence of the engineers towards making a cheaper and convenient availability of IT services tomorrow. Besides buying the extended capabilities as per the need, customers are also insistent on more transparency. They should and want to know where their data is being transported and how it being dealt with. The potential benefits and risks of Cloud Computing however are more deceptive. So, there is a need of assessing the all dimensions of risks; issues related to the qualifications of cloud architects, policy makers, coders and operators; risk-control methods and technical expertise; and the level of testing that's been done to verify that all the features and control processes are functioning as anticipated, and that vendors can identify unanticipated vulnerabilities.

The threat backdrop for a cloud computing environment is significantly different from long-established hosted web services in terms of mitigating tools and technologies. Cloud Computing providers are keenly being targeted, partially because

A. Mantri et al. (Eds.): HPAGC 2011, CCIS 169, pp. 84–91, 2011.

their relatively weak registration systems facilitate illusion, and providers' fraud detection capabilities are limited.

Instead of reaching directly to any conclusion or exposing a random weakness of public cloud, we must categorize the potential problems in all. A cloud may provide services at three levels such as HaaS, PaaS, SaaS. So, the associated risks may be categorized as:

# 2 Types of Risks

Practically, users get hold of computing platforms or IT infrastructures from Computing Clouds and then run their applications inside on adhoc basis. Therefore, Computing Clouds deliver the services for users to access hardware, software and data resources, subsequently an integrated computing platform as a service in a translucent manner. Here we argue a range of security [3] and privacy threats confronted by Cloud Computing at different layers of architecture (HaaS, PaaS and SaaS) and highlight the blend of resolutions 4 which developers and architects need to be alert of.

## 2.1 HaaS Based Risks

This layer is also known as Infrastructure layer. This layer being the bottom-most layer provides various resources such as Disk Space, CPU Cycles as infrastructure services as on demand basis. When these resources are offered in others hand may lead to bottleneck if not monitored appropriately. In other words, Cloud offers a flexible infrastructure of distributed data center services connected via Internet style networking hence requires due attention. As a user of HaaS (Amazon, Rackspace, GoGrid, Cloud) one would need to consider following aspects:

### 2.1.1 At Network Level

- **Connection pooling** is the exercise of creating and then reusing a connection resource. From security perspective, this can result in either the client or server using a connection for some unplanned purpose formerly being used by a privileged user. This can potentially expose vulnerability if the connection is not reauthorized when used by a new identity.
- **Liaison attacker** or eavesdropping intercepts both the client and server communications and then proceeds as man-in-the-middle between the two without their knowledge. This gives the ability to attacker to read and possibly modify the incoming and outgoing communication from whichever side.
- **Interruption of nodes** during the routing process to send packets from source to destination decreases total packet delivery i.e. increases the packet loss which in turn leads to lower throughput of connection.

### 2.1.2 At Host Level

- **Buffer overflows** [5] occurs when the attacker uses a practice to root insertion and execution of malicious code in such a way that the attacker obtains the control of the program and thus the most expensive resources like CPU may forcefully get involve into unplanned program processing.

- **Physical theft** of storage account information, code or other intellectual property is the most obvious concern since data is stored at third party site whose location at any moment is almost untraceable to client.

#### 2.1.3 At Application Level

- **Brute Force Attacks** again unintentionally exploits the CPU's raw processing power. For instance, trying different permutations of any character set to form a username and a password of certain length in order to attempt to gain access to a system. No intelligence is used in such kind of attacks to filter for likely combinations.

#### 2.1.4 At Data Communication and Storage Level

- **Data tampering** [11] is violating the integrity of data (at rest or in transit) by modifying it in local memory, in a data-store, or on the network.

### 2.2 PaaS Based Risks

This layer is also known as Platform layer. This layer being positioned in the middle of Infrastructure layer and application layer facilitates middleware providing application services and a platform-as-a-service (PaaS) runtime environment for cloud applications. The threats at this level tend to exploit existing software bugs and vulnerabilities with the intent of crashing a system. There may be cloud specific security threat like software bug leading to accidental exposure of information to other parties sharing the resources or leakage out of sensitive data from released resources. As user of PaaS (Azure, Salesforce, GAE), one would need to consider the following aspects:

#### 2.2.1 Code Vulnerabilities

- **Disclosure of sensitive/confidential data** [6] means unintended exposure of sensitive information usually achieved out of parameterized error messages, where an attacker will force an error and the program will pass sensitive information up through the layers of the program without filtering it.
- **File System or Registry tampering** may be achieved through some compromised web service. This issue is of even much higher risk in cloud based environment because of large emulation of VMs as compared to older web services.

#### 2.2.2 Data Storage and Access

- **Open ports** listening in insecure interfaces may allow communication to all open or externally addressable ports in case no proper packet filtering mechanism is used which in turn may lead to unauthorized data traffic. The attack can take various forms such as a distributed denial-of-service (DDOS) attack and the like.

### 2.2.3 Auditing and Logging [6]

- **Upgrade of a privilege** in which a user with limited privileges assumes the identity of a privileged user to gain advantaged access to an application to compromise and take control of a trusted process or account.

## 2.3 SaaS Based Risks

This layer is also known as application layer. This layer being close to end user provides applications delivered on demand. Here, threats like attempt to steal sensitive information etc. lies which many a times comes from inside by a discontented employee. Here cloud specific security threats may involve insecure interface and API [12] exposed by the cloud provider or losing control over their ability to ensure strong authentication at the user level. In spite of being least exposed as a user of a (SaaS), yet one would need to consider following aspects:

### 2.3.1 Secrecy / Privacy

- **Dictionary attacks** are quite cleverly focused and used in a try to access the passwords and other credentials and coding methods to access a system.
- **Weak Encryption** or disclosure of arbitrary secrets in a table or storage is equally dangerous to a scene if data is not pre-encrypted before uploading or if decryption keys are also stored in the same storage.

### 2.3.2 Identity and Access Management [7]

- **Cookie manipulation** scheme of an attacker will alter the cookies stored in the browser and then use the cookies to deceptively validate themselves to a service or Web site.
- **Cookie replay attack** happens when already valid cookies are used to cheat the server for something like passwords etc. by posing to server that a previously authenticated/valid session is still active.
- **Information disclosure** i.e. unwanted exposure of private data like Disclosure of Shared Access Signatures, data in transit between client and server, Disclosure of SSL Certificates/keys. Some other examples of information disclosure vulnerabilities include comments embedded in Web pages that contain database connection strings and connection details, the use of hidden form fields and weak exception handling that can lead to internal system-level details being revealed to the client.

### 2.3.3 Authentication and Authorization

- **Cross-Site Request Forgery (CSRF)** is interacting with a web site on behalf of another user to perform unkind operations. That site takes for granted that all requests it hears are not unintentional rather trustworthy.
- **Cross-site scripting (XSS)** may happen when an attacker succeeds in infusing a piece of executable script into a stream of data that will be delivered in a web browser. This malicious code will be executed in the user's current session [6] and will obtain all due privileges to the site and information.

### 2.3.4 At Application Level

- **Request flooding** may be a serious activity at the customer code or application level. Even though this concern is not only specific to cloud environment yet requires due attention.
- **Mis-configuration** of service settings or application settings may entail problems like scoping the cookies and other properties to the service sub-domain.

## 3 Methods to Minimize the Risks/ Strengthen the Assets of Cloud

Cloud computing have distinct security issues compared to traditional computing systems which may be stand-alone, or networked architecture based. To rely honestly on cloud, a user always requires the guarantee of integrity of user's own data and application etc on the remote machine. The similar kind of guarantee is also needed by remote machine about the user's process and data. While the safeguards of a traditional system aim at protecting the system and data from its users, the security orientation of cloud computing systems need to go a step ahead and also protect applications and data from the system where the computation takes place. The major security dilemma with cloud computing includes:

### 3.1 Network Security

The simplest and foremost idea to extend network security [13] is Port Scanning/ Service Enumeration which means that the only ports open and addressable (internally or externally) on a VM should be those which are explicitly defined in some kind of Service definition file coupled with firewall enabled on each VM to enhance VM switch packet filtering, which blocks unauthorized traffic. Whenever unauthorized port scanning is detected it should be stopped, blocked and this reported violation should be investigated seriously. Furthermore, to avoid spoofing VLANs may be used to partition the internal network and segment it in a way that prevents compromised nodes from impersonating trusted system, coupled with HTTPS connection. Packet sniffing by other tenants must not be allowed for example even two virtual instances that are owned by the same customer located on the same physical host should not be able to listen to each other's traffic. Encrypting the sensitive information over the network and securing the channel may also assist.

### 3.2 Application Security [8]

The security of cloud services is more or less dependent on the security of APIs. These interfaces must be designed to protect against both accidental and malicious attempts of illegal authentication, access control, encryption etc. Dependence on a fragile group of interfaces and APIs exposes the client's secrecy to a variety of security issues related to confidentiality, integrity, availability and accountability. The applications may be strengthened by developers to do application-level throttling of incoming requests for any kind of complex, time-intensive operation. Beyond all, client should select a Web interface wisely by inquiring from prospective Cloud provider if it offers a true web-interface, that is, one which allows secure access from anywhere, and have other features like platform independence (PC/Mac/Linux), and mobile device access. Exception shielding may help to build stronger applications where not all the exceptions are reported but only those exceptions are returned to the client that does not contain sensitive information in the exception message, and they do not contain a detailed stack trace, either of which might reveal sensitive information about the Web service's inner workings.

### 3.3 Data Security

The security measures to ensure data security are to meet the physical, transmission, storage, access, data, and application security needs of client organizations with strictest standards. Besides physical security of storage banks the filtering and auditing of data and erasure of data after use may improve to uplift the security mechanism.

Multiple storage keys design pattern can be used for one subscription to protect trusted storage data. This variety may be employed to diminish the publicity of a particular key to theft by placing lower-trust keys on lower-trust roles and higher-trust keys on higher-trust roles. Storage of data should provide a simple structured storage environment to avoid common SQL injection vulnerabilities [10]. To enhance the data protection over wire, all data in transit to Storage should be using HTTPS and data that won't be of interest to outside parties or eavesdroppers, HTTP can be used for a faster transfer. Moreover data origin authentication method may help to verify that data messages have not been tampered with in transit and they originate from expected (authentic) user.

### 3.4 Access and Identity Management

HaaS providers offer their clients the virtual environment of unlimited computation, network, and data storage capability, generally offered with very smooth registration process where a user with a valid credit card may get registered and may start using cloud services without delay. By abusing the relative mystery behind these registration and usage models, spammers, malicious coders, and other law breakers have been able to carry out their tricks with ease. The possible prevention to identity management may be strengthened with stricter initial registration and validation processes coupled with improved credit card usage model. Multi factor authentication may also be quite supportive in it where user has to enter a six digit code generated by an authentication device that valid user always keeps in his physical possession. Using encrypted credentials coupled with message signing which refers to signing a message with a digital signature using cryptographic methods, to confirm the source of the message and detect if the contents have been tampered with or not, and implement the schemes to protect sensitive data from being stolen from memory, from configuration files, or when transmitted over the network and cryptographic random number generators to generate session IDs, such methods can also build up even better security.

### 3.5 Monitoring and Reporting

Round the clock availability of general cloud services is reliant upon the capability of APIs. From access control to activity monitoring [9], these interfaces must be designed to log and report one and all deceptive attacks to circumvent policy. There should be reserved disk space for monitoring and logging capabilities. Auditing and logging is used to monitor and record important activities, such as transactions or user management event, logging each access request including request type, the requested resource, the requestor's IP, and the time and date of the request are recorded in log. Logged information enables efficient auditing of events in the case of an attack or a suspected attack. Using application instrumentation to expose behavior that can be monitored and stripping sensitive data [6] before logging may also assist in this direction. Besides Application reporting, regular system monitoring including resource usage, regular reviews with hosting partner and automated systems with live monitoring, encrypting the sensitive data in configuration files may prove to be a milestone in Cloud Computing.

## 4 Conclusion

Although security remains an important concern for many companies, Cloud computing solutions are very convincing to firms wishing to slim their capital investments. Software architects and developers must understand the vulnerabilities/ threats to software developed and use appropriate security design practices to counter threats in the cloud environment. Even though that few basic protection mechanisms are always well in place already for a client before moving to the cloud, yet Cloud services developers have to be enough mature and accountable for writing robust code and to ensure the security of their applications in different dimensions when dealing with malicious attacks/threats. The work required to develop secure web applications isn't new, revolutionary, or technically challenging; it simply expects the application designers and developers to recognize the potential threats to their applications in Cloud environment and apply the best possible practices available. Besides all these control and other security measures it is always better to be protective than to cure it later i.e. it is always advisable to keep the recovery plans accurate and compliant inside and outside the cloud.

## References

1. Weiss, A.: Computing in the Clouds. NetWorker 11(4), 16–25 (2007)
2. Buyya, R., Yeo, C.S., Venugopal, S., Broberg, J., Brandic, I.: Cloud Computing and Emerging IT Platforms: Vision, Hype, and Reality for Delivering Computing as the 5th Utility. Future Generation Computer Systems 25(6), 599–616 (2009)
3. Cloud Computing Security, A Trend Micro White Paper (August 2009), `http://emea.trendmicro.com/imperia/md/content/uk/solutions/enterprise/wp02_cloud-computing_090812us.pdf`
4. Cloud Expo::Attacks Cannot Be Prevented By Lori MacVittie, `http://cloudcomputing.sys-con.com/node/1668772`

5. white paper : On Verifying Stateful Dataflow Processing Services In Large-Scale Cloud Systems, http://www.techrepublic.com/whitepapers/on-verifying-stateful-dataflow-processing-services-in-large-scale-cloud-systems/2417851
6. Cloud Security Threats and Countermeasures at a GlanceJD Meier (July 8, 2010), http://blogs.msdn.com/b/jmeier/archive/2010/07/08/cloud-security-threats-and-countermeasures-at-a-glance.aspx
7. http://www.cloudsecurityalliance.org/guidance/csaguide.v2.1.pdf
8. http://www-07.ibm.com/in/ibm/cloud/?ca=googleaw&gclid=CPrsyYXb86cCFRFOgwod0kaacg
9. http://www.intel.com/itcenter/topics/cloud/security.htm?cid=apac:gglcloudsecurity_in_genti17A68s
10. The "Comparing Google App Engine, Amazon SimpleDB and Microsoft SQL Server Data Services" post of May 6, 2008 to the OakLeaf Systems blog, http://bit.ly/LtvxN,http://oakleafblog.blogspot.com/2008/04/comparing-google-app-engine-amazon.html
11. Hinchcliffe, Dion. Cloud computing: A new era of IT opportunity and challenges (March 03, 2009), http://blogs.zdnet.com/Hinchcliffe/?p=261&tag=rbxccnbzd1 (retrieved June 03, 2009)
12. Swaminathan, Kishore S., Daugherty, Paul, Tobolski, Joseph.: What the Enterprise needs to know about Cloud Computing. Accenture Technology Labs, p. 3-15 (2009)
13. white paper Securing the Cloud, A Review of Cloud Computing, Security Implications and Best Practices, http://www.vmware.com/files/pdf/cloud/VMware-Savvis-Cloud-WP-en.pdf

# Efficient Grid Scheduling with Clouds

L. Yamini, G. LathaSelvi, and Saswati Mukherjee

Department of Information Science and Technology,
College of Engineering, Guindy,
Anna University, Chennai- 600 025
yami.lp88@gmail.com, madam_selvi@rediffmail.com,
msaswati@yahoo.com

**Abstract.** An efficient technique for scheduling in grids is explored in this paper and is further extended it with clouds. Here, we consider bandwidth availability while selecting resources for job scheduling. Thus, this strategy selects the resource in such a manner that along with computational capability, the ability of the resource to quickly respond to a task is also taken into account by means of using available bandwidth. This is further extended with cloud in order to tackle non availability of resources in a grid environment. Thus, if peak demand arises in a grid environment, we instantiate an on demand cloud resource customized to meet the grid user requirements. The response time and thus the total completion time for the job is lowered as the waiting time of the jobs gets lowered, which is evident from the experimental results.

**Keywords:** Cloud computing, Grid Scheduling, Network information, on demand instance.

## 1 Introduction

Grid computing principles focus on large-scale resource sharing in distributed systems in a flexible, secure, and coordinated fashion. This coordinated sharing helps innovative applications to make use of high-throughput computing for dynamic problem solving. Grid computing (or the use of a computational grid) is defined as the application of the resources of many computers in a network to a single problem at the same time - usually to a scientific or technical problem requiring a great number of computer processing cycles or access to large amounts of data. Most of the application domains have in common is a need for coordinated resource sharing and problem solving in dynamic, multi-institutional virtual organizations [8].

Grid computing involves grouping and sharing of geographically distributed resources, for solving computational and data intensive applications. The term Grid Resource Management commonly describes all aspects of the process of identifying different types of resources, their availability, arranging for their use, utilizing them and monitoring their state. Management of these resources is very important in grid environment as the number of available resources keeps on changing over time. An important problem that arises here is optimal assignment of grid jobs to resources. A grid job is defined as anything that needs a resource – from a bandwidth request, to an application, to a set of applications (for example, a parameter sweep) [9]. The term

A. Mantri et al. (Eds.): HPAGC 2011, CCIS 169, pp. 92–102, 2011.

resource means anything that can be scheduled: a machine, disk space, a QoS network, and so forth [9]. As grid resources are heterogeneous and geographically distributed in nature, it is difficult to obtain a proper job to resource assignment schedule. Hence, the grid resource manager or scheduler should use some best resource selection strategy in terms of user requirements and should have the ability to adapt with dynamic resources.

Grid scheduling is defined as the process of making scheduling decisions involving resources over multiple administrative domains [10]. In this process, multiple administrative domains could be searched for using a single machine or a single job is scheduled to use multiple resources at a single site or multiple sites. A generic grid scheduler does the task of interacting with a local resource manager, external services like information, forecasting, submission, security or execution services, receiving a scheduling problem, calculating and returning a schedule.

Schedulers, in general, can be classified into various types. If the entire grid resources are managed to make their usage easier for the users, it is termed as a metascheduler or grid scheduler. A job template describing various requirements and directions are specified by the grid user to the grid scheduler entity and from that moment on, the grid scheduler takes control, performs matchmaking and identifies the best resource suitable for the job [1]. If, on the other hand, resources are managed at a single site or perhaps only for a single cluster or resource, it is termed as a local resource scheduler.

One of the primary differences between a Grid scheduler and a local resource scheduler is that the Grid scheduler does not "own" the resources at a site (unlike the local resource scheduler) and therefore does not have control over them [10]. Gridway, Condor are few examples of metaschedulers available for grid resource management.

Managing grid resources by using a scheduler that uses best resource selection strategy is very important. Matchmaking refers to the task of matching available resources to the incoming grid jobs as per the requirements of the grid user. This plays a major role in selecting resources thereby schedule generation for allocating and executing jobs. Current research trends in grid scheduling focuses on improving the efficiency of the scheduling algorithm so as to improve application performance. Most of the current metaschedulers take their scheduling decisions focusing just on the computing power (and utilization) of the available resources [1]. So, a metascheduler might decide that the most suitable resource to run a user's job is the most powerful or unloaded one [1].

To overcome this common resource selection strategy of the current metaschedulers and to make the resource selection and schedule generation more efficient, researchers have also focused on incorporating network related information in scheduling decisions [1]. In the current work, we use a modified resource selection strategy by making the scheduler consider the available bandwidth before generating a schedule and hence, submitting a job to the resource. We use a modified matchmaking that includes the bandwidth requirements consideration. Our objective is to select a resource that is less constrained with reasonable CPU power rather than selecting a resource which is highly constrained and is having more power in terms of computational capabilities.

Despite having a grid environment in place with an efficient scheduling technique, one can always benefit by extending them with on demand cloud resources. Organizations are more bend towards having a grid environment in place, as it provides a secure high performance environment. However, Grid is constrained by lack of resource during peak demand time. Availability of on demand resources to complete or continue with the grid job allocation or execution would help to reduce this constraint.

Today, Cloud Computing is emerging as the paradigm for the next generation of large-scale scientific computing, eliminating the need of hosting expensive computing hardware [4]. Since cloud environment abound with tightly coupled, homogeneous resources, these can be used in an on demand basis for meeting grid user requirements. We propose a set up, where in cloud resources can be utilised by a user (i.e., provided and dedicated to a user) on demand in a resource-as-a-service fashion.

The objective of this research is having an efficient grid scheduling strategy and usage of on demand cloud resources, thus extending grids with cloud resources. Scheduling focuses on including network information viz, bandwidth while choosing resources and extending with clouds focuses on providing grid users with on demand customized cloud resources in case of peak demands or heavy requirements.

This paper is organized as follows. In Section 2 we describe the related work. Section 3 explains our proposed method. In Section 4, we discuss our implementation methods. Section 5 describes the experimental results and Section 6 concludes the paper.

## 2 Related Work

In GNB-Grid Network Broker [2], an architecture that combines concepts from Grid scheduling with autonomic computing is proposed. An existing network QoS architecture is improved to consider the status of the network and this architecture is extended with autonomic behavior for adapting to changes in the system. They have used models for predicting latencies in a network and in CPU usage, which constitute the basis for the autonomic behavior of our architecture. Network status is considered when reacting to changes in the system—taking into account the workload on computing resources and the network links when making a scheduling decision.

The schedulers only paid attention to the load of the computing resource, thus a powerful unloaded computing resource with an overloaded network is chosen to run jobs, which decreases the performance received by users, especially when the job requires a high network I/O.

In [3], network information is used to perform the scheduling of jobs to computing resources, and it is done by extending gridway metascheduler. The network conditions are used both at rank and requirement sections in gridway, and they can also be combined with other parameters. This shows that including network related information improves performance.

In [4], authors have extended a grid workflow development and computing environment to use on-demand cloud resources in grid environments offering a limited amount of high performance resources. The extensions to the resource management architecture to consider cloud resources comprises of three new components: cloud management, image catalogue and security components.

In [5], architecture is proposed for building type of arbitrary complex grid infrastructures with a single point of access. This supports methods for extending grid infrastructures by requesting resources form other grid resource providers and also from cloud resource providers. This also supports dynamic addition of cloud resource for meeting grid user requirements. The virtualization overhead involved in booting the cloud instance up is analyzed in this work.

In [6], a straightforward deployment of virtual machines in a Grid infrastructure is achieved. This strategy does not require additional middleware to be installed and it is not bounded to a virtualization technology. Though the overall overhead induced by the virtualization technology decreases the application performance, it presents attractive benefits, like increasing software robustness or saving administration efforts, so improving the quality of life in the Grid.

In [7], DIET-Solve Grid middleware is used on top of the EUCALYPTUS Cloud system for demonstrating general purpose computing using Cloud platforms. This work proposes the use of a Cloud system as a raw computational on- demand resource for a Grid middleware. A proof of concept is illustrated by considering the DIET-Solve Grid middleware and the EUCALYPTUS open-source Cloud platform.

## 3 Proposed Method

This work brings in the usage of network related information in scheduling decisions, thus making the scheduler aware of the network and further, this infrastructure is extended with a cloud resource on demand for meeting various requirements of a grid user. Network aware scheduling exploits available bandwidth between the node submitting the job and the node to which the job is submitted. If the link is already constrained as a result of the node involved in various other computations, then the next node with less constrained link is opted for by the scheduler. Thus, higher ranks are assigned for resources with more available bandwidth and reasonable computing power rather than resources with higher computational capability and constrained bandwidth availability.

Further an instantiation of an on demand cloud resource aims at providing the grid users with additional resources for performing their computations rather than waiting for grid resources in case of high demands or non availability of resources. Also, since cloud provides homogeneous resources in a tightly coupled manner, user satisfaction becomes high with them.

### 3.1 System Architecture

Figure 1 sketches the architecture of the proposed system with the building block as gridway metascheduler performing various tasks.

**Scheduler.** In this block, we have mainly concentrated on improving the efficiency of the scheduler, by obtaining an allocation that corresponds to user's job requirements with network information added in scheduling decisions. Initially, RANK expression given by the user in the gridway job template is evaluated and REQUIREMENTS tag is parsed to find the available matching resources. Here, we have altered the match-making part in such a manner that higher ranks are assigned to resources with high bandwidth availability (less constrained links). We have also exploited the RANK policy of the gridway metascheduler while scheduling jobs and resources. Thus, we are assigning higher ranks to those resources meeting user requirements as well having higher available bandwidth. The scheduler thus makes a network aware selection.

**Metascheduler Core.** The metascheduler core (gridway core) consists of various middleware access drivers that are responsible for interfacing with the services provided by the underlying middleware (globus). The core is responsible for job execution management and resource brokering, providing advanced scheduling, and job failure and recovery capabilities.

**External Information Providers.** We have used these information providers for supplementing the middleware services with external information that are unavailable in general to them. They are basically other sources of data and they could be some executables or any other non–WS service. The important requirement is that they need to produce a valid XML document. We have used Ganglia for populating the information service of the underlying middleware with the resource information. We have also used Network weather service for obtaining bandwidth information. We have fed these details into the information service of the underlying middleware using XML documents.

**Extender.** We will monitor the number of pending jobs in grid environment waiting for allocation. Once this count exceeds a specified number (this depends on the availability of grid resources), we use extender component for sending a request to the cloud infrastructure for instantiating an on demand cloud resource.

**Cloud Infrastructure.** This component receives the request and retrieves the corresponding image to be used for instantiating the instance. This image used here is a customized one and the image manger component plays a major role in the creation and customization of this image. Upon, finding the corresponding image, an instance is started with that and is responsible for taking care of the grid user requirements. Whenever jobs are submitted to this on demand cloud instance, it tries to cater to the requirements and supports job allocation and execution. Eucalyptus is the cloud infrastructure used here for setting up the cloud and obtaining the virtual cloud instance.

**Image Manager.** We have designed this block for creating, customizing, bundling and uploading the image that will cater to the requirements of the grid user, to the storage component of the underlying cloud infrastructure

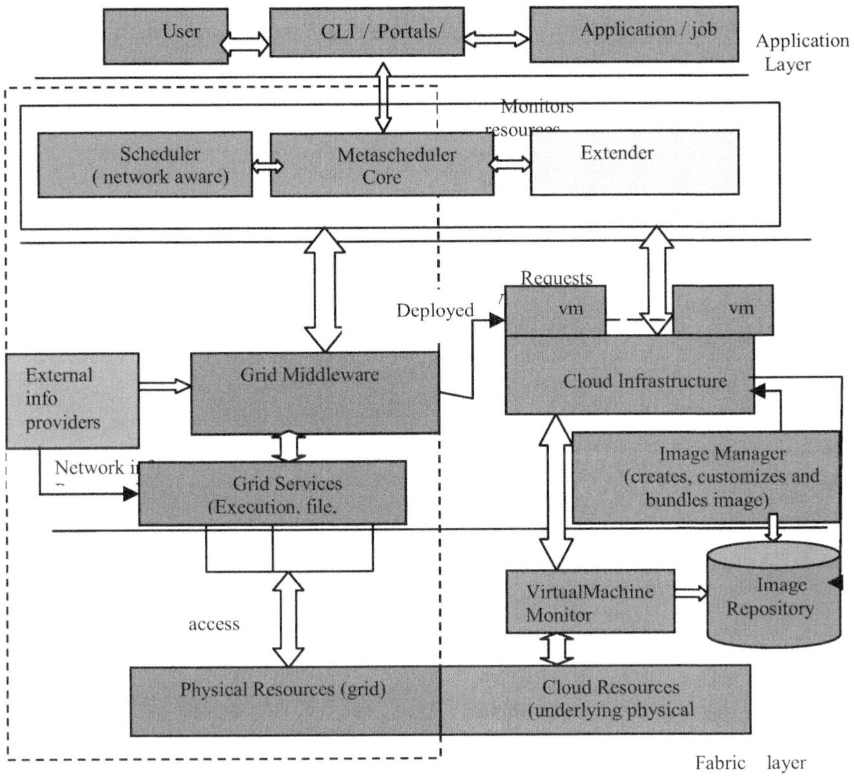


**Fig. 1.** System Architecture Diagram

# 4 Implementation

Gridway is the metascheduler used here for managing resources and the scheduling and matchmaking of gridway is altered to include network information in scheduling decisions. Globus toolkit is the underlying grid middleware used here for supporting the gridway metascheduler by providing various services like index, execution and file transfer for its working. Ganglia is used here for monitoring the availability of resources (free RAM, disk memory, load on the resource). Although Network Weather Service is another monitoring tool, here it is used here for obtaining bandwidth information.

## 4.1 Scheduling with Network Information

Gridway uses resource prioritization and job prioritization policies for scheduling. A priority is computed for the resource and schedules are generated based on the overall priority. We have exploited the RANK policy to obtain higher priority for the resource with higher ranks. This helped us in selecting the higher ranked resource as the most probable candidate to be considered for job allocation. Initially, we submitted jobs using the job templates meant for the metascheduler and noted down their respective completion and response times. Then, we incorporated this network aware scheduling and again we submitted jobs to the metascheduler. We observed that whenever network aware metascheduling is used and when the job does require transfer of heavy input and output files, the response and completion time becomes lower compared to the normal strategy.

If the transfer is not heavy, the response and completion time is still reasonably lower compared to the original strategy. Now we show the steps for implementing the concept of network related information in resource selection and thus in scheduling.

Input: Jobs and set of resources (S)
Output: Job to host assignment

1. Let u be a user submitting the job.
2. Let R be the set of resources available.
3. Compute available bandwidth between the node submitting job and resource

For each scheduling interval,
Begin
For the sets of jobs to be scheduled
for each job, use job policy and prioritize them.
Also, find the matching available resources for them.
Compute rank for the resources based on the obtained network information, combined with other job requirements parameters
Resource with high rank, thus a higher priority is chosen for allocation.
Keep on monitoring the available bandwidth
Allocate job to resources with higher ranks.
return job to host assignments.
End.

## 4.2 Cloud Extended Environment

We have used Eucalyptus - open source cloud computing tool, as the cloud infrastructure for providing us with on demand cloud instances. Whenever the number of pending jobs exceeds a specified limit, we send a request to the eucalyptus for running an on demand cloud instance. The image to be associated with the instance is carefully prepared by us (since it has to satisfy grid user requirements) and is uploaded into the storage component of the eucalyptus. Initially, an empty virtual disk file is created to hold the new custom image as required. Upon creation, this file could be used and the required operating system can be installed. The above created image can be customized to meet the grid user requirements. All required software supporting grid jobs are used here in this image. Also, the image is booted with the hypervisor for checking its proper working. Finally, this is bundled and uploaded into the eucalyptus storage component for running instances.

**Steps for booting up an on demand cloud instance are shown below:**

1. Use an empty virtual disk file to hold the custom image.
2. Install the required OS in the image.

3. Customize the created image can be to meet the grid user requirements by installing all software and mainly grid middleware.
4. Boot the image to check for its proper working via hypervisor.
5. Bundle and upload the customized image to eucalyptus.
6. Whenever request arises for cloud instance, associate this image with the instance to cater to the need of grid user requirements.
7. Thus, this additional on demand cloud resource supports grid users.

## 5 Experimental Results

Globus toolkit is the grid middleware used for providing services to the gridway metascheduler. Initially, individual jobs were submitted to the metascheduler and the scheduler was made to use network related information in scheduling decisions. Execution, start, end and transfer times of the jobs were noted down. Then a batch or array jobs (4 jobs) were submitted, all of them started at the same time and their execution time, transfer time, start time and end time were also found.

Transfer time is the time taken to transfer the input files to the resource and prepare them for execution. Also, outputs are transferred back to the original resource that came up with the request. Thus, the sum of transfer time (otherwise called as prolog here), waiting time in the queue for resources, execution time and epilog time (transferring the output back and clean up) constitutes the total completion time taken by a job.

**Table 1.** Time calculated with network info in scheduling

| Jobs | Completion time (in seconds) | Transfer time (in seconds) | Execution time (in seconds) |
|---|---|---|---|
| Job 1 | 76 | 5 | 5 |
| Job 2 | 103 | 5 | 4 |
| Job 3 | 122 | 6 | 6 |
| Array Job 1 (group of 4 tasks) | 30 (for each job) | 8 | 4 (for each job) |

Table 1 shows that the second job is waiting in the queue for the first job to complete and release the resource. Thus, the waiting time of the job in the queue for obtaining the local resource becomes high. But, the transfer time taken is somewhat nearly constant for the jobs, indicating that a resource with high network availability is chosen. Though the transfer time is less, still the waiting time of the job for obtaining the local resource in the queue increases in case of non availability of resources.

Fig.2. shows the total completion time of jobs (in percentage). The waiting time is high despite the transfer time being low. Thus, this set up is extended with a on – demand cloud resource, to which the forthcoming jobs are allocated to and executed. Thus, by doing so, these jobs are not forced to wait in the queue for obtaining local resource and they are sent to cloud resource for their allocation and execution. Table 2. shows the time taken for booting up an on demand cloud instance.

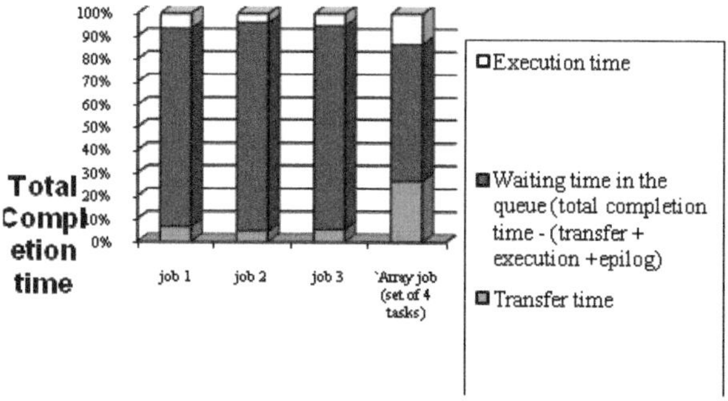


**Fig. 2.** Total completion time for jobs shown comprising of transfer, execution and waiting times in the queue

Now, if another set of jobs comes in (supposing jobs 4,5,6,7 comes in ) when there are already pending jobs in grid waiting for resources, in that case, we will move these set of jobs and submit them to a cloud resource. A request is sent for the instantiation of cloud resource and thereafter these jobs can be submitted to the cloud resource.

**Table 2.** Time taken for booting up a cloud instance

| Instance (from pending to running state) (in minutes) | Terminate running Instance (in seconds) | Total time approximately for having a running instance (in seconds) |
|---|---|---|
| 2 | 7 | 120 |

Table 2. shows the time taken for booting up an on demand cloud instance. Table 3. shows the time taken for completion when jobs are submitted to the obtained cloud resource. Thus, we are claiming that instead of making the jobs wait in the queue in grid, we provide them another on demand cloud resource and submit them there. This is certainly lowering the total completion time, as these jobs would have waited for a long time if they are submitted to grids. This cloud resource can always be made available by keeping it in running state. In that case, the overhead involved in instantiating the resource is also lowered. This cloud resource when not needed can

be terminated. Xen is the hypervisor used here by eucalyptus. Hence, the virtualization overhead involved becomes much lower as xen is a better performing hypervisor when compared to others.

**Table 3.** Time calculated after submitting jobs to a cloud resource

| Jobs | Completion time (in seconds) | Transfer time (in seconds) | Execution time (in seconds) |
|---|---|---|---|
| Job 1 | 25 | 4 | 4 |
| Job 2 | 54 | 5 | 5 |
| Job 3 | 116 | 28 | 16 |
| Job 4 | 150 | 15 | 6 |
| Array Job 1 (group of 4 tasks) | 15 | 4 | 5 |
| | 47 | 4 | 6 |
| | 71 | 5 | 5 |
| | 86 | 4 | 6 |

## 6 Conclusion

An efficient metascheduling technique that combines network related information in scheduling tasks or jobs is used here to have a better resource selection strategy. Resources are prioritized and ranked based on the requirements and thus, matchmaking is done. Despite using network information, this work also extends the setup with a cloud environment, by providing another on demand cloud instance meeting grid user requirements, in case of too many waiting jobs in the grid environment. By doing so, we are able to submit jobs to the cloud resource thereby reducing the waiting time of jobs in queue for obtaining local resource. This in turn ensures that the forthcoming jobs have lower response and completion times.

## References


1. Caminero, A., Caminero, B., Carrion, C., Tomas, L.: Improving GridWay with Network Information: Tuning the Monitoring Tool. In: IEEE International Symposium on Parallel and Distributed Processing, pp. 1–8 (2009)
2. Caminero, A., Rana, O., Caminero, B., Carrion, C.: Performance evaluation of network-aware Grid metaschedulers. In: International Conference on Parallel Processing Workshops, pp. 282–289 (2009)

3. Tomas, L., Caminero, A., Caminero, B., Carrion, C.: Studying the influence of network-aware grid scheduling on the performance received by users. In: Chung, S. (ed.) OTM 2008, Part I. LNCS, vol. 5331, pp. 726–743. Springer, Heidelberg (2008)
4. Ostermann, S., Prodan, R., Fahringer, T.: Extending Grids with cloud resource management for scientific computing. In: 10th IEEE/ACM International Conference on Grid Computing, October 13-15, pp. 42–49 (2009)
5. Blanco, C.V., Huedo, E., Montero, R.S., Llorente, I.M.: Dynamic Provision of Computing Resources from Grid Infrastructures and Cloud Providers. In: Workshops at the Grid and Pervasive Computing Conference, pp. 113–120 (2009)
6. Rubio-Montero, J., Huedo, E., Montero, R.S., Llorente, I.M.: Management of Virtual Machines on Globus Grids Using GridWay. In: IEEE International Parallel and Distributed Processing Symposium, p. 358 (2007)
7. Caron, E., Desprez, F., Loureiro, D., Muresan, A.: Cloud Computing Resource Management through a Grid Middleware: A Case Study with DIET and Eucalyptus. In: IEEE International Conference on Cloud Computing, pp. 151–154 (2009)
8. Foster, I., Kesselman, C., Tuecke, S.: The Anatomy of the Grid: Enabling Scalable Virtual Organizations. International Journal of High Performance Computing applications 15(3), 200–222 (2001)
9. Dong, F., Akl, S.G.: Scheduling Algorithms for Grid Computing: State of the Art and Open Problems, School of Computing, Queen's University Kingston, Ontario, Technical Report No. 2006-504 (January 2006)
10. Schopf, J.: Ten Actions When Grid Scheduling – They User as a Grid Scheduler. In: Grid Resource Management: State of the Art and Future Trends, pp. 15–23. Kluwer Academic Publishers, Dordrecht (2008)

# Security Concerns in Cloud Computing

Puneet Jai Kaur and Sakshi Kaushal

University Institute of Engineering and Technology,
Panjab University, Chandigarh
puneetkaur79@yahoo.co.in,
sakshi@pu.ac.in

**Abstract.** Since inception, the IT industry experienced a variety of natural evolution points, most marked with rapid change followed by years of internalization and consumption. According to most observers, the industry is rapidly evolving toward services as a core component of how consumers and business users interact with both software and one another The hype is deafening in places, and the key to success is recognizing that "cloud" adoption does not represent an all-or-nothing proposition. Organizations use cloud computing as a service infrastructure, critically like to examine the security and confidentiality issues for their business critical insensitive applications. Yet, guaranteeing the security of corporate data in the cloud is difficult, if not impossible, as they provide different services like SaaS, PaaS and IaaS. Each service has its own service issues. This paper discusses the security issues, requirements and challenges that cloud service providers face during the cloud engineering and the various deployment models for eliminating the security concerns.

**Keywords:** Cloud Computing, Public Cloud, Private Cloud, Cloud Security, Deployment Models.

## 1 Introduction

Cloud computing is a computing paradigm, where a large pool of systems are connected in private or public networks, to provide dynamically scalable infrastructure for application, data and file storage. With the advent of this technology, the cost of computation, application hosting, content storage and delivery is reduced significantly. Cloud computing is a practical approach to experience direct cost benefits and it has the potential to transform a data center from a capital-intensive set up to a variable priced environment. The idea of cloud computing is based on a very fundamental principal of reusability of IT capabilities. The difference that cloud computing brings compared to traditional concepts of grid computing or distributed computing is to broaden horizons across organizational boundaries. According to NIST - "Cloud Computing is a pay-per-use model for enabling available convenient, on-demand network access to a shared pool of configurable computing resources( e.g. networks, servers, storage, applications, services) that can be rapidly provisioned and released with minimal management effort or services provider interaction." Cloud Computing has given a boost to the IT industry by providing following benefits:

A. Mantri et al. (Eds.): HPAGC 2011, CCIS 169, pp. 103–112, 2011.

**1. Reduced Cost :** There are a number of reasons to attribute Cloud technology with lower costs. The billing model is pay as per usage; the infrastructure is not purchased thus lowering maintenance. Initial expense and recurring expenses are much lower than traditional computing.

**2. Increased Storage:** With the massive Infrastructure that is offered by Cloud providers today, storage & maintenance of large volumes of data is a reality. Sudden workload spikes are also managed effectively & efficiently, since the cloud can scale dynamically.

**3. Flexibility:** This is an extremely important characteristic. With enterprises having to adapt, even more rapidly, to changing business conditions, speed to deliver is critical. Cloud computing stresses on getting applications to market very quickly, by using the most appropriate building blocks necessary for deployment.

In this paper we have discussed the general concepts of the Cloud computing along with the various challenges faced. Along with we have also highlighted the security concerns of cloud computing . Section 2 provides an overview of the types of the cloud computing . Section 3 introduces the cloud computing security. Section 4 and 5 discusses the various security issues and concerns of Cloud computing and the deployment models for eliminating those security concerns.

## 2 Types of Cloud Computing

**Cloud** computing is typically classified in two ways:

**Location of the cloud computing** and the **type of services offered.**

### 2.1 Based on Location of the Cloud[1,2,3]

**Public cloud:** In Public cloud the computing infrastructure is hosted by the cloud vendor at the vendor's premises. The customer has no visibility and control over where the computing infrastructure is hosted. The computing infrastructure is shared between any organizations.

**Private cloud:** The computing infrastructure is dedicated to a particular organization and not shared with other organizations. Private clouds are more expensive and more secure when compared to public clouds.

**Hybrid cloud:** Organizations may host critical applications on private clouds and applications with relatively less security concerns on the public cloud. The usage of both private and public clouds together is called hybrid cloud. A related term is Cloud Bursting. In Cloud bursting organization use their own computing infrastructure for normal usage, but access the cloud for high/peak load requirements. This ensures that a sudden increase in computing requirement is handled gracefully.

### 2.2 Based upon the Services Offered [1,2,3]

**Infrastructure as a service (IaaS):** IaaS provides basic storage and computing capabilities as standardized services over the network. Servers, storage systems, networking equipment, data centre space etc. are pooled and made available to handle workloads. The customer would typically deploy his own software on the infrastructure. Leading vendors that provide Infrastructure as a service are Amazon EC2, Amazon S3, Rack space Cloud Servers and Flexi scale.

**Platform as a Service (PaaS ):** Here, a layer of software, or development environment is encapsulated & offered as a service, upon which other higher levels of service can be built. The customer has the freedom to build his own applications, which run on the providers infrastructure. Typical players in PaaS are Google's Application Engine, Microsofts Azure, Salesforce.com's force.com .

**Software as a service (SaaS):** In this model, a complete application is offered to the customer, as a service on demand. A single instance of the service runs on the cloud & multiple end users are serviced. On the customers side, there is no need for upfront investment in servers or software licenses, while for the provider, the costs are lowered, since only a single application needs to be hosted & maintained. Examples are Salesforce.coms ,Googles gmail and Microsofts hotmail, Google docs and Microsofts online version of office called BPOS (Business Productivity Online Standard Suite).

## 3 Cloud Computing Security

Cloud computing security is an evolving sub-domain of computer security, network security, and, more broadly, information security. It refers to a broad set of policies, technologies, and controls deployed to protect data, applications, and the associated infrastructure of cloud computing. There are a number of security issues associated with cloud computing but these issues fall into two broad categories: Security issues faced by cloud providers (organizations providing Software-, Platform-, or Infrastructure-as-a-Service via the cloud) and security issues faced by their customers. In most cases, the provider must ensure that their infrastructure is secure and that their clients' data and applications are protected while the customer must ensure that the provider has taken the proper security measures to protect their information. In cloud computing, end users' data is stored in the service provider's data centers rather than storing it on user's computer. This will make users concerned about their privacy. Moreover, moving to centralized cloud services will result in user's privacy and security breaches as discussed in. Security threats may occur during the deployment; also new threats are likely to come into view. Cloud environment should preserve data integrity and user privacy along with enhancing the interoperability across multiple cloud service providers. The security related to data distributed on three levels in [11]:

- *Network Level* The Cloud Service Provider (CSP) will monitor, maintain and collect information about the firewalls, Intrusion detection or/and prevention systems and data flow within the network.
- *Host Level* :It is very important to collect information about system log files., in order to know where and when applications have been logged.
- *Application Level:* Auditing application logs, which then can be required for incident response or digital forensics.

At each level, it is required to satisfy security requirements to preserve data security in the cloud such as confidentiality, integrity and availability as follows:

*1. Confidentiality:* Ensuring that user data which resides in the cloud cannot be accessed by unauthorized party. This can be achieved through proper encryption techniques taking into consideration the type of encryption: symmetric or asymmetric encryption algorithms, also key length and key management in case of the symmetric cipher.

*2. Integrity:* Cloud users should not only worry about the confidentiality of data stored in the cloud but also the data integrity. Data could be encrypted to provide confidentiality where it will not guarantee that the data has not been altered while it is residing in the cloud. Mainly, there are two approaches which provide integrity, using Message Authentication Code (MAC) and Digital Signature (DS). In MAC, it is based on symmetric key to provide a check sum that will be append to the data. On the other hand, in the DS algorithm it depends on the public key structure (Having public and private pair of keys). As symmetric algorithms are much faster than asymmetric algorithms, in this case, we believe that Message Authentication Code (MAC) will be the best solution to provide the integrity checking mechanism.

*3. Availability:* Another issue is availability of the data when it is requested via authorized users. The most powerful technique is prevention through avoiding threats affecting the availability of the service or data. It is very difficult to detect threats targeting the availability. Threats targeting availability can be either Network based attacks such as Distributed Denial of Service (DDoS) attacks or CSP availability.

# 4 Security Issues in Cloud

## 4.1 Various Security Issues

Security concerns have been raised due to the computing model introduced by cloud computing, which is characterized by off-premises computing, lost control of IT infrastructure, service-oriented computing, and virtualization, and so on. Here are seven of the specific security issues Gartner says customers should raise with vendors before selecting a cloud vendor.[5,10]

**1. Privileged user access.** Sensitive data processed outside the enterprise brings with it an inherent level of risk, because outsourced services bypass the "physical, logical and personnel controls" IT shops exert over in-house programs.

**2. Regulatory compliance.** Customers are ultimately responsible for the security and integrity of their own data, even when it is held by a service provider. Traditional service providers are subjected to external audits and security certifications. Cloud computing providers who refuse to undergo this scrutiny are "signaling that customers can only use them for the most trivial functions.

**3. Data location.** When we use the cloud, we probably won't know exactly where our data is hosted. In fact, we might not even know what country it will be stored in.

**4. Data segregation.** Data in the cloud is typically in a shared environment alongside data from other customers. Encryption is effective but isn't a cure-all. The cloud provider should provide evidence that encryption schemes were designed and tested by experienced specialists.

**5. Recovery.** Even if we don't know where our data is, a cloud provider should tell us what will happen to our data and service in case of a disaster.

**6. Investigative support.** Investigating inappropriate or illegal activity may be impossible in cloud computing. Cloud services are especially difficult to investigate, because logging and data for multiple customers may be co-located and may also be spread across an ever-changing set of hosts and data centers. If you cannot get a contractual commitment to support specific forms of investigation, along with evidence that the vendor has already successfully supported such activities, then your only safe assumption is that investigation and discovery requests will be impossible.

**7. Long-term viability.** Ideally, our cloud computing provider will never go broke or get acquired and swallowed up by a larger company. But we must be sure that our data will remain available even after such an event.

### 4.2 Security Policy in Cloud Computing Environment

In order to solve these problems, the security policy [9] should include the following points:

a) Divided into multiple security domains in the cloud computing environment, different security domain operation must be mutual authentication, each security domain internal should have main map between global and local.

b) Ensure that the user's connection and communications security with the SSL, VPN, PPTP, etc. Using license and allowing there are multiple authorizations among user, service owner and agents, to ensure user access to data securely.

c ) User data security assurance: according to the different user's requirements, different data storage protection should be provided. At the same time, the efficiency of data storage should be improving.

d) Using a series of measure to solve the user dynamic requirements, including a complete single sign-on authentication, proxy, collaborative certification, and certification between security domains.

e) Establishment of third-party monitoring mechanism to ensure that operation of cloud computing environment is safe and stable.

f) The computing requested by service requestor, should carry out the safety tests, it can check whether they contain malicious requests to undermine the security rules.

## 5 Security Models for Cloud Computing

### 5.1 Model for addressing Security Policies

The concept of Security Access Control Service (SACS). has been introduced [9] to address the above mentioned security policies in cloud computing environment.

Figure 1 represents the composition of its system modules. SACS includes Access Authorization, Security API, cloud connection Security. Access Authorization is used to authorize to users who want to request cloud service; Security API keeps users use specific services safely after accessing to the cloud; cloud connection security to ensure the bottom resource layer. Combining the SACS with the existing architecture of cloud computing, A security model of cloud computing is constituted, as shown in Fig.2.

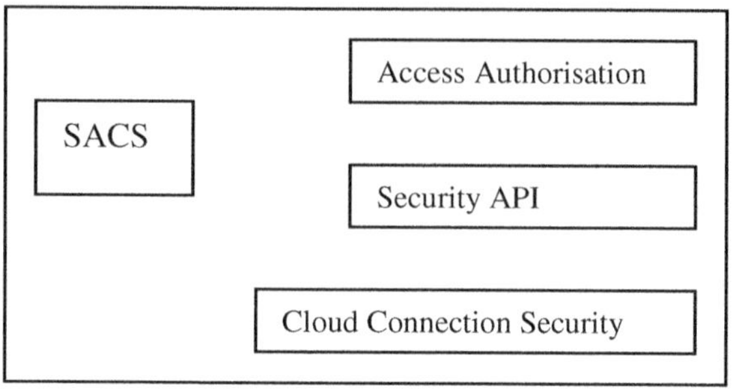


**Fig. 1.** The system modules of SACS

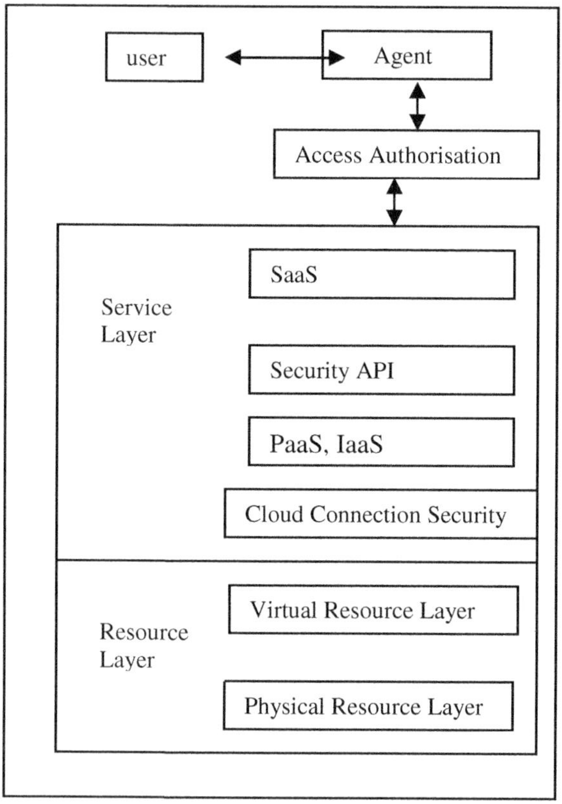


**Fig. 2.** Security Model

The process in the security model: First, the user creates a local user agent, and establish a temporary safety certificate, then user agent use this certificate for secure authentication in an effective period of time. This certificate, including the host name, user name, user ID, start time, end time, and security attributes, etc. the user's security access and authorization is complete. Second, when the user's task use the resource on the cloud service layer, mutual authentication take place between user agent and specific application, while the application check if the user agent's certificate is expired, a local security policy is mapped.

Third, according to user's requirements, cloud application will generate a list of service resource, and then pass it to the user agent. Through Security API, user agent connects specific services. And Cloud connection security ensures the safety of resource provided by the resource layer. The security API in this model should be achieved with SSL method, while the realization of cloud connection security uses SSL and VPN methods

## 5.2 Deployment Models for Eliminating Security Concerns

To address the above mentioned security issues various deployment models have been proposed In the following, we present five deployment models [8] that address users' security concerns with cloud computing.

**1. Separation Model:** The main idea is to have two independent services responsible for data processing and data storage. (Figure 3)

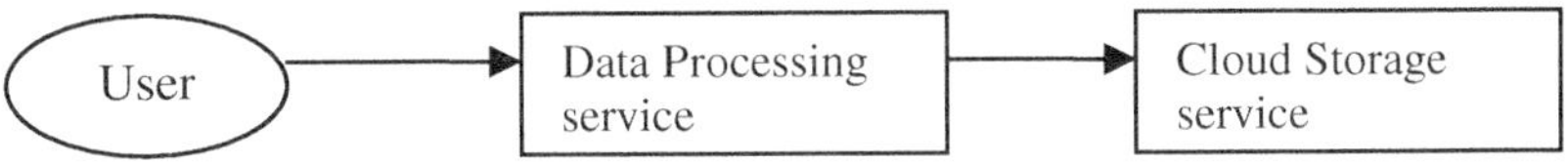


**Fig. 3.** Separation Model

Data are presented to users and are processed by the Data Processing Service. When the data need to be stored, they are handed over to the Cloud Storage Service, which will make the data persistent and ready for retrieval in the future. The Separation Model mandates at least two different cloud computing service providers be involved in a transaction. To some extent, this prevents some frauds and errors by preventing any single service provider from having excessive control over the transactions.

**2. Availability Model:** With the availability model, a user can work on her data via a data processing service, and the data will be kept on a cloud storage service.(figure 4).

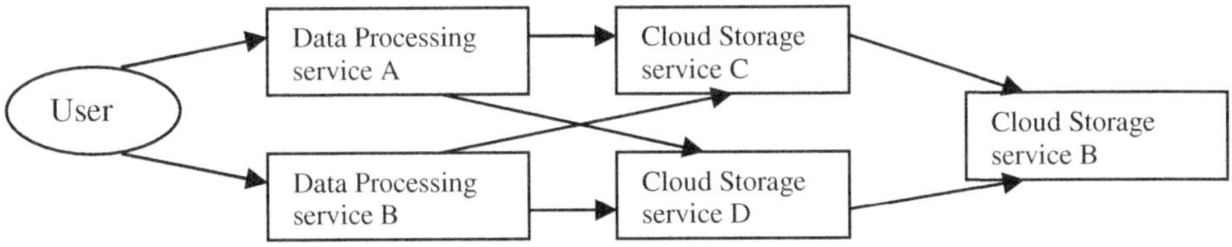


**Fig. 4.** The Availability Model

To ensure the availability of the services, there are at least two independent data processing services, Data Processing Service A and Data Processing Service B respectively, and two independent data storage services, Cloud Storage Service C and Cloud Storage Service D respectively. Either one of the data processing services can have access to the data on either one of the cloud storage services. Data are replicated and synchronized via a Replication Service. The Availability model imposes redundancy on both data processing and cloud storage. Hence there is no single point of failure with respect to data access. When a data processing service or a cloud storage service experiences failure, there is always a backup service present to ensure the availability of the data.

**3. Migration Model:** When data on clouds can only stay on the clouds where they are kept, users will be forced to stay with the clouds unless they decide to give up their data. This is not an acceptable situation.

**Fig. 5.** Migration Model

In this model (figure 5) users process their data via a Data Processing Service, where the data are kept on Cloud Storage Service A. The Cloud Data Migration Service can interact with Cloud Storage Service A and another cloud storage service, namely Cloud Storage Service B. The Cloud Data Migration Service can move data from Cloud Storage Service A to Cloud Storage Service B, and vice versa. By being able to move data from Cloud Storage Service A to Cloud Storage Service B, users need not worry about their data being excessively controlled by a cloud provider, knowing that they can switch to another service provider by moving the data out from the current cloud storage service provider to another.

**4. Tunnel Model:** The Tunnel model introduces a tunnel service located between the Data Processing Service and the Data Storage Service. (figure: 6)

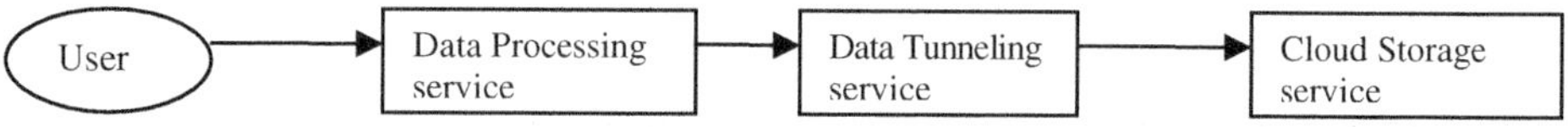


**Fig. 6.** Tunnel Model

The tunnel servers as a communication channel between the Data Processing Service and the Cloud Storage Service. It is responsible for providing an interface for the two services to interact with each other, for manipulating and retrieving data. The tunnel can in fact be implemented as a service as well. With the Tunnel Model, the Data Processing Service manipulates data based on the interface provided by the Data Tunneling Service. The Cloud Storage Service will not be able to relate the data it keeps with a specific data processing service. The Tunnel Model makes it extremely difficult for the Data Processing Service to collude with the Cloud Storage service for fraud.

**5. Cryptography Model:** For critical applications, the security of data, especially confidentiality and integrity, are key requirements. Data confidentiality and integrity are in most cases dependent on cryptography support. The Cryptography Model Figure 7, augments the Tunnel Model with a Cryptography Service, which provides support for cryptographic operations on data. The Data Processing Service feeds data to the Data Tunneling Service for persistence. The Data tunneling Service will invoke the Cryptography Service to perform a cryptographic operation on the data before handing the data over to the Cloud Storage Service. Thus the data kept by the Cloud Storage Service are cryptographically processed, meaning that they could be ciphertext that can only be read by those who have the decryption key, or they could be data augmented with digital signatures or message authentication codes, and so on, depending on the security requirements. With the Cryptography Model, data can be stored in their cryptographically processed form.

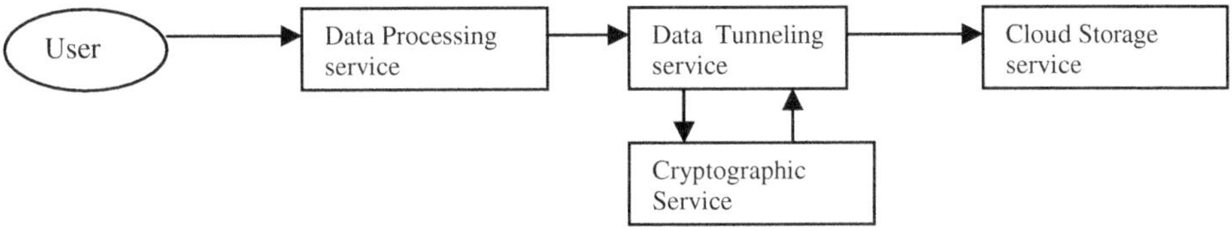


**Fig. 7.** Cryptography Model

# 6 Conclusion

With the continuous promotion of cloud computing, security has become one of the core issues In this paper security in cloud computing was elaborated in a way that covers security issues and challenges and the deployment models for eliminating Security concerns. These deployment models are developed to address the security issues raised by the identified security concerns. The proposed models are not without limitations. As the proposed models are at deployment architecture level, they do not include specific protocols and algorithms that can provide supports for confidentiality and integrity at cryptography level. Corresponding design patterns and interfaces should also be developed to allow cloud based applications be deployed on clouds in the manners specified by the proposed models. In all, Cloud computing platform need to provide some reliable security technology to prevent security attacks, as well as the destruction of infrastructure and services .

# References


1. Cloud Computing, Wikipedia, `http://en.wikipedia.org/wiki/CloudComputing`
2. Lenk, A., Klems, M., Nimis, J., Tai, S., Snadholm, T.: What's Inside the Cloud? An Architectural Map of the Cloud Landscape. In: IEEE Proceedings, ICSE 2009, May 23, pp. 23–31 (2009)

3. Harris, T.: Cloud Computing- An Overview., Whitepaper, Torry Harris Bussiness Solutins (January 2010)
4. Buyya, R., Pandey, S., Vecchiola, C.: Cloudbus Toolkit for Market-Oriented Cloud Computing. CloudCom, pp. 22-44 (2009)
5. Popovic, K., Hocenki, Z.: Cloud Computing security issues and challenges. In: MIPRO 2010, opatija, Croatia, May 24-28, pp. 344–349 (2010)
6. Jing, X., Zhang, J.-j.: A brief survey on the Security Model of Cloud Computing. In: Ninth International symposium on Distributed computing and Applications to Business, Engineering and Sciences, pp. 475–478. IEEE, Los Alamitos (2010)
7. Mukherjee, K., Sahoo, G.: A secure Cloud Computing. In: IEEE Proceedings of International Conference on Recent Trends in Information, Telecommunication and Computing, pp. 369–371 (2010)
8. Zhao, G., Rong, C.: Deployment models- towards eliminating Security Concerne from Cloud Computing. In: IEEE Proceedings of International Conference on High Performance Computing and Simulation (HPCS), pp. 189–195 (2010)
9. Jing, Y., Zhang, J.-j.: A brief survey on the Security model of Cloud Computing. In: Ninth International Symposium on Distributed Computing and Applications to Bussiness, IEEE Proceedings Engineering and Science, pp. 475–478 (2010)
10. Brodkan, J.: Gartner: Seven Cloud Computing Security Risks, `http://www.infoworld.com/.../security.../gartner-seven- cloud-computing-security-risks`
11. Almulla, S.A., Yeun, C.Y.: Cloud Computing Security Management. In: IEEE Proceedings of Second International Conference on Engineering Systems management and its Applications (ICESMA), pp. 1–7 (2010)

# Cloud Computing – The Future

Vinay Chawla and Prenul Sogani

Reliance Communications, DAKC, BHQ, Thane-Belapur Rd,
Navi Mumbai, Maharashtra, India
Vinay.Chawla@RelianceADA.com,
Prenul.Sogani@RelianceADA.com

**Abstract.** Cloud services are expected to become the driving force of IT innovation for the foreseeable future. Most companies are trying to be a part of the story either as enablers, vendors or service providers. With this, the cloud market is expected to grow at a phenomenal rate, with both large enterprises and SMEs going for it. Enterprise concerns over security, lock-in, etc. will be overcome by the benefits of the cloud. Large enterprises will prefer going for private or hybrid cloud deployment, while SMEs will prefer public clouds.

**Keywords:** cloud, computing, hybrid, IaaS, PaaS, SaaS, virtualization, VPC.

## 1 Introduction to the Cloud

Any application or service hosted at a remote location and accessed over the Internet or a private network is essentially a cloud service.

Cloud offerings can be classified into Software as a Service (SaaS), Platform as a Service (PaaS), and Infrastructure as a Service (IaaS). SaaS is any application service that is delivered over network on a subscription and on-demand basis. These apps are accessible through a thin client or even an Internet browser. PaaS involves provisioning of a platform (by way of APIs) for developers to create their applications, thereby providing them more control over the hosting environment. IaaS involves delivery of the basic infrastructure of network, storage and compute on which PaaS is hosted. These also include the monitoring and backup services.

## 2 Stakeholders in the Cloud System

Cloud players can be segregated on the basis of their presence in the cloud delivery value chain. However, the role of some mature players (such as IBM) can span across categories. Each player has its own preference of the cloud delivery model.

1. *Cloud Enablers:* These players provide the technology and infrastructure for the provisioning of cloud services. Examples are VMware, Citrix.
2. *Cloud Vendors:* These are players who actually provide the various types of cloud services such as SaaS, PaaS and IaaS. Some examples are Rackspace, Amazon.
3. *Service Providers:* Service providers provision the ancillary services that are required for the successful deployment of cloud systems.

A. Mantri et al. (Eds.): HPAGC 2011, CCIS 169, pp. 113–118, 2011.

## 3 Cloud Delivery Models

Cloud services can be provided primarily through three distinct models: private, public and hybrid clouds. Based on factors such as security, cost, a customer should determine the most suitable solution for cloud deployment.

*Private Cloud:* Private cloud delivers the low hanging fruits by allowing for operational and cost benefits while still preserving control over applications and information. In the private cloud delivery model, the entire IT service delivery infrastructure is owned and managed by the organization.

*Public Cloud:* In order to further enhance service capabilities and operational efficiency, organizations can have the services provided by their partners to be provisioned over the Internet. The same resources can be shared among multiple customers of the cloud service providers.

*Hybrid Cloud:* It is an amalgamation of private and public clouds. If an organization has regulatory or security concerns over cloud, they can host only the non-mission-critical data on the public cloud while critical data remains in-house.

*Virtual Private Cloud:* Virtual private clouds provide the cost benefits of the public cloud while providing security and reliability on par with private clouds. This model delivers services from a public cloud over MPLS based VPN networks.

## 4 Cloud Market and Economics

Given the positive sentiment towards deployment of cloud services, the cloud services market will exhibit significant growth. The overall market is expected to be a USD 44 billion market by 2013 [2] from the current USD 17.4 billion [3].

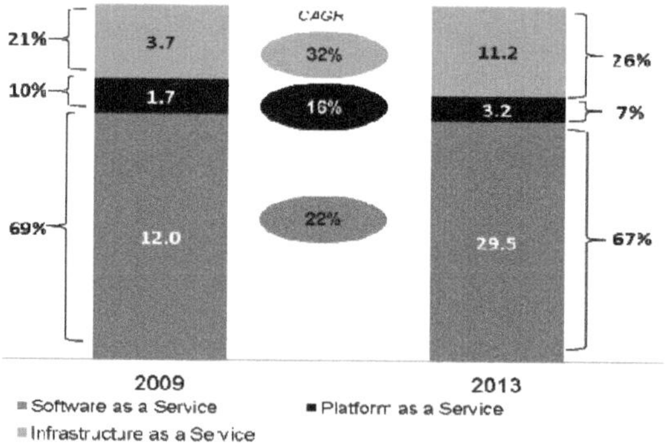


**Fig. 1.** Market Opportunity for Cloud Services [2], [3]

The Indian market for cloud services is expected to be around USD 1.1 billion by 2015, representing year-on-year CAGR of 29% during 2010-15 [4].

Although the industry is at a nascent stage, there have been some early estimates on the benefits accrued from implementing cloud services in totality. Typically, an enterprise can gain about 25% cost savings by using cloud services [5].

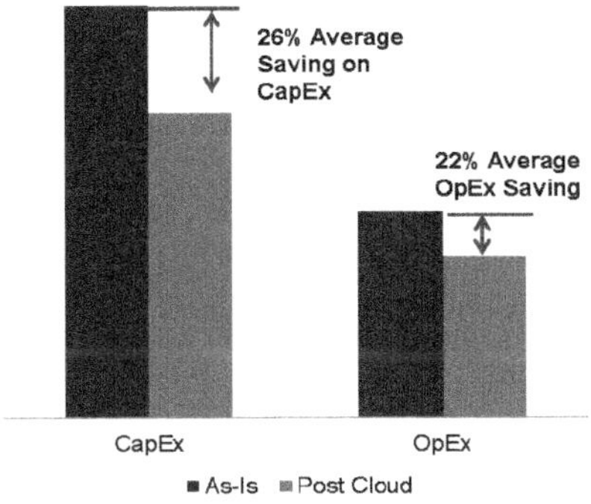


**Fig. 2.** Cost Savings from Cloud Services [5]

# 5 Cloud Opportunity

## 5.1 IaaS - Current Opportunity

Currently IaaS has the highest growth potential segment among cloud services (fig. 1). It has interested many players. Hosting and co-location vendors have realized that cloud services can help them improve margins [6] as compared to their current portfolio of services with minimal modifications in their current infrastructure.

But there is a chance that IaaS will get commoditized. In order to sustain themselves in the IaaS space, CSVs should look to provide more localized and customized IaaS. For example, in a newly developing market such as Africa, where the infrastructure is still being deployed, a CSV can project desktop as a service (DaaS) as an avenue to reduce costs.

## 5.2 SaaS – A Matured Opportunity

SaaS is the biggest cloud service category and will continue to remain so, in the near future (fig. 1). In 2013, it will be dominated by Content, Communication and Collaboration segment (44%) followed by CRM (32%), ERP (13%) & SCM (11%) [7].

CSVs who want to provide SaaS should go for partnerships with existing vendors as most of the SaaS segments have mature players. They can also reduce time to market of services by adopting a simple resale model. CSVs can even use the expertise of SaaS vendors to develop their IT systems to optimally serve cloud services.

### 5.3 PaaS – A Future Opportunity

PaaS is a relatively smaller cloud services market, but is expected to provide the differentiation to CSVs who have exhausted the IaaS and SaaS routes. A CSV needs to look at its technology readiness, ownership of content portfolio and its overall SaaS strategy before taking a decision on PaaS strategy. Based on these factors, a CSV can adopt a no PaaS strategy, a PaaS purchase strategy or a PaaS development strategy.

## 6 Target Customer Segments

Cloud services are expected to be important to both large enterprises as well as SMEs. Large enterprises have the majority share in this space, but the SMEs are expected to grow at a significantly higher rate in the future. Currently SMEs' share of cloud services is 25%, which is expected to rise to 40% by 2014 [8]. This will be because SMEs suffer from high IT costs and have the highest to gain from migrating to cloud services.

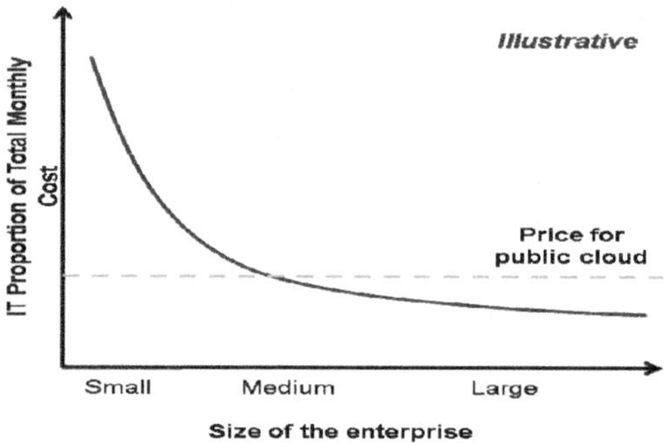


**Fig. 3.** IT Cost Comparison of Public Cloud for Enterprises and SMEs [9]

### 6.1 Strategies for Large Enterprises

Judging from fig. 3, large enterprises should ideally go for private clouds and should prefer public clouds only under the following conditions [10]:

1. *Frequency of Spikes:* If the frequency of spikes in demand is less than the inverse of public cloud cost to internal IT cost. For example, if the cost of public cloud services were four times as much as owned capacity, they still make sense if peak demand only occurs a quarter of the time or less.
2. *Magnitude of Spikes:* If the magnitude of spike in demand (from average) is higher than the public cloud cost to internal IT cost. For example, if public cloud services cost twice as much, it still makes sense for those demand curves where the peak-to-average ratio is two-to-one or higher.

Another factor which can help large enterprises decide whether to go for cloud services is the sophistication of cloud services they require. They can map their requirements on the following spider graph and decide to choose based on their requirement overlap with telcos or other CSVs.

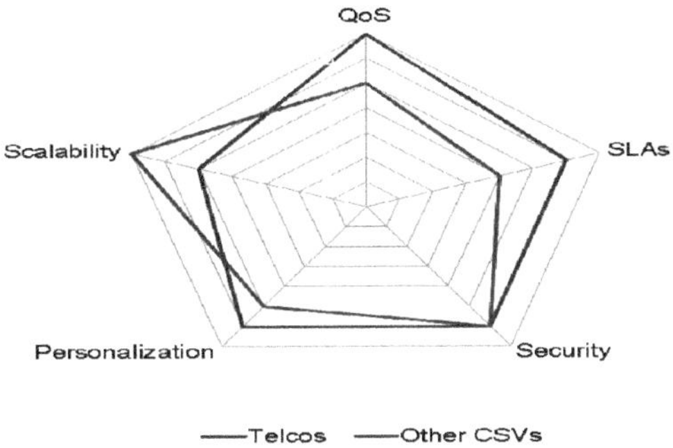


**Fig. 4.** Spider for deciding on the best CSV for a large enterprise

## 6.2 Strategies for SMEs

SMEs differ substantially in their requirements from large enterprises. Hence, a separate strategy has to be undertaken to target SMEs. Based on SME preferences, cloud solutions with the following features should be offered:

1. *Packaged Total Solutions:* SMEs prefer all their telecom needs to be catered to by a single vendor, thus preferring bundled services. Moreover, they want solutions which are easy to deploy and support.
2. *Standardized Solutions:* In the long run, SMEs look at the TCO of the solution rather than its price. TCO is the least for solutions which are off-the-shelf.
3. *Plug-and-Play Service Care:* SMEs lack in IT skills and resources to manage after sale of solutions. Thus they go with proven and tailored solutions that have good after-sales support.

## 6.3 Conclusion

Based on their inherent strengths and target customer segment, various CSVs will aim to provide specific cloud services.

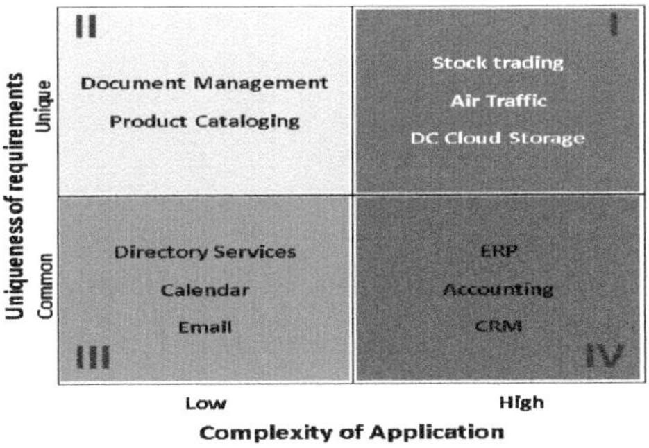


**Fig. 5.** Target customer applications for various CSVs

Stated below is the mapping of the best target segments for various CSVs:

1. *Target for Localized Telcos (Quadrant I in Fig. 5):* These customers require strong SLAs and proximity to the data center as they have customized and highly localized applications.
2. *Target for Global Telcos (Quadrant II in Fig. 5):* Global telcos can cater those applications which are highly specialized and have a small market in a single country but become profitable if catered globally.
3. *Target for Large CSVs (Quadrant III in Fig. 5):* Large CSVs such as Google and Amazon cannot provide stringent SLAs and hence target applications which are scale dependent with low margins.
4. *Target for ISVs (Quadrant IV in Fig 5):* Internet Service Providers (ISVs) such as Salesforce.com and NetSuite are best placed to provide highly complex applications to a large number of customers.

## References


1. AT&T Business Solutions: The Enterprise Cloud (2009)
2. Belmans, W., Puopolo, S., Yellumahanti, S.: Network service providers as cloud providers. In: Cisco IBSG Survey Results (2010)
3. Gens, F., Mahowald R., Villars, R.L., Bradshaw, D., Morris, C.: Cloud Computing 2010 - An IDC Update (September 2009)
4. Malini, N.: Third party data centers - crucial to market growth. In: CIOL.com (February 2011)
5. Drogseth, D.: The Responsible Cloud. In: Enterprise Management Associates (January 2010)
6. Cisco: Infrastructure as a Service: Accelerating Time to Profitable New Revenue Streams (2009)
7. Deloitte: Cloud Computing – Market Overview and Perspective (October 2009)
8. Hilton, S.: Seize the USD35.6 billion global market for enterprise cloud services. In: Analysis Mason (June 2010)
9. McKinsey: Clearing the Air on Cloud Computing (April 2009)
10. Weinman, J.: Mathematical Proof of the Inevitability of Cloud Computing. In: Cloudonomics.com (November 2009)

# Cloud Computing: A Need for a Regulatory Body

Bikramjit Singh[1], Rizul Khanna[2], and Dheeraj Gujral[2]

[1] BSNL, Nakodar, India
bikramjit.nitj@gmail.com
[2] Dept of Electronics and Comm Engineering, NIT Jalandhar, India
{rizulkhanna,dheeraj.nitj}@gmail.com

**Abstract.** There have been massive rise in spending on Cloud technologies from the past two years. Now, every IT Setup is expanding their horizons in the cloud services and related technologies. But turning to the issue of inexistence of a regulatory body or issues such as data protection, has kept many companies out of the cloud, particularly which are engaged in Financial Services, Health Care Services or Secret or Government Services, where data leakage or protection can't be compromised. In this paper, initially we are going to review generic cloud computing term, their types and services and later, we assert for the need of a regulatory body and its resolution in providing a model for this body, which looks after various aspects like protocols, securities, data interactions, etc.

**Keywords:** applications, cloud computing, data interaction, internet, regulatory body, security.

## 1 Introduction - Cloud Computing

Cloud computing broadly describes off-premise, on-demand computing where the end-user is provided applications, computing resources, and services (including operating systems and infrastructure) by clouds services provider via the Internet. Cloud computing offers computer application developers and users an abstract view of services that simplifies and ignores much of the details and inner workings. A provider's offering of abstracted Internet services is often called *The Cloud*. This frequently takes the form of web-based tools or applications that users can access and use through a web browser as if they were programs installed locally on their own computers [1].

Computing involving clouds consist of thousands of servers located at data centres running tens of thousands of application instances accessed by millions of users at the same time. Further, in order to provide secure access to computing resources for various user roles - cloud operators, service providers, resellers, IT administrators, application users - computing clouds need to have delegated administration and self-service capabilities [2].

In particular, five essential elements of cloud computing are clearly articulated –

**On-Demand Self-Service.** A consumer with an instantaneous need at a particular timeslot can avail computing resources in an automatic fashion without resorting to human interactions.

A. Mantri et al. (Eds.): HPAGC 2011, CCIS 169, pp. 119–125, 2011.

**Broad Network Access.** These computing resources are delivered over the network and used by various client applications with heterogeneous platforms situated at a consumer's site.

**Resource Pooling.** A cloud service provider's computing resources are pooled together in an effort to serve multiple consumers using either the multi-tenancy or the virtualization model.

**Rapid Elasticity.** For consumers, computing resources become immediate rather than persistent - there are no up-front commitment and contract as they can use them to scale up whenever they want, and release them once they finish scaling down.

**Measured Service.** Although computing resources are pooled and shared by multiple consumers (i.e. multi-tenancy), the cloud infrastructure is able to use appropriate mechanisms to measure the usage of these resources for each individual consumer through its metering capabilities [3].

### 1.1 Cloud Types

**Private Cloud.** Computing architecture of this cloud is dedicated to the customer and is not shared with other organisations and managed by the organization or a third party regardless whether it is located premise or off premise.

**Public cloud.** The public cloud is used by the general public cloud consumers and the cloud service provider has the full ownership of the public cloud with its own policy, value, and profit, costing, and charging model. The customer has no visibility over the location of the cloud computing infrastructure. The computing infrastructure is shared between organizations. Many popular cloud services are public clouds including Amazon EC2, S3, Google App Engine, Force.com, etc.

**Community Cloud.** Several organizations jointly construct and share the same cloud infrastructure as well as policies, requirements, values, and concerns. The cloud community forms into a degree of economic scalability and democratic equilibrium. For example, all the government agencies in a city can share the same cloud.

**Hybrid cloud.** The cloud infrastructure is a combination of two or more clouds (private, community, or public) that remain unique entities but are bound together by standardized or proprietary technology that enables data and application portability. Organizations use the hybrid cloud model in order to optimize their resources to increase their core competencies by margining out peripheral business functions onto the cloud while controlling core activities on-premise through private cloud [3].

### 1.2 Cloud as a Service

In practice, cloud service providers tend to offer range of services in a cloud computing environment. And these services are –

**Software as a Service (SaaS).** Cloud consumers release their applications on a hosting environment, which can be accessed through networks from various clients by application users to achieve economies of scale and optimization in terms of speed, security, availability, disaster recovery, and maintenance. It can be accessed by the customers on pay per use basis [4]. Examples of SaaS include SalesForce.com, Google Mail, Google Docs etc.

**Platform as a Service (PaaS).** It is a development platform supporting the full *Software Lifecycle* which allows cloud consumers to develop cloud applications (e.g. SaaS) directly on the PaaS cloud. Google App Engine is its famous known type.

**Infrastructure as a Service (IaaS).** Cloud consumers directly use IT infrastructures provided in the IaaS cloud. Virtualization is extensively used in IaaS cloud in order to integrate/decompose physical resources in an ad-hoc manner to meet growing or shrinking resource demand from cloud consumers. An example of IaaS is Amazon's EC2 which allows users to rent virtual computers on which to run their own computer applications [5].

**Data storage as a Service (DaaS).** The delivery of virtualized storage on demand becomes a separate Cloud service - data storage service called DaaS, could be seen as a special type IaaS. DaaS allows consumers to pay for what they are actually using rather than for the entire database. Some DaaS offerings provide table-style abstractions that are designed to scale out to store and retrieve a huge amount of data within a very compressed timeframe, often too large, too expensive or too slow for most RDBMS to cope with [6].

## 2 Regulation: A Dire Need for Future

One can consider regulation as actions of conduct imposing sanctions, so, in perspective cloud computing, what type of regulation can be here? Let's take a look in Fig. 1.

It shows that cloud 1 is connected to machine 1 and machine 2, and further connected to cloud 2 via machine 1. Such inter-connecting clouds are very much-existing in near future, when small companies or subsidiaries too get inclined to have such cloud technologies in their setups because of their more familiarization and foreseen cost reductions. Suppose a malicious application or intruder attacks machine 2 and possibility of theirs subsequently entering the cloud 1, as it is based on the theme 'everything is online', though machine 2 is also a part of cloud 1 and thus it too gets attacked by the same malicious code. Then this unwanted code tries to infect other interconnected machines or networks and thus, can easily enter into cloud 2 and affecting many of its systems. So we see here that a problem in one cloud can lead to a total collapse of a network. This is one of the key reasons behind its limited usage [7]. Some of the initials works identified the following threats in cloud computing - Abuse and Nefarious Use of Cloud Computing; Insecure Application Programming Interfaces; Malicious Insiders; Shared Technology Vulnerabilities; Data Loss/ Leakage; Account, Service & Traffic Hijacking; Unknown Risk Profile [8].

The extent of hacking rises proportionally with the advancements in security measures. Now days, Collection and analysis of data are now possible cheaply. So, what is the impact on privacy of abundant data and cheap data-mining? Because of the cloud, attackers potentially have massive, centralized databases available for analysis and also the raw computing power to mine these databases. For example, Google is essentially doing cheap data mining when it returns search results. How much more privacy did one have before one could be Googled [9]? Various law bodies and business setups like Microsoft Corporation also feel the need for a regulatory body. They argued that the U.S. government should create rules and regulations for cloud computing, a burgeoning technology that has gained traction among schools and colleges. As a growing number of businesses, governments, schools, and universities store sensitive data on off-site servers managed by third parties, lawmakers should draft legislation that would protect the integrity of this information [10]. Another example is of Google's myriad Cloud-based application offerings. A new complaint made by Epic (Electronic Privacy Information Center) to the US Federal Trade Commission urges the regulatory agency to "consider shutting down Google's services until it establishes safeguards for protecting confidential information." They blamed Google for revealing user's data in its Gmail webmail service, Google Docs online word processing and spreadsheets and in Google Desktop [11]. Along with it, the cloud that hosts Gmail, tens of thousands of Gmail accounts were lost, closed, reset and important data was lost [7]. Other sectors like financial services, who seem an unlikely customer of the cloud, are highly sensitized to the impact of latency and outages on their business.

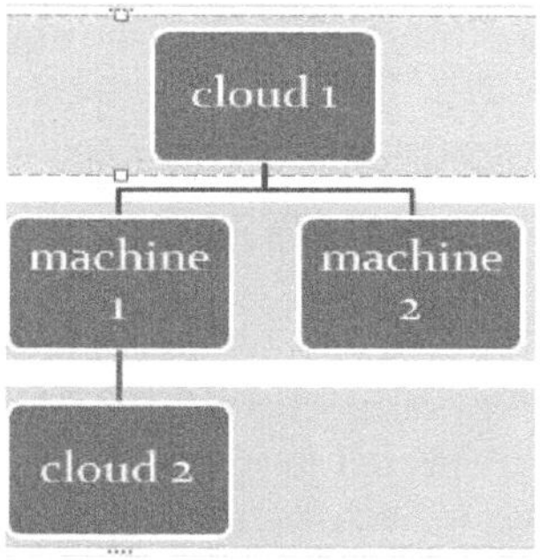


**Fig. 1.** Clouds and Machines Network

## 2.1 Regulatory Body

We propose to create a worldwide base regulatory authority to improve upon the aspects of regulations, future research and mainly security issues being faced by the clouds these days. The aim of this body is going to promote, provide, monitor, regulate the working and fix responsibilities/liabilities if something goes wrong or suffers at the hands of clouds, as this technology embedding more and more with the time, on this planet.

The reason for the assertion of this world wide authority is to fix standards or rights for the clouds. The authority will monitor the protocols dealing with cloud's inflow and outflow traffic. It will aim to create standards for the business clouds in particular, in order to check the un-authoritative access to their sensitive data. The need to create this single body network arises from the accelerating growth of clouds in the coming future, where it can help clear out the mess. The body is expected to work in the following pattern –

1. Every cloud needs to register themselves with this regulatory body.
2. A sort of nomenclature and standards should be defined in the naming of clouds for their easy identification by their administrators or regulators or intrusion monitors or security experts.
3. Each cloud should list them under the type category of cloud, and laying out standards for rights to data access/sharing among all these types.
4. Protocols should be defined for clouds data interaction, like of infusing of data packets with encrypted signatures/labels in their headers, so that malicious activities from clouds can be easily detected in the network.

The hierarchy of this regulatory body is shown pictorially in Fig. 2, demonstrating the working of the regulatory body being mentioned.

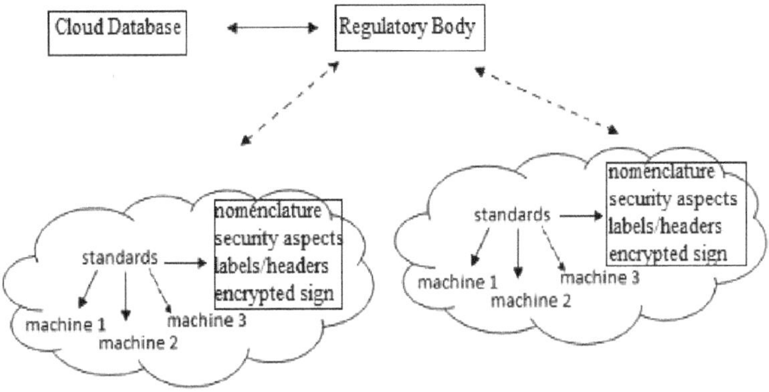


**Fig. 2.** Hierarchy of a Regulatory Body

How the regulatory authority is expected to execute its working in the system? We have opinionated an idea, which defines its role. Let us take an example of clouds network, shown in Fig. 3. In this figure, Cloud 1 is connected to machine 1 and 2, and also connected to Cloud 2. The identity of Cloud 2 is known to Cloud 1. As if Cloud 1 is being approached by unknown Cloud 3 for interaction, Cloud 1 would ask the regulatory body for security clearance or demanding standards laid by this body for interaction with other clouds. This regulatory body maintains the database containing cloud types, their rights, or even reputations apparent from indulging or reporting some malicious activities, or their security standards. After clarifying the instance, then it will proceed for the interaction upon achieving their required level of satisfaction. So, overall some standards for achieving clouds data interaction has been defined and given below -

1. Whenever a cloud needs to interact with the other cloud, the other cloud must check the encrypted signatures or labels of the cloud, as it carries the information regarding standards or protocols laid by the regulatory body.
2. In case a cloud carries out some illegal code like access to sensitive data of some other cloud, not deliberately, may be accidently, for example cloud 3 in Fig. 3 interacts with cloud 1 and tries to access its personal data, cloud 1 must report

or feedback to the regulatory authority and certain measures should be taken to subside this activity, and also may take some stringent actions if it's the clouds deliberate act and even bar from holding a cloud.

3. It should work in direction to strengthen people's faith in this cloud computing workings, and supporting this technology at every corner of this planet in a most efficient way. It should promote future research in upgrading the technologies, securities aspects, improvising various related models, etc.

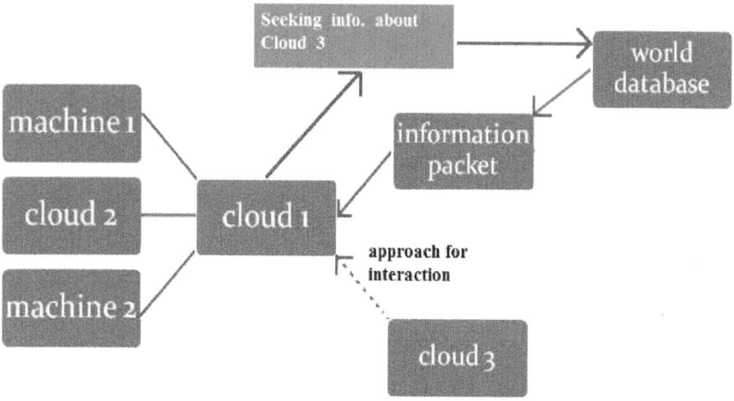


**Fig. 3.** A Network outline consisting of clouds, machines and a sole Regulatory Body

# 3 Conclusions

Cloud Computing holds a lot of promise and have wider influence on hosting and application development and this part going to play more dependent role in near future for IT setups. Amazon is currently the major player, although its competitors are offering compelling feature sets for researchers and enterprises looking to deploy existing applications with minimal changes. As this new technology is aggressively expanding, World should prepare themselves that the incumbent issues will be satisfactorily resolved, which greatly asserts the idea of having a unified collaborated body, a regulatory body, which looks or governs this technology's standards from various aspects. As, more and more we're interacting with this technology, more will be the need felt for this regulatory body. Also, there is no present collaboration on the part of enhancing security aspects for the clouds, further helps in demand or establishment of this concept.

# References

1. Cloud Computing, `http://searchcloudcomputing.techtarget.com/definition/cloud-computing`
2. Zhang, S., Zhang, S., Chen, X., Huo, X.: Cloud Computing Research and Development Trend. In: Proceedings of Second International Conference on Future Networks, pp. 93–97 (2010)

3. Dillon, T., Wu, C., Chang, E.: Cloud Computing: Issues and Challenges. In: Proceedings of 24th IEEE International Conference on Advanced Information Networking and Applications, pp. 27–33 (2010)
4. SaaS Introduction with Examples - Cloud Service Model, http://www.techno-pulse.com/2010/04/saas-introduction-example-cloud-service.html
5. Amazon Elastic Compute Cloud (Amazon EC2), http://aws.amazon.com/ec2/
6. Peng, J., Zhang, X., Lei, Z., Zhang, B., Zhang, W., Li, Q.: Comparison of Several Cloud Computing Platforms. In: Proceedings of Second International Symposium on Information Science and Engineering, pp. 23-27 (2009)
7. When The Cloud Hurts..., http://www.crazyengineers.com/tag/disadvantages-of-cloud-computing/
8. Cloud Security Alliance - Top Threats to Cloud Computing V1.0, http://www.cloudsecurityalliance.org/topthreats/csathreats.v1.0.pdf
9. Chow, R., Golle, P., Jakobsson, M., Shi, E., Staddon, J., Masuoka, R., Molina, J.: Controlling Data in the Cloud: Outsourcing Computation without Outsourcing Control. In: 16th ACM Conference Proceedings on Computer and Communications Security, pp. 85–90 (2009)
10. Microsoft calls for cloud-computing regulations, http://www.eschoolnews.com/2010/01/21/microsoft-calls-for-cloud-computing-regulations/
11. Can the Cloud survive regulation?, http://devcentral.f5.com/weblogs/macvittie/archive/2009/03/26/can-the-cloud-survive-regulation.aspx

# Clustering Dynamic Class Coupling Data to Measure Class Reusability Pattern

Anshu Parashar and Jitender Kumar Chhabra

Department of Computer Engineering,
National Institute of Technology, Kurukshetra,
Kurukshetra 136 119, India
conf.ppr@gmail.com

**Abstract.** Identification of reusable components during the process of software development is an essential activity. Data mining techniques can be applied for identifying set of software components having dependence amongst each other. In this paper an attempt has been made to identify the group of classes having dependence amongst each other existing in the same repository. We explore document clustering technique based on tf-idf weighing to cluster classes from vast collection of class coupling data for particular java project/program. For this purpose firstly dynamic analysis of java application is done using UML diagrams to collect class import coupling data. Then in second step, this coupling data of each class is treated as a document and represented using VSM (using TF and IDF). Then finally in the third step basic K-mean clustering technique is applied to find clusters of classes. Further each cluster is ranked for its goodness.

**Keywords:** Coupling, Data Mining, Software Reusability.

## 1 Introduction

Software Reuse is defined as the process of building or assembling software applications from previously developed software [16] The success of reusability is highly dependent on proper identification of whether a particular component is really reusable or not. Class coupling plays a vital role in measuring the reusability and selecting classes for reuse in combination because the highly coupled classes are required to be reused as a group [17] for ensuring the proper functioning of the application [8]. So for reuse, issues like maintaining class code repository, deciding what group of classes should be incorporated into repository and identifying exact set of classes to reuse, need to be addressed. By using clustering, one can find frequently used classes in the same cluster and can know their coupling with other classes in a particular application.

### 1.1 Data Clustering and Reusability

Data mining techniques can be used to analyze software engineering data to better understand the software and assist software engineering tasks. Clustering can be used

A. Mantri et al. (Eds.): HPAGC 2011, CCIS 169, pp. 126–130, 2011.

for document clustering in the information retrieval task. The Vector Space Model (VSM) is the basic model for document clustering. In this model, each document, $d_j$, can be represented as a term-frequency vector in the term-space:

$$d_{jtf} = (tf_{1j}, tf_{2j}, \ldots. tf_{vj}) \quad j=1,2,\ldots D$$

where $tf_{ij}$ is the frequency of the $i_{th}$ term in document $d_j$ , $V$ is the total number of the selected vocabulary, and $D$ is the total number of documents in the collection.[18]. One can weight each term based on its Inverse Document Frequency (IDF) [18,3]. After having VSM representation, K-mean algorithm can be applied to cluster the documents [15]. Clustering technique can be applied to cluster the Classes/components that may often be reused in combinations [19]. Due to the popularity of open source concept large amount of source code of classes is available on internet as software repositories. For this reason, it is desirable to have clustering mechanism that form cluster of classes based on their association or coupling patterns. In this paper, we explore document clustering technique based on tf-idf weighing [3] to cluster classes from vast collection of class coupling data for particular java project/program. For this purpose firstly dynamic analysis of java application is done using UML diagrams to collect class import coupling data. Then in second step, these collected coupling data of each class are treated as document and represented as VSM(using TF and IDF). Then finally in the third step basic K-mean clustering technique is applied to find cluster of classes. Further each cluster is ranked for their goodness based on some user specified threshold and clusters which are not satisfying threshold are discarded as bad cluster.

The rest of the paper is organized as follows. Section 2 discusses the related works. Section 3 describes the proposed methodology. Section 4 presents conclusion and future scope.

## 2 Related Works

For object-oriented development paradigm, class coupling has been used as an important parameter effecting reusability[7]. Arisholm et al[4] have provided a method for identifying import coupled classes with each class at design time using UML diagrams. Few algorithms[1,13] like CLARANS [14], BIRCH [21], UPGMA [11] have been proposed for clustering large data sets. K-Mean's and its family of algorithms have also been extensively used in document clustering [12]. Fung et al [9] proposed to use the notion of frequent itemsets, document clustering and they used an approach proposed by Agrawal et al [2]. There are some distance measures available in literature like Absolute distance, Euclidean distance and cosine distance [22,23,24,6]. Alzghool et al [3] also proposed a technique based on clustering the training topics according to their tf-idf (term frequency-inverse document frequency) properties. For Evaluation of cluster quality authors in [20,15] proposed cluster ranking to quickly single out the most significant clusters based on goodness and quality of cluster.

# 3 Proposed Methodology

In our approach we use document clustering technique to cluster classes from vast collection of class coupling data for particular java project/program. Dynamic analysis of java application is done using UML diagrams to collect class import coupling data and collected import coupling data of each class are treated as document and represented as VSM (using tf-idf weighing). Then basic K-mean clustering technique is applied to find cluster of classes. Our approach consists of three steps as described in section 3.1 to 3.3.

## 3.1 Collection of Class Import Coupling Data through UML

During this step, the existing software system is analyzed in order to extract import coupling of its classes. Dynamic analysis of programs can be done through UML diagrams [10] as described by Erik Arisholm[4]. IC_OC ($C_i$) counts the number of distinct classes that a method in a given object uses.

$$IC\_OC(c_1) = \{(m_1, c_1, c_2) \mid (\forall (o_1, c_1) \in R_{oc})(\exists (o_1, c_2) \in R_{oc} | \in N) c_1 \neq c_2 \wedge (o_1, m_1 \mid o_2, m_2) \in ME\}$$

## 3.2 Representation of Collected Data

Data collected in step one should be represented in some suitable intermediate representation, so that clustering algorithm can be applied easily to find the clusters for classes that can be reused together. In our approach we propose to use basic Vector Space Model commonly used for document clustering. For an application A, the class set of A is represented as Class_Set(A)={ $C_1$,$C_2$,$C_3$.....$C_n$} where n is total number of classes in an application A. Each class $C_i$ is represented as class-frequency vector CFV($C_i$) of length n The CFV($C_i$) contains import usage frequency of all classes in class $C_i$. CFV ($C_i$) = $\{cf_{1i}, cf_{2i}, \ldots, cf_{ni}\}$, where $cf_{1i}$ represent frequency of usage of class $C_1$ in $C_i$. Next inverse class frequency (ICF) weighing is used to weight each class $C_i$ based on idf. So we calculate ICF of each class using idf formula 1:

$$\text{ICF}(C_i)=\log\left(\frac{n}{ICoupF(C_i)}\right) \tag{1}$$

Where $n$ is total number of classes, $ICoupF(C_i)$ number of classes using $C_i$. Then finally import coupling ICoup($C_i$,$C_j$) of class $C_i$ with $C_j$ is represented as 2D-point ($cf_{ji}$, $ICoupF(C_i)*cf_{ji}$).

## 3.3 Clustering of Class Import Coupling Data

After representing the class import coupling for each pair of classes ($C_i$ , $C_j$ ) as 2D points the clustering process begins. Here aim is to obtain K clusters of classes and value of K input by the user. So for this purpose K-mean algorithm are to be applied as described in [15]. Once we decide on what are the clusters into which we arrange the classes then K-mean algorithm starts. In K-mean algorithm initial centroids are often chosen randomly. The centroid is the mean of the points in the cluster. The absolute distance function(formula 2) is used to measure the closeness.

$$d_A(x,y)=\sum_{i=1}^{n}|x_i - y_i| \tag{2}$$

In every iteration centroids are recalculated. Each class pair ($C_i$ , $C_j$ ) is assigned to the cluster with the nearest seed point. By iterating K-mean algorithm (until there is no movement of points), it will discover clusters of the form {(C1, C3), (C2, C3)}. Their union {C1, C2, C3} will form the final cluster. We interpret this as classes in a cluster are coupled with each other and will be reused together. After having K cluster each cluster is ranked by taking average and sum of its *x,y* points , e.g. Cluster I ={(a,b), (c,d)} then RankC(Cluster I)= (a+c/2)+(b+d/2). This RankC(k) should pass the threshold $th_c$ specified by the user. Threshold is the lowest possible permissible rank that will be used to classify cluster as good or bad. If RankC(k) < $th_c$ then cluster k is discarded otherwise k retained.

## 4 Conclusions and Future Work

In this paper, an approach has been proposed to determine class reusability pattern from dynamically collected class import coupling data of java application. We have explored the idea of document clustering and represented coupling between two classes using tf-idf weighting scheme. Our initial study indicates that basic technique of K-mean clustering can be constructive to find out cluster of most reusable classes by clustering class import coupling behaviour. Currently, we have applied our approach on a simple example. However in future we are planning to experiment our approach on larger java applications. Moreover, other mining and clustering algorithms will be explored to apply on dynamic class coupling data for finding class reusability patterns.

## References


1. Abrantesy, A.J., Marquesz, J.S.: A Method for Dynamic Clustering of Data. In: Proceedings of the British Machine Vision Conference, pp. 154–163 (1998)
2. Agrawal, R., Imielinski, T., Swami, A.: Mining Association Rules between Sets of Items in Large Databases. In: ACM, SIGMOD, pp. 207–216 (1993)
3. Alzghool, M., Inkpen, D.: Clustering the Topics using TF-IDF for Model Fusion. In: ACM Proceeding of the 2nd PhD Workshop on Information and Knowledge Management, pp. 97–100 (2008)
4. Arisholm, E.: Dynamic Coupling Measurement for Object-Oriented Software. IEEE Transactions on Software Engineering 30(8), 491–506 (2004)
5. Bhatia, P.K., Mann, R.: An Approach to Measure Software Reusability of OO Design. In: Proceedings of the 2nd National Conference on Challenges & Opportunities in Information Technology, pp. 26–30 (2008)
6. Cosine Similarity, `http://en.wikipedia.org/wiki/Cosine_similarity`
7. Czibula, I.G., Serban, G.: Hierarchical Clustering Based Design Patterns Identification. Int. J. of Computers Communications & Control 3, 248–252 (2008)

8. Eickhoff, F. Ellis, J., Demurjian, S., Needham, D.: A Reuse Definition, Assessment, and Analysis Framework for UML. In: International Conference on Software Engineering (2003), http://www.engr.uconn.edu/~steve/Cse298300/eickhofficse2003submit.pdf
9. Fung, B.C.M., Wang, K., Esterz, M.: Hierarchical Document Clustering Using Frequent Itemsets. In: Proceedings of the third SIAM International Conference on Data Mining (2003)
10. Gupta, V., Chhabra, J.K.: Measurement of Dynamic Metrics Using Dynamic Analysis of Programs. In: Proceedings of the Applied Computing Conference, pp. 81–86 (2008)
11. Kaufman, L., Rousseeuw, P.J.: Finding Groups in Data, An introduction to Cluster Analysis. John Wiley & Sons, Inc., Chichester (1990)
12. Kiran, G.V.R., Shankar, K.R., Pudi, V.: Frequent Itemset based Hierarchical Document Clustering using Wikipedia as External Knowledge. In: Proceeding pf Intl Conference on Knowledge-Based and Intelligent Information Engineering Systems (2010)
13. Li, W., Chen, C., Wang, J.: PCS: An Efficient Clustering Method for High-Dimensional Data. In: Proceedings of the 4th International Conference on Data Mining (DMIN 2008), July 14-17 (2008)
14. Ng, R.T., Han, J.: Efficient and effective clustering methods or spatial data mining. In: Proceeding of VLDB conference, pp. 144–155 (1994)
15. Rao, I.K.R.: Data Mining and Clustering Techniques. In: Proceeding of DRTC Workshop on Semantic Web (2003)
16. Shiva, S.J., Shala, L.: Software Reuse: Research and Practice. In: Proceedings of the IEEE International Conference on Information Technology, pp. 603–609 (2007)
17. Taha, W., Crosby, S., Swadi, K.: A New Approach to Data Mining for Software Design. In: 3rd International Conference on Computer Science, Software Engineering, Information Technology, e-Business, and Applications (2004)
18. Xiao, Y.: A Survey of Document Clustering Techniques & Comparison of LDA and moVMF. In: CS 229 Machine Learning Final Projects (2010)
19. Xie, T., Pei, J.: Data mining for Software Engineering, http://ase.csc.ncsu.edu/dmse/dmse.pdf
20. Yossef, Z.B., Guy, I.: Cluster Ranking with an Application to Mining Mailbox Networks. In: ACM Proceedings of the Sixth International Conference on Data Mining (2006)
21. Zhang, T., Ramakrishnan, R., Birch, L.M.: An efficient data clustering method for very large data-bases. In: ACM SIGMOD, pp. 103–114 (1996)
22. http://en.wikipedia.org/wiki/Distance
23. http://en.wikipedia.org/wiki/Euclidean_distance
24. http://en.wikipedia.org/wiki/Metric_mathematics

# Cloud Computing in Education: Make India Better with the Emerging Trends

Sunita Manro[1], Jagmohan Singh[2], and Rajan Manro[3]

[1] Computer Science Department, PIMT,
Mandi Gobindgarh, India
[2] Computer Science Department,
SBBSIET VPO Padhiana,
Jalandhar, India
[3] Computer Science Department, DBIMCS,
Mandi Gobindgarh, India
{sunitamanro,jagmohan08,rajanmanro}@gmail.com

**Abstract.** The objective of this paper is to study the impact of cloud computing on the modern education. Further, the study also attempts to answer whether the services of cloud computing are significant in the education sector. Education institutions are under increasing pressure to deliver more for less, and they need to find ways to offer rich, affordable services and tools. Both public and private institutions can use the cloud to deliver better services, even as they work with fewer resources. By sharing IT services in the cloud, your educational institution can outsource noncore services and better concentrate on offering students, teachers, faculty, and staff the essential tools to help them succeed.

**Keywords:** Cloud, Education, Public, Service, Virtualization.

## 1 Overview

Cloud Computing is the use of common software, functionality or business applications from a remote server that is accessed via the Internet. Basically, the Internet is the "cloud" of applications and services that are available for access by subscribers utilizing a modem from their computer. With Cloud Computing, one simply logs into desired computer applications - such as sales force or office automation programs, web services, data storage services, spam filtering, or even blog sites. Generally, access to such programs is by monthly or annual paid subscription. Through Cloud Computing, businesses may prevent financial waste, better track employee activities, and avert technological headaches such as computer viruses, system crashes, and loss of data. When computers are used in **education**, this will be another medium of teaching other than chalkboard. The integration of computers change the whole ecology of a school. The typical school has 1 computer per 20 students, a ratio that computer educators feel is still not high enough to affect classroom learning as much as books and classroom conversation.

A. Mantri et al. (Eds.): HPAGC 2011, CCIS 169, pp. 131–139, 2011.

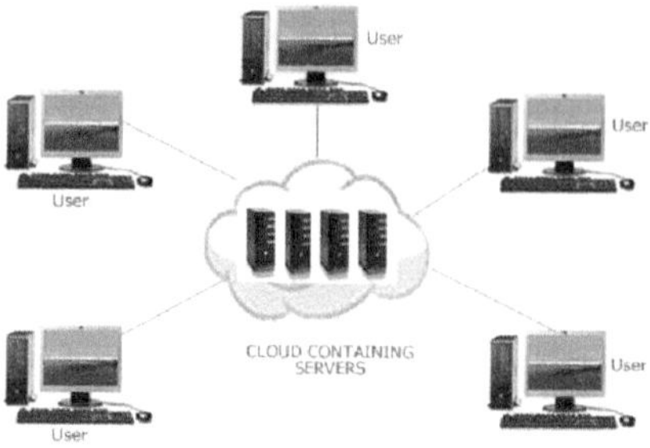


**Fig. 1.**

Computers are a new and exciting part of education and learning. They have changed the how student learn, study, and do assignments. Furthermore they have changed the way teachers teach. Every day in computers innovations are made that will improve how computers can be used by educators and students alike. The most basic way that computers help students is through word processing. It also gives students the ability to be creative and add pictures, highlight, underline, and use different fonts. In some classrooms the teachers use computers to compound what they teach. Computers can be used as projectors, to run programs, or simply to print out information quickly. Use of the internet is also now part of the modern classroom. There are many tutorial programs available. They are excellent in helping students hone their skills at home. These programs are for the most part affordable and have a wide range of topics.

Internet access is arguably the best form of computer innovations. Students and teachers alike can use the internet to do research. Furthermore teachers and students can use it to communicate or to send papers. The use of cloud computing in case of education has certain features like:

1. Computers improve both teaching and student achievement.
2. Computer literacy should be taught as early as possible; otherwise students will be left behind.
3. Technology programs leverage support from the business community - badly needed today because schools are increasingly starved for funds.
4. To make tomorrow's work force competitive in an increasingly high-tech world, learning computer skills must be a priority.
5. Work with computers - particularly using the Internet - brings students valuable connections with teachers, other schools and students, and a wide network of professionals around the globe. Those connections spice the school day with a sense of real-world relevance, and broaden the educational community.

## 2 Categories of Cloud Computing

The rapid improvement of the capacity of online connectivity gave birth to cloud computing. Although the term was already used since the 90's, the actual adoption of cloud computing in relation to online computing started in the 21st century. Cloud computing is a general term for anything that involves delivering hosted services over the Internet. Cloud Computing can be broadly classified into three *aaS, i.e., three layers of Cloud Stack, also known as Cloud Service Models or SPI Service Model. These services are broadly divided into three categories:

a) ***Infrastructure-as-a-Service (IaaS):*** This is the base layer of the cloud stack. It serves as a foundation for the other two layers, for their execution. The keyword behind this stack is *Virtualization*. Amazon EC2 is a good example of an IaaS. In Amazon EC2 (Elastic Compute Cloud) your application will be executed on a virtual computer (also known as an instance). You have your choice of virtual computer, meaning that you can select a configuration of CPU, memory and storage that is optimal for your application. The IaaS provider supplies the whole cloud infrastructure viz. servers, routers, hardware based load-balancing, firewalls, storage and other network equipment. The customer buys these resources as a service on an as needed basis. Infrastructure as a Service is a provision model in which an organization outsources the equipment used to support operations, including storage, hardware, servers and networking components. The client typically pays on a per-use basis. There are following characteristics and components of IaaS:

   - Utility computing service and billing model: *(charges per usage)*
   - Automation of administrative tasks.
   - Dynamic scaling.
   - Desktop virtualization: *(multiple networks, centrally located server.)*
   - Policy-based services.
   - Internet connectivity.

   The whole cloud infrastructure, including servers, routers, hardware based load-balancing, firewalls, storage and other network equipment, is provided by the IaaS provider. IaaS delivers computer and web infrastructure through virtualization.

b) ***Platform-as-a-Service (PaaS):*** IaaS delivers computer and web infrastructure through virtualization. But all this infrastructure is of no use without a platform. This post continues the series. Here we'll discuss the middle layer of Cloud Stack, i.e., PaaS (Platform as a Service).This middle layer of cloud is consumed mainly by developers or tech savvy individuals. Platform as a Service (PaaS) is a way to rent hardware, operating systems, storage and network capacity over the Internet. There are following Characteristics of Paas include:

   - Operating system features can be changed and upgraded frequently. Geographically distributed development teams can work together on software development projects.
   - Services can be obtained from diverse sources that cross international boundaries. Initial and ongoing costs can be reduced by the use of infrastructure services from a single vendor rather than

maintaining multiple hardware facilities that often perform duplicate functions or suffer from incompatibility problems.
- Overall expenses can also be minimized by unification of programming development efforts.

c) ***Software-as-a-Service (SaaS):*** Software as a Service (SaaS) is a software distribution model in which applications are hosted by a vendor or service provider and made available to customers over a network, typically the Internet. SaaS is becoming an increasingly prevalent delivery model as underlying technologies that support Web services and service-oriented architecture (SOA) mature and new developmental approaches, such as Ajax, become popular. SaaS is closely related to the ASP (application service provider) and On Demand Computing software delivery model.There are following Characteristics of the SaaS :

- easier administration
- automatic updates and patch management *(acquiring, testing, and installing multiple patches (code))*
- compatibility: All users will have the same version of software.
- easier collaboration
- global accessibility.

The traditional model of software distribution, in which software is purchased for and installed on personal computers, is sometimes referred to as *software as a product.*

## 3 Services of Cloud Computing in India: *21st Century Power House*

India, the 2nd fastest growing economy has mesmerized the world with its stunningly high economic growth rate since last 2 decades. A World Bank report has projected that in 2010, the rate of growth of India's economy would be faster than the currently fastest growing economy (i.e. China). What does this mean for SMBs, PSUs, MNCs or any business entity based in India? Last 5 years have seen Indian companies on a buying spree, resulting in acquiring many big & small overseas companies. In brief, Indian companies, sitting on a huge cash piles, are ready to rapidly scale up in their niche.

Interestingly, India is a global leader in providing IT services but the implementation of IT in its burgeoning domestic market is still lagging. This may be a boon in disguise as they have an opportunity to lap up the latest Cloud Computing technologies.

Currently, most of the companies are start-ups and are nowhere near the global giants like Amazon, Google, Salesforce or Microsoft but they have the potentials to compete with these giants in near future. Here's the list of India based Cloud Computing Service Providers, in random order.

***a) Zenith InfoTech*** (Location: Mumbai, India; Cloud Offering: PROUD; Type: IaaS): An IT product development and innovation company. With an investment of INR 175-crore, this is considered as India's one of the most ambitious R&D efforts in IT. The company is expecting 2,000 - 3,000 clients of Proud in next 2 years.

***b) Wolf Frameworks*** (Location: Bangaluru, India; Cloud Offering: Wolf PaaS;Type: PaaS): Founded in 2006, it provides affordable cloud service with 99.97% Service Level Assurance. Wolf is a browser based On Demand Platform-as-a-Service (PaaS) for rapidly designing and delivering database driven multi-tenant Software-as-a-Service (SaaS) applications.

***c) OrangeScape*** (Location: Chennai, India; Cloud Offering: OrangeScape; Type: PaaS): The experience of building business applications of varying complexity across industries has made OrangeScape the most comprehensive PaaS (Platform as a Service) offering in the market. You can transform your idea into a SaaS application and can showcase them to your investors, partners and potential customers. It has an impressive list of customers viz. Ford, Pfizer, Geojit, Sify etc.

***d) TCS*** (Location: India; Cloud Offering: ITaaS; Type: IaaS+SaaS): ITaaS is a Nano in software. ITaaS framework is a one-stop shop for total end-to-end IT and hardware solutions. It includes hardware, network, bandwidth & business software. Currently ITaaS is available for 5 sectors: Manufacturing, Retail, Healthcare, Education and Professional Services.

***e) Cynapse India*** (Location: Mumbai;Cloud Offering: Cyn.in;Type: IaaS + on Demand SaaS): Cyn.in on demand is a cloud hosted service and is the quickest way to get your own cyn.in server, without the hassles of having to set it up. With a Cyn.in on-demand system, you get a dedicated virtual server running a Cyn.in appliance that is maintained & updated by Cynapse and hosted by Amazon, ensuring an infrastructure-free and worry-free Cyn.in experience.

***f) Wipro Technologies*** (Location: India;Cloud Offering: Wipro w-SaaS; Type: SaaS): Wipro has built w-SaaS, a platform for rapid SaaS enablement and deployment on cloud, using some of the commonly accepted trends in software engineering and open standards. Wipro chose Oracle (Oracle Database, Oracle WebLogic Application Server and Oracle VM) as the deployment platform for w-SaaS enabled applications. The software vendor can deploy the same application on-premise or on the cloud using w-SaaS and Oracle.

***g) Netmagic Solutions*** (Location: Mumbai, India;Cloud Offering: CloudNet, CloudServe, PrivateCloud; Type: IaaS): Netmagic looks like a dedicated cloud provider in Indian market with a potential to become a big player in near future.

***h) Reliance Data Center-(a division of Reliance Communications)*** (Location: India; Cloud Offering: Reliance Cloud Computing Services;Cloud Type: IaaS+SaaS+PaaS):

A hosted infrastructure service based on the Microsoft platform for Enterprises and SMBs - geared to deliver India's largest cloud infrastructure.

*i) **Infosys Technologies*** (Base Location: Bangalore, India;Cloud Offering: Cloud based Solution for Auto Sector; Cloud Type: SaaS) : Infosys' Cloud Computing Consulting and Service offerings enable organizations to adopt the Cloud Computing platform selectively and effectively. But brand Infosys, the most recognized IT brand from India has to put significant efforts to catch up with other cloud providers.

Though the companies listed here have a long way to go before they can be compared with the best in the world yet they have the potential to grow big with the growing Indian economy.

## 4 Benefits of Cloud Computing for Institutions and Students

With cloud-based education tools, the whole world can learn from the best. The service provider will take care of all the nitty-gritty, leaving schools free to devote resources towards what they do best – teach our children. Also, think how convenient homework assignments will become. The students can work on the cloud, cooperate with team members and share knowledge, and be sure that they won't leave behind their homework assignments when they go to school. Since they are on the cloud, they can access them anywhere, be it home or school.

From schools, let's move to colleges. Many colleges do not have sufficient hardware or software to give students a complete learning experience. This problem is especially pronounced in the technical fields. However, with SaaS and IaaS, a limited budget will still allow students access to the latest technologies on offer. There are certain benefits of cloud computing like:

- Free of cost, robust service.
- Branding of your institute (school, college or university) as you've a custom domain e-Mail Ids with your school/college/university name as suffix (Say your institute's domain name is abcd.edu then your students will have e-Mail Ids like Student_Name@abcd.edu)
- Enterprise class hosted e-Mail: the quality of e-Mail service, collaboration tools & storage services are better than any of the available paid on-premise services. (I guess!)
- Quick & Effective Communication with Anytime Anywhere Access.
- Collaborate globally: Collaboration tools leads to collective intelligence & creativity as students may work on their project document at the same time.
- Help teachers (& students) in organizing their classroom presentations and schedules.
- No maintenance cost.
- Security will be taken care of by the provider.
- Privacy: Google Apps is in compliance with FERPA (Family Educational Rights and Privacy Act). I am not so sure about Live@Edu.

- Go Green: You'll save on notebooks, papers, printing etc.
- Easy to deploy.
- Finally, you choose any one of them & your institute is associating itself with one of the most respected global IT brands.

However, our schools need more than just extraordinary teachers and innovative educational programs to overcome today's challenges. They need more efficient and cost-effective systems to permit the teachers, administrators and even parents to focus more of their time and energies on these exciting initiatives. The good news is that an increasing number of secondary schools and universities are turning to SaaS and Cloud Computing alternatives to meet their escalating needs. Nearly half of the schools who participated in a 2009 survey conducted by SchoolDude.com and eSchool News were already using one or more SaaS solutions.

It is challenged at how he will be able to aggregate it, security, and make it accessible post-graduation to the state Department of Education, the student or anyone the student wishes to grant access, such as an employer or higher education institution.

The key advantages are:

- No additional cost for procurement of external hardware / software.
- No burden of paying an enormous amount at one time for procuring of the software. Work on a "Pay-as-you go" model.
- No need to employ any technical person at your Institution as all the technical aspects will be handled by us.
- Activities that are managed from central locations rather than at each School's/ Institution's site, enabling students/ parents/ faculty and the management to access applications remotely via the web application delivery that typically is closer to a one-to-many model (single instance, multi-tenant architecture) than to a one-to-one model.
- Data is highly secured and strong encryption techniques like Asymmetric key based encryption algorithms are used.
- Scalability becomes extremely simple and does not involve much additional costs.

Cloud computing is filling the business world with all its hype. It's a very fast emerging computing technology designed to help improve the efficiency of your computing needs, both personal and business.

## 5 Initial Results and Future Scenario of Cloud Computing in Education

While cloud computing is about a very simple idea—consuming and/or delivering services from 'the cloud' —there are many issues regarding the types of cloud computing and the scope of deployment. This makes the idea of cloud computing not nearly so simple and it has become successful in many countries. There are certain future scenarios regarding cloud computing:

**Scenario 1:** Through 2012, Global 1000 IT organizations will spend more money building private cloud computing services instead of offerings from public cloud computing service providers.

**Scenario 2:** By 2012, enterprise concerns over lock-in and standards will supplant security as the biggest objections to cloud computing.

**Scenario 3:** By 2013, at least two of the top three providers of SaaS and IaaS services will each offer a PaaS as a strategic offering.

Institutions are very comfortable with using Software-as-a-Service. Below is a graph showing SaaS usage among respondents. Facebook is of course the leader in the SaaS cloud race, with Twitter and Google Docs coming in right behind them.

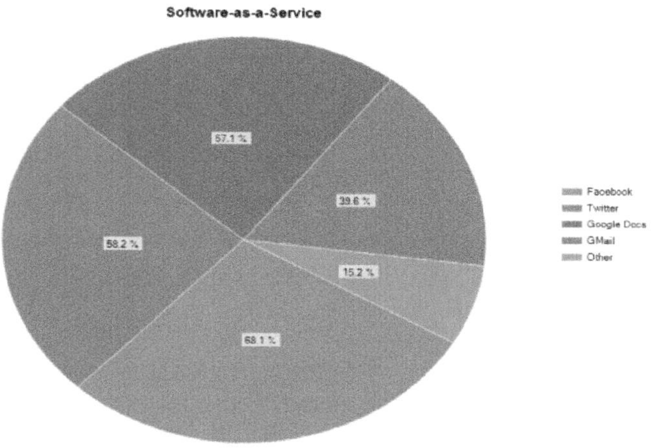


**Fig. 2.**

As far as PaaS and IaaS, most institutions are not using these services. A Brief List of Schools, Colleges, Universities & State Education Departments already on Cloud Applications:

- Oregon Department of Education (USA)
- University of Southern California (USA)
- Arizona State University (USA)
- FMS, the University of Delhi (India)
- New South Wales Department of Education and Training (Australia)
- Open University Malaysia (Malaysia)
- University of Cape Town - Graduate School of Business (South Africa)
- Alexandria University (Egypt)
- Riyadh College of Technology (Saudi Arabia)
- Universidad Nacional Autónoma de México, Facultad de Ciencias (Mexico)
- HCMC University of Technology (Vietnam)
- University of Batangas (Philippines)
- Politeknik Kesehatan Yogyakarta (Indonesia)

- University of Aberdeen (UK)
- Shree Chanakya Education Society (SCES), Pune *(India)*
- University of Queensland (Australia)
- The Universidad Europea de Madrid (UEM) (Spain)
- Government of the state of São Paulo, (Brazil)
- Ionis Education Group (France)
- Canadore College (Canada)
- Università della Calabria (Italy)
- Naresuan University (Thailand).

## 6 Conclusion

We see, in this paper, 40% of online Indians use webmail services, store data online, or use software programs such as word processing applications whose functionality is located on the web. Online users who take advantage of cloud applications say they like the convenience of having access to data and applications from any Web-connected device. But cloud computing can be used to address tactical problems with which IT continually deals, like resource availability. We hope this problem will be recovered soon.

## References


1. Armbrust, Michael, et al.: Above the Clouds: A Berkeley View of Cloud Computing, Technical Report No. UCB/EECS-2009028 (February 10, 2009), `http://www.eecs.berkeley.edu/Pubs/TechRpts/2009/EECS-2009.html`
2. Babcock, Charles.: Why 'Private Cloud' Computing is Real – and Worth Considering", in Information Week (April 11, 2009), `http://www.informationweek.com/story/showArticle.jhtml?articleID=216500083`
3. Gens, Frank.: Clouds and Beyond: Positioning for the Next 20 Years in Enterprise IT, presentation by Senior VP and Chief Analyst, IDC, March 5, San Jose, CA (2009)
4. Howard, Chris.: Cloud Computing: An Executive Primer, Burton Group Executive Advisory Program (April 20, 2009)
5. Manes, Thomas, A.: Cloud Computing: The Gap Between Hype and Reality. In: Presentation by VP and Research Director, Burton Group, ECAR Symposium, Boca Raton, FL, December 5 (2008)
6. McKinsey & Company, Clearing the air on cloud computing, Discussion document (March 2009), `http://www.slideshare.net/kvjacksn/mckinsey-co-clearing-the-air-on-cloud-compuitng`
7. McKinsey and Company, Enterprise Software Customer Survey, results of a survey of 850 enterprise software customers (2008)
8. Natis, Yefim, et al.: Key Issues for Cloud-Enabled Application Infrastructure, 2008, Gartner Research Number: G00155751 (April 21, 2008)
9. Plummer, Daryl C., et al.: Gartner's Top Predictions for IT Organizations and Users, 2008 and Beyond, Gartner Research Number: G00154035 (January 8, 2008)
10. Plummer, Daryl C., et al.: Cloud Computing: Defining and Describing an Emerging Phenomenon, Gartner Research Number: G00156220 (June 17, 2008)

# Enhancing Grid Resource Scheduling Algorithms for Cloud Environments

Pankaj Deep Kaur and Inderveer Chana

Computer Science and Engineering Department,
Thapar University, Patiala, India
{pankajdeep.kaur,inderveer}@thapar.edu

**Abstract.** Cloud computing is the latest evolution in the distributed computing paradigm and is being widely adopted by enterprises and organizations. The inherent benefits like instant scalability, pay for use, rapid elasticity, cost effectiveness, self-manageable service delivery and broader network access make cloud computing 'the preferred platform' for deploying applications and services. However, the technology being in nascent stage needs to be proven. The biggest challenge confronting service providers is effective provisioning and scheduling of cloud services to consumers leveraging the cost benefits of this computing paradigm. This paper attempts to investigate the key concerns for cloud resource management and explores possible alternatives that can be adapted from the existing Grid technology.

**Keywords:** Grid computing, Cloud computing, Virtualization, Resource management.

## 1 Introduction

In this rapidly evolving technology market place, Cloud computing is seen as the most attractive alternative to accommodate the growing information needs of businesses and organizations. It is described as a distributed computing paradigm that allows virtualized applications, software, platforms, computation and storage to be rapidly provisioned, scaled and released instantly through the use of self manageable services that are delivered over the web in a pay-as-you-go manner 3.

Cloud ecosystem comprises of three main entities: Cloud consumers, Cloud service providers, and the cloud services. *Cloud consumers* consume *cloud services* provided by the *cloud service providers*. These services may be hosted on the service provider's own infrastructure or on third party cloud infrastructure providers. The notion of service in cloud environments has been borrowed from the Service Oriented Architectures (SOA) where the services publish themselves in public registries, discover peer services and bind to other services using standardized protocols 1.

Cloud services are fundamentally categorized as *Infrastructure as a Service (IaaS)*, *Platform as a service (PaaS)* and *Software as a Service (SaaS)*. Cloud IaaS providers utilize virtualization techniques to provide dynamic infrastructure availability through standard interfaces. Consumer's specified hardware (number of CPU cores, physical memory size etc.) and software stack (operating systems, middleware and application

A. Mantri et al. (Eds.): HPAGC 2011, CCIS 169, pp. 140–144, 2011.

software) is immediately made available by the cloud IaaS providers. Cloud PaaS service facilitates developers with provider specific programming language and tools to develop cloud applications. Cloud SaaS offering allows end users to use cloud application without worrying about the underlying infrastructure 2. Businesses and enterprises can thus exploit the potentials of Cloud computing to reduce costs and increase business agility. Hence efficient provisioning and scheduling of cloud resources is must to leverage the cost benefits of this computing paradigm 45.

## 2 Clouds and GRIDs - A Close Alliance

Cloud computing is closely related to its predecessor Grid computing. Grids emerged to exploit the massive amount of available resources (compute, data, I/O, etc.) to solve a single, large problem that could not be performed on any one resource 67. However, as opposed to the declared target capabilities of Grid 8, the complex interfaces and lack of transparency hindered the adoption of Grid computing among business and consumer users 11. The use of Grid was thus confined to a limited number of applications and usage scenarios. Some representative applications depicting the Grid usage scenarios include compute-intensive, data-intensive, knowledge-intensive, and collaboration-intensive scenarios that address problems ranging from multiplayer video gaming and earthquake engineering to bioinformatics, biomedical imaging, and astrophysics 9.

Clouds, in contrast, rather than being resource specific, are more close to users and applications. Service providers currently use their existing datacenter infrastructures to host cloud services. However, a large number of public cloud providers have also spurred up to provide cloud services for commercial purposes presenting a limited user interface to allow consumers to access cloud capabilities from anywhere in the world through their web browsers using easy web interfaces 1011.

Applications in the Cloud environments belong to a broader distributed application class. Cloud platforms are best utilized for hosting traditional web applications and interactive applications that fully exploit the rapid scalability of clouds 12. Apart from that, parallel computing applications executing for short time intervals and utilizing enormous computational resources are hosted on cloud platforms 13. Furthermore, compute intensive analytical applications executing various analytical and data mining algorithms over the same data repeatedly are also targeted towards cloud environments.

## 3 Resource Management Problem for Cloud Computing

Clouds being an outgrowth of previous distributed systems require novelty in resource management and capacity planning capabilities 5. Resource management includes services to launch a job on a particular resource, check its status and retrieve results when the job is complete 14. The resource management mechanisms include:

i. Resource Information Dissemination: It involves collection of information about all the resources required for the execution of application.

ii. Resource Discovery: Resource discovery is the process of matching a query for resources, described in terms of required characteristics, to a set of resources that meet the expressed requirements 15.
iii. Resource Provisioning: Resource provisioning allows users and providers to access the specified resources as per their availability in the virtual environments created in the distributed environments.
iv. Resource Scheduling/Job execution: Resource scheduling comprises decisions for allocating jobs on the provisioned resources and performing the task of job execution.
v. Resource Monitoring and Re-Scheduling: Resources must be monitored to track the status of allocated resources, available resources and required resources for application execution. In case a job fails to perform or a resource bottleneck occurs, resources need to be rescheduled.

The virtual environments created in Cloud infrastructures provide a large pool of resources as compared to the limited number of actual physical resources. The two aspects to cloud resource provisioning include Host Level Provisioning and VM Level Provisioning. The former is concerned towards service provider's policies and the associated resource management cost while the latter takes into account the user QoS requirements 16. Thus, the traditional application provisioning models that assign individual application elements to computing nodes do not accurately represent the computational abstraction, which is commonly associated with Cloud resources 17.

## 4 Potential Grid Scheduling Techniques Suitable for Cloud

This section discusses few of the Grid scheduling techniques that can be possibly enhanced for cloud environments.

*Genetic algorithms (GA):* GA is a typical branch of evolutionary algorithms inspired by evolutionary biology such as inheritance, mutation, selection, and crossover 18. It operates on a population of potential solutions, applying the principle of survival of the fittest to produce exact or approximate solutions to the given problems. GA at first randomly selects an initial population of chromosomes on which genetic operators (selection, crossover and mutation) are applied to generate new offspring. Each of the chromosomes in the population is evaluated in terms of fitness expressed by the fitness function to carry over the selected fittest individuals over to the next generation. The algorithm terminates after some pre-specified stopping criterion is reached.

*Particle Swarm Optimization (PSO):* PSO is a swarm based intelligence algorithm influenced by the social behavior of animals such as flock of birds looking for a food source. A particle in PSO is analogous to a bird fling through a search space. The movement of each particle is coordinated by a velocity which has both magnitude and direction. Each particle position at any instance of time is influenced by its best position and the position of best particle in the problem space. The performance of particle is measured by a fitness value which is problem specific 19 23.

*Simulated annealing algorithm (SA):* SA is a probabilistic heuristic for the optimization problems 20. It aims to merely find an acceptably good solution in a fixed amount of time rather than the best possible solution. The input of the algorithm is an initial solution which is constructed by assigning a resource to each task at random. For solving a minimization problem, in each iteration the current solution X is given a small randomly generated perturbation, yielding a new solution X'. The resulting change in the objective function value say $\Delta f=f(X')-f(X)$ is calculated. If $\Delta f<=0$, the resulting change is accepted but if $\Delta f>=0$ the new solution is not straight away rejected but accepted with probability $P=\exp(-\Delta f/Kf)$. This acceptance criterion implies that uphill moves are occasionally acceptable, small uphill excursions are more likely to be accepted than larger ones. When f is large i.e. objective value is away from the optimal value, most of the uphill moves are accepted and as f approaches zero i.e. objective function approaches optimality, most of the uphill moves are rejected.

*Tabu search algorithm (TS)*: Tabu uses a local or neighbourhood search procedure to iteratively move from a solution to another solution in the neighborhood of until some stopping criterions have been satisfied. To explore regions of the search space that would be left unexplored by the local search procedure, tabu modifies the neighborhood structure of each solution as the search progresses. The new neighborhoods are determined through the use of memory structures. The most important type of memory structure used to determine the solutions admitted to the neighborhood of is the tabu list. In its simplest form, a tabu list is a short-term memory which contains the solutions that have been visited recently 21 22.

## 5 Conclusion

Reduced costs and increased business agility are considered as the key drivers of cloud computing. Today every small and medium business enterprises, academia, scientific and research organizations, government and technology institutes are trying to associate their services with cloud service model. The inherent benefits like fast deployment, pay-for-use, lower costs, scalability, rapid provisioning, instant elasticity, ubiquitous network access, greater resiliency, rapid re-constitution of services, low-cost disaster recovery and data storage solutions enabled cloud computing to be ranked as the second most emergent technology of 20103. This new technology infrastructure leveraging the decades of effort in Grid Computing is sure to accommodate the growing future needs of the society.

## References

1. Mei, L., Chan, W.K., Tse, T.H.: A Tale of Clouds: Paradigm Comparisons and Some Thoughts on Research Issues. In: APSCC 2008, pp. 464–469 (2008)
2. Mell, P., Grance, T.: The NIST Definition of Cloud Computing. Version 15, 10-7-09. National Institute of Standards and Technology, Information Technology Laboratory (2009)
3. Kaur, P.D., Chana, I.: Unfolding the Distributed Computing Paradigms. In: International Conference on Advances in Computer Engineering, pp. 339–342 (2010)

4. Armbrust, M., Fox, A., Griffith, R., Joseph, A., Katz, R., Konwinski, A., Lee, G., Patterson, D., Rabkin, A., Stoica, I., Zaharia, M.: Above the Clouds: A Berkeley View of Cloud Computing. UCB/EECS-2009-28 (February 10, 2009)
5. Jha, S., Katz, D.S., Luckow, A., Merzky, A., Stamou, K.: Cloud Book Chapter, Understanding Scientific Applications for Cloud Environments. John Wiley & Sons, Chichester (2010)
6. Foster, I., Kesselman, C., Tuecke, S.: The Anatomy of the Grid: Enabling Scalable Virtual Organizations. International Journal of Supercomputer Applications 15(3) (2001)
7. Foster, I., Kesselman, C.: The Grid 2: Blueprint for a New Computing Infrastructure. Morgan Kaufmann Publishers Inc., San Francisco (2003)
8. Foster, I., Kesselman, C., Nick, J., Tuecke, S.: The physiology of the Grid. Grid Computing: Making the Global Infrastructure a Reality (2003)
9. Nabrzyski, J., Schopf, J., Weglarz, J.: Grid Resource Management, State of the Art and Future Trends. Kluwer Academic Publishers, Dordrecht (2003)
10. Buyya, R., Yeo, C.S., Venugopal, S., Broberg, J., Brandic, I.: Cloud computing and emerging IT platforms: Vision, hype, and reality for delivering computing as the 5th utility. Future Generation Computer Systems 25(6), 599–616 (2009)
11. Jha, S., Merzky, A., Fox, G.: Using Clouds to Provide Grids Higher- Levels of Abstraction and Explicit Support for Usage Modes, `http://www.ogf.org/OGF_Special_Issue/cloud-grid-saga.pdf`
12. Varia J.: Architecting for the Cloud: Best Practices", Amazon Web Services (January 2010), `http://aws.typepad.com/aws/2010/01/new-whitepaper-architecting-for-the-cloud-best-practices.html`
13. Joseph J.: Patterns for high availability, scalability and Computing Power with Windows Azure. MSDN Magazine (May 2009), `http://msdn.microsoft.com/en-us/magazine/dd727504.aspx`
14. Li, M., Baker, M.: The Grid core grid Technologies. John Wiley & Sons Ltd., Chichester (2005)
15. Chana, I.: A framework for resource management in grid environment. Phd thesis, Thapar University (2009)
16. Buyya, R., Ranjan, R., Calheiros, R.N.: Modeling and simulation of scalable Cloud computing environments and the CloudSim toolkit: Challenges and opportunities. In: Proc. of the 7th High Performance Computing and Simulation (HPCS 2009), Leipzig, Germany (June 2009)
17. Deb, K.: Solving Goal Programming Problems Using Multi-Objective Genetic Algorithms. In: 1999 Congress on Evolutionary Computation, pp. 77–84. IEEE Service Center, Washington, D.C (1999)
18. Kennedy, J., Eberhart, R.: Particle Swarm Optimization. In: IEEE International Conference on Neural Networks, vol. 4, pp. 1942–1948 (1995)
19. Bertsimas, D., Tsitsiklis, J.: Simmulated Annealing. In: Probability and Algorithms, pp. 17–29. National Academy Press, Washington D.C
20. Ma, T., Yan, Q., Liu, W., Guan, D., Lee, S.: Grid Task Scheduling: Algorithm Review. IETE Technical Review, 158–167 (2011)
21. Glover, F.: Tabu search: a tutorial. Interfaces 20, 74–94 (1990)
22. Pandey, S., Wu, L., Guru, S., Buyya, R.: A Particle Swarm Optimization (PSO)-based Heuristic for Scheduling Workflow Applications in Cloud Computing Environments. In: AINA 2010, Perth, Australia, April 20-23 (2010)

# Development of Efficient Artificial Neural Network and Statistical Models for Forecasting Shelf Life of Cow Milk Khoa – A Comparative Study

Sumit Goyal[1], A.K. Sharma[2], and R.K.Sharma[1]

[1] School of Mathematics & Computer Applications,
Thapar University, Patiala, 147004, Punjab, India
[2] Computer Centre, DES&M Division,
National Dairy Research Institute (Deemed University),
Karnal-132001, Haryana, India
Tel.: +919896391267 (Mobile);
Fax: +911842250042
adesh_ndri@rediffmail.com,
adeshkumar.sharma@gmail.com

**Abstract.** *Khoa* is very popular milk product used to make variety of sweets in India. *Khoa* is made by milk thickening and heating it in an open iron pan. In this study, feedforward Backpropagation Neural Network (BPNN), Radial Basis Function Neural Network (RBFNN) and Multiple Linear Regression (MLR) models have been developed to predict shelf life of cow milk *khoa* stored at 37°C. Five input parameters, *viz.*, moisture, titratable acidity, free fatty acids, tyrosine and peroxide value are considered to predict sensory score. The dataset comprised of 48 observations. The accuracy of these models was judged with percent Root Mean Square Error (%RMSE). The BPNN model with Bayesian regularization algorithm provided static and consistent results. The residual shelf life of *khoa* was also computed using regression equations based on sensory scores. The BPNN model exhibited the best fit (%RMSE, 4.38) followed by MLR model (%RMSE, 9.27) and RBFNN model (%RMSE, 10.84).

**Keywords:** Backpropagation, Bayesian Regularization, Khoa, Model, Multiple Linear Regression, Neural Network, Prediction, Radial Basis Function, Shelf Life.

## 1 Introduction

*Khoa* is very popular Indian milk product used as a base material for the preparation of variety of sweets like burfi, milk cake, carrot halwa, *etc*. *Khoa* is being prepared for centuries in India for use as the base material for many sweets. *Khoa* is made by milk thickening and heating it in an open iron pan. Sweets made from *khoa* are in great demand today not only in India but also in foreign countries. Nutrient composition of *khoa* is important because it goes into the preparation of several Indian sweet meats. The consumer demands food products, under the legal standards, at low cost with high standards of nutritional, sensory and health benefits. To accommodate the new

A. Mantri et al. (Eds.): HPAGC 2011, CCIS 169, pp. 145–149, 2011.

paradigm, Martins *et al.* [1] recommended shelf life dating approaches with special emphasis on computational systems and future trends on complex systems methodologies applied to the prediction of food quality and safety. Shelf life of *khoa* can be estimated by sensory evaluation, but it is expensive, very time consuming and does not fit well with the dairy factories manufacturing it. Sensory analyses may not reflect the full quality spectra of the product. Moreover, traditional methods for shelf life dating and small-scale distribution chain tests cannot reproduce in a laboratory the real conditions of storage, distribution, and consumption on food quality [2]. In dairy products, Artificial Neural Network (ANN) model has been successfully used in various applications.

## 2 Materials and Methods

*Dataset:* The experimental data on quality parameters, *viz.*, moisture, free fatty acids, tyrosine, titratable acidity, and peroxide value of stored *khoa* at 37°C were taken as input parameters. The sensory score was used as output parameter for developing the artificial neural network model. Laboratory determined optimum dataset, which comprised of 48 observations were obtained from National Dairy Research Institute, Karnal for each input and output parameters, and were used for developing the models.

### 2.1 Feedforward Backpropagation Neural Network Model

Feedforward neural networks are the most common types of neurocomputing networks. The BPNN model consists of one input layer, one or several hidden layers and one output layer. In the development of BPNN model for shelf life prediction of *khoa* at 37°C, different combinations of several internal parameters, *i.e.,* number of hidden layers, data pre-processing, data partitioning approach, number of neurons in each hidden layer, transfer function, error goal, *etc.*, along with backpropagation algorithm based on Bayesian regularization mechanism as training function, were empirically explored in order to optimize the prediction ability of the model. The trial and error approach was tried to decide the optimum architectural parameters. BPNN models with single hidden layer as well as with two hidden layers were explored. The number of neurons in each hidden layer varied from 1 to 30. The dataset was randomly divided into two disjoint subsets, namely, training set containing 38 observations (79% of total observations) and testing set comprising of 10 observations, *i.e.,* 21% of total observations.

The feedforward backpropagation was used as training algorithm. The network was trained with training set after getting optimum values for architectural parameters. The sum square errors were performance function used during training of feedforward neural network. Weights and biases were randomly initialized. The network was trained with 100 epochs. The transfer function for each hidden layer was tangent sigmoid while for the output layer, it was linear function. Tangent sigmoid transfer function maps the input to the interval [-1, 1]. Neural network was simulated with the test dataset in order to validate the proposed BPNN model. In order to solve the overfitting problem, a variant of the backpropagation method based on Bayesian regularization technique was used, which determines the optimal regularization parameters in an automated fashion [3, 4].

### 2.2 Radial Basis Function Neural Network Model

Feedforward supervised RBFNN model was developed to predict shelf life of *khoa* stored at 37°C by using the same dataset. Various different combinations were tried to train the network like 'spread constant' by using different data partitioning approaches and spread constant 371, having 38 neurons was found fit. The dataset was randomly divided into two disjoint subsets namely, training set comprising 38 observations (79% of total observations) and testing set containing of 10 observations (21% of total observations).

### 2.3 Multiple Linear Regression Model

Using the same dataset (with same data partitioning scheme) as above, MLR model was developed for predicting shelf life of *khoa* and compared it with BPNN and RBFNN models. In this study, sensory score is dependent and moisture, titratable acidity, free fatty acids, tyrosine, and peroxide value are independent variables used to predict value of dependent variable *i.e.*, sensory score.

## 3 Results and Discussion

BPNN and RBFNN model performance matrices for predicting sensory scores are presented in Table 1 and Table 2, respectively.

**Table 1.** Performance of BPNN model for predicting sensory score

| Number of Hidden Layer (s) | Number of Neurons in Hidden Layer(s) | | Root Mean Square Error % |
|---|---|---|---|
| | I | II | |
| 1 | 3 | - | 9.72 |
| 1 | 6 | - | 13.47 |
| 1 | 10 | - | 9.61 |
| 1 | 12 | - | 14.76 |
| 1 | 15 | - | 14.36 |
| 1 | 17 | - | 12.71 |
| 1 | 18 | - | 13.39 |
| **1** | **19** | **-** | **4.38** |
| 1 | 21 | - | 11.66 |
| 2 | 24 | - | 10.18 |
| 2 | 30 | - | 12.04 |
| 2 | 5 | 5 | 11.99 |
| 2 | 7 | 7 | 9.69 |
| 2 | 8 | 8 | 12.12 |
| 2 | 14 | 14 | 14.30 |
| 2 | 17 | 17 | 11.45 |
| 2 | 20 | 20 | 9.10 |

**Table 2.** Performance of RBFNN model for predicting sensory score

| Number of Neurons in the hidden layer | RBFNN Model | |
|---|---|---|
| | Spread Constant | Root Mean Square Error % |
| 38 | 240 | 12.17 |
| 38 | 245 | 12.16 |
| 38 | 250 | 11.98 |
| 38 | 260 | 12.06 |
| 38 | 300 | 12.52 |
| 38 | 346 | 11.35 |
| 38 | 350 | 11.39 |
| 38 | 360 | 11.37 |
| 38 | 370 | 10.85 |
| **38** | **371** | **10.84** |
| 38 | 375 | 10.87 |
| 38 | 380 | 11.21 |
| 38 | 400 | 11.08 |
| 38 | 430 | 11.36 |

BPNN, RBFNN and MLR models were developed and compared with each other. Comparison of the three developed models is graphically shown in the Fig.1.

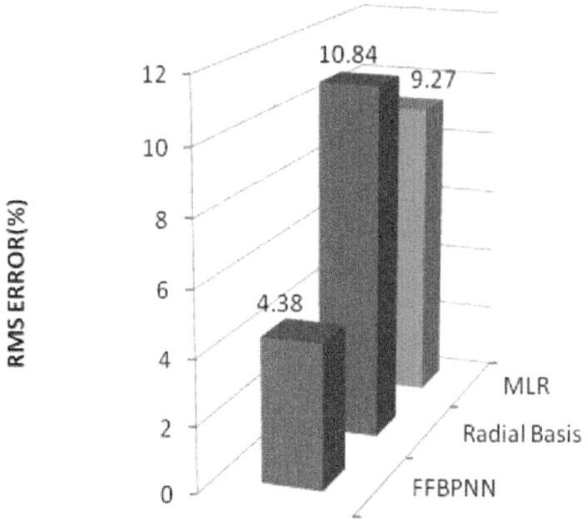


**Fig. 1.** Comparison of BPNN, RBFNN and MLR Models

Evidently, BPNN model seems to exhibit better results than MLR and RBFNN models for predicting shelf life of *khoa* stored at 37°C.

### 3.1 Prediction of Shelf Life

The regression equations were developed to estimate shelf life of the product, *i.e.*, in days (*d*) for which product has been in the shelf, based on sensory score. The *khoa* was stored at 37°C taking storage intervals (in days) as dependent variable and sensory score as independent variable.

| Sensory Score | Constant | Regression Coefficient | Coefficient of Determination |
|---|---|---|---|
| | 20.64 | -4.23 | 0.983 |

Coefficient of determination ($R^2$) was found to be 98% of the total variation as explained by sensory scores. For instance, the time period (in days) for which the product has been in the shelf can be predicted for an arbitrary sensory score of 4.98 for cow milk *khoa* stored at 37 °C: $d$=20.64-4.23 × 4.98 = -0.47.

Now, the shelf life is computed by subtracting the above obtained value of $d$ from experimentally determined shelf life, which was found to be 15 days. Hence, the residual shelf life comes out to be: 15-(-0.47) = 15.47 days. Since, this value is exceeding experimentally obtained shelf life of 15 days; hence, the product should be discarded.

## 4 Conclusion

Over the years, there has been research to improve food quality, which led to increased demand for healthy and delightful natural taste of the food products. To obtain better quality of food products, shelf life prediction is recommended. Therefore, artificial neural network models namely Backpropagation Neural Network (BPNN) and Radial Basis Function Neural Network (RBFNN) models have been developed in this paper to predict the shelf life of cow milk *khoa* stored at 37°C. To compare prediction potential between neurocomputing models and conventional regression models, a conventional Multiple Linear Regression (MLR) model was also developed. The results of the three developed models were compared with each other. The BPNN model (%RMSE, 4.38) exhibited the best fit followed by MLR (%RMSE, 9.27) and RBFNN (%RMSE, 10.84) models. Hence, the results of this study allow us to conclude that feedforward backpropagation neural network model has supremacy over MLR and RBFNN models in predicting shelf life of cow milk *khoa* stored at 37°C.

## References

1. Martins, R.C., Lopes, V.V., Vicente, A.A., Teixeira, J.A.: Computational Shelf-life Dating: Complex Systems Approaches to Food Quality and Safety. Food Biopro. Tech. 1, 207–222 (2008)
2. Goyal, S., Sharma, A.K., Sharma, R.K.: Comparison of Neurocomputing and Conventional Regression Models for Predicting Shelf Life of Khoa. Int. J. Comput. Intelli. Res. 6, 561–565 (2010)
3. Forsee, F.D., Hagan, M.T.: Gauss-Newton Approximation with Bayesian Regularization. In: IEEE International Joint Conference on Neural Networks, vol. 3, pp. 1930–1935. IEEE Press, New York (1997)
4. Mackay, D.J.C.: Bayesian Interpolation. Neu. Comput. 4, 415–447 (1992)

# QoS for Grid Systems

Vandana[1] and Tamanna Sehgal[2]

[1] Chitkara Unversity, Punjab
vandana.bajaj@chitkara.edu.in
[2] World Institute of Technology, Sohna, Gurgaon
tamannapuri@gmail.com

**Abstract.** QoS is important to the adoption of Grid Technologies. Grid Computing makes it possible for users to participate in distributed applications requiring data to be stored and delivered in timely manner. Users may wish to have control over Quality of Service (QoS) so that data is transferred on time in a distributed environment. Large scale Grids is composed of huge number of components from different sites and this requires efficient workflow management and Quality of Service (QoS).All the important components of this framework, integrated services and how workflows are managed with QoS are covered in this paper.

**Keywords:** Grid, Quality of Service (QoS), workflow management.

## 1 Introduction

In general Quality of Service refers to the ability to provide different treatment to different classes of traffic. The main goal of QoS is to increase the overall utility of the network by granting priority to higher valued data or data which is more sensitive. A very important factor is a kind of applications that are designed and run over network. Since networks are ultimately used by users running applications, it is imperative that the designers of networks and Internet service providers consider the effect of those applications operating over the network and also the effect of the network capabilities or service model on the usability and quality of applications. Applications need to consider the capabilities and limitations of the network that are used to transmit their data. Applications that are unresponsive to network conditions can cause network congestion or even congestion collapse, reduce network utilization, and suffer the consequences of their own behavior.

The purpose of Grid Systems is to build not only intensive computing but also huge scientific databases exploration and also many data transfer patterns occurs in this environment. Consequently, in Grids the heterogeneous mix of flows receives may affect the overall grid performance, as well as each individual application performance. So, QoS support gives the service instance which offers a performance level that satisfies the requirements of a given client. In this environment, the best effort usage of Grid services can result in denial of service problems as well as problems while configuring an instrument, collecting and storing relevant data from instrument, processing of this information and all such tasks which interact with each other which constitute a workflow.

A. Mantri et al. (Eds.): HPAGC 2011, CCIS 169, pp. 150–153, 2011.

## 2 Terminology in QoS

QoS can be of two types: *Qualitative and Quantitative* characteristics of Grid infrastructure. Qualitative characteristics highlight the attributes related to user satisfaction for service and service reliability. Quantitative characteristics gives the attributes for networks, CPU's or storage like the parameters such as *Delay* for network QoS , the time it takes a packet to travel from sender to receiver. *Delay Jitter*, the variation in the delay of packets taking the same route. *Throughput*, the rate at which packets go through the network (i.e., bandwidth). *Packet-loss rate'*, the rate at which packets are dropped, lost, or corrupted. Together, these four parameters form the QoS network measurement parameters.

The availability of Grid middleware tools facilitates persistent access to Grid services. The use of Grid middleware has expanded from scientific applications to business oriented disciplines while envisioning a service-oriented architecture to build sophisticated Grid applications with complex Grid resources requirements.

Large scale Grids, supporting the remote control and monitoring of distributed scientific and general-purpose equipments, need to allow exclusive access to instruments by users or groups of users, and to guarantee low delay and high responsiveness, given the interactive nature of many operations exposed by components. The support of QoS requirements needs to adopt various techniques in the services such as resource locking techniques, techniques dependent on physical location, communication infrastructure, the types of operations, number of clients etc.

There is a support of two types of guarantees: *Strict and Loose* [2].

*Strict:* These mechanisms are service-specific and depend on the type of agreement established. The service provider that is in charge of delivering guarantees offers a certain specific QoS profile for a given time duration on a contractual basis to the service consumers involved in the corresponding agreement.

*Loose*: Loose guarantees are delivered on a best-effort basis, and for this reason, they do not require the prior establishment of a Service Level Agreement (SLA).Loose guarantees consists of the capability of clients to select service providers that, on a statistical basis, exhibit a best-effort QoS profile which meets the client's requirements

## 3 QoS Service and Framework

QoS Service interacts with various modules to deliver QoS guarantees. These modules are QoS Handler, reservation manager, allocation manager, QoS registry service

**Reservation Manager:** Reservation manager checks the feasibility to grant a request whenever reservation request is received and if possible the requested resources are reserved and reservation table is updated and this agreement is generated and returned to the client. The resource manager doesn't have direct interaction with underlying resources.

**Allocation Manager:** Allocation manager allocates particular fraction of resource based on resource allocation request received and verifies that user has made a reservation based on supplied Service Level Agreement (SLA) if it passes, then other parameters are also passed from Allocation manager to Resource manager. The role of Allocation manager is to interact with underlying resource managers for allocation and de-allocation of resources. It is the bridge between QoS enabled Service (QeS) and resource manager.

**QoS registry Service:** It is a Web Service Registry providing users a means to publish and search for services with QoS properties. For example in case of QoS enabled services it offers allocation strategies, classes of network QoS, performance characteristics etc.

**QoS Handler:** Handler helps the service provider to publish properties of a service within the registry and can alter any parameters associated with it.

## 4 Workflow Management

Workflow management has been used to support various types of workflows on Grid. The development and execution of workflow within the Grid is a complex task. QoS restriction for interactions between the different Grid components thus become crucial to enable resource oriented flow and analysts are not concerned about the dynamic nature of resource allocation and de-allocation. In such a workflow process, the agreement between the two parties as discussed above and its success is based on management of QoS which further leads to quality products and its services. Therefore, workflow management system (WMS) should be able to monitor and control the QoS given by users whenever services are created and managed using workflow process.

Many workflow systems exists which highlights classification and characterization of workflow systems. One of the important features is where users are allowed to make reservation of resources, storage elements and instruments. Also some workflow languages also exist based on Web services.

Various components which are normally available in workflow management system are:

**WF Editor:** In order to facilitate the production of input documents related to QoS WF editor is required. Users can edit and save the workflow on the server and can access them from anywhere. Other facilities like dragging of different components on pane or creating QoS instances are also available. The architecture, different types and styles of Web Services and various complexities are hidden by automating different tasks.

**WF Mediator:** Mediator can be used to communicate between various Grid services such as storage elements, instrument elements etc.The workflow mediator is used and is responsible for ordered execution of the workflow task with analyzer monitoring progress.

**Analyzer:** The Analyzer is able to monitor progress of the executing workflow through status calls. If the workflow deviates from the expected plan the analyzer is able to invoke Builder to recompute the workflow in order to achieve the desired QoS requirements.

**WF Builder:** The function of Builder if given a QoS document which demands resource reservation will query the Repository to select a resource to make its reservation on. Then Agreement Service is contacted in order to make these reservations. The QoS document is then updated to indicate that the reservation has been made and records the unique token used to access the reservation. All the requests are processed here in one of the component of Builder. The QoS requirements and workflow is converted into a set of constraint equations. So one of the components in Builder provides the scheduling functionality.

**Valuable Repository:** It contains information about previous executions of tasks on the Grid. Based on repository workflow system decides if the submitted request can be achieved within QoS constraints. The place where information is stored is called Repository. It serves the data of *static* information like knowing how a particular service scales or its scope depending upon the resource on which it is running and also dynamic information like that of resources, current network status between two sites.

## 5 Conclusion

In this paper we have discussed how Quality of services are used for merging Grid Technology with that of instruments. High performance grid applications may need to use QoS control mechanisms to guarantee that data is available at remote systems instruments when required. An important limitation in network QoS work is the ability to control and manage workflow especially in Grid applications. Workflows are defined through editor and Mediator which provides ways to keep QoS on top. Enforcing the instruments to conform to the defined policy is the difficult task to achieve in reality. We believe however that resource agreement, their allocation is a work necessary to enable full QoS on Grids.

## References

1. Bhatti, S.N., Sørensen, S.-A., Clark, P., Crowcroft, J.: Network QoS for Grid Systems (2003)
2. Colling, D., Ferrari, T., Hassoun, Y.: Chenxi Huang, C. Kotsokalis, A. S. McGough, Y. Patel, E. Ronchieri, P. Tsanakas: On Quality of Service Support for Grid Computing
3. Guo, L., McGough, A.S., Akram, A., Colling, D., Martyniak, J., Krznaric, M.: QoS for Service Based Workflow on Grid. Imperial college, London
4. Foster, I., Kesselman, C. (eds.): The Grid: Blueprint for a New Computing Infrastructure. Morgan Kaufmann, San Francisco (1998)
5. Foster, I., Kesselman, C., Nick, J., Tuecke, S.: The Physiology of the Grid: an Open Grid Services Architecture for Distributed Systems Integration. Technical report, Global Grid Forum (2002)
6. W3C. Web Service, `http://www.w3.org/TR/ws-arch/`

# Creating Information Advantage in Cloudy World

Chahar Ravita and Mangla Vikram

Assistant Professor, Chitkara University, Punjab, India
ravitachahar@gmail.com, mangla.vikram@gmail.com

**Abstract.** To create knowledge of data management in this cloudy world we must have consistent, available and scalable data management systems which are capable of serving a billion of bytes of data to a number of users as well as large internet enterprises. One of the main security issue what all are facing is complications in data security in spite of providing different tools and security services. The security of cloud computing services is a contentious issue which may be delaying its adoption. The security depends on the methods adopting for the data management. In this paper we have analyzed the design choices and recommended approaches that allowed modern data management systems to achieve goals as compared to traditional databases.

**Keywords:** Approaches, Consistency, Data management, security.

## 1 Introduction to Cloud

Cloud computing is a general concept that unite software as a service(SaaS).web2.0 and other recent, well-know technology trends in which the common theme is reliance on the for satisfying the computing needs of users. For example Google Apps, provides common business applications online that are accessed from a web browser, while the software and date are stored on the servers. It describes different consumption and models deliver by information technologies. It's working is taking a user IP address which direct to cluster of servers which is nearest to it so that it can access rapidly actually user is taking the service obtain by cookie which is stored in the browser. This technology allows for much more efficient computing by centralizing storage, memory, processing and bandwidth.

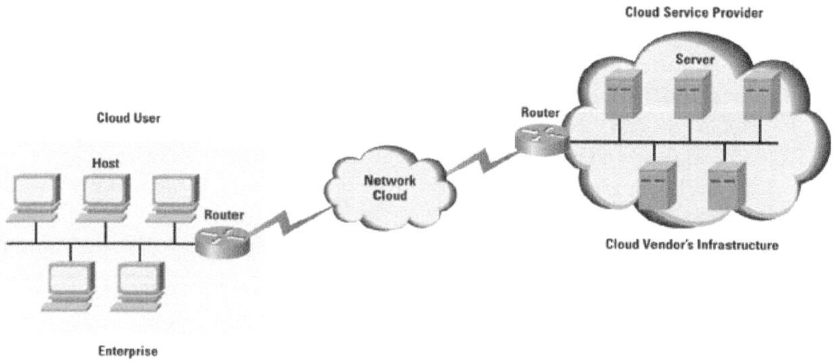


A. Mantri et al. (Eds.): HPAGC 2011, CCIS 169, pp. 154–158, 2011.

Cloud Computing

**ARCHITECTURE:** Cloud architecture is the systems architecture of the software systems involved in the delivery of cloud computing e.g hardware, software as designed by a cloud architect who typically works for a cloud integrator it typically involves multiple cloud components communicating with each other over application programming interfaces, usually web serves. The majority of cloud computing infrastructure currently consists of reliable service currently consist of reliable services delivered through data centers that are built on servers with different levels of virtualization technologies. The services are accessible anywhere in the world, with the cloud appearing as a single point of access for all the computing needs of consumers. **Reliability** is enhanced by way of multiple redundant sites .which makes it suitable for business continuity and disaster recovery, however IT and business for managers are able to do little when an outage hits them. **Security** typically improves due to centralization of DATA, increase security-focused resources, etc, but raises concerns about loss of control over certain sensitive data. **Sustainability** is achieved through improved resource utilization, more efficient systems and carbon neutrality. Nonetheless, computers and associated infrastructure are major consumers of energy.

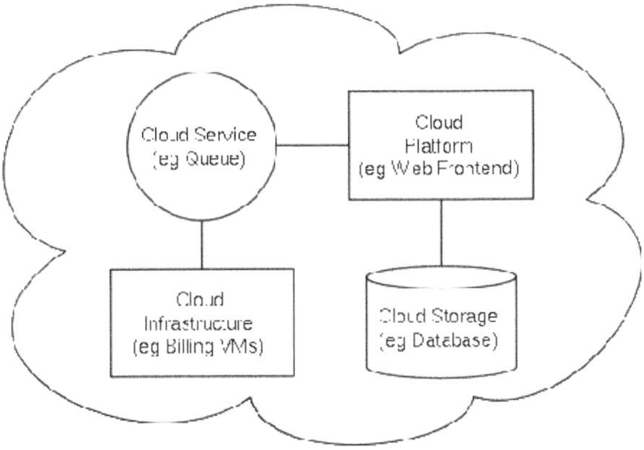


## 2 Data Management: Cloud

Scalable and consistent data management is a challenge that has confronted the database research community for more than two decades. Historically, distributed database systems were the first generic solution that dealt with data not bounded to the confines of a single machine while ensuring global serializability. This design was not sustainable beyond a few machines due to the crippling effect on performance caused by partial failures and synchronization overhead. As a result, most of these systems were never extensively used in industry. Recent years have therefore seen the emergence of a different class of scalable data management systems such Google's Bigtable], PNUTS from Yahoo!, Amazon's Dynamoand other similar but undocumented systems. All of these systems deal with petabytes of data, serve online requests with stringent latency and availability requirements, accommodate erratic workloads, and run on cluster computing architectures; staking claims to the territories used to be occupied by database systems. One of the more uncomfortable things about cloud computing is that it shines a spotlight on data management and governance. The data management skills of most IT organizations are sorely lacking. So when someone suggests that they need to start using cloud computing services to save money, the first thing that comes to mind is how they will manage which data is stored where. Of course, nobody wants to admit that they might have a data management issue. It's much easier to object to cloud computing on the grounds of security issues. But like it or not, cloud computing is coming, and with it a need to focus on data management and governance.

## 3 Data Management Applications

There are some applications on which the data management performs its functions. Some of which are discussed below:

### 1. Transactional data management

These applications typically rely on the ACID guarantees that databases provide, and tend to be fairly write-intensive. We speculate that transactional data management applications are *not* likely to be deployed in the cloud, at least in the near future, for the following reasons: Transactional data management systems do not typically use a shared-nothing architecture and it is hard to maintain ACID guarantees in the face of data replication over large geographic distances. Implementing a transactional database system using a shared-nothing architecture is non-trivial, since data is partitioned across sites and, in general, transactions can not be restricted to accessing data from a single site. This results in complex distributed locking and commit protocols, and in data being shipped over the network leading to increased latency and potential network bandwidth bottlenecks.

### 2. Analytical data management

By "analytical data management", we refer to applications that query a data store for use in business planning, problem solving, and decision support. Historical data along with data from multiple operational databases are all typically involved in the analysis. Consequently, the scale of analytical data management systems is generally larger than transactional systems (whereas 1TB is large for transactional systems, analytical systems are increasingly crossing the petabyte barrier). Furthermore, analytical systems tend to be read-mostly (or read-only), with occasional batch inserts. Analytical data management consists of $3.98 billion of the $14.6 billion database market (27%) and is growing at a rate of 10.3% annually. We speculate that analytical data management systems are well-suited to run in a cloud environment, and will be among the first data management applications to be deployed in the cloud, for the following reasons:

a) Shared-nothing architecture is a good match for analytical data management.
b) ACID guarantees are typically not needed.
c) Particularly sensitive data can often be left out of the analysis.

We conclude that the characteristics of the data and workloads of typical analytical data management applications are well-suited for cloud deployment. The elastic compute and storage resource availability of the cloud is easily leveraged by a shared-nothing architecture, while the security risks can be somewhat alleviated.

**4. Recommended approaches to managing data in the age of the cloud**

1) Enterprises moving to cloud computing are looking to move to the private cloud first. Private cloud can provide better control and more secure access to corporate information. Information governance policies in place today can be leveraged in private cloud environments, where information and applications are still under direct control of IT.
2) Global IT leaders identified four emerging conditions that can complicate the flow and value of information in hybrid and public clouds. These include:

- --Unchecked proliferation of incompatible cloud platforms and services
- --Fragmentation of an enterprise's information architecture through isolated data and content within the cloud
- -- Escalating potential for vendor lock-in
- --Complex chains of custody for information management and security

Different councils has a lot of good thoughts on this subject of data management, the one that stands out most is "the need to own your information." Far too many cloud computing providers, in the name of simplicity, skimp when it comes to giving customers the management tools they need to govern their data. So it's little wonder that there is a lot of resistance to public cloud computing.

The second one advice that vendors will get better at delivering data management and governance tools that customers can federate across private and public cloud infrastructure. When all is said and done, this data management and governance issue will slow the migration to public cloud infrastructure more than any other.

## Concluding Remarks

Among the primary reasons for the success of the cloud computing paradigm for utility computing are elasticity, pay as you go model of payment, and use of commodity hardware in a large scale to exploit the economies of scale. Therefore, the continued success of the paradigm necessitates the design of a scalable and elastic system that can provide data management as a service. In this paper, our goal was to lay the foundations of the design of such a system for managing "clouded data" and to present the same in terms of Data protection. Cloud computing is no exception. In this paper challenge, Approaches, Deployment models and key security issues which are currently faced by cloud computing are highlighted. We mentioned the how to manage data in point of security at different service models which easily give the view where we require more security and concentrate our focus to under developed areas.

## References

[1] Chow, R., Golle, P., Jakobsson, M., Shi, E., Staddon, J.: Controlling Data in the Cloud Outsourcing Computation without Outsourcing Control. In: CCSW, Chicago, Illinois, USA (2009)
[2] http://www.technologyreview.com/computing/23951/?a=f
[3] http://www.privatecloud.com/2010/12/14/data-management-in-the-cloud-computing-era/?fbid=V3F_TsOelAh
[4] Ramgovind, S., Eloff, M., Smith, E.: The Management of Security in Cloud Computing. In: IEEE International Conference on Service Computing, pp. 126–130 (2010)
[5] Krautheim, F.J., Phatak, D.S.: LoBot: Locator Bot for Securing Cloud Compupting Environments. Submitted 2009 ACM Cloud Computing Security Workshop, Chicago, IL (2009)

# Design of CMOS Energy Efficient Single Bit Full Adders

Manoj Kumar[1], Sujata Pandey[2], and Sandeep K. Arya[1]

[1] Department of Electronics & Communication Engineering
Guru Jambheshwar University of Science & Technology,
Hisar, 125 001, India
manojtaleja@gjust.org, arya1sandeep@rediffmail.com
[2] Department of Electronics & Communication Engineering
Amity University, Noida, 201303, India
spandey@amity.edu

**Abstract.** Here, three new low power single bit full adders using 9 and 10 transistor have been presented. The proposed adders have the advantage of low power consumption with small area requirements due less number of transistors. Low power objective has been achieved at circuit level by designing the adder with optimized XOR gates and multiplexer approach. Direct path between supply voltage and ground have been minimized in these designs. The circuits have been simulated in 0.18μm CMOS technology using SPICE. The first adder shows power dissipation of 23.8595pW with maximum output delay of 67.5566fs at supply voltage of 1.8V. The second adder shows power dissipation of 43.1258pW with maximum output delay of 58.9935fs. Third adder shows power dissipation of 33.5163pW with delay of 62.065fs. Further, simulations have been carried out with different supply voltage [1.8 - 3.3] V. Power consumption of proposed full adders have been compared with earlier reported circuits and proposed circuit's shows better results.

**Keywords:** CMOS, exclusive-OR (XOR), full adder, power consumption and power delay product.

## 1 Introduction

In recent years, rapid growth in mobile communication and other handheld portable devices added rapid research efforts in the field of low power CMOS circuit design. Low power design also increases the operation time of battery operated devices. With added functionality and complexity, numbers of components on integrated circuits are increases and power consumption of VLSI (Very large scale integration) circuits is rising exponentially. With increase in power consumption temperature of circuit rises, this further creates reliability problems and performance degradation of the system. Packaging and cooling costs of system also goes high with the rise in temperature and power consumption. Three major sources of power consumption exist in CMOS circuits: 1) dynamic power due to output switching 2) short circuit power due to current between supply voltage and ground during transition 3) static power due to leakage and static currents. Despite the scaling of device dimension and supply voltage, the total power consumption of VLSI circuits is going up due to increase in operating

A. Mantri et al. (Eds.): HPAGC 2011, CCIS 169, pp. 159–168, 2011.

frequency and rise in number of components [1].One of the most important operations in computer arithmetic is addition and efficient adders are highly desirable in computer arithmetic. Full adders are the core element of various VLSI circuits like comparators, parity checkers, multiplier and compressors [2]. Improvement in performance of full adder circuit in terms of power consumption, delay and other performance parameters will affect system performance as a whole. Design optimizations at circuit level are highly attractive to improve the performance in terms of power dissipation, delay and output logic level. A variety of full adder circuits has been reported in literature with diverse techniques and numbers of transistors. Conventional static CMOS full adder implemented with pull up and pull down with NMOS and PMOS networks uses 28 transistors [2],[3]. Complementary pass-transistor logic (CPL) adder with 32 transistors with superior driving capability at the cost of large power dissipation has been presented [4]. Transmission gate CMOS adder (TGA) based on transmission gates with 20 transistors has been reported [5]. Major drawback of TGA is that it employs twice the number of transistors that of PTL (pass transistor logic) for implementations of same logic functionality. Another drawback of CMOS transmission gate is that it needs complementary signals to control gates of PMOS and NMOS transistors. A full adder cell implemented with 14 transistors using XOR and transmission gates has been reported in [6]. A transmission function full adder (TFA) with 16 transistors based on transmission function theory has been reported in [7]. Multiplexer based adder (MBA) with elimination of direct path to power supply with 12 transistors has been reported [8]. Static energy recovery full (SERF) adder with 10 transistors gives reduced power consumption at the cost of large propagation delay is reported in [9]. Performance analysis of various arithmetic circuits also has been presented [10]. Another design for full adder with XOR/XNOR having 10 transistors has been reported in [11]. Full adder circuits using 22 transistors based on hybrid logic has been presented [12], [13]. In [14] a 16 transistor full adder cell with XOR/XNOR, pass transistors and transmission gate has been reported. Structured approach with disintegration of full adder cell into small modules using XOR/XNOR gates [15] is shown in figure.1. First stage is to generate intermediate XNOR/XOR functions and output of first stage is fed to second stage which generates Sum and Cout. With partitioning the full adder cell into small sub-module, Sum and Cout (Carry out) signals are obtained as

$$\text{Sum} = H \text{ xor } Cin = H.\ Cin' + H'\ Cin \quad (1)$$

$$\text{Cout} = A.\ H' + Cin.\ H \quad (2)$$

Where H is half sum (A xor B) and H' is complement of H.

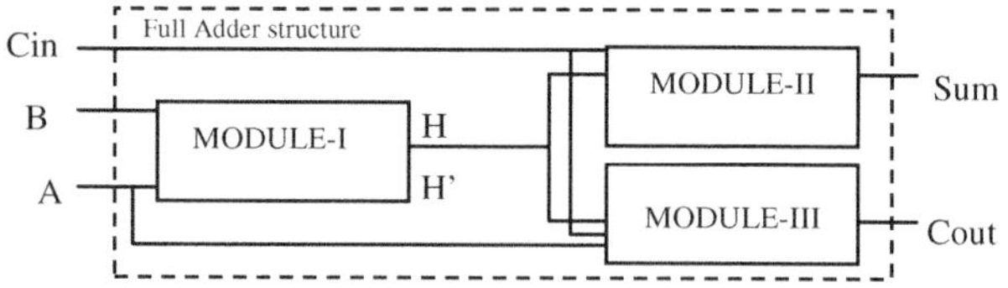


**Fig. 1.** Structure of single bit full adder

Improvement in design at circuit level can results in reduced power consumption, propagation delay, area with adequate output voltage level. Here, in current work energy efficient single bit full adders using nine and ten transistors have been presented. Full adder cell has been anatomized into smaller modules. Optimized XOR gates and multiplexer have been used for implementation of proposed adders. Rest of paper is organized as follows: In Section 2, three novel designs for single bit full adder have been presented. In Section 3 results of power consumption, maximum output delay and power delay product (PDP) have been obtained. Finally conclusions have been drawn in Section 4.

## 2 System Description

The first single bit full adder circuit having 10 transistors and implemented with two optimized XOR gates, one inverter and one multiplexer block has been shown in figure 2 (a). Sum is generated by combination of XOR-I, inverter & XOR-II. Cout is generated by two transistors multiplexer with inputs signal A, H' and Cin. Inverted output from XOR gate has been used to control the gate of two transistor multiplexer section.

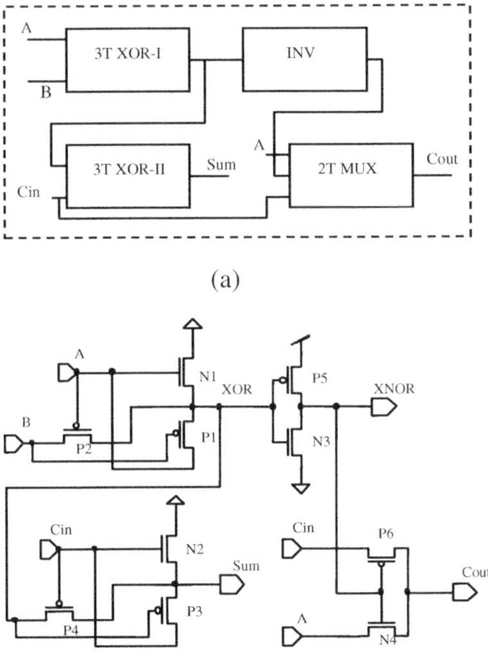


(b)

**Fig. 2.** (a) Block diagram (b) Circuit diagram of full adder using 10 transistors

XOR-I operation has been performed by two PMOS transistors (P1& P2) and one NMOS (N1) transistors connected as shown in figure 2 (b). XOR-II operation has been performed by another pair of PMOS (P3& P4) and NMOS (N2) transistors. One inverter for generation the complemented XOR and to restore the degraded logic is used with XOR-I. Gate lengths of all transistors have been taken as 0.18μm. Width of transistors P1, P2, P3 and P4 have been taken as 2.2μm. Widths of N1 and N2 have been taken as 0.22μm. Width of PMOS transistors P5 & P6 have been taken as 1.25μm whereas widths of NMOS transistors N3 & N4 taken as 0.50μm.

In the second proposed adder one 3T XOR gate, inverter and two multiplexers have been utilized as shown in block diagram of figure 3(a). Gate lengths of all transistors have been taken as 0.18μm. Width of transistors P1, P2, have been taken as 2.2μm whereas widths of N1 have been taken as 0.22μm. Width of all other PMOS transistors P3-P5 have been taken as 1.25μm whereas widths of NMOS transistors N2-N4 taken as 0.50μm.

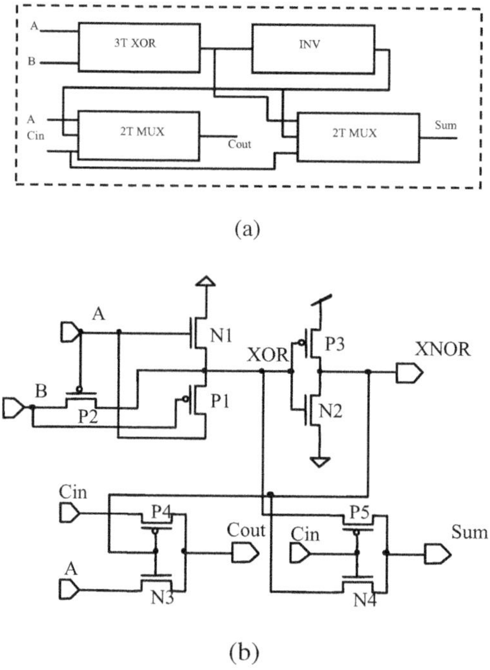


**Fig. 3.** (a) Block diagram (b) Circuit diagram of 9 T adder-I

In third proposed adder, a complemented Cin signal is generated which is used for generating the sum output in multiplexer section. Carry out (Cout) is generated by another 2 T multiplexer as shown in figure 4(a).Complete adder circuit implemented with 3T XOR, inverter & two multiplexers is shown in figure 4 (b).

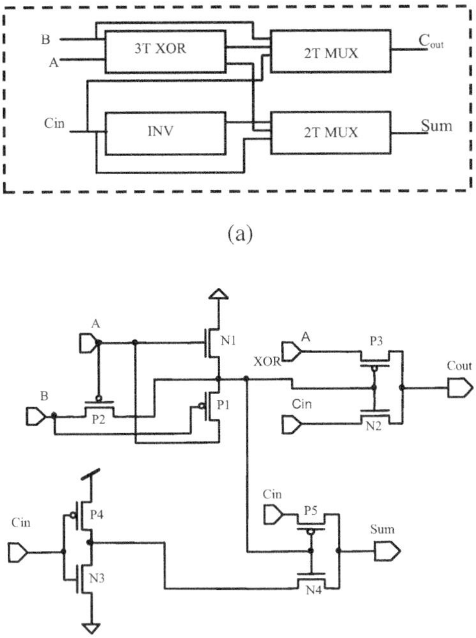


(b)

**Fig. 4.** (a) Block diagram (b) Circuit diagram of 9 T adder-II

## 3 Results and Discussions

Table 1 shows the results for power consumption, maximum output delay and power delay product (PDP) of single bit full adder with 10 transistors. Simulations have been performed using an input pattern that contains all possible input combination. Input and output waveforms at supply voltage of 3.3V are depicted in figure 5. Circuit of adder is simulated in SPICE using CMOS TSMC 0.18μm technology with supply voltage of [3.3 -1.8]V. Power consumption of adder varies from [103.6808 - 23.8595]

**Table 1.** Power consumption, delay and PDP of 10 T adder (Figure 2)

| Supply voltage (V) | Power consumption (pW) | Maximum output delay(ps) | Power delay product(PDP)x$10^{-24}$ |
|---|---|---|---|
| 3.3 | 103.6808 | 36.3781 | 3771.71 |
| 3.0 | 80.2657 | 40.8819 | 3281.41 |
| 2.7 | 61.2255 | 45.2928 | 2773.07 |
| 2.4 | 45.8559 | 50.4466 | 2313.27 |
| 2.1 | 33.5692 | 58.7628 | 1972.62 |
| 1.8 | 23.8595 | 67.5566 | 1611.86 |

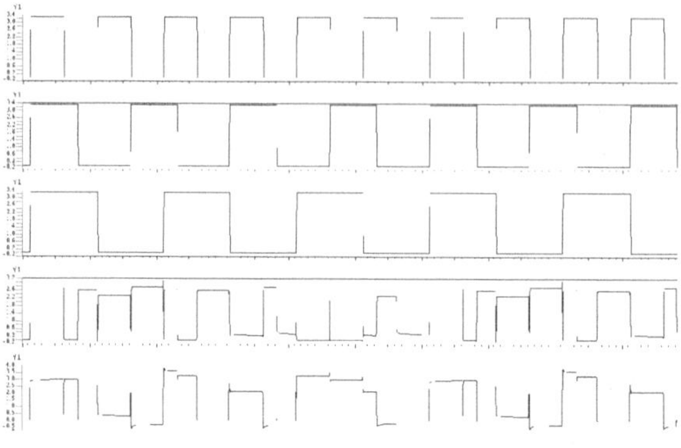

**Fig. 5.** Input and output waveforms for 10T adder at 3.3 V

pW with variation of supply voltage [3.3 - 1.8] V. Maximum output delay and power delay product(PDP) shows variation of [36.3781 - 67.5566] fs and [3771.71 to 1611.86] *10-24 J respectively.

Table 2 shows the results for power consumption, maximum output delay and power delay product (PDP) of single bit full adder with nine transistors (figure 3). Input and output waveforms at 3.3 V supply voltage are shown in figure 6. Power consumption of adder varies from [215.7757 - 43.1258] pW with variation of supply voltage [3.3 - 1.8] V. Maximum output delay and power delay product (PDP) shows variation of [53.9937 - 58.9935] fs and [11650.52 to 2544.14] $*10^{-24}$ J respectively.

Table 3 shows the results for power consumption, maximum output delay and power delay product (PDP) of single bit full adder with 9 transistors (figure 4). Input and output waveforms with supply voltage of 3.3V are shown in figure 7. Power consumption of adder varies from [134.6232 - 33.5163] pW with variation of supply voltage [3.3 - 1.8] V. Maximum output delay and power delay product (PDP) shows variation of [17.872 - 62.065] fs and [2405.98 - 2080.18] *10-24 J respectively.

**Table 2.** Power consumption, delay and PDP of 9T adder-I (Figure 3)

| Supply voltage (V) | Power consumption (pW) | Maximum output delay(ps) | Power delay product(PDP) x$10^{-24}$ |
|---|---|---|---|
| 3.3 | 215.7757 | 53.9937 | 11650.52 |
| 3.0 | 162.0149 | 52.6371 | 8527.99 |
| 2.7 | 120.1644 | 55.6838 | 6691.21 |
| 2.4 | 87.7247 | 56.2358 | 4933.26 |
| 2.1 | 62.7360 | 57.2426 | 3591.17 |
| 1.8 | 43.1258 | 58.9935 | 2544.14 |

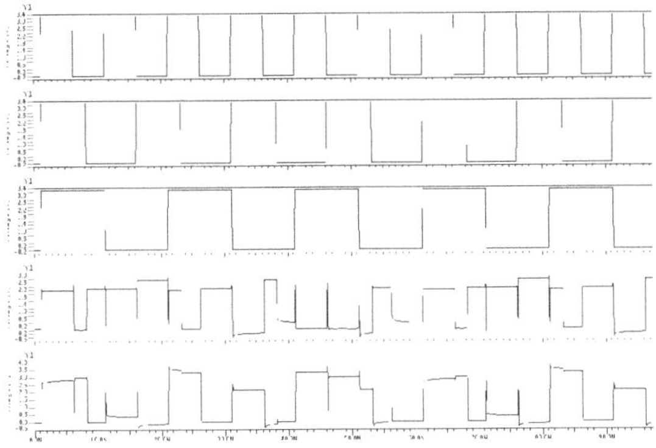

**Fig. 6.** Input and output waveforms for 9T adder-I at 3.3 V

**Table 3.** Power consumption, delay and PDP of 9T adder-II (Figure 4)

| Supply voltage (V) | Power consumption (pW) | Maximum output delay(fs) | Power delay product(PDP) x$10^{-24}$ |
|---|---|---|---|
| 3.3 | 134.6232 | 17.872 | 2405.98 |
| 3.0 | 105.7485 | 21.166 | 2238.27 |
| 2.7 | 81.8489 | 34.981 | 2863.15 |
| 2.4 | 62.2506 | 40.434 | 2517.04 |
| 2.1 | 46.3420 | 45.962 | 2129.97 |
| 1.8 | 33.5163 | 62.065 | 2080.18 |

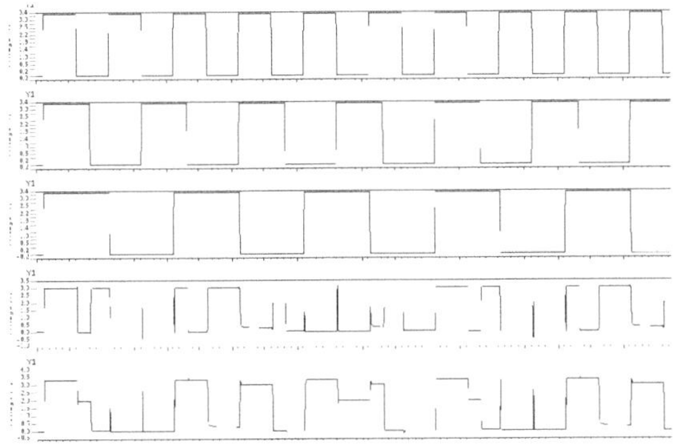

**Fig. 7.** Input and output waveforms for 9T adder-II at 3.3 V

Figure 8 show that power consumption variation for three proposed adder with supply voltage variations. 10T adder has minimum power consumption among the proposed adders. Figure 9 shows the variation of output delay of proposed circuit with power supply variations. Finally the figure 1) shows PDP variation of circuits with supply voltage. The designed circuits have the advantage of low power consumption due to less number of transistors and elimination of direct path to ground from supply voltage. With reduced numbers of transistors, magnitudes and numbers of internal node capacitances reduces and have great effect on power consumption. Power consumption during charging and discharging of nodes also decreases due to less capacitance.

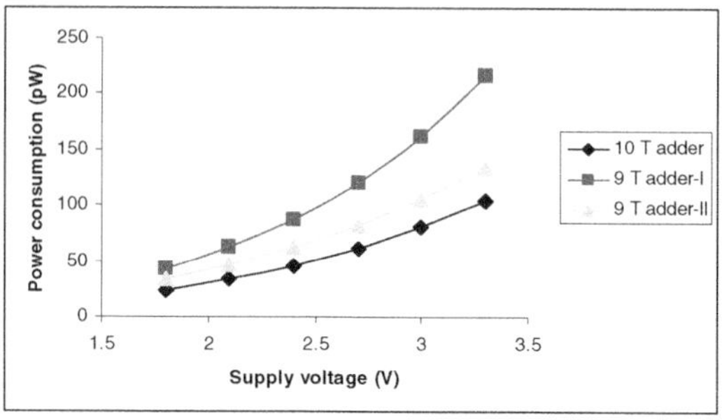


**Fig. 8.** Power consumption variations of proposed adders with supply voltage

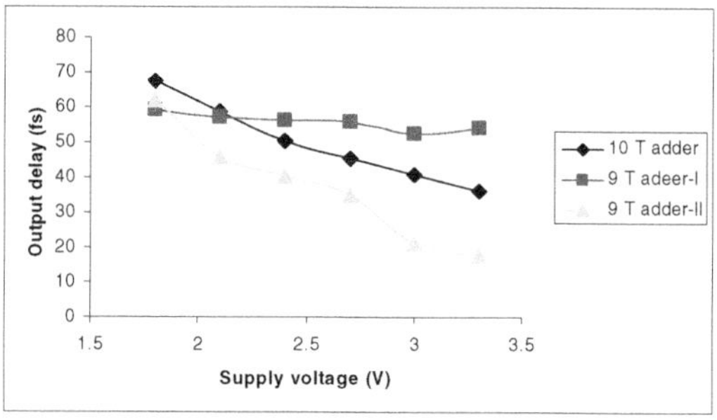


**Fig. 9.** Output delay variations of proposed adders with supply voltage

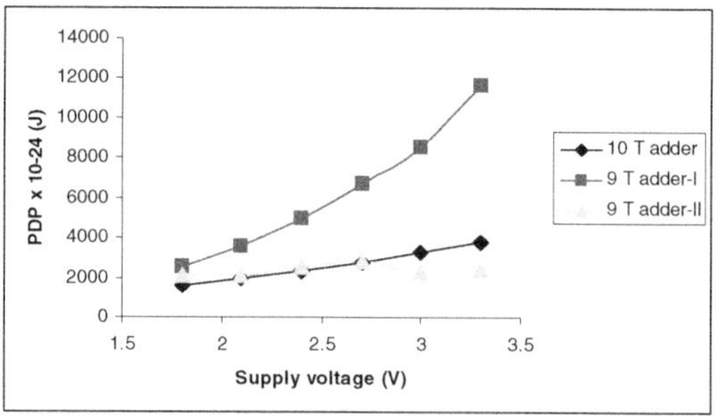


**Fig. 10.** Power delay product variations of proposed adders with supply voltage

Earlier reported adder circuits namely TGA, 16T, 22T, 18T, 10T have been prototyped in 0.18μm technology and simulated in SPICE with same input pattern as for proposed adders. Table 4 shows comparisons of power consumption of proposed circuit with earlier reported circuits.

**Table 4.** Power consumptions comparisons with earlier reported circuits

| Adder configuration | Power consumption | Numbers of transistors |
|---|---|---|
| 20T TGA[7] | 357.5880μW | 20 |
| 16T [14] | 167.2136μW | 16 |
| 22T [12] | 533.9007μW | 22 |
| 18T [3] | 548.2570 μW | 18 |
| 10T SERF[9] | 152.9083μW | 10 |
| Present work [10T] | 23.8595pW | 10 |
| Present work [9T adder-I] | 43.1258pW | 9 |
| Present work [9T adder-II] | 33.5163pW | 9 |

## 4 Conclusions

Three new designs for low power single bit full adder cell have been presented. The adder designs are based on optimized XOR circuit with combinations of multiplexer blocks. First proposed circuit shows power consumption of 23.8595pW with maximum output delay of 67.5566fs with supply voltage of 1.8V. Second adder shows power consumption of 43.1258pW with delay of 58.9935fs. Finally the third proposed circuit shows power consumption of 33.5163pW with delay of 62.065fs. Overall power delay product (PDP) also has been computed for proposed adders. Comparisons with earlier reported circuits have been made and the new cell outperforms the earlier reported circuits in terms of power consumption.

## References


1. Ekekwe, N., Cummings, R.E.: Power Dissipation Sources and Possible Control Techniques in Deep Ultra Submicron CMOS Technologies. Microelectronics Journals 37, 851–860 (2006)
2. Leblebici, Y., Kang, S.M.: CMOS Digital Integrated Circuits, 2nd edn. Mc Graw Hill, Singapore (1999)
3. Weste, N., Eshraghian, K.: Principles of CMOS VLSI Design, A System Perspective. Addison-Wesley, Reading (1993)
4. Zimmermann, R., Fichtner, W.: Low-Power Logic Styles: CMOS versus Pass-Transistor Logic. IEEE J. Solid State Circuits 32(7), 1079–1090 (1997)
5. Shams, A.M., Darwish, T.K., Bayoumi, M.A.: Performance Analysis of Low-Power 1-Bit CMOS Full Adder Cells. IEEE Transactions on Very Large Scale Integrations (VLSI) Systems 10(1), 20–29 (2002)

6. Adu-Shama, E., Bayoumi, M.: A New Cell for Low Power Adders. In: IEEE International Symposium on Circuits and Systems, pp. 1014–1017 (1996)
7. Zhuang, N., Wu, H.: A new design of the CMOS full adder. IEEE J. Solid-State Circuits 27(5), 840–844 (1992)
8. Al-Sheraidah, Y.J., Yuke Wang Sha, A., Jin-Gyun Chung, E.: A Novel Multiplexer-Based Low-Power Full Adder. IEEE Transactions on Circuits and Systems: Express Briefs 51(7), 345–348 (2004)
9. Shalem, R., John, E., John, L.K.: A Novel Low-Power Energy Recovery Full Adder Cell. In: Proc. Great Lakes Symp. VLSI, pp. 380–383 (1999)
10. Chang, C.H., Gu, J., Zhang, M.: A Review of 0.18μm Full Adder Performances for Tree Structured Arithmetic Circuits. IEEE Transactions on Very Large Scale Integration (VLSI) Systems 13(6), 686–695 (2005)
11. Bui, H.T., Wang, Y., Jiang, Y.: Design and Analysis of Low-Power 10-Transistor Full Adders Using XOR-XNOR Gates. IEEE Trans. Circuits Syst. II, Analog Digital Signal Process 49(1), 25–30 (2002)
12. Zhang, M., Gu, J., Chang, C.H.: A Novel Hybrid Pass Logic with Static CMOS Output Drive Full-Adder Cell. In: Proc. IEEE Int. Symp. Circuits Systems, pp. 317–320 (2003)
13. Goel, S., Kumar, A., Bayoumi, M.A.: Design of Robust, Energy Efficient Full Adders for Deep Sub Micrometer Design Using Hybrid-CMOS Logic Style. IEEE Transactions on Very Large Scale Integration (VLSI) Systems 14(12), 1309–1321 (2006)
14. Shams, A.M., Bayoumi, M.: A Novel High-Performance CMOS 1-Bit Full Adder Cell. IEEE Trans. Circuits Syst. II, Analog Digital Signal Process 47(5), 478–481 (2000)
15. Shams, A.M., Magdy, A.: A Structured Approach for Designing Low Power Adders. In: Conference Record of the Thirty-First Asilomar Conference on Signals, Systems & Computers, vol. 1, pp. 757–761 (1997)

# Exploring Associative Classification Technique Using Weighted Utility Association Rules for Predictive Analytics

Mamta Punjabi, Vineet Kushwaha, and Rashmi Ranjan

Indian Institute Of Information technology, Allahabad
Room No 209 Girls Hostel 1
IIIT Allahabad
Tel.: +91-9918809057
mamta.chgh@gmail.com,
vineet.kushwaha@gmail.com,
rromi.123@gmail.com

**Abstract.** Association rule discovery determine the "inter-dependence" among various items in a transactional database. Data mining researchers have augmented upon the quality of association rule discovery for business development by integrating the influential factors like quantity of items sold (weight), profit (utility), for extracting the association patterns. This paper proposes a new model (associative classifier) based on weightage and utility for useful mining of substantial class association rules. In process of predicting the class lables, all attributes do not have same importance. So our framework considers the different frequencies of individual items as their weights and varied significance can be assigned to different attributes as their utilities according to their predicting capability. Initially, the proposed framework uses the CBA-RG algorithm to produce a set of class association rules from a database and as well as exploits the downward closure property of the apriori algorithm. Subsequently, the set of class association rules mined are subjected to weightage and utility constraints like W-gain, U-gain and a combined Utility Weighted Score (UW-Score) is calculated for the mining of class association rules. We purport a theoretical model to innovate new associative classifier that takes vantage of valuable Class association rules based on the UW-Score.

**Keywords:** Associative Classifiers, Association Rule Mining, Apriori, Accuracy, Classifiers, Prediction, Utility, Utility gain (U-gain), Utility factor (U-factor), Utility Weighted Score (UWscore), Weightage, Weighted gain (W-gain).

## 1 Introduction

Employing the Association rule discovery for classification, improves the predictive accuracy of classification system. Associative Classification integrates association rule discovery problem into classification problem. Association rule discovery determine the "inter-dependence" among various items in a transactional database.

A. Mantri et al. (Eds.): HPAGC 2011, CCIS 169, pp. 169–178, 2011.

Classification rule mining takes a training data set (an object set whose class labels are already known) and generates a small set of rules to classify future data. To build a classifier based on association, we use a special subset of association rules known as Class Association rules (CARs) which perform association rule discovery techniques applicable to classification tasks. There are many associative classifiers that have been proposed such as CBA, CMAR, CPAR, MCAR and MMAC. There are many real applications such as retail marketing, financial analysis and business decision making, where to solve data mining problems association rule discovery has been widely used. The problem with mining association rules can be operated stepwise. The first step is to notice all the frequent itemsets in databases. The next step is to the generation of association rules. Generating association rules can be accomplished in linear time. Traditional ARM algorithms was designed assuming that all items have same frequency of occurrence (weight=1) and significance (utility=1) in a record, which is not always the case. However, in predictive modeling system where, attributes have different prediction capability it does not make sense to designate equal importance to each item. In order to get the better of the weakness of the conventional association rule mining, utility mining model [5] [6] and weighted association rule mining [8] have been proposed. Utility mining is a new research area and focused at integrating utility factors in data mining tasks. Utility of an item could be measured in terms of profit, value, cost, risk etc and it holds across the dataset. Utility of an item is dependent on users and applications. Given a transaction database, item utility, a utility table and a minimum utility threshold, the goal of utility based data mining is to find all itemsets with high utility. The total cost of stocking or the profitability an itemset cannot be decided using the support value alone. Thus practically, utility based data mining can be more useful than conventional association rule discovery. In weighted association rule mining (WARM) [8], itemsets are no longer simply calculated as they appear in a transaction. This change of calculating mechanism makes it necessary to accommodate traditional support to weighted support [13].

Weighted Association Rules can amend the confidence in the rules, as well as also render a mechanism to do more effectual target marketing by describing or sectioning customers based on volume of purchases or their potential degree of loyalty [7]. For example, a customer may purchase 13 coke bottles and 6 snacks bags and another may purchase 4 coke bottles and 1 snacks bag at a time. The traditional association rule mining approach treats the above two transactions in the similar manner, which could direct to the loss of some vital information [7]. So, Weighted ARM trades with the value of individual items in a database [9, 10, 11]. For example, some products are more beneficial or may be under promotion, therefore more concerning as compared to others, and hence rules interesting them are of greater value [12].

Recently, researchers are more interested at integrating both the weightage and utility for mining of valuable association rules. The incorporation, Weighted Utility association rule mining (WUARM) is the extension of weighted association rule mining can be ensured in the sense that it ensures items weights as their importance in the dataset and also trades with the number of occurrences of items in transactions. Thus, weighted utility association rule mining is interested with both the frequency and significance of itemsets and is also helpful in determining the most valuable and high selling items which contribute more to the company's profit [14].

Here, we propose a new model (associative classifier) based on weightage and utility for useful mining of substantial class association rules. In process of predicting the class labels, all attributes do not have same importance. So our framework considers the different frequencies of individual items as their weights and varied significance can be assigned to different attributes as their utilities according to their predicting capability [4]. Initially, the proposed framework uses the CBA-RG algorithm to produce a set of class association rules from a database and as well as exploits the downward closure property of the apriori algorithm. Subsequently, the set of class association rules mined are subjected to weightage and utility constraints like W-gain, U-gain and a combined Utility Weighted Score (UW-Score) is calculated for the mining of class association rules. Ultimately, we determine a subset of valuable class association rules based on the UW-Score estimated. We propose a theoretical model to innovate new associative classifier that takes vantage of valuable Class association rules based on the UW-Score. The model can generate high utility class association rules that can be lucratively applied in any domain such as business development to improve the prediction accuracy.

## 2 Related Work

Association Rule Mining (ARM), one of most commonly used descriptive data mining task, is the process of determining the "inter-dependence" among various items in a transactional database. Apriori has been recognized as the most renowned algorithm for ARM. Most algorithms for ARM including Apriori, mine potential data patterns chiefly based on frequency. The data patterns thus extracted based only on frequency would not be of the highest value for decision makers in business development. Hence in recent times, the incorporation of interestingness, utility and item weightage into the standard association rule mining algorithms have attracted voluminous research. In the proposed approach, we have incorporated two of the aforesaid measures together with Apriori for effectual mining of association rules. The two attribute measures chosen in the proposed research are:

**A. Weightage:** Generally in a transaction database, attributes comprise of numerical assets that gives the actual quantity of the attribute (count of the items) involved in the transaction. But traditional algorithms like Apriori mine association rules from a binary mapped database that only depicts the presence of the item in a transaction or not. So, standard ARM algorithms possibly overlook the quantitative information associated with an attribute, leading to frequent but less-weightage rules. In most cases of a customer transaction, some of the attributes may be actually more weighted in one transaction but it may not have occurred frequently in the database. However, in the business point of view, these set of attributes would be of significant value and should have been included in the frequent itemset. Hence, the proposed research will consider the weightage measure (W-gain) of the individual items in every transaction, for mining a subset of most significant rules from the set of frequent association rules mined.

**B. Utility:** The second measure employed in the proposed research to improve the quality of ARM is the individual utility (Gain) of the attributes. In general, a supermarket is likely to consist of attributes (items) that will yield different margins of profit. Hence, the rules mined without considering those utility values (profit margin) will lead to a probable loss of profitable rules. So, to attain a subset of high utility rules from the Apriori mined rules, the proposed approach makes use of a utility measure (U-gain). But, possibly the incorporation of both the measures together into ARM will enable more potential utility-oriented association rules. In this research, we incorporate the two measures weightage (W-gain) and utility (U-gain) to mine the association rules from a transaction database.

## 3 Problem Definition

Let $D$ be a database consisting of $n$ number of transactions $T$ and $m$ number of attributes $I = [i_1, i2,......,i_m]$ with positive real number weights $W_i$. A utility table $U$ comprising of $m$ number of utility values $U_i$, where $U_i$ denotes the profit associated with the $i$ attribute.

The major steps involved in the proposed approach for Associative Classifier based on weightage and utility are:

Step 1: Mining of class association rules from Database using CBA-RG algorithm.
Step 2: Computation of the measure W-gain.
Step 3: Computation of the measure U-gain.
Step 4: Computation of UW-score from W-gain and U-gain.
Step 5: Determination of significant class association rules based on UW-score.
Step 6: The selection of one subset of CARs to form the classifier from those generated at Step 5.
Step 7: Measuring the quality of the derived classifier on test data objects.

Figure 1 shows the major steps used in propose AC approach.

Apriori, a standard ARM algorithm is used in the proposed approach to mine the association rules. Classical Apriori generally processes on a binary mapped database

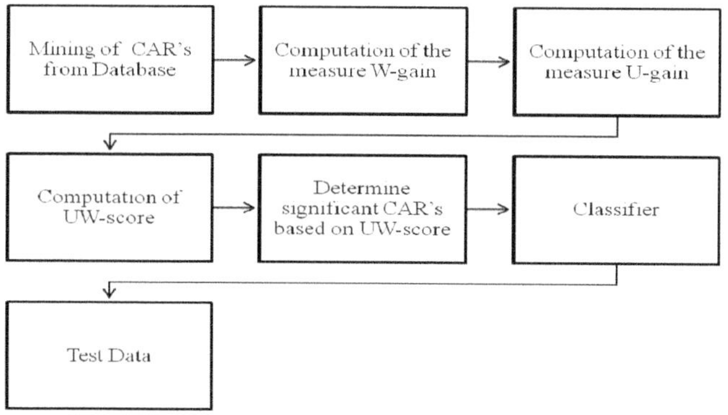


**Fig. 1.** Steps for Proposed Associative Classifier

$B_T$ for mining association rules. Hence, the input database $D$ is transformed to a binary mapped database $B_T$ that consists of binary values 0 and 1 denoting the non existence and existence of attributes in the transactions respectively. The weights $W_i$ associated with the individual attributes in the database $D$ is mapped onto the binary values using the following equation:

$$B_T = \begin{cases} 0 \ if \ W_i = 0 \ \forall \ T_k \\ 1 \ if \ W_i \neq 1 \ \forall \ T_k \end{cases} \quad (1)$$

Subsequently, the binary mapped database $B_T$ is given as an input to the Apriori algorithm [2] for mining of association rules.

### A. *Mining of class association rules using CBA-RG:*

Let $D$ be the dataset. Let $V$ *be* the set of class labels and $I$ *be* the set of all items in $D$. We say that a data case $d \in D$ contains $W \subseteq I$, a subset of items, if $W \subseteq d$. A class association rule (CAR) is an entailment of the form $W \rightarrow v$, where $W \subseteq I$, and $v \in V$.

Our targets are (1) to produce the complete set of CARs that gratify the user-defined minimum support is called *minsup* and minimum confidence is called *minconf* constraints and (2) to form a classifier from the CARs [1].

The pseudo code for the CBA-RG algorithm [1] is:

Let *k-ruleitem* denote a *ruleitem* whose *condset* has $k$ items. Let $F_k$ denote the set of frequent *k-ruleitems*. Each element of this set is of the following form: < (*condset*, *condsupCount*), (*y*, *rulesupCount*)>. Let $C_k$ be the set of candidate *k-ruleitems*.

```
F1 = {large 1-ruleitems};
CAR1 = genRules(F1);
prCAR1 = pruneRules(CAR1);
for (k = 2; Fk-1 ≠ ϕ ; k++) do begin
 Ck = candidateGen (Fk-1);
 for all data case d ∈ D do begin
  Cd = ruleSubset(Ck, d);
  for all candidates c ∈ Cd do
   c.condsupCount++;
   if d.class = c.class then
    c.rulesupCount++
  end
 end
 Fk = {c ∈ Ck | c.rulesupCount ≥
  minsup};
 CARk = genRules(Fk);
 prCARk= prune Rules(CARk);
end
CARs = ∪kCARk;
prCARs = ∪k prCARk;
```

Here, Pruning operation can be optional. The CBA-RG algorithm generates a *k* number of Class association rules (CARs) R = {$R_1$ , $R_2$...., $R_k$}. The set of CARs R is entered to the next phase of the suggested research, weightage and utility computation. The measures W-gain (weightage) and U-gain (utility) are calculated for every attribute present in the *k* Class association rules of *R*. For example, say an association rule $R_i$ of the form, [A, B] $\rightarrow$ C, where, C,A and B are the attributes in the rule $R_i$ , the measures U-gain, W-gain and UW-score are calculated for every attribute A, B and C individually. The sorted list of association rules is given by $S$ = {$R'_1$ ,$R'_2$,......, $R'_k$}, S $\in$ $R$ where conf ($R'_1$) $\geq$conf ($R'_2$) $\geq$conf ($R'_3$) ....... $\geq$conf ($R'_k$).

*B. Computation of W- gain:*

From the sorted list *S*, the first rule $R'_1$ is selected and the individual attributes of $R'_1$ are determined. Subsequently, the measure W-gain is calculated for every attribute in the rule $R'_1$.

*Definition 1*: Item weight ($W_i$): Item weight is the quantitative measure of the attribute contained in the transaction database *D*. Item weight value $W_i$ is a nonnegative integer.

*Definition 2:* Weighted Gain (*W*-gain): W-gain is defined as the sum of item weights $W_i$ of an attribute contained in every transaction of the database D as shown in the following equation:

$$\mathrm{W-gain} = \sum_{i=0}^{|T|} W_i \tag{2}$$

Where, *Wi* is the item weight of an attribute and *T* is the number of transactions in the database *D*.

*C. Computation of U-gain:*

Similarly, for *U*-gain computation, the first rule $R'_1$ from the sorted list *S* is selected and the individual attributes of $R'_1$ are determined. Subsequently, the *U*-gain measure is calculated for every individual attribute present in the rule $R'_1$, based on the U-factor and the utility value $U_i$ of the attribute.

*Definition 3:* Item Utility ($U_i$): The Item utility is generally defined as the margins of profit associated with that particular attribute. It is denoted as $U_i$ .

*Definition 4*: Utility table *U*: The utility table *U* comprises of 'm' utility values $U_i$ associated with the attributes present in the transaction database D. The utility table is signified by:

$$U = \begin{bmatrix} U_1 \\ U_2 \\ \vdots \\ U_k \end{bmatrix} \tag{3}$$

*Definition 5*: Utility factor ($U_{factor}$): The utility factor ($U_{factor}$) is a constant that is determined by the sum of the all items utility ($U_i$) contained in the utility table *U*. It is defined as:

$$U_{factor} = \frac{1}{\sum_{i=1}^{n} U_i} \tag{4}$$

Where, n is the number of attributes exhibit in the transaction database.

*Definition 6:* Utility Gain (U-gain): Utility Gain refers to the measure of an attribute's actual utility based on the $U_{factor}$ .

$$U - gain = U_i \quad * \; U_{factor} \tag{5}$$

The measure U-gain is computed for every attribute in the association rule $R'_1$.

*D. Computation of UW-score from W-gain and U-gain:*

Based on the calculated W-gain and U-gain measures for the individual attributes of an association rule, a single consolidated value termed UW-score is computed for every individual association rule.

*Definition 7*: Utility Weighted Score (UW-score): UW-score is outlined as the ratio between the sum of products of W-gain and U-gain for every attribute in the association rule to the number of attributes exhibit in the rule.

$$UW - score = \frac{\sum_{i=1}^{|r|}(W-gain_i)*(U-gain)}{|R|} \tag{6}$$

Where, $|R|$ represents the numbers of attributes in the Class association rule.

The aforesaid processes of W-gain, U-gain and UW-score computation are reiterated for the rest of the class association rules $R'_2$ to $R'_k$ present in the sorted list *S*. Now, all 'k' number of association rules in the sorted list *S* possess a UW-Score associated with it. Subsequently, the class association rules in the sorted list *S* are sorted based on the *UW-score* to get $S' = \{ R''_1, R''_2, \ldots\ldots, R''_k \}$ where *UW-score* $(R''_1) \geq$ *UW-score*$(R''_2) \geq$ *UW-score*$(R''_3)\ldots\ldots \geq$ UW-score$(R''_k)$.

*E. Determination of significant class association rules based on UW-score:*

A set of significant weighted utility class association rules $R_{WU}$ whose the UW-Score is above a predefined threshold are selected from the sorted list $S'$. The resultant weighted and utility based class association rules is given by $R_{WU} = \{R_{WU1}, R_{WU2}, \dots R_{WUl}\}$ where, $k \geq l$ and $R_{WU} \subseteq S'$ [3].

*F. Building the Classifier:*

Finally, one subset of CARs is selected from the set of significant weighted utility class association rules $R_{WU}$ to build the Classifier and measuring the quality of the derived classifier on test data objects.

## 4 Example

Given a training data set T and utility table U as shown in Table 1 and Table 2.

**Table 1.** Training Data Set

| Row-id | A | B | C | Class label |
|---|---|---|---|---|
| T100 | $i_1$ | $i_2$ | $i_5$ | 10 |
| T101 | $i_2$ | $i_4$ | $i_8$ | 20 |
| T102 | $i_2$ | $i_3$ | $i_6$ | 30 |
| T103 | $i_1$ | $i_2$ | $i_4$ | 10 |
| T104 | $i_1$ | $i_3$ | $i_5$ | 50 |
| T105 | $i_2$ | $i_3$ | $i_9$ | 60 |
| T106 | $i_1$ | $i_3$ | $i_5$ | 50 |
| T107 | $i_1$ | $i_2$ | $i_3$ | 80 |
| T108 | $i_1$ | $i_2$ | $i_3$ | 80 |
| T109 | $i_5$ | $i_6$ | $i_7$ | 100 |
| T110 | $i_5$ | $i_1$ | $i_2$ | 100 |
| T111 | $i_4$ | $i_5$ | $i_7$ | 100 |
| T112 | $i_1$ | $i_5$ | $i_2$ | 100 |
| T113 | $i_7$ | $i_8$ | $i_9$ | 140 |
| T114 | $i_3$ | $i_8$ | $i_1$ | 50 |
| T115 | $i_3$ | $i_1$ | $i_5$ | 50 |
| T116 | $i_1$ | $i_2$ | $i_3$ | 50 |
| T117 | $i_4$ | $i_1$ | $i_2$ | 10 |
| T118 | $i_3$ | $i_4$ | $i_5$ | 50 |
| T119 | $i_1$ | $i_2$ | $i_7$ | 10 |
| T120 | $i_5$ | $i_7$ | $i_8$ | 100 |
| T121 | $i_6$ | $i_7$ | $i_8$ | 100 |

**Table 2.** Utility Table

| Item no. | Items | Item value |
|---|---|---|
| 1 | I1 | 0.9 |
| 2 | I2 | 0.3 |
| 3 | I3 | 0.7 |
| 4 | I4 | 0.6 |
| 5 | I5 | 0.8 |
| 6 | I6 | 0.5 |
| 6 | I7 | 0.4 |
| 8 | I8 | 0.56 |
| 9 | I9 | 0.85 |

**Table 3.** Expected Result

| Min Support | Min Confi-dence | No. of rules in Classifier | | Classifier Accuracy | |
|---|---|---|---|---|---|
| | | CBA without UW-Score | CBA with UW-Score | CBA without UW-Score | CBA with UW-Score |
| 10 | 21 | 3 | 4 | 19.6 | 22.9 |
| 5 | 11 | 3 | 4 | 16.8 | 21.9 |

In this example we compare both techniques CBA without UW-Score and CBA with UW-Score based on two factors no of rules in classifier and average accuracy of the classifier. With the help of this implemented example we can say that our suggested approach can generate high utility class association rules based on UW-Score to improve the prediction accuracy.

## 5 Conclusion

We have proposed an effective approach based on utility and weight factor for efficient mining of high utility class association rules.

Initially, the proposed approach has made use of the traditional CBA-RG algorithm to produce a set of class association rules from a database. A combined Utility

Weighted Score (*UW-Score*) is calculated for every mined class association rule based on weightage (*W-gain*) and utility (*U-gain*) constraints. Ultimately, we have determined a subset of significant association rules based on the UW-Score computed. We propose a theoretical model to innovate new associative classifier that takes vantage of valuable Class association rules based on the UW-Score. The model can generate high utility class association rules that can be lucratively applied in any domain such as business development to improve the prediction accuracy.

## References


1. Liu, B., Hsu, W., Ma, Y.: Integrating classification and association rule mining. In: 4th Intl. Conf. on KDD (1998)
2. Agrawal, R., Imielinski, T., Swami, A.: Mining association rules between sets of items in large databases. In: Proceedings of the international Conference on Management of Data, ACM SIGMOD, Washington, DC, pp. 207–216 (1993)
3. Sandhu, S., Dhaliwal, S., Bisht, P.: An Improvement in Apriori algorithm Using Profit and Quantity. In: Proceedings of the 2nd International Conference on Computer and Network Technology, pp. 49–61 (2010)
4. Soni, S., Pillai, J., Vyas, O.P.: An Associative Classifier Using Weighted Association Rule. World Congress on Nature & Biologically Inspired Computing (NaBIC), 270–274 (2009)
5. Yao, H., Hamilton, H.J., Butz, C.J.: A Foundational Approach to Mining Itemset Utilities from Databases. In: Proceedings of the Third SIAM International Conference on Data Mining, Orlando, Florida, pp. 482–486 (2004)
6. Wang, J., Liu, Y., Zhou, L., Shi, Y., Zhu, X.: Pushing Frequency Constraint to Utility Mining Model. In: Proceedings of the 7th international conference on Computational Science, Beijing, China, pp. 685–692 (2007)
7. Zubair Rahman, A.M.J., Balasubram, P.: Weighted Support Association Rule Mining using Closed Itemset Lattices in Parallel. International Journal of Computer Science and Network security 9(3), 247–253 (2009)
8. Sun, K., Bai, F.: Mining Weighted Association Rules without Preassigned Weights. IEEE Transactions on Knowledge and Data Engineering 20(4) (2008)
9. Cai, C.H., Fu, A.W.C., Cheng, C.H.K., Wong, W.W.: Mining Association Rules with Weighted Items. In: Proceedings of the International Symposium on Database Engineering and Applications, Cardiff, Wales, UK, pp. 68–77 (1998)
10. Wang, W., Yang, J., Yu, P.S.: Efficient Mining of Weighted Association Rules (WAR). In: Proceedings of the KDD, Boston, MA, pp. 270–274 (2000)
11. Lu, S., Hu, H., Li, F.: Mining Weighted Association Rules. Intelligent Data Analysis 5(3), 211–225 (2001)
12. Sulaiman Khan, M., Muyeba, M., Coenen, F.: Fuzzy Weighted Association Rule Mining with Weighted Support and Confidence Framework. In: International Workshops on New Frontiers in Applied Data Mining, Osaka, Japan, pp. 49–61 (2009)
13. Tao, F., Murtagh, F., Farid, M.: Weighted Association Rule Mining using Weighted Support and Significance Framework. In: Proceedings of the International Conference on Knowledge Discovery and Data Mining, Washington, pp. 661–666 (2003)
14. Sulaiman Khan, M., Muyeba, M., Coenen, F.: A Weighted Utility Framework for Mining Association Rules. In: Proceedings of European Symposium on Computer Modeling and Simulation, Liverpool, pp. 87–92 (2008)

# Bio-enable Security for Operating System by Customizing Gina

Swapnaja A. Ubale[1] and S.S. Apte[2]

[1] Assistant Professor-CSE Dept.
SVERI'S CoE Pandharpur-India
Swapnaja_2283@yahoo.co.in
[2] HOD-CSE Dept.
WIT CoE Solapur-India

**Abstract.** Security is the core part of computer system and based applications. Gina DLL can be treated as the heart of security for windows operating system. User can customize Gina DLL for security to operating system. Paper briefly summarizes customization of Gina DLL for providing password and fingerprint security by considering biometrics as a main tool. The principles behind biometrics are common and used in everyday life. Hamster device is connected with the system for fingerprint recognition and security to operating system is provided at the starting level of the operating system by customizing Gina DLL.

**Keywords:** Bio-enable Security, Gina.

## 1 Introduction

People recognize family members by their faces, and individuals know friends by their voices and even their smell. Although human beings are excellent at doing this complex job, even they are not perfect – it may be very difficult to distinguish between identical twins, for example. The challenge for biometrics lies in the measurement and decision of what exactly is similar. There's no arbitrariness in matching a password – it either matches or it doesn't. And while biometric technology is advancing rapidly, it is not yet 100% accurate in matching a previously enrolled biometric feature to a present feature. For this reason, biometrics is still not quite as natural as human beings recognizing each other. As a field of analytic technique, biometrics uses physical and behavioral characteristics such as fingerprints, voice, face, handwriting and hand geometry to verify authorized users.

Biometrics devices use some measurable feature of an individual to authenticate their identity. The devices are built on the premise that physical human characteristics are unique and cannot be borrowed, misplaced, forged, stolen, duplicated, or forgotten. There are a number of different human characteristics that can be used in biometrics recognition like Fingerprints and other too. Biometrics identification such as fingerprint recognition can eliminate problems of forgotten passwords or lost cards and is currently becoming more popular for convenient and secure authentication. For that it is beneficial to implement bio-enable security for Operating System.

A. Mantri et al. (Eds.): HPAGC 2011, CCIS 169, pp. 179–185, 2011.

## 2 Interactive Logon

The interactive logon process is the first step in user authentication and authorization. Interactive logon is mandatory in the Microsoft Windows Server 2003, Windows XP, Windows 2000, and Windows NT 4.0 operating systems. Interactive logon provides a way to identify authorized users and determine whether they are allowed to log on and access the system. There are two ways for this Logon as –

Windows Server 2003 interactive logons begin with the user pressing CTRL+ALT+DEL to initiate the logon process. The CTRL+ALT+DEL keystroke is called a secure attention sequence (SAS); Winlogon registers this sequence during the boot process to keep other programs and processes from using it. The GINA generates the logon dialog box. The following figure shows the Windows Server 2003 logon dialog box.

**Fig. 1.** Windows Server 2003 Logon Dialog Box

A user who logs on to a computer using either a local or domain account must enter a user name and password, which form the user's credentials and are used to verify the user's identity. In the case of smart card logons, however, a user's credentials are contained on the card's security chip, which is read by an external device, a smart card reader. During a smart card logon, a user enters a personal identification number (PIN) instead of a user name, domain, and password.

## 3 Interactive Logon Architecture

Windows Server 2003 interactive logon architecture includes the following components:

- Winlogon
- Graphical Identification and Authentication (GINA) dynamic-link library (DLL)

- Local Security Authority (LSA)
- Authentication packages (NTLM and Kerberos).

### 3.1 Winlogon

Winlogon(windir%\System32\Winlogon.exe) is the executable file responsible for managing secure user interactions. Winlogon initiates the logon process for Windows Server 2003, Windows 2000, Windows XP, and Windows NT 4.0. Winlogon is responsible for the following tasks.

- Desktop lockdown
- Standard SAS recognition
- SAS routine dispatching
- User profile loading
- Screen saver control
- Multiple network provider support

Desktop Lockdown

Winlogon helps prevent unauthorized user's from gaining access to system resources by locking down the computer desktop. At any time, Winlogon is in one of three possible states: logged on, logged off, or workstation locked, as shown in the following figure.

Winlogon States

Winlogon switches between three desktops — Winlogon, Screen-saver, and Default — depending on its state and user activity. The following table lists and describes each of these desktops. Winlogon interacts with Gina manytimes when systrm is running.

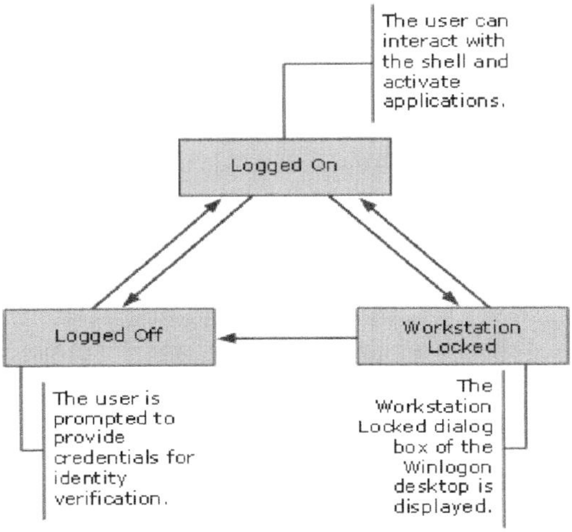


**Fig. 2.** Winlogon States

### 3.2 GINA

The GINA—a DLL component loaded by Winlogon—implements the authentication policy of the interactive logon model. It performs all user identification and authentication interactions. Msgina.dll, the standard GINA provided by Microsoft and loaded by Winlogon, can be replaced by a GINA that is custom-built by a third party.

GINA is the pluggable part of WinLogon that third parties may replace in order to customize the functionality or the UI of the logon experience in Windows®. By replacing GINA, you can choose the authentication mechanism Windows will use for interactive users. This is often useful for smartcard or biometric logons.

## 4 Proposed Work

Proposed work implement new MsGina.DLL with all its functions so that it can interact with Hamster Device for fingerprint Recognition. Here when System Boots firstly it loads winlogon. Then it calls our replaced Gina and then according to our replaced Gina it will ask for pressing Ctrl+Alt+Del. And then it will ask for password if password is correct then control gets transferred to Hamster Device where we are going to perform fingerprint matching. If matching is more than 95% (threashold) then it allows user to access the Operating System. If some succeeds to steal password then also due to biometric security he/she may not enter in Operating system. Thus operating system will become more secure.

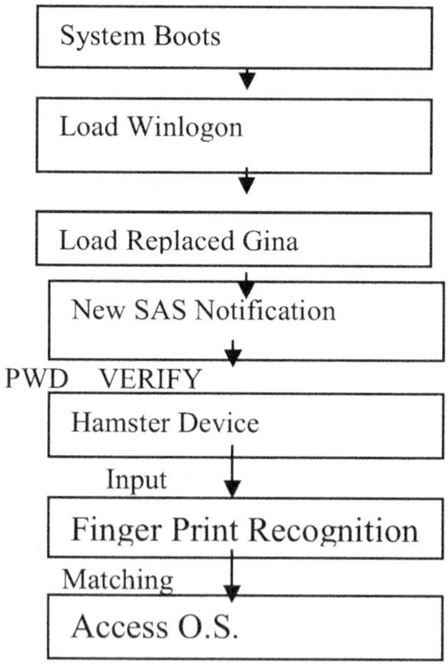


**Fig. 3.** Proposed work

## 5 Some Necessary Functions of GINA

Following are some of the necessary functions of GINA . Proposed work change there implementation by keeping prototype same.

- WlxNegotiate
- WlxInitialize
- WlxDisplaySASNotice
- WlxLoggedOutSAS
- WlxActiveUserShell
- WlxLoggedOnSAS
- WlxNetworkProviderLoad

WlxNegotiate allows the GINA to verify that it supports the installed version of Winlogon. Winlogon calls this function once for each window station present on the computer. Currently, the operating system supports one window station per workstation.

The context returned by this function will be passed back to the GINA in all subsequent calls. WlxInitialize also allocates space for the credentials UserName, Domain, Password, and OldPassword. Winlogon calls WlxDisplaySASNotice to display a notice when no users are logged on. Winlogon calls this function when it receives an SAS event while no users are logged on. This routine clears out the user process information and then calls Logon. Winlogon calls this function when it receives an SAS event while the user is logged on and the workstation is not locked. Winlogon calls WlxLoggedOnSAS when the logged-on user wants to shut down, log off, or lock the

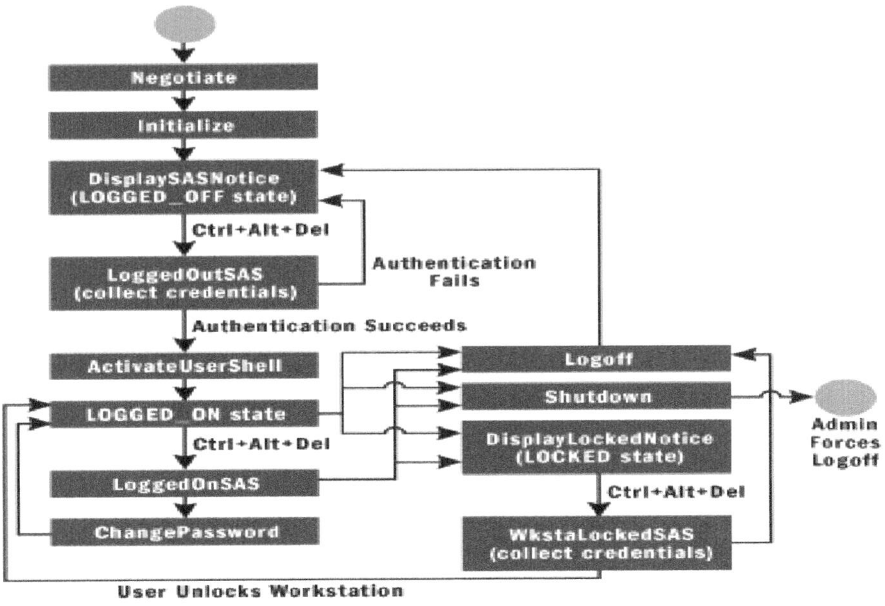


**Fig. 4.** Some necessary functions of Gina DLL

workstation. WlxDisplayLockedNotice allows the GINA to display information about the lock, such as who locked the workstation and when it was locked. Winlogon calls WlxWkstaLockedSAS function when it receives an SAS event and the workstation is locked. The GINA should interact with winlogon continuously for the operating system proper working.

# 6 Loading New GINA

After implementation and execution, copy created gina dll and place it at the given location as C:\WINNT\system32\msgina.dll

This is standard location for gina dll. Otherwise user can store it at any place and give path while creating string value as given below.

Activating new Gina Dll.

- Follow Path as –

HKEY_LOCAL_MACHINE\SOFTWARE\Microsoft\WindowsNT\CurrentVersion\Winlogon

- Add new String Value
- For that String Value Give Value Name : GinaDll
- Value Data : Path of msgina.dll
- Gina of type REG_SZ

After Loading Reboot System – Effect can take place.

# 7 Warning: Potential System Failure

Attempting to use a new GINA can cause a system to fail to boot correctly. If a new GINA is created, the registry is changed to use this new GINA, and attempts to restart the system are met with a Failed to Load xyz.dll error message.

## 7.1 Recovering from a GINA-Related System Failure

Recovery is possible without reinstalling the system. This requires having a boot disk, or equivalent workaround, that will boot the system without using a normal hard disk boot. To recover from this situation:

1. Boot the auxiliary system—potentially by using a boot disk.
2. Go to the System32 folder where the new, failed GINA DLL is located.
3. Move the failed GINA to a different folder.
4. Copy the original MSGINA DLL in the System32 folder and rename it with the name of the failed GINA.
5. Restart the computer and the original MSGINA DLL will load under the new name. At this point fixes can be made in the modified GINA.

## 8 Sample Results

Experimental results will be as given in following steps not shown diagrammatically here:

1.Booting of the system
2.Load replaced gina dll
3.Generatation of new SAS value as

    3.1 Demand for user name and password(if correct go on with further steps otherwise deny access and again ask for user name and password)
    3.2 Connect with database of Hamster Device

4. If match of fingerprints allow access of operating system if no deny access
5.While working of system provision of different options as lock, shut down
5.On lock again it will ask for same procedure as from step 3.1to4.

## References

1. Park, B.: One Touch Logon Replacing Multiple Passwords with Single Fingerprint Recognition. In: The sixth IEEE Conference on Computer and I.T(CIT 2006) (2006)
2. Frank Graf, A.: Capability Based Transparent Cryptographic File System CW 2005
3. Itoi, N.: NI_Pluggable Authentication Module for Windows NT Center For I.T. Integration. University of Michigan (January 1988)
4. Wolthusen, S.D.: Distributed Multipurpose Mail Guard. In: Proceedings of 2003 IEEE Workshop on Information Assurance United States Military

# A Destination Capability Aware Dynamic Load Balancing Algorithm for Heterogeneous Environment

Sharma Rajkumar[1], Kanungo Priyesh[2], and Chandwani Manohar[3]

[1] Computer Centre, Vikram University, Ujjain, India
rksujn@rediffmail.com
[2] Patel College of Science & Technology, Indore, India
[3] Instt. of Engg. & Tech., Devi Ahilya University, Indore, India

**Abstract.** Complexity of both data- intensive problems and computation-intensive problems has been increased in computational world. These problems require sophisticated mathematical & statistical techniques as well as knowledge creation and involve complex computations to be performed on huge amount of data. A number of large computational problems now work in terms of Petabytes of data rather than limited volume comprising of only Mbytes or Gbytes of data in olden days. In the present scenario, efficient utilization of system resources is necessary to solve large computational problems within desirable time constraint. In this paper, we present a dynamic load balancing algorithm for a cluster created by using heterogeneous commodity hardware. The algorithm uses system resources efficiently & effectively depending on the nature of the applications. Using this algorithm we observed an improvement in execution time of parallel applications if tasks are processed on favorable nodes instead of randomly assigned nodes.

**Keywords:** Computation-Intensive Problems, Data-Intensive Problems, Dynamic Load Balancing, Hardware Heterogeneity.

## 1 Introduction

A network created from commodity hardware can effectively and affordably be used for high performance computing. Major attraction from such networks is cost-effectiveness due to efficient utilization of existing hardware. By combining processing powers of many scattered heterogeneous workstations, a large computational problem can be solved within desirable time constraint, which is comparable to that of a proprietary or dedicated parallel computer. Popularly such systems are known as Cluster or Network of Workstations (NOWs)[3][4]. Among the state-of-art computers in the world, many have been developed by using commodity hardware clusters.

In this research work, we address the following three questions: (1) What is the impact on execution time of a parallel application due to heterogeneous system hardware ? (2) What are the effects of system heterogeneity such as variation in CPU power, cache memory, primary memory, size of hard drive etc. on data & computation-intensive problems ? (3) What should be the design of an efficient

A. Mantri et al. (Eds.): HPAGC 2011, CCIS 169, pp. 186–191, 2011.

dynamic load balancing algorithm in which processes are aware of the capabilities of destination nodes and get processed at the most favorable nodes ? We evaluate performance of computation-intensive & data-intensive problems on cluster of nodes having different computing powers.

### 1.1 Load Balancing

In Network of Workstations, load balancing plays a crucial role by migrating load from highly loaded node to a lightly loaded or idle node, thereby keeping system approximately in a balanced state. An effective load balancing algorithm should exploit all resources of a workstation efficiently[8].

### 1.2 Data and Computation Complexities

Computation-intensive problem requires more CPU cycles instead of memory usage and data transfers. On the other hand, data-intensive problem requires more memory usage and data transfers than CPU cycles. In earlier days, computational problems required moderate calculations and working on megabytes or gigabytes of data, but large computational problems involve complex computations based on advanced mathematical & statistical methods and work on voluminous data, may be in terms of Petabytes. Beginning from uniprocessor and homogeneous environment, most of the state-of-art applications are now based on distributed computing model comprising of heterogeneous environment as indicated in Fig. 1. Heterogeneity may exist due to different operating systems, architecture and data formats.

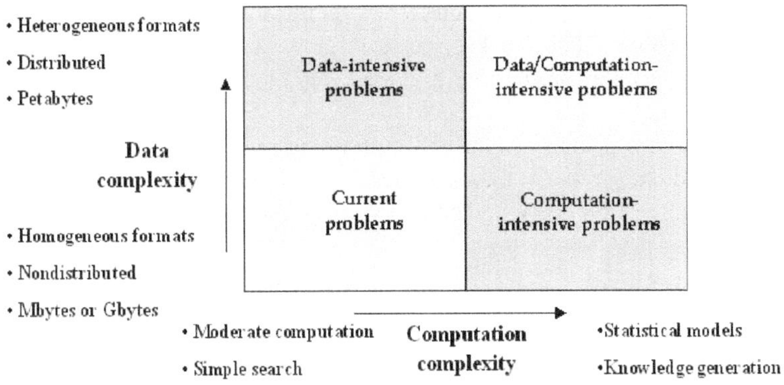


**Fig. 1.** Data and Computation Complexities

## 2 Proposed Method

In this research work, we propose a dynamic load balancing algorithm for heterogeneous computing environment which allocates workstations to the processes depending on the nature of the problem whether it is a data-intensive or a computation-intensive problem. The proposed algorithm considers heterogeneity of several hardware parameters as compared to few parameters considered by other researchers [1][2]. We use assignment factor and priority of nodes which assign processing load to nodes based on their computing performance. We use node index $j$ to refer one of the total nodes $N$ in a heterogeneous environment. If $I_j$ is the number of iteration performed by $j^{th}$ node of an application $P$ in a given period of time, then the assignment factor of $j^{th}$ node is given by:

$$AF_j = \frac{I_j(P)}{\sum_{j=1}^{N} I_j(P)} \tag{1}$$

We assign priority to nodes in decreasing order of their assignment factor. Nodes having higher assignment factor will get higher priority and will receive more work to process as compared to nodes with lower priority. A sample of assignment factor and priority is shown in Table 1. The system heterogeneity can be characterized as the variance of assignment factor among the nodes. By using standard deviation the system heterogeneity $H_{sys}$ can be defined as follows:

$$H_{sys} = \sqrt{\frac{\sum_{j=1}^{N} (\overline{AF} - AF_j)^2}{N}} \tag{2}$$

where, $\overline{AF} = \frac{\sum_{j=1}^{N} AF_j}{N}$ is the average of assignment factor of nodes in the system.

**Table 1.** Sample of Assignment Factor and Priority

| Node No. | Group Id. | Supports Problem of Type | Assignment Factor | Priority |
|---|---|---|---|---|
| 1 | G1 | Data-Intensive | 0.27 | 2 |
| 2 | G1 | Data-Intensive | 0.39 | 1 |
| 3 | G1 | Data-Intensive | 0.18 | 3 |
| 4 | G1 | Data-Intensive | 0.16 | 4 |
| 5 | G2 | Computation-Intensive | 0.25 | 2 |
| 6 | G2 | Computation-Intensive | 0.17 | 4 |
| 7 | G2 | Computation-Intensive | 0.22 | 3 |
| 8 | G2 | Computation-Intensive | 0.36 | 1 |

The basic idea is to enroute the incoming process to a favorable group for efficient use of hardware resources. We evaluate assignment factor and assign priority to nodes within each group respectively.

## 3 Experimental Results and Performance Comparison

To evaluate and analyze the performance of a parallel application, we performed our experiments on a real heterogeneous cluster of eight nodes having different computing powers, connected with a Ethernet 10/100 IEEE 802.3 switch. A Message Passing Interface (MPI) Library is needed to establish point-to-point and collective communication among the processors to run parallel application on the cluster [5][7].

Based on our previous studies [6], higher processor speed, CPU cores, motherboard FSB are favorable for computation-intensive problems, on the other hand higher amount of RAM, size of hard drive and its RPM improve the performance of data-intensive problems, therefore we formed two groups G1 and G2 supporting both types of problems respectively. We determine assignment factor for nodes within each group by running a benchmark application several times and averaging the results. Therefore, for our experiments we finally prepared a cluster of heterogeneous nodes divided into two groups, one group favors computation-intensive problems while another favors data-intensive problems. With the help of assignment factor and priority we distributed load among all nodes within a group. We also form groups G3 to G10 consist of random selection of nodes. We executed each type of problem by distributing load on group G1 to G10 and by repeating the experiments several times we computed the mean execution time as shown in Table 2.

**Table 2.** Comparative Execution Time

| Problem Type | Execution Time (ms) | | | | | | | | | |
|---|---|---|---|---|---|---|---|---|---|---|
| | G1 | G2 | G3 | G4 | G5 | G6 | G7 | G8 | G9 | G10 |
| Computation-Intensive Problem | 305 | 254 | 320 | 296 | 309 | 317 | 288 | 310 | 302 | 294 |
| Data-Intensive Problem | 342 | 422 | 418 | 425 | 402 | 409 | 380 | 411 | 420 | 416 |
| Mixed Problem | 360 | 345 | 378 | 412 | 407 | 380 | 417 | 410 | 399 | 385 |

We observed a significant improvement in execution time and CPU utilisation as shown in Fig. 2 and Fig. 3. Computation-intensive problem executed in improved time on group G1 which is intended for processing of computation-intensive problems. Similarly for data-intensive problem, there is minimum execution time on group G2. Group G2 comprised of nodes that are having hardware parameters supporting data-intensive problems. There is no noticeable performance gain for mixed types of problems running on either G1, G2 or groups from G3 to G10 which consist of random selection of nodes.

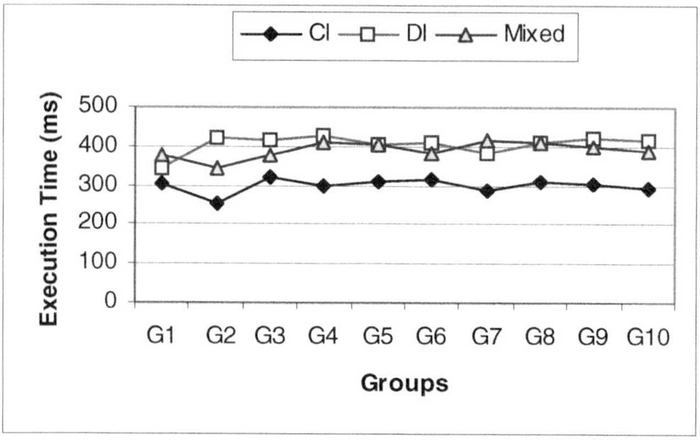


**Fig. 2.** Comparative Execution Time

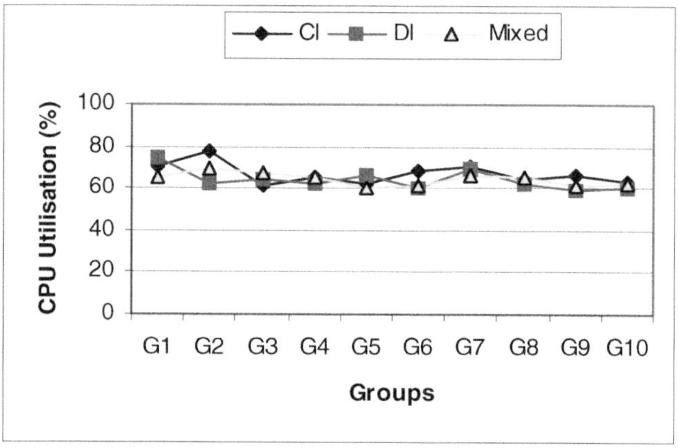


**Fig. 3.** Comparative CPU Utilisation

# 4 Conclusion

Heterogeneous network of workstations are becoming popular for high performance computing. Cost-effectiveness is major advantage of such networks due to efficient and effective use of existing and scattered hardware. In this paper, we have experimentally evaluated performance of the proposed dynamic load balancing algorithm. Based on our experiments and analysis we observed the impact on execution time of a parallel application due to heterogeneous system hardware. In addition to system heterogeneity due to CPU power and memory, in our study, we considered other important systems hardware parameter also to improve the performance of a parallel application. We have designed and experimented an effective dynamic load balancing algorithm in which processes are aware of the capabilities of destination nodes by using hardware profile of the nodes. By diverting load to the best supporting nodes depending on the nature of the problem, computation-intensive or data-intensive problem, we found a substantial improvement in execution time in our experiments.

## References


1. Li, X., Xiaodong, Z., Yanxia, Q.: Effective Load Sharing on Heterogeneous Networks of Workstations. In: 14th IEEE International Parallel & Distributed Processing Symposium, Mexico, pp. 431–438 (2000)
2. Xing, D., Xiaodong, Z.: Coordinating Parallel Processes on Networks of Workstation. J. Parallel and Distributed Computing 4(8), 125–135 (1997)
3. Huajie, Z.: On Load Balancing Model for Cluster Computers. International J. Computer Science and Network Security 8(10), 263–269 (2008)
4. Hong, J., et al.: Modeling Parallel Applications Performance on Heterogeneous Systems. In: 17th IEEE International Parallel & Distributed Processing Conference, France, pp. 201–205 (2003)
5. William, G., Ewing, L., Thakur, R.: Using MPI-2: Advanced Features of the Message-Passing Interface, `http://www.mcs.anl.gov/research/projects/mpi`
6. Sharma, R., Kanungo, P., Chandwani, M.: A Comparison of Computation Intensive and Data Intensive Problems on Heterogeneous Nodes in Network of Workstations. In: IEEE National Conference on Education & Research in Information Tecnology, Guna, pp. 291–298 (2010)
7. Xiaodong, Z., Yanxia, Q., Li, X.: Improving Distributed Workload Performance by Sharing Both CPU and Memory Resources. In: 20th International Conference on Distributed Computing Systems, Taiwan, pp. 233–241 (2000)
8. Maheshwari, P.: A Dynamic Load Balancing Algorithm for a Heterogeneous Computing Environment. In: 29th International Conference on System Sciences, Hawaii, pp. 338–346 (1996)
9. Post, E., Goosen, H.: Evaluating the Parallel Performance of a Heterogeneous System. In: 5th International Conference on High Performance Computing, Canberra, pp. 310–323 (2001)

# Reliable Mobile Agent in Multi – Region Environment with Fault Tolerance for E-Service Applications

M. Vigilson Prem[1] and S. Swamynathan[2]

[1] Assistant Professor,
Department of Information Technology,
R.M.D.Engineering College,
Kavaraipettai – 601206, Tamil Nadu, India
vigiprem@gmail.com
[2] Associate Professor,
Department of Information Science and Technology,
College of Engineering, Anna University Chennai,
Chennai – 600 025, Tamil Nadu, India
swamyns@annauniv.edu

**Abstract.** Mobile agent technology is emerging fast on web today that can be applied for developing distributed applications like information retrieval. Mobile agents travel through servers to perform their tasks. Ensuring the availability of mobile agents in the presence of agent server's failure, during their travel, is a challenging issue. Communication among mobile agents in multi-region environment, when they are in move, makes the issue more complex. Considering this scenario, we provide a reliable mobile agent model with fault tolerance which ensures that the Mobile agent and the information collected by the mobile agent is available even in the case of server failure. This model works for the complete itinerary failure of a region also. The model is experimented in Internet using IBM Aglets, a tool for Mobile agents. The experimental results appear to be more encouraging in claiming this mobile agent model is more reliable.

**Keywords:** Agent Tracking, Fault Tolerance, Mobile Agents, Location Management, Witness Agents.

## 1 Introduction

### 1.1 Mobile Agents

A mobile agent is a program that represents a user in a computer network. When injected into the network, by applications, it autonomously migrates from one node to another on a predetermined path or one that the agent themselves determined, based on dynamically gathered information, to perform a task given by creator. After accomplishing their goals, they can return to their owner (home server) to report their results [1]. A data processing application can move to its data, avoiding the overhead of transferring the data over the network [2]. The main difference between mobile agents and client server architecture lies in network load reduction, dynamic adaptation, synchronous and autonomous execution, robust and fault tolerance etc. [3].

A. Mantri et al. (Eds.): HPAGC 2011, CCIS 169, pp. 192–200, 2011.

### 1.2 Fault Tolerance

Fault Tolerance means guaranteeing the continuous execution of mobile agents even in the case of system failure. The components of the network like machine, link or agent may fail at any time. Eventually, these situations may lead to blocking of agents. Fault Tolerance is achieved normally through Replication or Checkpointing mechanisms. Replication resolves the problem of blocking [4][5]. But maintaining many physical servers and consistency of data among them is difficult. Also, in an asynchronous system like Internet, it is very difficult to detect whether the agent has actually failed or whether it is slow [6]. When replication prevents blocking, it may lead to multiple execution of agents (i.e.) violation of exactly-once execution property. Checkpointing is an act of saving a programs' state on stable storage. In this scheme, the state of an agent is saved into stable storage, upon the arrival of agent at agent server [7][8].

## 2 Related Work

S. Pears et al. proposed two exception handling designs for information retrieval applications: a mobile time out design - tolerate any number of agent server crash failures with no additional overheads imposed on mobile agents and servers, but in the event of agent server crash all agent servers are revisited; and a mobile shadow design - all agent servers are not revisited, but, the overhead is, it must spawn a replica before migration and it can solve the failure of one server at a time only. [9]. Witnessing Agent scheme was introduced by D. Johansen and co. where they used two different agents, an actual agent and a witness agent - to monitor the aliveness of its master, detects and recovers the master when it lost. A peer-to-peer message passing mechanism, where all actions of actual agent are logged, which is useful to recover committed actions, during failure is proposed in [10]. Assuming no 'k' or more servers fail at same time preserves witness dependency that is simplified by the witness length <= k. It also assumes that the host in the itinerary will never fall into long time un-repairable server failure [11]. Making the witness dependency to start from head node and some other related works can be found in [12][13]. To allow communication among mobile agents, finding the location of mobile agent is an important requirement. Tracking of mobile agent's location in an efficient manner is a challenging issue. The location server approach [15] maintains the information of location in a centralized way to keep track of mobile agents. An approach of broadcasting [16], broadcasts a message contains the identifier of a mobile agent to group of nodes. A group communication system for mobile agents is proposed in [17].

### 2.1 Motivation

From the existing work it is clear that the witnessing agent concept is used to ensure the reliability of Mobile Agents. In addition the agent tracking system is provided for mobile agents to communicate with each other. The major draw back of these schemes is (i) In the witness agent proposal, it is necessary to maintain n-1 witness agents in n-1 hosts to provide the reliability. In addition, these entire hosts should allocate considerable storage to maintain mobile agent clone, address table, etc. until the agent completes its itinerary. (ii) The necessary frequent communication between agents that checks the agents' availability, results in considerable network traffic. (iii) There is no provision to recover an agent from complete itinerary failure, as the first witness agent does not have its own witness agent to monitor it. These problems are addressed in [18][19]. In this paper a reliable mobile agent model is presented to overcome the specified limitations. This model, first, allows the mobile agents in multi region environment to communicate each other in a reliable way [20]. Then they exchange the information collected so far and continue their itinerary. After exchanging, the witness dependency is eliminated. It also ensures that the mobile agent reaches the sender successfully with the data collected by the agents.

## 3 Notations

| | | |
|---|---|---|
| $A_i$ | - | Agent Identification number i |
| $Ct_i$ | - | Cloning time at node i |
| $D_i$ | - | Data collected from node i |
| $DT_t$ | - | Data Transmission time |
| $Mt_{i,i+1}$ | - | Migration time from node i to i+1 |
| $Pt_i$ | - | Processing time at node i |
| $R_{i,j}$ | - | Response from node i to node j |
| $RS_i$ | - | Region Server i |
| $Rt_{i,i-1}$ | - | Response time from node i to i-1 |
| $S_i$ | - | Server i |
| $T_t$ | - | Total traverse time of mobile agent |
| $W_t$ | - | Waiting time for response |

## 4 Model for Mobile Agent Failure Recovery

Consider a client dispatches his mobile agent to collect necessary information from the bulk data available in N number of hosts. The mobile agent starts its itinerary, visits each host as per the predefined order. During this travel, if this agent fails due to host failure, it is not known to the sender. To overcome this problem response model is used [21][22]. This model requires every host that receives mobile agent to send a response message to the preceding host. By receiving the response, the preceding host (sender, not owner) can confirm the successful dispatch of agent to the succeeding host.

### 4.1 Simple Model

To begin with the working of a simple mobile agent is considered. The state transition for this model is given as,

$S_1 \rightarrow S_2 : D_1, A_1$
$S_2 \rightarrow S_1 : R_{2,1}, A_1$
$S_2 \rightarrow \ldots : D_1, D_2, A_1$

For simplicity, the first transition is taken into account. The transition indicates that, the mobile agent with identity $A_1$ is migrating from Server $S_1$ to $S_2$. In addition, the data $D_1$ collected from Server $S_1$ associates with the agent $A_1$.

### 4.2 K – Way Response Model

By keeping the mobile agent migration model discussed in section 4.1 in mind, the next level of migration between multiple servers is given as K-way response model.

The state transition for this model is given as, Migration without failure or attack

$S_1 \rightarrow S_2 : D_1, A_1$
$S_2 \rightarrow S_1 : R_{(2,1)}, A_1, S_3$

..............................

$S_{k-1} \rightarrow S_k : D_1, D_2, D_3, D_4, D_5, ...D_{(k-2)}, D_{(k-1)}, A_1$
$S_k \rightarrow S_{(k-1)} : R_{(k,k-1)}, A_1, S_{(k+1)}$
$S_k \rightarrow S_{(k-2)} : R_{(k,k-2)}), A_1, S_{(k+1)}$

.............................................

$S_k \rightarrow S_1 : R_{(k,1)}, A_1, S_{(k+1)}$

In this model, suppose the agent needs to visit 8 nodes (k=7), it is necessary for all the previous hosts (i.e. 7 hosts) to keep the clone of actual agent. This is because, suppose in this example, consider nodes $S_2$ & $S_3$ are compromised as a result of colluded attack. That is, $S_2$ forwards the agent to $S_3$ and sending response to $S_1$. But $S_3$ sends the response to $S_2$ but not forwards the agent to $S_4$. In this situation it is necessary for $S_1$ to get the response from a host where the agent has successfully completed its tasks. Here, applying K-Response Model, $S_4$ should send the response to $S_3$, $S_2$ and $S_1$. So, $S_1$ can easily notify that $S_3$ did not forward the agent to $S_4$. But the real problem lies in this model is, fixing up the number of response needed for successful delivery of agents. Especially, if the number of nodes to be visited by the mobile agent is very large (i.e. in large networks) it will become more complex. The number of response needs to be increased as long as the number of nodes involved in colluded attack increases. This will result in overhead in (i) space management i.e. until successful response is received the clones has to be maintained in all preceding servers; (ii) waiting time for response i.e. the waiting time for response increases as long as the number of node failure increases; (iii) mobile agent availability i.e. in complete itinerary failure, the agent and data within the agent is lost forever.

### 4.3 Multi – Region Information Sharing Model

To overcome the problem stated in previous section, multi-region environment is applied for free roaming mobile agent. Here the nodes to be visited by mobile agent are divided into various regions spatially (here in Fig. 1, it is 2 regions). Home server initiates the itinerary by creating required number of agents (for simplicity, one agent per region) and allots the work for them. Each agent starts their itinerary in each region as per the k-way response model. In Fig. 1, agent for Region 1 starts with node 'a' and travels through 'b','c','d','e' and 'f'. Same itinerary is followed in Region 2 also. During their travel, after visiting predefined number of servers (in Fig. 1 it is 3 servers) the information collected so far by agent in region '$R_1$' is shared with the agent in region $R_2$. This is done by contacting the lookup server.

That is, in Fig. 1, agent resides in node 'c' of Region 1 requests the Lookup server, the current location of agent in Region 2. The function of lookup server is to monitor the movement of mobile agents in each server. After getting the location of agent 2 (here it is node 'r'), agent 1 will directly exchange the information collected so far with agent 2. After this exchange both the agents continue their itinerary. So that the data collected so far by the agents $A_i$ and $A_j$, is now available in two different regions $R_i$ and $R_j$. This process will continue with other regions also. As a result, at any given time the data collected by each agent is available with all other agents belongs to all regions. On failure of one or more nodes, the corresponding agent may fail and data may lose. But, that data can be recovered, as they are available in other regions' mobile agent. Also, a new agent can be dispatched to this region (where mobile agent failed/killed), if there is no response or data sharing for a time period by other regions.

State Transition for this model is given by, Region 1.

$S_a \rightarrow S_b : D_a, A_1$
$S_b \rightarrow S_a : R_{(b,a)}, A_1, S_c$
$S_b \rightarrow S_c : D_a, D_b, A_1$
$S_c \rightarrow S_b : R_{(c,b)}, A_1, RS_2$
$S_c \rightarrow S_a : R_{(c,a)}, A_1, RS_2$
$S_c \rightarrow RS_2 : D_a, D_b, D_c, A_1$
$RS_2 \rightarrow S_c : D_p, D_q, D_r, A_2$
$S_c \rightarrow S_d : D_{(RS2)}, A_1$
$S_d \rightarrow S_c : R_{(d,c)}, A_1, S_e$

This will clearly increase the data availability even in the case of complete itinerary failure in a region. Also the number of nodes maintaining the clone is limited to a constant (here it is 3) as per 3-way response model. Further, the waiting time for response is reduced because it is enough here to wait for response from 3rd server at maximum. The merits of this model costs time and communication overhead.

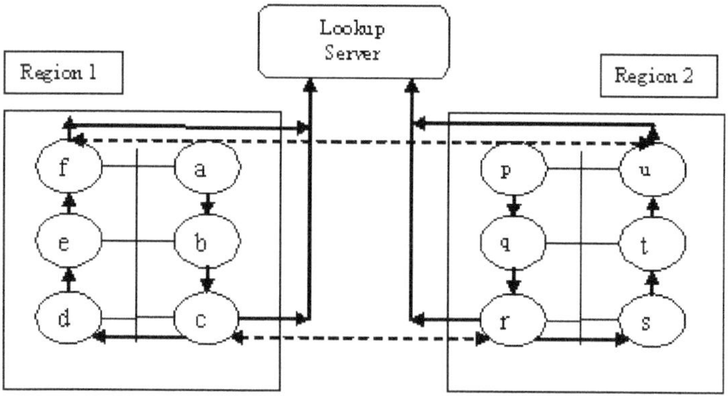


**Fig. 1.** Data sharing between mobile agents in multi-region environment

## 5 Cost Evaluation

In this environment, for normal operation, the sender will wait for the response from receiver for a time period. This waiting time includes, migration time from sender to receiver, processing time at receiver, response time from receiver to sender.

The waiting time for 1-response model is evaluated from the following equation.

$$Wt\ (1) = [\ (2 * Mt) + x\ ] + [\ Pt + y\ ] + Ct\ . \tag{1}$$

Where, x indicates the additional time taken for transmission due to network traffic and y indicates the waiting time for process allocation. Using equation (1), the waiting time for k-response model with no node failure can be calculated as,

$$Wt\ (k) = Wt(1) + Wt(2) + ... + Wt(k-1)\ . \tag{2}$$

The total agent traversal time without agent or node failure is,

$$Tt = \sum (\ Mt_{i,i+1} + Ct_i\ )\ ,\ 1 \leq\ i \leq n\ . \tag{3}$$

To calculate the total traversal time with 1-failure, additional clone migration time and response time should be considered. Now, equation (3) becomes,

$$Tt = \sum (Mti,i+1\ + Cti) + Wt\ (1)\ . \tag{4}$$

F0or k failures, equation (4) can be written as,

$$Tt = \sum (Mti,i+1\ + Cti) + Wt\ (k)\ . \tag{5}$$

Now consider the total traversal time for Multi-Region Data Sharing model. The total time given in equation (5) is for one region without data sharing. If we also include the data sharing time between regions, for example region 1 and 2, then the new Tt is,

$$Tt = Tt_{R1} + Tt_{R2} + (\ DTt * No.\ of\ DT\ )\ . \tag{6}$$

Agent will not wait for the response time between the hosts. It carried outs its process until there is no attack or failure. If there is an attack or failure, then the waiting time Rt needs to be included; otherwise the time taken by the agent to roam in the network is same as (1). The given equations meant for the one node failure

## 6 Experimental Design

The model is tested in internet using 3 machines, located at a distance of around 30 Km, of configuration 1GB RAM and 3.2GHz processor. IBM Aglet is used as implementation tool which can act as agent server. This server holds the agent and controls the migration of agent. The results obtained are given in following tables.

**Table 1.** Processing time at each host

| Host Number | IP Address | Processing Time (milli seconds) |
|---|---|---|
| 1 | 112.113.196.248 | 3842 |
| 2 | 192.168.1.10 | 4027 |
| 3 | 112.113.196.238 | 3783 |
| 4 | 192.168.1.24 | 3998 |

**Table 2.** Migration time between hosts

| Source Host | Destination Host | Migration Time (milli seconds) |
|---|---|---|
| 1 | 112.113.196.248 | 3842 |
| 2 | 192.168.1.10 | 4027 |
| 3 | 112.113.196.238 | 3783 |
| 4 | 192.168.1.24 | 3998 |

From this the waiting time of the sender to receive the acknowledgement can be calculated as follows,

Waiting time = ( Travel time * delay ) + ( Processing time * load )

Cloning time to be included if there is any failure in the host. Cloning time at node 192.168.1.10 is 3453 milli seconds.

Total turn around time = ( Total travelling time + Total processing time + Failure recovery time ) * ( delay in travel + delay in processing + response time [if node fails])

Total turn around time of agent in multi – region.

| | | |
|---|---|---|
| TAT | = | ( TAT (R) * Rn )+ ( DST * DSn ) |
| TAT | = | Total turn around time |
| TAT (R) | = | Turn around time of single region |
| Rn | = | No. of regions<br>(not included if agents run in parallel) |
| DST | = | Time taken for data sharing |
| DSn | = | No. of times data being shared. |

## 6 Conclusion and Future Work

In this paper a recovery model for continuous and free roaming mobile agent after the node failure is presented. Applying information sharing technique in K-response model, it is guaranteed that at least one of the hosts in each region will have the data collected by the agent of that region. In addition, the client can surely get the required information even in the case of node failure but may be with time delay (because of resending the agent in the middle after failure). This model will promote the mobile agent to incorporate in more number of the e-service applications. As a future work, a comparative analysis to be made for (i) state of agent recovery in various number of failures (ii) difference in transmission time with and without failure against different number of hosts in each region (iii) delay in communication with and without lookup server against number of regions.

## References


1. Karnik, N.M., Tripathi, A.R.: Design issues in mobile agent programming systems. IEEE Concurrency 6(3), 52–61 (1998)
2. Spalink, T., Hartman, J.H., Gibson, G.A.: A mobile agent's effects on file service. IEEE Concurrency 8(2), 62–69 (2000)
3. Lange, D.B., Oshima, M.: Seven good reasons for mobile agents. Communications of the ACM 42(3), 88–89 (1999)
4. Pleish, S., Schiper, A.: Modeling fault-tolerant mobile agent execution as a sequence of agreement problems. In: Proceedings of the 19th IEEE Symposium on Reliable Distributed Systems, pp. 11–20. IEEE Press, Germany (2000)
5. Pleisch, S., Schiper, A.: FATOMAS – A fault-tolerant mobile agent system based on the agent-dependent approach. In: Proceedings of International Conference on Dependable System and Networks, Sweden, pp. 215–224 (2001)
6. Fischer, M.J., Lynch, N.A., Paterson, M.S.: Impossibility of distributed consensus with one faulty process. Journal of the ACM 32(2), 374–382 (1985)
7. Chen, X., Lyu, M.R.: Performance and effectiveness analysis of checkpointing in mobile environments. In: Proceedings of 22nd International Symposium on Reliable Distributed System, Italy, pp. 131–140 (2003)
8. Park, T., Byum, I., Kim, H., Yeom, H.Y.: The performance of checkpointing and replication schemes for fault tolerant mobile agent systems. In: Proceedings of 21st IEEE Symposium Reliable Distributed System, Japan, pp. 256–261 (2002)
9. Pears, S., Xu, J., Boldyreff, C.: Mobile agent fault tolerance for information retrieval applications: an exception handling approach. In: Proceedings of 6th International Symposium on Autonomous Decentralized System, Italy, pp. 115–122 (2003)
10. Lyu, M.R., Chen, X.: Design and evaluation of a fault-tolerant mobile agent system. IEEE Intelligent System 19(5), 32–38 (2004)
11. Behesthi, S., Movaghar, A.: Fault tolerance in mobile agent system by cooperating the witness agents. In: Proceedings of 2nd International Conference Information and Communication Technologies: From Theory to Applications, vol. 2, pp. 3018–3021 (2006)
12. Summiya, K., Ijaz, U.M., Arshad Ail, S.: A fault tolerant infrastructure for mobile agents. In: Proceedings of International Conference on Computational Intelligence on Modeling Control and Automation and International Conference on IEEE Intelligent Agents, USA, p. 235 (2006)

13. Umar, M., Ijaz, K., Shamim, W.: Arshad Ali S.: Ensuring data security and consistency in FTIMA – a fault tolerant infrastructure for mobile agents. Proceedings of World Academy of Science, Engineering and Technology, (27), 343–348 (2007)
14. Baumann, J.: A comparison of mechanisms for locating mobile agents. IBM Research Report No.3333 (1999)
15. Deugol, D.: Mobile agent messaging models. In: Proceedings of the $5^{th}$ International Symposium on Autonomous Decentralized Systems, Texas, pp. 278–286 (2001)
16. Xu, W., Cao, J., Jin, B., Li, J., Zhang, L.: GCS-MA: A group communication system for mobile agents. Journal of Network and Computer Applications (30), 1153–1172 (2007)
17. Vigilson Prem, M., Swamynathan, S.: A reliable mobile agent based approach for information retrieval. In: Proceedings of International Conference on Information System and Software Engineering, Chennai, pp. 204–207 (2009)
18. Vigilson Prem, M., Swamynathan, S.: Group communication based approach for reliable mobile agent in information retrieval applications. In: Proceedings of International Conference on Advances in Computing, Control and Telecommunication, Thiruvananthapuram, pp. 464–467 (2009)
19. Choi, S., Baik, M., Kim, H., Byun, E., Choo, H.: A Reliable Communication Protocol for Multi-Region Mobile Agent Environment. IEEE Transactions on Parallel and Distributed Systems 21(1), 72–85 (2010)
20. Venkatesan, S., Chellappan, C.: Free-Roaming Mobile Agent (FRoMA) protection against multiple attacks. International Journal of Communication Networks and Distributed Systems 3(4), 362–383 (2009)
21. Venkatesan, S., Chellappan, C.: Recovery model for Free Roaming Mobile Agent against Multiple Attacks. In: ACM Proceedings of International Conference on Mobile Multimedia Communication, Greece, pp. 275–279 (2007)

# Composition of Composite Semantic Web Services Using Abductive Event Calculus

D. Paulraj[1] and S. Swamynathan[2]

[1] Department of Information Technology,
R.M.D. Engineering College,
Chennai, India
kingrajpaul@gmail.com
[2] Department of Information Science and Technology,
Anna University,
Chennai, India
swamyns@annauniv.edu

**Abstract.** Web Service composition is necessary when a single Web Service cannot satisfy the complex functional requirements. One of the key challenges in Composite Semantic Web Services is the composition of its atomic processes. In this work a novel architecture is proposed for the composition of composite semantic web services. An algorithm is used for the service discovery, which performs a fine grained match at the level of atomic process, rather than at the level of entire service. The architecture takes the advantage of abductive event calculus that uses abductive theorem prover to generate a plan for the composition of the atomic services.

**Keywords:** Semantic Web Service, Composition, Event Calculus.

## 1 Introduction

Web services are platform-neutral, language neutral, heterogeneous software entities that can be described, published and discovered in the internet. Web services are designed to provide interoperability between diverse applications. The main problem with the web services is the gap between the concepts people use and the information computers interpret. Semantic Web technologies are introduced to overcome this barrier. In this direction, OWL-S coalition has promoted Ontology Web Languages for Services (OWL-S)[5], which enrich WSDL and BPEL4WS with rich semantic annotations. Using OWL-S, web services are extended with an unambiguous description by relating the input and output parameters of the service to common concepts defined in *Web Ontology* [18], which serve as the key mechanism to globally define and reference concepts. In Semantic Web Services (SWS), service discovery is generally achieved by using service profile of OWL-S, and indeed the clients rely on the details provided by the service providers of the services. Profile of a service is a derived, concise description and not a functional part of the service. The information contained in the service profile is sufficient for atomic service discovery, but it is not sufficient for the discovery of composite services. Also there is no schema present in the

A. Mantri et al. (Eds.): HPAGC 2011, CCIS 169, pp. 201–213, 2011.

service profile to describe the input, output, precondition and effect (IOPE), and are not always annotated with ontology concepts. In order to overcome these limitations, a complementary composition approach is proposed in this paper that uses process model (the subclass of service model) of OWL-S instead of service profile.

The proposed service composition approach uses an algorithm which performs a fine grained match at the level of atomic process rather than at the level of the entire service in a composite semantic web service. The rest of the paper is organized as follows. Event Calculus is briefed in section 2, which is followed by the related work in section 3. The system architecture is explained with a motivating example in section 4. Section 5 explains the abductive event calculus based service composition and execution and section 6 highlights the findings in the conclusion.

## 2 Event Calculus – An Introduction

The event calculus (*EC*) was introduced by Kowalski and Sergot [10] as a logic programming formalism for representing events and their effects. In *EC* it is possible to infer what is true when, given a set of events at certain time points and their effects. The ontology of the *EC* comprises of *fluents*(F), *events(E)* (or actions), and *time-points(T)*. Fluents are properties that may have different values at different time points. Events manipulate fluents. Fluents are initiated and terminated by events, and that a fluent was true at the beginning of time. Given an event narrative (a set of events), the *EC* theorem and domain-specific axioms together define which fluents hold at each time. The event calculus used in this paper is a subset of Shanahan's circumscriptive event calculus [15]. It is based on many-sorted first-order predicate calculus with predicates to reason about events. The predicates are:

*Initiates(E,F,T)* expresses that fluent *F* holds after timepoint *T* if event *E* happens at T.
*Terminates(E,F,T)* expresses that fluent *F* does not hold after timepoint *T* if event *E* happens at T.
*Releases(E,F,T)* expresses that fluent F is not subject to the common sense law of inertia after event E at time T.
*InitiallyTrue(F)* define if *F* holds at timepoint 0.
*InitiallyFalse(F)* define if *F* not holds at timepoint 0.
*Happens(E,T)* is true iff event *E* happens at *T*.
*HoldsAt(F,T)* is true iff fluent F holds at T.
*Clipped*($T_1$,*F*,$T_2$) expresses if fluent *F* was terminated during time interval [$T_1$,$T_2$].
*Declipped*($T_1$,*F*,$T_2$) expresses if fluent *F* was initiated during time interval [$T_1$,$T_2$].

*EC* is a logical mechanism that infers "what's true when", given "what happens when" and "what actions do". The "*what happens when*" part is a *narrative* of events, and the "*what actions do*" part describes the *effects* of actions. Literature survey reveals that the abductive event calculus (a subset of EC) is the best suitable formalism for the web service composition.

In an abductive task, "*what actions do*" and "*what's true when*" are supplied, and "*what happens when*" is expected [19]. Traditionally, abductive means inference to a best explanation and is a pattern of reasoning. It proceeds from an outcome to a hypothesis that explains for the outcome. For instance, O ← E concludes E as an explanation for the outcome O. Abduction is handled by a second order abductive theorem prover. Shanahan [21] explained that, when *EC* formulae are submitted to a suitably tailored resolution based abductive theorem prover, the result is purely logical planning system whose computations mirror closely those of a hand-coded planning algorithm.

## 3 Related Work

There is a considerable amount of research done in the field of web service composition. Some of the most significant contributions related to this work are cited in this section.

An OWL-S service profile ontology based framework is used, for the retrieval of web services based on subsumption relation and structural case-based reasoning which performs domain-dependant discovery [11]. Brogi, Corfini and Popescu [1] have presented an algorithm called Service Aggregation Matchmaking (SAM), for composition-oriented discovery of web services. A context-based approach to the problem of matching and ranking semantic web services for composition has been proposed by Segev and Toch [13].

Automatically composing composite semantic web services is a challenging and critical task. One of the techniques that has been proposed for this task is the AI planning. In AI planning method, generally the users specify the necessary inputs and required outputs. A plan is generated automatically by an AI planner [6,7,16,20]. In [17], the fluent calculus is used for the AI planning for the composition of web services. This approach is a domain specific scenario that uses the domain ontologies and fluent calculus formalisms together for the planning and composition of web services. Like AI planning, the EC is also a very suitable formalism for the problem of web service composition. The use of EC has been proposed by Aydin et al., [4], as a solution for the web service composition problem. It is shown that when a goal situation is given, the EC can generate several plans as web service composition by using an abductive planning technique. The execution engine selects the best plan by compiling it in to a graph. A monolithic approach for web service composition and execution is proposed by Okutan and Cicekli [12]. In this work several plans are generated by the planner from which the best solution is selected according to some ranking. Ozorhan, Kuban and Cicekli [9], have used EC for the solution of automated web service composition problem using the interleaved and the template-based approaches.

In order to ensure the correct and reliable composite service execution, an event driven approach has been proposed [14]. This approach analyzes and checks the web service's transactional behavior consistency at design time and report recovery mechanisms deviations after runtime. Chifu et al. [3] have described how the planning capabilities of the fluent calculus can be used to automatically generate the web service composition. BPEL based semantics for a new specification language based on the Π-calculus is used for interacting web services [2]. Some of the limitations found in the AI based planning approaches are its lack of scalability, extendibility, concurrency and its tendency to generate more plans. Almost all of the existing works in the area of composition of the semantic web services have concentrated only on the composition of atomic services. But the problem of composing atomic processes from a composite semantic web services have received a lot of attention.

# 4 System Architecture

The proposed architecture shown in Fig. 1. is having two distinct phases. The first phase focuses on the discovery of atomic processes and the second phase mainly focuses on the composition of the discovered atomic processes and its execution. The output of the first phase of the proposed architecture is a list of atomic processes that are required for the composition.

## 4.1 A Motivating Example

This work has taken up two composite semantic web services namely, *Book_store* service and an *Online_banking* service, having several atomic processes as example. According to David et al.[5], any composite semantic web service can be considered as a tree whose non-terminal nodes labeled with control constructs. The leaves of the tree are atomic processes which are directly invoked by a client or an agent. Based on this hypothesis, the composite semantic web services *Book_store* and *Online_banking* services are represented as a tree and are shown in Fig. 2. and Fig. 3. respectively. The *Book_Store* service sells all type of books like, engineering, drawing and novels etc.

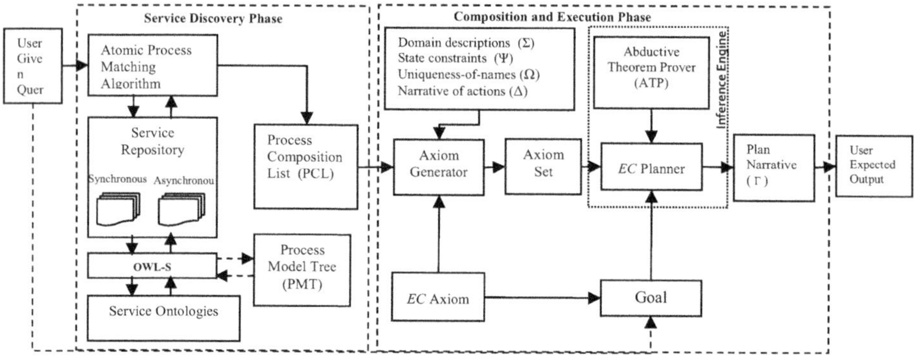


**Fig. 1.** Architecture for the discovery, composition and execution with abductive reasoner

*Online_Banking* service is able to create virtual credit cards. A client can obtain a credit card number, credit card type and the PIN number through a bank transfer. Assume that, a client is travelling and wants to purchase an engineering book and a novel. The client specifies the necessary input and the requested output from the service as a query. In this example the input and required output are specified as follows.

> ***Inputs:*** *username, password, isbn, author,publication, acc_no, amount*
> ***Outputs:*** *engg_book_purchase_receipt, novel_purchase_receipt*

According to this query, neither of these two services alone can fulfill the request by itself. For example *cr_card_type*, *cr_card_num* and *pin_num* are the three inputs required by both *engg_book_payment* and *novel_payment* atomic processes of *Book_store* service. But these three inputs are not provided in the given query. So, the *Book_store* service requires another service to generate these three parameters. However, these inputs can be obtained by executing *Online_banking* service, as the output of its atomic process *Virtual_Credit_Card* generates *cr_card_num, cr_card_type* and *pin_num* as the output.

**Listing 1.** Pseudocode for the atomic process discovery.

```
Input
      Q_input - User given input.
Output
      CL  - Composition List consists of a set of
relevant atomic processes.
Local Resources
      SR - Service Registry
      R_Ai - Root node of atomic node A^i
      A_i-input - Input of Atomic Process A_i
      A_i-output - Output of Atomic Process A_i
CreateCL(Service Registry SR, Query Q_input, Composition
List CL)
repeat
   forall Services S_i in {S_1,S_2,…,S_n} ∈ SR do
      forall P_i in {P_1,P_2,…,P_n} do
         forall A_i in {A_1,A_2,…,A_n} do
           if ( (∃A_i-input ∈ Q_input ) V (∃A_i-output ∈ Q_input ))
then
             add A_i into CL
          next  A_i
      next  P_i
   next  S_i
until no more processes  node is added into CL
```

The pseudo code shown in Listing 1 first invokes the root of the tree, and descends towards the atomic node and matches its inputs and outputs with the given query Q. If a match is found, then that atomic service is added into the Composition List (CL). CL is a temporary list used to store the discovered atomic processes in a specific order according to the control constructs present in the root node of the corresponding atomic node. The final PCL list produced by the architecture is shown in Fig. 4.

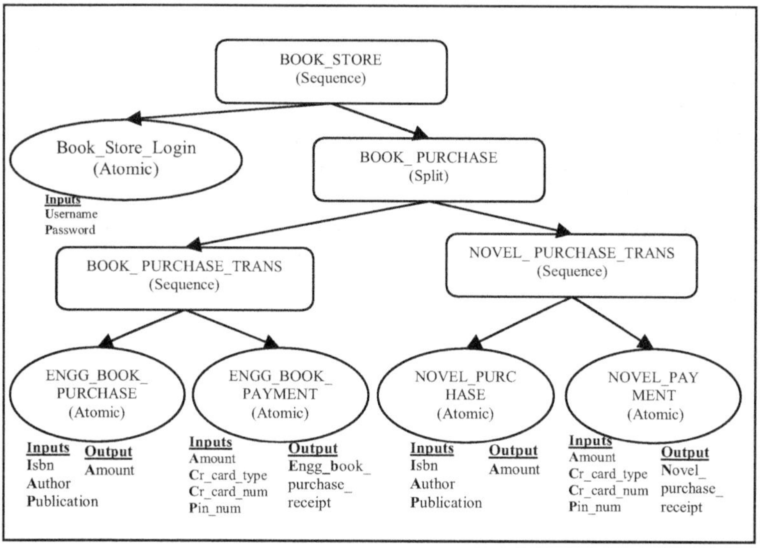


**Fig. 2.** Tree Representation of Book Store Service

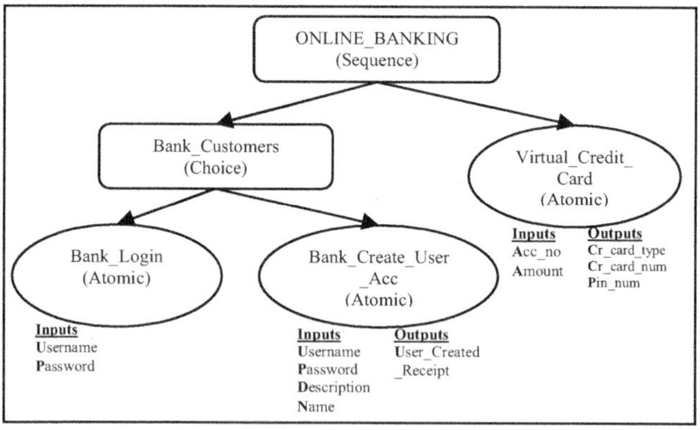


**Fig. 3.** Tree Representation of Online Banking Service

```
Book_Store_Login(Username, Password),
Engg_Book_Purchase(Isbn,Author,Publication,Amount)
Novel_Purchase(Isbn,Author,Publication,Amount)
Bank_Login(Username, Password),
Virtual_Credit_Card(Acc_num, amount, cr_card_num,
cr_card_type,pin_num),Engg_Book_Payment(Amount,
Cr_Card_Type,Cr_Card_num,Pin_num,Engg_Book_Purchase_R
eceipt)
Novel_Payment(Amount,Cr_Card_Type,Cr_Card_num,
```

**Fig. 4.** Composition List (CL)

## 5 Services Composition Using Abductive Event Calculus

The second phase of the proposed architecture is mainly focused on the composition and execution of the discovered services in a specific order. Hence a plan has to be generated to compose and execute the candidate services present in the CL. In order to generate the plan, the discovered atomic services must be translated into event descriptions [8] and the user required output must be translated into a goal state. The inference engine in the architecture shown in Fig.1. generates always a unique plan which is domain independent,and extendible to meet the desired goal set by the user.A set of EC axioms comprising of *domain dependent descriptions, state constraints* and *uniqueness-of-names axioms* are used for the planning. The axioms are proved by the abductive theorem prover coded as a Prolog meta-interpreter. Abductive theorem prover is tailored for the event calculus by compiling the EC axioms into the meta-level, resulting in an efficient implementation. The job of an abductive meta-interpreter is to construct a *residue* of *abducible* literals that cannot be proved from the object-level program. To construct a plan for the given goal, the effects of events are to be described using EC *initiates,* and *terminates* clause. A list of *holds_at* predicates that represents the goal is presented to the abductive theorem prover. It generates the *residue* (plan), comprising of *happens* and *before* literals. The EC axiom set generated by the axiom generator for the candidate services present in the CL is shown in Listing 2.This axiom set along with a list of *holds_at* predicates that represents the goal (user required output) is presented to the EC planner as input and the plan is generated. The user goal of the given example is fed into the inference engine as follows:

`abdemo([holds_at(receipt(engineering_book_purchase_receipt,novel_purchase_receipt),t)],R).` In this work, Shanahan's(2000) abductive event calculus planner is used for the inference engine.

**Listing 2.** Axiom set generated by the axiom generator.

```
axiom(initiates(bookstore_login(X,Y),f_logged_in,T),[bookstoreLogin(X,Y)])
.
axiom(initiates(engineering_book_purchase(A,B,C,D),f_engg_book_purchase,T)
,[pur(A,B,C,D),holds_at(f_logged_in,T)]).
axiom(initiates(novel_purchase(A,B,C,D),f_novel_purchase,T),[pur(A,B,C,D)]
).
axiom(terminates(engineering_book_purchase(A,B,C,D),f_engg_book_purchase,T
),[holds_at(f_engg_book_purchase,T)]).
axiom(terminates(novel_purchase(A,B,C,D),f_novel_purchase,T),[holds_at(f_n
ovel_purchase,T)]).
axiom(initiates(bank_login(A,B),f_bankLogin,T),[banklogin(A,B),holds_at(f_
purchaseboth,T)]).
axiom(initiates(virtual_credit_card(A,B,C,D,E),f_vcard,T),[creditcard(A,B,
C,D,E),holds_at(f_bankLogin,T)]).
axiom(initiates(engineering_book_payment(A,B,C,D,E),f_bookreceipt,T),[paye
ngg(A,B,C,D,E),holds_at(f_vcard,T)]).
axiom(initiates(novel_payment(A1,B1,C1,D1,E1),f_novelreceipt,T),[paynovel(
A1,B1,C1,D1,E1)]).
axiom(terminates(engineering_book_payment(A,B,C,D,E),f_bookreceipt,T),[hol
ds_at(f_bookreceipt,T)]).
axiom(terminates(novel_payment(A,B,C,D,E),f_novelreceipt,T),[holds_at(f_no
velreceipt,T)]).

/* state constraints */
axiom(holds_at(f_purchaseboth,T),[holds_at(f_engg_book_purchase,T),holds_a
t(f_novel_purchase,T)]).
axiom(holds_at(receipt(engineering_book_purchase_receipt,novel_purchase_re
ceipt),T),[holds_at(f_bookreceipt,T),holds_at(f_novelreceipt,T)]).

/* contextual conditions */
axiom(bookstoreLogin(username, password),[]).
axiom(pur(isbn,author,publication,amount),[]).
axiom(banklogin(username, password),[]).
axiom(creditcard(acc_no,amount,cr_card_num,cr_card_type,pin_num),[]).
axiom(payengg(cr_card_num,cr_card_type,pin_num,amount,engineering_book_pur
chase_receipt),[]).
axiom(paynovel(cr_card_num,cr_card_type,pin_num,amount,novel_purchase_rece
ipt),[]).
```

The abductive theorem prover present in the inference engine will populate the plan narrative into the residue R given in the goal. The residue R populated by the inference engine consists of a set of *happens* and *before* literals as shown in Fig. 5. Services are the events in the domain and they are to be defined in the axiom definition through the *executable* clause. The **`executable`**`(bookstore_login (X,Y)` clause in Listing 2 means that the `bookstore_login` event is used in constructing a plan. The EC predicates such as *initiates* and *terminates* are used to specify how the execution can evolve. In addition to defining the fluent they initiate or terminate, the required preconditions for activating these predicates must be specified. In the initiates axiom, the first argument is the *initiates* clause, and the second argument is a set of preconditions necessary for the *initiates* clause to be applicable. This fact is illustrated in the following *initiates* axiom:

```
axiom(initiates(engineering_book_purchase(A,B,C,D),f_engg_
book_purchase,T),[pur(A,B,C,D),holds_at(f_logged_in,T)]).
```

Evidently, the event engineering_book_purchase initiates the fluent f_engg_book_purchase with the precondition holds_at(f_logged_in,T). That is, the requirement to execute the event `engineering_book_purchase` is that, the fluent `f_logged_in` of `bookstore_login` event must hold. The argument `engineering_book_purchase(A,B,C,D)` is known to be the *contextual condition*. Contextual condition can also be represented by using the *holds_at* predicates. But then there is no distinction between the precondition and the contextual condition, in this formalism. So the *domain constraints* or the *state constraints* (Ψ) are used only to represent the preconditions, which give rise to actions with indirect effects. This expresses a logical relationship held between fluents at all times. In the event calculus, state constraints are *holds_at* formulae with a universally quantified time argument as the one given below:

```
axiom(holds_at(receipt(engineering_book_purchase_receipt,
novel_purchase_receipt),T),[holds_at(f_bookreceipt,T),hol
ds_at(f_novelreceipt,T)]).
```

R = [[happens(novel_payment(cr_card_num, cr_card_type, pin_num, amount, novel_purchase_receipt), t7, t7),
happens(novel_purchase(isbn, author, publication, amount), t6, t6),
happens(bookstore_login(username, password), t5, t5),
happens(engineering_book_purchase(isbn, author, publication, amount), t4, t4),
happens(bank_login(username, password), t3, t3),
happens(virtual_credit_card(acc_no, amount, cr_card_num, cr_card_type, pin_num), t2, t2), happens(engineering_book_payment(cr_card_num, cr_card_type, pin_num, amount, engineering_book_purchase_receipt), t1, t1)],
[before(t7, t),
before(t6, t3),
before(t5, t4),
before(t4, t3),
before(t3, t2),
before(t2, t1),

**Fig. 5.** The plan narrative generated by Prolog inference engine

### 5.1 Concurrent Execution of Events in the Plan

The most significant advantage of the event calculus is the inherent support for concurrency. The events h*appens($e_1$,$t_1$), happens($e_2$,$t_2$), happens($e_3$,$t_3$), happens($e_4$,$t_4$)*, $t_1 < t_2 < t_4$, $t_1 < t_3 < t_4$ are examined. Since there is no relative ordering between $e_2$ and $e_3$ they are assumed to be concurrent as shown in Fig. 6. It is to be observed that the process model tree of the *Book_store* service having two atomic processes namely, *Engg_Book_Payment* and *Novel_Payment* that are connected under a *split* control construct. According to the OWL-S formalism, these two atomic processes are to be executed concurrently.

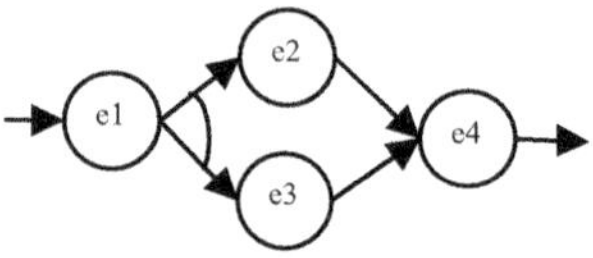


**Fig. 6.** Concurrency of events

The axiom generator in the proposed architecture is designed in such a way that, it generates the axiom sets that are proved by the abductive theorem prover and the inference engine generates the plan with simultaneous occurrence of two events. Following is the events generated by the inference engine for the two atomic processes:

```
happens(novel_payment(cr_card_num, cr_card_type, pin_num,
amount, novel_purchase_receipt), t7, t7),
happens(engineering_book_payment(cr_card_num,
cr_card_type, pin_num, amount,
engineering_book_purchase_receipt), t1, t1)],

. . . .
before(t7, t),
before(t1,t)
```

The literal `before(t7,t)` means that, t7 < t. Here, the events *novel_payment* and *engineering_book_payment* are to be executed at time t7 and t1 respectively. But both should be executed just before t and since there is no relative ordering between t7 and t1, it is assumed that these two events are to be executed concurrently, as shown in the execution order of the plan in Fig. 7.

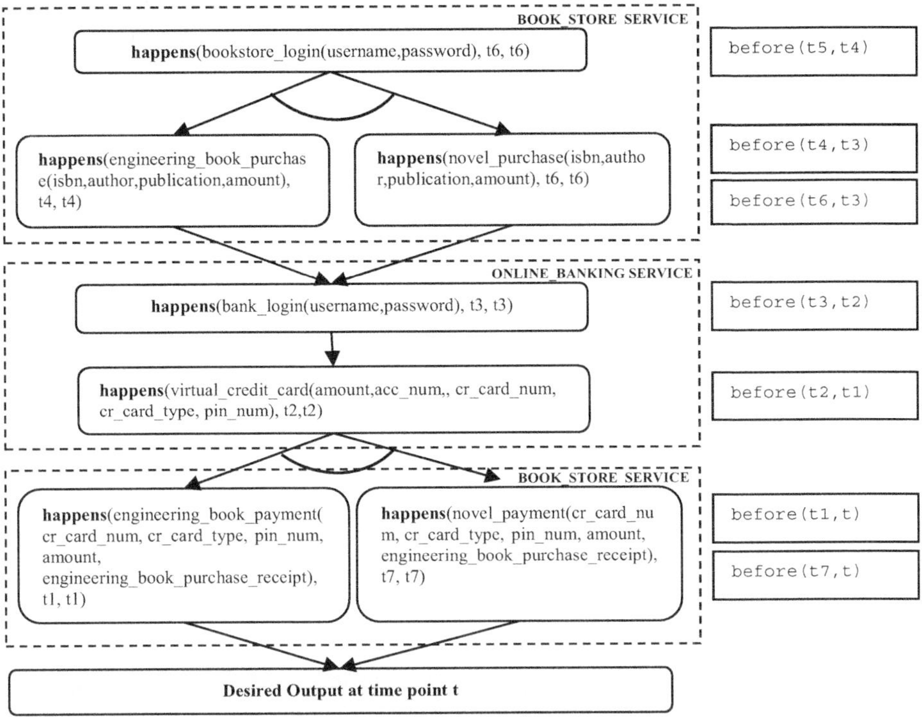


**Fig. 7.** Execution path generated by the EC planner

## 6 Conclusion and Future Work

A new novel architecture is proposed in this work for atomic service discovery and composition of composite semantic web services. It is proved that the process model ontology is effectively used to discover the atomic services. In the architecture, the second phase takes the advantage of the abductive event calculus. The inference engine in the planner uses the second order abductive theorem prover as its main inference method. All the time the planner generates a unique and correct plan, which is domain independent, and extendible to meet the desired goal set by the user. The plan is scalable, sound and complete. Since the planner generates a unique plan, the proposed work totally avoids any manual intervention to select a best plan. Other works in this area have proposed solutions for the composition of *atomic services* only, but this work has proposed a solution for the composition of *composite semantic web services*.

## References


1. Brogi, A., Corfini, S., Popescu, R.: Semantic-Based Composition-Oriented Discovery of Web Services. ACM Transactions on Internet Technology 8(4), 19:1-19:33 (2008)
2. Abouzaid, F., Mullins, J.: A Calculus for Generation, Verification and Refinement of BPEL Specifications. Electronic Notes in Theoretical Computer Science 200(3), 43–65 (2008)
3. Chifu, V., Salomie, I., Chifu, E.: Fluent Calculus-Based Web Service Composition - From OWL-S to Fluent Calculus. In: Intelligent Computer Communication and Processing, Cluj-Napoca, Romania, August 28-30, pp. 161–168. IEEE Explore Digital Library, Washington, USA (2008)
4. Aydin, O., Kesim Cicekli, N., Cicekli, I.: Automated Web Service Composition with the event calculus. In: 8th International Workshop on Engineering Societies in the Agents World, pp. 142–157. Springer, Heidelberg (2007)
5. David, M., Mark, B., Jerry, H., Ora, L., Drew, M., Sheila, M., Srini, N., Massimo, P., Bijan, P.,Terry, P., Evren, S., Naveen, S., Katia, S.: OWL-S Semantic Markup for Web Services, `http://www.w3.org/submission/owl-s`
6. Hoffmann, J., Bertoli, P., Helmert, M., Pistore, M.: Message-Based Web Service Composition, Integrity Constraints, and Planning under Uncertainty: A New Connection. Journal of Artificial Intelligence Research 35(1), 49–117 (2009)
7. Seog-Chan, O., Lee, D., Kumara, S.: A Comparative Illustration of AI Planning-based Web Services Composition. ACM SIGecom Exchanges 5(5), 1–10 (2005)
8. Paulraj, D., Swamynathan, S.: Dynamic Discovery and Composition of Semantic Web Services Using Abductive Event Calculus. In: Proceedings of ITC 2010- International Conference on Recent Trends in Information, Telecommunication, and Computing, Kochi, India, March 12-13, pp. 70–74. IEEE Computer Society, Washington, USA (2010)
9. Ozorhan, E., Kuban, E., Cicekli, N.: Automated composition of web services with the abductive event calculus. Elsevier Information Sciences 180(19), 3589–3613 (2010)
10. Kowalski, R.A., Sergot, M.J.: A Logic–based calculus of events. New Generation Computing 4(1), 67–95 (1986)
11. Meditskos, G., Bassiliades, N.: Structural and Role-Oriented Web Service Discovery with Taxonomies in OWL-S. IEEE Transactions On Knowledge And Data Engineering 22(2), 278–290 (2010)
12. Okutan, C., Cicekli, N.: A monolithic approach to automated composition of semantic web services with event calculus. Elsevier Knowledge-Based Systems 23(5), 440–454 (2010)
13. Segev, A., Toch, E.: Context-Based Matching and Ranking of Web Services for Composition. IEEE Transactions On Services Computing 2(3), 210–222 (2009)
14. Gaaloul, W., Rouached, M., Godart, C., Hauswirth, M.: Verifying composite service transactional behavior using event calculus. In: Chung, S. (ed.) OTM 2007, Part I. LNCS, vol. 4803, pp. 353–370. Springer, Heidelberg (2007)
15. Shanahan, M.: A Circumscriptive Calculus of Events. Elsevier Artificial Intelligence 77(2), 249–284 (1995)
16. Petrie, C.: Planning Process Instances with Web Services. In: Proceedings of the International Conference on Enterprise Information Systems AT4WS 2009, Milan, Italy (pp, May 6-7, pp. 31–35. IGI Publishing (2009)
17. Salomie, I., Viotica, R., Harsa, I., Gherga, M.: Towards Automated Web Service Composition with Fluent Calculus and Domain Ontologies. In: Proceedings of the iiWAS2008, Linz, Austria, November 24-26, pp. 201–207. ACM, New York (2008)

18. Sean, B., Frank, H., Jim, H., Ian, H., Deborah, L., Peter, F. and Lynn, A.: OWL Web Ontology Language (2004), `http://www.w3.org/TR/owl-ref/`
19. Shanahan, M.: The Event Calculus Explained. In: Veloso, M.M., Wooldridge, M.J. (eds.) Artificial Intelligence Today. LNCS (LNAI), vol. 1600, pp. 409–430. Springer, Heidelberg (1999)
20. Seog-Chan, O., Lee, D., Kumara, S.: Web Service Planner (WSPR): An Effective and Scalable Web Service Composition Algorithm. International Journal of Web Services Research 4(1), 1–22 (2007)
21. Shanahan, M.: An Abductive Event Calculus Planner. Journal of Logic Programming 44(1-3), 207–240 (2000)

# Ant Colony Optimization Based Congestion Control Algorithm for MPLS Network

S. Rajagopalan[1], E.R. Naganathan[2], and P. Herbert Raj[3]

[1] Dept of CSE, Alagappa University,
Karaikudi, India
Tel.: 9443978509
raja_04@hotmail.com
2 Professor & Head, Dept. of Computer Applications,
Velammal Engineering College,
Chennai, India
naganathaner@hotmail.com
[3] Department of Technical Education, Brunei
herbi_raj@hotmail.com

**Abstract.** Multi-Protocol Label Switching (MPLS) is a mechanism in high-performance telecommunications networks which directs and carries data from one network node to the next with the help of labels. MPLS makes it easy to create "virtual links" between distant nodes. It can encapsulate packets of various network protocols. MPLS is a highly scalable, protocol agnostic, data-carrying mechanism. Packet-forwarding decisions are made solely on the contents of this label, without the need to examine the packet itself. This allows one to create end-to-end circuits across any type of transport medium, using any protocol. There are high traffics when transmitting data in the MPLS Network due to emerging requirements of MPLS and associated internet usage. This paper proposes an Ant Colony Optimization (ACO) technique for traffic management in MPLS Network. ACO is a swarm intelligence methodology which offers highly optimized technique for dozen of engineering problems. In our proposed work, the ACO provides optimal value than existing algorithms.

**Keywords:** Ant Colony Optimization, MPLS Network, Traffic Management.

## 1 Introduction

MPLS operates at an OSI Model layer that is generally considered to lie between traditional definitions of Layer 2 and Layer 3, and thus is often referred to as a "Layer 2.5" protocol. It was designed to provide a unified data-carrying service for both circuit-based clients and packet-switching clients which provide a datagram service model. It can be used to carry many different kinds of traffic, including IP packets, as well as native ATM, SONET, and Ethernet frames. MPLS is currently replacing some of these technologies in the marketplace. It is highly possible that MPLS will completely replace these technologies in the future, thus aligning these technologies with current and future technology needs.

A. Mantri et al. (Eds.): HPAGC 2011, CCIS 169, pp. 214–223, 2011.

In particular, MPLS dispenses with the cell-switching and signaling-protocol baggage of ATM. MPLS recognizes that small ATM cells are not needed in the core of modern networks, since modern optical networks (as of 2008) are so fast (at 40 Gbit/s and beyond) that even full-length 1500 byte packets do not incur significant real-time queuing delays (the need to reduce such delays — e.g., to support voice traffic was the motivation for the cell nature of ATM). MPLS was originally proposed by a group of engineers from IPSILON Networks, but their "IP Switching" technology, which was defined only to work over ATM, did not achieve market dominance. Cisco Systems, Inc., introduced a related proposal, not restricted to ATM transmission, called "Tag Switching". It was a Cisco proprietary proposal, and was renamed "Label Switching". It was handed over to the IETF for open standardization. The IETF work involved proposals from other vendors, and development of a consensus protocol that combined features from several vendors' work.

One original motivation was to allow the creation of simple high-speed switches, since for a significant length of time it was impossible to forward IP packets entirely in hardware. However, advances in VLSI have made such devices possible. Therefore the advantages of MPLS primarily revolve around the ability to support multiple service models and perform traffic management. MPLS also offers a robust recovery framework that goes beyond the simple protection rings of synchronous optical networking (SONET/SDH). In 2000, the first iteration of pure IP-MPLS was implemented by a project team led by Preston Poole of Schlumberger NIS. Through a series of research-joint ventures, this team successfully engineered, deployed, and commissioned the world's first commercial IP-MPLS network. Originally consisting of 35 Points of Presence (PoP) around the globe, this network was first purposed to serve the Oil and Gas community by delivering the DeXa suite of services. Later iterations of this commercial IP-MPLS network included VSAT Satellite access via strategic teleport connections, access to finance and banking applications, and Drilling Collaboration centres. Further developments in the IP-MPLS field deployed by Mr. Poole's team included mathematical conception and development of the most commonly used algorithms for what is known today as Bandwidth on Demand (BoD), Video on Demand (VoD), and Differentiated Services for IP MPLS.

## 2 MPLS Working Model

This ensures end-to-end circuits over ANY type of transport medium using ANY network layer protocol. In view of the fact that MPLS supports Internet Protocol revised versions (IPv4 and IPv6), IPX, AppleTalk at Layer3; Ethernet, Token Ring, Fiber Distributed Data Interface (FDDI), Asynchronous Transfer Mode (ATM), Frame Relay, and PPP (Point to Point Protocol) at Layer 2, it is referred as 'Layer 2.5 protocol'.

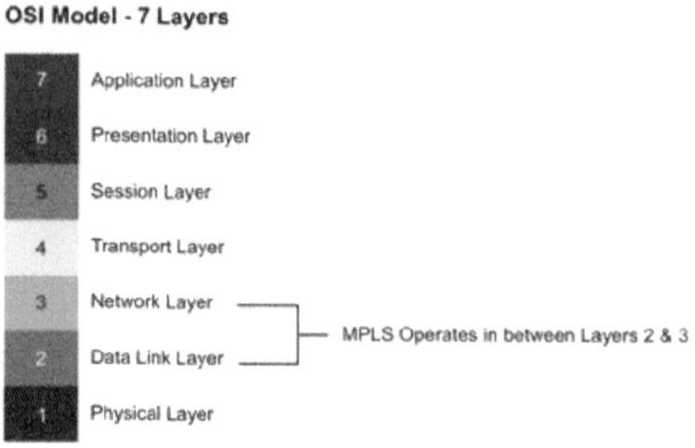


**Fig. 1.** Operation of MPLS in OSI Layer

MPLS works by prefixing packets with an MPLS header, containing one or more "labels". This is called a label stack. Each label stack entry contains four fields:

- A 20-bit label value.
- a 3-bit Traffic Class field for QoS (quality of service) priority (experimental) and ECN (Explicit Congestion Notification).
- a 1-bit bottom of stack flag. If this is set, it signifies that the current label is the last in the stack.
- an 8-bit TTL (time to live) field.

These MPLS-labeled packets are switched after a label lookup/switch instead of a lookup into the IP table. As mentioned above, when MPLS was conceived, label lookup and label switching were faster than a routing table or RIB (Routing Information Base) lookup because they could take place directly within the switched fabric and not the CPU. The entry and exit points of an MPLS network are called label edge routers (LER), which, respectively, push an MPLS label onto an incoming packet and pop it off the outgoing packet. Routers that perform routing based only on the label are called label switch routers (LSR). In some applications, the packet presented to the LER already may have a label, so that the new LER pushes a second label onto the packet. For more information see penultimate hop popping.

Labels are distributed between LERs and LSRs using the "Label Distribution Protocol" (LDP). Label Switch Routers in an MPLS network regularly exchange label and reachability information with each other using standardized procedures in order to build a complete picture of the network they can then use to forward packets. Label Switch Paths (LSPs) are established by the network operator for a variety of purposes, such as to create network-based IP virtual private networks or to route traffic along specified paths through the network. In many respects, LSPs are not different from PVCs in ATM or Frame Relay networks, except that they are not dependent on a particular Layer 2 technology. When an unlabeled packet enters the ingress router and needs to be passed on to an MPLS tunnel, the router first determines the forwarding equivalence class (FEC) the packet should be in, and then inserts one or more labels in the packet's newly-created MPLS header. The packet is then passed on to the next hop router for this tunnel. When a labeled packet is received by an MPLS router, the topmost label is examined. Based on the contents of the label a swap, push (impose) or pop (dispose) operation can be performed on the packet's label stack. Routers can have prebuilt lookup tables that tell them which kind of operation to do based on the topmost label of the incoming packet so they can process the packet very quickly.

## 3 Traffic Engineering and Congestion Control in MPLS

Traffic engineering is a method of optimizing the performance of a telecommunications network by dynamically analyzing, predicting and regulating the behavior of data transmitted over that network. Traffic engineering is also known as tele-traffic engineering and traffic management. The techniques of traffic engineering can be applied to networks of all kinds, including the PSTN (public switched telephone network), LANs (local area networks), WANs (wide area networks), cellular telephone networks, proprietary business and the Internet.

The theory of traffic engineering was originally conceived by A.K. Erlang, a Danish mathematician who developed methods of signal traffic measurement in the early 1900s. Traffic engineering makes use of a statistical concept known as the law of large numbers (LLN), which states that as an experiment is repeated, the observed frequency of a specific outcome approaches the theoretical frequency of that outcome over an entire population. In telecommunications terms, the LLN says that the overall behavior of a large network can be predicted with reasonable certainty even if the behavior of any single packet cannot be predicted.

When the level of network traffic nears, reaches or exceeds the design maximum, the network is said to be congested. In a telephone network, traffic is measured in call seconds (CCS) or erlangs. One CCS is equal to 100 seconds of telephone time. One erlang is equal to one hour or 36 CCS of telephone time. In a congested network, one of three things can happen when a subscriber attempts to send a message or place a call:

- The user receives a busy signal or other indication that the network cannot carry out a call at that time.
- A message is placed in a queue and is eventually delivered according to specified parameters.
- A message is rejected, returned or lost.

When message queues become unacceptably long or the frequency of busy signals becomes unacceptably high, the network is said to be in a high-loss condition. A major objective of traffic engineering is to minimize or eliminate high-loss situations. In particular, the number of rejected messages or failed call attempts should be as close to zero as possible. Another goal of traffic engineering is to balance the QoS (Quality of Service) against the cost of operating and maintaining the network.

Most of the previous works on two-layer models focus on the optimization of flow aggregation and routing and in particular on Wavelength Division Multiplexing (WDM) networks. In this context, the problem considered is usually referred to as grooming problem since the main goal is to aggregate flows in order to better exploit the large capacity available on each wavelength. In order to define the logical topology of the WDM network, also wavelengths must be assigned to light paths and converters located where needed. Different variants of the problem can be considered including hierarchical flow aggregation, but it has been shown that even for simple network topologies where the routing is trivial, the grooming problem is inherently difficult.

In WDM networks, resilience to failures is also an important feature and protection and restoration techniques at different layers can be jointly applied and optimized. From the network technology perspective, the integration of the optical layer with electronic layers within a converged data-optical infrastructure, based on classical IP or modern GMPLS (Generalized MPLS) architectures, is a key element in the current trend in broadband network evolution. Two-layer network design problems, where also link and node dimensioning is included in the model, have been considered only quite recently. Some works specifically consider MPLS technology and some of them address the problem of MPLS node location. Given the complexity of the optimization models, several authors rely on path formulations and column generation coupled with branch & bound, joint column and row generation methods, branch & cut with cut-set inequalities or LP-based decomposition approaches. For mid-to-large networks, the solution of choice remains heuristic algorithms, which provide a feasible solution in limited time. However, to the best of our knowledge, the effect of statistical multiplexing has not been previously considered in such network design and routing models.

Chun Tung Chou [1] proposed a virtual private network architecture using MPLS, which allows granularity and load balancing. This paper shows feasible result in the view of link utilization, but the multi objective function proposed by the author not optimal in other aspects like response time and packet loss. Shekhar et al [2] introduced a distortion factor for heterogeneous streams in traffic engineering of MPLS backbone networks in the presence of tunnelling and capacity constraints by formulating a distortion-aware non-linear discrete optimization problem. The author presented a two-phase heuristic approach to solve this formulation efficiently.

Francesco Palmieri [3] explains the MPLS hierarchical architecture for label-switched networks can be used to address all required functions of converged/unified networks, from initial IP level authentication and configuration, security, session control, resource reservation, admission control, to quality of service and policy management, enhanced only where necessary to address the idiosyncrasies of the mobile wireless environment. This architecture encompassing mostly IETF (Internet Engineering Task Force) standardized protocols, takes advantage of MPLS flexibility to address wireless-specific requirements such as micro mobility as well as non-wireless specific requirements, such as traffic engineering and quality of service and does not involve specific requirements in the mobile terminal for initiating label-switched paths over the wireless interface and allowing end to end interconnection to the backbone network.

Bosco et al [4] analysed the performance of a traffic engineering (TE) strategy for MPLS based network, described in [5] is carried out. Specifically the implementation based on a distributed control plane (Internet-like) has been investigated and realized by means of a test bed where real signalling protocol (RSVP-TE) and routing protocols (OSPF-TE) have been implemented. All these previous works are provided feasibility but optimal result based on the current requirement of the internet users such as lower response time and lesser packet loss.

## 4 Proposed Work

The proposed system involves swarm intelligence. The swarm intelligence – Ant colony is used for optimal congestion control. Ant colony algorithms [6], [7] have been inspired by the behavior of the real ant colony. The algorithm can find the optimum solution by generating artificial ants. As the real ants search their environment for food, the artificial ants search the solution space. The probabilistic movement of ants in the system allows the ants to explore new paths and to re-explore the old visited paths. The strength of the pheromone deposit directs the artificial ants toward the best paths and the pheromone evaporation allows the system to forget old information and avoid quick convergence to suboptimal solutions. The probabilistic selection of the paths allows searching large number of solutions. ACO has been applied successfully to discrete optimization problems such as the traveling salesman problem [8], routing [9], and load balancing [10]. A number of proofs for the convergence to the optimum path of the ACO can be found in [11] and [12]. The implementation of the proposed system [13] [14] in the wired environment which provides optimum result and suggested traffic free routing.

The ant agents move in the network randomly to scan large number of network nodes. While it is moving, it collects information about the network and delivers it to the network nodes. The algorithms of this category are not using the agents to optimize the paths as in S-ACO or S-ACO meta-heuristic [6]. It is just used to deliver more updated information about the network to the network nodes, which speeds up the optimization process. This category of ant like mobile agent algorithm is already successfully implemented in the GPS / ant-like routing algorithm for mobile ad hoc networks and ant-adhoc on-demand distance vector routing (AODV) hybrid routing protocol.

Every node in the network can function as a source node, destination node, and/or intermediate node. Every node has a pheromone table and a routing table. The routing table can be constructed based on the state transition rule and pheromone update policy. The following random proportional rule is applied as State transition rule: for destination D, at node i, the probability of selecting a neighbor j is

$$prob(D,i,j) = Fun(TD,i,j,\eta) ----if, j \in N \quad --- \tag{1}$$

Where TD is the pheromone value corresponding to neighbor j at node i and $0<TD<1$ is the local heuristic value of the link (i, j) / or node j. $0 < \eta < 1$. is the value can represent neighbor's information (i.e., neighbors queue delay, battery's remaining energy, processing power, link's signal-to-noise ratio, link's bandwidth, bit-error rate, etc.).

Fun (TD, i, j, η) is a function in TD and η (this function value is high when TD and η are high). N is the set of all feasible neighbor nodes defined by the ant's information and the routing constraints (i.e., the guarantee of loop free). Assuming that at a given moment in time m1 ants have used the first bridge and m2 the second one, the probability p1 for an ant to choose the first bridge is:

$$Fun(TD, r, s) = \begin{Bmatrix} \dfrac{T(r,s) \bullet [\eta(r,s)]^{\beta}}{\sum T(r,s) \bullet [\eta(r,s)]^{\beta}} \rightarrow if \ldots route \ldots found \\ 0 \rightarrow otherwise \end{Bmatrix} \text{---} \quad (2)$$

where T(r, s) is the pheromone deposited in the path between 'r' and 's', η(r, s) is the corresponding heuristic value which is the inverse of length of the particular path. β is a parameter which determines the relative importance of pheromone versus distance (β > 0). The pheromone update policy is as follows:

$$T(r,s) \leftarrow (1-\alpha) \bullet T(r,s) + \sum (1-\alpha) \bullet T(r,s) \text{---} \quad (3)$$

$$\Delta T_K(r,s) = \begin{Bmatrix} \dfrac{1}{L_K} \rightarrow if \ldots route \ldots found \\ 0 \rightarrow otherwise \end{Bmatrix} \text{---} \quad (4)$$

Where L is the length of tour, α is the pheromone decay parameter which is lies between '0 to 1'. The pheromone values of each entry in the table can be initialized to equal values, thus providing nonbiased search for the best path. If some information about the best path is available, the pheromone values of the entry can be set to closer values to the optimum, thus, speed up the algorithm.

***The Ant Colony Optimization Algorithm:***

1) //Initialization Phase

    For each pair *(r, s), the value of* $\tau(r, s) := \tau_0$ End-for

    For $k := 1$ to $m$ do

    Let $(r, k_1)$ be the starting city for an ant $k$

    $J_k(r_{k1}) := \{1, \ldots, n\} - r_{k1}$

    // $J_k(r_{k1})$ is the set of yet to be visited cities for //ant $k$ in city $r_{k1}$

    $r_k := r_{k1}$

    // $r_k$ is the city where ant $k$ is located

    End-for

2) //This is the phase in which ants build their tours. The tour of ant $k$ is stored in //$Tour_k$

    For $i := 1$ *to* $n$ do

    If $i < n$

```
            Then
              For k := 1 to m do
                      Choose the next city S_k
                  J_k(S_k) := J_k(r_k) - S_k
                  Tour_k(i) := (r_k, S_k)
              End-for
            Else
              For k := 1 to m do
                      //In this cycle all the ants go back to the initial city r_k1
                      S_k := r_k1
                      Tour_k (i) := (r_k , S_k )
              End-for
            End-if
        //In this phase local updating occurs and pheromone is updated
      For k := 1 to m do
        T(r,s) ← (1−α)•T(r,s)+∑(1−α)•T(r,s)
        r_k := s_k // New city for ant k
      End-for
      End-for
3)    //In this phase global updating occurs and pheromone is updated
      For k := 1 to m do
        Compute L_k // L_k is the length of the tour done by ant k
      End-for
      Compute L_best
      /*Update edges belonging to L_best
      For each edge (r, s)
      T(r,s) ← (1−α)•T(r,s)+∑(1−α)•T(r,s)
      End-for
4)    //In this phase assigns priority
        For k := 1 to m do
          Sort the routing table based on pheromone values
          {
                Assigns high priority to higher pheromone density path
                Choose the best path based on priority in the wired network and
             availability & priority in the case of wireless network
          }
        End-for
```

This proposed routing is simulated in NS2. The result and performance analysis is described in the coming section.

## 5 Result and Performance Analysis

**Table 1.** Packet Loss in Single Path Routing under normal load (number of packets)

| | OSPF | RIP | ACO |
|---|---|---|---|
| 8Nodes | 1 | 1 | 0 |
| 20Nodes | 1 | 1 | 0 |
| 50Nodes | 1 | 2 | 1 |
| 75Nodes | 1 | 19 | 1 |
| 100Nodes | 1 | 22 | 1 |
| 200Nodes | 1 | 30 | 1 |

**Table 2.** Packet Loss in Single Path Routing (number of packets) under medium load

| | OSPF | RIP | ACO |
|---|---|---|---|
| 8Nodes | 2 | 1 | 1 |
| 20Nodes | 2 | 1 | 1 |
| 50Nodes | 2 | 4 | 1 |
| 75Nodes | 2 | 35 | 1 |
| 100Nodes | 2 | 40 | 1 |
| 200Nodes | 3 | 53 | 1 |

**Table 3.** Packet Loss in Single Path Routing under heavy load (number of packets)

| | OSPF | RIP | ACO |
|---|---|---|---|
| 8Nodes | 3 | 2 | 1 |
| 20Nodes | 3 | 2 | 1 |
| 50Nodes | 4 | 6 | 1 |
| 75Nodes | 4 | 63 | 2 |
| 100Nodes | 4 | 72 | 2 |
| 200Nodes | 5 | 96 | 2 |

**Table 4.** Routing Packet Size (in Bytes)

| OSPF | RIP | ACO |
|---|---|---|
| 44 | 44 | 16 |

## References

[1] Chou, C.T.: Traffic engineering for MPLS-based virtual private networks. Computer Networks 44, 319–333 (2004)
[2] Srivastava, S., van de Liefvoort, A., Medhi, D.: Traffic engineering of MPLS backbone networks in the presence of heterogeneous streams. Computer Networks 53, 2688–2702 (2009)
[3] Palmieri, F.: An MPLS-based architecture for scalable QoS and traffic engineering in converged multiservice mobile IP networks. Computer Networks 47, 257–269 (2005)
[4] Boscoa, A., Bottab, A., Conteb, G., Iovannaa, P., Sabellaa, R., Salsanoc, S.: Internet like control for MPLS based traffic engineering: performance evaluation. Performance Evaluation 59, 121–136 (2005)
[5] Iovanna, P., Sabella, R., Settembre, M.: Traffic engineering strategy for multi-layer networks based on the GMPLS paradigm. IEEE Netw. 17(2), 28–37 (2003)
[6] Di Caro, G., Dorigo, M.: AntNet: A Mobile Agents Approach to Adaptive Routing. Tech. Rep. IRIDIA/97-12, Univ. Libre de Bruxelles, Brussels, Belgium (1997)
[7] Schoonderwoerd, R., Holland, O., Bruten, J.: Ant like agents for load balancing in telecommunication networks. In: Proceedings of the First Int. Conf. on Autonomous Agents, pp. 9–216. ACM Press, New York (1997)
[8] Duan, H., Yu, X.: Hybrid Ant Colony Optimization Using Memetic Algorithm for Traveling Salesman Problem. In: Proceedings of the IEEE Symposium on Approximate Dynamic Programming and Reinforcement Learning, pp. 92–95 (2007)
[9] Subramanian, D., Druschel, P., Chen, J.: Ants and reinforcement learning: A case study in routing in dynamic networks. In: Proceedings of the 15th Int. Joint Conf. on Artificial Intelligence, pp. 823–838. Morgan Kaufmann, San Francisco (1997)
[10] Sim, K.M., Sun, W.H.: Ant Colony Optimization for Routing and Load-Balancing: Survey and New Directions. IEEE Transactions on Systems, Man, and Cybernetics 33(5), 560–572 (2003)
[11] Xing, L.-N., Chen, Y.-W., Wang, P., Zhao, Q.-S., Xiong, J.: A Knowledge-Based Ant Colony Optimization for Flexible Job Shop Scheduling Problems. Applied Soft Computing 10, 888–896 (2010)
[12] Lopez-Ibanez, M., Blum, C.: Beam ACO for the traveling sales man problem with time windows. Computers & Operations Research 37, 1570–1583 (2010)
[13] Chandra Mohan, B., Sandeep, R., Sridharan, D.: A Data Mining Approach for Predicting Reliable Path for Congestion Free Routing Using Self-motivated Neural Network. SCI, vol. 149, pp. 237–246. Springer, Heidelberg (2008)
[14] Chandra Mohan, B., Baskaran, R.: Redundant Link Avoidance Algorithm for improving Network Efficiency. International Journal of Computer Science Issues 7(3) (May 2010)

# Low Power Optimized Array Multiplier with Reduced Area

Padma Devi[1], Gurinder Pal Singh[2], and Balwinder Singh[2]

[1] CT Institute of Engineering Management and Technology, Shahpur, Jalandhar, India
padma.devi6@gmail.com
[2] Academics and consultancy Services Division, Centre for Development of Advanced Computing, Mohali, India
gurinderpalsingh900@gmail.com, balwinder_cdacmohali@yahoo.com

**Abstract.** Multiplication is a fundamental operation in most arithmetic computing systems. Multipliers are indispensable part of DSP processing, FFT, convolution and many more areas where computation is required. In this paper an improved optimized design of 32-bit unsigned array multiplier with low power and reduced area is proposed. The power dissipation of optimized multiplier design is reduced by 3.82 percent and more than 30 percent as compared to multipliers using ripple carry and carry select adders. The area reduction is highly achieved by reducing the gate count.

**Keywords:** Multiplier, carry select adder, Partial product generation/ reduction/ accumulation.

## 1 Introduction

Great computation capacity and large amount of energy consumption is required by the signal processing and computation applications. Along with performance and area, power consumption has become a critical concern because of steady growth of operating frequency and processing capacity per chip, large current has to be delivered and the heat generated due to large power consumption must be removed by proper cooling techniques. Also battery life in portable electronic devices is limited. Multiplication is tremendously used in number of processes such as finite impulse response filters, fast Fourier transforms convolution, discrete cosine transforms and other DSP applications. Multipliers have large area, long latency and consume considerable power. The desired characteristics of a good multiplier are low power, less area and high speed. Different algorithms have been proposed for low-power, high speed multipliers at technology, physical, circuit and logic levels. These algorithms place two constraints on multiplication: latency and throughput.

A. Mantri et al. (Eds.): HPAGC 2011, CCIS 169, pp. 224–232, 2011.

## 2 Related Work

A more efficient method to increase the number of computations is to increase the throughput. In [1] a multiplier is proposed which uses modified booth encoding MBE along with a technique called spurious power suppression technique to reduce the switching power dissipation by 22%. In [20] frequency of transitions is observed in high versus low order bits of a multiplier.

At the algorithm and architecture level [9] addresses low-power multiplier design systematically from two aspects: internal efforts and external efforts. Left -to-right algorithms without carry propagation are proposed in [7] which reduce power consumption. In [5] structure optimization techniques are used to reduce power and delay in Left-right array multipliers. Carry save adders are used in [6] multiplier, to speed up the carry propagation along array multipliers. A simple modification in carry save adder [11] along with booth encoding reduces total switching activity inside the adder array which further leads to reduction in power dissipation. Operand transformation in a MAC design [8] reduces transitions and therefore power dissipation. On the basis of performance; multiplier [2] with carry lookahead adder is double in speed than ripple carry adder multiplier. Heterogeneous adder architecture is proposed for efficient delay area which consists of sub-adders of various sizes and different carry propagation schemes [3]. To improve the speed of a multiplier by optimizing the PPR can also be achieved by TDM three dimensional multiplication technique [12]. Wallace tree introduced in [22] is no doubt a fast design but it possesses irregularities in the layout, consumes more area due to complex structure which results in wiring overhead.

In section 2 we discussed about power consumption in multipliers. Proposed multiplier architecture is discussed in section 3. In section 4 results are provided. Sections 5 concludes the paper and at the last references are given.

## 3 Power Consumption in Multiplier

There are two modes of power dissipation in integrated circuits:

Static power – due to leakage current
Dynamic power

- Due to switching transient current
- Due to charging and discharging of load capacitances.

**Total power can be summarized as:**

$$P_{avg} = P_{switching} + P_{short\ circuit} + P_{leakage}$$

Switching component of power is mainly responsible for power dissipation in multipliers. The switching component of power is

$$P_{switching} = \alpha\, C_L\, V_{dd}\, 2\, f_{clk}$$

$C_L$ is the total capacitance of the circuit, $V_{dd}$ is the supply voltage, $f_{clk}$ is the clock frequency and $\alpha$ is the activity factor that corresponds to the number of transitions occurring per second. The supply voltage of a CMOS circuit is decided by the characteristics of the CMOS technology. Reducing the clock frequency reduces power and performance in proportion, so has no impact on power efficiency. The two other techniques are: minimizing the physical capacitance; minimizing the switching activity. Adder blocks in multipliers generate both carry and sum bits. These two signals have different delay properties, so that both array and Wallace tree reduction networks experience glitching due to these unequal delays and hence lead to power consumption. The best approach to reduce the power in multiplier designs is to reduce the dynamic power which is the dominant factor and contributes around 80 % of the system power.

## 4 Multiplier Architecture

Multiplication can be considered as the process which consists of sequence of addition and shift operations. Array multipliers and tree multipliers are two of the most popular kinds of multiplier.

Multiplication consists of three steps: 1) generation of partial products, 2) accumulation of all partial production to get only two rows at the end and last two rows of partial products are added by using a carry propagation adder. 3) last step is to efficiently add the two rows of partial products which can be fastened by using appropriate adder structure. In one method of partial product generation AND gate is used to generate partial products when these partial products are arranged by bit position then a trapezoidal structure comes in to existence as shown in figure 1. The resulting structure is called a partial product array.

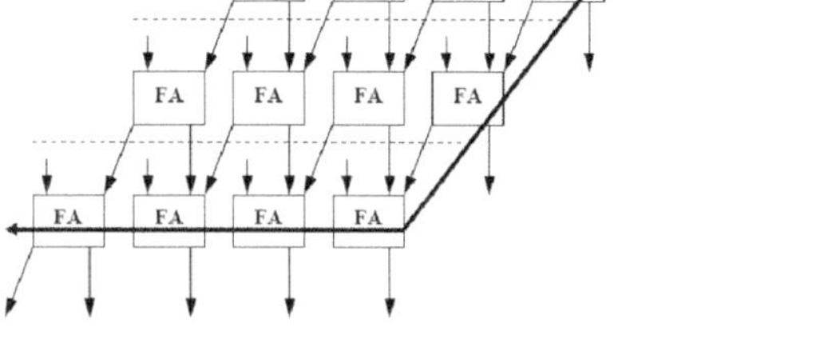


**Fig. 1.** Example of Array multiplier

This step is called partial product array generation. The dimensions of the array are functions of the size of the multiplier and multiplicand. The second method uses Booth encoded schemes popular as radix-2, radix-4 (modified booth encoding (MBE)) to generate partial products. The advantage of array multiplier comes from its regular structure. Since it is regular, it is easy to layout and has a compact size. Advantages of array multipliers are; easy to layout, due to compact size, and easy to design for pipelined structure. But tree structure is faster than this having complex layout and irregular structure which causes more power consumption.

## 5 Proposed Multiplier

In this paper we propose 32-bit array multipliers which use carry select adders as the base adders to perform the partial product additions.

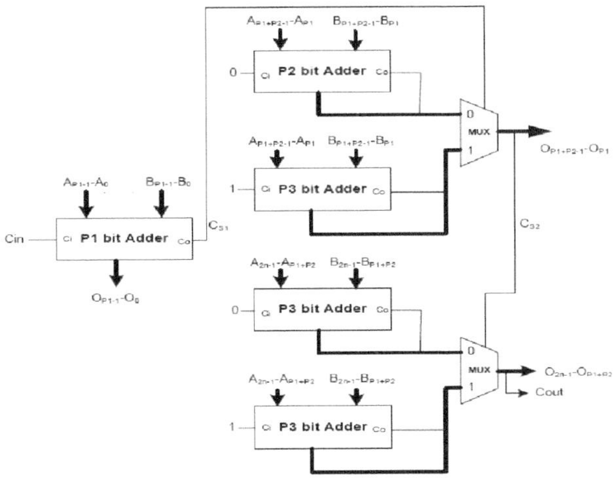


**Fig. 2.** Three stages carry select adder [4].

The conventional k-bit Carry select adder (CSA) consists of two k/2 adders, one for least significant bits and the other for most significant bits. The adder for higher bits consists of two adders one considering input carry as '1' and other as'0'. Using multiplexer, final carry out and sum are selected taking carry out from the k/2 bit least significant adder as a control signal. This adder increases in area but delay is reduced. Further multiple stage carry select adders proposed in [4] are used to develop optimized 32-bit array multipliers. Figure 2 shows the 3 stage CSA. Where Cs1 is the carry out of P1 bit adder. Cs2 is the carry propagated from the other part of adder. Cout is the final carry output of the adder. Similarly 4 stage, 5 stage CSA based 32-bit array multipliers are designed.

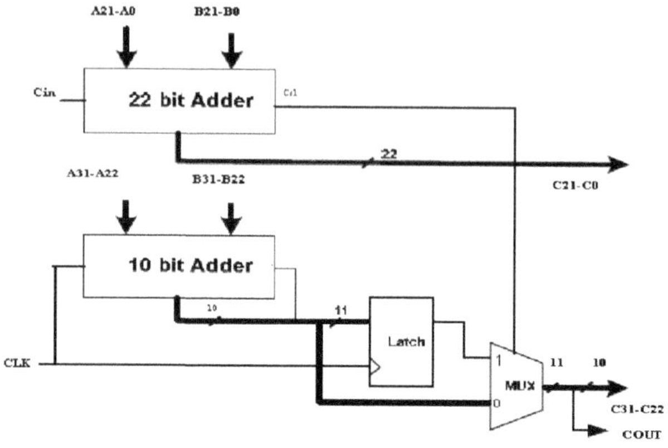


**Fig. 3.** CSAS architecture [4]

One more design of CSA named as clock select adder with sharing (CSAS) [4] is proposed shown in figure 3. Instead of using two separate adders in conventional CSA, one for the Cs1=1 and another for the Cs1=0 one adder will be used to reduce the area and power dissipation. A clock signal shared by the adder circuit hence named so. Multiplier using this adder structure is optimized in area and power consumption to conventional multipliers.

Ripple carry adders (RCAs) are most compact in area (O (n) ) among all types of adders, they are the slowest types of adders (O(n) time). On the other hand, Carry Look-ahead Adders (CLAs) are the fastest adders (O(log(n) time), but they are the worst from the area point of view (O(nlog(n)) area). Carry Select Adders (CSAs) have been considered as a compromise solution between RCAs and CLAs (O(n) time and O(2n) area) [4] because they offer a good tradeoff between the compact area of RCAs and the short delay of CLAs. Carry skip adders (CSKAs) also show results between RCA and CLA. Even variable block carry skip adders (VBCSKAs) are better in performance than CSKA.

To design 32-bit array multipliers using different adder architectures algorithm is: Algorithm for array Multiplier [2].

Begin Program
Multiplier = 32 bits, Multiplicand = 32 bits
Register = 64 bits
Put the multiplier in the least significant half, and clear the most significant half
If the least significant bit of the 64-bit register contains binary '1'
Add the Multiplicand to the Most Significant Half using the adder.
Shift the 64-bit Register one bit to the right throwing away the least significant bit
Else; Shift the 64-bit Register one bit to the right throwing away the least significant bit
End If; End Loop
Register = Sum of Partial Products
End Program

## 6 Results

All array multipliers are designed in VHDL and synthesized using Xilinx project navigator ISE 9.1i. Simulation results are obtained through Model Sim XE III 6.2g and designs are implemented on Spartan 3, XC3S400 kit.

Table 1 and 2 provide the results obtained for the optimized and conventional 32-bit array multipliers respectively. Table 3 shows the percent gain of reduction in dynamic power dissipation and reduced gate count of CSAS multiplier over RCA, CSKA, and VBCSKA multipliers and in table 4; carry select adder (CSA), 3 stage CSA , 4 stage CSA and 5 stage CSA based multipliers are compared for the same parameters. As lowering the gate count directly leads to the reduction in area consumption, CSAS multiplier showed tremendous decrease in gate count as compared to other designed multipliers.

**Table 1.** Results obtained for 3 stage, 4 stage, 5 stage and CSAS based optimized 32-bit array multipliers

| | **3 stage carry select multiplier** | **4 stage carry select multiplier** | **5 stage carry select multiplier** | **Clock Select Adder with Sharing multiplier** |
|---|---|---|---|---|
| **Delay (ns)** | 44.982 | 43.982 | 43.234 | 65.109 |
| **LUT's** | 3807 | 3784 | 3704 | 3149 |
| **Bonded IO/B's** | 128 | 128 | 128 | 129 |
| **Gate count** | 23,449 | 23,203 | 22,762 | 20,319 |
| **Dynamic Power (mw)** | 161.50 | 157.74 | 159.67 | 106.87 |

**Table 2.** Results obtained for RCA, CSKA, VBCSKA and CSA based multipliers

| | **Ripple carry multiplier** | **Carry skip multiplier** | **Variable carry skip multiplier** | **Carry select multiplier** |
|---|---|---|---|---|
| **Delay(ns)** | 47.234 | 42.175 | 40.605 | 46.234 |
| **LUT's** | 4118 | 4669 | 3970 | 4034 |
| **Bonded IO/B's** | 128 | 128 | 128 | 128 |
| **Gate count** | 26,065 | 28,669 | 24,496 | 24,802 |
| **Dynamic power(mw)** | 111.12 | 147.7 | 153.77 | 155.45 |

**Table 3.** Comparison of Power reduction and gate count of CSAS multiplier with RCA, CSKA and VBCSKA multiplier

| | Ripple Carry Multiplier | Carry Skip Multiplier | Variable Carry Skip Multiplier | Parameter |
|---|---|---|---|---|
| CSAS Multiplier | +03.82% | +27.65 | +30.50 | Reduction in Power Dissipation (%) |
| CSAS Multiplier | -5,746 | -8,350 | -4,177 | Gate Count Difference |

**Table 4.** Comparison of power reduction and gate count of CSAS multiplier with carry select adder, 3 stage carry select, 4 stage carry select, and 5 stage carry select multipliers

| | Carry Select Multiplier | 3 Stage Carry Select Multiplier | 4 Stage Carry Select Multiplier | 5 Stage Carry Select Multiplier | Parameter |
|---|---|---|---|---|---|
| CSAS Multiplier | +31.25 | +33.82 | +32.24 | +33.06 | Reduction in Power Dissipation (%) |
| CSAS Multiplier | -4,483 | -3,130 | -2,884 | -2,443 | Gate Count Difference |

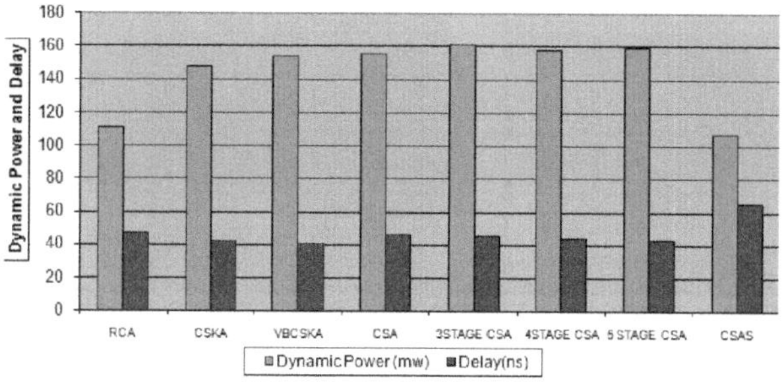


**Fig. 4.** Dynamic Power and delay of designed multipliers

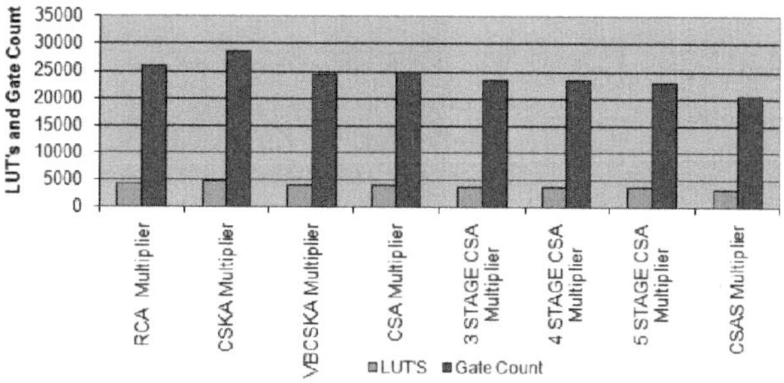


**Fig. 5.** Area utilization of designed multipliers

Figure 4 and 5 depicts the graphical representation of dynamic power and delay and area consumption of 32-bit designed array multipliers respectively.

## 7 Conclusion

The optimized multiplier design consumes low power and least area as compared with designed conventional multipliers. The power reduction obtained in case of CSAS multipliers as compared to carry select multiplier designs is more than 30 %. With respect to RCA multiplier 03.82% reduction in dynamic power is achieved. Reduction in power and area consumption is also improved as compared to carry skip and variable carry skip multipliers. Gate count of optimized multiplier is reduced by 22.044% as compared to RCA multiplier, 29.12% with carry skip and 18.75 % with carry select multiplier.

## References

1. Marimuthu, C.N., Thangarj, P.: Low power high performance multiplier. ICGST International Journal on Programmable Devices, Circuits and Systems, PDCS 8(1), 31–38 (2008)
2. Krat, H., Al-Taie, A.Y.: Performance analysis of a 32-bit multiplier with a carry-look-ahead adder and a 32-bit multiplier with a ripple adder using VHDL. Journal of Computer Science 4, 305–308 (2008)
3. Lee, J.-G., Lee, J.-A., Lee, B.-S., Ercegovac, M.D.: A design method for heterogeneous adders. In: Lee, Y.-H., Kim, H.-N., Kim, J., Park, Y.W., Yang, L.T., Kim, S.W. (eds.) ICESS 2007. LNCS, vol. 4523, pp. 121–132. Springer, Heidelberg (2007)
4. Amelifard, B., Fallah, F., Pedram, M.: Closing the gap between carry select adder and ripple carry adder: a new class of low-power high-performance adders. In: Proc. of IEEE International Symposium on Quality Electronic Design, ISQED (2005)
5. Zhijun, Huang: High-performance low-power left-to-right array multiplier design. IEEE Trans. Comput. 54(3), 272–283 (2005)
6. Fonseca, M.R., da Costa, E.A.C., Bampi, S.: Performance optimization of radix-2^m multipliers using carry save adders. In: XI Workshop Iberchip, Salvador-BA, pp. 126–131 (2005)
7. Wang, J.-S., Kuo., C.-N., Yang, T.H.: Low power fixed width multiplier. In: Proc. of the International Symposium on Low Power Electronics and Design, pp. 307–312 (2004)
8. Fujino, M., Moshnyaga, V.G.: Dynamic operand transformation for low power multiplier accumulator design. In: Proc. of Circuit and Systems, vol. 5, pp. 345–348 (2003)
9. Huang, Z.: High-Level optimization techniques for low-power multiplier design. Ph.D dissertation (2003)
10. Rabaey, J.M., Chandrakasan, A., Nikolic, B.: Digital integrated circuit. Electronics and VLSI Series. Prentice Hall, Englewood Cliffs (2003)
11. Yu, Z., Wasserman, L.: A painless way to reduce power dissipation by over 18% in Booth-encoded carry-save array multipliers for DSP. In: IEEE Workshop on Signal Processing Systems, pp. 571–580 (2000)
12. Oklobdzija, V.G.: High-Speed VLSI Arithmetic Units: Adders and Multipliers. In: Chandrakasan, A. (ed.) Design of High-Performance Microprocessor Circuits. IEEE Press, Los Alamitos (2000)

13. Parhami, B.: Computer Arithmetic, Algorithm and Hardware Design, pp. 91–119. Oxford University Press, New York (2000)
14. Meier, P.C.H.: Analysis and design of low power digital multipliers. Ph.D dissertation, Carnegie Mellon University,Pittsburgh, Pennsylvania (August 1999)
15. Keane, G., Spanier, J., Woods, R.: The impact of data Characteristics and hardware topology on hardware selection for low power DSP. In: Proc. Low Power Electronics and Design, pp. 94–96 (1998)
16. Farooqui, A.A., Oklobdzia, V.G.: General data-path organization of a MAC unit for VLSI implementation of DSP processors. In: Proc. IEEE Circuits and Systems Conf., pp. 260–263 (1998)
17. Oklobdzjia, V.G., Villeger, D., Liu, S.S.: A Method for Speed Optimized Partial Product Reduction and Generation of Fast Parallel Multipliers Using an Algorithmic Approach. IEEE Transactions on Computers 45(3) (March 1996)
18. Wallace, C.S.: A suggestion for a fast multiplier. IEEE Trans. Electron. Computing EC-13, 14–17 (1964)

# Simulink Library Development and Implementation for VLSI Testing in Matlab

Gurinder Pal Singh and Balwinder Singh

Acadamics and consultancy Services Division
Centre for Development of Advanced Computing, Mohali, India
Gurinderpalsingh900@gmail.com, balwinder_cdacmohali@yahoo.com

**Abstract.** In ATPG, faults in the VLSI circuits are detected with D-algorithm, SPODEM and FAN algorithms. This paper gives the emphasis on presenting first two algorithms in MATLAB. Implementation of these algorithms for complex VLSI circuits is very tedious job. So an environment in Simulink is presented here, which is further verified on some benchmark circuits. Simulink provides an environment for intellectual properties (IP) building block based circuit engineering design as well as project simulation environment. In PODEM the requirement is the exact values of Controllability and Observability. For effective and fast calculation of COM, Simulink based designed models are used.

**Keywords:** ATPG (Automatic Test Pattern Generator), COM (Controllability and Observability Measures), CUT (Circuit Under Test), S-PODEM (Sequential-Path Oriented Decision Making), VLSI (Very Large Scale Integrated) Circuits, GUI (Graphic User Interface), CC (Controlled Controllability), CO (Controlled Observability).

## 1 Introduction

Selection of proper test vector is essential, after discarding equivalent faults. The test vectors are generated manually or by special software tool which automates the process, that is Automatic Test pattern Generator (ATPG) [6]. The ATPG is fault oriented; it selects a fault from list of candidate faults, and attempts to create a test for fault.

There are well-known three algorithms for the ATPG for digital circuits. Each algorithm require some information related to the DUT (Device under Test) i.e, information related to the circuit design, the fault list information, information of the individual components and their fault free behavior, the way through which fault effects are propagates to primary outputs and the fault equivalence list. Each algorithm operates with fault generator, to minimum collapsed fault list. ATPG algorithms are multi-purpose that means they are able to; generate circuit test-patterns, to find the redundant circuit logic, to match circuit implementation with another circuit. Those three algorithms are: D-Algorithm, PODEM (Path Oriented Decision Making) Algorithm and Fan algorithm.

A. Mantri et al. (Eds.): HPAGC 2011, CCIS 169, pp. 233–240, 2011.

The functioning of D-Algorithm is mainly based on the Roth's five valued logic shown in table1 [10], Developed by Roth in 1960s. This algorithm uses a logical value for representation of "good" and "faulty" circuits. The method is based on the intersection of the D-cubes. The supported 5 values are D,$\overline{D}$, 1, 0, X. PODEM algorithm applied of combinational circuits is also based on this algebra but the difference is that the through which the values of the primary input for the objective defect is decided that is the backtracing applied in PODEM is totally based on the COM [14, 15].

Controllability is VLSI circuit is the difficulty of setting a particular logic signal to 1 or 0. Observability is the difficulty of observing the state of a logic signal [17]. For DUT, includes the sequential elements that are flip flops, this 5-valued logic algebra is not able to detect the fault. For that the algorithm used is sequential PODEM (S-PODEM). This approach uses the concept of time compression and synthesizes in one time frame a single test vector representing the compressed form of multiple time frames. The single vector is then expanded into a test sequence [21].

For sequential circuit testing, when a test pattern is applied for the fault propagation, then each signal line can be static or pulsating. So for these pulsating lines a 'P- model' value is used to reflect the behavior of the pulsating line model is used for CUT. So for the S-PODEM a '11-value' logic model is used to reflect the behavior of each and every gate and flip flop in the CUT. These Values of 11-value logic that are: D, $\overline{D}$, 0, 1, X, P, P0, P1, 1P, 0P and PP.

Where D: denotes logic value 1 in the good circuit and 0 in faulty circuit.

$\overline{D}$: indicates logic value 0 in the good circuit and 1 in faulty circuit.

P: value used for any pulsating signal.

P0 and P1: indicates a pulsating signal in the good circuit and static value 0(1) in the faulty circuit respectively.

0P and 1P: indicates 0 in the good circuit and P in the faulty circuit respectively.

PP: represents a signal pulsating in the good machine with another signal pulsating in the faulty machine [21].

In sequential circuits the primary inputs remain at some static values during test generation for a particular fault. Only a few primary inputs exhibit pulsating behavior. Some common and important terms in S-PODEM are explained below:

Static Lines: A line which remains at a static value for both faulty and faulty free circuits, during the entire test operation is regarded as a static line, and is designated by SLS.

Pulsating Lines: If line faces a change in its logic value in any one of faulty or faulty free circuits during the test experiment, then that line is known as pulsating line.

In section 2 includes the related work. Section 3 explains the way to model D-algorithm and PODEM in MATLAB/Simulink and shows the modified library of Simulink representation. Designing of SPODEM is represented on benchmark circuit S27 and Serial Multiplier is shown in section 4. At the last in section 5, paper is concluded by presenting all results of designed circuit for purposed environment in tabular form, followed by future work.

## 2 Related Work

Tommo Inoue et. al suggest the fast detection of the faults in ATPG by following 5-valued algebra instead of following 16-valued logic[2]. Jozesf Sziary et. al presents two algorithm that are D-algorithm and composite justification are analyzed on the basis of computational complexity, for calculating fault detection for digital circuits. From analyzing these two it is concluded that composite justification requires less computational steps [3]. Bushnell M. et. al in his book tells about ATPG algorithms that are POPDEM and D algorithm described. Benefits of using PODEM over D algorithm are defined by considering the speed of testing as the main factor [10]. Wen-Ben Jone ,et. al gives idea on Testing of sequential element that is flip flop. A method for sequential element testing is suggested which is based a new PODEM technique which depends on 11- value logic [21]. Mohamed Lamoussi et. al presents all the basic formulas for the calculation of controllability and observability based on Rutman's system model are used, which are further based on Reduced Ordered binary Decision Diagram (ROBDD) [23]. Noatake Kamuira et. al introduces the use of COM to guide D algorithm for multi valued logic, and used for determining the execution of the D-drive at the fan out[17].

## 3 Simulink Based Implementation for Testing for Stuck-at Faults Using D-Algorithm and PODEM

### 3.1 Testing for Combinational Circuits

D-cubes are the collapsed truth table that can be used to characterize an arbitrary logic block. D-cubes are utilized to describe both the faulty and normal functions simultaneously. Conceptual purpose of the D-cubes is the path sensitization. Notation D represents logic value "1" in the good circuit or fault free and "0" in the faulty circuit. Compliment of D represents "0" in the fault free circuit and "1" in the faulty circuit, shown in table1. So In D-Cubes main parameters are D, $\overline{D}$ and X. But these are not understandable to Simulink. To interpret these to Simulink some notations are used that are; For interpretation of 'D' it is replaced with the numeric value 2, similarly '$\overline{D}$' is replaced with 3 and parameter unknown ('X') is replaced with 4 illustrated in table 1.

**Table 1.** Interpretation values for the D-Logic

| | | | | |
|---|---|---|---|---|
| 0 | 1 | 2 | 3 | 4 |
| 0 | 1 | D | $\overline{D}$ | X |

Now based on these values the behavior of all logic gates is defined in Simulink so that it is understandable to user and Simulink both. D-cube is generated for fault propagation. For any fault to be detected, condition is that its effect should be propagated to the output [9]. By keeping this in view logic gates are designed by transporting the translated values of D-cubes into 2-D lookup tables of Simulink library.

Its rows act as first primary input and column as second primary input. All logic gates are designed and properly masked and added to Simulink library (as shown in figure 1) permanently by modifying "slblocks.m", to have user friendly environment.

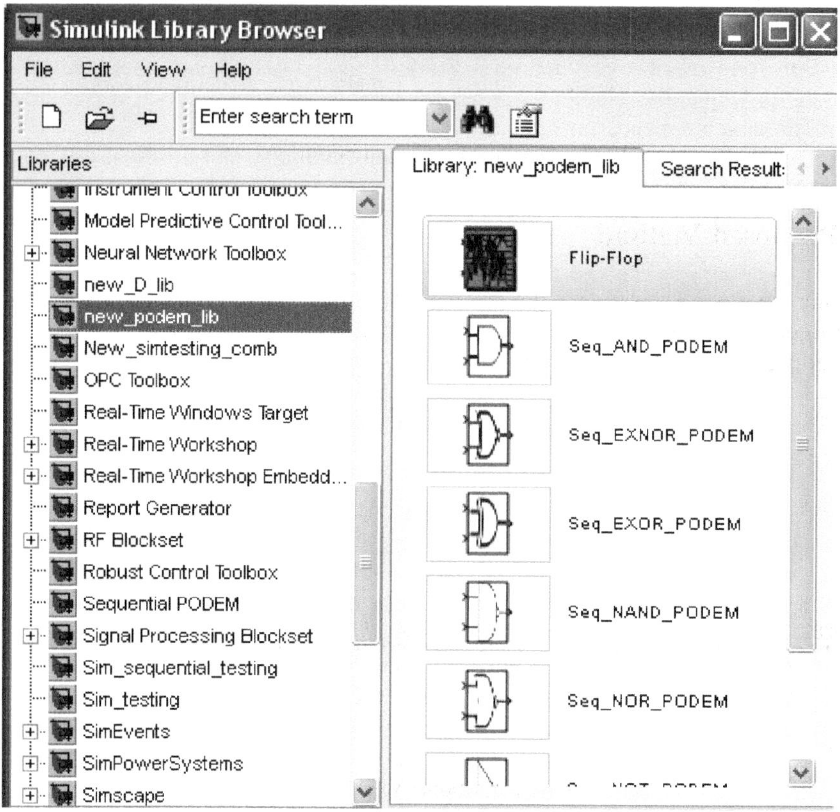


**Fig. 1.** Updated Simulink library with D-cube based user designed logic gates

By using this newly designed logic gate library, 74181-4-bit ALU is designed in Simulink as shown in figure 2. To make it utilized for testing purpose some random faults are forced in the circuit shown in testing model, which are also mentioned in table 2 and this table also includes their respective test patterns. Randomly forced fault and their respective patterns for fast adder 74823 is given in table 3

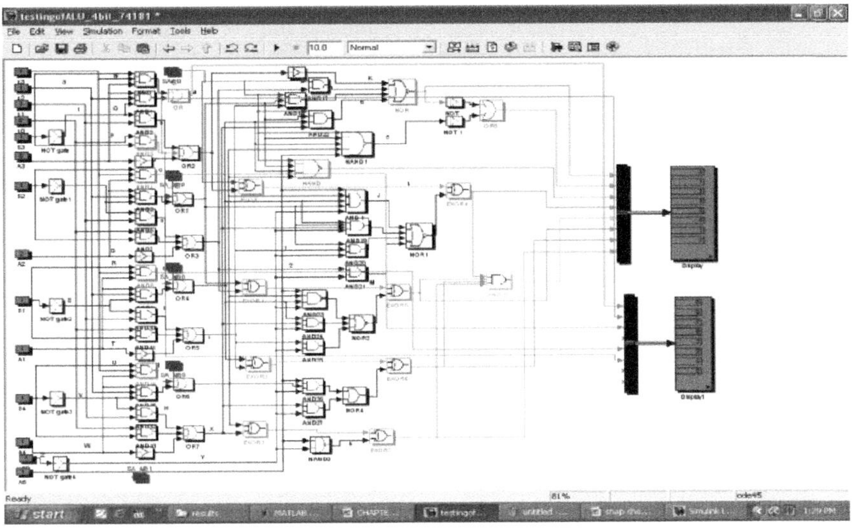

**Fig. 2.** D-algorithm based testing model for *74181 4bit ALU* using library shown in figure 3.1

**Table 2.** Test vector based on D-algorithm for forced SA-0,1 for *ALU- 74181 4-bit*

| S.NO. | Nodes SA-0, SA-1 | Simulated test vectors for SA-0 | Simulated test vectors for SA-1 |
|---|---|---|---|
| 1. | 1c10a | 01011011011101111 | 11001110111110110 |
| 2. | ce32j | 11100110111101101 | 01111001101101111 |
| 3. | Bd6j | 11011001111010111 | 00100110011010001 |
| 4. | 45fg | 11110110001101111 | 11011001110010010 |
| 5. | 7891 | 11110111101111111 | 11010001010011000 |
| 6. | kmnpqrs | 11010011110111101 | 00111001110011100 |

**Table 3.** Test vector based on D-algorithm for forced SA-0,1 for fast adder 74283

| S.NO. | Nodes forced for SA-0 | Simulated test vectors for SA-0 | Nodes forced for SA-1 | Simulated test vectors for SA-1 |
|---|---|---|---|---|
| 1. | aceg | 11101011111111 | dfhi | 11010011000111 |
| 2. | klmp | 11000110101101 | ronq | 01110110011101 |
| 3. | suvy | 11011100110110 | twxz | 11011001111000 |
| 4. | b145 | 11101001101101 | 236 | 10010110011101 |

## 3.2 Testing for Sequential Circuits

As mentioned in [20] testing of sequential circuits is depends on 11 valued logic called sequential PODEM, discussed earlier. As for 5-valued D-logic these 11-valued logic is also interpreted in to numeric form to make it understandable by the Simulink environment. This is represented in table 4.

**Table 4.** Interpretation values for the D&P-Logic

| 0 | 1 | 2 | 3 | 4 | 5 | 6 | 7 | 8 | 9 | 10 |
|---|---|---|---|---|---|---|---|---|---|---|
| 0 | 1 | D | $\overline{D}$ | P | 0P | 1P | P | P0 | P1 | PP |

By using these values, a noble Simulink library model is developed for sequential testing, as done in pervious section 3.1. But in sequential based PODEM testing techniques COM calculations are essential [13]. Need of these depend upon the complexity of any circuit, because at certain levels in a CUT to propagate faults to PO it is not clear which input node is set for fault propagation. In that case controllability and observability measures are helpful. Here these are also calculated by using COM user Library model for Simulink and presented in table 6. To present sequential testing CUT taken is S27. Primary input values taken are;CC0a, CC1a=1 for sequential_NOR_Gate_Sys, sequential_Not_Gate_Sys,sequential_OR_Gate. CC0b, CC0b, CC1b=1 for sequential_NOR_Gate_Sys3.

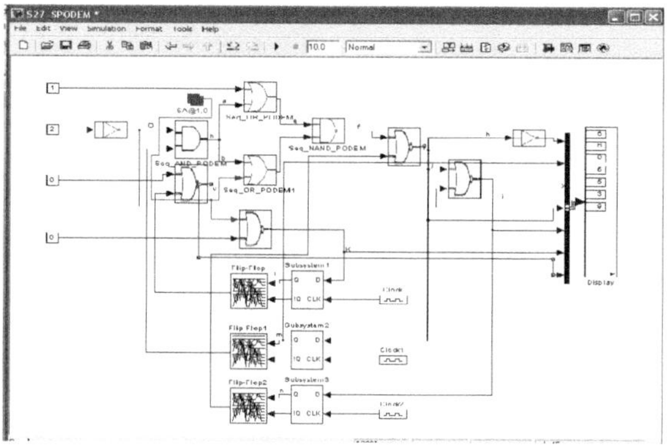

**Fig. 3.** S27 testing for SA faults with SPODEM

To test this circuit for stuck-at0-1, faults are forced randomly as shown in figure 5.

**Table 5.** COM Values for S27

| Gates | CC0Z | CC1Z | CC0Q | CC1Q |
|---|---|---|---|---|
| sequential_OR_Gate | 2 | 1 | x | x |
| sequential_OR_Gate1 | 2 | 2 | x | x |
| sequential_Not_Gate_Sys1 | 2 | 10 | x | x |
| sequential_NAND_Gate | 6 | 2 | x | x |
| D flip-flop1 | | | 4 | 6 |
| D flip-flop2 | | | 4 | 14 |
| D Flip-flop3 | | | 4 | 7 |

**Table 6.** Test vectors for S27

| S.NO. | Nodes forced for SA-0,SA-1 | Simulated test vectors for SA-0 | Simulated test vectors for SA-1 |
|---|---|---|---|
| 1. | abf | 010x | 11x0 |
| 2. | oec | 01x0 | 1001 |
| 3. | knf | 1101 | 0010 |
| 4. | ghj | 011x | 100x |
| 5. | ilm | 1100 | 0011 |

## 4 Conclusion

In this paper a new approach for VLSI testing in MATLAB is proposed and implemented. In this technique we have designed a user library for Simulink by using 2-D logic array block, for D-algorithm and SPODEM. By using that libraries benchmark circuits are implemented for random testing. Results are presented in tabular form. Controllability and Observability measures required in PODEM are calculated by using Simulink model for CUT. This environment is useful for VLSI testing to make understand the concept of testing to students and can be used as teaching tool.

## References


1. Sziray, J.: Computational Complexity in Logic testing. Intelligent Engineering system (INES). In: 14th international Conference, Las Palmas of Grain Canaria, Spain, May 4-7, pp. 97–102 (2010)
2. Inoue, T., Izumi, N., Yoshikawa, Y., Ichhihra, H.: A Fast Threshold Generation Algorithm Based on 5-Valued Logic. In: 5th IEEE International Symposium on Electronic Design, Test & Applications, pp. 345–349 (2010)
3. Jayanthy, S., Bhuvaneswari, M.C.: Simulation Based ATPG for Crosstalk Delay Faults in VLSI Circuits using Genetic Algorithm. ICGST-AIM Journal 9(2), 11–17 (2009)
4. Efthymioy, A.: Redundancy and Test-Pattern Generation for Asynchronous Quasi-delay-Insensitive Combinational Circuits. Design & Diagnostics of Electronic Circuits and Systems (2007)
5. Perelroyzen, E.: Digital Integrated Circuits Design for test Using Simulink and State flow, pp. 143–193. CRC Press Taylor&Francis Group (2007)
6. Wang, L.-T., Wu, C.-W.: VLSI test Principles and Architectures design for testability., pp. 11–21, 37-48, 161-244. Margen Kaufmann Publishers, San Francisco (2006)
7. Grout, I.A.: Integrated circuit test engineering, 3rd edn., pp. 41–85. Springer, London (2006)
8. Vaseekar Kumar, M.M., Padmanaban, S., Tragoudas, S.: Low Power ATPG for Path Delay Faults. In: Proceeding of the 14th ACM Great Lakes Symposium on VLSI, April 26-28, pp. 398–392 (2004)
9. Bushnell, M.L., Agrawal, V.D.: Essential of Electronic Testing for Digital Memory & Mixed-Signal VLSI Circuits, pp. 81–206. Kluwer Acadmic Publishers, Dordrecht (2004)

10. Using Simulink: Dynamic System Simulation for MATLAB. The Mathworks, Inc., Natick, MA (2004)
11. Marchand, P., Thomas Holland, O.: Graphics and GUIS with Matlab, 3rd edn., ch. 10. Chapman and Hall/CRC (2003)
12. Jha, N.K., Gupta, S.: Testing of Digital system, 1st edn., pp. 266–306. Cambridge university Press, Cambridge (2003)
13. Miczo, A.: Digital Logic testing and Simulation, 2nd edn., pp. 119–128, 165-201. A John Willey & Sons, West Sussex (2003)
14. Kamiura, N., Isokawa, T., Mastsui, N.: PODEM based on Static Testability Measures and Dynamic Testability Measures for Multiple-Valued Logic Circuits. In: Proceedings of the 32nd IEEE International Symposium on Multiple-Valued Logic (2002)
15. Chapman, S.J.: MATLAB programming for Engineers, a division of Thomson learning USA, 2nd edn (2002)
16. Kamiura, N., Hata, Y., Matsui, N.: Controllability /Observability Measures for multiple-valued test generation based on D-algorithm. In: 30th IEEE international Proceeding, ISMVL (2000)
17. Kirovski, D., Potkonjank, M., Guerra, L.M.: Improving the Observability and Controllability of Data paths for Emulation-Based Debugging. IEEE Transaction on Computer-Aided Design Of Integrated Circuits and Systems 18(11), 1529–1536 (1999)
18. Hamzaoglu, I., Patel, J.H.: New Techniques for Deterministic Test Pattern Generation. In: 16th IEE Proceedings VLSI test Symposium (1998)
19. Shmerko, V.P., Yanushkevich, S., Levashenko, V.: Test Pattren Generation for Combinational Multi-Valued Networks based on Generalized D-algorithm. In: 27th international symposium preceding, pp. 139–144 (1997)
20. Jone, W.-B., Shah, N., Gleason, A., Das, S.R.: PGEN: A Novel Approach to Sequential Circuit Test Generation. VLSI design. OPA (oversea publishers association) 4(3), 149–165 (1996)
21. Hong, S.J.: A 15-valued Fast Test Generation for Combinational Circuits. In: IEEE Test Symposium, pp. 113–118 (1993)
22. Jamoussi, M., Kamonska, B.: Controllability and Observability Measures For functional-Level Testability Evalution. In: IEEE VLSI Test Symposium, pp. 211–216 (1992)
23. Butler, K.M., Kapur, R., Ray Mercer, M., Ross, D.E.: The Role Of Controllability and Observability in Design for Test. In: IEEE VLSI Test Symposium, pp. 154–157 (1992)

# Processing of Image Data Using FPGA-Based MicroBlaze Core

Swagata Samanta[1], Soumi Paik[1],
Shreedeep Gangopadhyay[1], and Amlan Chakrabarti[2]

[1] Dept. of Electronics and Communication Techno India, Kolkata, India
friend_swagata@yahoo.co.in, imsoumi@gmail.com,
deep_mistu@yahoo.com
[2] A.K. Choudhury School of Information Technology,
University of Calcutta, India
acakcs@caluniv.ac.in

**Abstract.** This paper proposes the technique of storing of image data in the FPGA memory and subsequent processing of the stored image data using MicroBlaze processor core of the Xilinx FPGA device. Though related research work exists in processing image data using DSP blocks available in FPGA devices but very little work exists in processing the same using FPGA based processor core. This type of design is extremely important for real time embedded system design for image processing applications. Our work deals with the inversion of a binary image in the FPGA memory and the recovery of the inverted image into its original form for verification of the inversion process. We have used the Xilinx EDK 11.1 tool, Spartan 3E FPGA kit and MATLAB is used for pre and post processing of the image data.

**Keywords:** EDK, FPGA, image inversion, image processing, System C.

## 1 Introduction

The current available technologies like Microcontroller-based systems, DSP processor-based systems possess programmable processors, which are based on the Von-Neumann architectural concept, where a given application (program) is stored in the memory in sequential fashion. So, these Von-Neumann processors fail to exploit the concurrency in the algorithm even though they offer flexibility. Thus performance is not achieved for many applications especially where high speed is required. This leads to an alternative solution where, speed of the hardware can be achieved retaining the flexibility of the software (programmability).

Field programmable gate array (FPGA) represents an evolutionary improvement in gate array technology, which offers potential reductions in prototype system costs and product time-to-market, and simplifies debugging and verification. Re-programmable FPGAs are capable of dynamically changing their logic and interconnect structure to adapt to changing system requirements. Hardware design techniques such as parallelism and pipelining techniques can be developed on a FPGA, which is not possible in dedicated DSP designs 3.

A. Mantri et al. (Eds.): HPAGC 2011, CCIS 169, pp. 241–246, 2011.

During the last few years' researchers made an unprecedented effort in the field of image processing in hardware. Prior research may be categorized based on the types of hardware and the image processing algorithms implemented. The types of hardware considered for image processing include Application Specific Integrated Chips (ASIC), Digital Signal Processors (DSP) and Reconfigurable Logic Devices (FPGA).The image processing algorithms considered for hardware implementation include: convolution, image filtering and edge detection (Sobel's, Prewitt's and Canny's edge detection). Some researchers have also considered hardware implementations specific to FPGA vendors like Xilinx, Actel and Altera. Most of the system level hardware programming languages introduced and commonly used in the industry are highly hardware specific and requires intermediate to advance hardware knowledge to design and implement the system. In order to overcome this bottleneck various C-based hardware descriptive languages have been proposed over the past decade 3, 4.C-based hardware descriptive languages have been proposed and developed since the late 1980s. Some of the C-based hardware descriptive languages include: Cones, HardwareC, TransmogrifierC, SystemC, OCAPI, C2Verilog, Cyber, SpecC, NachC, CASH.

This paper deals with FPGA technology and the development environment used is Xilinx embedded development kit (EDK) 1, 2. EDK is a Xilinx software suite for designing complete embedded programmable systems and this enables the integration of both hardware and software components of an embedded system. The soft-core processor that is used is MicroBlaze. The MicroBlaze has a versatile interconnect system to support a variety of embedded applications. It consists of about 900 LUTs having RISC architecture 2. The programming platform that is required here is SystemC 7. Another feature of our design is that in our FPGA system (Spartan 3E) there was no dedicated graphics port but still we were successful in transmitting the image data using the serial RS232 interface.

Image processing involves processing or altering an existing image in a desired manner. It is a computational process used to convert an image data into a desired image representation. Image information demands visual perception. Therefore, displaying image information in various forms is a necessity 6.The first step is obtaining an image in a readable format. Once the image is in a readable format, image processing software needs to read it so that it can be processed and written back as an image file.

The organization of the paper is as follows: Section 2 shows the image data basics. Section 3 describes the proposed methodology. Section 4 deals with the results. Applications, limitations and suggestion for further improvement are presented in sections 5 and 6 respectively. In Section 7 we discuss on some important issues of our work.

## 2 Image Data Basics

An image consists of a two-dimensional array of numbers. The color or gray shade displayed for a given picture element (pixel) depends on the number stored in the array for that pixel. In binary image, image data is black and white. Here each pixel is either 0 or 1.Gray-scale images appear like common black-and-white photographs;

they are black, white, and shades of gray. This paper discusses methods of viewing an image into a desired form. This paper presents a program to display the raw pixel data of an image in text format and the recovery of the image in its original form.

**Fig. 1.** Gray-scale image

## 3 Proposed Methodology

From the 2-D image the raw pixel data is transferred into an environment where by taking the pixel data our desired operation is accomplished. Here the environment is text environment. Inversion logic 5 is applied on the raw readable image data through MicroBlaze processor and EDK tool. The inverted image is shown in Figure 5.

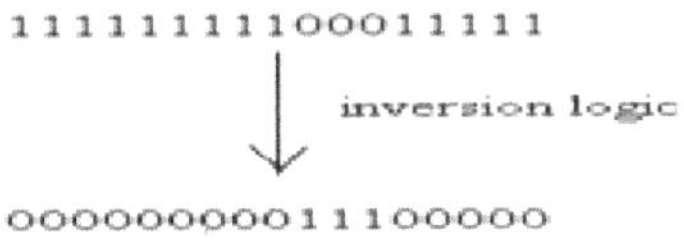


**Fig. 2.** Inverted pixel data obtained by applying inversion logic

Inverted text raw image data is obtained. After obtaining the inverted data reverse process is applied in order to get the original image. The pixel data of the inverted image is viewed at hyper terminal through RS-232 serial port which is shown in Figure 8.

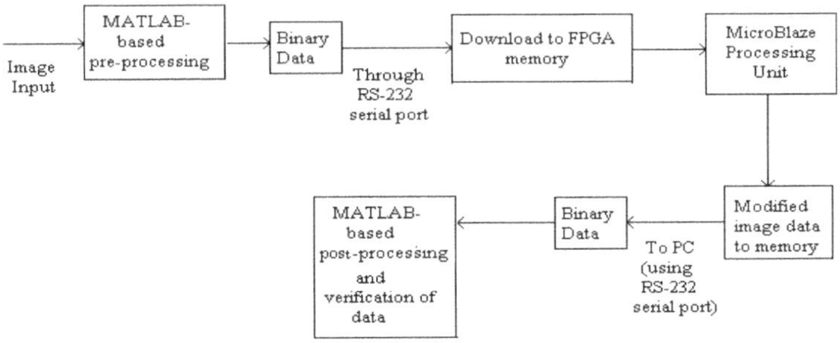


**Fig. 3.** Flow diagram of the proposed methodology

Figure 1 is the example gray scale image, which is converted into binary image. Figure 3 shows the Flow diagram of the proposed methodology. First block represents the MATLAB-based pre-processing of the image. MATLAB-based pre-processing generates the required binary data of the image which is downloaded to the FPGA memory through RS-232 serial port. Next the MicroBlaze processing unit processes the image data and stores the modified data back to the memory. Through the RS-232 serial port the binary data of the image is transmitted to the computer for subsequent MATLAB-based post-processing and verification of the processed image data. Ultimately we get the inverted image of the initial input image.

## 4 Results

The IO devices that are needed to perform the entire process are: DDR_SDRAM, RS232_DCE, LEDs, DIP_Switches_4bit, Buttons_4bit. Table1 shows the device utilization summary.

**Table 1.** Device utilization summary

| **logic utilization** | **used** | **available** | **utilization** |
|---|---|---|---|
| no. of slice flip flops | 2,561 | 9,312 | 27% |
| no. of 4 input LUTs | 2,625 | 9,312 | 28% |
| no. of occupied slices | 2,429 | 4,656 | 52% |
| no. of slices containing only related logic | 2,429 | 2,429 | 100% |
| no. of slices containing unrelated logic | 0 | 2,429 | 0% |
| **Total no. of 4 input LUTs** | 2,747 | 9,312 | 29% |

Figure 4 shows the binary data which is being converted from gray scale image (Figure 1) using MATLAB. The binary image is then processed in EDK and the corresponding pixel data is viewed at hyper terminal (Figure 5). After processing in EDK the inverted image data is obtained. Next, the inverted image data is transferred to the computer from the SDRAM memory of the FPGA kit through the RS232 serial link. The image data is recovered into its binary form by the use of MATLAB based post processing. Figure 6 and 7 shows the pixel data of the binary image and the inverted image respectively. The final inverted image is obtained as shown in Figure 8. This experimental result verifies the inversion processing using the Microblaze processor core.

**Fig. 4.** Binary image

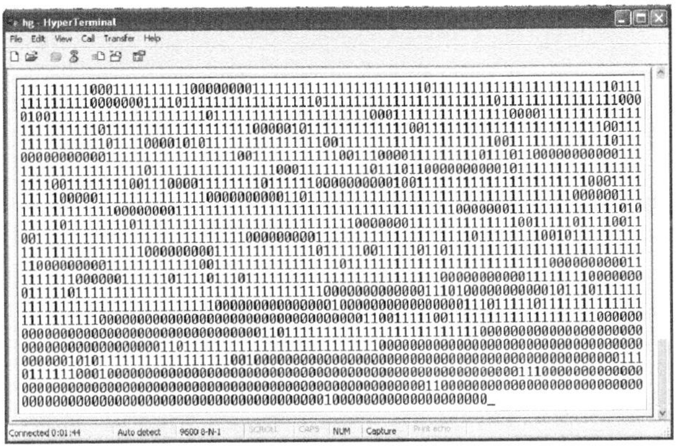


**Fig. 5.** Pixel data viewed at hyper terminal

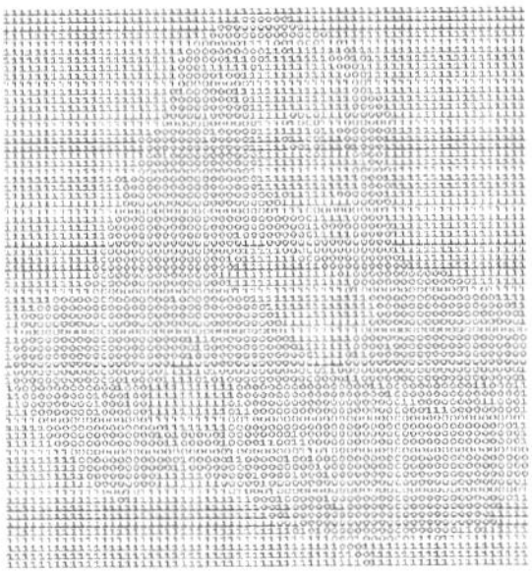

**Fig. 6.** Pixel data of binary image

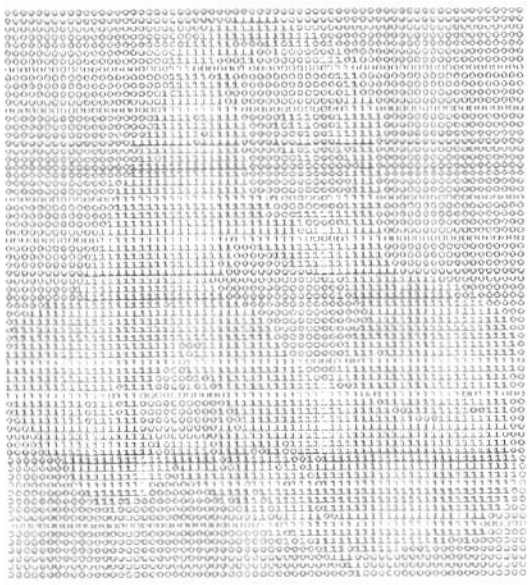

**Fig. 7.** Pixel data of inverted image

**Fig. 8.** Inverted image

## 5 Applications

This proposed development work is useful for the real time embedded applications where processing of image data is needed. In surveillance applications there is a need of processing data from visual sensors in real time, for effective decision making and hence embedded systems can give a better solution. Our work addresses this problem and we have tried to develop FPGA based embedded system for processing image data. We have adopted the processor core based design which will be effective for interfacing the system with different data communication protocols easily compared to the DSP based image processing existing in the related research works

## 6 Limitations and Suggestions for Improvement

We have implemented our design on Spartan3E, which has small RAM size so our test image sizes are small. In higher versions of FPGA i.e. Vertex II or Vertex IV we expect to implement the design for larger image sizes and also aim for video data processing.

## 7 Discussion

As EDK is a new development tool in embedded system design domain, inverting a binary image is really a challenging job. In Spartan 3E device, there is no graphics port to download an image directly so the image data is converted to raw binary data and is transferred as a file to the FPGA using the RS232 link. So it can be concluded that this is a useful design for devices which don't have a dedicated graphics port but still we can transfer the image data if any other communication link is available.

## References

1. Introduction to EDK 10.1 and the Digilent V2Pro Board" – online tutorial
2. EDK Concepts, Tools and techniques Version EDK 9.2i, http://www.xilinx.com
3. Rao, D.V., Patil, S., Babu, N.A., Muthukumar, V.: Implementation and Evaluation of Image Processing Algorithms on Reconfigurable Architecture using C-based Hardware Descriptive Languages. International Journal of Theoretical and Applied Computer Sciences 1(1), 9–34 (2006)
4. Edwards, S.A.: The challenges of hardware synthesis from C-like languages. International Journal of Theoretical and Applied Computer Science (June 2004); Proc. IWLS, Temecula, California
5. Image arithmetic-invert/Logical NOT –online tutorial
6. Displaying different image types, http://www.mathworks.com/help/toolbox/images/f10-30847.html
7. Grotker, T., Liao, S., Martin, G., Swan, S.: System Design with System. Kluwer, Dordrecht (2002)

# Parametric Analysis of Zone Routing Protocol

Rani Astya[1], Parma Nand[2], and S.C. Sharma[3]

[1]IILM-CET, Greater Noida India (91)
astya2005@gmail.com
[2] Wireless Computing Research Lab,
Indian Institute of Technology, Roorkee, India (91)
{astyadpt,scs60fpt}@iitr.ernet.in

**Abstract.** A Mobile Ad-hoc Network (MANET) consists of a set of autonomous, self-configuring, decentralized, power constrained mobile hosts that may communicate with one another from time to time without any base station support. Each host is equipped with a CSMA/CA (carrier sense multiple access with collision avoidance) transceiver. Routing is the process of finding a path from a source to some arbitrary destination on the network. In this paper Zone Routing Protocol (ZRP), which is hybrid routing protocol in MANET, is evaluated. Zone Routing Protocol (ZRP) is a most promising and widely used in scenario where nodes are placed in zone structure. The performance of this protocol is analyzed using performance metrics throughput, packet delivery ratio and average end-to-end delay. The total packets routed through Interzone Routing Protocol (IERP) are also analyzed using network simulator Qualnet 5.0.2.

**Keywords:** Adhoc networks; wireless networks; CBR, routing protocols; route discovery; MAC; IEEE 802.11.

## 1 Introduction

The adhoc network is a multihop infrastructure less network topology. Mobile adhoc networks (MANETs) are created dynamically. The dynamic nature of MANETs is more challenging the wired network topology. In such networks, each mobile node operates not only as a host but also as a router [1]. The nodes, in the network, forward packets to other mobile nodes. These nodes may not be within direct wireless transmission range of each other. Each node must cooperate dynamically to establish routing among them. Each node participates in an adhoc routing protocol that allows it to discover "multi-hop" paths through the network to any other node. Some applications of the ad hoc networking are: students using laptop to participate in an interactive lecture, business associates sharing information during a meeting, soldiers relaying information for situational awareness on the battlefield, and emergency disaster relief personnel coordinating efforts after a hurricane or earthquake etc.

In this paper hybrid routing protocol called Zone Routing Protocol (ZRP) is discussed briefly. Zone Routing Protocol (ZRP) is a most promising and widely accepted and well proved hybrid routing protocol in Mobile Ad-hoc Networks (MANETs) for its performance. The performance of this protocol and Interzone Routing Protocol (IERP) is analyzed using performance metrics throughput, packet delivery ratio and average end-to-end delay using Qualnet 5.0.2[2].

A. Mantri et al. (Eds.): HPAGC 2011, CCIS 169, pp. 247–251, 2011.

## 2 Routing Protocol: Classification in Brief

Routing is the process of finding a path from a source to some arbitrary destination on the network. The broadcasting [16,17,18] is inevitable and a common operation in ad-hoc network. It consists of diffusing a message from a source node to all the nodes in the network. Broadcast can be used to diffuse information to the whole network. It is also used for route discovery protocols in ad-hoc networks. The routing protocols are classified on the basis of the way the network information is obtained in these routing protocols.

*Proactive or Table-driven routing protocol:* For example

1. Destination sequenced Distance vector routing (DSDV)[5]
2. Source Tree Adaptive Routing (STAR) [8]

*Reactive or On-demand routing protocol:* For example

1. Ad-Hoc On-demand Distance Vector (AODV) [6]
2. Dynamic Source Routing (DSR) [3]

*Hybrid Protocols:* For example

1. Temporally ordered routing algorithm (TORA) [7]
2. Zone Routing Protocol (ZRP) [9]

These classes of routing protocols are reported but choosing best out of them is very difficult as one may be performing well in one type of scenario and the other may work in other type of scenario [13, 14, 15].

## 3 Zone Routing Protocol (ZRP)

Zone Routing Protocol (ZRP), a hybrid routing, is suitable for a wide variety of mobile ad-hoc networks, especially those with large network spans and diverse mobility patterns. Each node proactively maintains routes within a local region (referred to as the routing zone). Knowledge of the routing zone topology is leveraged by the ZRP to improve the efficiency of a globally reactive route query/reply mechanism. The proactive maintenance of routing zones also helps in improving the quality of discovered routes, by making them more robust to changes in network topology. The ZRP can be configured for a particular network by proper selection of a single parameter, the routing zone radius.

Choosing an appropriate routing zone is very important for this protocol. Large routing zones are preferred when demand for routes is high and/or the network consists of many slowly moving nodes. In the extreme case of a network with fixed topology, the ideal routing zone radius would be infinitely large. On the other hand, smaller routing zones are appropriate in situations where route demand is low and/or the network consists of a small number of nodes that move fast relative to one another. In the "worst case", a routing zone radius of one hop is best, and the ZRP defaults to a traditional reactive flooding protocol.

The neighbor discovery is done through the Neighbor Discovery Protocol (NDP). Periodically network information is broadcasted to all the neighbors. The periodic broadcasting of "hello" beacons is used for this purpose and successful reception of this indicates the connectivity with the neighbors.

It uses sequence number for messages to discover loopfree routes. ZRP uses two different route discovery protocols

1. Intrazone Routing Protocol (IARP)[10] and
2. Interzone Routing Protocol (IERP)[11].

The IARP is derived from globally proactive link state routing protocols that provide a complete view of network connectivity.

Interzone Routing Protocol (IERP) is very similar to classical route discovery protocols. An IERP route discovery is initiated when no route is locally available to the destination of an outgoing data packet. The source generates a route query packet, which is uniquely identified by a combination of the source node's address and request number. The query is then relayed to a subset of neighbors as determined by the bordercast algorithm. Upon receipt of a route query packet, a node checks if the destination lies in its zone or if a valid route to it is available in its route cache

# 4 Simulation Setup

The Qualnet 5.0.2 network simulator is used for the analysis. The IEEE 802.11[12] for wireless LANs is used as the MAC layer protocol with omni directional antenna. The channel bandwidth 2.4 GHz and data rate of 2 Mbps is used in it. The 54 nodes are placed uniformly over the region of 1500mx1500m. The random waypoint mobility model is used in the rectangular field. The multiple data traffic of Constant bit rate (CBR) application is applied over different source and destination nodes.

## 4.1 Performance Metrics

*Throughput:* Throughput is the average rate of successful data packets received at destination. It is usually measured in bits per second (bit/s or bps), and sometimes in data packets per second.

*End-to-End Delay:* A specific packet is transmitting from source to destination and calculates the difference between send times and received times. Delays due to route discovery, queuing, propagation and transfer time are included in the delay metric.

*Packet Deliver Ratio* (PDR): The (PDR) is defined as the ratio between the amount of packets sent by the source and received by the destination.

*Interzone Routing Protocol (IERP)*: Interzone Routing Protocol is initiated route discovery when no route is locally available to the destination of an outgoing data packet.

# 5 Result and Discussion

The Qualnet 5.0.2 network simulator is used to analyze the parametric performance of Zone Routing Protocol (ZRP).

*Throughput:* With the varying CBR and FTP data traffic the throughput is analyzed. The successful packet delivery in an adhoc network is observed with increasing MAC based traffic load. It is found that ZRP performs at nodes having number (ID) more than 31 as route discovery for them is easy being close to source. The performance is shown in figure 1.

*End-to-End Delay:* When a packet is transmitted from source to destination it takes time to reach. This time includes different delay as described in its definition above. In this analysis it is found as expected the delays are more for the nodes 4,18,25,31 as they are far from the source and ZRP takes time to send the packet successfully. The performance is shown in figure 2.

*Packet Deliver Ratio:* Performance is analyzed on this parameter and it is observed that ZRP performs better at the node 43 as it is very close to the source. The PDR is also better at node 18 and 36 but it is poorer for rest of the nodes as they are not easily traced by ZRP as shown in figure 3.

*Interzone Routing Protocol (IERP)*: The performance is analyzed on this metrics to observe the how many times and how many zones of nodes are being formed. It is observed that mostly source and destinations are found to be in different zones and packets are routed through interzone routes to improve network performance as shown in figure 4.

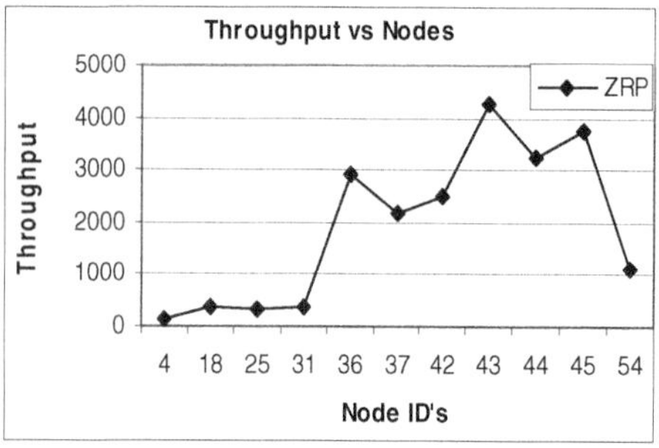


**Fig. 1.** Throughput vs Nodes

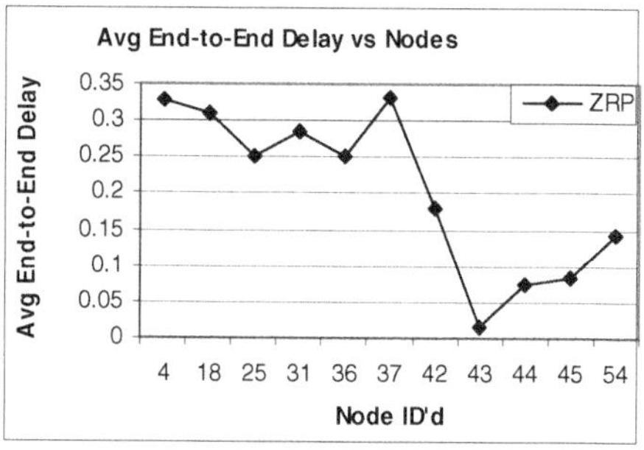


**Fig. 2.** Avg. End-to-End Delay vs Node

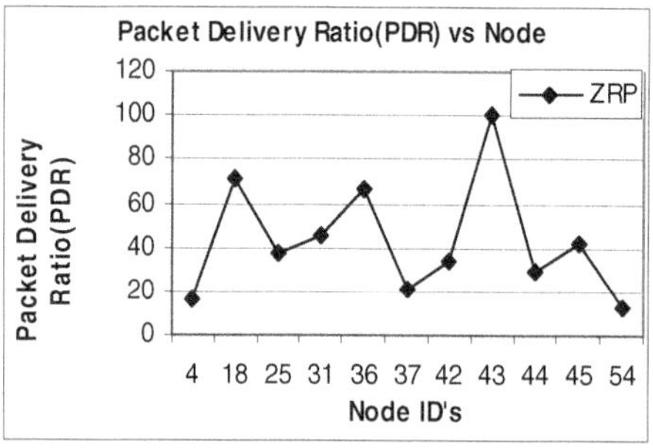


**Fig. 3.** Packet Delay Ratio vs Node

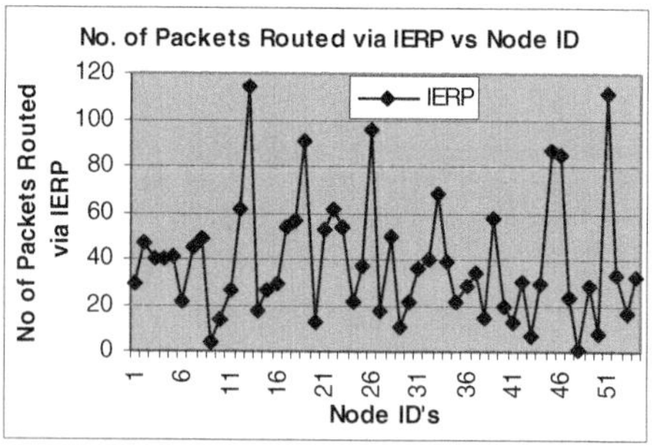


**Fig. 4.** No. Pkts Routed via IERP vs Node

## 6 Conclusion

The ZRP performs better for the scenarios having low mobility. It does not perform when the mobility is very high because of its Zone wise route discovery. The poor performance of ZRP is also because it doesn't have suitable mechanism to expire the expired or stale routes and therefore the average end-to-end delay is also very high. It is also observed that the throughput is average with the mobility of nodes because of the reasons mentioned above. One of the future research works is to develop an efficient and optimized routing protocol with heavy mobility and routing overheads.

## References

1. Toh, C.-K.: Ad hoc Mobile Wireless Networks: Protocols and Systems, pp. 55–77. Prentice Hall PTR, Englewood Cliffs (2002)
2. Qualnet Simulator, http://www.scalable-networks.com
3. Broch, J., Johnson, D., Maltz, D.: The dynamic source routing protocol for mobile adhoc networks for IPv4 IETF RFC 4728 (February 2007)
4. Perkins, C., Bhagwat, P.: Highly Dynamic Destination-Sequenced Distance-Vector Routing(DSDV) Routing. In: SIGCOMM 1994 Computer Communication Rewiev, vol. 24(4), pp. 234–244 (October 1994)
5. Perkins, C., Royer, E., Das, S.: Ad hoc on demand distance vector (AODV) routing. IETF RFC No. 3561 (July 2003)
6. Park, V., Corson, S.: Temporally-Ordered Routing Algorithm (TORA) Version 1 Functional Specification. draft-ietf-manet-tora-spec-04.txt (July 2001)
7. Garcia-Luna-Aceves, J.J., Spohn, M.: Source-Tree Adaptive Routing in Wireless Networks. In: Proceedings of the 7th Annual IEEE international conference on Network Protocols, Toronto, Canada, October 31- November 3 (1999), http://www.ieee-icnp.org/1999/papers/1999-29.pdf
8. Haas, Z.J., Pearlman, M.R., Samar, P.: The Zone Routing Protocol(ZRP) for Adhoc Networks. draft-ietf-manet-zone-zrp-04.txt (July 2002)
9. Haas, Z.J., Pearlman, M.R., Samar, P.: Intrazone Routing Protocol (IARP), IETF Internet Draft, draft-ietf-manet-iarp-02.txt (July 2002)
10. Haas, Z.J., Pearlman, M.R., Samar, P.: Interzone Routing Protocol (IERP), IETF Internet Draft, draft-ietf-manet-ierp-02.txt (July 2002)
11. IEEE, Wireless LAN Medium Access Control(MAC) and Physical layer PHY) Specifications, IEEE Std. 802.11 (1997)
12. Nand, P., Sharma, S.C., Astya, R.: Traffic Load based Performance Analysis of DSR, STAR & AODV Adhoc Routing Protocol. International Journal of Advanced Computer Science and Applications 1(4), 58–62
13. Pirzada, A.A., McDonald, C., Datta, A.: Performance Comparison of Trust-Based Reactive Routing Protocols. IEEE Transactions on Mobile Computing 5(6), 695–710 (2006)
14. Ni, S.Y., Tseng, Y.C., Chen, Y.S., Sheu, J.P.: The broadcast storm problem in a mobile ad hoc network. In: Proceedings of the 1999 Fifth Annual ACM/IEEE International Conference on Mobile Computing and Networking, pp. 151–162. IEEE Computer Society, New York (1999)

# Vision of 5G Communication

Mohd. Maroof Siddiqui

(MISECE)
Department of Electronics and Communication Engineering,
Integral University, Lucknow, India
maroofsiddiqui@yahoo.com

**Abstract.** This paper is designed to introduce the reader to the fundamental information for future or next Generation Technology .Forth Generation system implement in few countries & 4G systems is still a predominantly research and development and make an overview of next generation system i.e.5G communication. In this Paper presents an overview of the fifth generation mobile networks with emphasis on current and future trends in the areas of wireless networking, multimedia technology, network architecture, and network services. Related research in the development of future mobile systems has been highlighted. Beginning with prognoses of 5G Communication, architecture model of wireless communication & comparison with all the generation technology

**Keywords:** Prognoses, Architecture network, Comparison of all generation technology.

## 1 Introduction

**5G** technology will change the manner in which cellular are offered worldwide(5th generation mobile networks or 5th generation wireless systems) is a name used in some research papers and projects to denote the next major phase of mobile telecommunications standards beyond the upcoming 4G standards (expected to be finalized between approximately 2011 and 2013). Currently, 5G is not a term officially used for any particular specification or in any official document yet made public by telecommunication companies or standardization bodies such as 3GPP, WiMAX Forum or ITU-R. New standard releases beyond 4G are in progress by standardization bodies, but at this time are not considered as new mobile generations since implementation and rollout of systems compliant with 4G is still under way; the goals of a 5G-based telecommunications network would ideally answer the challenges that a 4G model would present once it has entered widespread use. The fifth generation communication system is envisioned as the real wireless network, capable of supporting Wireless World Wide Web (wwww) applications in 2010 to 2015 time frame. There are two views of 5G systems: evolutionary and revolutionary. In evolutionary view the 5G (or beyond 4G) systems will be capable of supporting wwww allowing a highly flexible network such as a Dynamic Ad-hoc Wireless Network (DAWN). In this view advanced technologies including intelligent antenna and flexible modulation are keys to optimize the ad-hoc wireless networks. In revolutionary view, 5G systems

A. Mantri et al. (Eds.): HPAGC 2011, CCIS 169, pp. 252–256, 2011.

should be an intelligent technology capable of interconnecting the entire world without limits. An example application could be a robot with built-in wireless communication with artificial intelligence. The 4G system is still a predominantly research and development

## 2 Prognoses

The implementation of standards under a 5G umbrella would likely be around the year of 2020. A new mobile generation has appeared every 10th year since the first 1G system (NMT) was introduced in 1981, including the 2G (GSM) system that started to roll out in 1992, and 3G (W-CDMA/FOMA), which appeared in 2001. The development of the 2G (GSM) and 3G (IMT-2000 and UMTS) standards took about 10 years from the official start of the R&D projects, and development of 4G systems started in 2001 or 2002.It is expected that in terms of data streams, a 5G standard would have peak download and upload speeds of more than the 1 Gbps to be offered by ITU-R's definition of 4G systems. The development of the bit rates offered by cellular systems is however hard to predict, since the historical bit rate development has shown very little resemblance with a simple exponential function of time (as opposed to for example Moore's law for computing capacity). The data rate increased by a factor 8 from 1G (NMT 1.2 kbps) to 2G (GSM 9.6 kbps). The peak bit rate increased by a factor 40 from 2G to 3G for mobile users (384 kbps), and by a factor of 200 from 2G to 3G for stationary users (2 Mbps). The peak bit rates are expected to increase by a factor 260 from 3G to 4G for mobile users (100 Mbps) and by a factor 500 from 3G to 4G for stationary users (1 Gbps).

## 3 Network Architecture

The basic block of wireless mobile communication system consists of a mobile phone connected to the wired world via a single hop wireless connection to a Base Station (BS), which is responsible for carrying the calls within its region called cell (figure 1).

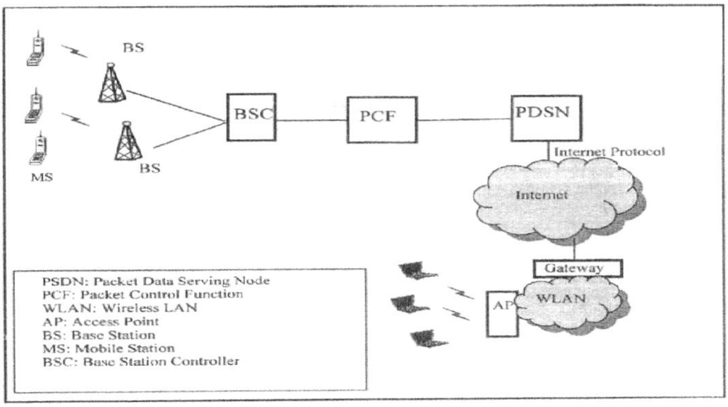


**Fig. 1.** Wireless Mobile System Network Architecture

In general, 4G architecture includes three basic areas of connectivity; PANs (such as Bluetooth), WANs (such as IEEE 802.11), and cellular connectivity. Under this umbrella, 4G will provide a wide range of mobile devices that support global roaming. Each device will be able to interact with Internet-based information that will be modified on the fly for the network being used by the device at that moment (Figure 2).

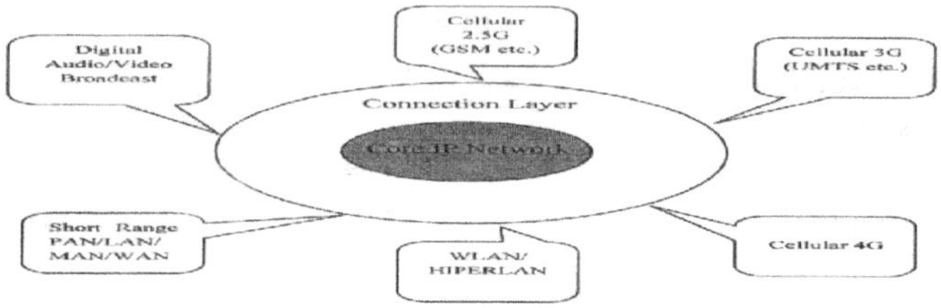


**Fig. 2.** Seamless Connected of network in 4G

In 5G mobile IP, each cell phone is expected to have a permanent "home" IP address, along with a "care-of" address that represents its actual location. When a computer somewhere on the Internet needs to communicate with the cell phone, it first sends a packet to the phone's home address. A directory server on the home network forwards this to the care-of address via a tunnel, as in regular mobile IP. However, the directory server also sends a message to the computer informing it of the correct care-of address, so future packets can be sent directly. This should enable TCP sessions and HTTP downloads to be maintained as users move between different types of networks. Because of the many addresses and the multiple layers of sub netting, IPv6 is needed for this type of mobility. For instance, 128 bits (4 times more than current 32 bit IPv4 address) may be divided into four parts (I thru IV) for supporting different functions. The first 32-bit part (I) may be defined as the home address of a device while the second part (II) may be declared as the care-of address allowing communication between cell phones and personal computers. So once the communication path between cell and PC is established, care-of address will be used instead of home address thus using the second part of IPv6 address. The third part (III) of IPv6 address may be used for tunneling to establish a connection between wire line and wireless network. In this case an agent (a directory server) will use the mobile IP address to establish a channel to cell phones. The fourth and last part (IV) of IPv6 address may be used for local address for VPN sharing. Figure 3 illustrates the concept.

| I | II | III | IV |
|---|---|---|---|
| | | | |

**Fig. 3.** IPv6 address divided into four IPv4 addresses for

## 4 Comparison of 5G Technology with Other Technologies

The goal of 4G and 5G is to replace the current proliferation of core mobile networks with a single worldwide core network standard, based on IPv6 for control, video, packet data, and voice. This will provide uniform video, voice, and data services to the mobile host, based entirely on IPv6. The objective is to offer seamless multimedia services to users accessing an all IP-based infrastructure through heterogeneous access technologies. IPv6 is assumed to act as an adhesive for providing global connectivity and mobility among networks.

| Technology/ Features | 1G | 2G/2.5G | 3G | 4G | 5G |
|---|---|---|---|---|---|
| Start/ Deployment | 1970/ 1984 | 1980/ 1999 | 1990/ 2002 | 2000/ 2010 | 2010/ 2015-2020 |
| Data Bandwidth | 2 kbps | 14.4-64 kbps | 2 Mbps | 200 Mbps to 1 Gbps | 1 Gbps and higher |
| Standards | AMPS | 2G: TDMA, GSM 2.5G: GPRS, EDGE, | WCDMA, CDMA-2000 | Single unified standard | Single unified standard |
| WEB Standard | ---- | www | www (IPv4) | www (IPv4) | wwww (IPv6) |
| Technology | Analog cellular technology | Digital cellular technology | Broad bandwidth CDMA, IP technology | Unified IP and seamless combination of broadband, LAN/WAN/ PAN and WLAN | Unified IP and seamless combination of broadband, LAN/WAN/ PAN/WLAN and wwww |
| Service | Mobile telephony (voice) | 2G: Digital voice, short messaging 2.5G: Higher capacity packetized data | Integrated high quality audio, video and data | Dynamic information access, wearable devices | Dynamic information access, wearable devices with AI capabilities |
| Multiplexing | FDMA | TDMA, CDMA | CDMA | CDMA | CDMA |
| Switching | Circuit | 2G: Circuit 2.5G: Circuit for access network & air interface; Packet for core network | Packet except circuit for air interface | All packet | All packet |
| Core Network | PSTN | PSTN | Packet network | Internet | Internet |
| Handoff | Horizontal | Horizontal | Horizontal | Horizontal and Vertical | Horizontal and Vertical |

## 5 5G Research

Key concepts suggested in research papers discussing 5G and beyond 4Gwireless communications are:

- Real wireless world with no more limitation with access and zone issues.
- Wearable devices with AI capabilities.
- Internet protocol version 6 (IPv6), where a visiting care-of mobile IP address is assigned according to location and connected network.
- One unified global standard.
- Pervasive networks providing ubiquitous computing: The user can simultaneously be connected to several wireless access technologies and seamlessly move between them These access technologies can be a 2.5G, 3G, 4G, or 5G mobile networks, Wi-Fi, WPAN, or any other future access technology. In 5G, the concept may be further developed into multiple concurrent data transfer paths.
- Cognitive radio technology, also known as smart-radio: allowing different radio technologies to share the same spectrum efficiently by adaptively finding unused spectrum and adapting the transmission scheme to the requirements of the technologies currently sharing the spectrum.
- This dynamic radio resource management is achieved in a distributed fashion, and relies on software defined radio. See also the IEEE 802.22 standard for Wireless Regional Area Networks.
- High altitude stratospheric platform station (HAPS) systems.
- Supporting 1Gbps downloading speed

## 6 Conclusion

This paper has presented a unique vision of the 5G Communication in next generation technology. It aims to promote technology innovate to achieve a reliable & efficient outcomes from next generation technology. With world wide wireless web (wwww) platform this latest technology will enable increased flexibility in control and reconfigurable dynamic. 5G system should be an intelligent technology capable of interconnecting the entire world without limits. Since this initial work cannot address everything within the proposed framework and vision, more research and development efforts are needed to fully implement the proposed framework through a joint effort of various entities.

## References

1. Lee, W.C.Y.: Mobile cellular telecommunications Analog & Digital systems. Mc Graw Hill, New York
2. Feher/ Wireless Digital communications: Modulation & spread spectrum Applications/PHI
3. Forouzan, B.A.: Data Communication & Networking, 4th edn. TMH
4. Encyclopedia of multimedia technology and networking, Global snippet (2009)
5. Encyclopedia of 5G Communication

# Secure Satellite Images Transmission Scheme Based on Chaos and Discrete Wavelet Transform

Musheer Ahmad[1] and Omar Farooq[2]

[1] Department of Computer Engineering,
Faculty of Engineering and Technology,
Jamia Millia Islamia, New Delhi 110025, India
[2] Department of Electronics Engineering,
ZH College of Engineering and Technology,
AMU, Aligarh 202002, India

**Abstract.** Many applications based on satellite communication like national defence and security rely on the satellite images as an important source of information. It is therefore, mandatory to secure satellite imagery while transmitting them over communication channels to protect from unauthorized access and usage. In this paper, chaotic logistic map based satellite image encryption scheme is proposed to meet the requirement of secure satellite-based communication. The algorithm is based on the concept of permuting the pixels of satellite image, then improving the pixels gray value distribution from cryptographic viewpoint. The permutation of image pixels is carried out in discrete wavelet domain and the relationship between the encrypted and the original satellite image is confused using chaotic-state modulation technique, thereby significantly increasing the resistance to statistical attacks. Experimental results demonstrate that the scheme has the advantage of high sensitivity to secret key and large key space. Moreover, the encrypted satellite images have uniform gray level distributions, high entropies and low correlation coefficients. Hence, the theoretical and experimental analyses confirm that the proposed scheme has high security and can be applied for practical satellite image protection.

**Keywords:** Satellite communication, satellite images, chaotic logistic map, security, image encryption, discrete wavelet transform.

## 1 Introduction

Recent advancement of satellite and network communication technologies have made the satellite communication crucial in application areas of defence and security, multimedia broadcasting, tele-medicine, tele-education and training, weather forecasting, disaster management and support, managing and monitoring the land and ocean resources [1]. Satellite images are the most preferred source for satellite-based communications [2]. For example, the national defence and military extensively relies on the satellite imagery as an important source of information. These application areas require reliable and secure mechanism to store and transmit satellite images. It is unsafe to directly transmit sensitive satellite images over various communication channels.

A. Mantri et al. (Eds.): HPAGC 2011, CCIS 169, pp. 257–264, 2011.

To fulfill the requirement of privacy and secrecy of satellite images, cryptographic techniques need to be applied. Therefore, a lot of encryption algorithms based on different techniques have been proposed to protect images [3-14]. Among them, chaos-based cryptographic techniques have suggested some new and efficient ways to develop secure image encryption schemes to meet the demand for reliable and secure image transmission. Chaotic signals have cryptographically desirable properties such as high sensitivity to initial conditions/parameters, long periods, high randomness and mixing. These properties make chaos-based image cryptosystems excellent and robust against statistical attacks.

In this paper, the cryptographic characteristics of chaos system are utilized to develop a reliable and secure encryption scheme to protect satellite images. The proposed encryption scheme performs two level of encryption. At first level, the original image is permuted in discrete wavelet domain using multi-resolution analysis to have visual effect of disorder so that image becomes meaningless or noise-like for a casual observer. The permutation matrix is generated using sequence generated from two chaotic logistic maps with different secret initial seeds and parameters. At second level, the gray value distribution of the permuted image is enforced by chaotically modulating the states of the logistic map. The modulated states are then used to extract keys to encrypt the permuted image by simple XOR operation. Rest of the paper is organized as follows: The multi-resolution wavelet transform and Chaos are briefly discussed in Section 2. Section 3 discusses the proposed satellite image encryption scheme in detail. The experimental results are discussed in section 4 followed by conclusion.

## 2 The DWT and Chaos

The discrete wavelet transform (DWT) is becoming popular in image/video applications due to its multi-resolution representation feature. The idea of multi-resolution DWT for an image is to decompose it at level $R$=1 into four sub-band images with different space and frequency: vertical and horizontal direction of low frequency sub-band $LL_1$, $LH_1$, $HL_1$ and $HH_1$ sub-bands. In the low frequency sub-band signal $LL_1$, the main energy is concentrated. The remaining three sub-bands signals are known as the details of signal. The low frequency sub-band $LL_1$ can be decomposed again at level $R$=2 into four sub-bands of $LL_2$, $LH_2$, $HL_2$ and $HH_2$. Similarly, the low frequency sub-band can be further decomposed on and on at multiple resolutions as desired for the application.

The 1D chaotic logistic map is one of the simplest discrete nonlinear dynamical systems that exhibit chaotic behavior and is defined by the following equation:

$$x(n+1) = \lambda x(n)(1-x(n)) \ . \tag{1}$$

Where $x(0)$ is initial condition, $\lambda$ is the system parameter, $n \geq 0$ is the number of iterations and $0 < x(n) < 1$ for all $n$. The research shows that the logistic map has good chaotic properties if $3.9 < \lambda < 4$.

## 3 Proposed Satellite Image Encryption Scheme

Shannon has suggested two basic properties for the design of practical cryptosystems: confusion and diffusion. To achieve desired confusion and diffusion, the proposed satellite image encryption scheme makes use of permutation and diffusion architecture. In the permutation phase, the pixels of original image are permuted to transform the original image into a meaningless or distorted image. In the diffusion phase, the pixels gray values are modified sequentially to further improve the security of the encryption scheme. A lot of permutation schemes are suggested to permute the image in spatial domain. The spatial domain permutation has disadvantage that it keeps the statistical characteristics like gray value distribution, entropy, etc of image same after permutation phase. Therefore, it is not safe to perform image permutation in spatial domain as the attacker can utilize these characteristics of permuted image to recover the original image through regression analysis. Image pixels permutation in transform domain has advantage that the above mentioned characteristics of image are altered and make the work of attacker more difficult. In the proposed scheme, the permutation of image is performed in DWT domain using multi-resolution analysis. The image is decomposed to higher levels and LL sub-band of that level is permuted using two chaotic logistic maps given in Eq. 2 and Eq. 3. To get cumulative effect, the image obtained is further decomposed into lower level and permutation of LL sub-band of this level is carried out, this process is repeated till the permutation of LL sub-band of first level decomposition is achieved.

$$x_1(n+1) = \lambda_1 x_1(n)(1 - x_1(n)) \tag{2}$$

$$x_2(n+1) = \lambda_2 x_2(n)(1 - x_2(n)) \tag{3}$$

$$x_3(n+1) = \lambda_3 x_3(n)(1 - x_3(n)) \tag{4}$$

$$x_4(n+1) = \lambda_4 x_3(n+1)(1 - x_3(n+1)) \tag{5}$$

In diffusion phase, similar logistic maps given in Eq. 4 and Eq. 5 with different initial condition and parameters are utilized. The gray values of permuted image are modified to enforce the secrecy of the proposed encryption scheme by chaotically modulating the states of logistic map given in Eq. 5 through the states of the map given in Eq. 4. The input state of Eq. 5 is the output state of Eq. 4 and the output state of Eq. 5 are used to extract the keys for encrypting gray values of image pixels and at the same time; it also acts as input state to Eq. 4, i.e. for beginning iteration $i+1$, where $i \geq 1$, we have $x_3(i) = x_4(i)$.

The proposed encryption scheme is symmetric; therefore, the decryption can be performed by executing the encryption procedure in reverse order. The block diagram of the proposed satellite image encryption scheme is depicted in Fig. 1. The complete proposed satellite image encryption scheme is illustrated below:

***Step 1.*** Let $O(i, j)$ be the original satellite image. Now, Set $I(i, j) = O(i, j)$.
***Step 2.*** Evaluate DWT decomposition of $I(i, j)$ at level $R$. Let $N_R$x$M_R$ be the size of LL sub-band coefficient matrix $C_R(i, j)$ of $I(i, j)$at level $R$.
***Step 3.*** Take $x_1(0)$ & $x_2(0)$ as initial conditions and evaluate new initial conditions for level $R$.

$y_1=x_1(0)*R*3;\ x'_1(0)= y_1 - floor(y_1)$ ;
$y_2=x_2(0)*R*7;\ x'_2(0)= y_2 - floor(y_2)$;

***Step 4.*** Iterate maps of Eq. 2 and 3 with $x'_1(0)$ & $x'_2(0)$ as initial conditions for $N_R*M_R$ & $N_R*M_R$ times, respectively. We get sequences $x_1(n)$ & $x_2(m)$, $1 \le n \le N_R*M_R$, $1 \le m \le N_R*M_R$. Determine $\delta(x)=max(x(n))-min(x(n))$
***Step 5.*** Generate permutation matrix as: Take array $x=[]$ and $k=1$.

*for* $k_1$=1 to $N_R*M_R$
$t=floor[[\{x_1(k_1) - min(x_1(n))\}*(N_R - 1)]/\delta(x_1)]$
$t=t+1$
*if*($t$ is not a member of array $x$)
{ Set $x(k)=t$ and $k=k+1$}
*if*($length(x)=N_R$) then *break*
*end*

Do the same to determine array $y$ of length $M_R$.
***Step 6.*** Move coefficient of LL band matrix $C_R(i, j)$ from position $(i, j)$ to $(x(i), y(j))$, where $1 \le i, x(i) \le N_R$, $1 \le j, y(j) \le M_R$.
***Step 7.*** Reconstruct $I(i, j)$ with shuffled LL using IDWT from level $R$. Set $R=R-1$.
***Step 8.*** Repeat Steps 2 to 7 while $R \ge 1$.

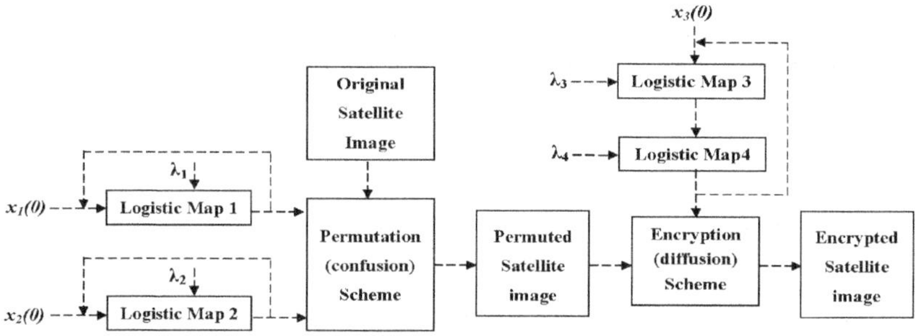


**Fig. 1.** Proposed satellite image encryption scheme

***Step 9.*** Set $P(i, j)= I(i, j)$ as permuted satellite image.
***Step 10.*** Arrange the pixels of permuted satellite image $P(i, j)$ from left to right and then top to bottom. Take $x_3(0)$ as initial condition.
***Step 11.*** Iterate maps given in Eq. 4 and 5 and extract the key $\Phi(i)$ from the current state of Eq. 5 as:

$$\Phi(i)=[x_4(i)*10^{14}]mod(256)$$

*Note*: $x_3(i)$ $(=x_4(i))$ will be the input state of map given in Eq. 4 for next iteration.

***Step 12.*** Calculate the pixels gray value $E(i)$ of encrypted image as:

$$E(i)=P(i) \oplus \Phi(i)$$

***Step 13.*** Repeat Steps 11 to 12 to encrypt all pixels of permuted satellite image. Arrange the encrypted pixels $E(i)$ into 2D encrypted satellite image $E(i, j)$.

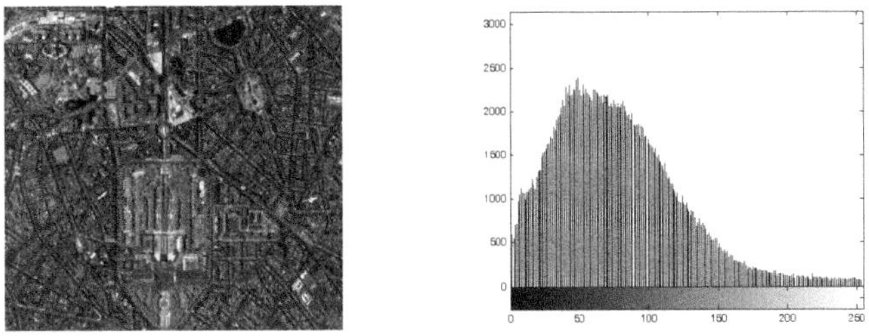


(a) (b)

**Fig. 2.** Original satellite image *Brussels* and its pixels gray value distribution

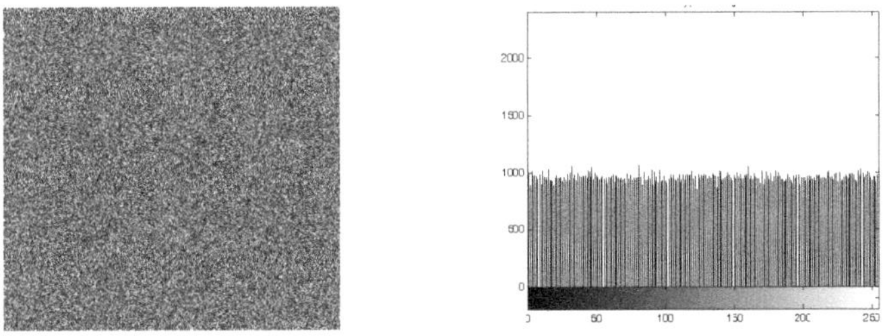


(a) (b)

**Fig. 3.** Encrypted satellite image of *Brussels* and its pixels gray value distribution

## 4 Experimental Results

The proposed encryption scheme is experimented and tested with a number of satellite images. However, only the *Brussels* satellite image of size 512x480 is shown in Fig. 2. The initial conditions and parameters taken for experimentation and simulation are: $x_1(0)$ = 0.67234, $x_2(0)$ = 0.31578, $x_3(0)$ = 0.57691, $\lambda_1$ = 3.995, $\lambda_2$ = 3.985, $\lambda_3$ = 3.9975, $\lambda_4$ = 3.9995. The mother wavelet filter used is *Daubechies* 'db4' and resolution level $R$ = 4. The *Brussels* satellite image is encrypted using proposed encryption scheme; encrypted image obtained is shown in Fig. 3. It can be seen that the encrypted image is very much indistinguishable and noise-like. Moreover, the statistical characteristics of original Brussels satellite image are changed absolutely in encrypted image in such a manner that encrypted images have good balance property. Because the gray value distributions in encrypted images are fairly uniform and much different from the gray value distributions of original satellite images. It shows that the encrypted image doesn't provide any information regarding the distribution of gray values to the attacker.

### 4.1 Key Space

Key space is the total number of different keys that can be used in the cryptographic system. The secret key of proposed encryption scheme is: $\boldsymbol{K}(x_1(0), x_2(0), x_3(0), \lambda_1, \lambda_2, \lambda_3, \lambda_4)$. The initial conditions for $x_1(0), x_2(0), x_3(0), \lambda_1, \lambda_2, \lambda_3, \lambda_4$ constitute secret key of encryption and decryption. In the proposed cryptosystem, all the variables are declared as type double which has a 15-digit precision, therefore the key space of proposed scheme is $[(4\text{-}3.9)(10^{14})]^4(10^{14})^3$ i.e. $10^{94} \approx 2^{312}$. Thus, the key space of the scheme is extensively large enough to resist the exhaustive attack.

### 4.2 Pixels Correlation

Confusion and diffusion properties when incorporated in image encryption schemes reduce the correlations among adjacent pixels in encrypted images. The correlation between two adjacent pixels of encrypted images is calculated as:

$$\rho = \frac{N\sum_{i=1}^{N}(x_i \times y_i) - \sum_{i=1}^{N} x_i \times \sum_{i=1}^{N} y_i}{\sqrt{(N\sum_{i=1}^{N} x_i^2 - (\sum_{i=1}^{N} x_i)^2) \times (N\sum_{i=1}^{N} y_i^2 - (\sum_{i=1}^{N} y_i)^2)}} \tag{6}$$

where $x$ and $y$ are gray values of two adjacent pixels in an image. We randomly select 1000 pairs of vertically, horizontally and diagonally adjacent pixels and calculate the correlation coefficients in three directions separately. Correlation coefficients of adjacent pixels in original satellite images in three directions are listed in Table 1. The values of correlation coefficients obtained for encrypted images are listed in Table 2. These correlation coefficients show that the two adjacent pixels in original images are highly correlated to each other as the correlation coefficients are almost 1, whereas in case of encrypted images, coefficients are close to 0, this means that the proposed scheme highly de-correlate the adjacent pixels in encrypted images and satisfy the confusion and diffusion properties.

**Table 1.** Correlation of adjacent pixels in Original satellite images

| Satellite Image | Vertical | Horizontal | Diagonal |
|---|---|---|---|
| *Brussels* | 0.74143 | 0.67129 | 0.62339 |
| *Andes* | 0.97260 | 0.97131 | 0.95409 |
| *Pearl Harbor* | 0.82669 | 0.80461 | 0.75545 |

**Table 2.** Correlation of adjacent pixels in Encrypted satellite images

| Satellite Image | Vertical | Horizontal | Diagonal |
|---|---|---|---|
| *Brussels* | 0.00054 | 0.00145 | 0.00091 |
| *Andes* | 0.00005 | 0.00088 | 0.00042 |
| *Pearl Harbor* | 0.00034 | 0.00017 | 0.00259 |

### 4.3 Information Entropy

The entropy $H$ of a symbol source $S$ can be calculated by following equation.

$$H(S) = -\sum_{i=0}^{255} p(s_i)\log_2(p(s_i)) \tag{7}$$

where $p(s_i)$ represents the probability of symbol $s_i$ and the entropy is expressed in bits. If the source $S$ emits $2^8$ symbols with equal probability, i.e. $S = \{s_0, s_1, \ldots, s_{255}\}$ then the entropy is $H(S) = 8$, which corresponds to a true random source and represents ideal value of entropy for message source $S$. If the entropy of an encrypted image is significantly less than the ideal value 8, then, there would be a possibility of predictability which threatens the image security. The values of information entropy obtained for original and encrypted satellite images are given in Table 3. The values obtained for encrypted satellite images imply that the information leakage in the proposed encryption scheme is negligible and is secure against the entropy attack.

**Table 3.** Information entropies of Original and Encrypted satellite images

| Satellite Image | Original Image | Encrypted Image |
|---|---|---|
| *Brussels* | 7.43289 | 7.99913 |
| *Andes* | 7.34712 | 7.99934 |
| *Pearl Harbor* | 7.37640 | 7.99928 |

### 4.4 Sensitivity to Secret Key

A good cryptosystem should be sensitive to secret key i.e. a small change in secret key during decryption process should results into a completely different decrypted image. In proposed scheme, a tiny change in key even of the order of ($\Delta$=) $10^{-14}$, results into noise-like decrypted image. The *Brussels* encrypted image of Fig. 3 is decrypted using correct key and then using $x_1(0)+\Delta$, $x_3(0)+\Delta$, $\lambda_4+\Delta$ separately and shown in Fig. 4, the images decrypted with wrong keys are completely unrecognizable and appears like noise. Similar sensitivity is noticed for the case of wrong $x_2(0)$, $\lambda_1$, and $\lambda_3$. Hence, it can be said that the proposed satellite image encryption scheme has high sensitivity to secret key.

(a) (b) (c)

**Fig. 4.** Brussells satellite image decrypted with (a) correct key (b) x1(0)+Δand (c) λ4+Δ

## 5 Conclusion

In this paper, a chaos-based encryption scheme is proposed for reliable and secure transmission and storage of sensitive satellite images. To satisfy the property of Shannon's confusion and diffusion, the scheme is based on permutation-diffusion architecture. The pixels permutation is performed using multi-resolution wavelet transform by permuting the coefficients of low frequency LL sub-bands at multiple resolution levels. The statistical characteristics of permuted image are further enforced by extracting encryption keys from chaotically modulated states of chaotic logistic map and then performing simple XOR operation with the pixels of permuted image. All the experimental and theoretical analyses prove that the proposed encryption scheme is robust, reliable, has high performance and can be applied practically for secure transmission and storage of satellite imagery to have secure satellite-based communication.

## References

1. Nair, G.M.: Role of Communications Satellites in National Development. IETE Technical Review 25(1), 3–8 (2008)
2. Haller, L. L., Sakazaki, M. S.: Commercial Space and United States National Security, `http://www.fas.org/spp/eprint/article06.html` (accessed on December 4, 2010)
3. Bourbakis, N., Alexopoulos, C.: Picture data encryption using SCAN patterns. Pattern Recognition 25(6), 567–581 (1992)
4. Chang, C.C., Hwang, M.S., Chen, T.S.: A new encryption algorithm for image cryptosystems. Journal of Systems Software 58, 83–91 (2001)
5. Fridrich, J.: Symmetric ciphers based on two-dimensional chaotic maps. International Journal of Bifurcation and Chaos 8(6), 1259–1284 (1998)
6. Cheng, H., Li, X.B.: Partial encryption of compressed images and videos. IEEE Transaction on Signal Processing 48(8), 2439–2451 (2000)
7. Chen, G.Y., Mao, Y.B., Chui, C.K.: A Symmetric Image Encryption Scheme based on 3D Chaotic Cat maps. Chaos Solitons and Fractals 21(3), 749–761 (2004)
8. Guan, Z., Huang, F., Guan, W.: Chaos-based image encryption algorithm. Physics Letters A 346(1-3), 153–157 (2005)
9. Mao, Y., Lian, S., Chen, G.: A Novel Fast Image Encryption Scheme based on 3D Chaotic Baker maps. International Journal of Bifurcation and Chaos 14(10), 3616–3624 (2004)
10. Lian, S., Sun, J., Wang, Z.: A Block Cipher based on a Suitable use of Chaotic Standard map. Chaos, Solitons and Fractals 26(1), 117–129 (2005)
11. Zhang, L., Liao, X., Wang, X.: An Image Encryption Approach based on Chaotic maps. Chaos, Solitons and Fractals 24(3), 759–765 (2005)
12. Pareek, N.K., Patidar, V., Sud, K.K.: Image Encryption using Chaotic Logistic map. Image and Vision Computing 24(9), 926–934 (2006)
13. Tong, X., Cui, M.: Image encryption scheme based on 3D baker with dynamical compound chaotic sequence cipher generator. Signal Processing 89(4), 480–491 (2008)
14. Tang, Y., Wang, Z., Fang, J.: Image encryption using chaotic coupled map lattices with time varying delays. Communication in Nonlinear Science and Numerical Simulation 15(9), 2456–2468 (2009)

# Computational Analysis of Availability of Process Industry for High Performance

Shakuntla, A.K. Lal, and S.S. Bhatia

School of Mathematics and Computer Application,
T.I.E.T. Patiala,
Punjab, India

**Abstract.** This paper aims to proposed methodology to evaluate the availability of the rice plant in a realistic environment. A complex mechanical system consisting five subsystems is considered .The failure of the parallel unit of any subsystems, the system remains operative for a short period of time. The paper discussed the three states of system: good state reduced state and failed state. The problem is formulated using supplementary variable technique .Lagrange's Method is used to solve the governing equations. Availability of the system is evaluated followed by behavior analysis of subsystems.

**Keywords:** Availability, Supplementary variable technique, Lagrange's Method, Mat lab.

## 1 Introduction

In the present era of rapid technology evolution, modern technology and integrated automation of manufacturing has developed a tendency to design and manufacture equipments of greater capital cost, sophistication, complexity and capacity. High productivity and high payback ratios have become essential for the survival of these systems. Availability analysis of complex system has assumed important issue in recent years.Availabilty analysis of a system can benefit the industry in terms of higher productivity and low maintenance cost. It is possible to improve the availability of the plant with proper maintenance planning and monitoring. Singh(1976) developed a mathematical model on mission reliability for a complex system with priority repair policy having imperfect switch. Earlier Kodama et al (1973) considered a simple mission reliability model. In the last four decades, several research paper and books have been published that discuss various techniques of reliability modeling in different areas of application. A number of researchers have attempted to analyze the performance of real industrial system using reliability engineering principles. Cherry et al (1978) performed reliability analysis of a chemical industry by calculating long run availability of the plant. Dhillon and Rayapati (1988) emphasized the need of reliability engineering principles in the chemical industry. Singh (1989) discussed application of reliability technology in a bio gas plant. Gupta et al (2005) evaluated reliability

A. Mantri et al. (Eds.): HPAGC 2011, CCIS 169, pp. 265–274, 2011.

parameters of a butter manufacturing system in a dairy plant considering exponential distributed failure rates of various components. Priel (1974) developed methodology for failure analysis in process plants. Most of the work carried out is related to the numerical analysis of the steady state of the various system .In this paper we made an attempt to analyze availability by assuming failure and repair rates are variable.

## 2.1 System Description

### 2.1.1 Sub-system E (Elevator)

Elevator (or lift) is vertical transport equipment that efficiently moves people or goods between floors (levels, decks of a building, vessel or other structure. Elevators are generally powered by electric motors that either drive traction cables and counterweight systems like a hoist, or pump hydraulic fluid to raise a cylindrical piston like a jack. These are five units $(E_i, i1,3,5,7,9)$.Failure of any one unit causes complete failure of the system.

### 2.1.2 Sub-system C (Cleaning)

Removing all impurities and unfilled grains from paddy. These are two identical units $(C_i, i = 1,2\,)$working in parallel.This unit can work with one unit in reduced capacity.

### 2.1.3 Sub-system H (Husking)

Removing husk from paddy. There is one unit subjected to major failure.

### 2.1.4 Sub-system S (Separation)

Separating the unhusked paddy from brown rice there is one unit subjected to major failure.

### 2.1.5 Sub-system W (Whitening)

Removing all or part of the bran layer and germ from brown rice. These are two identical units $(W_i, i = 1,2)$ working in parallel. This unit can work with one unit in reduced capacity.

### 2.1.6 Sub-system L (Polishing)

Improving the appearance of milled rice by removing the remaining bran particles and by polishing the exterior of the milled kernel.

These are two units $(L_i, i = 10,11)$.Failure of any one unit causes complete failure of the system.

## 2.2 Notations

−: The Sub-system/unit is running without any failure.: g: Unit is good state but not operative.

m : Unit is under preventive maintenance.: r: unit is under repair or repair continued.

r: unit is under repair or repair continued.

$B^z$: $(B = H, S)$ indicate the working state of husking separation machine w.r.t z,(z=-, g, m, r).

$E_{i\ j}^{x\ y}$ : indicates the working state of the sub-system $E_i$ and $E_j$ w.r.t , $y, (x, y = -, g, r)$ : : $i = 1,3,5,7,9$:: $j = i + 2, i + 4, i + 6\ \ if\ i = 1;\ \ j = i - 2, i + 2, i + 4$ :$if\ i = 3; j = i - 2, i - 4, i + 2, i + 4\ if\ i = 5; j = i - 2, i - 4, i - 6, i + 2\ if\ i = 7;$

$P_{k\ l}^{x\ y}$ : indicates the working state of the sub-system $P_k$ and $P_l$ w.r.t $x, y, (x, y = -, g, r)$ : : $k = 10,11$ :: $l = 11\ if\ k = 10; l = 10\ if\ k = 11.$

$L_{u\ 3-u}^{t\ n}$ : indicates the working states of the subsystem L the order pair $\binom{t}{u}$ and $\binom{t}{3-u}$ represents the functioning of the sub-system L w. r. t to "t" and "n"$(u = 1,2; t, = \ -, r)$.

$W_{v\ 3-v}^{t\ n}$ : indicates the working states of the subsystem W the order pair $\binom{t}{v}$ and $\binom{t}{3-v}$ represents the functioning of the sub-system W w. r. t to "t" and "n"$(v = 1,2; t, = -, r)$.

$\lambda_i(y)$: refers failure rate of the sub-system $E_1, E_3, E_5, E_7, E_9, L_{10}, L_{11}, H, S$and $W$ from normal to failed state $(i = 1,3,5,7,9,10,11,12,13)$.

$\lambda_k(y)$: refers failure rate of the sub-system $C$ and $W$ from normal to reduced state $(k = 2,4)$.

$\lambda_l$: refers constant transition state of the subsystem $H$ and $S$ which transits the system into the 6 and 7respectively on reaching to these state preventing maintenance of H and S states start immediately, $(l = 6,8)$.

$\mu_i(x)$: Time dependent repair rates of the subsystem $E_1, E_2, E_3, E_5, E_7, E_9, P_{10}, P_{11}, H, S$ and $W$ to return it from failed to normal state and elapsed repair time x,$(i = 1,3,5,7,9,10,11,12,13)$

$\mu_k(x)$: Time dependent repair rates of the subsystem $C$and $W$ to return it from reduced to normal state and elapsed repair time "x"$(k = 2,4)$.

$P_k(x, y, t)$: Probability that the system is in state k at time t and has an elapsed failure time 'y' and elapsed repair time 'x'$(k = 1, \ldots 13)$.

## 2.3 Assumption

The assumptions, on which the present analysis is based on, are as follows:(i) Repair and failure rates are independent of each other.(ii) Failure and Repair rates of the subsystems are taken as variable.(iii) Performance wise, a repaired unit is as good as new one for a specified duration.(iv) Sufficient repair facilities are provided.(v) System can work at reduced capacity also.

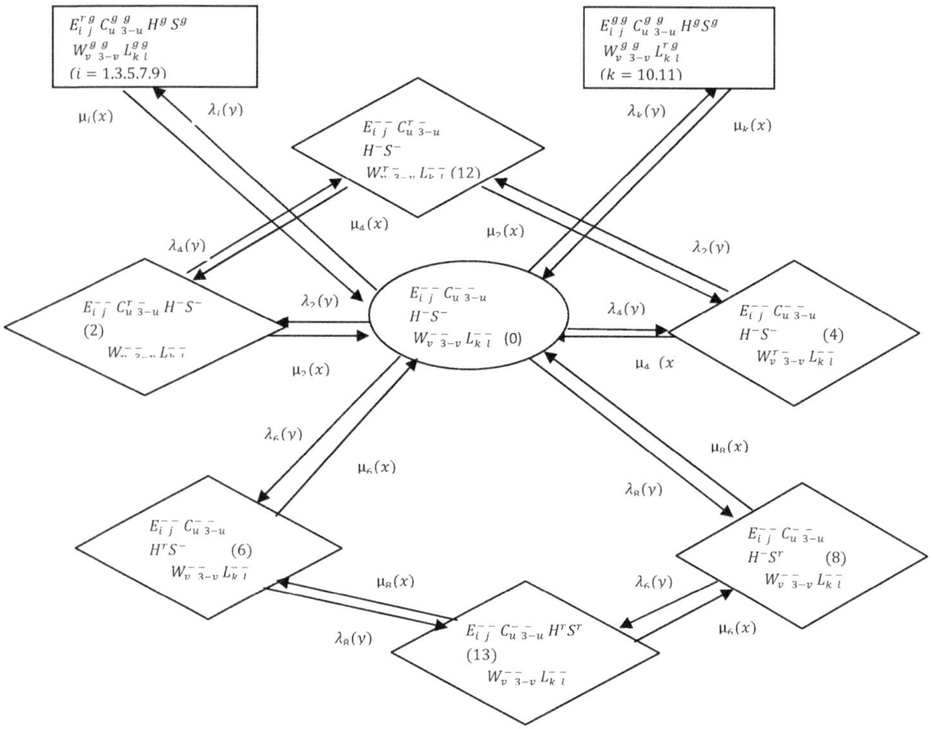


**Fig. 1.** Transition diagram of rice industry

# 3 Mathematical Modeling of the System in Transient State

## 3.1 When Both Failure and Repair Rates Are Variable

In this section we develop the Chapman-Kolmogorov differential equation assuming variable failure and repair rates of the subsystems by applying supplementary variable technique. In the transient state, Probability considerations give the following system of differential difference equation associated with the state transition diagram (fig. 1) of the system at time $(t + \Delta t)$. Using mnemonic rule, we have

$$\left[\frac{d}{dt} + \int \sum_{i=1}^{11} \lambda_i(y)dy\right] P_0(t) = \int \sum_{i=1}^{11} \mu_i(x)\, P_i(x, y, t)dxdy \tag{1}$$

Similarly, for the other states, we can write the differential equation as:

$$\left[\frac{\partial}{\partial t} + \frac{\partial}{\partial x} + \frac{\partial}{\partial y} + \lambda_4(y) + \mu_2(x)\right] P_2(x, y, t) = \lambda_2(y)P_0(t) + \mu_4(x)P_{12}(x, y, t) \tag{2}$$

$$\left[\frac{\partial}{\partial t} + \frac{\partial}{\partial x} + \frac{\partial}{\partial y} + \lambda_2(y) + \mu_4(x)\right] P_4(x, y, t) = \lambda_4(y)P_0(t) + \mu_2(x)P_{12}(x, y, t) \tag{3}$$

$$\left[\frac{\partial}{\partial t}+\frac{\partial}{\partial x}+\frac{\partial}{\partial y}+\mu_i(x)\right]P_i(x,y,t)=\lambda_i(y)P_0(t)i=1,3,5,7,9,10,11. \quad (4)$$

$$\left[\frac{\partial}{\partial t}+\frac{\partial}{\partial x}+\frac{\partial}{\partial y}+\mu_6(x)+\lambda_8(y)\right]P_6(x,y,t)=\lambda_6P_0(t)\ \mu_8(x)P_{13}(x,y,t) \quad (5)$$

$$\left[\frac{\partial}{\partial t}+\frac{\partial}{\partial x}+\frac{\partial}{\partial y}+\mu_8(x)+\lambda_6(y)\right]P_8(x,y,t)=\lambda_8P_0(t)+\mu_6(x)P_{13}(x,y,t) \quad (6)$$

$$\left[\frac{\partial}{\partial t}+\frac{\partial}{\partial x}+\frac{\partial}{\partial y}+\mu_4(x)+\mu_2(x)\right]P_{12}(x,y,t)=\lambda_2(y)P_4(x,y,t)+\lambda_4(y)P_2(x,y,t) \quad (7)$$

$$\left[\frac{\partial}{\partial t}+\frac{\partial}{\partial x}+\frac{\partial}{\partial y}+\mu_6(x)+\mu_8(x)\right]P_{13}(x,y,t)=\lambda_8(y)P_6(x,y,t)+\lambda_6(y)P_8(x,y,t) \quad (8)$$

**Boundary Condition**

$$P_i(0,y,t)=\lambda_i(y)P_0(t)\ ;\ i=1,2,3,4,5,7,9,10,11 \quad (9)$$
$$P_{12}(0,y,t)=\int\lambda_4(y)P_2(x,y,t)dx+\int\lambda_2(y)P_4(x,y,t)\,dx \quad (10)$$
$$P_{13}(0,y,t)=\int\lambda_8(y)P_6(x,y,t)dx+\int\lambda_6(y)P_8(x,y,t)\,dx \quad (11)$$

**Initial Conditions**

$$P_i(x,y,0)=0\ ;\ i=1,\ldots.13 \quad (12)$$

Solving these equations (1-9) together with initial and boundary conditions (10-14) using Lagrange's method, we get the state probabilities as given below:

$$P_0(t)=e^{-T_0t}[1+\int S_0(t)e^{T_0t}dt] \quad (13)$$

$$P_i(x,y,t)=e^{-\int\mu_i(x)dx}[\lambda_i(y-x)P_0(t-x)+\int\lambda_i(y)P_0(t)e^{\int\mu_i(x)dx}dx]\ i=1,3,5,7,9,10,11, \quad (14)$$

$$P_2(x,y,t)=e^{-\int T_1(x,y)dx}[\lambda_2(y-x)P_0(t-x)+\int S_1(x,y,t)e^{\int T_1(x,y)dx}dx] \quad (15)$$

$$P_4(x,y,t)=e^{-\int T_2(x,y)dx}[\lambda_4(y-x)P_0(t-x)+\int S_2(x,y,t)e^{\int T_2(x,y)dx}dx] \quad (16)$$

$$P_6(x,y,t)=e^{-\int T_3(x,y)dx}[\lambda_6(y-x)P_0(t-x)+\int S_3(x,y,t)e^{\int T_3(x,y)dx}dx] \quad (17)$$

$$P_8(x,y,t)=e^{-\int T_4(x,y)dx}[\lambda_8(y-x)P_0(t-x)+\int S_4(x,y,t)e^{\int T_4(x,y)dx}dx \quad (18)$$

$$P_{12}(x,y,t)=e^{-\int T_5(x)dx}\int[\lambda_4(y-x)P_2(x,y-x,t-x)+\lambda_2(y-x)P_4(x,y-x,t-x)+ S_5(x,y,t)e^{\int T_5(x)dx}]dx \quad (19)$$

$$P_{13}(x,y,t)=e^{-\int T_6(x)dx}\int[\lambda_8(y-x)P_6(x,y-x,t-x)+\lambda_6(y-x)P_8(x,y-x,t-x)+ S_6(x,y,t)e^{\int T_6(x)dx}]dx \quad (20)$$

Where

$T_0 = \int \sum_{i=1}^{11} \lambda_i(y)dy$

$T_1(x,y) = \lambda_4(y)+\mu_2(x)$ ;$T_2(x,y) = \lambda_2(y)+\mu_4(x)$;$T_3(x,y) = \lambda_8(y)+\mu_6(x)$

$T_4(x,y) = \lambda_6(y)+\mu_8(x)$;

$T_5(x) = \mu_4(x)+\mu_2(x)$; $T_6(x) = \mu_6(x)+\mu_8(x)$:$S_0(t) = \int \sum_{i=1}^{11} \mu_i(x)\, P_i(x,y,t)dxdy$

$S_1(x,y,t) = \lambda_2(y)P_0(t)+\mu_4(x)P_{12}(x,y,t)$; $S_2(x,y,t) = \lambda_4(y)P_0(t)+\mu_2(x)P_{12}(x,y,t)$

$S_3(x,y,t) = \lambda_6(y)P_0(t)+\mu_8(x)P_{13}(x,y,t)$; $S_4(x,y,t) = \lambda_8(y)P_0(t)+\mu_6(x)P_{13}(x,y,t)$

$S_5(x,y,t) = \lambda_4(y)P_2(x,y,t) + \lambda_2(y)P_4(x,y,t)$;

$S_6(x,y,t) = \lambda_8(y)P_6(x,y,t) + \lambda_6(y)P_8(x,y,t)$

If the industry provide the failure and repair rates, one can calculate the availability $A_v(t)$ in term of the probability $P_0(t)$, which is shown in equitation (1), thus, the time dependent availability $A_v(t)$ of the system is given by

$$A_v(t) = P_0(t) + \int \sum_{i=2,4,6,8,12,13} P_i(x,y,t)\, dxdy \tag{21}$$

### 3.2 When Both Failure and Repair Rates Are Constant

To find the availability of the system, when both failure and repair rates are constant, system of equitation's (1-9) reduces to more simplified form, which are given below:

$$\left[\frac{d}{dt} + \sum_{i=1}^{13} \lambda_i\right] P_0(t) = \sum_{i=1}^{11} \mu_i P_i(t) \tag{22}$$

$$\left[\frac{d}{dt} + \lambda_4+\mu_2\right] P_2(t) = \lambda_2 P_0(t)+\mu_4 P_{12}(t) \tag{23}$$

$$\left[\frac{d}{dt} +\lambda_2+\mu_4\right] P_4(t) = \lambda_4 P_0(t)+\mu_2 P_{12}(t) \tag{24}$$

$$\left[\frac{d}{dt} +\mu_i(x)\right] P_i(t) = \lambda_i P_0(t)\ i = 1,3,5,6,7,8,9,10,11 \tag{25}$$

$$\left[\frac{d}{dt} + \mu_6+\lambda_8\right] P_6(t) = \lambda_6 P_0(t)\ +\mu_8(x) P_{13}(t) \tag{26}$$

$$\left[\frac{d}{dt} +\mu_8+\lambda_6\right] P_8(t) = \lambda_8 P_0(t) + \mu_6(x) P_{13}(t) \tag{27}$$

$$\left[\frac{d}{dt} +\mu_4+\mu_2\right] P_{12}(t) = \lambda_2 P_4(t) + \lambda_4 P_2(t) \tag{28}$$

$$\left[\frac{d}{dt} +\mu_6+\mu_8\right] P_{13}(t) = \lambda_8 P_6(t) + \lambda_6 P_8(t) \tag{29}$$

**Initial conditions**

$$P_i(0) = \begin{Bmatrix} 1, i = 0 \\ otherwise\ 0 \end{Bmatrix} \tag{30}$$

Most of the authors have used Laplace transformation and matrix method to solve the reliability function. But it is difficult to find Laplace inverse since expression for probability transforms are in very complicated form and the complexity increase with the increase in number of equation. To overcome such type of problems the system of differential equation (22-29) with initial conditions (30) has been solved numerically following the approach earlier used by Gupta et. al. (2007). The numerically computation have been carried out starting from $t = 0$ to $t = 360$ days assuming $t = 0.005$ assuming as one day. The availability of rice mill has been obtained by taking different combination of the constant failure and repair rates of the subsystems collected from the concerned industry. It is evident that availability$A_v(t)$= of the system can be computed as,

$$A_v(t) = \sum_{i=0,2,4,6,8,14} P_i(t) \tag{31}$$

**Table 1.** Effect of Failure Rate of elevator on rice plant

| $\lambda_6$ days | .04 | .14 | .24 | .34 |
|---|---|---|---|---|
| 30 | 0.8249 | 0.8051 | 0.7629 | 0.7188 |
| 60 | 0.7993 | 0.7688 | 0.7056 | 0.6420 |
| 90 | 0.7696 | 0.7296 | 0.6486 | 0.5698 |
| 120 | 0.7410 | 0.6923 | 0.5961 | 0.5057 |
| 150 | 0.7136 | 0.6570 | 0.5479 | 0.4489 |
| 180 | 0.6871 | 0.6235 | 0.5036 | 0.3985 |
| 210 | 0.6617 | 0.5918 | 0.4629 | 0.3537 |
| 240 | 0.6372 | 0.5616 | 0.4256 | 0.3140 |
| 270 | 0.6137 | 0.5330 | 0.3912 | 0.2788 |
| 300 | 0.5910 | 0.5059 | 0.3596 | 0.2475 |
| 330 | 0.5691 | 0.4802 | 0.3306 | 0.2197 |
| 360 | 0.5481 | 0.4557 | 0.3039 | 0.2050 |

## 4 Result Analyses

Table 1 shows that behavior of availability of the sub-system husking for a period of 360 days divided over an interval of 30 days. It seems in the first row with the increase in failure rate ($\lambda_6$) of husking from0.04 (once in 25 hrs.) t0 0.34(once in 2.9 hrs.) effect the availability of the system by 12% where as it affect in the first column by 33.5% with the increase in time from 3o days to 360 days.

**Table 2.** Effect of Failure Rate of separation machine on rice plant

| $\lambda_8$ days | .002 | .006 | .010 | .014 |
|---|---|---|---|---|
| 30 | 0.8249 | 0.7922 | 0.7907 | 0.7893 |
| 60 | 0.7993 | 0.7697 | 0.7675 | 0.7653 |
| 90 | 0.7696 | 0.7422 | 0.7393 | 0.7365 |
| 120 | 0.7410 | 0.7157 | 0.7123 | 0.7090 |
| 150 | 0.7136 | 0.6903 | 0.6864 | 0.6826 |
| 180 | 0.6871 | 0.6658 | 0.6615 | 0.6573 |
| 210 | 0.6617 | 0.6423 | 0.6376 | 0.6330 |
| 240 | 0.6372 | 0.6195 | 0.6145 | 0.6096 |
| 270 | 0.6137 | 0.5976 | 0.5923 | 0.5871 |
| 300 | 0.5910 | 0.5765 | 0.5709 | 0.5654 |
| 330 | 0.5691 | 0.5561 | 0.5503 | 0.5446 |
| 360 | 0.5481 | 0.5386 | 0.5304 | 0.5245 |

**Table 3.** Effect of Failure Rate of polishing machine on rice plant

| $\lambda_{10}$ days | 0.03 | 0.06 | 0.09 | 0.12 |
|---|---|---|---|---|
| 30 | 0.8249 | 0.7937 | 0.7686 | 0.7450 |
| 60 | 0.7993 | 0.7660 | 0.7277 | 0.6921 |
| 90 | 0.7696 | 0.7339 | 0.6838 | 0.6382 |
| 120 | 0.7410 | 0.7031 | 0.6425 | 0.5884 |
| 150 | 07136 | 0.6738 | 0.6039 | 0.5427 |
| 180 | 0.6871 | 0.6457 | 0.5676 | 0.5005 |
| 210 | 0.6617 | 0.6188 | 0.5335 | 0.4617 |
| 240 | 0.6372 | 0.5930 | 0.5015 | 0.4217 |
| 270 | 0.6137 | 0.5683 | 0.4713 | 0.4259 |
| 300 | 0.5910 | 0.5446 | 0.4430 | 0.3928 |
| 330 | 0.5691 | 0.5219 | 0.4164 | 0.3623 |
| 360 | 0.5481 | 0.5002 | 0.3914 | 0.3983 |

Table 2 shows that the availability decreases in first row by 4.3% with the increase in the failure rate $\lambda_8$ of separating machine from .002(once in 500 hrs.) to .014(once in 70 hrs.) Further we also find in the first column that when the time increases from 30 days to 36o days the availability decreases by 33.5%.

Table 3 shows that in the first and sixth row ,the availability of the system affects 9.6% to 27.3% approximately with increase in failure rate ($\lambda_{10}$)of polishing machine from(.03 to .12) whereas increase in time affect it app. (33.5% to 46.5%) with the increase in time from 30 days to 360 days.

Table 4 shows affect of repair rate($\mu_6$) of husking machine on the availability of the system. We can see that increase in time from 30 days to 360 days decrease the availability of the system in the first and forth column by33.5% and 8.0% however increase in repair rate ($\mu_6$) of husking machine is increased from ($3.8$ $to$ $4.4$ ) increase in app. By .0001 and 39.8% in the first and sixth row respectively.

**Table 4.** Effect of Repair Rate of elevator on rice plant

| $\mu_6$ | 3.8 | 4.0 | 4.2 | 4.4 |
|---|---|---|---|---|
| 30 | 0.8249 | 0.8331 | 0.8337 | 0.8337 |
| 60 | 0.7993 | 0.8331 | 0.8342 | 0.8342 |
| 90 | 0.7696 | 0.8259 | 0.8274 | 0.8274 |
| 120 | 0.7410 | 0.8185 | 0.8204 | 0.8204 |
| 150 | 0.7136 | 0.8111 | 0.8134 | 0.8134 |
| 180 | 0.6871 | 0.8038 | 0.8065 | 0.8065 |
| 210 | 0.6617 | 0.7966 | 0.7997 | 0.7997 |
| 240 | 0.6372 | 0.7894 | 0.7929 | 0.7929 |
| 270 | 0.6137 | 0.7823 | 0.7862 | 0.7862 |
| 300 | 0.5910 | 0.7752 | 0.7795 | 0.7795 |
| 330 | 0.5691 | 0.7683 | 0.7729 | 0.7729 |
| 360 | 0.5481 | 0.7613 | 0.7663 | 0.7663 |

**Table 5.** Effect of Repair Rate of separation machine on rice plant

| $\mu_8$ days | .03 | 1.03 | 2.03 | 3.03 |
|---|---|---|---|---|
| 30 | 0.8249 | 0.8273 | 0.8293 | 0.8300 |
| 60 | 0.7993 | 0.8073 | 0.8131 | 0.8154 |
| 90 | 0.7696 | 0.7891 | 0.7957 | 0.7984 |
| 120 | 0.7410 | 0.7713 | 0.7785 | 0.7814 |
| 150 | 0.7136 | 0.7540 | 0.7615 | 0.7648 |
| 180 | 0.6871 | 0.7512 | 0.7450 | 0.7486 |
| 210 | 0.6617 | 0.7028 | 0.7288 | 0.7486 |
| 240 | 0.6372 | 0.7048 | 0.7129 | 0.7327 |
| 270 | 0.6137 | 0.6891 | 0.6974 | 0.7171 |
| 300 | 0.5910 | 0.6738 | 0.6823 | 0.7018 |
| 330 | 0.5691 | 0.6588 | 0.6675 | 0.6869 |
| 360 | 0.5481 | 0.6441 | 0.6529 | 0.6723 |

**Table 6.** Effect of Repair Rate of polishing machine on rice plant

| $\mu_{10}$ days | .80 | 1.80 | 2.80 | 3.80 |
|---|---|---|---|---|
| 30 | 0.8249 | 0.8302 | 0.8337 | 0.8337 |
| 60 | 0.7993 | 0.8072 | 0.8118 | 0.8118 |
| 90 | 0.7696 | 0.7886 | 0.7936 | 0.7936 |
| 120 | 0.7410 | 0.7704 | 0.7755 | 0.7755 |
| 150 | 0.7136 | 0.7528 | 0.7577 | 0.7577 |
| 180 | 0.6871 | 0.7356 | 0.7404 | 0.7404 |
| 210 | 0.6617 | 0.7188 | 0.7235 | 0.7235 |
| 240 | 0.6372 | 0.7024 | 0.7070 | 0.7070 |
| 270 | 0.6137 | 0.6863 | 0.6908 | 0.6908 |
| 300 | 0.5910 | 0.6707 | 0.6750 | 0.6750 |
| 330 | 0.5691 | 0.6554 | 0.6596 | 0.6596 |
| 360 | 0.5481 | 0.6404 | 0.6445 | 0.6445 |

Table 5 shows that availability increase in first and sixth row by.61% to 22.8% resp.with increase in repair rate ($\mu_8$) of separation machine from (0.03 to 3.03) Further we also find that in the first and fifth column that by an increase time from 30 days to 360 days decreases the availability of the system by app.33.5% to 19%.

Table 6 shows that behavior of availability of the system with repair rate ($\mu_{10}$) of polishing machine for a period of 360 days divided over an interval of 30 days. It seems that in the first and sixth row increase in repair rate ($\mu_{10}$) of polishing machine from( .80 to 3.80) affect the availability of the system by app.(1.06% to 17.5%) whereas it affect the first and fifth column (33.5 to 22.5%) with the increase in time from 30 days to 360 days.

## 5 Conclusions

The reliability management of complex industrial system is highly difficult for reliability analysis due to difficulties in modeling and evaluating the performance of the system, especially during strategic maintenance planning. Through rigorous efforts have been made by researchers to evolve methods to study the effect of subsystem conditions and maintenance policies on system performance, these methods involve complex computations and the computations grow tremendously with further growth in number of subsystems.The performance analysis of rice mill system can help in increasing the production and quality of rice. Detailed study reveals that the polishing subsystem is critical part of the system and needs utmost care of management. Thus, the concerned managers can plan and adapt suitable maintenance practices/strategies for improving the system performance. Apart from these advantages the system performance analysis may help to conduct cost benefit analysis, operational capability studies, inventory spare parts management and replacement decisions.

## References

1. Singh, J.: Some problem on queues and reliabilityPh.D.thesis.Kurukshetr University Kurukshetra,India (1976)
2. Kodam, M., Fukuta, J., Takamatsu, S.: Mission relabilty for 1-unitsystem with allowed down time. IEEETransictions on reliability 22(5), 268–270 (1973)
3. Cherry, D., Grogan, J.C., Holmes, W., Perris, F.A.: Availability analysis for chemical plants. Chemical Engineering Progress 74, 55–60 (1978)
4. Dhillon, B., Rayapati, S.: Chemical system reliability a review. IEEE transictions on Reliability 37(2), 21–26 (1988)
5. Singh, J.: Reliability analysis of a bio gas plant having two dissimilar units. Microelectronic Relaibilty 29(5), 779–781 (1989)
6. Gupta, P., Singh, J., Singh, I.P.: Numerical analysis of reliability and availability of the series processes in buter oil processing plant. International Journal of Quality and Reliability Management 22(3), 303–316 (2005)
7. Priel, V.Z.: Twenty ways to track maintenance performance.Facotry, pp. 81–91. Mc Graw-Hill, New York (1974)

# A Preprocessing Technique for Recognition of Online Handwritten Gurmukhi Numerals

Rajesh Kumar Bawa and Rekha Rani

Punjabi University Patiala
rajesh_k_bawa@yahoo.com, errekhagoyal@yahoo.com

**Abstract.** In this paper, a preprocessing technique involving removal of duplicate points, normalization, interpolation of missing points, sharp point detection, removing hook and smoothing is applied for recognition of online handwritten Gurmukhi numerals. Above stages are performed on the data collected from different persons. It is observed that our preprocessing technique improves feature extraction rate by increasing the accuracy in recognition of some features like hole and junction.

**Keywords:** Smoothing, Sharp point detection, Preprocessing, Interpolation of missing points, Feature extraction, Accuracy.

## 1 Introduction

Online Handwriting recognition is the recognition of handwriting that is obtained from the movements of the pen or digital tablets. Most digital tablets have low-pass hardware filter. Thus, when such devices are used to capture handwriting strokes, the shapes of the strokes present the jagged forms. Due to the hesitated writing, most handwriting strokes contain some duplicate points and hooks. Furthermore, some points of a stroke may be missing, and some wild points might exist in a stroke etc. Generally, such noise information influences the exploration of the profile of the handwriting in such a way as to influence further processes, such as feature extraction and classification.Many techniques have been applied to preprocess online handwritten data. However, most of them may not efficiently deal with the variety of the writing styles (such as different writing speeds from different writers). Huang et al. [3] proposed a new preprocessing technique for online handwriting. The goal of the preprocessing was to remove some of the variations of handwriting styles and to simplify the shapes of the symbol. This proposed preprocessing technique has improved the recognition rates by at least 10%. Deepu et al. [7] used Gaussian technique to filter handwritten data. However, the Gaussian can not interpolate the points. Sharma et al. [1] described an elastic matching technique to recognize online handwritten Gurmukhi characters. Recognition of online handwritten Gurmukhi character was done in two stages. First stage recognizes the stroke and second stage evaluate the character on the basis of recognized strokes. Sharma et al. [2] presented a system to recognize online handwritten Gurmukhi words. It proposed a new step as rearrangement of recognized strokes in online handwriting recognition procedure. Prasanth et al. [4] compared the results of DTW with four different sets of features for

A. Mantri et al. (Eds.): HPAGC 2011, CCIS 169, pp. 275–281, 2011.

Tamil. It was observed that the generalized shape context feature was better than shape context + tangent context features in terms of accuracy. Joshi et al. [5] described a system for the automatic recognition of isolated writer dependent handwritten Devanagari characters obtained by linearizing consonant conjuncts. It proposed a structural recognition technique to reduce some characters to others and residual characters were then classified using the subspace method. Finally the results of structural recognition and feature-based matching were mapped to give final output. Parui et al. [6] described a novel scheme for recognition of online handwritten basic characters of Bangla by using hidden markov method. Database consist 24,500 online handwritten isolated character samples composed of one or more strokes written by 70 persons. These strokes were manually grouped into 54 classes based on the shape similarity of the graphemes that constitute the ideal character shapes. One HMM was constructed for each stroke class and second stage of classification was used for recognition of characters using stroke classification results along with 50 look-up-tables. Swethalakshmi et al. [8] proposed a system for recognition of online handwritten characters for Indian Devanagari and Telugu scripts. A handwritten character was represented as a sequence of strokes whose features were extracted and classified by using support vector machine. Santosh et al. [9] focused on a survey on online handwriting recognition system along with the real application by taking Nepali natural handwriting. The survey mainly included preprocessing, feature vector and similarity measures in between the non-linear 2D sequences of coordinates. In order to avoid distortion of the handwriting and solve the above limitations, we propose a new preprocessing technique. The approach, firstly remove the duplicate points in the strokes and then normalize the stroke. After normalization interpolation of missing points is done by using Bezier interpolation and then sharp points are detected to remove the hook from the stroke. Finally, the online handwritten symbol is smoothened by using K-neighbors method.

## 2 Data Collection

Online handwriting recognition works with handwritten data that is inputted with a digitizing tablet and special stylus. This data is a sequence of (x, y) coordinates. An (x, y) coordinate is called a point.

## 3 Preprocessing

The goal of preprocessing is to discard irrelevant information that can negatively affect the recognition. Generally, the irrelevant information includes duplicate points, wild points, hooks, noise etc. The approach is summarized in Figure 1. Given an online handwritten symbol or expression, the approach contains of removing duplicate points, normalization, interpolation of missing points, sharp point detection, hook removal and smoothing data. As a result, the features of online handwritten symbols can be easily and accurately detected. In the following we describe each of these steps.

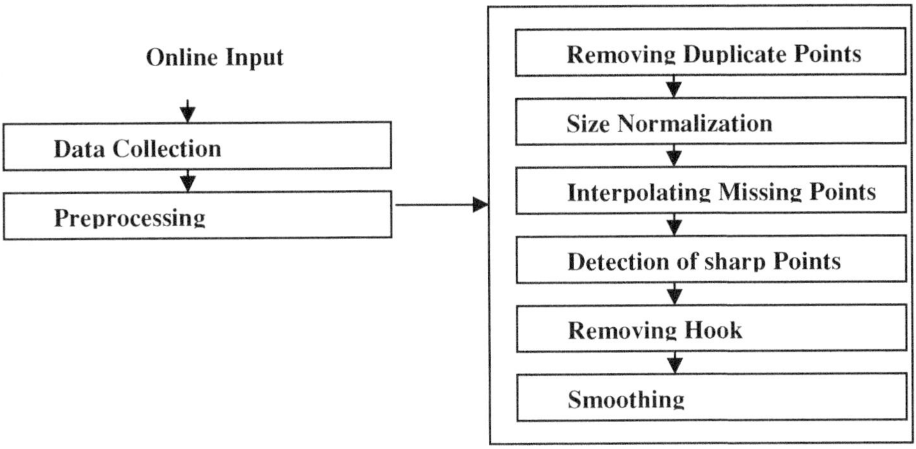


**Fig. 1.** Preprocessing Steps

### 3.1 Removing Duplicate Points

In online handwriting recognition data is collected using the electronic tablet or digitizer. Digitizer is a sensitive device and captures every slight movement of the pen, even when the pen tip is not quite touching the digitizer but is over the plane. This will cause co-occurrence of the coordinates at a point. Slow handwriting is also a cause for the repetition of coordinates at the same position, usually at corners. All these wild points are removed by checking whether the coordinates of any two points are the same. If they are at the same position, one of them is kept and other are removed.

### 3.2 Size Normalization

Size normalization of a stroke plays an important role in online handwriting recognition. A nominal sized numeral is recognized much faster than a varying sized numeral. Normalization includes basic techniques like scaling, translation and rotation. Normalization is done to remove some of the variations of online handwriting style and to simplify the shape of numeral. In Figure 2: (a) represents online handwritten input and (b) represents a normalized online handwriting.

**Fig. 2.** (a) Raw Data (b) Normalized Data

### 3.3 Interpolating Missing Points

In online handwriting recognition user directly interact with the computer. While writing variability in hand movements and fast speed of writer will skip many points between the two consecutive points. This incomplete data is not sufficient for correct recognition. These missing points can be calculated using Bezier interpolation. Bezier interpolation helps to interpolate points among fixed number of points. Figure 3 shows the online handwriting after interpolating missing points.

**Fig. 3.** Interpolation of Missing Points

### 3.4 Sharp Point Detection

A sharp point is at the peak of the wavelet where the writing direction has changed. The algorithm [3] for exploring sharp points has been developed based on the changed-angles of pen motion. Firstly, according to the given stroke S, the algorithm calculates the slope angles of the lines that are defined by any two consecutive points. Secondly, the algorithm calculates the changed angles based on these slopes. Note that a changed-angle is the angle between two consecutive lines.In the values of these changed-angles, a turn (or joint) values between the subset of the consecutive increasing changed-angle values and the subset of the consecutive decreasing changed-angle values can be identified, which represents a sharp point. In addition, the beginning point and end point of a stroke are sharp points. Therefore, the sharp points in a stroke can be calculated by the turn values with the start and end points.

### 3.5 Removing Hooks

If there are more than two sharp points, the hook of the stroke can be found based on two segments: one segment $seg_b$ is between the beginnings of two sharp points; the other segment $seg_e$ is between the two end sharp points. Let their slope angles be $ang_b$ and $ang_e$ respectively. Let the length of the segment $seg_b$ or $seg_e$ be $l_{seg}$ .
In addition, the two segments have to be defined: one segment $seg_{b+1}$ is between the first second and third sharp points, the other segment $seb_{e-1}$ is between the last second and third sharp points. Let the angle between segments $seg_b$ and $seg_{b+1}$ or between segments $seg_e$ and $seg_{e-1}$ be QQ [3].

Thus, if the length $l_{seg}$ and the angle QQ of segment $seg_b$ or $seg_e$ match the following condition, the segment is a hook. Figure 4 represents the online handwriting after removing hook from it.

$$QQ <= threshold_{ang} \ \&\& \ l_{seg} <= threshold_{len}$$

Here, $threshold_{ang}$ is set to 90 degree and $threshold_{len}$ is set to 3% of the diagonal line.

**Fig. 4.** Removal of Hook

### 3.6 Smoothing Data

Smoothing of a stroke is done to eliminate the errors due to the erratic hand motion during the writing. Some time handwritten data contains flickers due to the fast writing of writer or any hardware and software problem. These flickers affect the accuracy of online handwriting recognition system. These flickers can be smoothened by k-neighbors. In k-neighbors method each point is modified with the mean value of its k-neighbors. Figure 5 shows the result after smoothing online handwriting.

**Fig. 5.** Smoothing of Online Handwriting

## 4 Results and Discussion

As explained in section 3 the preprocessing approach first removes the duplicate points of a stroke and normalize it, then Bezier interpolation is applied to interpolate missing points and the technique of detecting sharp points is applied to eliminate hooks. They also are coupled with K-neighbors to efficiently smooth handwritten data. Figure 6 shows the shape of Gurmukhi Numeral 'two'before and after the preprocessing. Preprocessing of online handwriting is done before the feature extraction and classification.

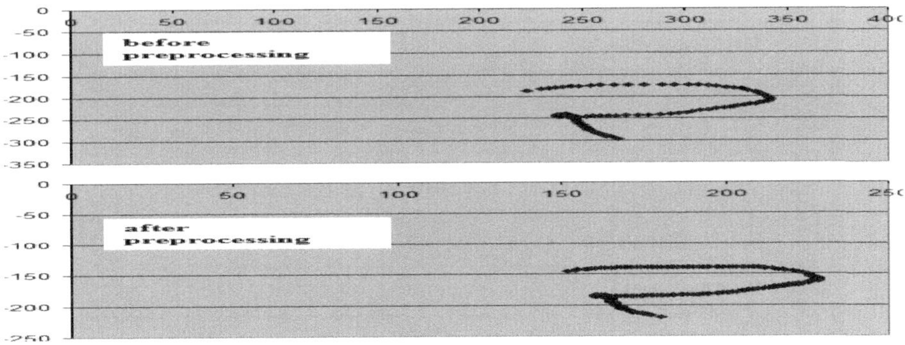


**Fig. 6.** Shows Gurmukhi Numeral 'Two'before and after the Preprocessing

The evaluation of the preprocessing methods is carried out by extracting the features which are already used. To evaluate the preprocessing method, a set of 4 writers were told to write Gurmukhi Numerals in their natural handwriting. We collected five samples of each Gurmukhi Numeral from each writer. In the experiment, features like hole and junction were firstly extracted from the un-preprocessed Gurmukhi Numerals and then from the preprocessed Gurmukhi Numerals. In contrast, obviously, the feature extraction rate of Gurmukhi Numerals with preprocessing is higher than the Gurmukhi Numerals without preprocessing. Table 1 shows the feature extraction rates of Gurmukhi Numerals with preprocessing and without preprocessing.

**Table 1.** Feature Extraction Rate before and after Preprocessing

| Sno | Numeral | Correctly recognized hole | Correctly recognized junction |
|---|---|---|---|
| 1 | Un-Preprocessed ੦ | 20% | 100% |
| | Preprocessed ੦ | 90% | 100% |
| 2 | Un-Preprocessed ੧ | 30% | 30% |
| | Preprocessed ੧ | 60% | 70% |
| 3 | Un-Preprocessed ੨ | 35% | 80% |
| | Preprocessed ੨ | 50% | 85% |
| 4 | Un-Preprocessed ੩ | 5% | 85% |
| | Preprocessed ੩ | 20% | 100% |
| 5 | Un-Preprocessed ੪ | 45% | 80% |
| | Preprocessed ੪ | 65% | 100% |
| 6 | Un-Preprocessed ੫ | 100% | 30% |
| | Preprocessed ੫ | 100% | 90% |
| 7 | Un-Preprocessed ੬ | 100% | 85% |
| | Preprocessed ੬ | 5% | 95% |
| 8 | Un-Preprocessed ੭ | 100% | 100% |
| | Preprocessed ੭ | 100% | 100% |
| 9 | Un-Preprocessed ੮ | 15% | 65% |
| | Preprocessed ੮ | 35% | 100% |
| 10 | Un-Preprocessed ੯ | 10% | 70% |
| | Preprocessed ੯ | 15% | 95% |

## 5 Conclusions and Future Work

A new efficient preprocessing approach for online handwriting recognition has been applied for recognition of online handwritten Gurmukhi numerals. Pre-processing refines the incomplete, noisy and inconsistent data, collected directly from user to make it suitable for analyzing. Preprocessing enhances the accuracy of feature extraction phase in online handwriting recognition. Features like hole and junction are more easily and accurately extracted from a preprocessed handwritten stroke than an un-preprocessed handwritten stroke. The current work can be used for feature extraction and recognition of preprocessed online handwritten Gurmukhi numerals more efficiently.

## References

[1] Sharma, A., Sharma, R.K., Kumar, R.: Recognizing Online Handwritten Gurmukhi Characters Using Elastic Matching. In: International conference on Image and Signal Processing. IEEE, Los Alamitos (2008)

[2] Sharma, A.: Rearrangement of Recognized Strokes in Online Handwritten Gurmukhi Words Recognition. In: International Conference on Document Analysis and Recognition. IEEE, Los Alamitos (2009)

[3] Huang, B.Q., Zhang, Y.B., Kechadi, M.-T.: Preprocessing Techniques for Online Handwritten Recognition. In: International conference on Intelligent System Design and Applications. IEEE, Los Alamitos (2007)

[4] Prasanth, L., Babu, V.J., Sharma, R.R., Rao, G.V.P.: Elastic Matching of Online Handwritten Tamil and Telgu Scripts Using Local Features. In: Proceeding of the 4th International conference on Document Analysis and Recognition (2006)

[5] Joshi, N., Sita, G., Ramakrishnan, A.G.: Matchine Recognition of Online Handwritten Devanagari Characters. In: Proceeding. of the 8th International conference on Document Analysis and Recognition (2005)

[6] Parui, S.K., Guin, K., Bhattacharya, U., Chaudhuri, B.B.: Online Handwritten Bangla Character Recognition Using HMM. In: Proceeding of the 19th International Conference on Pattern Recognition (2008)

[7] Deepu, V., Madhvanath, S., Ramakrishnan, A.: Principal Component Analysis for Online Handwritten Character Recognition. In: International conference on Pattern Recognition. IEEE, Los Alamitos (2004)

[8] Swethalakshmi, H., Jayaraman, A., Chakravarthy, V.S., Sekhar, C.C.: Online Handwritten Character Recognition of Devanagari and Telugu Characters Using Support Vector Machines. In: International workshop on Frontiers in Handwriting Recognitipon (2006)

[9] Santosh, K.C., Nattee, C.: A Comprehensive Survey on Online Handwriting Recognition Technology and Its Real Application to the Natural Handwriting. Kathmandu University Journal of Science, Engineering and Technology (2009)

# A Framework for Vulnerability Analysis during Software Maintenance

Jitender Kumar Chhabra and Amarjeet Prajapati

Department of Computer Engineering
National Institute of Technology, Kurukshetra-136119 India
jitenderchhabra@rediffmail.com,
amarjeetnitkkr@gmail.com

**Abstract.** Need of vulnerability analysis during software maintenance has been highly stressed by many vulnerability response experts. An analysis of why and how vulnerability happened is crucial for developing appropriate countermeasures to prevent recurrence of the vulnerabilities. In this paper, we present a framework for vulnerability analysis which needs to be applied during software maintenance. The framework helps in better and efficient cause-detection, identification of reasons of the breaches and development of countermeasures for already existing as well as new vulnerabilities.

**Keywords:** Vulnerability analysis, vulnerability modeling, incident analysis.

## 1 Introduction to Vulnerability

Vulnerabilities are weaknesses in a system which can be exploited at any point of time and lead to undesired consequences [1]. Analysis of vulnerability during security incident handling provides crucial information for countermeasure. For the last decade, a lot of research has been done in analysis of detected vulnerability for different purposes. Various software companies and organizations have started analyzing and reporting detected vulnerabilities of software and services. For example, the National Vulnerability Database [2], SANS top-20 annual security risks [3], Common Weakness Enumeration (CWE) [4],and Common Vulnerability Scoring System (CVSS [5]. But there has been a lack of framework and approaches for vulnerability analysis during software maintenance time.

## 2 Related Works

Various solutions have been proposed in the literature to address security vulnerability problems related to software. During software development, various modeling approaches such as Attack Trees [6], SAGs [7], SGITs [8], and SGMs [9]. After the deployment of the system if vulnerability is detected VCGs [10] and ICGs which model the incidents and related vulnerabilities [11] are being used. However there are no methodologies in literature for analysis and modeling of new version of vulnerability.

A. Mantri et al. (Eds.): HPAGC 2011, CCIS 169, pp. 282–287, 2011.

## 3 Proposed Framework

In this section, we propose a framework for vulnerability analysis during software maintenance. There are four major components of the framework as shown in fig. 1 i.e. security incident handling, vulnerability handling, analysis, modeling of new version of vulnerability and patch development & installation. The framework starts with security incident analysis in which various analysis activities related to vulnerability is performed and later various types of data is collected regarding vulnerabilities which serve as a basis for rapid and cost effective development of patch file for vulnerability mitigation.

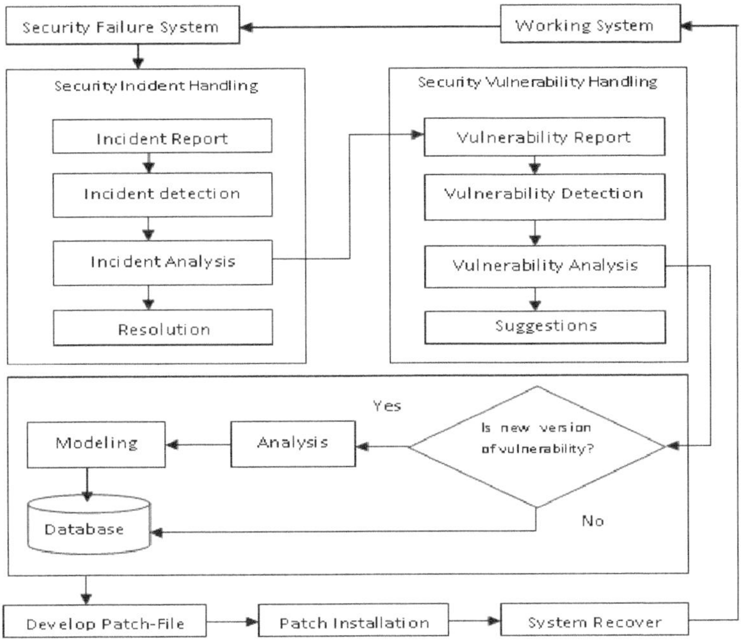


**Fig. 1.** Components of proposed Framework

## 4 Working and Methodology

The step-by-step working details and methodology of the proposed framework is explained below:

### 4.1 Security Incident Handling

Security incident handling process involves: 1) incident reporting, it includes a manual description of the incident or event, the appropriate taxonomy and some additional information such as incident category, date & time, and method used to identify the incident etc. 2) Detection of the Incident, it involves observation of signs of incident events which occur in system, that recorded mainly by logging and computer security software. 3) Analysis of the Incident, it starts with collecting the evidence about the incident. Collect a set of incident hypotheses, conducting interviews of all involved parties and accepting or rejecting the hypotheses. The next step is to build the incident cause graph (ICG) [11]. During this step, possible hypotheses will be identified, refined and finalized when constructing the ICG. 4) After the detection and analysis of the incident the suitable strategy is applied for resolution of the incident.

### 4.2 Security Vulnerability Handling

Vulnerability handling process starts with vulnerability report which includes various evidence of the failure component. It is generated during incident analysis. Next the vulnerability is detected by intensive research and detail information related to vulnerability is collected. In the last, the detected vulnerabilities are classified along six axes which are nature of the flaw, time of introduction, exploitation domain of the vulnerability, the effect domain, the minimum number of components needed to exploit the vulnerability and the source of the identification of the vulnerability. Find out the direct and indirect causes. Use CVSS (Common Vulnerability Scoring System) to measure the severity of vulnerability [14].

### 4.3 Security Vulnerability Modeling

For modeling of discovered vulnerability use Vulnerability Cause Graph (VCG) [10]. On the basis of the analysis performed, suggestions are extracted.

### 4.4 Patch Development and Installation

After the validation of vulnerabilities corresponding patch files are developed (if new version of vulnerability) or downloaded (if known vulnerability) for mitigation and then it is installed on the failure system.

## 5 Case Study

We have applied our method to an example, denial of service incident in Valve software's Half-Life server. Valve Software's Half-Life Server is a game server for the online multi-player game. It allows up to four players to connect to the server, per copy of the game [14].

### 5.1 Security Incident Handling

The reported security incident is Off-by-one error in the GeoIP module in the AMX Mod X1.76d plug-in for Half-Life server. On the basis of reports we initiate surveillance over the reported module GeoIP and find out that it allows execution of arbitrary code which is compromised by attacker to prevent some services. After that we start with creating a base graph with an exit node. The direct predecessor of the exit node is the answer to the question "what has caused the incident?" The incident is the result of a denial of service. Then we continue building the graph by analyzing the node "Denial of service". Based on our hypothesis, the denial of service is cause of weak detection measures of input data validation and buffer overflow vulnerability. (see Figure 2).

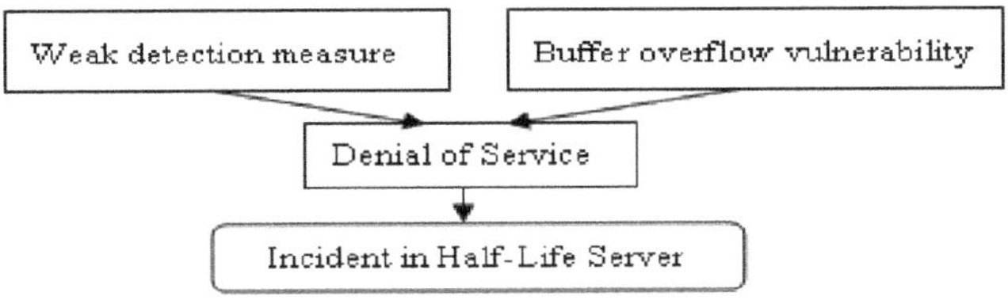


**Fig. 2.** ICG construction

## 5.2 Security Vulnerability Handling

During security incident analysis the affected modules such as geoip_code2 (), geoip_code3 () and error messages is reported. We detect the vulnerability as exhausting of memory by error messages and collect associated information so the vulnerability type is-Buffer overflows. Causes of vulnerability is geoip_code () function incorrectly triggers error message and execution of arbitrary code in geoip_code () function. It's CVSS score is calculate as 7.5[13], so it severity is labeled "high" severity.

## 5.3 Security Vulnerability Modeling

Buffer overflow can be modeled with vulnerability cause graphs (VCG). (see Fig 3).

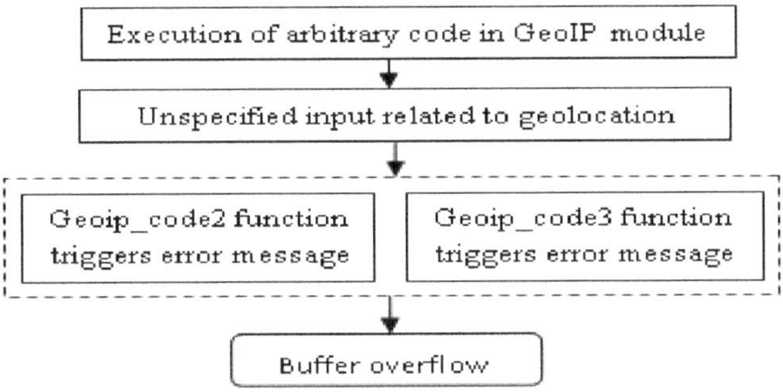


**Fig. 3.** The vulnerability model of Buffer overflows

We further develop a complete security incident cause graph which also includes buffer overflow vulnerability as shown in fig. 5. The resulting model shows that in order to prevent the incident and vulnerability in the future, the denial of service must be prevented. This can be accomplished by addressing the causes "Weak detection measure" and by preventing the vulnerability Buffer overflow. Addressing its causes can prevent the vulnerability.

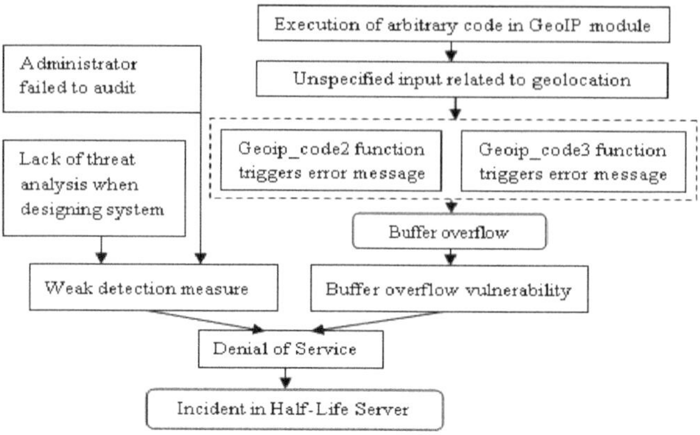


**Fig. 4.** ICG for incident of Half-Life Server

### 5.4 Patch Development and Installation

A patch file for the Buffer overflow vulnerability is developed and install on the Half-Life Server to recover from failures.

## Conclusion

This paper has proposed a new framework for the analysis of vulnerability during software maintenance. The approach involves four major steps- security incident handling, vulnerability handling, analysis and modeling of new version of vulnerability and patch development and installation. The proposed framework's usefulness and working has been demonstrated using a case study of Denial of Service caused due to exploitation of buffer overflow vulnerability. The study indicated that all unknown and partially known breaches of security could be identified easily and systematically through this framework and the results can be used to plug-in these vulnerabilities in an efficient way through patches and other mechanisms.

## References

1. Viega, J., McGraw, G.: Building Secure Software. Addition Wesley, Reading (2005)
2. SANS, `http://www.sans.org/`
3. Common Weakness Enumeration, `http://www.cwe.mitre.org`
4. Common Vulnerability Scoring System, `http://www.first.org.cvss`
5. Ray, I., Nayot, P.: Using Attack Trees to Identify Malicious Attacks from Authorized Insiders. In: di Vimercati, S.C., Syverson, P.F., Gollmann, D. (eds.) ESORICS 2005. LNCS, vol. 3679, pp. 231–246. Springer, Heidelberg (2005)
6. Byers, D., Shahmehri, N.: A Cause-Based Approach to Preventing Software Vulnerabilities. In: The Third International Conference on Availability, Reliability and Security, pp. 276–283 (2008)

7. Kloos, J., Elberzgager, F., Eschbach, R.: Systmatic Construction of Goal Indicator Trees for Indicator-Based Dependability Inspections. In: 36th Conference on Software Engineering and Advanced Applications (SEAA), pp. 279–282 (2010)
8. Byers, D., Shahmehri, N.: Unified modeling of attacks, vulnerabilities and security Activities. In: Workshop on Software Engineering for Secure Systems (SESS). ACM, Cape Town (2010)
9. Incident handling analysis CERT/CSIRT style Hand-on training. Cyber Security Malaysia (2010)
10. Byers, D., Ardi, S., Shahmehri, N., Duma, C.: Modeling Software Vulnerabilities With Vulnerability Cause Graphs. In: 22nd IEEE International Conference on Software Maintenance, pp. 411–422 (2006)
11. Ardi, S., Shahmehri, N.: A Post-Mortem Incident Modeling Method. International Conference on Availability, Reliability and Security, 1018–1023 (2009)
12. `http://www.valvesoftware.com/games/`
13. `http://www.cvedetails.com/cve-details/`
14. Mell, M., Scarfone, K., Romanosky, S.: Common Vulnerability Scoring System. IEEE Security & Privacy 4, 85–89 (2006)

# Performance Optimization for Logs of Servers

M. Vedaprakash[1], Ramakrishna Alavala[2], and Veena R. Desai[3]

[1] Department of Information Science and Engineering,
RV College Of Engineering, Bangalore, India
mvedprakasha@gmail.com
[2] Performance Architect, Lab Services,
IBM, Bangalore, India
rkalavala@in.ibm.com
[3] Assistant Professor, Department of ISE,
RVCE, Bangalore, India
desaiveena@gmail.com

**Abstract.** Performance of any entity is said to be high, if it does greater units of work in lesser amount of time. However tuning can make a significant amount of change in the efficiency of a system. Servers are of important considerations in these days with its greater potential to serve more number of clients at different levels of architectural tiers. Hence there is need to enhance the performance of these Logs of servers so that errors and related administrative tasks can more accurately known and updated to required levels so as to minimize the impact of the performance. There are a number of parameters that influences the performance of these servers, which upon automated optimization shall result in, greater help to the analysts and further use of the servers to persistently provide the services to the clients.

**Keywords:** Optimization, Tuning, Performance.

## 1 Introduction

Internet is accepted as the at most requirement for the E-commerce transactions. These applications traverse across three tiers. The request – response unit in the network is basically a three layered architecture. The request made by any client shall first reach the web server, followed by application server and then the required data is retrieved from the database. The diagram shows the flow of the request across the three tier architecture.

The web servers are the servers which accept the incoming requests form the client browsers. These servers have certain load limits because of the constraint on the limited concurrent connections. This is based further on its own settings, HTTP request type etc [3]. The application server is the primary runtime component in all configurations. It is where an application executes. All WebSphere Application Server configurations can have one or more application servers [4]. Any errors, administrative tasks are written into the logs. All products produce problem determination data, such as trace records, log records and messages. The activity of writing these details to a log has certain performance impact and can be enhanced

A. Mantri et al. (Eds.): HPAGC 2011, CCIS 169, pp. 288–292, 2011.

further by appropriately optimizing the parameters that influences the write activity of log. This paper discusses such parameters and the optimization of these parameters across all the three tiers of web server, application server, and database.

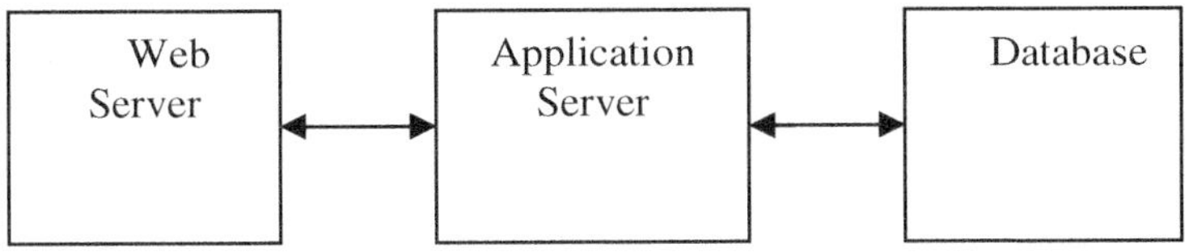


**Fig. 1.** Flow of the request-response across the three tier architecture

The three tier architecture of the network is as represented in the figure 1. The request from the client reaches the web server; this is redirected to the application server from where it reaches the database. The requested data is retrieved from the database and this response travels in the exactly reversed direction to reach the client.

## 2 Related Work

In order to hike the performance of the servers, duplicate servers or Content Distribution Networks (CDN) can be used. Here the replicas of the servers are maintained, which shall contain the copies of the content. As soon as the request form the client is arrived, the CDN server responses it back. But it suffers from the minimum p median problem. So the paper proposes the placement algorithm for solving it and hike the performance of the entire system by enhancing the performance of each server. Here the two basic considerations that have been made are client demand traffic and distance between each client and related replica [1].

The performance of the servers can also be enriched by load balancing. A linear quadratic regulator is used to control such as the overheads of changing resource allocations and the performance degradation of transient load server imbalances. The paper also presents the empirical data model response time benefits of IBM DB2 data pools. It shall also provide with the equations of the trade offs between the transient load imbalances and the cost of changing resource allocations. Resources such as memory pool is maintained by the cases such as memory pool sizes are constant and response time benefits are close to each other, memory pool sizes are constant and response time benefits are differ form one to other, memory pool sizes oscillate [2].

## 3 Approach

An observation is made as to what parameters influences the logs of the three tier architecture. Those parameters are altered in order to meet the optimized requirements. It is important to note that, rather than analyzing the content of the Logs, it is better to control what needs to be incorporated into the log.

### 3.1 IBM HTTP Server

The technical details can be extracted as the performance optimization needs certain inputs such as weather the user needs a single log for all his administrative tasks, weather the Domain name is needed in the Log, weather the importance has to be given to session affinity or file existence or hardware interrupts etc. Based on these inputs the analysis is made as to enabling what parameters will satisfy these requirements of the user. Those corresponding parameters are enabled and rest is disabled. Thus the optimized traces for the log are made by updating the correct values for the administrative tasks of log.

The HTTP server has the properties file – ‘httpd.conf’. The settings of this file shall alter the log details of the server. The Domain name is influenced by the parameter called “HostNameLookups”. This directive enables DNS lookups so that host names can be logged. By default, the HostNameLookups directive is set to Off. The server writes the IP addresses of incoming requests to the log files. When HostNameLookups is set to on, the server queries the DNS system on the Internet to find the host name associated with the IP address of each request, then writes the host names to the log.

Log levels of the HTTP Server can be set to emerg, alert, critical, error, warning, audit, info, debug, which represents the decreasing order of significance. Upon choosing the highest level the lower leveled details are also logged in. So choosing the correct value is very much required, else causes the Performance impact. The tool shall find the current log level and gives the recommendations as to what details are logged in excess and how to rectify it.

If there is a need for single log, then all the parameters such as error maintenance, access details, information regarding the client, from where the reference of client is taken etc has to be put into the single log. If there is need for the separate logs then, the parameter “CustomLog logs/access.log combined” has to be enabled else this parameter can be documented. An agent log is a log which provides the details regarding the identifying information that the client browser reports about itself. If these details are needed in the separate log then “CustomLog logs/agent.log agent” can be enabled, if not it can be commented. A referrer log is one that has the information related to site that the client reports having been referred from. This can be enabled by the parameter ” CustomLog logs/referer.log referrer”.

### 3.2 WebSphere Application Server

As we have seen in the case of the IHS there are several levels of logging even in the case of the WAS. There are two groups of packages in WAS. First one is those that change at configuration and the second one are those that change at runtime. Upon selecting the Configuration tab, a static list of well-known components, packages, and groups is displayed. This list is not exhaustive. So choosing Runtime tab for all purposes is preferable. If we select the Runtime tab, the list of components, packages, and group are displayed with all the components that are registered on the running application server and in the static list. * Specifies all traceable code running in the application server, including the product system code and customer code.

For convenience the packages are classified as Java related packages, IBM related packages, Apache related packages, Security oriented packages, and others. For instance if the entry exit status of the java packages is on, then all the java grouped files will be set to "Finer", which has an additional impact of calculating the configuration status, sub task progress, and event details. These details are given as the recommendations by the tool and upon choosing the update, the tool optimizes the requirements to the recommended value.

### 3.3 DB2

The parameter that influences the log of the Database2 is diaglevel. This parameter specifies the type of diagnostic errors that will be recorded in the db2diag.log file. The requirements of the database log can be either to capture the errors, warnings, severe errors, or all information. Capturing all information if the requirement is just to capture errors shall result in the low performance, which has to be avoided. The error levels are specified here in terms of the numerical values such as 0 to 4 respectively for no capturing of data, severe errors, all errors, warnings, and all information. Traces of lesser numerical values are also captured in the higher numerical valued traces.

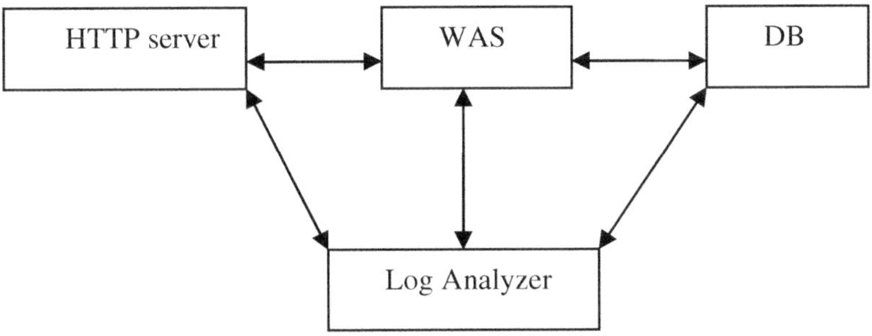


**Fig. 2.** Flow of Control

## 4 Advantages of Analyzer

- Most of the Log Analyzers are based on log's contents and none has the Log Level analysis that is "what details have to be logged into the log".
- Greater the level of log greater is the computation needed and lower the level minimum requirements are not met. So there is a need for the average level (optimized) to be established, which the tool performs.
- Log analysis, which shall keep track of the tracing levels and inform as the recommendations as to what logs are on, and give the recommendations as to what is needed to be disabled for a given log.
- Easy-to-use, intuitive viewing. Data can be viewed in real time using easy-to-read textual views conveniently organized into logical groups.

## 5 Conclusion

This process shall perform the analysis of all these three tier architecture and give the recommendations as to which values are more suited for the requirements specified. Also, it shall have the option of automatically updating the required values to the respective architectural groups. As a future work it can be extended to different operating systems, and of different configurations. For application server, we can have the greater classification level of the packages used for the server, which shall give the better results.

## References


[1] Li, Y., Liu, M.T.: Optimization of Performance Gain in Content Distribution Networks with Server Replicas

[2] Diao, Y., Hellerstein, J.L., Storm, A., Surendra, M., Lightstone, S., Parekh, S., Garcia-Arellano, C., Thomas, I.B.M., Watson, J.: Research Center, Hawthorne, New York, USA IBM Toronto Lab, Markham, Ontario, Canada Incorporating Cost of Control into the Design of a Load Balancing Controller

[3] Akyildiz, I.F.: Senior Member IEEE Georgia Institute of Technology, Atlanta, W. Liu, Student Member IEEE, Georgia Institute of Technology, Atlanta Performance Optimization of distributed-System Models with Unreliable Servers

[4] http://www.redbooks.ibm.com/redbooks/pdfs/sg246451.pdf

[5] Nahum, E., Barzilai, T., Kandlur, D.D.: Performance Issues in WWW Servers

[6] Mu rye1 a Weihru u ch DB2 For OS1390 V5 VS. V6 Outer Join Perfromance

[7] Yu, H.,Moreira, J.E., Dube, P., Chung, I., Zhang, L.: Performance Studies of a WebSphere Application, Trade, in Scale-out and Scale-up Environments

[8] Marsal, G., Denis, B., Faure, J.-M., Frey, G.: Evaluation of Response Time in Ethernet-based Automation Systems

[9] Li, W.-S., Zilio, D.C., Batra, V.S., Subramanian, M., Zuzarte, C., Narang, I.: Load Balancing for Multi-tiered Database Systems through Autonomic Placement of Materialized Views

[10] Max Felser Ethernet TCP/IP in Automation A short Introduction to real-time Requirements

# Ontology Based Information Retrieval for Learning Styles of Autistic People

Sanchika Gupta and Deepak Garg

Department of Computer Science,
Thapar University, Patiala, Punjab
sanchigr8@gmail.com, dgarg@thapar.edu

**Abstract.** In this paper an ontology based prototype system for information retrieval on the Internet is described. User is interested in the focused results regarding a product with some specific characteristics. A product may have different characteristics like size, length, color, functionality based parameters etc. It is, however, difficult for autistic people to identify appropriate keywords due to their lack of ability to process and retain the information. Therefore, a large amount of unwanted and irrelevant data is included in the outcome. In this proposal user may type the search queries using some words. The objective is to find the right set of keywords from the search paragraph and retrieval of correct patterns or products from the web. This is based on memories of such people and their learning styles that help them find the desired result.

**Keywords:** autistic, information retrieval, ontology, semantic web.

## 1 Introduction

Information retrieval (IR) is the area of study concerned with searching for documents, for information within documents, and for metadata about documents and finding material of an unstructured nature that satisfies an information need from within large storage [1]. IR is also used to facilitate semi structured search such as finding a document where the title contains Java and the body contains threading and clustering which is the task of coming up with a good grouping of the documents based on their contents. An information retrieval process begins when a user enters a query into the system. Queries are formal statements of information needs, for example search strings in web search engines. In information retrieval a query does not uniquely identify a single object in the collection. Instead, several objects may match the query, perhaps with different degrees of relevancy. Today's technologies are advancing the way we learn and communicate. With innovative tools, autistic people can read books, play games, and communicate with computers. New technologies are being developed to understand the special needs of such people. The motive here is to look for easy ways for these people to access electronic information and more importantly information retrieval on the web.

'Learning styles' is a concept which attempts to describe the methods by which people gain information about their environment and that help someone learn and retain a skill or fact. Some individuals believe that spatial methods are the only way to

A. Mantri et al. (Eds.): HPAGC 2011, CCIS 169, pp. 293–298, 2011.

go with autistic children whereas some believe that the only way to deliver skills training is by utilizing words. The predominantly used learning styles are: Visual (spatial), Aural (auditory-musical), Verbal (linguistic), Physical (kinaesthetic), Logical (mathematical), Social (interpersonal), Solitary (by self). For example, viewing a book of pictures or reading a textbook involves learning through vision, listening to a lecture in a classroom or on tape involves learning through hearing, and pressing buttons to determine how to operate a computer involves hands-on learning [2].

Various user friendly and efficient tools were developed for the assistance of disabled people such as Graphical User Interface (GUI), which was developed by Douglas Engelbart [8], a talking typewriter for blind people, a talking display terminal, a screen reader for the sight impaired, a speech recognizer (Via Voice), a talking web browser (Home Page Reader) which were developed by IBM and many assistive tools were developed by Microsoft [9], [11].

User is interested in the focused results regarding a product with some specific characteristics. It is, however, too difficult for autistic people to identify appropriate keywords due to their lack of ability to process and retain the information. Therefore, a large amount of unwanted and irrelevant data is included in the outcome. The proposal handles the search queries related to the autistic people based on their learning styles. The information system can identify desired keywords for the query that have strong relationships with ontology in their learning styles.

## 2 Information System for Autistic People

Autism is a lifelong developmental disability that first appears during infancy or childhood, and generally follows a steady course without remission. It affects how a person communicates with, and relates to, other people. It also affects how they make sense of the world around them. Overt symptoms gradually begin after the age of six months, become established by age two or three years, and tend to continue through adulthood, although often in more muted form. Autism interferes with the normal development of the brain. The three main areas of difficulty which all people with autism share are sometimes known as the 'triad of impairments'. They are: difficulty with social communication, difficulty with social interaction, difficulty with social imagination.

Difficulties concerning communication are like immature language skills, communication with gestures, not able to understand the facial expressions or tone of voice, inability to attach meaning to words etc.

As for difficulty with social interaction, individuals with autism prefer to spend time alone rather than with others, doesn't understand the unwritten social rules, are less responsible to social gestures such as eye contact, are insensitive to others feeling and are anxious.

As for difficulty with social imagination is concerned it is hard for autistic people to comprehend abstract ideas, understand the concept of danger, solve problems on their own, distinguish between right and wrong, imagine and plan for future.

Many problems that autistic children experience can be alleviated through special education. The main goals when treating children with autism are to lessen associated difficulties and to increase quality of life and functional independence. No single treatment is adequate and is typically tailored to the child's unique needs [3].

Based on above difficulties it is realized that autistic individuals are more likely to rely on only one style of learning. One way to improve these children's lives is to introduce learning tools called Augmentative Communication. This technology provides visual information in a creative format which helps a child in completing activities of daily living both in the home and at school. For example, an Activity Schedule can be created to remind the child what they have to do when they wake up to get ready for school. The morning routine can be represented using various photographs, drawings or words placed sequentially. Eg: Page 1: a toothbrush and toothpaste; Page 2: a washroom; Page 3: a hairbrush; Page 4: a shirt, pants, socks, and shoes; Page 5: a bowl of cereal and milk; Page 6: a backpack, lunch box, bottle. After completing each action it's important for the child to cross off the item or place the picture in an "all done" envelope. When the information is presented in this way it helps the child understand the sequence of daily events.

As can be inferred from the work of many researchers (such as [4, 5, 6]), individuals with autism think in pictures, not words, and play a video in their mind when reasoning. Below is the pictorial representation of phases of an eating event.

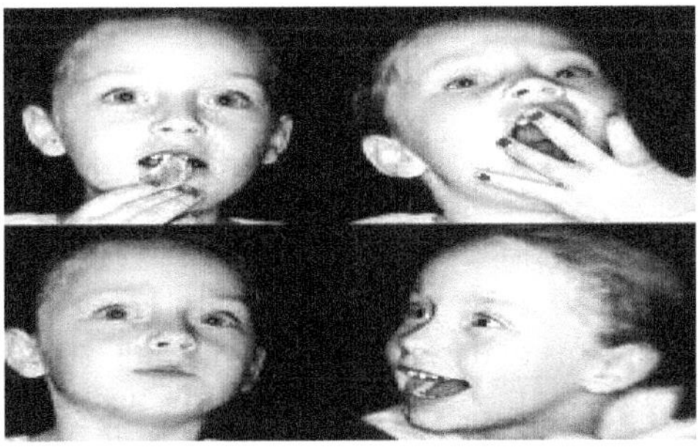

**Fig. 1.** Phases of an eating event

One of the most remarkable mysteries of autism has been the magical ability of most autistic people to perform outstandingly at visual spatial skills while performing so poorly at verbal skills. Visual thinking enables them to build entire systems in their imagination. They prefer to use diagrams, pictures and movies to see the information they are learning. Temple Grandin, an architect with autism, often says that she can translate both spoken and written words into full colour movies, which run like a movie in her mind [3].

From the above explanation we can see that, for autistic people to remember something, they must first recall when and where the event happened, and then identify it. When they are searching for something with Web information retrieval systems, they need to recall memories of the events already occurred in the form of sequence of images or a short movie and use them for their search.

Autistic children are motivated by imagination and consistency which is why computers are the perfect assistive technology tool for learning. They put the child in the driver's seat and make them learn to function independently. Research has found that students with Autism who use computers have increased attention spans, can stay in their seats longer, develop improved fine motor skills, and show greater ability to generalize skills across environments by repeating a wanted behaviour at home that was learned at school. Moreover computer games can help children with autism in educational areas such as learning new vocabulary, practicing math skills or improving eye-hand coordination. Due to the enormous benefits computers can provide, they should be an integral part of a special education student's daily curriculum.

The "Autistic Learning style Ontology" consists of various events in their lives in the form of entities and information regarding those events as annotations. The retrieved events that occur in autistic people's daily lives are stored as subclass information of the event class. Therefore, by asking the child when and where the event happened, and by showing images related to the events the system can infer which kind of event happened to the child. Such questions asked of the child can help them to discover the queries and get the desired results. Below (Fig 2.) is a section of the "Autistic learning style Ontology" which presents various games for autistic people.

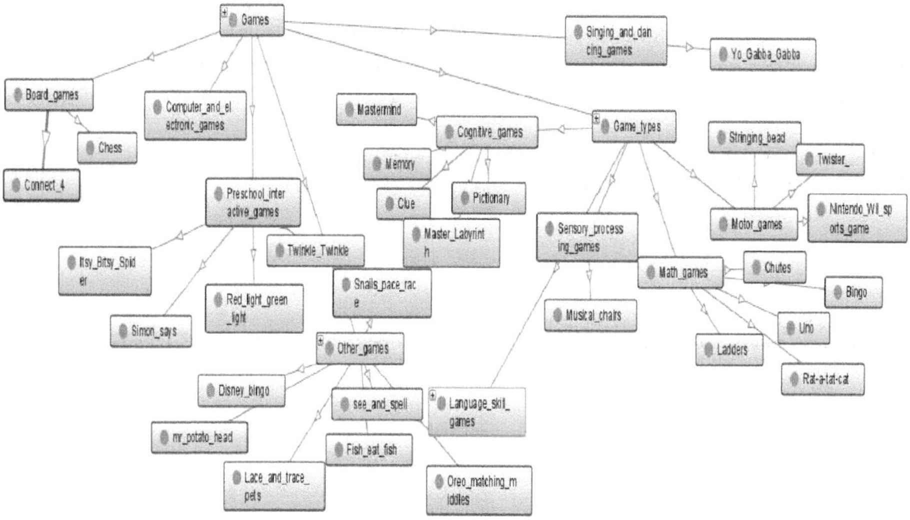


**Fig. 2.** Autistic learning style Ontology graph

## 3 Semantic Information Retrieval Process

The Semantic Web combined with ontology can be used for visualization techniques in several different ways, but the visualization is dependent on characteristics of the ontology used. Ontology helps both people and machines communicate more effectively by providing a common definition of a domain. The GUI serves as an interface between the user and the system.

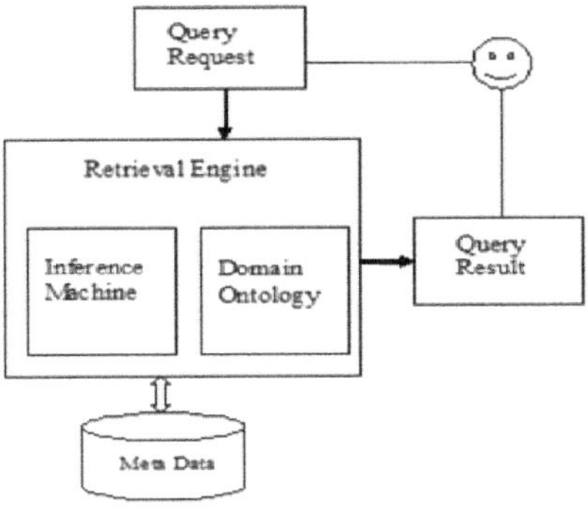


**Fig. 3.** Semantic Information Retrieval system

A web retrieval process is discussed in this section. Imagine this scenario. An autistic child Bunny aged 9 wants to play a game "reader rabbit" which he saw his friend playing on his computer a day before. He doesn't know the name of the game but know the name of the character "rabbit". If he search that term in any search engine, the results presented are hardly helpful. There are listings for animals, Chinese zodiac sign rabbit, web proxy mixed into the results. Only after sifting through multiple listings and reading through the linked pages will he be able to find the exact thing he wants. But being an autistic child he is not able to do this.

However, in a Semantic Web-enabled environment, you could use a Semantic Web agent to search the Web for "reader rabbit". The following are interactions between bunny and the web Agent:

1. Bunny inputs "rabbit" as a search keyword. The resulting data include many irrelevant pages such as animal, zodiac etc.
2. Agent: "Where did you see rabbit?"
3. Bunny inputs "game".
4. The system collects a set of character names in popular games on the Web.
5. Agent: "You are searching for Reader Rabbit in which a rabbit with blue and red striped sweater plays?
6. Bunny: inputs "Yes."
7. Agent: The system dynamically finds the keyword "Reader rabbit" in the "Autistic Learning style Ontology." The subclass hierarchy consists of "Games," "Game types," "Language skill games" and "Reader Rabbit".
   Agent: "Would you like to play the game Reader Rabbit?"
8. Bunny: inputs "Yes"
9. The agent starts the game on the computer and tells him about the rules of the game.

## 4 Conclusion

The Semantic Web provides a futuristic approach to the web. In this information is given explicit meaning, making it easier for machines to automatically process and integrate information available on the Web. The main problem with the existing ontology is that it does not provide the latest information because it is static and fixed, no new data can be added to it. Whereas this "autistic learning style Ontology", can be modified and enhanced taking the latest information from various sources related to the autistic people. E.g. Blog's written by mothers, teachers of autistic children and from the web. This ontology is to help autistic people retrieve information about the various events happening in their lives and their learning styles. When children want to search for something, they use memories and sequence of images in the form of a video or a short story to express what they want. To identify the right query, their unique and special ways to express things needs to be understood. By using this "autistic learning style ontology", it is expected to better understand their needs and learning styles.

## References


1. Manning, C.D., Raghavan, P., Schutze, H.: An Introduction to information retrieval, pp. 26–569. Cambridge University Press, Cambridge (2008)
2. Sumiyoshi, H., Yamada, I., Murasaki, Y., Kim, Y.B., Yagi, N., Shibata, M.: Agent Search System for A New Interactive Education Broadcast Service, Nhk Strl R&D No.84 (2004)
3. Kilicaslan, Y., Ucar, O., Guner, E.S., Bal, K.: An NLP-Based Assistive Tool For Autistic And Mentally Retarded Children: An Initial Attempt, Trakya University, Faculty of Engineering and Architecture, Department of Computer Engineering, 1–8 (2006)
4. Gradin, T.: Thinking in Pictures. Vintage Press (2006)
5. Williams, D.: Autism: An Inside-Out Approach. Jessica Kingsley Publishers (1996)
6. Williams, D.: Autism and Sensing: The Unlost Instinct. Jessica Kingsley Publishers (1998)
7. Vallet, D., Fernandez, M., Castells, P.: An Ontology Based Information Retrieval Model, Universidad Autonoma de Madrid
8. Poole, J.B., Skymcilvain, E., Jackson, L., Singer, Y.: Education for an Information Age: Teaching in the Computerized Classroom, 5th edn., p. 401 (2005)
9. Microsoft Corporation History of Microsoft's Commitment to Accessibility (2004), http://www.microsoft.com/enable/microsoft/history.aspx
10. Griswold, D.E., Barnhill, G.P., Myles, B.S., Hagiwara, T., Simpson, R.L.: Asperger Syndrome and academic achievement. Focus on Autism and Other Developmental Disabilities, summer 17, 94–102 (2002)
11. World Wide Web Consortium Web Content Accessibility Guidelines 2.0, http://www.w3.org/TR/2005/WDWCAG20050630/checklist

# Analyze the Performance of New Edge Web Application's over N-Tiers Layer Architecture

Pushpendra Kumar Singh [1], Prabhakar Gupta [1], S.S. Bedi[2], and Krishna Singh [1]

[1] Sshri Ram Murti Smarak College of Engineering & Technology, Ram Murti Puram, 13 Km. Barelly Nainital Road, Bareilly, (U.P.), India
[2] M.J.P. Rohilkhand University, Bareilly, (U.P.), India

**Abstract.** This paper is an empirical case study to predict or estimate the performance and variability of same type software frameworks used for web application development. First we explore and analyze (Web Applications) PHP and ASP.NET considering quality attributes. Second we develop two same web applications i.e. Online Book's Mart (Web applications to purchase books online) in PHP and ASP.NET. Finally we conduct automated testing to determine and analyze application's performance. Software architecture, CSS, database design, database constraints were tried to keep simple and same for both applications i.e. Applications developed in PHP and ASP.NET. This similarity helps to establish realistic comparison of applications performance and variability. The application's performance and variability is measured with help of automated scripts.

**Keywords:** ASP.NET, Autometed Testing / Scripts, CSS, PHP Software Engineering, Software Rchitecture

## 1 Introduction

Today, the Internet has grown to be a million-network network, and internet dependence on our life is increasing day by day. We need internet in almost every field of life. We use internet mostly in form of web applications. We use web applications for paying utility bills, social networking, email, online transactions, etc. There are three main frameworks for interactive application development. These are EJB (Enterprise JavaBeans), PHP (Hypertext Preprocessor) and ASP.NET. In this paper, we try to compare these frameworks in terms of performance and quality attributes.

## 2 Quality Attributes Analysis

When we are starting to develop a website using .Net framework 3.5 / PHP, then we should consider about important quality attributes which may be crucial for both business and end user perspective.

A. Mantri et al. (Eds.): HPAGC 2011, CCIS 169, pp. 299–305, 2011.

### 2.1 ASP.NET 3.5

- **USABILITY:** we used Visual Studio 2008 for online book's mart application development. Visual Studio is a powerful IDE for application development in ASP.NET.

- **PERFORMANCE:** ASP.NET is power full framework among others. We can easily check the performance of our website, to calculate the response time of a request or action. If our requests take minimum time to perform any action then we can say our website has good performance.

- **AVAILABILITY:** of ASP.NET website is based on different quality attributes such as performance, maintainability, supportability, testability, security and scalability [4]. Right now ASP.NET has great availability for small scale and large scale web applications.

- **TESTABILITY:** Usually ASP.NET web applications have n-tiers layer architecture and each layer is totally independent to other layers. We can easily test each layer separately using Unit testing. Visual Studio 2008 provides powerful unit testing feature to test developed components.

### 2.2 PHP (Hypertext Preprocessor Version)

- **USABILITY:** PHP fully supports Object Oriented Programming (OOP) which makes PHP more powerful and usable language. PHP is an open source programming language, it means developer easily develop their application using PHP without any cost [1].

- **PERFORMANCE:** PHP is an interpreter language; it does not need to go into memory for compiling the code like others languages e.g. C#, VB, etc. Due to this mechanism PHP has better speed and is efficient on any type of browser [1]. PHP performance is based on PHP version; in which you are developing your application, web server environment and your system/coding complexity. To increase website performance, we can use PHP caching module and by buffer flushing [2].

- **AVAILABILITY:** As we know PHP is an interpreter scripting language [1], it does not need to compile inside memory that's way its response back time is very good.

- **TESTABILITY:** The syntax and logic of PHP applications are very straight forward and easy to understand. Those applications developed inside Content Management Systems (CMS) provide drag and drop mechanism e.g. WordPress and Joomla. These applications require not much effort for testing because most of testing part is done by CMS development side [3].

# 3 Requirements Definition

## 3.1 General Requirements

Main objective is to develop online book's mart for professional book's and latest arrivals; i.e. customers should be able to shop new and latest books. The application shall be developed using ASP.NET 3.5 Framework and PHP 5.3.0. The testing of the application should be accomplished to measure and compare the framework's performance. Following are the core application requirements:

- Customer registration and log in/out
- Product browsing and shopping cart
- Checkout of free components and services (not requiring payment)
- Download of purchased books
- Order history / Product search
- Product management.

# 4 Implementation

Implementation environment for application developed in ASP.NET and PHP require the operating system is Windows Server 2003 for both and the implementation tools, software and technologies are Visual Studio 2008, ASP.NET 3.5, SQL Server 2005 and WAMP Server 2.0.

# 5 Performance Testing

## 5.1 Test Methodology

We decided response time as key criteria for measuring the speed and variability. "Response time" here refers to the time interval between client sending the request and receiving the response. We chose response time, because we are comparing the same application developed using two different technologies i.e. ASP.NET and PHP. PHP and ASP.NET application's architecture, database design, CSS and database records were kept same during application development. We request same page(s) of different applications with same data from database and measure the response time.

## 5.2 Small Scale Performance Testing

Small scale testing involves application testing with small amount of data and requests. Web Pages used for testing are: Page1. Default.(aspx/php), Page2. SearchProduct. (aspx/php), Page3. Products. (aspx/ php), Page4. SignIn. (aspx/ php), Page5. UserOrders. (aspx/ php), Page6. ProductDetail. (aspx/ php), Page7. UserOrdersDetail. (aspx/ php), Page8. ProductDownload. (aspx/ php), Page9. AddEditProduct. (aspx/ php), Page10. ProductDetail. (aspx/ php)

Web pages mentioned above are part of application developed in PHP and ASP.NET. We performed 3 tests in small scale testing.

- In test 1, 10000 page requests were generated from one client. Table 1, 2 shows the response time results of 10000 requests for ASP.NET and PHP clients respectively.

**Table 1.** Test 1 result for ASP.NET application

| Client # | Page # | No. of Req. | Avg. Time(s) | Max. Time(s) | Min. Time(s) | Med. Time(s) | Total Time(s) |
|---|---|---|---|---|---|---|---|
| 1 | 2 | 10000 | 0.006 | 0.563 | 0.005 | 0.005 | 59.11 |

**Table 2.** Test 1 result for PHP application

| Client # | Page # | No. of Req. | Avg. Time(s) | Max. Time(s) | Min. Time(s) | Med. Time(s) | Total Time(s) |
|---|---|---|---|---|---|---|---|
| 1 | 2 | 10000 | 0.007 | 1.463 | 0.005 | 0.005 | 65.52 |

- In test 2, 10000 requests of page #6 are sent by two clients simultaneously.

**Table 3.** Test 2 result for ASP.NET application

| Client # | Page # | No. of Req. | Avg. Time(s) | Max. Time(s) | Min. Time(s) | Med. Time(s) | Total Time(s) |
|---|---|---|---|---|---|---|---|
| 1 | 6 | 10000 | 0.013 | 18.204 | 0.003 | 0.003 | 127.82 |
| 2 | 6 | 10000 | 0.010 | 01.157 | 0.003 | 0.003 | 99.93 |

**Table 4.** Test 2 result for PHP application

| Client # | Page # | No. of Req. | Avg. Time(s) | Max. Time(s) | Min. Time(s) | Med. Time(s) | Total Time(s) |
|---|---|---|---|---|---|---|---|
| 1 | 6 | 10000 | 0.012 | 4.791 | .005 | 0.006 | 113.49 |
| 2 | 6 | 10000 | 0.012 | 4.783 | .005 | 0.006 | 127.22 |

- Test 3 results shows poor performance by ASP.NET clients. In Test 3, 10000 requests of page #9 and page #10 were generated by two clients simultaneously.

**Table 5.** Test 3 result for ASP.NET application

| Client # | Page # | No. of Req. | Avg. Time(s) | Max. Time(s) | Min. Time(s) | Med. Time(s) | Total Time(s) |
|---|---|---|---|---|---|---|---|
| 1 | 9 | 10000 | 0.256 | 04.531 | 0.008 | 0.302 | 2561.47 |
| 2 | 10 | 10000 | 0.015 | 0.905 | 0.003 | 0.003 | 146.28 |

**Table 6.** Test 3 result for PHP application

| Client # | Page # | No. of Req. | Avg. Time(s) | Max. Time(s) | Min. Time(s) | Med. Time(s) | Total Time(s) |
|---|---|---|---|---|---|---|---|
| 1 | 9 | 10000 | 0.016 | 8.4 | 0.006 | 0.008 | 160.64 |
| 2 | 10 | 10000 | 0.016 | 8.676 | 0.005 | 0.006 | 166.06 |

### 5.3 Small Scale Testing Results Analysis

Table 7 shows the average response time and median response time of ASP.NET application versus PHP application for all clients for each test. We can see that except for test #3, ASP.NET clients average response time is faster than PHP clients. But in test 3, ASP.NET application is slower than the PHP application.

**Table 7.** ASP.NET and PHP Performance

| Test # | Average ASP.NET response time(s) | Average PHP response time(s) | Average ASP.NET median time(s) | Average PHP median time(s) |
|---|---|---|---|---|
| 1 | 0.006 | 0.007 | 0.005 | 0.005 |
| 2 | 0.023 | 0.024 | 0.003 | 0.006 |
| 3 | 0.271 | 0.033 | 0.153 | 0.007 |

### 5.4 Large Scale Performance Testing

In large scale performance testing, we tried to test applications with large amounts of data i.e. more database records. Two web pages were used for testing in this phase are Page6. ProductDetail. (aspx/ php), and Page7. UserOrdersDetail. (aspx/ php)

- In test 1, clients request 1000 products details. Results show that ASP.NET clients are more than twice as faster than PHP clients.

**Table 8.** Test 1 results for ASP.NET application

| Client # | Page # | No. of Req. | Avg. Time(s) | Max. Time(s) | Min. Time(s) | Med. Time(s) | Total Time(s) |
|---|---|---|---|---|---|---|---|
| 1 | 6 | 1000 | 0.007 | 3.382 | 0.003 | 0.003 | 6.98 |

**Table 9.** Test 1 results for PHP application

| Client # | Page # | No. of Req. | Avg. Time(s) | Max. Time(s) | Min. Time(s) | Med. Time(s) | Total Time(s) |
|---|---|---|---|---|---|---|---|
| 1 | 6 | 1000 | 0.018 | 0.918 | 0.005 | 0.015 | 17.68 |

- In test 2, each client requests 1000 distinct products detail pages each. There are total 10 clients sending requests simultaneously.

**Table 10.** Test 2 results for ASP.NET application

| Client # | Page # | No. of Req. | Avg. Time(s) | Max. Time(s) | Min. Time(s) | Med. Time(s) | Total Time(s) |
|---|---|---|---|---|---|---|---|
| 1 | 6 | 1000 | 0.013 | 0.044 | 0.004 | 0.013 | 12.89 |
| 2 | 6 | 1000 | 0.013 | 0.043 | 0.005 | 0.013 | 13.35 |
| 3 | 6 | 1000 | 0.013 | 0.046 | 0.004 | 0.013 | 13.18 |
| 4 | 6 | 1000 | 0.013 | 0.042 | 0.005 | 0.013 | 13.41 |
| 5 | 6 | 1000 | 0.013 | 0.044 | 0.003 | 0.013 | 12.78 |

**Table 11.** Test 2 results for PHP application

| Client # | Page # | No. of Req. | Avg. Time(s) | Max. Time(s) | Min. Time(s) | Med. Time(s) | Total Time(s) |
|---|---|---|---|---|---|---|---|
| 1 | 6 | 1000 | 0.532 | 433.503 | 0.005 | 0.037 | 531.64 |
| 2 | 6 | 1000 | 0.540 | 473.141 | 0.005 | 0.016 | 539.51 |
| 3 | 6 | 1000 | 0.530 | 438.180 | 0.006 | 0.034 | 529.65 |
| 4 | 6 | 1000 | 0.545 | 482.036 | 0.005 | 0.009 | 544.92 |
| 5 | 6 | 1000 | 0.102 | 19.071 | 0.007 | 0.048 | 101.47 |

- In test 3, each client requests 10000 pages each. Each page contains distinct product detail.

**Table 12.** Test 3 results for ASP.NET application

| Client # | Page # | No. of Req. | Avg. Time(s) | Max. Time(s) | Min. Time(s) | Med. Time(s) | Total Time(s) |
|---|---|---|---|---|---|---|---|
| 1 | 6 | 10000 | 0.014 | 0.134 | 0.003 | 0.013 | 144.36 |
| 2 | 6 | 10000 | 0.014 | 0.083 | 0.006 | 0.013 | 144.69 |
| 3 | 6 | 10000 | 0.014 | 0.079 | 0.004 | 0.013 | 144.72 |
| 4 | 6 | 10000 | 0.014 | 0.137 | 0.003 | 0.013 | 144.72 |
| 5 | 6 | 10000 | 0.014 | 0.082 | 0.003 | 0.013 | 144.37 |

**Table 13.** Test 3 results for PHP application

| Client # | Page # | No. of Req. | Avg. Time(s) | Max. Time(s) | Min. Time(s) | Med. Time(s) | Total Time(s) |
|---|---|---|---|---|---|---|---|
| 1 | 6 | 10000 | 0.089 | 97.738 | 0.006 | 0.018 | 890.17 |
| 2 | 6 | 10000 | 0.071 | 102.220 | 0.006 | 0.008 | 705.19 |
| 3 | 6 | 10000 | 0.074 | 94.826 | 0.006 | 0.010 | 743.30 |
| 4 | 6 | 10000 | 0.086 | 109.728 | 0.006 | 0.021 | 861.47 |
| 5 | 6 | 10000 | 0.071 | 107.370 | 0.005 | 0.017 | 705.53 |

### 5.5 Large Scale Performance Testing Results Analysis

Table 14 shows the total response time and average median response time of all the clients in each of 3 tests. We can see that in all 3 tests ASP.NET clients average response time is faster than PHP clients. However, ASP.NET and PHP clients median response times are faster in 3 tests each.

**Table 14.** ASP.NET vs. PHP Performance comparison

| Test # | Average ASP.NET response time(s) | Average PHP response time(s) | Average ASP.NET median response time(s) | Average PHP median response time(s) |
|---|---|---|---|---|
| 1 | 0.007 | 0.018 | 0.003 | 0.015 |
| 2 | 0.012 | 0.531 | 0.012 | 0.030 |
| 3 | 0.015 | 0.801 | 0.013 | 0.015 |

## 6 Conclusion

The paper aimed at calculating and analyzing the performance and variability of ASP.NET 3.5 and PHP 5.3.0 frameworks. We developed web application i.e. online books mart using these frameworks. We found that Visual studio 2008 along with ASP.NET provide a strong IDE (Integrated Development Environment) for web application development.

We performed a number of automated tests to measure the performance of these applications. We performed small scale, and large scale tests. In terms of performance, ASP.NET performed slightly better than PHP in small scale tests. However, ASP.NET and PHP showed mixed results in large scale tests. PHP performed better in database access from multiple tables.

## References

[1] Biz Five. PHP Vs ASP.NET, http://www.bizfive.com/articles/web-design/comparing-PHP-and-ASP.NET/ (accessed on March 08, 2010)
[2] Google Code. Let's make the web faster, http://code.google.com/speed/articles/optimizing-PHP.html (accessed on March 08 , 2010)
[3] Joomla. Unit Testing, http://docs.joomla.org/Unit_Testing (Accessed on March 26 , 2010)
[4] Chapter 16: Quality Attributes, http://msdn.microsoft.com/en-us/library/ee658094.aspx (accessed on July 6 , 2010)

# Self-Configurable Scheduling Algorithm for Heterogeneous Computing Systems

A. Prashanth Rao[1] and A. Govardhan[2]

[1] Reasearch Scholar,
JNTU College of Engineering
KarimNagar (Dt), A.P, India
adirajupppy@yahoo.com
[2] Professor of CSE & Principal,
JNTU College of Engineering,
KarimNagar (Dt), A.P, India
govardhan_cse@yahoo.co.in

**Abstract.** Research in real time task scheduling algorithm is a challenging problem for high performance computing system; especially achieving mapping of tasks to processors is a key design issue in the heterogeneous computing environment. There are many existing scheduling algorithms covered in the literature but none of them mapped the specific task to the corresponding processor. In this paper, we build a new scheduler model that is suitable to provide timing requirement constraints, (specific processor, specific task) pair and load balancing taken into account. This paper also addresses percentage of tasks missing their deadline due to the running out of resources of the system. When scheduler makes use of cloud of resources, the missing tasks can be scheduled properly. The new scheduler combines all types of real time and non real-time tasks and schedules them. Hereafter, we call this algorithm as Self-Configurable Scheduling (SCS) Algorithm. This algorithm automatically adjusts the jobs among processing elements and there will be a feedback loop from each processing element to the load balancer. We will build a simulation that can estimate shortage of resources in order to add resources and also allocate missed tasks to the cloud of resources.

**Keywords:** Cloud Computing, Deadline, Feedback Loop, Scheduling, Utilization Value.

## 1 Introduction

A Distributed Heterogeneous Real Time Computing System consists of diversely capable machines harnessed together to execute a set of tasks that vary in their computing requirements. The computational requirements of real time tasks may be varying due to many reasons such as Heterogeneity of Processor or I/O operation or speed of the processor. Due to these reasons the task execution cannot be deterministic which violates timing requirements and leads to severe problems. Over the last decade, heterogeneous systems have been widely used for scientific and commercial applications [1].There are

A. Mantri et al. (Eds.): HPAGC 2011, CCIS 169, pp. 306–315, 2011.

wide variety of scheduling algorithms [2],[3],[4],[5] which were developed to improve the better performance of heterogeneous computing environment. Ranaweera and Agrawal proposed a scalable scheduling scheme called STDP for heterogeneous systems [6].

The main objective of scheduling theory is that required number of processing elements make available and need to minimize that number such that timing requirement met. The availability is defined as the ratio of the total computing time for processing elements is said to be functional to given interval of time. There are many scheduling algorithms were developed to enhance the performance metrics such as availability, average response time, QoS, processor utilization of the system [7],[8],[9],[10],[11]. The performance of a heterogeneous system will be degraded if one or multiple processing elements are overloaded and out of order due to random breakdown or preventive maintenance. Many of high-performance applications such as Military applications , 24×7 healthcare applications and international business applications require computing platforms with high availability [12][13][14][15][16]. In all these type of applications there is need to schedule both real time and non-real time tasks. Our objective is to schedule all those type of tasks together for better resource utilization.

The execution time of a task on the different processors can vary dramatically change due various resons. The total execution time units allocated in the heterogeneous computing environment can be minimized if (processor, task) pair considered together. This paper addresses the problem of dynamic scheduling algorithm of set of tasks on heterogeneous computing environment, where execution time of tasks diversely changes. All types of tasks whether static or dynamic are scheduled together, balance the load among the processing elements, (task, processor) pair mapped appropriately and also added with timing requirement guaranteed.

**Motivation:** The main objective of this paper to make use small computing power devices such as mobile phones, a laptop, a PC, or a microwave oven. Some of these machines may have a microcontroller and some other may have a processor with operating system. We can expect all other machines to have operating system along with a processor or a microcontroller in future. All devices surround us will make use of computing elements to solve a complex problem.

This paper is organized as follows. Section 2 describes Basic Terminology and Definitions. Model Description and Problem Formulation is presented in Section 3. Section 4 describes Self Configurable Scheduler and implementation. Sections 5 demonstrate Results and Discussions. Finally, the Conclusion and future scope are given in Section 6.

## 2 Basic Terminology and Definitions

In this section we first define some terms and then present SCS algorithm which uses these terms.

**Definition 1:** Each $task_i$ characterized by four tipple form phase (©), periodicity or deadline ($d_i$) and execution time ($e_i$).If deadline parameter not mentioned in the task system, it simply periodic set otherwise set will be treated as non-periodic set.

**Definition 2:** The Feasible Schedule for a given processor defined such that it's total execution time units ($T_{jk}$) less than or equal to Feasibility Check Parameter($T_{jk} \leq d_{lk}$) where $T_{jk}$ is total time execution units allocated to processor j in cluster k and $d_{lk}$ is maximum deadline of a task on processor j.

**Definition 3:** The Expected Time Matrix ETM(i,j) can defined as the execution time the task i on machine j and ETM matrix contains all n tasks execution times among m machines.

**Definition 4:** The Threshold Value (TL) defined the load on each processor which can be stated if

a) $T_{jk} < d_{lk}$ Lightly Loaded
b) $T_{jk} = d_{lk}$ Load balanced
c) $T_{jk} > d_{lk}$ Heavily Loaded.

**Definition 5:** k cluster are formed from ETM (i,j) such that lowest cluster index have minimum task-processor pair combinations and so on. The lowest cluster index have more dedicated or specialized pair than higher index cluster.

Using above definitions we can develop a self-configurable scheduling algorithm which discussed in the section (4).Before that we need define problem formulation and model description to be discussed in next section.

## 3 Model Description and Problem Formulation

We consider a queuing architecture of a Real Time Heterogeneous Computing System in which real time application composed of n tasks needs to be scheduled on m heterogeneous processing elements. The heterogeneous multiprocessor system contains m dedicated processors $p_1$ ,$p_2$ , $p_3$, $p_4$ , ............$p_m$ and one centralized scheduler which is fault free. Both m and n are finite integers that are equal or greater than 1. The task execution time may vary in computing elements and this variation is due to differencein processing speeds, processor architecture, specialized processor which are suitable to particular domain of application and also other constraints like memory, communication among tasks. Initially all those types of tasks are executed sequentially in all different types of processors which are configured to our application and task information recorded in expected time matrix (ETM)[14]. Similarly task information under different load conditions are tested and updated in the ETM matrix (Def. 3). This matrix is used for modeling of given Heterogeneity Environment (HC) System. The communication between processors and centralized scheduler is through dispatch queues. The system architecture model, depicted in Fig1, is known as self-configurable scheduler (SCS).

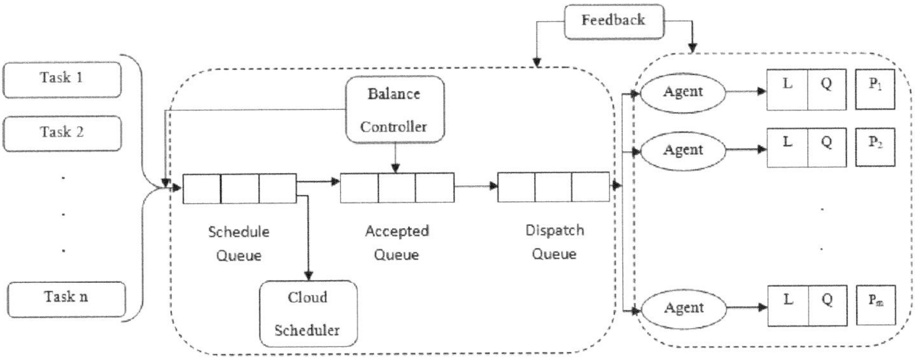


**Fig. 1.** Self Configurable Scheduler

### 3.1 MajorScheduler Components

The scheduler in this context works for balancing the load among different heterogeneous processing elements and other details such ETM (i,j), QoS, and dynamically adding resources are not subject of this paper. The major components required to understand the SCS algorithm depicted below.

#### 3.1.1 Balancing Controller

The load balance algorithm controls the load on each processor with proper communication between an agent on each processor and balancing controller. If load exceeds a certain threshold value (Def.4) then it automatically detaches from one of the tasks from local queue and adds to waiting queue. The task which is removed from local queue is mostly the last task or higher cluster index task and it adjusts automatically the feasibility check parameter. So we need to remove one of the tasks from local queue such that both the remaining tasks in the local queue and the removed task need to meet the deadline. The following are some of the reasons for the overloading of the processor but not limited.

a) Unknown and unpredictable tasks and its behavior.
b) Delay for getting I/O data
c) Context switch among tasks

Load balancing is an important issue considered in the scheduling strategy and allocation of feasible execution time units to the processor (Def.2). The purpose of load balancing is to distribute the work equally among all processing elements. When the work is distributed equally, then at the end of a time interval all processing elements will be free further allocation. Hence, the degree of the load balancing is able to be evaluated by the standard deviation of the processing elements.

Let $d_{ij}$ deadline of task i on processing element j and this task deadline will setting as feasibility check parameter and $TE_{tj}$is total execution units allocated to processor j.

*Ld* be the load balancing parameter which is given by

$$Ld = \sqrt{\sum_{j=1}^{m} \left(TE_j - d_{lj}\right)^2 / m}$$

When Ld is equal to zero, then load is balanced between the processing elements.

### 3.1.2 Local Agent

Local agent responsible monitoring tasks locally and it updates information to load balancer whenever the following problems are encountered.

1. The total execution units of priority queue exceed its feasibility check parameter.
2. The task which is being executed may take longer time then that was expected earlier.
3. It removes one of the tasks from local queue if there is imbalance in the local queue.

### 3.1.3 Cloud Scheduler

The cloud scheduler is responsible for adding/deleting resources to existing system whenever resources available are not sufficient. The cloud scheduler identifies the required resource based on the behavior of the task and its importance. The cloud computing connects many computing elements which surround us. These computing devices canbe alaptop, a mobile device, processing element or any device having computing capabilities. The cloud scheduler is invoked whenever tasks are not scheduled using available resources. The load balancer identifies tasks which are not urgent and some soft real time tasks are allocated to cloud of resources. The clouds of resources are generally configured with our system initially. The main components in cloud scheduler(Fig2) are

a. Cloud Manager: The cloud manager is responsible to register new computing elements or to remove existing elements. Specifically we need to configure certain computing devices.
b. Task Queue: The tasks which are not scheduled in the real time environment are kept in this queue and are simply known as non-real time tasks. There are certain tasks which may have deadline but not lead to severe problems if they are not meeting timing constraints. All those are scheduled using cloud computing.
c. Registered Computing Element: The registered computing elements or devises, randomly pick up processing elements which surround you. The primary use of these processing elements may entirely different. But whenever an application is required we can make use of those resources.
d. Mapping Process: The mapping process defines suitability of the task where it executes and sends results back to global scheduler.

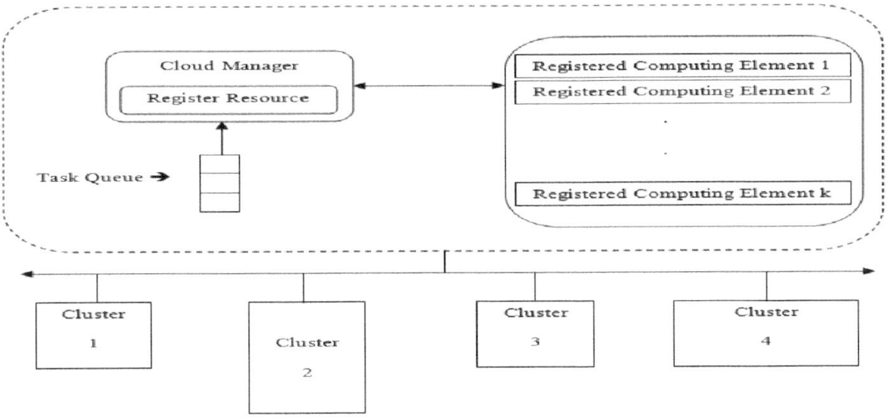


**Fig. 2.** Cloud Scheduler

Using above terminology we need develop a Self-Configurable Scheduler (SCS) discussed in next section.

# 4 Self-Configurable Scheduler Algorithm

To balance load among the processing elements is the primary objective of minimizing the total execution time of a set of tasks on heterogeneous processor architecture. We are aware that tasks can be categorized as either static or dynamic based on the information available. The task scheduling problem addressed in this paper is that of assigning the independent tasks to processor such that all types of tasks meet timing requirements. We propose a novel algorithm referred to as self-configurable scheduling to addresses the following issues simultaneously.

1. On line branch and bound search algorithm performs scheduling periodically at regular interval of time. This means the algorithm is self-configurable.
2. When (Processor, Task) pair is mapped appropriately, overallefficiency is improved. This means the specific domain task mapped to corresponding dedicated or embedded processor.
3. Every processor maintains priority local queue and agent .The priority queue length is equal to its feasibility check parameter and the agent monitors tasks locally.If the total execution units are allocated to processor exceeds its feasibility check parameter then agent notifies load balancer and receives communication from it for further action.

This technique invokes partial task scheduling in regular interval of time and places dynamic arrived tasks on already allocated processors.SCSAlgorithm uses online tuning parameter (def.2) which determines the feasibility condition and also queue length.

SCS is an on-line branch and algorithm that searches through space of all possible partial and complete schedules. In our task scheduling problem, a space of all partial and complete schedules can be represented as a tree G (V, E) consisting of aset of processing elements$v_{jk} \varepsilon V$ and set of edges $(v_{jk},v_{jk'}) \varepsilon E$ which need to get task from cluster k' to cluster k on the processor j. An edge $(v_{jk},v_{jk'}) \varepsilon E$ in the graphG represents a transform function in the state space that transformsprocessing element $v_{jk}$ to another processing element$v_{jk'}$. Each processing element known as cluster element $v_j$ represents a partial schedule$S_j = \{( T_{lj}, p_j)/ T_{lj} \varepsilon T , p_j \varepsilon P\}$ that assigns a set of tasks whose total execution units $T_{lj}$ at scheduling phase $\emptyset$ to corresponding processor j. An edge $(v_{jk},v_{jk'})$ in G represents the extension of the partial schedule of node $v_{jk}$by adding one more task to processor j from cluster k'.The partial schedule of a successor node $v_{jk'}$ of $v_{jk}$is thus $S_{k'} = S_j U\{(e_{ij}, p_j)\}$ ,where $P_j \varepsilon P$ and $e_{ij}$ execution time of task $t_i$ in processor $P_j$ that has not yet been scheduled as part of $S_j$. The total execution cost of processor $P_j$ can be calculated as ( $T_{lj} + e_{ij}$ ) known as CPU bound. The deadline of newly arrived task from cluster k' is $d'_{ij}$ and it is compared with Feasibility check parameter ($d_{lj}$ ).The greater among the two will be set as Feasibility Check parameter($d_{lj}$ ).The Algorithm 1 describes SCS which maps tasks to processing elements and dynamically adjusts load.

**Algorithm 1.** Self-Configurable Scheduler

1. Initially at Phase =$\emptyset$
   a. SCS select of one schedule and assign the tasks processors based on certain policy.
   b. Load on processor

$$T_{ij} = \sum_{i=1}^{n_{ij}} e_{ij} \quad 1 < i < n ,1 < j \le m$$

2. Invoke cluster algorithm at Phase $\emptyset'$ and newly arrived task with its execution time ($e_{ij'}$), deadline ($d_{lj}$') (Def.5)
3. Compute maximum deadline $d_{max} = \max (d_{lj} , d_{lj}' )$ (Def.2)
4. Load at Phase $\emptyset'$=(load at $\emptyset$-{($\emptyset'$) + ($e_{ij}$)} )

If (Load at Phase $\emptyset' \le d_{max}$ ) then

{

Allocate task processor j

Assign Load at $\emptyset$ = Load at $\emptyset'$ and push current load onto stack.

Else if (Load at Phase $\emptyset' \le d_{lj}$ ) then

{

Allocate non-real time task to j

Assign Load at $\emptyset$ = Load at $\emptyset'$ and push current load onto stack.

}

Else if {

Select another processor j+1 ;

Repeat steps 2 to 5 till all processor are visited. }

Else {

Search in the cloud for suitable processor which are already configured with our system}

Step1: Initially at Phase zero, the SCS scheduler map tasks to the processor based certain policy and $n_{ij}$ be the number of tasks allocated to the processor j and estimate load, check feasibility.
Step2: Monitors incoming tasks at Phaseare equal to Cand invokes cluster algorithm (Def.5), note deadline, execution time of incoming tasks.
Step3: Compare deadline or periodicity of incoming tasks with feasibility check parameter of given processor.
Step4: Estimate load on each processor and assign task with less loaded processor otherwise search suitable processor in the cloud. Best way remove low priority task from any of local queue and added to cloud of resources if incoming task may be tight deadline.

## 5 Results and Discussions

We evaluate in this section the performance of SCS algorithm using simulations. The two important parameters task arrival rate and executions time units are considered. The execution time $e_{ij}\varepsilon$ ETM (i.j) where$e_{ij}$ execution time of task i on processor j.We compare SCS with well-known algorithm SASC scheduling to reveal the strength of

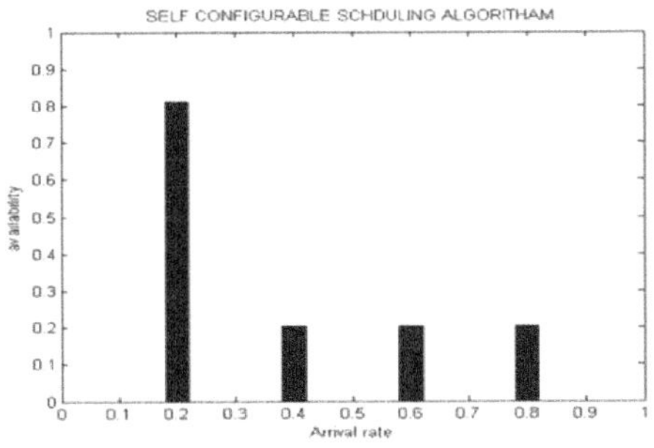


**Fig. 3.** Graph between availability Arrival Rate

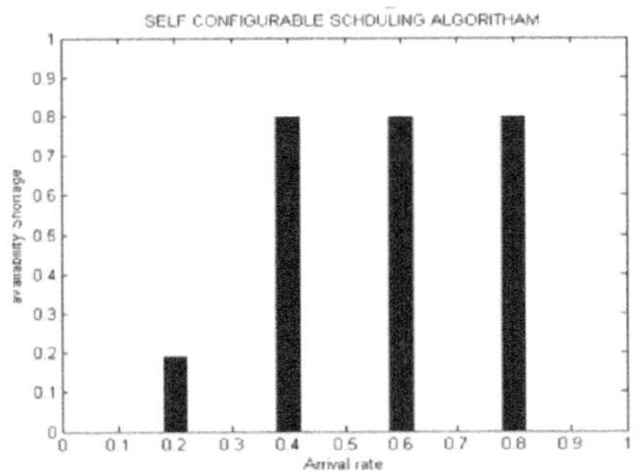


**Fig. 4.** Graph between availability shortage vs arrival rate

proposed scheduling strategy. The SASC algorithm usesthreshold load (TL) parameter to balance load among the processing elements which is static for given application. But SCS algorithm uses dynamically computed threshold load (TL) suitable for any type of real time applications. Thethreshold load (TL) parameter for processing element is equal to its feasibility check parameter and its value will be different for each processing element. In this simulation we have taken different sets of values to execute our proposed algorithm. But SASC uses static value whereas SCS uses dynamically adjusted parameter. Here wepresentperformance results of our proposed algorithm whose main metric is the availability of the resources and task arrival rate. When task arrival rate increases and available resources are not adequate then firstly we try to schedule the real time tasks. The remaining non-real time tasks schedulable tasks are directed to the cloud scheduler. From Fig 3 we can conclude that utilization of all processing elements is the percentage of total tasks units running out of total required execution time units. We can vary the mean arrival rate from 0.2 to 1.0 with an increment 0.2.From Fig 4 can be concludes that shortage of the system quantifies the discrepancy between availability demands and actual availability offered by the system.

## 6 Conclusions

This paper provides a better algorithm called SCS for load balancing, among processing elements with each processor will have a feedback loop to the balancer. The rejected tasks will be submitted to cloud of resources and they will be allocated to properly configure computing elements. Our future work is to design specialized device which connects any type of task to the respective processing elements. New learning capabilities will be added to the scheduler to make it work under heterogeneous computing environment.

## References

1. Dogen, A., Özgüner, F.: LDBS: A Duplicate Based Scheduling Algorithms for Heterogeneous Computing Systems. In: Proc. Int'l Conf. Parallel Processing (ICPP), B.C., Canada, pp. 352–359 (2002)
2. Dogan, A., Özgüner, F.: Reliable Matching and Scheduling of Precedence-Constrained Tasks in Heterogeneous distributed computing. In: Proc. Int'l Conf. Parallel Processing, pp. 307–314 (2000)
3. Topcuoglu, H., Hariri, S., Wu, M.-Y.: Performance-effective and Low-complexity Task Scheduling for Heterogeneous Computing. IEEE Trans. Parallel and Distributed Sys. 13(3) (March 2002)
4. Kuo, C.-F., Hai, Y.-C.: Real-Time Task Scheduling on Heterogeneous Two-Processor Systems. In: Hsu, C.-H., Yang, L.T., Park, J.H., Yeo, S.-S. (eds.) ICA3PP 2010. LNCS, vol. 6082, pp. 68–78. Springer, Heidelberg (2010)
5. Kim, K., Kim, D., Park, C.: Real-time scheduling in heterogeneous dual-core architectures. In: Proc.12th International Conference on Parallel and Distributed Systems, ICPADS 2006, July 12-15, vol. 2, p. 6 (2006)

6. Ranaweera, S., Agrawal, D.P.: Scheduling of Periodic Time Critical Applications for Pipelined Execution on Heterogeneous Systems. In: Proc. Int'l Conf. Parallel Processing (ICPP), pp. 131–138 (September 2001)
7. Qin, X., Xie, T.: An Availability-Aware Task Scheduling Strategy for Heterogeneous Systems. IEEE Transactions on Computers 57(2), 188–199 (2008)
8. Qin, X., Xie, T.: Security-Aware Resource Allocation for Real-Time Parallel Jobs on Homogeneous and Heterogeneous Clusters. IEEE Transactions on Parallel and Distributed Systems 19(5) (May 2008)
9. Zhu, X.M., Lu, P.: Multi-dimensional scheduling for real-time tasks on heterogeneous clusters. Journal of Computer Science and Technology 24(3), 434–446 (2009)
10. Lau, H.C., Zhang, C.: Job Scheduling with Unfixed Availability Constraints. In: Proc. 35th Meeting of the Decision Sciences Institute (DSI), Boston, USA, pp. 4401–4406 (November 2004)
11. Apon, A., Wilbur, L.: AmpNet - a highly available cluster interconnection network. In: Proceedings IEEE Intl' Symp. Parallel and Distributed Processing, April 22-26 (2003)
12. Sadfi, C., Ouarda, Y.: Parallel Machines Scheduling Problem with Availability Constraints. In: Proc. Int'l Workshop Project Management and Scheduling (2004)
13. Sanlaville, E., Schmidt, G.: Machine scheduling with availability constraints. Acta Informatica 35(9), 795–811 (1998)
14. Smith, S.P.: An Efficient Method to Maintain Resource Availability Information for Scheduling Applications. In: Proc. IEEE Int'l Conf. Robotics and Automation, vol. 2, pp. 1214–1219 (May 1992)
15. Sethuraman, J., Squillante, M.S.: Optimal Stochastic Scheduling in Multicalss Parallel Queues. In: Proc. ACM Sigmetric Conf. (May 1999)
16. Braun, T.D., Siegel, H.J., Beck, N., Bölöni, L., Maheswaran, M., Reuther, A.I., Robertson, J.P., Theys, M.D., Yao, B., Freund, R.F., Hensgen, D.: A comparison study of static mapping heuristics for a classify meta-tasks on heterogeneous computing systems. In: 8th IEEE Heterogeneous Computing Workshop (HCW 1999), pp. 15–29 (April 1999)

# Performance Analysis of Proposed MAES Cryptographic Techniques

Richa Kalra, Ankur Singhal, Rajneesh Kaler, and Promila Singhal

Haryana Engineering College, Jagadhri, MMEC, Mullana,
Rayat Bahara Group of Institutions, Patiala Campus, GIMT, Kanipala
kalraricha51@gmail.com, asking2K@gmail.com,
Rajneesh@gmail.com

**Abstract.** Cryptography is an emerging technology, which is important for network security. Research on cryptography is still in its developing stages and a considerable research effort is still required for secured communication. This paper is devoted to the security and attack aspects of cryptographic techniques. The simulation based information content test such as Entropy, Floating Frequency, Histogram, N-gram, Autocorrelation and Periodicity on ciphers is done. The simulation based Randomness test such as Frequency test, Pokers test, Serial test, Long run test on ciphers are done using CrypTool. Finally, we have benchmarked proposed MAES cryptographic algorithm in search for the best compromise in security.

**Keywords:** Cryptography, Ciphers, CrypTool, Decryption, Encryption, Performance, Security.

## 1 Introduction

The Network Security is the most vital component in information security because it is responsible for securing all information passed through networked computers. There are four fundamental precepts for preventing the information loss: Confidentiality specifies that only sender and intended receiver must be able to access the contents of a message and interception causes loss of confidentiality. Authentications mechanisms help establish proof of identities and Fabrication is possible in the absence of proper authentication mechanisms. Integrity assures that received message has not been altered in any way after sender sends it i.e. modification can takes place in original data. Availability means message must be available to authorized parties at all the times otherwise interruption takes place. In other words, only one particular element underlies many of the security mechanisms in use: Cryptographic techniques; hence our focus is on this area. The goal of Cryptography is to transform the original information into some other form. This transformation can takes place the form of encoding messages that make them Non-Readable. This art and science of achieving security is known as Cryptography [1]. Every encryption and Decryption process has two aspects. The algorithm and the key used for encryption and decryption. Broadly, there are two cryptographic mechanisms depending on what keys are used. If same key or one key is used for encryption and decryption we call the mechanism as symmetric key cryptography. If two different

A. Mantri et al. (Eds.): HPAGC 2011, CCIS 169, pp. 316–321, 2011.

key is used for decryption on we call the mechanism as asymmetric key Cryptography. Both ideas are illustrated in the figure1 and 2.

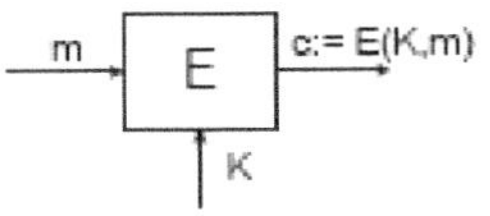


Fig. 1.

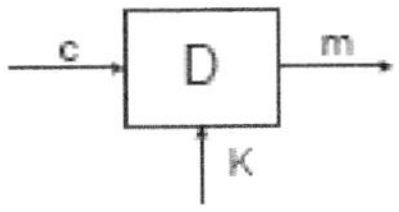


Fig. 2.

An efficient algorithm E, called encryption (algorithm), takes the key and a plaintext and outputs a cipher text, and an efficient algorithm D, called decryption (algorithm), takes the key and a cipher text and outputs a plaintext. Usually it denotes symmetric keys by K; $K_{1;}$ : :$K_n$, messages by m; $m_1$; : :$m_n$, and cipher texts by c; $c_1$ : :$c_n$.

## 2 Simulation Environment

The cryptanalyst main goal is to break the cryptographic system in every possible way, with his existing knowledge and available infrastructure. A part from this there are possibilities a cryptanalyst might exploit for different information from an encryption scheme e.g. information theory tests & randomness tests. To analyze the performance of Existing and new cipher is done using CRYPTOOL v1.4.30. Developing new Cipher "MAES" abbreviated as Modified Advance Encryption Standard and comparison of "MAES" with existing Ciphers is done using C will be the next future work. Where as in AES look up table is fetched at the time of Encryption & Decryption; MAES need not to be fetched as it is in-built in programming.

**Information Theory Tests:** The information theory tests mechanisms for cryptographic techniques are:

| | |
|---|---|
| Entropy | Calculate the entropy of a document. |
| Floating Frequency | Calculate the floating frequency of a document. |
| Histogram | Calculate the character frequency of a document. |
| N-Gram | Analyze the frequency of N-Grams of a document. |
| Autocorrelation | Perform autocorrelation of characters in a document. |
| Periodicity | Analyze the periodicity of a document. |

These parameters are evaluated with simulator CrypTool.

**Randomness Analysis Tests:** Following Parameters are taken into account:

**Significant level (α):** is the test of Hypothesis H0 gives the probability of incorrectly rejecting Hypothesis H0 although it is correct..

**Default offset & Test length:** decides whether default values are test/ user defined values. Generally, offset=0 & test length= file length.

**Use randomly chosen test block:** user can use its own default value & test length of file.

**Max. Test value:** is the statistical value that is dependent on α.

**Test outcome:** is a statistical value generated by test which is compared with max. test value.

**Tuple:** is an ordered ensemble of K elements.

The Randomness analysis contains the following tests

| | |
|---|---|
| Frequency Test | Checks the random quality of the active document with the Frequency test. |
| Poker Test | Checks the random quality of the active document with the Poker test. |
| Runs Test | Checks the random quality of the active document with the Run and Long-Run test. |
| Serial Test | Checks the random quality of the active document with the Serial Test. |

## 3 Simulation Results

Table 1 indicates the results obtained by applying information theory tests on different algorithms. When the entropy of cipher text is analyzed AES & MAES are better then other encryption algorithms. When Correlation is analyzed MAES is more effective and AES is the worst. Based on periodicity all are better then vernam ciphers. Figure 3 shows graphically.

**Table 1.** Results of Information Theory Tests

| Encryption Algorithm | Key length | Entropy | Autocorrelation | Periodicity |
|---|---|---|---|---|
| **Plain text** | ------ | 4.17/4.70 | 1-9,Z | No |
| **Vernam** | 11 | 0.00/4.70 | ----- | 1,1,8666,00 |
| **Hill** | 2*2 matrix | 2.25/4.70 | 0-1,S | No |
| **IDEA** | 128 bits | 5.49/8.00 | 1-2,Z | No |
| **DES(ECB)** | 64 bits | 5.66/8.00 | 1-2,Z | No |
| **DES(CBC)** | 64 bits | 5.45/8.00 | 1-2,Z | No |
| **3-DES(CBC)** | 128 bits | 5.70/8.00 | 1-2,Z | No |
| **AES** | 256 bits | 5.78/8.00 | 1-3,Z | No |
| **MAES** | 256 bits | 5.78/8.00 | 0-1,Z | No |
| **Digital signature** | Same as MD5 length | 4.17/8.00 | 1-5,Z | No |

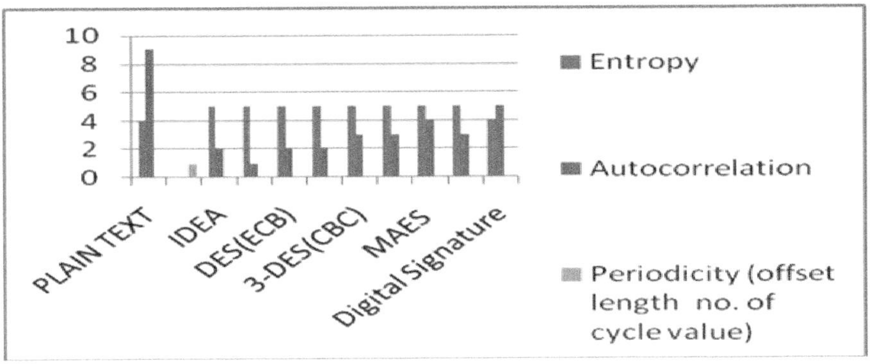


**Fig. 3.**

Further simulation is repeated for Randomness Tests are shown in table 2. Table 2 indicates the results obtained by applying randomness tests on different algorithms and results of MAES are better than others. Figure 4 shows graphically.

**Table 2.** Randomness Analysis of Encryption Techniques

| \Parameters Encryption Algorithm | Frequency Test A=0.05 MTV= 3.841 | Poker Test MTV=14.07 | Run Test MTV= 9.488 | Serial Test MTV=5.99 | Long Run Test MTV=34 |
|---|---|---|---|---|---|
| Plain Text | 455.58, F | 1031.63, F | 1382.04, F | 481.98, F | 9, P |
| Vernam | 69328.58, F | 16176.11, F | 30331.2, F | 13865.32, F | 0, P |
| Hill | 26.28, F | 66.70, F | 27.66, P | 25.79, F | 7, P |
| IDEA | 0.166, P | 7.875,P | 3.718, P | 0.1758,P | 6, P |
| DES(ECB) | 0.260,P | 18.06, F | 5.32,P | 33.63, F | 7, P |
| DES(CBC) | 0.161, P | 8.250,P | 3.32, F | 40.20, F | 9, P |
| 3-DES(CBC) | 3.760,P | 5.75,P | 1.180, F | 23.67, F | 10, P |
| AES | 0.66667,P | 0.52,P | 48.23, P | 14.43, F | 17, P |
| MAES | 3.4453,P | 6.75, P | ,4.06,P | 157.14, F | 21, P |
| Digital signature | 37.51, F | 536.87, F | 117.12, F | 101.12, F | 5, P |

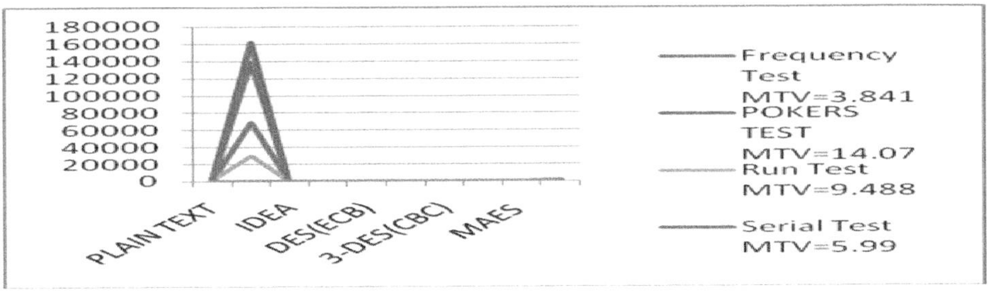


**Fig. 4.**

## 4 Proposed MAES Cryptographic Technique

The proposed work is the modification in existing AES algorithm named as MAES abbreviated as Modified AES. The steps of algorithm are as follows.

1. Initialize values.
2. Define S-box.
3. Expand Key.
4. Define xtime.
5. Perform Rounds.

## 5 Conclusion

The simulation based information content tests such as Entropy, Floating Frequency, Histogram, N-gram, Autocorrelation and Periodicity on ciphers are done. The simulation based Randomness test such as Frequency test, Pokers test, Serial test, Long run test on ciphers are done using CrypTool. Finally MAES cryptographic algorithm is designed for the best compromise in security.

## References

[1] Kahate, A.: Cryptography and Network security. Tata Mc Graw Hill Publishing Company Limited (2003)
[2] Lincke, S.J., Hollan, A.: Network Security: Focus on Security, Skills, and Stability. In: Proceedings of 37th ASEE/IEEE Frontiers in Education Conference
[3] Zhang, L., White, G.B.: Anomaly Detection for Application Level Network Attacks Using Payload Keywords. In: Proceedings of the 2007 IEEE Symposium on Computational Intelligence in Security and Defense Applications, CISDA 2007 (2007)
[4] Khalifa, O.O., Rafiqul Islam, M., Khan, S., Shebani, M.S.: Communications Cryptography. In: Proceedings RF and Microwave Conference, RFM 2004 (2004)
[5] Nadeem, A., Younus Javed, M.: A performance comparison of data Encryption Algorithm. In: Global Telecommunications Conference Workshops, Globe COM Workshops 2004. IEEE, Los Alamitos (2004)
[6] Wang, Y., Hu, M.: Timing Evaluation of known cryptographic Algorithm. In: International Conference on Computational Intelligence and security (2009)
[7] Singh, Y., Choba, Y.: Information Theory Tests based Performance Evaluation of Cryptographic Techniques. In: Proceedings of International Journal of Information Theory and Knowledge Management, vol. 1(2), pp. 475–483.
[8] Khaddour, M., Wang, Z., Hammami, O.: Student Member, IEEE, Student Member, IEEE and Member, IEEE , Performance Evaluation and Analysis of Parallel Software Implementations of TDES on a16-PE Embedded Multiprocessor Platform. In: Proceedings of 2009 ESRGroups, France (2009)

[9] Ahmad, M., Izharuddin: Randomness Evaluation of Stream Cipher for Secure Mobile Communication. In: Proceedings in 1st International Conference on Parallel, Distributed and Grid Computing, PDGC - 2010 (2010)
[10] Orner, S., Mansoor, S.S.P.: Performance analysis of Stream and Block cipher algorithms. In: 2010 3rd International Conference on Advanced Computer Theory and Engineering, ICACTE (2010)
[11] Ammar, A., El Sherbini, A., Ashour, I., Shiple, M.: Random Data Encryption Algorithm. In: Twenty Second National Radio Science Conference (NRSC 2005), Cairo-Egypt, March 15-17 (2005)

# Analysis of Core-Level Scale-Out Efficiency for OpenMP Programs on Multi-core Processors

Sareh Doroodian, Nima Ghaemian, and Mohsen Sharifi

Distributed Systems Research Laboratory, School of Computer Engineering
Iran University of Science and Technology,
Hengam St., Resalat Sq., Tehran, Iran
{sdoroodian,nima}@comp.iust.ac.ir,
msharifi@iust.ac.ir

**Abstract.** The majority of existing OpenMP compilers select the maximum number of available processing cores on a multi-core machine at runtime to execute a parallelized program on that machine. In this paper, we show that the use of maximum number of available cores does not necessarily result in speedup or efficiency. We show that in a considerable number of cases the use of more cores results in diminishing returns on execution time and efficiency. To help in choosing the proper number of cores, we propose an analytical method to estimate the execution times of OpenMP programs using different numbers of cores while considering the synchronization, excess computation and load imbalance overheads caused by the chosen number of cores. We validate our proposed method through a case study covering most recurrent and important structures and constructs of OpenMP.

**Keywords:** Modeling, Efficiency, HPC, OpenMP, Performance.

## 1 Introduction

Nowadays, the new generation of processors with more than one processing cores on a chip are found in almost every desktop computer and the number of cores on a chip seems to be following the Moore's law. This architectural evolution has affected the desktop computing paradigms and has made the multithreaded and concurrent programming a dominant paradigm. Predicting the adequate number of cores needed to execute a program plays an important role in multi-core platforms and affects the performance. This is also important in Cloud computing where one has to pay for resource usage as well as in real-time systems where knowing the worst-case runtime of a program is essential [1] [2].

In this paper, we propose an analytical modeling approach to estimate the execution time of OpenMP programs on multi-core processors. To run an OpenMP program, majority of compilers use the maximum number of cores available on the system. We show that for some programs, adding more cores will not increase the speedup and even for some applications, the overhead caused by parallelization can decrease the speedup. We have categorized the overheads of OpenMP programs into three classes of excess computations, load-imbalance, and synchronization. The

A. Mantri et al. (Eds.): HPAGC 2011, CCIS 169, pp. 322–326, 2011.

excess computations overhead relates to the extra instructions executed for compiler implementations of OpenMP constructs. The load-imbalance overhead relates to the maximum number of instructions a thread must execute in order to get to the next barrier. The synchronization overhead relates to the time spent on a section of code that executes in critical or atomic regions.

To measure the performance of our case study we use two general metrics, speedup and efficiency [3]. We run our tests on Nahelam, an 8 core Intel processor with each core supporting up to two threads. The rest of the paper is organized as follows. Section 2 presents notable related works. Section 3 presents our proposed mechanism and Section 4 concludes the paper.

## 2 Related Work

Adhianto et al [4] proposed an analytical approach to predict the performance of hybrid OpenMP and MPI applications. The approach is for clustered symmetric multiprocessors platforms. To evaluate the model they have used a matrix multiplication sample and the error rate of the method is reported about 5% for computation and 12-32% for communication.

Lobachev et al [5] have proposed a skeleton-based approach to estimate parallel performance of applications; it is assumed that programs are described using algorithmic skeletons. They have measured the serial run-time and the parallel overhead for different problem sizes and numbers of processors. Using the results, they have estimated the execution time of unknown inputs and non-available number of processors.

Furlinger et al [6] have presented a methodology to analyze the scalability behavior of OpenMP programs. Extending ompP, a profiler for OpenMP programs, they determine execution counts and times for all OpenMP constructs in target application. Depending on the region, different timing and count categories have been reported.

## 3 Proposed Method

To estimate the execution time of an OpenMP program, we divide each program into parts between barriers, applying all the implicit and explicit barriers of a program. Using p number of executing cores, we estimated the longest path between each two subsequent barriers, Total execution time of a program is the sum of longest paths between barriers as in relation (1).

$$T_p = \sum_{i=0}^{n} T_{i,i+1} \tag{1}$$

To estimate the longest path between each two subsequent barriers, first we count the maximum instructions assigned to each thread, then we add up the overhead caused by them. Using this approach all three kinds of overhead are taken into account. Using EPCC micro-benchmark we know the overhead of extra instructions of OpenMP directives. The longest path between each two barrier points takes the load-imbalance and communication overheads into account. In the rest of this section we have formulated the execution time of the most important constructs of OpenMP.

### 3.1 Work-Sharing Constructs

**Sections construct.** The Sections construct is suitable when there are different units of woks for different cores. The number of sections limits the parallelism, because each section is assigned to only one thread. We formulate the maximum time spent to execute sections construct by relation (2). When the number of cores is more than the number of sections, the execution time is bounded to the largest section time. In addition, when the number of cores is less than sections, sections are assigned to threads in the order of their appearance in the sections construct. Therefore, the execution time is the maximum sum of sections assigned to threads.

$$T_{section} = T_{extra\ instructions} + \begin{cases} T_{section} = max\, T_{section} & : p \geq n_{section} \\ T_{section} = max \sum_{k=0}^{n} T_{section\frac{n}{p}+k} & : p < n_{section} \end{cases} \quad (2)$$

**For construct.** In OpenMP programs a For construct is placed before a For loop. The iterations of the loop are distributed in a balanced way. The execution time of a For loop is equal to the execution time of one single thread in a team. This can be formulated as relation (3). However the total execution time is increased by overheads such as job distribution, thread creation and destruction.

$$T_{for} = \frac{W_{for}}{p} + overhead(p) \quad (3)$$

**Single construct.** The basic block guarded by a Single construct is executed by only one thread. No matter what the number of threads is, the execution time is the same and added to the total execution time of the program.

### 3.2 Synchronization Constructs

OpenMP has a shared memory programming model. The synchronization constructs are used to control the access to the shared data. In OpenMP each thread can have a temporal view of shared data until it reaches a barrier point. At that point the shared data are flushed and each thread has to check the memory to get the new value of the shared data. Therefore each time a barrier point is reached and overhead of memory latency is added to the run-time of the program. In our case according to Intel reports memory latency is about 32 nanoseconds [7]. For critical construct the flush function is called twice, one at the start of the region and one at the exit. We used relation (4) to estimate the execution time of Critical region. Due to the similarities, we used the same relation for the atomic construct.

$$T_{Critical} = \left(T_{basic-block} + T_{extera-inst.} + 2 * T_{shared\ memory\ access}\right) * p \quad (4)$$

### 3.3 Case Study

The program in Fig. 1 has 4 implicit barriers and its execution is the sums of $T_{i,i+1}$.

```
#pragma omp single
      time1 = omp_get_wtime();
#pragma omp for reduction(+:sum)
firstprivate(x)
for (j=1;j<= 1000000000; j++){
      x = (j-0.5)*step;
      sum = sum + 4.0/(1.0+x*x); }
#pragma omp critical
{     printf(…);
      time2 = omp_get_wtime();}
#pragma omp single
{     pi = step * sum;
      printf(…);
      x = 0, k =0;}
#pragma omp sections
firstprivate(x,k)
```

```
#pragma omp section{
for (k=1;k<= 312500000; k++){
   x = (k-0.5)*step;
   sum1 = sum1 + 4.0/(1.0+x*x);
#pragma omp section
for(k=312500000;k<=687500000;k++)
{     x = (k-0.5)*step;
      sum2 = sum2 + 4.0/(1.0+x*x);
}
#pragma omp section
for(k=687500000;k<=1000000000;k++)
      x = (k-0.5)*step;
      sum3 = sum3 + 4.0/(1.0+x*x);
#pragma omp master
      sum = sum1+ sum2+sum3;
      time3 = omp_get_wtime();
      pi = step * sum;
```

**Fig. 1.** A general program

Using instruction counting, the total number of instructions in the program shown in Fig. 1 was 3600 million instructions. This means that on our system the serial version of the program runs in 34.99 seconds. The left side of Fig. 2 shows the graphical model of the program. Using instruction counting, 50% of the instructions were in Sections. If we run this program using two cores, results the Sections construct to be executed with two workloads a 69% and a 32%. This means that the execution time of Sections using two threads will be 12.18 seconds and using three cores will be about 6 seconds. Using more threads will not reduce the execution time of sections.

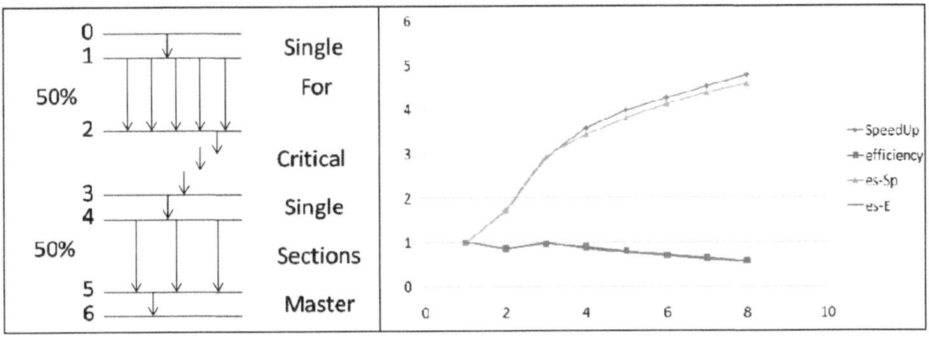


**Fig. 2.** Left Diagram is the graphical model of the program in Fig. 1, The Right side is the real and estimated speedup and efficiency of the program in Fig. 1; the horizontal axis represents the number of cores

The For construct has the other 50% of the program workload. The execution time of the serial version is about 17 seconds. Therefore, using more cores will reduce the execution time of the program. Relation (5) estimates the execution time of the program for more than three cores.

$$T(p) = 6 + \frac{17}{p} + overhead(p) \tag{5}$$

The real speedup and efficiency of this program is shown in Fig. 2. The right side of Fig. 2, the es-Sp and es-E are the speedup and efficiency estimated using relation (5) and the other two are real speedup and efficiency. Our approach shows the maximum error rate of 5% for the case studies. In comparison to [4] the error rates are the same but our approach has a better estimation in case on load-imbalance.

## 4 Conclusion

The ambition of parallelizing a program is to achieve more performance. The majority of existing OpenMP compilers uses the maximum processing elements available. We showed that scaling out the numbers of processors does not necessarily result in speedup or higher efficiency. We proposed a method to estimate the proper number of cores to execute a program on multi-core machines. The case study used in this paper includes the most useful constructs of OpenMP. Due to space constraints we could only use one case study in the paper but we examined these constructs one by one as well as their combinations and the results validated our method.

## References

1. Mancini, E., Rak, M., Villano, U.: PerfCloud: GRID Services for Performance-oriented Development of Cloud Computing Applications. In: 18th IEEE International Workshops on Enabling Technologies: Infrastructures for Collaborative Enterprises, Groningen (2009)
2. Abhijit, R., Wu, J., Srikanthan, T.: Estimating Processor Performance of Library Function. In: Yang, L.T., Zhou, X.-s., Zhao, W., Wu, Z., Zhu, Y., Lin, M. (eds.) ICESS 2005. LNCS, vol. 3820, Springer, Heidelberg (2005)
3. Grama, A., Gupta, A., Karupis, G., Kumar, V.,, T.: Introduction to Parallel Computing, 2nd edn. Addison Wesley, London (2003)
4. Adhianto, L., Chapman, B.: Performance Modeling of Communication and Computation in Hybrid MPI and OpenMP Applications. In: Proceedings of the Twelfth International Conference on Parallel and Distributed Systems, Minneapolis, Minnesota (2006)
5. Lobachev, O., Loogen, R.: Estimating Parallel Performance A Skeleton-Based Approach. In: Proceedings of the Fourth International Workshop on High-Level Parallel Programming and Applications HLPP 2010, Baltimore (2010)
6. Fürlinger, K., Gerndt, M.: Analyzing overheads and scalability characteristics of openMP applications. In: Daydé, M., Palma, J.M.L.M., Coutinho, Á.L.G.A., Pacitti, E., Lopes, J.C. (eds.) VECPAR 2006. LNCS, vol. 4395, pp. 39–51. Springer, Heidelberg (2007)
7. The Tech Report PC Hardware Report, `http://techreport.com/articles.x/19196/4`

# SQLIVD - AOP: Preventing SQL Injection Vulnerabilities Using Aspect Oriented Programming through Web Services

V. Shanmughaneethi, Ra. Yagna Pravin, C. Emilin Shyni, and S. Swamynathan

Department of Computer Science and Engineering
College of Engineering Guindy Campus
Anna university Chennai, India
shanneethi@nitttrc.ac.in, yagnapravin@gmail.com,
shyniedwin@yahoo.co.in, swamyns@annauniv.edu

**Abstract.** Security remains a major threat to the entire Web for many kinds of transactions. Most of the threats are created through application level vulnerabilities and have been exploited with serious consequences. Among the various types of application level vulnerabilities, command injection is the most common type of threat in web applications. Among command injection attack, SQL injection type of attacks are extremely prevalent, and ranked as the second most common form of attack on web. SQL injection attacks involve the construction of application's input data that will result in the execution of malicious SQL statements. Hence, this paper (SQLIVD-AOP) proposes a mechanism to intercept SQL statements without any modification of an application using Aspect Oriented Programming and to analyze the query for its legitimacy, and to customize the errors. This mechanism is different from others by query interception and separation of the main scripting code with SQL injection code. The SQL validations and injection detections code are implemented by means of web services.

**Keywords:** Web security, SQL injection, Web Service, Tautology, Query Engine, XML Schema, Piggybacking.

## 1 Introduction

Web applications are complex entities and it is designed to perform a specific function directly for the user or for another application program. Securing a Web application is difficult, not only because of the cross-departmental coordination involved, but because most security tools are not designed to address the Web application as a whole, including how the different pieces of the application interact with each other [1]. Traditional threats such as by pass authentication and authorization, privilege escalation and challenging the integrity of the data are all application level attacks. There are many tools are available to detect some of the traditional threats, but these tools are useful for their specific functions, they do not address all of the issues. Threat modeling is recommended to be part of the design stages in web application. The purpose of threat modeling is to analyze the

A. Mantri et al. (Eds.): HPAGC 2011, CCIS 169, pp. 327–337, 2011.

application's architecture and identify the potentially vulnerable areas. Weak input validation is an example of an application layer vulnerability, which can result in SQL injection attack. SQL injection is a technique for exploiting web applications that uses client-supplied data in SQL queries without stripping potentially harmful characters [9]. However, SQL injection bugs lead to disclosing sensitive information, tampering the data, running SQL commands with an elevated privilege.

## 1.1 Command Injection

Web application normally a three-tiered architecture, consisting of a web-browser, an application server, and a back-end database server. A web application retrieves and presents data to the user based on the user's input as part of the application's functionality; it is not intended to be simply an interface for arbitrary interaction with the database. However, if the user's input is not handled properly, serious security problems can occur. This is because queries are constructed dynamically in an ad hoc manner through low-level string manipulations [2]. Handling of user input, makes web applications susceptible to a large class of malicious attacks known as command injection attacks. An SQL command injection attack occurs when a malicious user, through specially crafted input, causes a web application to generate and send a query that functions differently than the programmer intended. There are various types of SQL Injection are possible in web applications.[4][8]

## 1.2 Variants of SQL Injection

### 1.2.1 Tautologies

Tautology-based attacks are among the simplest and best known types of SQL Injection. The general goal of a tautology based attack is to inject SQL tokens that cause the query's conditional statement to always evaluate to be true. For example *sql = SELECT * FROM login WHERE log_id='xxx' AND log_pwd= '' or '1'='1'*

The WHERE clause of this query is always evaluated to be true for *log_pwd= '' or '1'='1'*, thus an attacker can bypass the authentication, regardless of the data provided in the name field.

### 1.2.2 Union Queries

SELECT queries retrieve information from a database. If the part of the query in WHERE clause can be manipulated to return records other than what the programmer intended. This allows multiple SELECT queries to be specified in one statement.

For example, consider the following query:

*SQLString = SELECT LastName FROM Employees WHERE City = '' UNION ALL SELECT OtherField FROM OtherTable WHERE ''=''*

Here, *' UNION ALL SELECT OtherField FROM OtherTable WHERE ''='* is the intended input appended with the legitimate query by the attacker to create injection.

The database engine inspects the Employees table, looking for a row where City is set to "nothing." The database engine does not find this row, so no records are returned. The only records that are returned are from the injected query.

### 1.2.3 Query Enumeration with Syntax Errors

Attacker must require some prior knowledge of the database schema to create threat to database through web application, which is often unknown. But, malformed queries allows the attacker for overcoming this problem by taking advantage of overly descriptive error messages that are generated by the database when a malformed query is rejected. When these messages are directly returned to the user of the web application, it can be make use to identify vulnerable parameters and infer the schema of the underlying database.[3]

### 1.2.4 Piggybacked Queries

Similar to union queries, this kind of attack appends additional queries to the original query string. If the attack is successful, the database receives and executes a query string that contains multiple distinct queries. In following example, an attacker could inject the text "chennai; drop table employee" into the pin input field and have the application generate the following query:

SQLString = SELECT * FROM employee WHERE city='chennai'; DROP TABLE employee

The database treats this query string as two queries separated by the query delimiter (";") and executes both. The second malicious query causes the database to drop the users table in the database, which would have the catastrophic consequence of deleting all user information.[4]

### 1.2.5 Insert Command Queries

Attacker can use INSERT statements for vulnerabilities in the same way as the WHERE statements. An INSERT injection differs from SELECT injection. To take advantage of INSERT vulnerability, attacker can be able to view the information that was submitted, regardless of where the information is located.

For example, If an attacker wants to manipulate the arguments in the VALUES clause to make them retrieve other data and can do so using subselects.

*SQLString = "INSERT INTO TableName VALUES ('" &strValueOne & "', '" & strValueTwo & "', '" & strValueThree & "')"*

Fill out the form:

*Name: ' + (SELECT TOP 1 FieldName FROM TableName) + '*

*Email: blah@blah.com*

*Phone: 333-333-3333*

This modifies the SQL statement to say:

*SQLString = INSERT INTO TableName VALUES ('' + (SELECT TOP 1 FieldName FROM TableName) + '', 'blah@blah.com','333-333-3333')*

Attacker can see the first value in FieldName where the user name is normally located. Unless use TOP 1 in the subselect, attacker may receive an error message saying that the subselect returned too many records. Hence, all the above threats can be prevented from the web applications by performing sanitizing of user inputs and validating the query structure.

## 2 Aspect Oriented Programming (AOP)

Aspect-oriented programming (AOP) makes possible to clearly express programs involving such aspects, including appropriate isolation, composition and reuse of the aspect code. It could be applied during runtime or compile time to adding security to the applications without the need to change the source code of the project. It performs *Joinpoint* which is used to identify the defined points in code, *Pointcut* is a way of specifying a Joinpoint by some means of configuration or code, *Advice* is a form of expressing a cross cutting action that needs to occur and *Mixin* is an instance of a class to be mixed in with the target instance of a class to introduce new behavior. In our approach, an AOP module is designed to detect the vulnerabilities that arise through the dynamically generated query. This module consists of different modules that are used for detecting vulnerabilities present in the web application code. Using AOP to intercept SQL string is much better than some of the existing approaches.

## 3 Literature Survey

A lot of works have been done in this area of SQL Injection and SQL related problems. David Morgan [6] describes the concepts of SQL injection, explores the attack vectors, and cites examples of preventative best practice in a work published in Web application. Welty [7] describes the nature of SQL errors, various frequencies in which they occur and methods of detecting and correcting them. Johannes B. Ullrich, Jason Lam [8] discuss how a website can be defaced through SQL Injection. Zhendong and Gary [9] have presented an overview of command injection attacks and their consequences. Halfond, Orso and Manolios [10] define the different types of SQL injection attacks that are possible..Gregory et al [11] present a parse tree mechanism for preventing SQL injection attacks. The technique involves comparing, at runtime, the parse tree of the SQL query before inclusion of the user input with that resulting after the inclusion of the input query. Gabriel et al [12] presents a framework named AProSec, which is a security aspect for detecting SQL injections and Cross Scripting Site (XSS).The authors define clearly the need for AOP for providing security to web applications. This approach provides a framework for identifying vulnerabilities than previous works, but suffers from the critical detection mechanisms where it uses a Document Type Definition (DTD) for defining the possible vulnerabilities.

## 4 Proposed System Architecture

This proposed system architecture consists of three major modules namely Query interceptor, syntactic verification query and error customization as shown in figure 1.

The first module intercepting the dynamic SQL query generated after including in the user inputs from the web applications. The dynamically generated query can be used to detect any type of SQL injection, since injection is not only possible through user inputs. It may be from HTTP header and server variables. The intercepted query would be passed to second module for checking its legitimacy and further the query passed to error customization module. When the user provides the required inputs intothe web forms, they are then placed into the SQL string in an appropriate place of the application. Finally, the complete SQL string is generated for processing data transaction. The generated ( or framed) SQL string may cause SQL injection in a webapplication.SQL Injection attack is possible only when the SQL Query is directly processed by the SQL processing engine without any validation. The invalidated SQL Query may lead to different type of SQL injection attacking mechanism like tautology, logically incorrect query and other type of SQL injection.

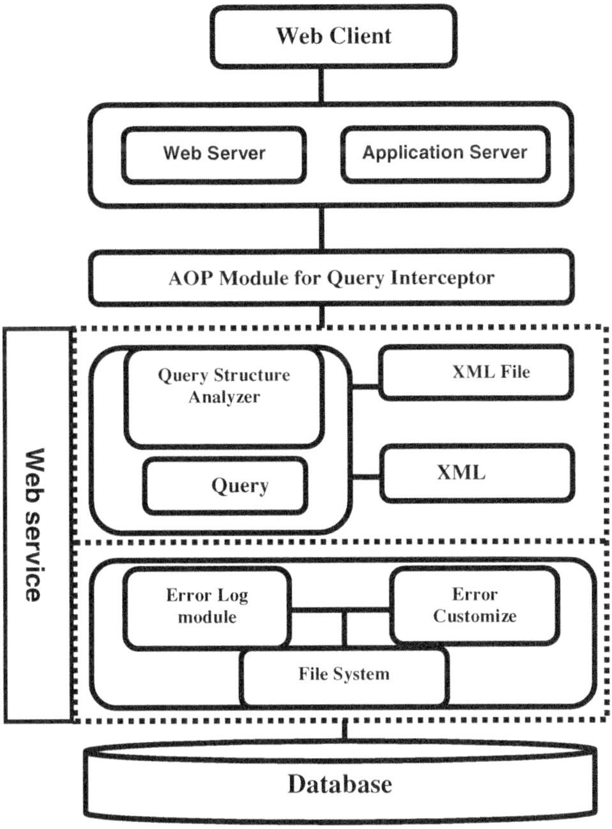


**Fig. 1.** Proposed SQLIVD - AOP System architecture

### 4.1 Interception of SQL Statement

Query interception involves intercepting the SQL query that is generated at run time. This query that is generated is executed in the database server and results are obtained. In order to detect injection vulnerabilities, this dynamically generated query has to be intercepted. Also intercepting the run time query can be used to detect any type of injection, since the sink points for causing the injections are these queries that are generated at run time. AOP is the best technique in the case of intercepting functions without affecting the business logic. Using AOP, these run time generated queries can be intercepted before they are executed in the database server. Intercepting these run time queries is done by writing point cuts which match the functions that execute these queries. Once a point cut is matched, advices are defined to obtain the parameters of the functions intercepted which are the run time queries.

The point cuts defined are functions that are used for executing a dynamically generated query in the database server. If any of these point cuts are matched then the corresponding query can be obtained. Using AOP for intercepting these queries can be very useful and powerful, since it does not involve any changes in the business logic.

## 4.2 Query Analyzer and XML Generation

Once, the intercepted query string received from the query interceptor module, the query has to be parsed to generate XML file to validate the legitimacy of the user input in the query string. The words in SQL query is divided into types such as SQL key words and non - SQL words based on the SQL grammar[13]. The SQL keywords form the names of the tags and the non - SQL keywords form the values inside the tag. Once the XML structure is generated using the above rules, the XML so formed from the given SQL statement is validated against the meta definition to find whether the SQL is valid or not. Another module in our approach creates the meta definition as a meta XML Schema as shown in figure 4. The XML Schema structure is developed for validating the XML file which is generated from the SQL query. This is created in such a way that it validates almost any type of SQL statement. The basic structure of XML Schema is divided into five types of statements as select, insert, update, delete and create. The schema developed must consider all the aspects such as the child elements pertaining to a root tag. For Instance a select statement must have the tags <select>, <from>, <where>, <and>, <or> and so on as shown in figure 4. The proposed XML generation module generates the XML structure from the SQL query. The intercepted SQL query is converted into a XML structure and subsequently stored as a XML file. The XML file is taken as an input file for validation against XML schema. The query is scanned for the list of SQL keywords and tokens.

Two arrays, one containing SQL keywords and the other containing the non-SQL keywords are formed. Using the keywords the XML element nodes are then created with the SQL keywords as the names of the element and their corresponding non-SQL text as their value. The following example describes this feature

SQL String: *Select roll_no, grade, addr from student*

For a given SQL string *<sql>* is the root tag as per our definition. In this statement, the verb of the SQL statement is under the classification select. Hence, the tag used is *<selects>*. The SQL query returns the value of roll_no, grade and address. These are non-SQL keywords. Therefore, all the non SQL keywords are to be placed as value for the element.

*<select> roll_no </select>*
*<select> grade </select>*
*<select> addr </select>*

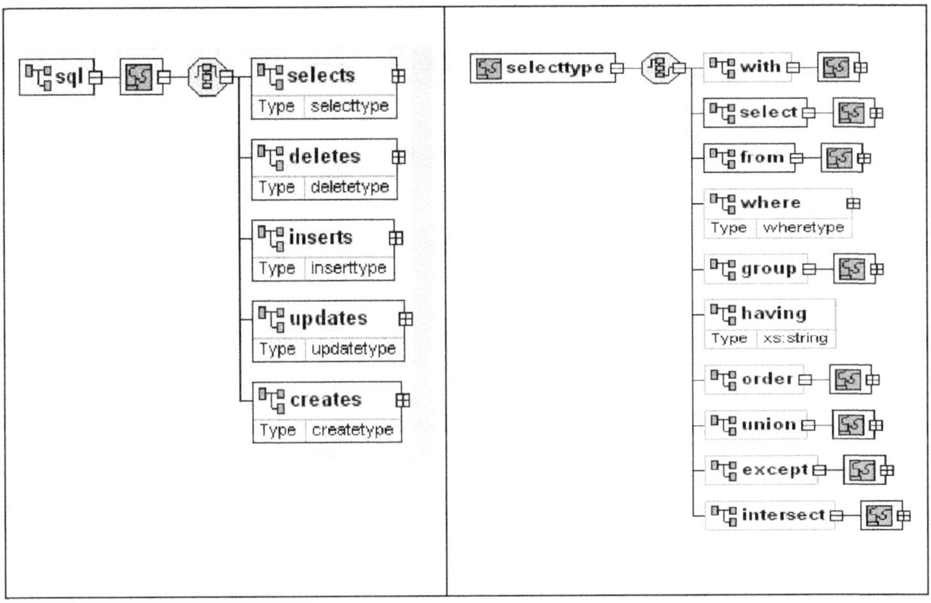


**Fig. 2.** Internal SQL Statement Schema Structure

Further, another SQL keyword is *'from'* and the non-SQL keyword is *'student'*. So, the tag is
*<from> student </from>*.

Hence, the XML file will be as follows

```
<?xml version ="1.0">
<sql>
    <selects>
        <select> roll_no </select>
        <select> grade </select>
        <select> addr </select>
        <from> student </from>
    </selects>
</sql>
```

The non-SQL keywords validation module demands for the validation of the non-SQL keywords in the input SQL query. This mainly includes the table names, column names and values. An array of non-SQL keywords is considered for this function. In a SQL query, a comment may cause serious threat to web application through SQL injection. To analyze the comment, the comment analyzer module scans the entire SQL query for any comments. If any comments are found in the SQL query, then it is rejected and error is returned from the error generated service. However, tautology based SQL injection causes major threat to web applications. The tautology analyzer performs the protection of verifying the tautology in the SQL String. From the non-SQL keywords the column names and table names are identified. This information is then used to check whether the column belongs to their respective database table as mentioned in the query. If a mismatch is found, then a generalized error is returned to the web client.

### 4.3 Query Validation

The Query validation validates the structure of the input SQL query using the generated XML by validating it against to our XML schema. It takes the generated XML file and schema as input and does XML schema validation. If the validation passes, then the query structure is considered to be correct by returning true. This module returns true if and only if; there will not be a suspected input in the query string. If any suspected inputs are placed in the SQL query, then the generated XML file doesn't parsed with our internal schema. Some time the correct syntactic query may logically incorrect. Such queries are leads to reveal the database schema to the attacker. By injecting illegal / logically incorrect requests, an attacker may gain knowledge about the injectable parameters, data types of columns within the table, name of tables, etc[14]. Although every database management system in the commercial market supports ANSI/ISO standard Structured Query Language, each vendor also develops a proprietary SQL language dialect. Almost every SQL injection attack within a web application targets a specific database. Therefore, there is a need for general solution to commonly targeted databases like MS SQL Server, MySQL, Oracle, DB2, Sybase, Informix and MS Access. Most of the available solutions are specific to commercial database software. Through our customized error generation module, the threat agent cannot deduce specific details such as injectable parameters. The framed query returns a valuable result set, then the result set will be returned as a dataset to the client via web server as shown in figure 3.


```
<NewDataSet xmlns="">
- <Table diffgr:id="Table1" msdata:rowOrder="0">
    <stu_id>2009501</stu_id>
    <stu_name>neethi's</stu_name>
    <stu_gender>male</stu_gender>
    <stu_dob>1975-01-13T00:00:00+04:30</stu_dob>
    <stu_dept>computer centre</stu_dept>
    <stu_degree>M.E</stu_degree>
    <stu_addr>trichy</stu_addr>
  </Table>
</NewDataSet>
```


**Fig. 3.** Valuable Result Set

If the processed query returns the specific error messages, then the error message is customized by an error customizer module as SQL error and this will be returned to the web server for client information. If any intended query framed by the attacker, to know about schema that can be protected by returning the customize error to the user. If the error is not customized then, the error set reveals some of the information about the database schema. So, whenever error returned to the client, the error has to be generalized shown in figure 7.

```
- <NewDataSet xmlns="">
  - <Table1 diffgr:id="Table11" msdata:rowOrder="0" diffgr:hasChanges="inserted">
      <Error_Message>SQL Error</Error_Message>
    </Table1>
```

**Fig. 4.** Customize Error message

This module will not reveal the real SQL error message or error code to the client to achieve SQL Injection. From the generalized error message as SQL Error shown in figure 4, the attacker will not be able to understand the table schema or any type of database information. Moreover, a log file has maintained to monitor the injected query and their domain.

```
127.0.0.1 -System.Data.Odbc.OdbcException: ERROR [42000] [Microsoft][O
Microsoft Access Driver] Syntax error (missing operator) in query expressic
and stu_name='neethi''.

127.0.0.1 -System.Data.Odbc.OdbcException: ERROR [42000] [Microsoft][O
Microsoft Access Driver] Syntax error (missing operator) in query expressic
and stu_name='''.

127.0.0.1 -System.Data.Odbc.OdbcException: ERROR [42000] [Microsoft][O
Microsoft Access Driver] The SELECT statement includes a reserved word (
argument name that is misspelled or missing, or the punctuation is incorrec
```

**Fig. 5.** Log File

This log file shown in figure 6 maintains the SQL injection request towards to our testing web server. This log file helpful for further understanding of command injection created by the hackers

## 5 Result and Discussions

In this paper, we have presented the formal definition and variation of SQL injection attacks in web applications. Based on this attack, we have developed a secure and complete runtime checking strategy for preventing SQL injection. Our tool tested with different types of code injection queries. The result proves that, there is no false negative reply from our SQLIVD-AOP tool. When this tool tested in the real time environment, the response time is increased in few milli seconds shown in table-1 and the graph shows that the difference is very minimal is shown in the following graph. In the following graph X – axis represents number of sample test performed in real-time environment and the Y- axis represents response time in milli seconds.

The time difference between typical web environment and inclusive with our approach is shown in figure 6. When compare to the consequences of the injection, this time delay difference with our approach is negligible.

**Table 1.** Response Time assessment of SQLIVD - AOP tool

| No. of Test | Response Time Without SQLVID - AOP (milli seconds) | Response Time With SQLVID - AOP (milli seconds) | Response Time Difference (milli seconds) |
|---|---|---|---|
| 1 | 83.83 | 130.89 | 47.06 |
| 2 | 92.88 | 169.76 | 76.88 |
| 3 | 72.98 | 131.07 | 58.09 |
| 4 | 81.89 | 133.42 | 51.53 |
| 5 | 82.21 | 150.12 | 67.91 |

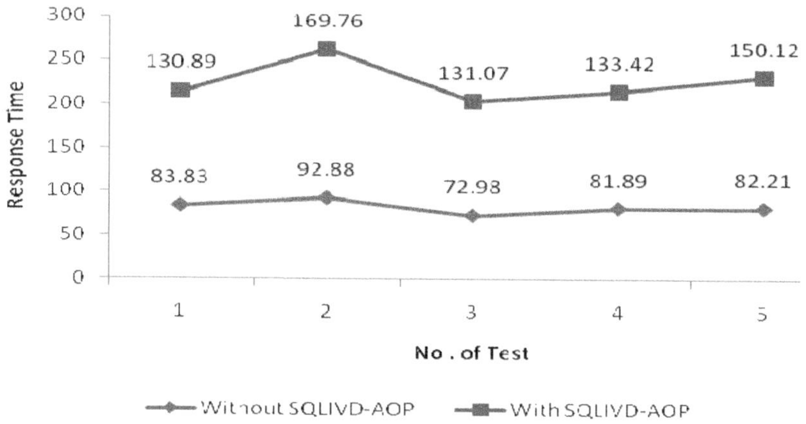


**Fig. 6.** Response Time (ms) versus No. of Test assessment of SQLIVD - AOP tool

But, this way of preventing SQL injection, completely free from application layer command injection attacks.

## 6 Conclusion and Future Work

Many web sites in the world has vulnerable, which can be hacked by such SQL injection technique. In this paper, we proposed an AOP module for intercepting SQL string in web application. The intercepted SQL string send to the web service for detection of Injection by tautology, illegal/ logically incorrect queries and piggy pack. Comparing with previous approaches, this approach is independent to the platform and work on any type of back end database. We analyzed the web application with the web service and found that the response time of the web application. In future, we intend to analyze the input string which is given as input to the web form by a user. The independent analysis of the input string will give the greater performance to protect SQL Injection. This strategy can be also followed for detection of XPath injection in web services.

## References


1. Security Focus, http://www.securityfocus.com
2. Anley, C.: Advanced SQL Injectio. In: SQL Server Applications. white paper, Next Generation Security Software (2002)
3. SQL Injection: are your web applications vulnerable? White Paper, HP (October 2007)
4. OWASP, Category: OWASP Top Ten Project, Vol. 2007 (2007)
5. Web application security- SQL injection attacks. Network Security 2006(4), 4–5 (2006)
6. Correcting user errors in SQL. International Journal of Man-Machine Studies 22(4), 463–477 (1985)
7. Ullrich, J.B.: Chief research officer and Jason Lam Defacing websites via SQL injection. Journal of Network Security 2008(1), 9–10 (2008)
8. Halfond, W.G.J., Orso, A.: Member and Panagiotis Manolios, " WASP: Protecting Web Applications Using Positive Tainting and Syntax-Aware Evaluation. IEEE Transactions on Software Engineering 34(1), 65–81 (2008)
9. Su, Z., Wassermann, G.: The Essence of Command Injection Attacks in Web Applications. In: Proceedings of the Thirty Third ACM Symposium on Principles of Programming Languages, South Carolina, pp. 372–382 (2006)
10. Buehrer, G.T., Weide, B.W., Sivilotti, P.A.G.: Using Parse Tree Validation to Prevent SQL Injection Attacks. In: SEM 2005: Proceedings of the Fifth International Workshop on Software Engineering and Middleware, New York, pp. 106–113 (2005)
11. Hermosillo, G., Gomez, R., Seinturier, L., Duchien, L.: Using Aspect Programming to Secure Web Applications. Journal of Software 6(2), 53–63 (2008)
12. SQL Grammar, http://msdn.microsoft.com/en-us/library/ms709391VS.85.aspx
13. CVE. Common Vulnerabilities and Exposures (April 2008), http://cve.mitre.org/

# Analysis and Study of Incremental K-Means Clustering Algorithm

Sanjay Chakraborty and N.K. Nagwani

National Institute of Technology (NIT)
Raipur, CG,
India
sanjay_ciem@yahoo.com,
nknagwani@gmail.com

**Abstract.** Study of this paper describes the incremental behaviours of partitioning based K-means clustering. This incremental clustering is designed using the cluster's metadata captured from the K-Means results. Experimental studies shows that this clustering outperformed when the number of clusters increased, number of objects increased, length of the cluster radius decreased, while the incremental clustering outperformed when the number of new data objects are inserted into the existing database. In incremental approach, the K-means clustering algorithm is applied to a dynamic database where the data may be frequently updated. And this approach measure the new cluster centers by directly computes the new data from the means of the existing clusters instead of rerunning the K-means algorithm. Thus it describes, at what percent of delta change in the original database up to which incremental K-means clustering behaves better than actual K-means. It can be also used for large multidimensional dataset.

**Keywords:** Clustering, Data mining, Incremental, K-Means.

## 1 Introduction

Data mining is a method to extract novel, useful, hidden knowledge from massive data sets. Data clustering is a popular unsupervised data mining technique for automatically finding classes or there must be group of similar data whereas dissimilar data should belong to different clusters. K-means clustering algorithm takes the input parameter, K, and partitions a set of n objects into K clusters. K-means has several limitations. Basically the actual K-means algorithm takes lot of time when it is applied on a large database. That's why the incremental clustering concept comes into the picture to provide quick and efficient clustering technique on large dataset. The overall concept of the paper is shown by the figure below.

A. Mantri et al. (Eds.): HPAGC 2011, CCIS 169, pp. 338–341, 2011.

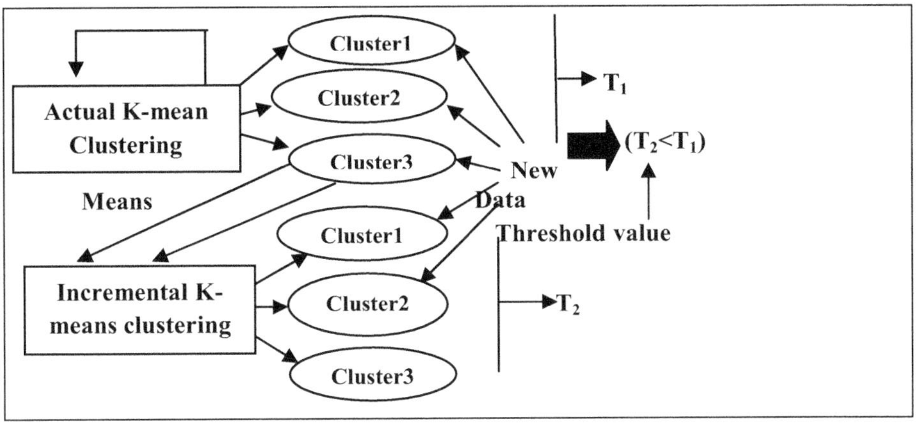


**Fig. 1.** Methodology of Incremental K-Mean algorithms

## 2 Literature Survey

Some work is done on the performance of partitional and incremental models which are developed based on the number of clusters and threshold values 1.To improve the efficiency, speed of clustering in data mining applications and in machine learning fields a fast and stable incremental clustering technique comes into the picture 2.In some cases, a fast incremental clustering algorithm has the ability to changing the radius threshold value dynamically. The algorithm restricts the number of the final clusters and reads the original dataset only once3.A new clustering algorithm which is to rigorously derive the updating formula of the k-modes clustering algorithm with the new dissimilarity measure 4.A slightly different approach is proposed on a weighted email attribute similarity based data mining model is proposed to discover email groups 4.

## 3 Incremental K-means Clustering

The term incremental means "% of $\delta$ change in the original database" i.e. insertion of some new data items into the already existing clusters. Such as,

$$\%\delta \text{ change in DB} = \frac{(\text{NEW DATA}-\text{OLD DATA})}{\text{OLD DATA}} \times 100. \quad (1)$$

An incremental clustering approach is the way to solve the problems that arise from partitional clustering.

### 3.1 Proposed Algorithm

The following are the steps of the proposed clustering algorithm

```
Input
D: A dataset containing n objects{X1,X2, X3, …, Xn} and n:
number of data items.
Output
K1: A Set of clusters.
Algorithm
Let,Ci(where i=1, 2, 3 …)is the new data item.
1. Run the actual k-means algorithm and clustered the new
data item Ci properly. Repeat till all data items are
clustered.
 Actual K-means ——▶ T1(Processing Time).
Incremental K-means Pseudo-code:
Start
2.  a>Let, K represents the already existing clusters.
    b>Compute the means (M) of existing clusters.And
directly clustered the new item Ci.
for i = 1 to n do
find some mean M in some cluster Kp in K such that dis (
Ci, M ) is the  smallest;
if dis ( Ci, M )=min then
Kp = Kp U Ci ;
Recomputed the mean M and compare it again.
Else If dis(Ci!= min) ——▶then Ci will be treated as
outliers or noisy data.
Update the existing cluster.
c > Repeat step b till all the data samples are
clustered.
Incremental K-means ——▶ T2(Processing Time).
End ;
 3.Compare(T1,T2),Result ◀—— (T2<T1)Upto some certain
Threshold value in the database.
 4.Compute the actual threshold value. Finds the
incremental clustering as a better time
Efficient.
```

## 4 Conclusion and Future Scope

This paper propose an improve K-means clustering approach which provides better and fastest result compare to the existing K-means clustering algorithm up to some certain point of change in the original database. The future scope of the work could be analyzing the other popular clustering techniques in incremental fashion.

## References


1. Bakar, Z.A., Deris, M.M., Alhadi, A.C.: Performance analysis of partitional and incremental clustering. In: SNATI 2005 (2005); ISBN-979-756-061—6
2. Young, S., Arel, I., Karnowski, T.P., Rose, D.: A Fast and Stable incremental clustering Algorithm. In: 7th International (2010) TN 37996

3. Su, X., Lan, Y., Wan, R., Yuming.: A Fast Incremental Clustering Algorithm. In: International Symposium on Information Processing (ISIP 2009), August-21-23, pp. 175–178 (2009)
4. Ng, M.K., Li, M.J., Huang, J.Z., He, Z.: On the Impact of Dissimilarity Measure in k-Modes Clustering Algorithm. IEEE Transaction on Pattern Analysis and Machine Intelligence 29(3) (March 2007)
5. Nagwani, N.K., Bhansali, A.: An Object Oriented Email Clustering Model Using Weighted Similarities between Emails Attributes. IJRRCS 1(2) (June 2010)

# Computational Model for Prokaryotic and Eukaryotic Gene Prediction

Sandeep Kaur[1], Anu Sheetal[2], and Preetkanwal Singh[3]

[1] SSIET, Dinanagar,
Punjab, India
sandeep.kaur18@gmail.com
[2] GNDU, Regional Campus,
Gurdaspur, Punjab, India
[3] BCET, Gurdaspur, Punjab, India

**Abstract.** In this paper we have design a computational model for prokaryotic and eukaryotic gene prediction by using the clustering algorithm. The input DNA (Deoxyribonucleic Acid) sequence is spliced and the open reading frames are identified. For identification of consensus sequences various data mining algorithm is applied for creation of clusters. This model saves the implementation time, as whole of the database is present online so the sequence to be predicted is just taken from any one of the available database. Several experiments have been done where the parameters of gene prediction are changed manually. The performance has been tested on different unknown DNA sequences found on the internet. The sequences having score greater than or equal to the threshold value are entered into one cluster and rest of the sequences having score less than the given threshold are entered into second cluster and GC (Guanine and cytosine)-content percentage is calculated.

**Keywords:** DNA, FASTA, GC, RNA.

## 1 Introduction

Bioinformatics is the science of managing, mining and interpreting information from biological sequences and structures. In this area of science, biology, computer science and information technology, all the three merge into a single discipline. During the last few years, bioinformatics has been overwhelmed with increasing floods of data, both in terms of volume and in terms of new databases and new types of data. Basic types of data that can be analyzed in bioinformatics are (a) Raw Deoxyribonucleic Acid sequences (b) Protein sequence (c) Macromolecular structure (d) Genome (e) Gene Expression [1].

### 1.1 Introduction to Gene Expression

While the specific sequence of nucleotides in a deoxyribonucleic acid molecule can have important information content for a cell, it is actually proteins that do the work of altering a cell's chemistry by acting as biological catalysts called enzymes. In

A. Mantri et al. (Eds.): HPAGC 2011, CCIS 169, pp. 342–348, 2011.

chemistry catalysts are neither consumed nor altered in the course of such a chemical process and can be used to catalyze the same reaction many times. The term gene is used in many different ways, but one of its narrowest and simplest definitions is that genes spell out the instruction needed to make the enzyme catalysts produced by cells. In the world of cells there are two major groups of cells: the prokaryotes and eukaryotes. The major similarities between prokaryotes and eukaryotes are that they both have deoxyribonucleic acid as their genetic material, both have ribosomes and open reading frames. The major differences between gene expression of prokaryotes and eukaryotes are that prokaryotes have single circular chromosome, whereas eukaryotes have many bar shaped chromosomes. The gene expression for eukaryotes completes in two different steps, one with transcription than with translation. In eukaryotes the open reading frames are longer as compared to open reading frames of prokaryotes.

## 1.2 Genetic or Evolutionary Algorithms

Genetic or evolutionary algorithms borrow their inspiration from the process of evolution by natural selection found in nature. They start with a population of possible hypotheses, and evaluate them on some training data. The best hypotheses are kept and used to create a new generation of hypotheses. Neural Networks is used for eukaryotic gene prediction.

Gene prediction by computational methods for finding the location of protein coding regions is one of the essential issues in bioinformatics. Two classes of methods are generally adopted: similarity based searches and *ab initio* prediction. Various gene prediction algorithms were developed both for prokaryotic and eukaryotic genomes, few of them are HMM, Glimmer, GenScan, BLAST, FASTA. These algorithms are compared on the availability and accuracy as the primary criteria. Accuracy means consistent performance of a program on a set of standard test cases. An evaluation is made on the number of true positives (TP), where the length and end sequence positions are correctly predicted and the number of over predicted positive predictions or false positives (FP), true negatives (TN) and unpredicted residues as false negative (FN) predictions. Based upon this (TP, TN, FP, FN) specificity and sensitivity is measured [7].

Lakshmi et al. (2004), in their paper "Bioinformatics tools and applications", discussed that the biological research in the twenty-first century is primarily driven by high precision instrumentation for exploring the complexity of biological systems in greater detail. Very large datasets are generated from these instruments that require efficient computational tools for data mining and analysis.

Valdimir (2001) proposed new promoter finding method, known as Dragon Promoter Finder (DPF) which locates RNA polymerase II promoter in DNA sequences of vertebrates by predicting transcriptional start site (TSS) positions. DPF uses sensors for three functional regions (promoters, introns, exons) and an artificial neural network. Results are better as compared to previous promoter-finding algorithms.

It is found from the literature survey that yet there is a requirement to design new model which predicts the gene in lesser time and with less complexity both for prokaryotes and eukaryotes. Our paper has been separated into three sections, section 2 is theory, section 3 is results and discussion and last section 4 includes conclusion.

## 2 Theory

The main focus of this work is to design and implement a new model for gene prediction. To achieve this objective, the proposed work will involve the use of data mining techniques to analyze the given database and retrieve the desired information. The data will be taken from online resources and web servers. A new model is created which computes the given DNA sequence based upon its GC content, and open reading frames. Connectivity is done between front-end and back-end and an interface is created. DNA sequence is entered to this interface. Algorithm searches all the possible reading frames from six reading frames. It calculates the resultant vector for each of the open reading frames. Than classify open reading frames as genes and non-genes. After this the GC content is computed. Based on parameters if ratio count>= 6, than it is eukaryotes otherwise prokaryotes. Algorithm includes both transcription and translation for predicting the gene [2 , 7].

### 2.1 Gene Prediction

The process by which information is extracted from the nucleotide sequence of a gene and then used to make a protein is essentially the same for all living things on Earth and is described by the grandly named central dogma of molecular biology shown in figure 1. Quite simply, information stored in DNA (Deoxyribonucleic Acid) is used to make a more transient, single stranded polynucleotide called RNA (Ribonucleic Acid) that is in turn used to make proteins. The process of making a ribosomal ribonucleic acid copy of a gene is called transcription and is accomplished through the enzymatic activity of an RNA polymerase.

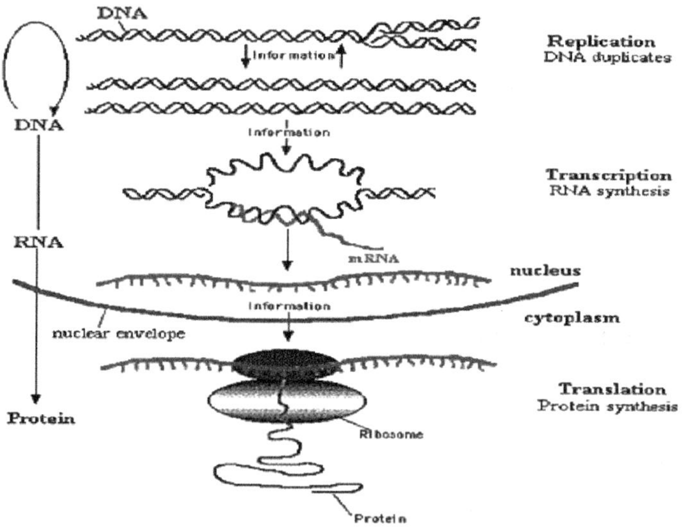


**Fig. 1.** Gene expression

There is one-to-one correspondence between the nucleotides used to make ribonucleic acid (G, A, U and C where —U is an abbreviation for Uracil) and the nucleotide sequences in deoxyribonucleic acid (G, A, T, and C, respectively). The process of converting that information from nucleotide sequences in ribonucleic acid to the amino acid sequences that make a *protein* is called translation and is performed by a complex of proteins and called ribonucleic acid ribosomes. Finding the particular beginning of genes for transcription is done by ribonucleic acid polymerase and that beginning sequence is known as Promoter Sequences. In case of prokaryotic genomes the promoter sequences are easy to find as compared to those in eukaryotic genomes. The problem of recognizing eukaryotic genes in genomic sequences data is a major challenge for bioinformatics. The best methods used are neural network and dynamic programming techniques. Eukaryotes have large genomes but low gene density. Some genes have strong and others have weak promoters. Strong promoters have sequences close to the ideal consensus sequences TTGACA (-35 box) or TATAAT (-10 box) shown in figure 2. So at the least there must be one promoter region upstream of TSS for the polymerase to bind [6].

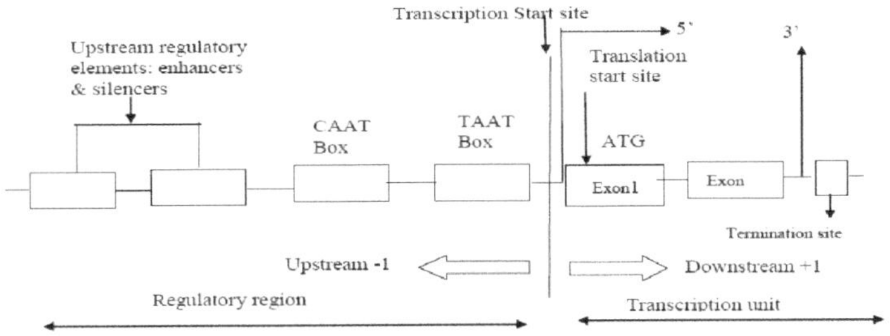


**Fig. 2.** A generalized structure of genes transcribed by RNA polymerase II displaying various structural and functional domains[6]

This computational model implements in two main module (1) applying the algorithm for ORF Prediction and (2) applying the model for finding the GC content.

## 3 Results and Discussion

The user inputs the DNA sequence. The extra spaces present in the input sequence are also trimmed. Local Alignment score is calculated with the sequences having greatest similarity calculated though the data mining algorithm applied on the consensus sequences. If score >=8 then score of entered DNA sequence is matched with the items of the cluster having similarity>=8 and consensus sequence having greatest score is found out otherwise best similarity with the second cluster is calculated. Then the percentage match of the entered DNA sequence with the consensus sequence is calculated and if that match is greater than or equal to the entered threshold value then various outputs are displayed.

The clustering algorithm that works at back end is similar to the single link technique. In this algorithm a threshold, is used to determine if items will be added to existing clusters or if a new cluster is created. The basis for making clusters is local alignment between the sequences, as larger the score more the sequences will be similar. So the sequences having score greater than or equal to the threshold value are entered into one cluster and rest of the sequences having score less than the given threshold are entered into second cluster and GC-content percentage is calculated as: ((G+C) / (A+T+G+C))*100. It is observed that GC content can vary so dramatically across prokaryotic species with values ranging from 25% to 75% GC. The proposed algorithm works with large data sets and the classification of the sequences is done on the basis of (1) Consensus Sequence, (2) Open Reading frame and (3) GC Content ratio

The complexity of this algorithm actually depends on the number of items. For each loop, items must be compared to each item already in a cluster. This is n in the worst case. Thus, the time complexity is O $(n^2)$. Space requirement is assumed to be also $O(n^2)$. Proposed algorithm is changed form of nearest neighbor algorithm. Changes are based on seeing the characteristics of input data.

**Table 1.** Comparison of clustering algorithms [5]

| | | | | |
|---|---|---|---|---|
| Nearest neighbour | Partitional | $O(n^2)$ | $O(n^2)$ | Iterative |
| PAM | Partitional | $O(n^2)$ | O(tk(n-k)2) | Iterative; Adapted agglomerative; Outliers |
| K-means | Partitional | O(n) | O(tkn) | Iterative , not categorical |
| DBSCAN | Fixed | $O(n^2)$ | $O(n^2)$ | Sampling, out-liners |

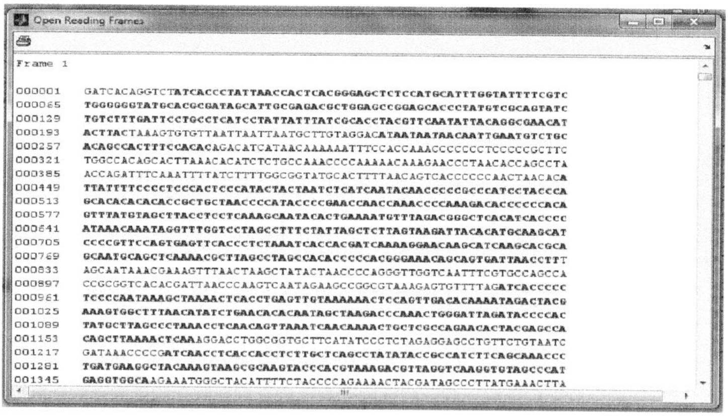


**Fig. 3.** Screen Shots for detecting open reading frames

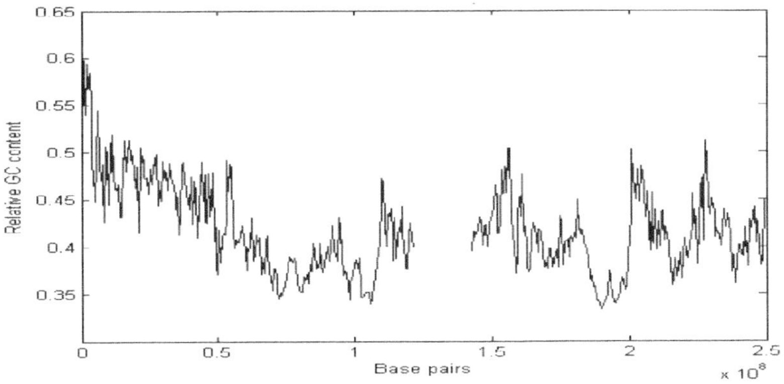


**Fig. 4.** GC Density

## Conclusions

In the field of Gene Prediction, the sequences of nucleotide in DNA molecules have important information contents of a cell. The information in DNA sequences is used to make single-stranded RNA sequence which in turn will further convert into Protein sequence. This designed computational model first find the open reading frames, based upon which the model distinguish between gene and non-genes and GC density is calculated, based upon these parameters the sequence is classified. This model saves the implementation time, as whole of the database is present online; the sequence to be predicted is just taken from any of the online available databases. Interface is opened and deoxyribonucleic acid sequence is entered in its FASTA format. All the complexities such as calculating GC content and locating open reading frames are computed by algorithm. Several experiments have been done where the parameters selected from classification changed manually. The global error was then estimated to about 10%. In general this error is too high. The performance has been tested on different unknown DNA sequences found on the internet.

## References

[1] Al Shahib, A., Rainer, B., Gilbert, D.R.: Predicting protein function by machine learning on amino acid sequences – a critical Evaluation. BMC Genomics 10, 1–10 (2007)
[2] Au, W.H., Chan, K.C.C., Yao, X.: A Novel Evolutionary Data Mining Algorithm with Applications to Churn Prediction. IEEE Trans. Evolutionary Computation 7(6), 532–545 (2003)
[3] Baker, D., Sali, A.: Protein structure prediction and structural genomics. Nucleic Acids Research 294(5540), 93–96 (2001)
[4] Brunak, S., Engelbrecht, J., Knudsen, S.: Prediction of human mRNA donor and acceptor sites from the DNA sequence. Journal of Molecular Biology 220, 49–65 (1991)

[5] Lakshmi, K.M., Steven, G.S.: Department of Bioinformatics and Computational Biology George Manson university, Lecture- Bioinformatics tools and applications, book reference, Vol. 21 (2004)
[6] Myburgh, G.: Euokaryotic RNA Polymerase II start site detection using artifical neural networks, M.Tech thesis, University of Pretoria (2005)
[7] Vladimir, Makarov: Computer programs for eukaryotic gene prediction, vol. 3(2), pp. 195–199. Henary Stewart Publications 1467-5463 (2002)

# Detection of Malicious Node in Ad Hoc Networks Using Statistical Technique Based on CPU Utilization


Deepak Sharma[1], Deepak Prashar[2],
Dalwinder Singh Salaria[2], and G. Geetha[2]

[1] Shri Mata Vaishno Devi University, Katra,
Jammu and Kashmir, India – 182320
deepak.sharma@smvdu.ac.in
[2] CSE/IT Department, Lovely Professional University,
Phagwara Punjab, India
deepakprash@gmail.com, dalwinder.singh@lpu.co.in,
gitaskumar@yahoo.com



**Abstract.** Proposing a strategy based on statistical value provided by each node of the network for detecting their malicious activity by comparing the node's present characteristic value with the old estimated value .If the difference between the two values is higher than expected value then that particular node become suspicions, a knowledge based system can take decision to expel the malicious node from the network topology.



**Keywords:** Malicious node, Node profile, Threshold value, Regression analysis, Auto correlation coefficient.


## 1 Introduction

Now a days, handled devices like laptops, mobile phones and PDAs has take an important place in everyone everyday life because of wide range of powerful leading application , including mobile conferencing, home networking, emergency/disaster services, Personal Area Networks (PANs). Depending on the device's Architecture the applications to be able or not able to read and rewrite code .If the reading and rewriting of the software is not needed for application, then security for that network could be set very high using tamper proof hardware, leaving no chance for the node usage malicious purposes in case of capturing by an attacker. Keeping the cost effective in mind most of the time, nodes are used on which read /rewrite of the codes can be formed. Imagine if an important application installed on the node is hacked by some hacker and then changed are made in the application for malicious purpose i.e. with wicked or mischievous intentions or motive or some other node is deployed in place of the original node in a mobile Ad hoc network having the same hardware ,same id ,also having same features i.e. duplicate node of the original authenticate node but having altered application for mischievous intentions or motives and this leads to corruption of network. To avoid the corruption of the network through captured node or duplicate node, immediate detection of the malicious nodes should be done and then immediately it must be expelled from the Ad hoc network.


A. Mantri et al. (Eds.): HPAGC 2011, CCIS 169, pp. 349–356, 2011.

## 2 Related Work

There are various types of malicious attack performed by the captured node and their intension is to disrupt the network .To avoid the disruption caused by malicious node various techniques have been proposed for the detection of malicious node in the Ad Hoc network, based on the reputation-based scheme ,a node may drops some or all packets forwarded to him ,this was solved by Reputation -based Scheme which uses both self-observation and second hand information to establish compressive reputation of a node. Node with bad compressive reputation will be excluded from the network. The local reputation is not only related to the node's packet-forwarding ratio (the proportion of correct forwarded packets with respect to the total number of packets to be forwarded during a fixed time), but also related to the busy state of the nodes. The reputation is calculated by R(a,b)=(1-α)* Rold(a ,b) + α* Rcur(a ,b),where $R_{old}$ = Old reputation and $R_{cur}$ = New reputation [1]. Another technique for the detection of malicious node for HELLO flood attack and wormhole attack, in which a malicious node may try to transmit a message with an abnormally high power so as to make all nodes believe that it is their neighbor. The Purposed mechanism was based on values of signal strength and geographical information for detecting malicious nodes staging HELLO flood and wormhole attacks. The idea was to compare the signal strength of a reception with its expected value, calculated using geographical information and the pre-defined transceiver specification of the model. As each node can hear both the transmission it compares the expected and the actual signal strength of the received signal, if the ratio of Expected value / Actual is greater than the threshold value, then it is said to be malicious message. All the nodes are uniquely identified, and know their own geographical position, which can be obtained using a positioning system such as GPS. The value of a node's geographical position as well as its identifier are included in each of the message it sends and the message is protected against tempering using some cryptographic mechanism[20]. (Expected value) $P_r = Pt \times G_t \times G_r\, h^2t\, h^2r \,/\, d^4 \times L$. ,***where*** $P_r$ *= is the received signal power in watts,* $Pt$ *= is the transmission power in watts,* $G_t$ = is the transmission gain , $G_r$ = *is the receiver antenna gain,* $h^2t$ *= is the transmitter antenna height in meters,* $h^2r$ *= is the receiver antenna height in meters,* **L**= *is the system Losses,* **d** *= is the distance between the transmitter and receiver distance.* In Blackhole attack [2], a malicious node sends fake routing information, claiming that it has an optimum route and causes other good nodes to route data packets through the malicious one. One of the method for detecting Black hole attack (The route confirmation request (CREQ) and route confirmation reply (CREP) to avoid the black hole attack as proposed by S. Lee, B. Han, and M. Shin, The black hole attack is able to inject a RREP message that is faked by changing the SN in the message and to deceive the source node in order to make the source node send its data packet to the attacker .The goal of method is to protect the network from the attack by detecting the malicious events related to attack during the route setting up phase .When an intermediate node unicasts a RREP a message ,the node also unicasts a newly defined control message to the destination node to request for the up-to-date SN. Then the destination node unicasts a reply message to inform the source node of the up-to-date SN after receiving the request message sent by the intermediate node. This reply from the destination node enables the source to verify if the intermediate node has sent a faked RREP message by checking if the SN in the RREP message is larger than the up-to-date SN .Further, this reply can also be used to confirm whether the intermediate node really has a route to the destination node.[3].Another malicious attack is a flooding attack, in this the attacker exhausts the network resources, such as bandwidth and to consume a node's resources, such as computational and battery power or to disrupt the routing operation to cause severe degradation in network performance. A simple mechanism was proposed to prevent the flooding attack in the AODV protocol. In this approach, each node monitors and calculates the rate of its neighbors' RREQ. If the RREQ rate of any neighbor exceeds the predefined threshold, the node records the ID of this neighbour in a blacklist. Then, the node drops any future RREQs from nodes that are listed in the blacklist. The limitation of this approach is that it cannot prevent against the flooding attack in which the flooding rate is below the threshold. Another drawback of this approach is that if a malicious node impersonates the ID of a legitimate node and broadcasts a large number of RREQs, other nodes might put the ID of this legitimate node on the blacklist by mistake. In the authors show that a flooding attack can decrease throughput by 84 percent. The authors proposed an adaptive technique to mitigate the effect of a flooding attack in the AODV protocol. This technique is based on statistical analysis to detect malicious RREQ floods and avoid the forwarding of such packets. As proposed P. Yi et al, in this approach, each node monitors the RREQ it receives and maintains a count of RREQs received from each sender during the preset time period. The RREQs from a sender whose RREQ rate is above the threshold will be dropped without forwarding. Unlike the method proposed in [4], where the threshold is set to be fixed, this approach determines the threshold based on a statistical analysis of RREQs. The key advantage of this approach is that it can reduce the impact of the attack for varying flooding rates.

In a link spoofing attack, a malicious node advertises fake links with non-neighbours to disrupt routing operations. A location information-based detection method is proposed [5] to detect link spoofing attack by using cryptography with a GPS and a time stamp. This approach requires each node to advertise its position obtained by the GPS and the time stamp to enable each node to obtain the location information of the other nodes. This approach detects the link spoofing by calculating the distance between two nodes that claim to be neighbors and checking the likelihood that the link is based on a maximum transmission range. The main drawback of this approach is that it might not in a situation where all MANET nodes are not equipped with a GPS. Furthermore, attackers can still advertise false information and make it hard for other nodes to detect the attack. In [6], the authors show that a malicious node that advertises fake links with a target's two-hop neighbors can successfully make the target choose it as the only MPR. Through simulations, the authors show that link spoofing can have a devastating impact on the target node. Then, the authors present a technique to detect the link spoofing attack by adding two-hop information to a HELLO message. In particular, the proposed solution requires each node to advertise its two-hop neighbors to enable each node to learn complete topology up to three hops and detect the inconsistency when the link spoofing attack is launched. The main advantage of this approach is that it can detect the link spoofing attack without using special hardware such as a GPS or requiring time synchronization. One limitation of this approach is that it might not detect link spoofing with nodes further away than three hops. Daniel-Ioan Curiac,Ovidiu Banias,Octavian

Dranga proposed a for the detection of malicious node if the application on the captured node is altered, this strategy based on the past and present values provided by each sensor of a network for detecting their malicious activity. Basically, every moment it compare sensor's output with estimated value computed by an auto regression predictor. If the difference between the two values is higher than a chosen threshold, the sensor node becomes suspicious and a decision block is activated. These solutions can also a way to discover the malfunctioning nodes.

The prediction value can be obtained from the following equation

yA(t)=node1(t) . node1 (t-1) + node2 (t). node2 (t-1) +..............+node n(t) . node n(t-1)

(error) eA(t)=xA(t)- yA(t). {Comparing with the present (xA(t)) and estimated value (yA(t))}

If the error is greater than the threshold, then sensor node becomes suspicious and a decision block is activated-[7].application.

## 3 Proposed Model

Statistical modeling is among the earliest methods used for detecting malicious activity in electronic information systems. It is assumed that an intruder's behavior is noticeably different from that of a normal behavior, and this statistical model is used to aggregate the node's behavior which distinguishes an attacker from a normal node behavior. Our statistical techniques is applicable to program or application running on any node. The observed behavior of a node is flagged as a potential malicious if it deviates significantly from the node's expected behavior or from different nodes in the same Ad hoc network. The expected behavior of a node is stored in the profile of the server node of the Ad hoc network. Statistical Mean measures are used to measure for detecting malicious activity of the node.

This algorithm analyzes a node's activities according to a four-step process.

First, the algorithm generates different data collected vectors to represent the activities of a particular node by monitoring the processor utilization for some time period after some interval of time. Let the different collected vectors generated represented by X1, X2 at different time T = <t1, t2, …, tn>.The session vector Xi =<x1, x2, …, xn > represents the data's collected from a single session.

Second, A threshold value range is calculated from different X1,X2……where X1 = <x11, x12, …, xmn >, X2 = <x21, x22, …, xpn >, XN , at different interval of time T 1= <t1, t2, …, tn>, T 2= <t1, t2, …, tn>, ………. Tn where <ti is min, ti is max. > by calculating means of acquired different set of data vectors at different interval of time T 1= <t1, t2, …, tn>, T 2= <t1, t2, …, tn> . The threshold value range is formed from different X1,X2…… . The threshold value range is then stored for a particular's node profile at the server of the network. Let the generated threshold value for a particular node is represented by Vn.. Same process is repeated for each node in the network and for each node a threshold value range is made and then stored for a particular's node profile at the server of the network, only if nodes differ in their application or architecture /manufacture.

Third, this step in the algorithm to detect the malicious activity of particular node. A session vector is formed which represent the activities of a particular node for the current a session by monitoring the processor utilization is acquired for a time period with some fixed interval T 1= <t1, t2, …, tn>, .The time interval and the size of the vector should be same as adjusted during the formation of the threshold range. The already calculated threshold value formed by acquiring different set of data vectors at different interval of time T = <t1, t2, …, tn>, where <ti, min, ti, max> is compared with the current threshold for this particular's node at the server of the network, if it falls outside the range then it is represent a malicious node /corrupted node or otherwise not a malicious node.

Fourth, the final step, the algorithm generates a suspicion quotient to represent how suspicious this session is compared with all other sessions and a knowledge based system can take decision to expel the malicious node from the network topology. Our proposed model is shown in the Figure:1.

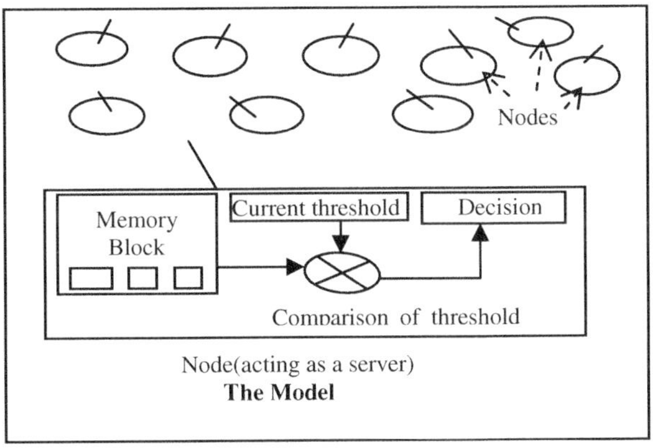


**Fig. 1.**

## 4 Experiment

As our statistical techniques is applicable to program or application running on any node. The observed behavior of a node is done by first creating vectors to represent the activities of a particular node by monitoring the processor utilization for a time period of 0-69 seconds with interval of time 3seconds between two readings and the same process is repeated till the formation of the three different collected vectors generated represented by X1, X2 at different time T = <t1, t2, …, tn>. Statistical mean is calculated for each vectors and a range is formed. Now this range describes the character of the processor for the particular application which we run. The range is stored in the profile of the node. We altered the code with repeating same of codes. Again we repeated the above described step and formed a vector, and mean is calculated and compared with the threshold range. The results are as shown in the graph Figure: 2, vector X5 and X6 shows a significant deviation from the node's expected behavior which flagged as a potential malicious. We again crosschecked our result by repeating the last step, which potentially confirm that the node is malicious the result are We also repeated the above steps by varying the time for monitoring the processor utilization from 0-10 sec with the interval of time 3 seconds and the end result shows a significant deviation from the node's expected behavior which flagged as a potential malicious.

Table 1 in the appendix shows the regression analysis & auto correlation coefficient of the data obtained from different observations by the use of SPSS version 17.0; as in our data we have taken the significance level to be equal to 10 % hence according to that the X5 & X6 are not found to be significant because of their respective values i.e. 0.510 & 0.607; this shows that observations obtained for these variables are significant at 51% & 60% respectively. And also it is important to note that these two variables are showing the problem of auto correlation as there coefficients i.e Durbin- Watson coefficient have values less than the permissible level which shows that the values of these two variables are themselves auto correlated. Hence from this regression analysis & D-W analysis it is evident that variables X5 & X6 are not showing the perfect behaviour so it can be considered as malicious.

Graphical representation v/s %CPU Utilization and Time as shown in Figure :2 .It shows the actual behavior of a node for vectors X1, X2, X3, X4 and X5,X6 shows the malicious behavior of the node after the modification in the code.

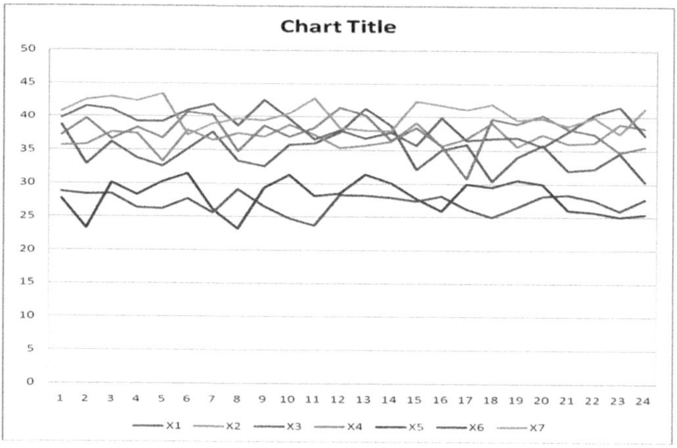


**Fig. 2.** Different collected vectors generated represented by X1, X2, X3, X4, X5, and X6

## 5 Conclusion

Statistical Approaches are very efficient in the detection of malicious node in early stage and helps to avoid the corruption of the network. The above statistical approach is excellent and fast in detecting the malicious node once the profile is created. As in the approaches other than Statistical Approach, the network has to first investigate then find the malicious node form the Ad hoc network and then removed from the network and then rewrite malicious nodes if conditions are favorable to so. If the detection is not made in appropriate time, which can leads to great loss of data or corruption of the network through captured node. But in our statistical approaches based on processor utilization technique we can easily detect malicious behavior in short span of time. Processor utilization technique can also be useful in detecting abnormal condition if any occur in the network other than capture node.

## References


[1] Hua, S.J., ChuanXiang, M.: A reputation-based Scheme against Malicious Packet Dropping for Mobile Ad Hoc Network

[2] Hu, Y.-C., Perrig, A., Johnson, D.: Wormhole Attacks in Wireless Networks. IEEE JSAC 24(2) (February 2006)

[3] Lee, S., Han, B., Shin, M.: Robust Routing in Wireless Ad Hoc Networks. In: 2002 Int'l. Conf.Parallel Processing Wksps., Vancouver, Canada, August 18–21 (2002)

[4] Yi, P., et al.: A New Routing Attack in Mobile Ad Hoc Networks. Int'l. J. Info. Tech. 11(2) (2005)

[5] Raffo, D., et al.: Securing OLSR Using Node Locations. In: Proc. 2005 Euro. Wireless, Nicosia, Cyprus, April 10–13 (2005)

[6] Kannhavong, B., et al.: A Collusion Attack Against OLSR-Based Mobile Ad Hoc Networks. In: IEEE GLOBECOM 2006 (2006)

[7] Junior, W.R.P., de Paula Figueiredo, T. H., Wong, H.C.: Malicious Node Detection in Wireless Sensor Networks

[8] Khokhar, R.H., Ngadi, A., Mandala, S.: A Review of Current Routing Attacks in Mobile Ad Hoc Networks

[9] Karlof, C., Wagner, D.: Secure routing in wireless sensor network: Attacks and countermeasures. In: First IEEE International Workshop on Sensor Network Protocols and Applications (May 2003)

[10] Curiac, D.-I., Banias, O., Dranga, O.: Malicious Node Detection in Wireless Sensor Network Using an Auto regression Technique

[11] Perkins, C., Belding-Royer, E., Das, S.: Ad Hoc On demand Distance Vector (AODV) Routing. IETF RFC 3561 (July 2003)

[12] Kannhavong, B., Nakayama, H., Nemoto, Y., Kato, N., Jamalipour, A.: A survey of routing attacks in mobile ad hoc networks. Security in Wireless Mobile ad Hoc and Sensor Networks, 85–91 (October 2007)

[13] Kannhavong, B., et al.: A Collusion Attack Against OLSR-Based Mobile Ad Hoc Networks. In: IEEE GLOBECOM 2006 (2006)

[14] Karakehayov, Z.: Using REWARD to Detect Team Black-Hole Attacks in Wireless Sensor Networks. In: Wksp. Real-World Wireless Sensor Networks, June 20-21 (2005)

[15] Kurosawa, S., et al.: Detecting Blackhole Attack on AODV-Based Mobile Ad Hoc Networks by Dynamic Learning Method. In: Proc. Int'l. J. Network Sec. (2006)

[16] Johnson, D., Maltz, D.: Dynamic Source Routing in Ad Hoc Wireless Networks. In: Imielinski, T., Korth, H. (eds.) Mobile Computing, p. 146. Kluwer, Dordrecht

[17] Raju, J., Garcia-Luna-Aceves, J.J.: A comparison of On-Demand and Table-Driven Routing for Ad Hoc Wireless etworks. In: Proceeding of IEEE ICC (June 2000)

[18] Hu, Y.-C., Perrig, A., Johnson, D.: Wormhole Attacks in Wireless Networks. IEEE JSAC 24(2) (February 2006)

[19] Perkins, C., Royer, E.: Ad Hoc On-Demand Distance Vector Routing. 2nd IEEE Wksp. Mobile Comp. Sys. and Apps., 149

[20] Qian, L., Song, N., Li, X.: Detecting and Locating Wormhole Attacks in Wireless Ad Hoc Networks Through Statistical Analysis of Multi-path. In: IEEE Wireless Commun. and Networking Conf. (2005)
[21] Raffo, D., et al.: Securing OLSR Using Node Locations. In: Proc. 2005 Euro. Wireless, Nicosia, Cyprus, April 10–13 (2005)
[22] Sanzgiri, K., et al.: A Secure Routing Protocol for Ad Hoc Networks. In: Proc. 2002 IEEE Int'l.Conf. Network Protocols (November 2002)
[23] Yi, P., et al.: A New Routing Attack in Mobile Ad Hoc Networks. Int'l. J. Info. Tech. 11(2) (2005)
[24] http://spie.org/x8693.xml?ArticleID=x8693 (Internet February 28, 2011)

# Appendix

**Table 1.** Regression analysis & auto correlation coefficient

| Model | 1 | 3 | 5 | 7 | | | | | | |
|---|---|---|---|---|---|---|---|---|---|---|
| | 2 | 4 | 6 | 8 | 9 | 10 | 4 | 6 | 8 | 11 |
| X1 | 0.309 | 0.095 | 0.054 | 20.63042 | 0.055 | 2.318 | 1 | 22 | 0.142 | 1.630 |
| X2 | 0.248 | 0.062 | 0.019 | 21.01148 | 0.062 | 1.444 | 1 | 22 | 0.242 | 1.670 |
| X3 | 0.462 | 0.213 | 0.178 | 19.23702 | 0.021 | 5.968 | 1 | 22 | 0.023 | 1.750 |
| X4 | 0.21 | 0.044 | 0.001 | 21.20674 | 0.044 | 1.014 | 1 | 22 | 0.325 | 1.890 |
| X5 | 0.105 | 0.011 | -0.034 | 21.56932 | 0.510 | 0.247 | 1 | 22 | 0.624 | 0.036 |
| X6 | 0.086 | 0.007 | -0.038 | 21.60939 | 0.607 | 0.164 | 1 | 22 | 0.689 | 0.031 |
| X7 | 0.302 | 0.091 | 0.05 | 20.68057 | 0.091 | 2.2 | 1 | 22 | 0.152 | 1.730 |

1 R
2 Rsquare Change
3 R square
4 F Change
5 Adjusted R square
6 df1
7 Standard error of the estimate
8 df2
9 Sig. F Change
10 R Square Change
11 Durbin-Watson

# Optimum Controller for Automatic Generation Control

Rahul Agnihotri[1], Gursewak Singh Brar[1], and Raju Sharma[2]

[1] Assistant Professor,
Electrical Engg. Department,
BBSBEC, FGS
[2] Assistant Professor,
Electronics and communication Engg. Department,
BBSBEC, FGS
Punjab Technical University (Pb.), India

**Abstract.** This paper deals with automatic generation control of area consisting of many generating sources i.e. hydro, thermal and gas. One percent load perturbation is given to each area considering combination of thermal, thermal hydro and thermal hydro gas generating station and the response of system frequency is analyzed. Accurate transfer function model is first required to analyze the system. To investigate the system dynamic performance, optimal control design is implemented in the wake of 1% step load disturbance.

**Keywords:** AGC, multi area generation, optimum controller.

## 1 Introduction

Continuously increasing demand of electricity leads to the increase in interconnected systems to transfer the electricity from far distant generating stations at the consumer end. Continuous and reliable power supply is dependent upon some power system parameters like frequency and tie-line power flow. Variations in these parameters arises due to continuously changing loading conditions i.e. change in real and reactive power demands. Change in real power leads to change in frequency of the power system which may lead to the system collapse if not controlled properly. So, the load frequency control (LFC) is the backbone of the stable power system operation. Large scale power systems are normally composed of control areas or regions representing coherent groups of generators. Area load changes and abnormal conditions lead to mismatches in frequency and scheduled power interchanges between areas. These mismatches have to be corrected by Automatic Generation Control (AGC), which is defined as the regulation of the power output of generators within a prescribed area. Each control area must meet its own demand and its scheduled interchange power. Any mismatch between the generation and load can be observed by means of a deviation in frequency [1]. This balancing between load and generation can be achieved by using Automatic Generation Control (AGC). Varieties of models have been

A. Mantri et al. (Eds.): HPAGC 2011, CCIS 169, pp. 357–363, 2011.

developed over the last few decades considering different types of generation in each area. In real situations, each control area may have various types of generation such as hydro, thermal, gas, nuclear etc. The results in [2] is an attempt to study the performance of AGC with thermal, hydro and gas generations in the same area. Work reported in literature on AGC pertains to either two-area thermal or hydro-hydro or combination of these two but there is no or very little work on AGC for multigenerational thermal system, hydro system and gas system. In a mixed power system [3] it is usual to find the area regulation for gas generation or thermal generation or in combination of thermal, hydro and gas station. In the present work, optimal control design are used to restore the frequency to its nominal value and their dynamic responses are compared for system consisting of thermal, hydro and gas based generation.

## 2 System Investigated

The prime sources of electrical energy supplied by utilities are the kinetic energy of water and thermal energy derived from fossil fuels. The prime movers convert these sources of energy into mechanical energy that is, in turn, converted to electrical energy by synchronous generators. The prime mover governing system provide a means of controlling power and frequency, a function commonly referred to as load frequency control[4]. To make more realistic studies, appropriate mathematical models of steam turbines e.g. reheat non reheat steam turbines, hydro turbines and gas turbine are to be considered for the dynamic simulation of the system behaviour. For AGC studies it is necessary to obtain appropriate models of the interconnected power systems. The important component for controlling the speed of the turbine is the governor. Governor for each system differs from other. Transfer function models [5] for these turbine governors of references are used for the studies undertaken in this paper. Automatic generation control basically is the load frequency control which controls the real power & frequency. As we know that our real power & reactive power both are not steady & changes with rising or falling trends, so we have to regulate steam input & excitation of generator to match the real & reactive power contentiously. The small change in real power is mainly dependent on change in rotor angle & thus the frequency and the reactive power is mainly dependent on voltage magnitude i.e. on the generator excitation. In particular the following have been investigated

1. The effect of change of load on each type of generating area and selection of the best combination of generators for good system response.
2. The effect on frequency response of an area while working with individual isolated system and when area is operated in combination of more than one generating system
3. The design of optimum controller based on transfer function model for automatic generation control using state equations.

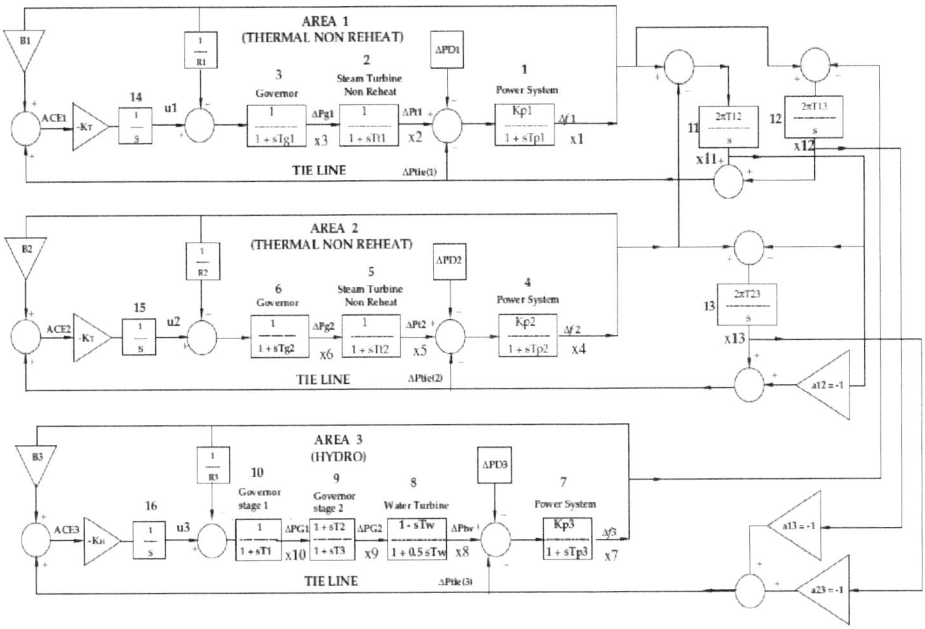


**Fig. 1.** Three area thermal-thermal-hydro power system

**Linear Optimal control: Three** area thermal, thermal and hydro power system transfer function model shown in Fig-1 is used to obtain state equations to obtain optimum control theory. The control is achieved by feeding back the state variables through a regulator with constant gains. Consider the control system presented in the state-variable form

$$\dot{x}(t) = Ax(t) + Bu(t) \tag{1}$$

$$y(t) = Cx(t)$$

For full state feedback, the control vector u is constructed by a linear combination of all states, i.e.

$$u(t) = -Kx(t) \tag{2}$$

Where K is a 1 X n vector of constant feedback gain.The purpose of this system is to return all state variables to values of zero when the states have been perturbed. In this section, the design of optimal controllers with linear systems with quadratic performance index, the so called linear quadratic regulator (LQR) has been discussed. The object of optimal regulator design is to determine the optimal control law u*(x, t) which can transfer the system from its initial state to final state such that a given performance index is minimized. The performance index is selected to give the best trade- off between performance and cost of control. The performance index that is widely used in optimal control design is known as quadratic performance index and is based on minimum- error and minimum energy criteria. Consider the plant described above

$$\dot{x}(t)=Ax(t)+Bu(t)$$

$$u(t)=-K(t)x(t)$$

Which minimizes the value of a quadratic performance index J of form?

$$J=\int_{t0}^{tf}(x'Qx+u'Ru)dt$$

Q is a positive semi- definite matrix and R is real symmetric matrix. Q is positive semi definite, if all its principal minors are non-negative. The choice of elements of Q and R allows the relative weighting of individual state variables and individual control inputs.

To obtain a formal solution, we can use the method of Lagrange multipliers. The constraint problem is solved by augmenting (1) into (2) using a non-vector of lag range's multiplier, λ. The problem reduces to the minimization of the following unconstrained function.

$$L(x, \lambda, u, t) = [x' Q x + u'R u] + \lambda' [Ax + Bu - \dot{x}]$$

The optimal values (denoted by the subscript *) are found by equating the partial derivatives to zero.

$$\frac{\partial L}{\partial \lambda} = AX^* + Bu^* - \dot{x}^*=0$$

$$\dot{x}^*=AX^* + Bu^*$$

$$\frac{\partial L}{\partial u}=2Ru^* + \lambda' B=0$$

$$u^*=\frac{-1}{2} R^{-1} \lambda' B \tag{3}$$

$$\frac{\partial L}{\partial x}=2x'^*Q + \dot{\lambda}' B=0$$

$$\dot{\lambda} = -2Qx^* - A'\lambda \tag{4}$$

Assume that there exists a symmetric, time-varying positive definite matrix p(t) satisfying

$$\lambda = 2p(t)\, x^*$$

Obtaining the derivative of λ, we have

$$\dot{\lambda} = 2(\dot{p}\, x^* + p\, \dot{x}^*) \qquad (5)$$

Substituting (3) into (4) gives the optimal control law

$$u^*(t) = -R^{-1} B' p(t) x^*$$

Finally, equating (4) with (5), we obtain

$$\dot{p}(t) = -p(t) A - A' p(t) - Q + p(t) B R^{-1} B' p(t)$$

The above equation is referred to as the matrix Ricatti equation.For linear time-invariant systems, since $\dot{p}=0$, when the process is of infinite duration, that is $t_f = \infty$, equation reduces to the algebraic Ricatti equation

$$pA + A'p + Q - pB R^{-1} B' p = 0$$

## 3 Result Discussion

The MATLAB Control System Toolbox function [K,p]=lqr2(A,B,Q,R) can be used for the solution of the algebraic Ricatti equation. The LQR design procedure is in stark contrast to classical control design, where the gain control matrix K is selected directly. To design the optimal LQR, the design engineer first selects the design parameter weight matrices Q and R. Then, the feedback gain K is automatically given by matrix design equations and the closed loop time responses are found by simulation. If these responses are unsuitable, new values of Q and R are selected and the design is repeated. This has the significant advantages of allowing all the control loops in a multiloop system to be closed simultaneously, while guaranteeing closed-loop stability. We have selected the state cost weighing matrix Q, k loops are opened as an identity matrix of proper dimension for each case study considered. Control cost weighing matrix R. For computation the control cost weighing matrix R is used as an identity matrix. For each of the case study its dimension is taken 3X3.System response with optimum parameters.

For three thermal generating stations connected in an area for step load perturbation frequency response is as shown above in Fig.-2.Similarly the transfer function model for two thermal one hydro station is drawn and state equations are written which are used to implement the optimum controller as in case of three thermal stations. . The graph showing frequency variation of two thermal and one hydro station is shown in Fig-3 and Fig-4. Finally we considered the effect of including a gas system along with combination of hydro and thermal station .from transfer function model state equations are written which are further used to implement optimum controller.

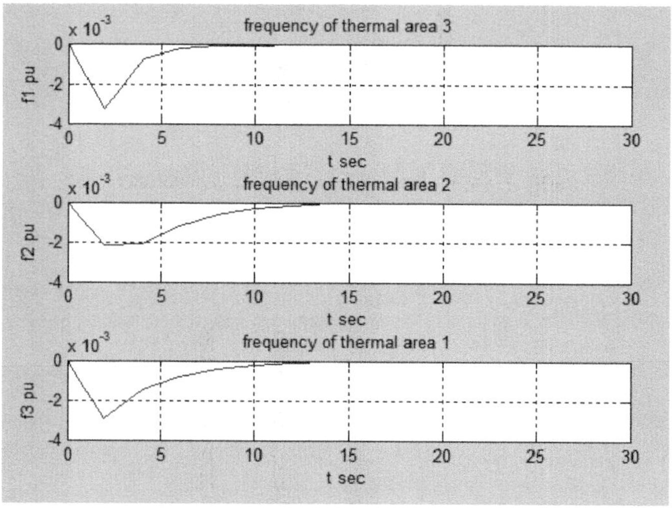


**Fig. 2.** Area with three thermal generating stations

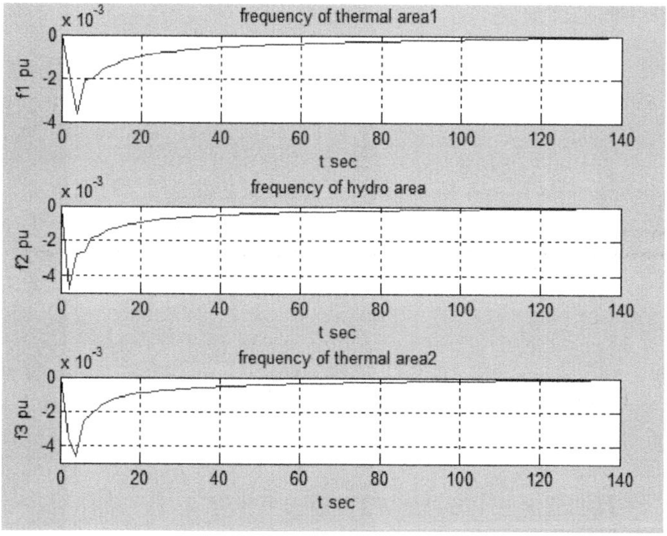


**Fig. 3.** Area with two thermal and one hydro generating station

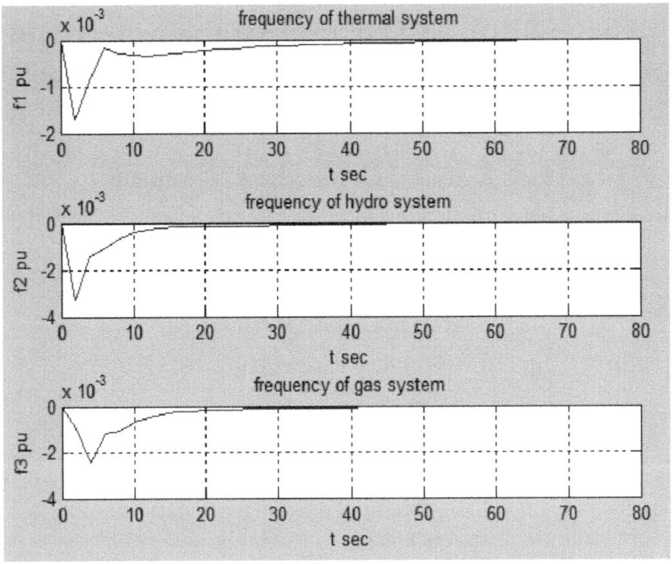


**Fig. 4.** Area with thermal, hydro and gas generating station

# 4 Conclusion

It has been found that if all the parameters are considered same, then frequency drop will be 1/3$^{rd}$ of that which would be experience if the control areas were operating alone. It has been found that as per unit disturbance in thermal area produces more oscillations in hydro and gas based system in comparison to thermal based systems. While a per unit disturbance in hydro or gas based system produces a large disturbance in hydro and gas based system rather frequency and tie line deviation for thermal system in all the three different interconnected systems.

# References

[1] Elgerd, Fosha, C.: Optimum megawatt-frequency control of multiarea electric energy systems. IEEE Trans. Power Apparatus &Systems PAS-89(4), 556–563 (1970)

[2] Ramakrishna, K.S.S., Bhatti, T.S.: Load frequency control of interconnected hydro-thermal power systems. In: International Conference on Energy and Environment 2006, ICEE 2006 (2006)

[3] Ramakrishna, K.S.S., Bhatti, T.S.: Automatic generation control of single area power system with multi-source power generation. In: Proc. IMechE vol. 222 Part A: J. Power and Energy (2008)

[4] Aldeen, M., Trinh, H.: Load frequency control of interconnected power systems Via constrained feedback control schemes. Computers & Electrical Engineering 20(1), 71–88 (1994)

[5] Chan, W.C., Hsu, Y.Y.: Automatic Generation of Interconnected Power Systems using Variable Structure Controllers. In: IEE Proc., pt. C, vol. 128(5), pp. 269–279 (September 1981)

# Abstraction of Design Information from Procedural Program

R.N. Kulkarni, T. Aruna, and N. Amrutha

Departure of Information Science & Engineering,
Ballari Institute of Technology & Management, Bellary
rn_kulkarni@rediffmail.com, arunatadapatri@gmail.com,
amrutha.n07@gmail.com

**Abstract.** In the past two decades there has been a continuous change in the software development. Organizations use different programming languages for developing different software applications. The applications which were developed earlier were based on procedural programming languages like 'C', FORTRAN, COBOL etc. The applications which are being developed now, may be based on object oriented languages or procedural languages or a mix of both. In order to understand how the information system is designed one may need to understand the behavior of the program. The behavior of the program can be understood with the help of design information. This design information about the application program can be abstracted the from data flow diagram.

In this paper we are proposing a methodology to abstract the behavior of the program and then representing this behavior in the form of a data flow diagram through a series of steps.

**Keywords:** Data flow diagram, Design information, Process, Program behavior.

## 1 Introduction

When we consider any system or a program there exist certain information which is helpful in understanding what the system or the program does. And also there is a continuous flow of data from one process to another process. This information cannot be understood or possibly difficult to interpret by just looking at the program. Therefore we need to abstract such information which gives a clear idea of what the program does, and also the flow of information from one process to another process. Such information can be represented by a data flow diagram (DFD). Data flow diagrams are very useful in understanding a system and can be effectively used during analysis as well as design of the system.

DFD shows the flow of data through a system and it views the system as a functional unit that transforms the input into the output that undergoes a series of transformation. These transformations are captured by DFD [5].

In the data flow diagram each process is represented by a symbol such as bubble or a circle. The input to the process may be either from an external entity or from a file or from another process. Similarly the output of the process may be either to an external entity or stored in a file or sent as a input to another process. Representing

A. Mantri et al. (Eds.): HPAGC 2011, CCIS 169, pp. 364–372, 2011.

the data flow diagram pictorially by making use of DFD symbols is difficult because it consumes more space and memory and also for representing the large applications pictorially is next to impossible. To avoid this complexity we are going for a tabular representation of data flow diagram.

Many works have been done in the field of decomposition of data flow diagrams. Data flow diagram process decomposition, as applied in the analysis phase of software engineering, is a top-down method that takes a process, and its input and output data flows, and logically implements the process as a network of smaller processes. The decomposition is generally performed in an ad hoc manner by an analyst applying heuristics, expertise, and knowledge to the problem [1]. In an algebra that formalizes process decomposition using the De Marco representation scheme. In this algebra, the analyst relates the disjoint input and output sets of a single process by specifying the elements of an input/output connectivity matrix. A directed acyclic graph is constructed from the matrix and is the decomposition of the process [2].

An approach for recovering data flow oriented design of a software system from its source code employs reverse engineering techniques that create hierarchical clusters of functions and procedures to identify the "bubbles" at various levels in the hierarchy of DFD's. It uses results from inter-procedural flow analysis to compute the "logical" flow of data between these bubbles. And it uses information about data types provided with the source code to create the data dictionary. The paper also identifies the open problems whose solutions would enable the recovery of data flow oriented designs [3].

Dragos Truscan, Joo M. Fernandes, Johan Lilius, Arndt T., Guercio A. stated that a set of quality measures described by Adler do not correspond to the intuitive notion of a good decomposition. The use Adler's algebra leads to an inefficient decomposition process, as well as one which is not guaranteed to find a good decomposition [4].These authors proposed an approach of automating the process of DFD design.

Inspite of all the above work done there exists a complexity in analyzing the flow of data in a program. Hence there is a need for better representation of DFD. Representing the processes present in a program and their inputs and outputs in the form of table provides a clear view of how data flows and what function is performed by a program. In this paper we identify all user defined functions and consider them as process and the interaction between the processes is depicted.

## 2 Taxonomy

**Data flow diagram:** The functions and the data items that are exchanged between different functions are represented in a diagram known as data flow diagram.

**Control flow graph:** A control flow graph captures the flow of control within a program.

**Context diagram:** A context diagram shows the system boundaries, external entities that interact with the system, and the relevant information flows between these external entities and the system [5].

**Process:** A process is a function which takes the valid inputs, apply certain computation and produce desired output.

**Data flow graph notations:**

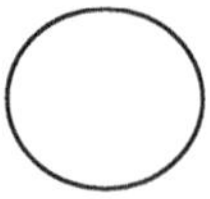 → Bubbles used to represent functions.

 → Arrows used to represent data flow.

 → Open boxes represented persistent

data storage

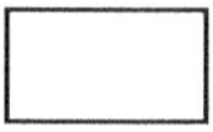 → Input/output boxes represent data

acquisition and production during

human computer interaction

**User defined functions:** These are the functions where the parameters passed, their data types, body of the function and the return value is specified by the user.

**Built-in functions:** These are the functions which are not defined by the user, their functionality cannot be changed.

## 3 Proposed Methodology

We are taking an executable 'C' program as input and storing it in a file. We read each and every line of the program to identify user defined processes and count is incremented by one after reading each line. Once the process is identified, determine the statement number at which process ends. Then for these numbers of statements identify the referenced and defined variables. The referenced variables are the inputs to the process and defined variables are the outputs. Then this information is represented in the form a table.

### 3.1 Algorithm

Defined: The variables in a program where a value is produced and in general at all statements that change the value of some or more variable[6].

Used: The variables whose value can be accessed and in general in all statements whose execution extracts a value from a variable[6].

//to abstract the data flow diagram from procedural program

//input: executable C program

```
//output: representation of data flow diagram in the form of a table
Step1: start
Step2: FSTREAM file
Step3: int count, i←0,j←0
Step4: char a[50],b[50],buf[1000],s1[100];
Step5: count←1, s1←NULL
Step6: file.open("stud.txt",iopen::out)
Step7: while (!feof) do
        file.getline(buf,1000,\n)
        while(buf[i]!=" ") do
          a[i]← buf[i]
          i←i+1
          flag←1
        end while
       if(flag) then
         while(buf[i]!="(") do
         b[j]←buf[i]
         i←i+1
        j←j+1
        end while
         end if
Step8: for j←0 to i
        strcat(s1,b[i])
Step9: if(strcmp(s1,"main")!=0) then
        Display all the referenced variables as inputs to that process
```

Display all the defined variables as outputs to that process

end while

Step10: end

## 4 Case Study

```
#include<stdio.h>
void insert();
void search();
void display();
struct student
{
char name[30];
int usn,m1,m2,m3;
};
struct student s1[100];
int num,i,n;
FILE *fp;
void main()
{
int ch;
for(;;)
{
printf("1.insert\n 2.search\n 3.display\n 4.exit\n");
printf("enter the choice\n");
scanf("%d",&ch);
switch(ch)
```

```
{
case 1: insert();
        break;
case 2: search();
        break;
case 3: display();
        break;
case 4: return;
}
}
}
void insert()
{
printf("enter the number of students\n");
scanf("%d",&n);
fp=fopen("stud.txt","a");
for(i=0;i<n;i++)
{
printf("enter the details of the student");
printf("enter student usn,name\n");
scanf("%d%s",&s1[i].usn,s1[i].name);
printf("enter the marks");
scanf("%d%d%d",&s1[i].m1,&s1[i].m2
&s1[i].m3);
if(fp!=NULL)
```

```
fprintf(fp,"usn=%d name=%s m1=%d m2=%d m3=%d",s1[i].usn,
s1[i].name, s1[i].m1, s1[i].m2,

s1[i].m3);

}

fclose(fp);

}

void search()

{

fp=fopen("stud.txt","r");

printf("enter usn to be searched\n");

scanf("%d",&num);

for(i=0;i!=EOF;i++)

{

fscanf(fp,"%d",s1[i].usn);

if(s1[i].usn==num)

{

printf("student details are\n");

printf("student name:%s\n",s1[i].name);

printf("marks:%d\n%d\n%d\n",s1[i].m1,

s1[i].m2,s1[i].m3);

}

}

fclose(fp);

}

void display()

{
```

```
int c;
fp=fopen("stud.txt","r");
c=getc(fp);
while(c!=EOF)
{
putchar(c);
c=getc(fp);
}
fclose(fp);
}
```

**Fig. 1.** A sample input 'C' program

The data flow diagram for the sample 'C' program shown in figure 1 is depicted in figure 2. This data flow diagram drawn here is a manual process.

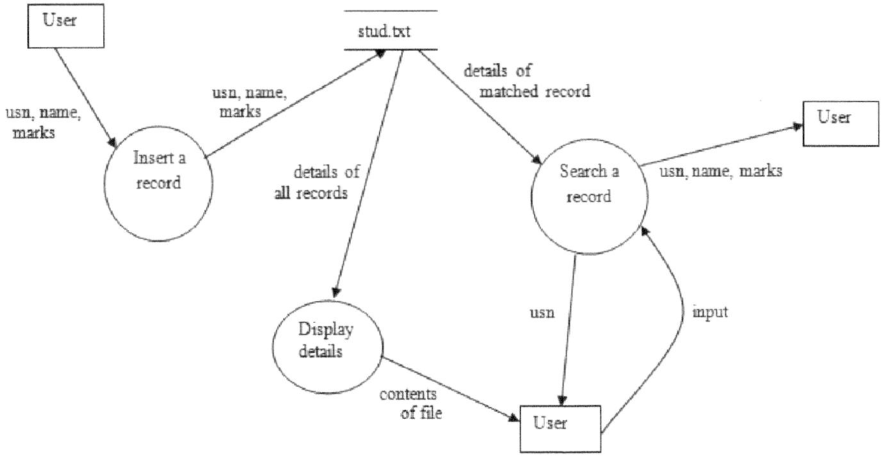


**Fig. 2.** Data Flow Diagram for the program shown in figure 1 (manual process)

The proposed tool developed here abstracts the information from the program shown in figure 1 and displays the output in the form of a table shown in table 1. The table provides the same information which is presented in the data flow diagram shown in figure 2.

**Table 1.** Data flow table for the above program

| Input | Process | output |
|---|---|---|
| Usn,name,m1,m2,m3 | Insert() | Stored in a file stud.txt |
| Usn,stud.txt | Search() | Message to user |
| Stud.txt | Display() | Details to the user |

## 5 Conclusion

This paper proposes a semi automated tool which abstracts the user-defined functions from a procedural program which are helpful in understanding the behavior of the program. This behavior of the program helps in understanding the flow of data or the design information of an information system. The proposed tool represents the inputs and outputs of a process and stores the abstracted information in the form of a table. The tool is verified for its completeness by giving different sized programs.

## References

1. Adler, M.: An Algebra for Data Flow Diagram Process Decomposition. IEEE Transactions on Software Engineering 14(2), 169 (1988)
2. Lakhotia, A.: An approach To Recovering Data Flow Oriented Design of a Software System, The Center for Advanced Computer Studies, University of Southwestern Louisiana Lafayette, LA 70504
3. Butler, G., Grogono, P., Shinghal, R., Tjandra, I.: Retrieving Information from Data Flow diagrams
4. Ardnt, T., Guercio, A.: Decomposition of Data Flow Diagrams
5. Jalote, P.: An Integrated Approach to Software Engineering
6. Pezze, M., Young, M.: Software Testing and Analysis: Process, Principles and Techniques. John Wiley & Sons, Inc., Chichester
7. Mall, R.: Fundamentals of Software Engineering, 3rd edn. PHI Learning Private Limited, New Delhi (2010)
8. Mathur, A.P.: Foundations of Software Testing, Dorling Kindersley (India) Pvt. Ltd

# Design of an Intelligent and Adaptive Mapping Mechanism for Multiagent Interface

Aarti Singh[1], Dimple Juneja[1], and A.K.Sharma[2]

[1] M. M. University, Mullana,
Ambala (Haryana), India
[2] Professor,
Y.M.C.A. University of Science & Technology,
Faridabad, Haryana, India
singh2208@gmail.com

**Abstract.** The main intent of this work is to propose an intelligent interface that facilitates agent interaction in homogenous as well as heterogeneous ontologies. Literature survey indicates that there exist mapping mechanisms serving well for homogenous domains but very few researchers have made an attempt to propose a mapping interface for heterogeneous domains that is possessed with learning abilities and is therefore adaptive by nature. This work uniquely contributes towards the future vision of an intelligent and adaptive mapping mechanism that not only overcomes the curse of already existing mapping mechanisms but also is time efficient. The performance of the proposed strategy has been evaluated and compared with the existing strategies in the related fields and the results are found be to be competitive.

**Keywords:** Agent-Communication, Multi-agent Systems, Ontologies, Ontology Mapping, Semantic Web.

## 1 Introduction

The advent of Semantic Web, comprising of multiples of agents, has given birth to ontology that facilitates agent interaction on the web. Basically ontology is vocabulary that describes objects and their relationships in a domain under consideration. Mathematically, an ontology can be defined as a pair O= {C, R} where C is a directed labeled graph describing the ontological concepts and R is set of rules specifying the intended interpretation of the vocabulary in concerned domain. The graph C={V,E} comprises a finite set of nodes V and a finite set of edges E, where nodes represent the concept name and edges represent the semantic relationship among the concepts[10].

However, in order to support the variety of tasks envisaged by the current and future scenario of semantic web, different applications demands for huge number of ontologies that must cooperate and coordinate in order to achieve the goal. Since large number of applications are executing in various domains which might or might not be compatible, therefore, the desire is to have a common layer through which various ontologies can be accessed and information can be exchanged semantically. This

A. Mantri et al. (Eds.): HPAGC 2011, CCIS 169, pp. 373–384, 2011.

work aims to propose such a common layer and henceforth, the layer shall be referred to as Intelligent and Adaptive Mapping Mechanism for Multiagent Interface (IAM3I) as illustrated in Fig. 1.

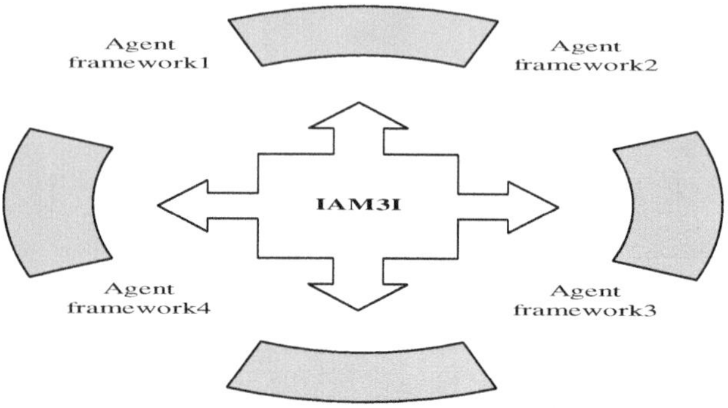


**Fig. 1.** Domain Model of Ontology Mapping System

This paper has been organized into four sections. Section 2 provides the related work in the area of ontology mapping for agents and elaborates the challenges that need to be focused on. Section 3 proposes the framework for mapping of ontologies through agents which is evaluated and a comparison is presented in section 4. Section 5 concludes the paper.

The upcoming section discusses the background of ontology mapping pertaining to agents participating in semantic web.

## 2 Related Work

Ontology mapping can be considered as translating every entity in ontology $O_1$ into some corresponding entity defined in ontology $O_2$. In fact, ontology mapping involves tasks concerning the alignment [4], articulation and merging of ontologies where ontology alignment refers to establishing a set of binary relations between the vocabularies of two ontologies and ontology articulation involves generation of rules through which fusion or merging of ontologies are carried out.

Many eminent researchers have put in efforts to resolve the issues pertaining to ontology mapping. Koes et. al [9] proposed augmenting the capabilities of multi-agent systems (MAS) for efficient transfer of low level information in multi-robot systems. Payne et. al [13] proposed a template based shallow parsing approach for message construction/decomposition improving the robustness of inter-agent communication but it allows exchange of simple messages only. Choi et. al [4] highlighted the need to improve methods for constructing global ontology using mapping techniques of local ontologies where as Mitra et. al [10] presented various automated methods to match terms used in different ontologies however, the methods still requires human intervention.

The algorithm [12] relies on the possibility to point to objects in the world but it failed to discuss the strategy used for pointing. The approach [16] overcomes interoperability problem in agents on a very small database thereby, lacking scalability. Juneja et. al [7] presented an agent based semantic match-making algorithm that exploits the concept of ontology at user end and semantically enhances the user-input without considering the domain compatibility.

As is evident from the background, a lot of work has been done in the area of ontology mapping but a complete solution for ontology mapping is still lacking. For instance, various techniques for mapping of ontologies [1,3, 10] have been proposed in literature, but there is no standard technique for mapping of ontologies that delegate mapping task to agents completely. Also most of the researchers have been focusing on communication scenario of agents in homogenous domains only while very few have diverted their attention towards their application in heterogeneous domains. Therefore, a mapping mechanism that is not only intelligent but also is adaptive is strongly recommended to establish a standard communication. Next section proposes the Intelligent and Adaptive Mapping Mechanism for Multiagent Interface (IAM3I).

## 3 Intelligent and Adaptive Ontological Interface (IAM3I)

The complete framework mainly comprises of three prime components *Source_Domain, Destination_Domain* and *IAM3I* wherein *Source_Domain* and *Destination_Domain* comprises of Personal task agents (PTAs), Ontology Database (ODB) and an Interface Agent (IFA) whereas *IAM3I* contains three agents namely, Information Agent (*IA*), Message Handler Agent (*MHA*) and Mapping Master Agent (*MMA*). Following are the few key assumptions accounting to simplicity and reusability:

### 3.1 Key Assumptions

- ***Support for IA:*** It gathers information from *Source_Domain* and passes mapped message to *Destination_Domain*. IA has access to ontologies of all participating domains. It can not only fetch RDF communication tags but also, can extract the definition of various terms from their respective URIs as and when required.
- ***Support for MMA:*** It is associated with two data structures namely, *Thesaurus_Based_Dictionary (TBD)* where *TBD* has been devised using available semantic networks such as wordnet [19] and a *Temporary Log (TL)* that holds the definitions of the most recently accessed keywords along with their URIs. TL has been implemented as two-dimensional array where rows represent $C_s$ as concept names and the $URI_s$ containing definitions for that $C_s$, form the columns. Thus there are multiple columns for each Cs. Initially matrix will be populated with null values but since intelligence and adaptivity have been incorporated as two important parameters, it is assumed that as the matrix will become more and more populated with $C_s$ and $URI_s$, it would behave intelligently and would definitely provide the desired definition.

- ***Support for IAM3I:*** The proposed framework makes use of two concepts namely Extension and Intension as discussed below:

  ***Extension:*** Extension of a concept is the set of instances that can be classified under that concept. The extension of the universal concept (everything) $C_u$ is the set of all instances while the extension of the absurd concept (nothing) $\perp C$ is the empty set [9]. Mathematically it can be explained considering: (Cs set of concepts in source ontology $O_s$) ($C_T$ is set of concepts in target ontology $O_T$.)

$$
\begin{array}{l}
C_S \in O_S \wedge C_T \in O_T \\
C_{S'} \supseteq C_S \wedge C_{T'} \supseteq C_T \text{ then} \\
\text{if } \left( (C_S \rightarrow \neg C_T) \vee (\neg C_S \rightarrow C_T) \right) \text{ then} \\
\text{extension relation holds.}
\end{array}
\tag{1}
$$

  ***Intension:*** It can be defined as the subsumption of sets of instances classified under the concept reflecting the $\sqsubset_c$ relation [9]. Mathematically it can be formulated as

$$
\begin{array}{l}
if \begin{pmatrix} (O_S \rightarrow O_T \; \forall \; O_S \supseteq O_T) \vee (O_{S'} \rightarrow O_T \forall O_{S'} \supseteq O_S) \\ \vee \, (O_S \rightarrow O_{T'} \forall O_{T'} \subseteq O_T) \end{pmatrix} \\
then \\
\text{intension } relation\ holds
\end{array}
\tag{2}
$$

The explanation of concepts of extension and intension is provided in the upcoming sections.

## 3.2 The High Level View

This section presents the ecology of agents included in the proposed work. Fig. 2 given below illustrates the coordination among the *source_domain*, *destination_domain* and *IAM3I*.

The high level view of Fig. 3 is explained as follows and the working algorithms are presented in the subsequent sections:

1. $PTA_{source\_domain}$ communicates the *input_phrase* to $IFA_{source\_domain}$ which provides interface necessary for communication with other agent based applications.
2. $IFA_{source\_domain}$ extracts the ontology $O_S$ from $ODB_{source\_domain}$ and forwards the clubbed message (*input_phrase and* $O_S$) to $IA_{IAM3I}$.
3. $IA_{IAM3I}$ hands over the message to $MHA_{IAM3I}$ for converting the message in standard format by eliminating stop words and grammatical words. $IA_{IAM3I}$ has access to both $O_S$ and $O_T$ and it passes them to $MMA_{IAM3I}$ for usage.

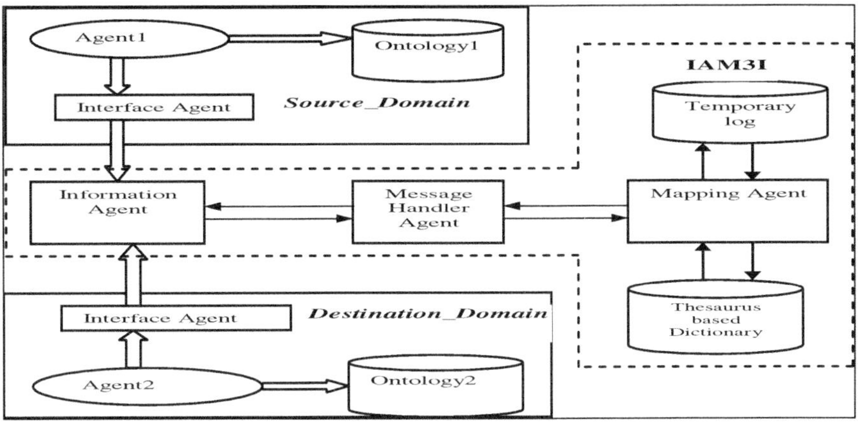


**Fig. 2.** Proposed Intelligent Ontological Interface

4. $MHA_{IAM3I}$ passes the message to $MMA_{IAM3I}$ .
5. Mapping algorithm firstly checks the TBD to check if the keywords in the source message are synonymous with the keywords in $O_T$.
6. If keywords are synonymous then $MMA_{IAM3I}$ replaces the source keywords with the keywords of $O_T$ and the mapped message is sent to $IFA_{destination\_domain}$ through $IA_{IAM3I}$.
7. In case if keywords are not synonymous then $MMA_{IAM3I}$ will explore definitions of the keywords, so as to be sure that same things are not referred using different names.
8. For increasing the recall rate and reducing the mapping time, $MMA_{IAM3I}$ is provided with TL, which keeps the definitions of most recently accessed keywords along with their tags and URIs. $MMA_{IAM3I}$ will first explore the TL before requesting $IA_{IAM3I}$ for fetching definitions of the keywords, if definitions containing same tags are found they can be used , leading to amplified recall rate.
9. In case definition is not available in the TL, $MMA_{IAM3I}$ will send request to $IA_{IAM3I}$ for fetching the required definitions from their respective URIs. $IA_{IAM3I}$ fetches and passes definitions to $MMA_{IAM3I}$, which explores definitions to decide whether intension holds with the classification given in $O_T$, so that mapping can be performed with some loss of information. If intension relation holds, message is mapped and sent to $IA_{IAM3I}$ .
10. If the source definitions are not subset of the classification used in $O_T$ then $O_S$ needs to be extended. Thus $MMA_{IAM3I}$ sends ontology extension request to $IA_{IAM3I}$ along with the class that needs to be added in $O_S$ .
11. $IA_{IAM3I}$ in turn passes the extension request to $IFA_{source\_domain}$ along with the class needed for the extension of the $O_S$ .
12. $IFA_{source\_domain}$ extends the $O_S$ by adding the class at appropriate place. This is possible because of the learning ability of the agents and is required due to the

fact that agent doesn't always need to communicate with other frameworks containing subset of their ontologies or working on similar kind of ontologies only. In case of frameworks working on similar domains but using different classification or working on entirely different domains, communication can be ensured only due to the extension feature of source_domain ontology, in our proposed framework. With the passage of time extension of ontologies will make them richer and easy to communicate with other frameworks. Now mapping becomes feasible and can be performed either through exact or intension relation of keywords mapping as discussed above.

13. Finally the mapped message is delivered by $IA_{IAM3I}$ to the $IFA_{destination\_domain}$, which finally passes it to the desired $PTA_{destination\_domain}$.

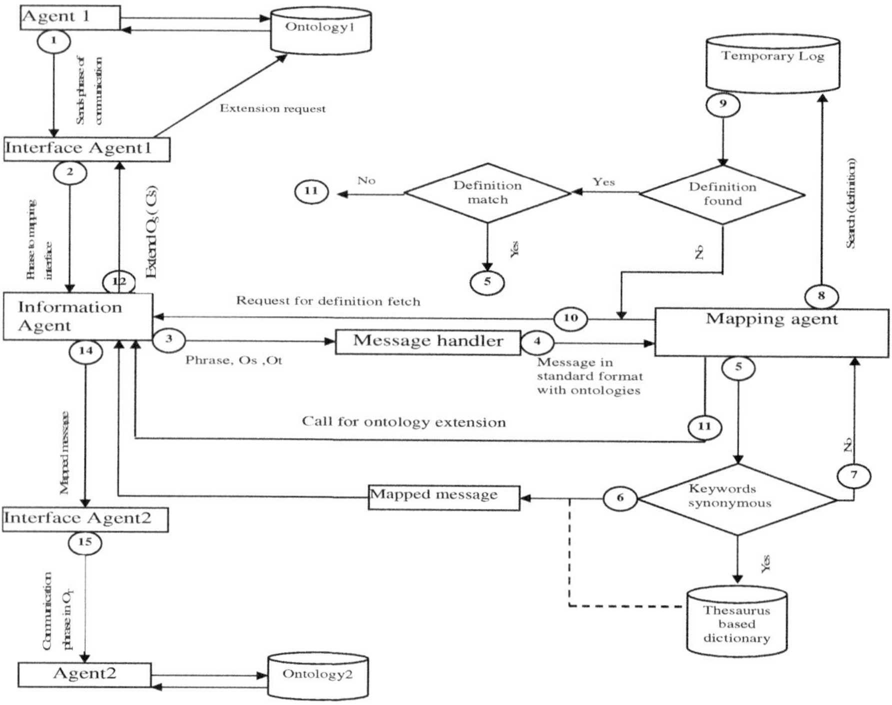


**Fig. 3.** High Level View of Ontological Mapping Framework

Algorithms for various agents are depicted in figures 4(a), 4(b), 4(c) and 4(d), along with these, algorithm for updating TL is also provided in fig. 4(e). Efficiency of TL will depend on its implementation, if implemented with appropriate amount of memory, it will result in increased precision and recall rate of the proposed system.

```
Interface  agent()
Input: message ,extension_request(Cs)
Output: message to IA
Action: activate, sleep, extension(Os)
{
 on(input) → activates;
I f(message)
 { pass to IA;}
if (extension_request(Cs))
{extend(Os) s.t. (Cs ⊂ Os' ∀ Os' ⊇ Os )}
if (message_from_IA)
{pass to desired agent in domain;}
call(sleep) }
```

**Fig. 4(a).** Interface Agent

```
Information agent()
Input: message,O_S,O_T, fetch(Dc), extension_request(Cs);
Output:  to_MHA(message, Os,Ot);
Action: activate,sleep;
{ on(input)  →  activates;
if(message)
{pass to MHA(message ,O_S,O_T);}
else
if( fetch(Dc))
{ fetches(Dc,URI);
Pass Dc to MMA;}
If(extension_request(Cs))
{  pass extend(Cs) to Interface Agent();}
call(sleep);}
```

**Fig. 4(b).** Information Agent

```
Mapping Master agent()
Input: message, O_S,O_T;
Output: message mapped in O_T, extend(Cs);
Action: activate, sleep, extension;
{
on(input) —activate;
Case 1: if( (∀ Cs ∈Os) ∃( C_T ∈O_T) s.t.C_S≡C_T)
        { return( mapped message);}
Case2: if(( O_S —O_T) s.t. O_S ⊒O_T) ∨(O_S —O_T'') s.t. (O_T'' ⊑O_T))
        { return(mapped message);}
Case 3: if (Os — —O_T) ∨(—Os —O_T))
        { return ( call(extension));}
    cal l(sleep); }
```

**Fig. 4(c).** Mapping Master Agent

```
Message handler Agent()
Input: message, O_S,O_T;
Output: message to MMA;
Action: activate, sleep;
{
 on(input)  →  acticate;
 If(message)
{
transform(message) →  standard_form;
 pass(message, O_S ,O_T) to MMA
}
  call(sleep);}
```

**Fig. 4(d).** Message Handler Agent

```
Temporary_Log()
TL is a two dimensional matrix of Cs and URIs;
{  for each (cs ∈ Cs: cs ⊇ Cs,a)
 {  if (not(TL[Cs,URI]=Null) )then
          return (Dc);
 else
    create TL[Cs,URI]=Dc; }
 return TL;  }
```

**Fig. 4(e).** Updation algorithm for Temporary Log

A case study is presented in the next section demonstrating the working of proposed framework.

### 3.3 Case Study

Suppose communication is desired between Source_domain: Student and Destination_domain: Hospitatal.

$PTA_{source\_domain}$ Input: A student needs medical services.
$IFA_{source\_domain}$ : <phrase, $O_S$ , $O_T$ >
$IA_{IAM3I}$: <phrase, $O_S$ , $O_T$ >
$MMA_{IAM3I}$ : (i) student $\neq$ Person
(ii) {student} $\subset$ {Person} so intension relation holds
$IA_{IAM3I}$ output: A person needs medical services.
$IFA_{destination\_domain}$ Input: A person needs medical services.

If communication is required in the reverse direction then {Person} $\supseteq$ {Student} extension is required. Thus $IA_{IAM3I}$ will send ontology extension request to $IFA_{source\_domain}$ with concept Person with all its attributes and its relationship with student (student $\subset$ person). Only in this way agent working in source_domain can understand what a person is and can participate in communication.

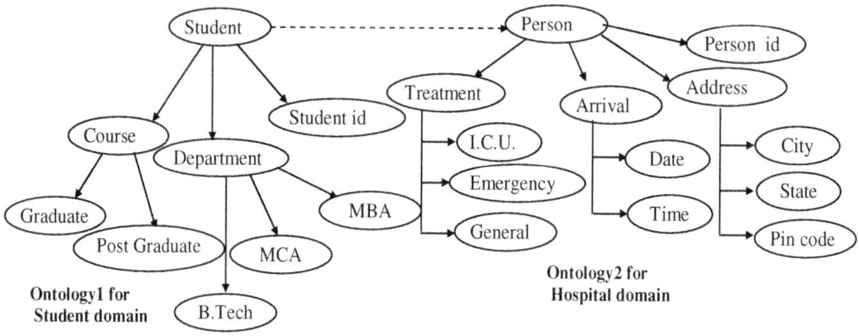


Extension of ontology is possible due to learning ability of agents and it will result in richness of domain ontology with passage of time.

The next example illustrates the usage of ontology mapping interface for homogenous domains. Both ontologies include vocabulary of animals but use different classifications.

source_domain: classification of living beings
destination_domain: classification of animals

Case 1:
$PTA_{source\_domain}$ Input: A snack is in the farm.
$IFA_{source\_domain}$ : <phrase, $O_S$ >
$IA_{IAM3I}$ : <phrase, $O_S$, $O_T$ >
$MMA_{IAM3I}$ : (i) snack $\neq$ animal
(ii) {snack} $\subset$ {Reptile} so intension relation holds
$IA_{IAM3I}$ output: A reptile is in the farm.
$IFA_{destination\_domain}$ Input: A reptile is in the farm.

Case 2:
$PTA_{source\_domain}$ Input: Lizard eats mosquitoes.
$IFA_{source\_domain}$ : <phrase, $O_S$ >
$IA_{IAM3I}$ : <phrase, $O_S$, $O_T$ >
$MMA_{IAM3I}$ : (i) Lizard $\neq$ animal
(ii) {lizard} $\subset$ {Reptile} so intension relation holds
$IA_{IAM3I}$ output: reptile eats mosquitoes.
$IFA_{destination\_domain}$ Input: reptile eats mosquitoes.

Thus in both cases, keywords Snack and Lizard will be mapped to reptile class in $O_T$ resulting in the loss of their specific attributes. However this kind of loss of information has to be tolerated due to the large and distributed nature of ontologies. We can never ensure similar classification to be used and similar attributes to be included while creating ontologies. Also difference in origin, application area and the thought process of the ontology developer leads to differences in ontologies defined even for similar domains.

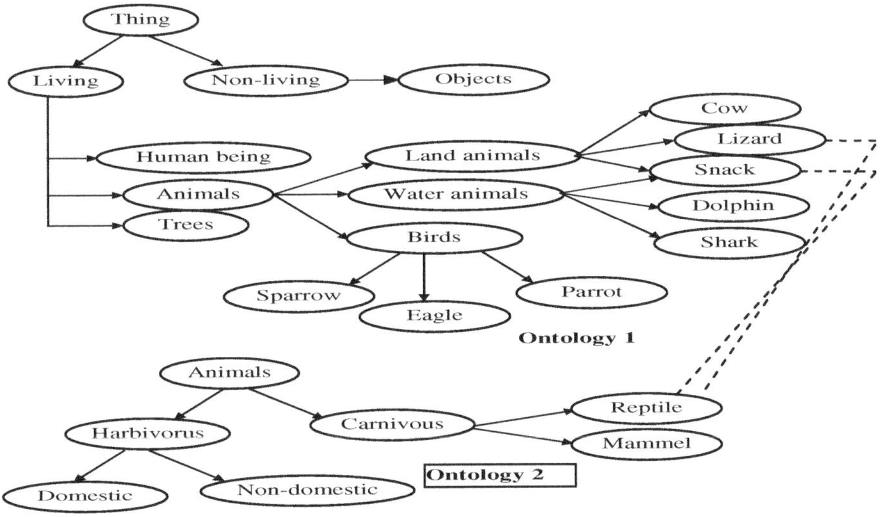


## 4 Evaluation

This section evaluates the proposed framework, with some existing ontology mapping mechanisms, as shown in Table 1. For evaluation purpose, measures available in literature for ontology mapping tools are used [11].

Graph shown in Fig. 5 illustrates that with increase in size of data set on X axis, number of matches on Y axis increase rapidly due to the learning ability of this system.

**Table 1.** Evaluating IAM3I with other Mapping Mechanisms

| Criterion | CTXMATCH[2] | GLUE[5] | ONION[10] | PROMPT[11] | IAM3I |
|---|---|---|---|---|---|
| **Input** | Concepts in concept hierarchy | Two taxonomies with their data instances in ontologies | Terms in two Ontologies | Two input ontologies | Communication phrase with Source & target ontologies |
| **Output** | Semantic relation between concepts | A set of pairs of similar concepts | Set of articulation rules between two ontologies | A merged ontology | Mapped communication phrase from source to target ontology |
| **User Interaction** | No, being an algorithm | User provides data for training and also provides similarity measures | A human expert chooses or discards or modifies suggested matches using a GUI tool | The user accepts, rejects or adjusts system's suggestion | No interaction is required as it is a layer of service and is hidden from user. |
| **Mapping strategy/Algorithm** | Logical deduction | Multi-strategy learning approach: Machine learning technique | Linguistic matcher, structure and inference based heuristics | Heuristic based analyzer | Lexical similarity, whole term, word constituent and type matching |
| **Scalability** | Poor, since it works more effectively when data is less | Good, works effectively when amount of data is large | Poor, for larger ontologies the algorithm doesn't scale well | Poor | Scalability is very Good due to composition being agent based |
| **Recall rate** | Poor | Poor | Good (nearly 70%) | Good | Very Good, due to the use of temporary log |
| **Precision** | Good | Good (66%-97%) | Good | Good | Very Good, since no suggestions are required from outside |

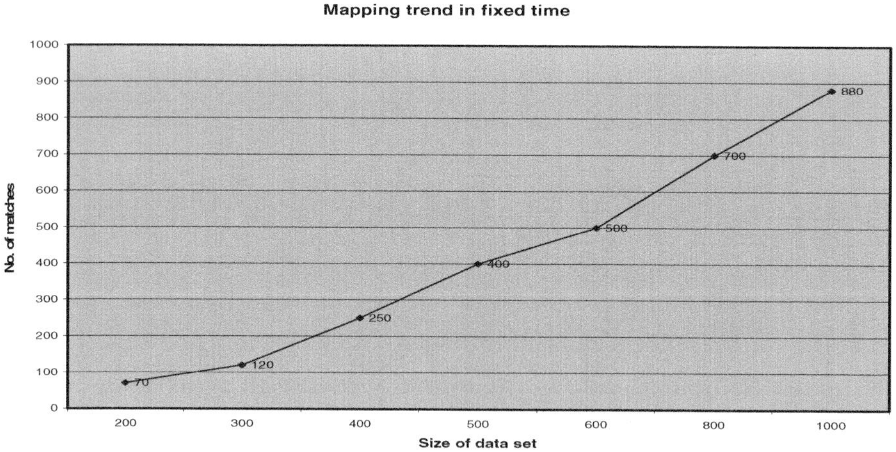


**Fig. 5.** Graph for Performance of Proposed System

## 5 Conclusions

This paper contributes towards meeting the challenge of providing an Intelligent and adaptive ontology mapping mechanism, which delegates the ontology mapping job to agents completely. This framework can provide faster mapping in homogenous as well as in heterogeneous ontologies, thus can improve communication efficiency of multiagent systems. Ontology extension feature can help make ontologies richer with the passage of time and can reduce communication delays occurring due to failures in mapping. Although we have evaluated the proposed framework on the available metrics and compared it with already existing mapping mechanisms, still the framework can be evaluated using fuzzy logic, which is left as part of future work.

## References

1. Aart, C.V., Caire, G., Pels, R., Bergenti, F.: Creating and Using Ontologies in Agent Communication. Telecom Italia EXP magazine 2(3) (September 2002)
2. Bouquet, P., Serafini, L., Zanobini, S.: Semantic coordination: A new approach and an application. In: Fensel, D., Sycara, K., Mylopoulos, J. (eds.) ISWC 2003. LNCS, vol. 2870, pp. 130–145. Springer, Heidelberg (2003)
3. Calvanese, D., Giacomo, G.D., Lenzerini, M.: A Framework for Ontology Integration. In: The Emerging Semantic Web, pp. 201–214. IOS Press, Amsterdam (2002)
4. Choi, N., Song, I.Y., Han, H.: A Survey on Ontology Mapping. SIGMOD Record 35(3) (September 2006)
5. Doan, A., Madhavan, J., Domingos, P., Halevy, A.: Learning to Map between Ontologies on the Semantic Web. VLDB Journal (2003); Special issue on the Semantic Web
6. Hideki, M., Sophia, A., Nenadie, G., Tsujii, J.: A Methodology for Terminology Based Knowledge Acquisition and Integration. In: Proceedings of COLING 2002, Tapai, Taiwan, pp. 667–673 (2002)
7. Juneja, D., Iyengar, S.S., Phoha, V.V.: Fuzzy Evaluation Of Agent Based Semantic Match Making Algorithm For Cyberspace. International Journal of Semantic Computing 3(1), 57–76 (2009)
8. Kalfoglou, Y., Schorlemmer, M.: Ontology Mapping: the State of the Art. The Knowledge Engineering Review 18(1), 1–31 (2003)
9. Koes, M.B., Nourbakhsh, I., Katia, S.: Communication Efficiency in Multi-Agent Systems. In: Proceedings of IEEE 2004 International Conference on Robotics and Automation, April 26 - May 1, pp. 2129–2134 (2004)
10. Mitra, P., Wiederhold, G.: Resolving Terminological Heterogeneity in Ontologies. In: Proceedings of the ECAI 2002 workshop on Ontologies and Semantic Interoperability (2002)
11. Natalya, F.N., Mark, A.M.: Evaluating Ontology Mapping Tools: Requirements and Experience. In: Proceedings of the Workshop on Evaluation of Ontology Tools at EKAW 2002 (EOEN2002), Siguenza, Spain (2002)
12. Obitko, M., Mařík, V.: Mapping between ontologies in agent communication. In: Mařík, V., Müller, J.P., Pěchouček, M. (eds.) CEEMAS 2003. LNCS (LNAI), vol. 2691, p. 191. Springer, Heidelberg (2003)
13. Payne, T.R., Paolucci, M., Singh, R., Sycara, K.: Communicating Agents in open Multi-Agent Systems. In: First GSFC/JPL Workshop on Radical Agent Concepts, WRAC (2002)

14. Sheremetov, L.B., Contreras, M., Smirnov, A.V.: Implementation of an ontology sharing mechanism for multiagent systems based on web services. In: Favela, J., Menasalvas, E., Chávez, E. (eds.) AWIC 2004. LNCS (LNAI), vol. 3034, pp. 54–63. Springer, Heidelberg (2004)
15. Wache, H., Vogele, T., Visser, U., Stuckenschmidt, H., Schuster, G., Neumann, H., And Hubner, S.: Ontology-Based Integration of Information-A Survey of Existing Approaches. In: Proceedings of IJCAI 2001 Workshop: Ontologies and Information Sharing, Seattle, WA, pp. 108–117 (2001)
16. Wiesman, F., Roos, N., Vogt, P.: Automatic Ontology Mapping for Agent Communication. In: Falcone, R., Barber, S.K., Korba, L., Singh, M.P. (eds.) AAMAS 2002. LNCS (LNAI), vol. 2631, Springer, Heidelberg (2003)
17. Wordnet-a lexical database for English, `http://www.cogsci.princeton.edu/wn/` Technical report, Princeton University

# Autonomous Robot Motion Control Using Fuzzy PID Controller

Vaishali Sood

Department of Electronics & Communication Engineering
Beant College of Engineering & Technology, Gurdaspur, India
G.N.D.U. Regional Campus, Gurdaspur, India
san1198@gmail.com

**Abstract.** Autonomous robots roles are increasing in different aspects of engineering and everyday life. This paper describes an autonomous robot motion control system based on fuzzy logic Proportional Integral Derivative (PID) controller. Fuzzy rules are embedded in the controller to tune the gain parameters of PID and to make them helpful in real time applications. This paper discusses the design aspects of fuzzy PID controller for mobile robot that decrease rise time, remove steady sate error quickly and avoids overshoot. The performance of robot design has been verified with rule based evaluation using Matlab and results obtained have been found to be robust. Overall, the performances criteria in terms of its response towards rise time, steady sate error and overshoot have been found to be good.

**Keywords:** Artificial intelligence, Robotics, Robot design, PID controller, Fuzzy logic, Rise time, Steady state error, Overshoot.

## 1 Introduction

Current research in robotics aims to build autonomous intelligent robot systems to meet the increasing industrial demand for automatic manufacturing systems. One of the most important features needed for autonomous robot is its capability of motion planning. Motion planning enables a robot to move in its surroundings steadily for executing a given task. The main design constraints for robot are cost, reliability and adaptability. The different performance objectives in robot design are insensitivity to parameter variations, distribution rejection properties and stability of the system.

A lot of research work has been carried out to develop techniques for obstacle-free motion planning for robots. Still, it requires a lot of attention because it is the primary requirement for robotics in real time motion. PID controller has been broadly used to control various engineering objects because of its simple configuration, better robustness and high consistency. However, the performance of a PID controller totally depends on the tuning of its gain parameters. Researchers have suggested many methods based on artificial intelligence to design PID controllers such as the differential evolution (DE) algorithm, genetic algorithm (GA) [1], simulated annealing (SA) algorithm and fuzzy logic control [2]. In these methods, the fuzzy logic control has a high

A. Mantri et al. (Eds.): HPAGC 2011, CCIS 169, pp. 385–390, 2011.

quality control effect particularly for the processes with nonlinear or uncertain properties or the processes whose modeling are very difficult to build with higher accurately.

In 2010, Zacharie [3] proposed a method consisted of two components: the process monitor that detects changes in the process characteristics and the adaptation mechanism that used information passed to it by the process monitor to update the controller parameters. They used Adaptive Fuzzy Knowledge Based Controller (AFKBC). The first part is the performance monitor that detects changes in the process characteristics by assessment of the controlled response of the process. The second part is the adaptation mechanism. It uses information passed to it by the performance monitor to update the controller parameters and so adapts the controller to the changing process characteristics.

The selection of the appropriate membership functions is very important for the design of controller. The important problem for fuzzy PID controller is lack of a very efficient and universal design method that is widely suitable to various kinds of processes. Till now, several methods have been developed for robot motion planning but each of these methods has its own limitations for time complexity and suitability. Thus, a more versatile and efficient method is desired. In the present work, a method has been developed by combining the fuzzy logic approach with PID controller to solve the robot motion problem and tested on a number of scenarios. This paper discusses the efficient design method for the optimal fuzzy PID controller.

This paper is organized as follows. Section 2 describes the design of robot with PID control loop and fuzzy inference mechanism. Section 3 discusses the design aspects based on different parameters and the results are presented. Section 4 concludes the paper.

## 2 Design of Robot

The basic idea behind our fuzzy control PID is to design a controller using fuzzy logic scheme on the PID controller to adjust its various parameters so that the robot motion can be controlled under various non-linear conditions. Based on the fuzzy logic control, a technique for fuzzy PID controller for adaptive robot motion is proposed. In this method, fuzzy control is used to optimize the input and output factors of controller so as to optimize the rise time (RT), to calculate the steady sate error (SSE) and control the overshoot (OS). If there is any variation in the dynamics of the robot motion then it will adapt the change automatically.

The robot has an on board computer (Pentium IV Quad Core Processor), with which a fuzzy logic PID controller is interfaced. The robot acquires the information from sensors and based on this, fuzzy control rules are activated. The outputs of the activated rules are combined by the fuzzy logic operations to increase the $k_p$ (proportional gain), $k_i$ (integral gain) and $k_d$ (derivative gain) of the PID controller so as to reduce the rise time, eliminate the steady state error (SSE) quickly and to decrease the overshoot (OS) respectively.

## 2.1 PID Control Loop

A PID control loop is useful in order to calculate whether it will actually reach a stable value. If the inputs are chosen incorrectly, the controlled process input oscillates and the output never stays at the set-point. The generic transfer function for PID controller is as shown ahead in Equation 1.

$$H(S) = P * \frac{(DS^2 + S + I)}{(S + C)} \quad \text{.............................(1)}$$

C is a constant which depends upon the bandwidth of the controlled system and S is the variable parameter.

The output of the controller i.e. the input to the robot is given as

Output (t) = $P_{contribution} + I_{contribution} + D_{contribution}$

$$\text{Output (t)} = k_p \,[\text{e (t)} + k_{ip} \int_0 \text{e (t) dt} + k_{dp} \,(\text{de/dt})] \text{........... (2)}$$

Where e (t) = set point – measurement (t) = error signal

$k_p$ = proportional gain, $k_{ip}$ = $k_i / k_p$, where $k_i$ is integral gain and $k_{dp}$ = $k_d / k_p$, where $k_d$ is derivative gain.

The controller is implemented with $k_p$ gain applied to the $I_{contribution}$, $D_{contribution}$ according to Equation 2. We tune the gain parameters using the standard Zeigler-Nicholas tuning method. To tune the above mentioned gain parameters, we have to increase the original value of gain parameters by very small incremental factor.

## 2.2 Fuzzy PID Controller

With input variation for each step, the fuzzy controller examines the variation of e, fuzzfy it, makes online adjustment by using IF-THEN rule for gain parameters, get the crisp value by centre of sums defuzzification method.

Figure 1 gives the structure of a fuzzy PID controller where set-point is the input of the system and e (t) is error of the system. $k_p$, $k_i$ and $k_d$ are the output of fuzzy controller and u is the control action generated by PID controller and y is the output of system.

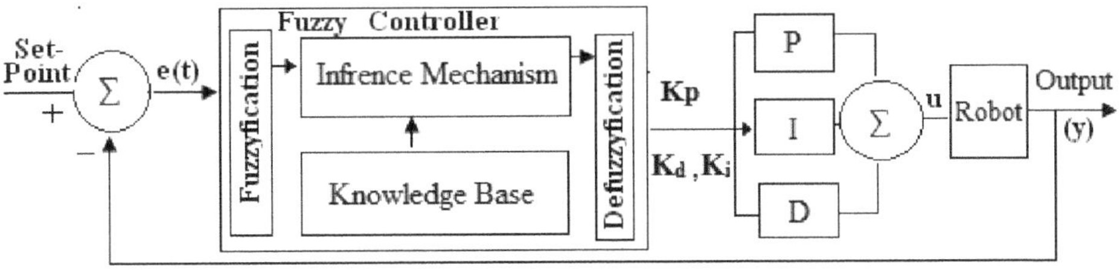


**Fig. 1.** Fuzzy PID Controller

### 2.3 Fuzzy Inference Mechanism

The gain parameters $k_p$, $k_i$ and $k_d$ of the PID controller must be real time, so as to cope with the real time practical applications of robot. Therefore, input of the robot must be real time so as to adjust with the changes. For this, a set of fuzzy IF-THEN rules is applied to the PID controller.

#### 2.3.1 Design of Knowledge Base

The knowledge base consist of two parts i.e. rule base and database. Rule based consist of fuzzy control IF- THEN rules and design of database consists of partition of variable space. Linguistic term such as fast, medium and slow are defined for robot motion (RM). Terms such as high, medium and low for gain parameters [rise time (RT), steady state error (SSE) and overshoot (OS)]. The membership functions are triangular or trapezoidal and inference mechanism used is Mamdani.

**Rules:** If $k_p$ is $k_{pi}$ and RT is $(RT)_j$ then RM is $RM_{ij}$. If $k_i$ is $k_{ii}$ and SSE is $SSE_j$ then RM is $RM_{ij}$. If $k_d$ is $k_{di}$ and OS is $OS_j$ then RM is $RM_{ij}$.

Where i, j are having values 1, 2, 3 because each $k_p$, $k_i$, $k_d$, RT, SSE, OS as well as RM has three membership functions. Using the center of sums defuzzification method, the crisp value of the $k_p$, $k_i$ and $k_d$ has been obtained as follow:

$$x^* = \frac{\sum_{i=1}^{N} x_i \sum_{k=1}^{n} \mu_{A_k}(x_i)}{\sum_{i=1}^{N} \sum_{k=1}^{n} \mu_{A_k}(x_i)}$$

We used this method of defuzzification because it leads to rather fast inference cycles and can be implemented easily. Fuzzy rules used for the adaptive robot motion are listed in Table 1, 2 and 3.

| Fuzzy Rule Number | Table 1 | | | Table 2 | | | Table 3 | | |
|---|---|---|---|---|---|---|---|---|---|
| | Kp | RT | RM | Ki | SSE | RM | Kd | OS | RM |
| 1 | High | High | Slow | High | High | Slow | High | High | Slow |
| 2 | High | Med | Slow | High | Med | Slow | High | Med | Fast |
| 3 | High | Low | Fast | High | Low | Fast | High | Low | Fast |
| 4 | Med | High | Slow | Med | High | Slow | Med | High | Slow |
| 5 | Med | Med | Slow | Med | Med | Slow | Med | Med | Slow |
| 6 | Med | Low | Fast | Med | Low | Fast | Med | Low | Fast |
| 7 | Low | High | Slow | Low | High | Slow | Low | High | Slow |
| 8 | Low | Med | Slow | Low | Med | Slow | Low | Med | Slow |
| 9 | Low | Low | Fast | Low | Low | Fast | Low | Low | Fast |

## 3 Discussion and Results

The adaptive robot motion controller presented in this paper is a fuzzy logic controller that combines non-linear fuzzy rules to control the gain parameters of the linear PID controller to control the robot motion in its domain. The rules embedded in the fuzzy logic controller have to be designed by the designer of the controller. When the robot is facing a change in speed, the PID controller must change its $k_p$, $k_i$ and $k_d$ parameters. The fuzzy rules for this are listed in Table 1, 2 and 3. e.g. according to rule 3, if the value of $k_p$ is high and rise time (RT) is low then the robot will move fast. An autonomous controller means a controller with adjustable parameter and a mechanism for adjusting parameter. Due to parameter adjustment, the controller becomes non-linear. In our purpose autonomous fuzzy PID controller, the adaptation is done by modifying the membership function in proportion to the undesired effect.

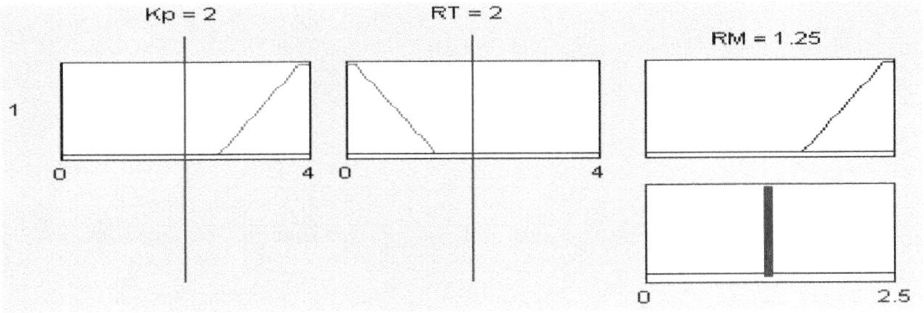


**Fig. 2.** For Rule Number 3 of Table 1

The values of $k_p$, $k_i$ and $k_d$ are incremented so as to control the rise time, eliminates the SSE quickly and to decrease the overshoot during robot motion. The system is more robust, faster and has a higher probability in obtaining the globally optimal solution. The results have been drawn from MatLab as shown in figure above.

## 4 Conclusion

This paper presents a novel autonomous robot motion controller system by taking the conceptual advantages of fuzzy control rules to control the gain parameter of PID controller. The proposed method is effective in terms of smooth response while considering overshoot, removal of steady state error quickly and its response towards rise time so that there is a faster and effective response. As compared with other methods based on fuzzy control rules, it has been found that the proposed PID controller has better performance in faster response, error removal and decrease in rise time. It has been tested in the MatLab and found that with change in operating point there is no need to retune it and results are found to be robust. The proposed method is used to deal with the rise time, steady sate error and overshoot problems efficiently.

## References

1. Krohling, R.A., Rey, J.P.: Design of Optimal Disturbance Rejection PID Controllers using Genetic Algorithms. IEEE Transactions on Evolutionary Computation 5(1), 78–82 (2001)
2. Khellaf, S.A., Leulmi, S.: Genetic Training of a Fuzzy PID. In: International Conference on Modeling and Simulation (ICMS 2004), Spain, pp. 185–186 (2004)
3. Zacharie, M.: Adaptive Fuzzy Knowledge based Controller for Autonomous Robot Motion Control. Journal of Computer Science 6(10), 1019–1026 (2010)

# A Multiresolution Technique to Despeckle Ultrasound Images

Parvinder Kaur[1] and Baljit Singh[2]

[1] Student (M.Tech)
[2] Assistant Professor
B.B.S.B.E.C,
Fatehgarh Sahib India

**Abstract.** Ultrasonography is a very prevailing technique for imaging soft tissue structures and organs of human body. But when an Ultrasound image is captured it gets noisy and this added noise is known as speckle noise which hinders the diagnostic process of the radiocologists and doctors. In this paper a method to remove speckle noise from ultrasound images is proposed. So many methods have been proposed in spatial, frequency and wavelet domains. Here new thresholding method in wavelet domain is proposed which takes into account statistical properties of the image using a weighted window. Performance of the proposed algorithm is compared with conventional methods based on Peak Signal to Noise Ratio (PSNR) and Mean Square error (MSE). Results show that proposed algorithm performs better than conventional methods.

**Keywords:** Despeckle, MSE, Multiresolution, PDF, PSNR, Speckle noise.

## 1 Introduction

Medical imaging is very dominant method for detection of diseases in human body. Ultrasound imaging has become popular due to its inexpensiveness and non-invasiveness and portability. It is a field of research because the presence of speckle noise makes it difficult to interpret the image. Sometimes a dark spot that is due to speckle noise can be considered as a cyst. Speckle is a dominant source of noise and should be filtered out 1-3. So Speckle is a random, deterministic, interference pattern in an image formed with coherent radiation of a medium containing many sub-resolution scatterers. It has been observed that speckle noise follows Rayleigh amplitude Probability Density Function PDF .

## 2 Existing Methods

The existence of speckle is unattractive since it disgraces image quality and it affects the tasks of individual interpretation and diagnosis.. Frost (1982) provided an adaptive filter for multiplicative noise. Kuan, Sawhauk and Strand (1987) provided adaptive restoration method for speckle noise removal. Adaptive filters have major limitations in preserving the sharp features of the original image. Mallet and Zhong (1992) used

A. Mantri et al. (Eds.): HPAGC 2011, CCIS 169, pp. 391–396, 2011.

median filter for speckle noise reduction 4. Solbo and Eltofit (2004) provided homomorphic filtering method in Fourier domain. The classical Wiener filter is not adequate for removing speckle since it is designed mainly for additive noise suppression. Uses of wavelet transform based techniques are the recent trends for speckle removal. Wavelet denoising attempts to remove the noise present in the signal while preserving the signal characteristics, regardless of its frequency content. Donoho (1995) provided a method for speckle noise reduction using soft thresholding. Gupta, Kaur and Chauhan (2003) proposed a method for enhancement of ultrasound images. They provided a wavelet based statistical approach for speckle noise reduction 5. Byung-Jun and Vaidyanathan (2004) proposed a wavelet based algorithm using customized thresholding 6. Sudha, Suresh and Sukanesh (2009) provided Speckle Noise Reduction method in Ultrasound Images by Wavelet Thresholding based on Weighted Variance. But they used hit and trial method to assign weights to the window used for calculating variance 7.

## 3 Proposed Method

Wavelet transform, due to its excellent localization property, has rapidly become an indispensable signal and image processing tool for a variety of applications like denoising. Problem can be formulated as

$$I(x, y) = S(x, y).\ \eta_m(x, y) + \eta_a(x, y) \tag{1}$$

Where I(x, y) be the recorded ultrasound image and S(x, y) be the noise-free image that has to be recovered, $\eta_m(x, y)$ and $\eta_a(x, y)$ the corrupting multiplicative and additive speckle noise components, respectively.

Step 1: As speckle is a multiplicative noise it needs to be converted into additive noise first. Transform the multiplicative noise model into an additive one by taking the logarithm of the original speckled data.

$$\log I(x, y) = \log S(x, y) + \log \eta(x, y). \tag{2}$$

Step2: Performing discrete wavelet transform: First sub step in performing the DWT is to choose a wavelet and number of levels for decomposition. We have chosen symlet wavelet. Symlet are compactly supported wavelets with least asymmetry and higher number of vanishing moments for a given support width. Associated scaling filters are near linear phase filters.

Step 3: Calculate the variance: First parameter that needs to be estimated is the noise variance. It is denoted by $\sigma^2$. It is estimated from sub band D by the robust median estimator,

$$\sigma^2 = \left[\frac{\text{median}|m,n|}{0.6745}\right]^2 \tag{3}$$

Step 4: Calculate the near optimal threshold value to threshold wavelet coefficients

Threshold selection is an important question when denoising an image. A small threshold may yield result close to the input image but the result may still be noisy. A large threshold on the other hand produces a signal with large number of zero coefficients. It leads to a smooth signal so details are destroyed and it may cause blur and artifacts. So the problem is to find the optimal threshold so that mean squared error between the original image and its estimate is minimized. Proposed method uses adaptive threshold based on local variance. We have used a 3x3 window to calculate the local weighted variance $\sigma_w$ m, $n^2$ of each wavelet coefficient Y m, n at level l.

| | | |
|---|---|---|
| W4 | W2 | W4 |
| W3 | W1 | W3 |
| W4 | W2 | W4 |

**Fig. 1.** A 3x3 window with different weights for calculating weighted variance

W1 the current coefficient is considered as least dominant and is given minimum weight. W2 corresponding to vertical neighbors of the current coefficient is most dominant one so given maximum weight. Now it is easy to distinguish between signal coefficients and noise coefficients. Local variance depicts correlation structure of the wavelet coefficients. W3 is given weight more than W1 but less than W2. W4 the diagonal coefficients are given less than W3 but more than W1.All the above assumptions are made on the basis of the fact that magnitudes of the wavelet coefficients show correlations which decay exponentially with the distance. The weighted variance of a coefficient Ym, n for a window size 3x3 and weights w= { $w_{i,j}$ , I, j□ N} is given by:

$$\delta = \sigma_w[m,n]^2 = \frac{\sum_{i,j\in N} w_{i,j}\, Y[m,n]^2}{\sum_{i,j\in N} w_{i,j}} \qquad (4)$$

Now the threshold value λ for each pixel can be given by:

$$\lambda(\mathrm{m,n}) = \frac{\sigma^2{}_n}{\delta} \qquad (5)$$

Step 5: Threshold all the coefficients using soft thresholding and using the threshold value obtained from previous step.

Step 6: Perform the inverse DWT to reconstruct the denoised image and take exponent.

## 4 Implementation and Results

All the simulations are done using MATLAB. Performance is compared with Kuan filter, Frost filter, Lee filter, soft thresholding, hard thresholding, custom thresholding. Speckle noise of variance .06, .07, .08, .09 is taken. Objective evaluation is done based on two parameters. These are Peak signal to noise ratio (PSNR) and mean square error (MSE).

$$MSE = \frac{1}{MN}\sum_{i=1}^{M}\sum_{j=1}^{N}\left(X(i,j) - Y(i,j)\right)^2 \tag{6}$$

$$PSNR = 10\, log_{10} \frac{255}{MSE} \tag{7}$$

Results show that by taking statistical properties of wavelet coefficients into account improved PSNR and minimized MSE are achieved. Hence proposed algorithm leads to better image enhancement.

**Table 1.** Comparison of PSNR of conventional filters with proposed algorithm

| Variance | 0.06 | 0.07 | 0. 08 | 0.09 |
|---|---|---|---|---|
| Kuan Filter | 33.3502 | 33.1935 | 33.1667 | 32.7881 |
| Frost Filter | 31.8250 | 31.6023 | 31.2022 | 30.8355 |
| Lee Filter | 33.1897 | 33.0732 | 32.8479 | 32.7357 |
| Soft thresholding | 35.2328 | 35.3277 | 35.1918 | 35.2930 |
| Hard thresholding | 35.0741 | 34.7970 | 35.5222 | 34.0764 |
| Custom Thresholding | 35.4600 | 34.9423 | 34.8034 | 34.6591 |
| Proposed method | 36.2553 | 36.4919 | 36.5933 | 36.4948 |

**Table 2.** Comparison of MSE of conventional filters with proposed algorithm

| Variance | 0.06 | 0.07 | 0. 08 | 0.09 |
|---|---|---|---|---|
| Kuan Filter | 5.4832 | 5.5830 | 5.6002 | 5.8497 |
| Frost Filter | 6.5357 | 6.7054 | 7.0215 | 7.3243 |
| Lee Filter | 5.5854 | 5.6609 | 5.8096 | 5.8851 |
| Soft thresholding | 4.4147 | 4.3667 | 4.4356 | 4.3842 |
| Hard Thresholding | 4.4961 | 4.6418 | 4.7911 | 5.0434 |
| Custom Thresholding | 4.3007 | 4.5648 | 4.6384 | 4.7161 |
| Proposed method | 3.9244 | 3.8189 | 3.7747 | 3.8177 |

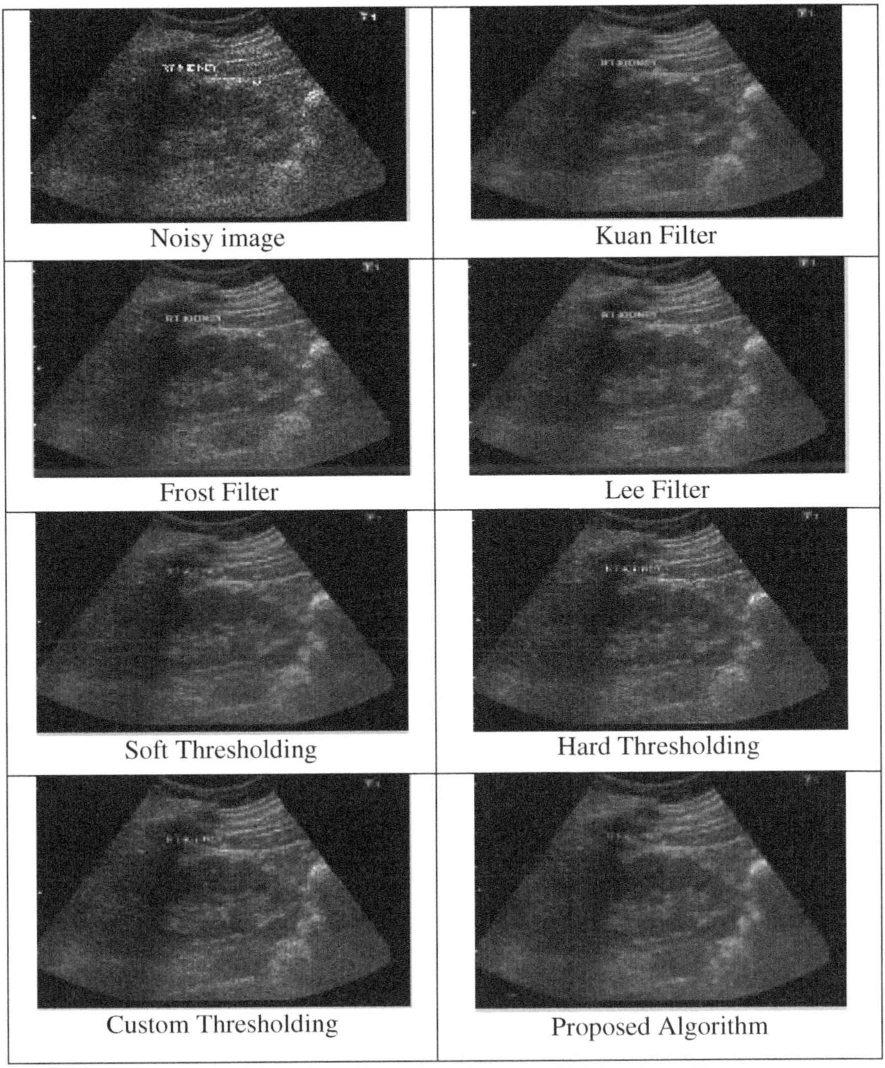


**Fig. 2.** Effect of different filters on an ultrasound image with noise variance 0.09

# References

1. Mastriani, Mario.: Denoising and Compression in Wavelet Domain via Projection onto Approximation Coefficients. International Journal of Signal Processing 5(1), 20–30 (2009)
2. Donoho, D.L.: De-noising by soft-thresholding. IEEE Trans. Inform. Theory 41(3), 613–627 (1995)
3. Gnanadurai, D., Sadasivam, V.: An Efficient Adaptive Thresholding Technique for Wavelet Based Image Denoising. International Journal of Signal Processing 2(2), 114–119 (2006)

4. Ashish, K., Khare, M., Jeong, Y., Kim, H., Jeon, M.: Despeckling of medical ultrasound images using Daubechies complex wavelet transform. Signal Processing 90, 428–439 (2010)
5. Gupta, S., Kaur, L., Chauhan, R.C., Saxena, S.: A versatile technique for visual enhancement of medical ultrasound images. Digital Signal Processing 17, 542–560 (2007)
6. Byung-Jun, Y., Vaidyanathan, P.P.: Wavelet based denoising by customized thresholding. IEEE Trans. ICASSP 2, 924–928 (2004)
7. Sudha, S., Suresh, G.R., Sukanesh, R.: Speckle Noise Reduction in Ultrasound Images by Wavelet Thresholding based on Weighted Variance. International Journal of Computer Theory and Engineering 1(1), 1793–8201 (2009)

# Design and Analysis of the Gateway Discovery Approaches in MANET

Koushik Majumder[1], Sudhabindu Ray[2], and Subir Kumar Sarkar[2]

[1] Department of Computer Science & Engineering,
West Bengal University of Technology, Kolkata, India
koushikmail@yahoo.com
[2] Department of Electronics and Telecommunication Engineering,
Jadavpur University, Kolkata, India

**Abstract.** The demand for any time anywhere connectivity has increased rapidly with the tremendous growth of the Internet in the past decade and due to the huge influx of highly portable devices such as laptops, PDAs etc. In order to provide the users with the huge pool of resources together with the global services available from the Internet and for widening the coverage area of the MANET, there is a growing need to integrate the ad hoc networks to the Internet. Due to the differences in the protocol architecture between MANET and Internet, we need gateways which act as bridges between them. The gateway discovery in hybrid network is considered as a critical and challenging task and with decreasing pause time and greater number of sources it becomes even more complex. Due to the scarcity of network resources in MANET, the efficient discovery of the gateway becomes a key issue in the design and development of future hybrid networks. In this paper we have described the design and implementation of the various gateway discovery approaches and carried out a systematic simulation based performance study of these approaches using NS2 under different network scenarios. The performance analysis has been done on the basis of three metrics - packet delivery fraction, average end-to-end delay and normalized routing load.

**Keywords:** Average end-to-end delay, gateway discovery approaches, Internet, Mobile ad hoc network, normalized routing load, packet delivery fraction, performance study.

## 1 Introduction

A group of mobile devices can form a self-organized and self-controlled network called a mobile ad hoc network (MANET) [1-6]. The main advantage of these networks is that they do not rely on any established infrastructure or centralized server. But due to the limited transmission range of the MANET nodes, the total area of coverage is often limited. Also due to the lack of connectivity to the fixed network, the users in the MANET work as an isolated group. In order to access the global services from the Internet and to widen the coverage area, there is a growing need to connect these ad hoc networks to the Internet. For this purpose we need Internet Gateways

A. Mantri et al. (Eds.): HPAGC 2011, CCIS 169, pp. 397–405, 2011.

(IGW). These gateways work as bridges between the different network architectures of MANET and the Internet and they need to understand the protocols of both the mobile ad hoc protocol stack and the TCP/IP protocol suite. The gateway discovery approaches can be broadly classified into three categories- proactive [7, 8], reactive [9, 10] and hybrid [11, 12].

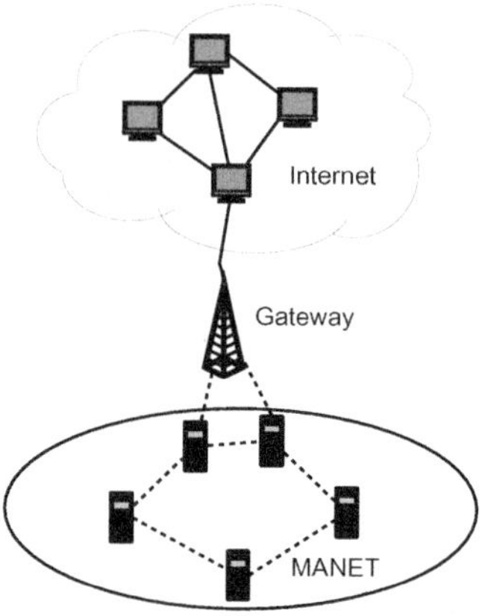


**Fig. 1.** Hybrid Network

Although a lot of research has been done on the mobile ad hoc routing protocols [13], the area of hybrid networking has remained less regarded. In this work we have used the extended AODV reactive routing protocol to support communication between the MANET and the Internet. In this paper we have described the design and implementation of various gateway discovery approaches and studied the performance differentials of these approaches under different scenarios using ns2 based simulation.

The rest of the paper is organized as follows. We investigate the different gateway discovery approaches in section 2. Section 3 and section 4 details the simulation model and the key performance metrics respectively. The simulation results are presented and analyzed in section 5. Finally section 6 concludes the paper and defines topics for future research.

## 2 Gateway Discovery Approaches

Depending on who initiates the gateway discovery, these approaches can be broadly classified into the following three categories.

### 2.1 Proactive Gateway Discovery

The gateway itself starts the proactive gateway discovery by periodically broadcasting the gateway advertisement (GWADV) message. This message is an extended version of the RREP_I message containing the additional RREQ ID field form the RREQ message and is transmitted at regular intervals after the expiration of the gateway's timer (ADVERTISEMENT_INTERVAL). The mobile nodes which are within the transmission range of the gateway, receive the advertisement and either create a new route entry or update the existing route entry for the gateway in their routing table. After this, a mobile node checks to find whether a GWADV message with the same originator IP address and same RREQ ID has already been received within the same time interval. If it is not so then the new advertisement is rebroadcasted, otherwise it is discarded. This solves the problem of duplicated advertisement messages and allows the flooding of the advertisement message through the whole network with controlled congestion.

## 2.2 Reactive Gateway Discovery

In this approach a mobile node that wants to find a new route or update an existing route to the gateway, initiates the gateway discovery. If a source mobile node wants to communicate with an Internet node, it first performs the expanding ring search technique to find the destination within the ad hoc network. When it obtains no corresponding route reply even after a network-wide search, the source mobile node broadcasts a RREQ_I message to the ALL_MANET_GW_MULTICAST address. This is the IP address for the group of all gateways. Thus only the gateways receive and reply to this message. The intermediate mobile nodes receiving this message simply rebroadcast it after checking the RREQ ID field, to avoid any kind of duplicate broadcast. After receiving the RREQ_I, the gateways unicast back RREP_I message to the source node. The source then selects one of the gateways based on the hop count and forwards the data packet to the selected gateway. Next, the gateway sends the data packet to the destination node in the Internet.

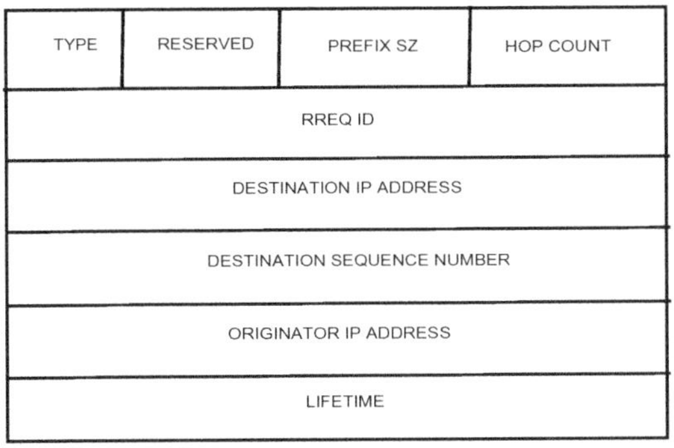


**Fig. 2.** Format of Gateway Advertisement (GWADV) Message

## 2.3 Hybrid Gateway Discovery

In the hybrid gateway discovery approach the gateway periodically broadcasts the GWADV message. The TTL is set to ADVERTISEMENT_ZONE so that the advertisement message can be forwarded only up to this maximal number of hops through the ad hoc network. The mobile nodes within this region receive this message and act according to the proactive approach. The nodes outside this region discover the default routes to the gateways using the reactive approach.

## 3 Simulation Model

We have done our simulation based on ns-2.34 [14, 15]. Our main goal was to measure the performance of the different gateway discovery approaches under a range of varying network conditions. We have used the Distributed Coordination Function (DCF) of IEEE 802.11[16] for wireless LANs as the MAC layer protocol. DCF uses RTS/CTS frame along with random back off mechanism to resolve the medium contention conflict. As buffering is needed for the data packets which are destined for a particular target node and for which the route discovery process is currently going on, the protocols have a send buffer of 64 packets. In order to prevent indefinite waiting for these data packets, the packets are dropped from the buffers when the waiting time exceeds 30 seconds. The interface queue has the capacity to hold 50 packets and it is maintained as a priority queue. In our simulation environment the MANET nodes use constant bit rate (CBR) traffic sources when they send data to the Internet domain. We have used the cbrgen traffic-scenario generator tool available in NS2 to generate the CBR traffic connections between the nodes. We have used two different communication patterns corresponding to 10 and 20 sources. The complete list of simulation parameters is shown in Table 1.

**Table 1.** Simulation Parameters

| Parameter | Value |
|---|---|
| Number of Mobile nodes | 50 |
| Number of sources | 10,20 |
| Number of gateways | 2 |
| Number of hosts | 2 |
| Transmission range | 250 m |
| Simulation time | 900 s |
| Topology size | 1200 m X 800 m |
| Source type | Constant bit rate |
| Packet rate | 5 packets/sec |
| Packet size | 512 bytes |
| Pause time | 0,100,200,300,400,500, 600,700,800,900 seconds |
| Maximum speed | 20 m/sec |
| Mobility model | Random way point |
| Gateway discovery approaches | Proactive, reactive and hybrid |

### 3.1 Hybrid Scenario

We have used a rectangular simulation area of 1200 m x 800 m. Our mixed scenario consists of a wireless and a wired domain. The simulation was performed with the first scenario of 50 mobile nodes among which 10 are sources, 2 gateways, 2 routers and 2 hosts and the second scenario of 50 mobile nodes among which 20 are sources, 2 gateways, 2 routers and 2 hosts. One of the two hosts in the wired domain is chosen randomly as the required destination for each data session. Each host is connected to the gateway through a router. For our hybrid network environment we have two gateways located at each side of the simulation area and running both extended AODV and fixed IP routing protocols. Their x,y-coordinates in meters are (200, 400) and (1000, 400). In our two simulation scenarios 10 and 20 mobile nodes respectively act as constant bit rate traffic sources. They are initially distributed randomly within the MANET. These sources start sending data packets after the first 10 seconds of simulation in order to ensure that the data packets are not dropped due to the lack of routes not yet established. They stop sending data packets 5 seconds before the end of the simulation so that the data packets sent late get enough time to reach their destinations.

## 4 Performance Metrics

We have primarily selected the following three parameters in order to study the performance comparison of the three gateway discovery approaches.

**Packet delivery fraction:** This is defined as the ratio between the number of delivered packets and those generated by the constant bit rate (CBR) traffic sources.

**Average end-to-end delay:** This is basically defined as the ratio between the summation of the time difference between the packet received time and the packet sent time and the summation of data packets received by all nodes.

**Normalized routing load:** This is defined as the number of routing packets transmitted per data packet delivered at the destination. Each hop-wise transmission of a routing packet is counted as one transmission.

## 5 Simulation Results and Analysis

In this section we have studied the effect of the three gateway discovery approaches under varying pause time and increasing number of sources, on the performance of the hybrid ad hoc network.

### 5.1 Packet Delivery Fraction (PDF) Comparison

From Figure 3 we see that the proactive approach has better packet delivery performance than the reactive approach. This happens because - due to the periodic update of route information form the gateway, routes form all the nodes to the gateway are always available. As a result majority of the packets are delivered smoothly. In case of reactive approach, a node wishing to send data to the destination needs to find the route to the gateway first. This takes a certain amount of time and no packet can be sent during this period due to the unavailability of routes.

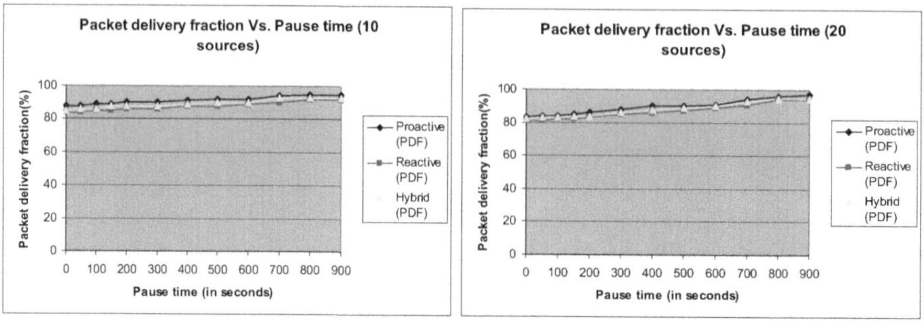


**Fig. 3.** Packet Delivery Fraction Vs. Pause Time for 10 and 20 sources

From the figure it is evident that the packet delivery performance deteriorates with decreasing pause time in all three approaches. Due to high mobility and frequent link breaks, nodes won't be able to send data packets to the gateway thereby reducing the packet delivery ratio. In the reactive approach, the routes are not optimized and nodes continue to maintain longer routes. As pause time decreases, the topology becomes highly dynamic. Due to the frequent link breaks, the older routes tend to become stale quickly. But the source node continues to send packets through these stale routes until it receives RERR message from a mobile node having a broken link. With longer routes it takes greater time for the source node to receive RERR. As a result, during this time greater numbers of packets are dropped. From the figure we also see that as the number of sources is increased, initially the packet delivery performance becomes better. This is due to the fact that with less number of sources, the channel capacity is not fully utilized. Therefore, increasing the number of sources also increases the packet delivery ratio. However, when the number of sources is increased more, there will be high volume of traffic in the network leading to congestion. Due to greater control traffic, less portion of the channel is left for the data. This ultimately reduces the packet delivery ratio.

## 5.2 Average End-to-End Delay Comparison

The average end-to-end delay with the proactive and hybrid gateway discovery approach is less in comparison to the reactive gateway discovery. In proactive approach, due to periodic route updates from the gateway, routes are optimized regularly and the nodes have fresher and shorter routes to the gateway. Moreover, all the routes are maintained all the time. This instant availability of the fresher and shorter routes enables the nodes to deliver packets to their destinations with less delay. In reactive approach, a node needs to find a route to the gateway first before sending the packet. This initial path setup delays the delivery of the packets.

The average end-to-end delay increases with decreasing pause time and increasing number of sources. As the nodes become more mobile, the links break more frequently. This together with the greater number of sources, necessitates the reactive route discovery process to be invoked more often thus causing huge amount of control traffic. The data traffic also increases with more number of sources. This results in

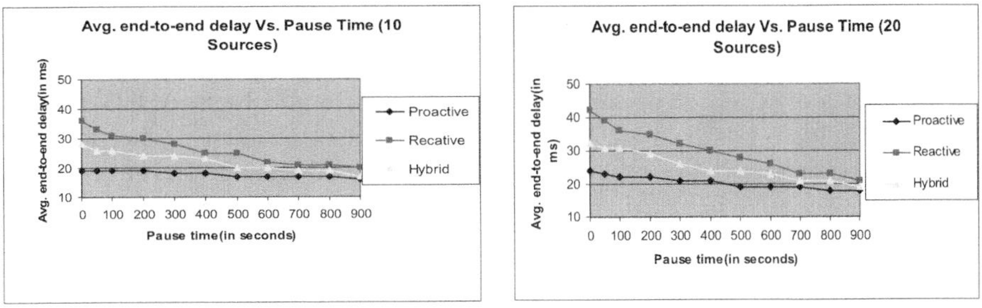


**Fig. 4.** Average End to End Delay Vs. Pause time for 10 and 20 Sources

more collisions, more retransmissions and further congestion in the network. Consequently the constrained channel increases the route discovery latency which in turn increases the average end-to-end delay. In the absence of any regular route update mechanism, reactive approach suffers from older and longer routes which increase the chances of link breaks, leading to further delay.

In case of hybrid approach, in the simulation done in our work, the gateways broadcast the gateway advertisement messages periodically up to three hops away and the nodes beyond that region follow the reactive gateway discovery approach. As a result the average end-to-end delay becomes less than that of the reactive approach but more than that of the proactive approach.

### 5.3 Normalized Routing Load Comparison

In terms of normalized routing load the reactive approach outperforms the proactive and hybrid approaches. In the reactive approach, the gateway discovery is initiated only when a mobile node needs to send a data packet which results in comparatively less routing overhead. As hybrid approach is a combination of proactive and reactive approaches, its normalized routing load lies between them.

The normalized routing overhead of the proactive approach remains almost constant for a particular advertisement interval irrespective of the pause time. Whereas in case of reactive approach with decreasing pause time, the gateway discoveries need to be invoked more often due to frequent link breaks. Moreover, as the reactive approach continues using longer and older routes and does not use route optimization until the route is broken, the chances of link breaks also increases. This further adds to the number of route discoveries. With this greater number of gateway discoveries, the control traffic also increases, which ultimately results in higher normalized routing load.

From the figure we see that the normalized routing load decreases for the proactive approach with more number of sources. The amount of control overhead remains almost same for a particular advertisement interval irrespective of the number of sources in case of the proactive gateway discovery mechanism. But with increasing number of sources the number of received data packets increases. This leads to the reduced normalized routing load of the proactive approach.

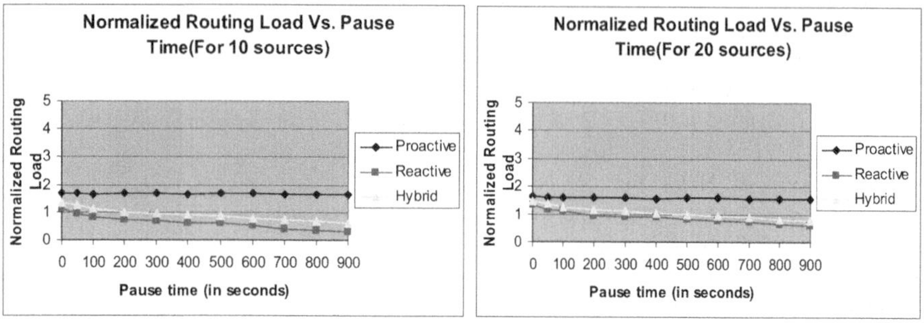


**Fig. 5.** Normalized Routing Load Vs. Pause Time for 10 and 20 Sources

In case of reactive approach, with greater number of source mobile nodes, the number of gateway discovery also increases. This causes higher volume of control overhead. More number of sources with higher volume of data traffic also creates congestion in the network which causes further collisions, more retransmissions and newer route discoveries. This further adds to the already increased control overhead resulting in higher normalized routing load.

## 6 Conclusion

In this paper we have described the design and implementation of the various gateway discovery approaches and carried out a detailed ns2 based simulation to study and analyse the performance differentials of these approaches under different scenarios. From the simulation results we see that the proactive approach shows better packet delivery performance than the reactive approach mainly due to the instant availability of fresher and newer routes to the gateway all the time. In terms of the average end-to-end delay, the proactive and hybrid gateway discovery approaches outperform the reactive gateway discovery. As we decrease the pause time and increase the number of sources, all the approaches suffer form greater average end-to-end delay. As far as normalized routing overhead is concerned, the reactive approach performs better than the proactive and hybrid approaches. In case of the proactive approach the normalized routing load remains almost constant for a particular advertisement interval irrespective of the pause time. With more number of sources, the number of received data packets increases for the proactive approach which accounts for its reduced normalized routing load. Whereas for the reactive approach, with decreasing pause time and increasing number of sources, the number of gateway discoveries and as a result the amount of control traffic also increases, which ultimately results in higher normalized routing load. The hybrid approach being a combination of proactive and reactive approaches, its normalized routing load lies between them.

In our future work, we plan to study the performance of these gateway discovery approaches under other network scenarios by varying the network size, the number of connections, distance between the gateways, the mobility models and the speed of the mobile nodes etc.

## References

1. Toh, C.K.: Ad-Hoc Mobile Wireless Networks. Prentice Hall, Englewood Cliffs (2002)
2. Corson, S., Macker, J.: Mobile Ad hoc Networking (MANET): Routing Protocol Performance Issues and Evaluation Considerations. IETF MANET Working Group RFC-2501 (January 1999)
3. Blum, J.I., Eskandarian, A., Hoffman, L.J.: Challenges of inter-vehicle Ad hoc Networks. IEEE Transactions on Intelligent Transportation Systems 5(4) (December 2004)
4. Royer, E.M., Toh, C.K.: A Review of Current Routing Protocols for Ad hoc Mobile Wireless Networks. IEEE Personal Communications Magazine, 46–55 (April 1999)
5. Dow, C.R.: A Study of Recent Research Trends and Experimental Guidelines in Mobile Ad-Hoc Networks. In: Proceedings of $19^{th}$ International Conference on Advanced Information Networking and Applications, vol. 1, pp. 72–77. IEEE, Los Alamitos (2005)
6. `http://www.ietf.org/html.charters/manet-charter.html`
7. Jonsson, U., Alriksson, F., Larsson, T., Johansson, P., Maguire Jr, G.Q.: MIPMANET – Mobile IP for Mobile Ad Hoc Networks. In: The First IEEE/ACM Annual Workshop on Mobile Ad Hoc Networking and Computing (MobiHOC 2000), Boston, Massachusetts, USA, August 11, pp. 75–85 (2000)
8. Sun, Y., Belding-Royer, E., Perkins, C.: Internet Connectivity for Ad hoc Mobile Networks. International Journal of Wireless Information Networks 9(2) (April 2002); Special Issue on Mobile Ad Hoc Networks (MANETs): Standards, Research, Applications
9. Broch, J., Maltz, D.A., Johnson, D.B.: Supporting Hierarchy and Heterogeneous Interfaces in Multi-Hop Wireless Ad Hoc Networks. In: Proceedings of the Workshop on Mobile Computing, Perth, Australia (June 1999)
10. Wakikawa, R., Malinen, J.T., Perkins, C.E., Nilsson, A., Tuominen, A.J.: Global connectivity for IPv6 mobile ad hoc networks. draft-wakikawa-MANET-globalv6-03.txt (October 23, 2003)
11. Ratanchandani, P., Kravets, R.: A Hybrid Approach to Internet Connectivity for Mobile Ad Hoc Networks. In: Proceedings of the IEEE Wireless Communications and Networking Conference (WCNC), New Orleans, Louisiana, USA, March 16-20 (2003)
12. Lee, J., et al.: Hybrid Gateway Advertisement Scheme for Connecting Mobile Ad Hoc Networks to the Internet. In: Proceedings of $57^{th}$ IEEEVTC 2003, Jeju, Korea, vol. 1, pp. 191–195 (April 2003)
13. Perkins, C.E.: Ad hoc networking. Addison Wesley, Reading (2001)
14. Fall, K., Vardhan, K., eds.: Ns notes and documentation (1999), `http://www.mash.cd.berkeley.edu/ns/`
15. Network Simulator-2 (NS2), `http://www.isi.edu/nsnam/ns`
16. IEEE Computer Society LAN MAN Standards Committee. Wireless LAN Medium Access Control (MAC) and Physical Layer (PHY) Specifications, IEEE Std 802.11-1997. The Institute of Electrical and Electronics Engineers, New York (1997)

# Wireless Sensor Network Security Research and Challenges: A Backdrop

Dimple Juneja[1], Atul Sharma[1,*], and A.K. Sharma[2]

[1] MM Institute of Computer Technology & Business Management, MM University, Mullana (Ambala), Haryana, India
[2] YMCA University of Science & Technology, Faridabad, Haryana, India
atul44885@gmail.com

**Abstract.** If sensor networks are to attain their potential, security is one of the most important aspects to be taken care of. The need for security in military applications is obvious, but even more benign uses, such as home health monitoring, habitat monitoring and sub-surface exploration require confidentiality. WSNs are perfect for detecting environmental, biological, or chemical threats over large scale areas, but maliciously induced false alarms could completely negate value of the system. The widespread deployment of sensor networks is directly related to their security strength. These stated facts form the basis for this survey paper. This paper present a brief overview of challenges in designing a security mechanism for WSN, classify different types of attacks and lists available protocols, while laying outline for proposed work.

**Keywords:** Wireless Sensor Networks, Security Protocols, Network Threats.

## 1 Introduction

Our previous work pertaining to use of Wireless Sensors in Subsurface exploration proposed novel and efficient deployment strategy [1], routing strategy [2], and information processing using Extended Kalman Filter [3]. Sensor network proponents predict a future in which numerous tiny sensor devices will be used in almost every aspect of life. The goal is to create smart environments capable of collecting massive amounts of information, recognizing significant events automatically, and responding appropriately. Sensor networks facilitate comprehensive, real-time data processing in complex environments. Typical applications of sensors include emergency response information, energy management, medical monitoring, inventory control, and battlefield management. If sensor networks are to attain their potential, secure communication techniques must be developed in order to protect the system and its users [4]. The need for security in military applications is obvious, but even more benign uses, such as home health monitoring, and sub-surface exploration require confidentiality. WSNs are perfect for detecting environmental, chemical, or biological threats over large scale areas, but maliciously induced false alarms are capable of negating value of the system. Widespread deployment of sensor networks is directly related to their security strength.

* Corresponding author.

A. Mantri et al. (Eds.): HPAGC 2011, CCIS 169, pp. 406–416, 2011.

These stated facts form the basis for this survey paper. Structure of the paper is as follows: Section 2 presents background and throws light on the work of researchers who proposed in-network security mechanisms. Section 3 presents attacks and defenses within WSN, while Section 4 outlines Sensor Security Challenges. Section 5 presents conclusion and proposed future work.

## 2 Related Work

Far-reaching research is being done in the area of Wireless Sensor Networks. Researchers have been concentrating on solving a variety of challenges ranging from limited resource capabilities to secure communication. Literature indicates that sensor networks are deployed in public or abandoned areas, over insecure wireless channels [5], [6], [7], [8]. It is therefore alluring for a malicious device / intruder to eavesdrop or inject messages into the network. The traditional solution to this problem has been to take up techniques such as message authentication codes, public key cryptography and symmetric key encryption schemes. However, since there are resource scarcities for motes, the major challenge is to devise these encryption techniques in an efficient way without sacrificing their scarce resources. One method of shielding any network against external attacks is to apply a straightforward key infrastructure. However, it is known that global keys do not provide network resilience and pair wise keys are not robust solution. A more intuitive solution is needed for WSNs.

TinySec [9] introduced security to the link layer of TinyOS suite [10] by incorporating software-based symmetric keying with low operating cost requirements. Not all vulnerabilities present in TinySec could be addressed for example techniques to avoid insider attacks. In contrast, Zigbee or the 802.15.4 standard [11] introduced hardware-based symmetric keying with success. However, in order to provide thorough security, use of public cryptography in order to create secure keys throughout network deployment and maintenance phases [12] is also being tested out. This concept has opened an unheard area for discussion of sensor network cryptographic infrastructure. Widespread research is also being carried out on topics such as key storage & key sharing [13], key preservation [14] and shared key pools [15]. Now, since sensor nodes need to cluster aiming to fulfill a particular task, it is desired that the group members' converse securing between each other, in spite of the actuality of global security also present. But contrary to this fact secure grouping has been researched to a very low extent in the past and only a few exhaustive solutions exist.

Further, although, data aggregation (sensor nodes aggregate sensed data from environment before finally transmitting it to the base station) is one of the promising strategies to reduce cost and network traffic but such data is always susceptible to attacks by intruders. A challenger with control over an aggregating node can choose to disregard reports or produce fake reports, affecting reliability of the generated data and at times whole network as well. The main aim in this area is to use flexible functions, which will be able to discover and report forged reports through demonstrating authenticity of the data somehow. Wagner [16] established a technique in which aggregator uses hash trees to create proof of its neighbors' data, which in turn is used to verify purity of collected data to the base station. Another approach [17], takes advantage of network density by using the aggregator's neighbors as witnesses. It is also possible to reduce amount of traffic heading to base station by using bloom filters to filter out false aggregations [18]. Latest research trends towards security measures indicate development of Secure Protocols. The main research challenge in this area is to discover new defense techniques to be applied to existing routing protocols, without compromising connectivity, coverage or scalability [19]. Perrig et al [20] made the first attempt to devise a secure protocol for sensor networks. Security Protocols in Sensor Networks (SPINS) provide data authentication, semantic security and low overhead, along with replay protection.

Fig 1 elaborates the energy cost of adding security protocols to sensor network. Majority of overhead arises from transmission of extra data rather than any computational costs. SPINS was later used to design a secure cluster based protocols such as LEACH. Karlof and Wagner [5] have provided an extensive analysis on the WSNs routing vulnerabilities and possible countermeasures. According to their study common sensor network protocols are generally vulnerable due to their simplicity and hence security should be incorporated into these protocols right from design time. In particular, their study targets TinyOs, directed diffusion and geographic routing.

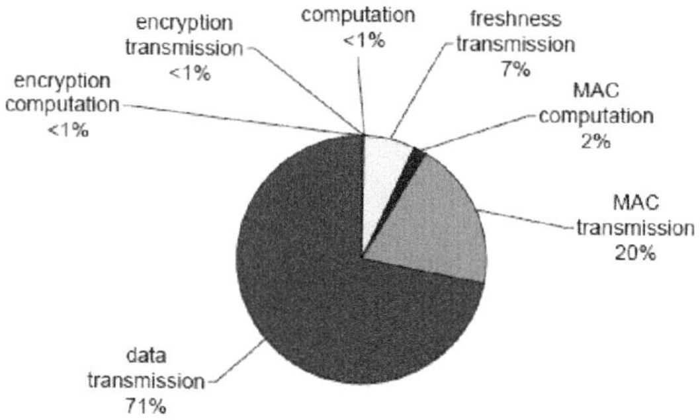


**Fig. 1.** Energy costs from SPINS [20]

## 3 Attacks and Defenses

Goals for security sensor networks include the same four primary objectives as conventional networks: availability, secrecy, integrity, and authentication. Though WSN security is characterized by the same properties as compared to traditional network security, but at the same time they are prone to new attacks. Attacks are made at several levels on the network, like Physical Layer, Link Layer or Network Layer.

Attacks at physical level include radio signal jamming as well as tampering with physical devices. One of the most prominent attacks at this layer is Jamming [21], a well-known attack on wireless communication. In jamming, intruder interferes with wireless frequencies on which the transceivers used by a device operates. It represents an attack on the network accessibility. Jamming is different from normal radio transmission in that it is redundant and disorderly, thus creating a denial-of-service condition. The degree of jamming is determined by physical properties such as available power, antenna design, obstacles, and height above ground. Jamming is extremely successful against single channel networks, i.e., when all nodes transmits in small band, single wireless spectrum.

Tampering [22] is the second security issue at physical layer. Sensor nodes are generally deployed in hostile environment, away from personal monitoring. These sensors are available for easy access to intruders, which can potentially harm these devices by tampering, duplicating or even destroying them. One available solution to this problem is manufacturing of tamper-proof sensor nodes. These nodes are smart enough to delete any cryptographic information available within them as soon as they sense some sort of tampering. But these are not economically viable since tamper-proof sensor nodes increase overall cost. Other solutions might be using of multi-key security algorithms. In these security algorithms intruders will not have access to complete data even if one of the key has been compromised upon.

Like the physical layer, link layer is particularly vulnerable to denial of service attacks. The link and media access control (MAC) layer handles neighbor-to-neighbor communication and channel arbitration. The first type of attack at this layer is known as Collision [23]. If a challenger is able to generate a collision of even part of a transmission, one can interrupt the entire packet. A single bit error will cause a Cyclic Redundancy Check (CRC) variance and would require retransmission. In some media access control protocols, a corrupted ACK (acknowledgment) may cause exponential back-off and pointlessly increase latency. Although error-correcting codes guard against some level of packet corruption, intentional corruption can occur at levels which are beyond the encoding scheme's capability to correct. The advantage, to the challenger, of this jamming at MAC level over physical layer jamming is that much less energy is required to achieve the same effect.

Another malicious goal of intruders is Exhaustion [24] of a sensor node's battery power resources. Exhaustion may be initiated by an interrogation attack. A compromised sensor node could repeatedly transmit RTS (Request To Send) packets in order to bring forth CTS (Clear To Send) packets from a uncompromised neighbor, eventually draining the battery power of both nodes. Still more damaging attack on Link Layer is Unfairness [25].

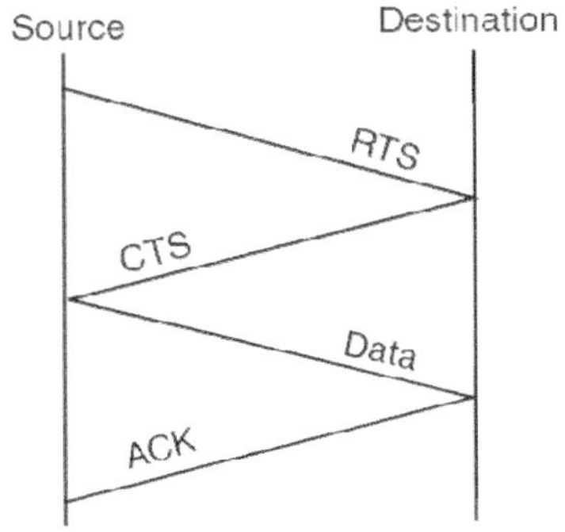


**Fig. 2.** A Four-Way Handshake ensures collision avoidance in 802.11 networks

In this type of attack at Link Layer, a compromised node can be misrepresented to sporadically attack the network in such a fashion which induces biasness in the priorities for granting of medium access. This fragile form of denial of service attack might, increase latency resulting in real-time protocols miss their deadlines. Another form of this attack generally target one particular flow of data in order to restrain recognition of some event. The use of tokens which avert a compromised node from capturing the channel for a long period of time has been proposed.

Due to the ad-hoc nature of sensor networks, each node eventually at some point of time assumes routing responsibilities. Since every node in a sensor network virtually enact as a router, hence WSN are highly susceptible to routing attacks at network layer. Researchers have identified a variety of routing attacks [26] and have shown them to be effective against major sensor network routing protocol. Various classifications of attacks are summarized below and followed by a general discussion of secure routing techniques.

The most prominent attack on routing is to alter, spoof, or just replay routing information. This type of attack is known as False Routing Information. The false information may allow intruder to attract or repel traffic, create routing loops, shorten or extend route lengths, increase latency, and even partition the network, as shown in Fig 3. Clearly, the distortion of routing information can cripple complete network. The standard solution is to require authentication for routing information, i.e., routers only accept routing information from valid routers encrypted with valid shared key information.

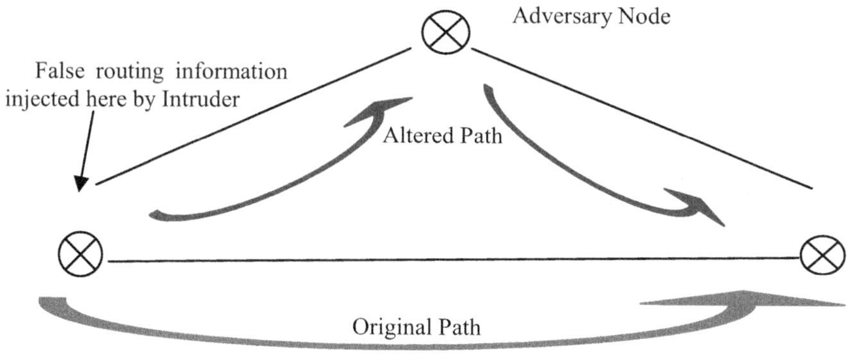


**Fig. 3.** Redirecting traffic through an adversary node via False Routing Information attack

Another attack, known as Selective Forwarding [27] is a more clever attack in which the compromised node is made to transmit forward only some of the packets correctly, while others are silently dropped. Smart networks are capable to routing data along another path, in case of a failure of a particular node. If all packets from a node are dropped, it will be considered as a dead network. Hence only selective packets are being forwarded by compromised node, creating an illusion that it is still active, and that data can be routed via it.

Routing decisions in network are based on distance between nodes. In Sinkhole Attack [28] a compromised node is made to advertise a luring route to the base station or sink. Thus all neighboring nodes are made to route their data towards the compromised node, as shown in Fig 4. The intruder at compromised node thus gains access to major data within its area, and might destroy, manipulate or even modify these packets.

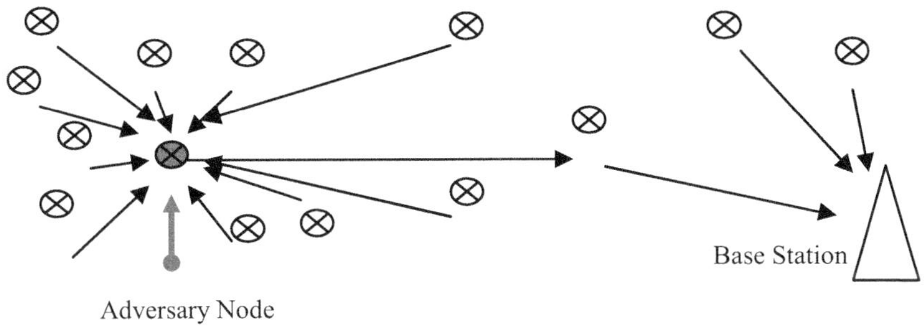


**Fig. 4.** Model of Sinkhole attack

In Sybil attack [29], the compromised node spoof neighboring nodes by broadcasting multiple identities. The compromised node claims to be other node present within the network, hence presenting a great threat to overall routing process [Fig 5]. The malicious effect aggravates as other nodes unknowingly further transmit routing data received from compromised node to their neighbors.

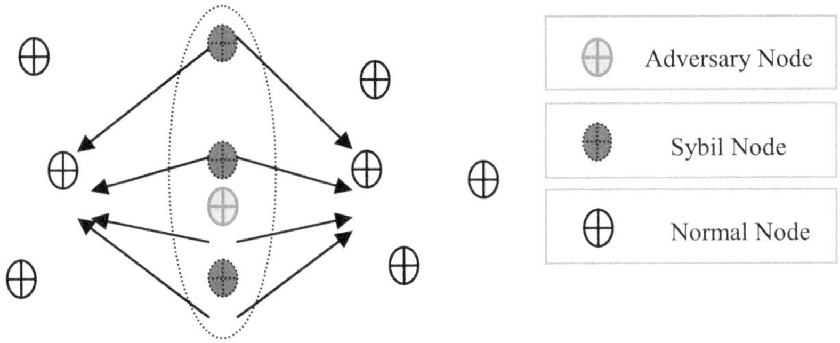


**Fig. 5.** Model of Sybil attack

In Wormhole Attack [30], two mutually understanding malicious nodes form an out-of-bound channel or transmission tunnel in between them. The end points of this tunnel are called as Start & End point. The compromised node at Start point transmits its data via tunnel to malicious node present at End point, as shown in Fig 6. The End point node then re-transmits the received data packets, hence creating an illusion that these distant nodes are neighbors. This sort of attack is likely to be used in arrangement with selective forwarding or eavesdropping.

Nodes present within a network rely on acknowledgment received from neighboring nodes. In Acknowledgment Spoofing attack [31], a malicious node may respond back to a transmitting node on behest of a weak or a non-active node, and thus deceiving sensor about strength of link. This way sender unknowingly keeps on transmitting to the non-active node and data is eventually lost or captured and destroyed by malicious node. There have been several approaches to defend against network layer attacks. Authentication and encryption may be initial steps, but more proactive techniques such as monitoring, probing, and transmitting redundant packets have also been suggested. Secure routing methods protect against some of previous attacks. Proposed techniques include Authentication & Encryption. Link layer authentication and encryption protect against most outsider attacks on sensor network routing protocol. Even a simple scheme which uses a globally shared key will prevent unauthorized nodes from joining topology of the network. In addition to preventing selective forwarding and sinkhole attacks, authentication and encryption make Sybil attack almost impossible because nodes will not accept even one identity from the malicious node.

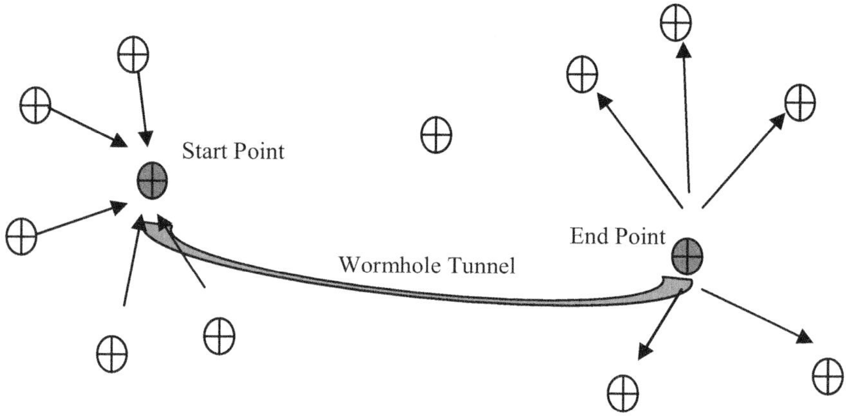


**Fig. 6.** Model of Wormhole attack

Another technique is Monitoring, which is a more active strategy for secure routing, where-in nodes monitor their peers and watch for suspicious behavior. In this approach, motes act as "watchdogs" to monitor next hop transmission of the packet. In event that misbehavior is detected, nodes will update routing information in order to avoid the compromised node. Another proactive defense against malicious routers is probing. This method periodically sends probing packets across the network to detect blackout regions. Since geographic routing protocols have knowledge of the physical topology of the network, probing is especially well-suited to their use. Probes must appear to be normal traffic, however, so that compromised nodes do not intentionally route them correctly in order to escape detection. Redundancy is another strategy for secure routing. Redundancy simply transmits a packet multiple times over different routes. Hopefully one of the routes remains uncompromised and will correctly deliver message to the destination. Despite its inefficiency, this method does increase the difficulty for an attacker to stop a data flow.

## 4 Challenges in Sensor Security

Five of the most looked for challenges in designing security schemes for large wireless sensor networks are Wireless Medium, Ad-Hoc Deployment, Hostile Surroundings, Resource Scarcity and Immense Scale. Applications proposed for sensor networks necessitate wireless communication links. The deployment scenarios for ad-hoc sensor motes renders use of wired media communication totally infeasible [32]. This leads to more security concerns in WSN, since wireless medium is always prone to security attacks since its method of operation / transmission makes it an easy prey for eavesdropping. Wireless communication can be easily trapped, modified or even replaced by intruders. The wireless media allows intruders to destroy genuine communication packets and inject deceptive data into network, with least of the efforts. Wireless media security problem has been intrinsic to traditional networks too, but enhanced and robust solutions are required for sensor networks, owing to their unpredictable deployment and ad-hoc arrangement.

Another challenge for WSN security is its ad-hoc deployment. Sensors may be required to deploy in deterministic or non-deterministic environments. In both cases no fixed topology can be framed in advance. Even the deployed network may have to change its topology every now and then, subject to addition of new nodes, node failures etc. [33]. Under such conditions, robust security protocols are required which can adapt dynamically as per changing configuration / topology of WSN. Hence in sensor networks traditional security mechanisms based on static configurations cannot be applied.

The environment within which sensor nodes operate, collect and transmit data is hostile. Intruders might have know-about the geographical locations of sensor motes, and subsequently reach them to capture / destroy them. No security protocol can fend WSN against such kind of physical attacks, but these needs to be kept in scenario while designing a security framework, in order to provide self-healing capabilities to network.

Another challenge in WSN is resource scarcity within sensor motes. Due to hostile conditions and non-predictable environment sensor nodes cannot be replenished in terms of battery power. In addition to battery, the memory size and computational powers too are low due to small size of nodes. These factors make efficient but resource extensive security mechanisms totally infeasible for WSN. A representative example of sensor device is Mica mote. It has a 4 MHz Atmel ATMEGA103 CPU with 128 KB of instruction memory, 512 KB of flash memory, and just 4 KB of RAM for data [34]. The radio operates at up to 40 Kbps bandwidth with a transmission range of a few dozen meters. Such constraints on resources demand extremely competent security algorithms in terms of computational complexity, memory as well as bandwidth. While energy is perhaps the most prized resource for sensor networks, earlier research work has given little to no attention to energy efficiency. Transmission is especially expensive in terms of power, as apparent from SPINS [Fig 1] too.

Another challenge for WSN security mechanism is its large scale deployment. Traditional networks might be limited to an office or to a bigger geographical location but in a controlled fashion. But in case of sensors, the area being covered may be large and un-predictable. In many cases sensors are even air-dropped and hence their exact geographical location may be different than what might have been thought of. In such cases providing security to all nodes present becomes a challenging task. Security mechanism needs to be developed which can cater to large number of nodes spread over a large scale, and at the same time maintaining computational and communication efficiency.

## 5 Conclusion and Future Work

The paper presented known threats and security protocols available for wired and wireless networks. Works of researchers in this field have been extensively studied. While many frameworks have been devised for WSN, but none were found for robust security mechanisms in subsurface exploration. Keeping in view the extreme harsh conditions prevailing in subsurface, the demand is to devise a novel security mechanism which will make communication within sensors more robust, scalable and efficient.

## References

1. Juneja, D., Sharma, A., Kumar, A.: A Novel and Efficient Algorithm for Deploying Mobile Sensors in Subsurface. Computer and Information Science 3(2), 94–105 (2010); ISSN 1913-8989 (Print), ISSN 1913-8997 (Online)
2. Juneja, D., Sharma, A., Kumar, A.: A Query Driven Routing Protocol for Wireless Sensor Nodes in Subsurface. International Journal of Engineering Science and Technology 2(6), 1836–1843; ISSN: 0975-5462
3. Juneja, D., Sharma, A., Kumar, A.: A Novel Application Of Extended Kalman Filter For Efficient Information Processing In Subsurfaces. International Journal of Computer Applications 17(2), 28–32 (2011); Published By FCS (Foundation of Computer Science, USA). ISSN: 0975-8887
4. Al-Sakib, Pathan, K., Lee, W., Hyung, Hong, S., Choong.: Security in Wireless Sensor Networks: Issues and Challenges. In: ICACT 2006 (2006)
5. Lu, B., Habetler, T.G., Harley, R.G., Gutiérrez, J.A.: Applying Wireless Sensor Networks in Industrial Plant Energy Management Systems – Part I: A Closed-Loop Scheme. In: Sensors, October 30 -November 3, pp. 145–150. IEEE, Los Alamitos (2005)
6. Virone, G., Wood, A., Selavo, L., Cao, Q., Fang, L., Doan, T., He, Z., Stankovic, J.A.: An Advanced Wireless Sensor Network for Health Monitoring. In: Transdisciplinary Conference on Distributed Diagnosis and Home Healthcare (D2H2), Arlington, VA, April 2-4 (2006)
7. Bokareva, T., Hu, W., Kanhere, S., Ristic, B., Gordon, N., Bessell, T., Rutten, M., Jha, S.: Wireless Sensor Networks for Battlefield Surveillance. In: Land Warfare Conference 2006, Brisbane, Australia (October 2006)
8. Mainwaring, A., Polastre, J., Szewczyk, R., Culler, D., Anderson, J.: Wireless Sensor Networks for Habitat Monitoring. In: ACM WSNA 2002, Atlanta, Georgia, USA, September 28, pp. 88–97 (2002)

9. Wireless Sensor Networks, http://en.wikipedia.org/wiki/Wireless_Sensor_Networks
10. Tiny Operating System, http://en.wikipedia.org/wiki/TinyOS
11. Sastry, N., Wagner, D.: Security considerations for IEEE 802.15.4 networks. In: Proceedings of the 2004 ACM Workshop on Wireless Security. ACM Press, New York (2004)
12. Malan, D., Welsh, M., Smith, M.: A public-key infrastructure for key distribution in TinyOS based on elliptic curve cryptography. In: Sensor and Ad Hoc Communications and Networks (2004)
13. Chan, H., Perrig, A., Song, D.: Random key predistribution schemes for sensor networks. In: Proceedings of the Symposium Security and Privacy (2003)
14. Du, W., Deng, J., Han, Y., Chen, S., Varshney, P.: A key management scheme for wireless sensor networks using deployment knowledge. In: INFOCOM 2004: Twenty-third AnnualJoint Conference of the IEEE Computer and Communications Societies (2004)
15. Eschenauer, L., Gligor, V.D.: A key-management scheme for distributed sensor networks. In: Proceedings of the 9th ACM Conference on Computer and Communications Security. ACM Press, New York (2002)
16. Wagner, D.: Resilient aggregation in sensor networks. In: SASN 2004: Proceedings of the 2004 ACM Workshop on Security of Ad Hoc and Sensor Networks (2004)
17. Du, W., Han, Y.S., Deng, J., Varshney, P.K.: A Pairwise key predistribution scheme for wireless sensor networks. In: Proceedings of the ACM Conference on Computer and Communications Security (2003)
18. Ye, F., Luo, H., Lu, S., Zhang, L.: Statistical en-route filtering of injected false data in Sensor Networks. In: Proceedings - IEEE INFOCOM (2004)
19. Hoger, K., Andreas, W.: Protocols and Architecture for Wireless Sensor Networks. John Wiley & Sons Ltd, Chichester (2005); ISBN: 0-470-09510-5
20. Perrig, R., Szewczyk, V., Wen, D., Culler, J.D.: SPINS: security protocols for sensor networks. In: Proceedings of ACM MobiCom 2001, Rome, Italy, pp. 189–199 (2001)
21. Raymond, D.R., Marchany, R.C., Brownfield, M.I., Midkiff, S.F.: Effects of Denial-of-Sleep Attacks on Wireless Sensor Network MAC Protocols. IEEE Transactions on Vehicular Technology 58(1), 367–380 (2009)
22. Wood, A.D., Stankovic, J.A.: Denial of Service in Sensor Networks. IEEE Computer 35(10), 48–56 (2002)
23. Brownfield, M., Gupta, Y., Davis, N.: Wireless sensor network denial of sleep attack. In: Proceedings from the Sixth Annual IEEE SMC Information Assurance Workshop, IAW 2005, pp. 356–364 (2005)
24. Wood, A.D., Stankovic, J.A.: Denial of Service in Sensor Networks. IEEE Computers, 54–62 (October 2002)
25. Padmavathi, G., Shanmugapriya, D.: A Survey of Attacks, Security Mechanisms and Challenges in Wireless Sensor Networks. International Journal of Computer Science and Information Security (IJCSIS) 4(1 & 2), 1–9 (2009)
26. Karlof, C., Wagner, D.: Secure routing in wireless sensor networks: attacks and countermeasures. In: Proceedings of the First IEEE International Workshop on Sensor Network Protocols and Applications, pp. 113–127 (2003)
27. Yu, B., Xiao, B.: Detecting selective forwarding attacks in wireless sensor networks. In: Proceedings of the Second International Workshop on Security in Systems and Networks (IPDPS 2006 Workshop), pp. 1–8 (2006)
28. Krontiris, I., Dimitriou, T.D., Giannetsos, T., Mpasoukos, M.: Intrusion detection of sinkhole attacks in wireless sensor networks. In: Kutyłowski, M., Cichoń, J., Kubiak, P. (eds.) ALGOSENSORS 2007. LNCS, vol. 4837, pp. 150–161. Springer, Heidelberg (2008)

29. Newsome, E., Song, S.D., Perrig, A.: The sybil attack in sensor networks: analysis & defenses. In: IPSN 2004: Proceedings of the Third International Symposium on Information Processing in Sensor Networks, pp. 259–268. ACM Press, New York (2004)
30. Hu, Y.-C., Perrig, A., Johnson, D.B.: Wormhole detection in wireless ad hoc networks. Department of Computer Science, Rice University, Tech. Rep. TR01-384 (June 2002)
31. Tumrongwittayapak, C., Varakulsiripunth, R.: Detecting Sinkhole Attacks In Wireless Sensor Networks. In: Proceedings of the IEEE ICROS-SICE International Joint Conference, pp. 1966–1971 (2009)
32. Feng, Z., Leonidas, G.: Wireless Sensor Networks (An Information Processing Approach). Morgan Kaufmann Publisher under Elsevier; ISBN:1-55860-914-8
33. Deepak, G., Alberto, C., Wei, Y., Yan, Y., Jerry, Z., Deborah, E.: Networking Issues in Wireless Sensor Networks. Elsevier Science, Amsterdam (2003)
34. CrossBow Technology Inc., `http://www.xbow.com/`

# Automated Test Case Generation for Object Oriented Systems Using UML Object Diagrams

M. Prasanna[1] and K.R. Chandran[2]

[1] Research Scholar, Dept. of CIS,
PSG College of Technology, Coimbatore, India
mp_psg@rediffmail.com
[2] Professor of IT & Head, Dept. of CIS,
PSG College of Technology, Coimbatore, India
chandran_k_r@yahoo.co.in

**Abstract.** To reduce the effort in identifying adequate test cases and to improve the effectiveness of testing process, a graph based method has been suggested to automate test case generation for Unified Modeling Language object diagram. The system files produced in the modeling exercise have been used to list all possible valid and invalid test cases that are required to validate the software. The diagrams are treated as graphs to generate the test cases. The effectiveness of the test cases has been evaluated using mutation testing.

**Keywords:** Object diagram, mutation testing, test case, uml, weighted graph.

## 1 Introduction

Software development life cycle is a model of a detailed plan on how to create, develop, implement and eventually fold the software. Among all stages, software testing [1] plays an important role, since it determines quality of the developed product. With the increasing complexity and size of software applications more emphasis has been placed on object oriented design strategy to reduce software cost and enhance software usability. However, object-oriented environment for design and implementation of software brings about new issues in software testing. This is because the important features of an object oriented program such as encapsulation, inheritance, polymorphism, dynamic binding etc create several testing problems and bug hazards. Most reported research proposed test case generation based on program source code. However, generating test cases from program source code, especially for the present day complex applications is very difficult and time consuming. One significant approach is the generation of test cases from UML models. The main advantage with this approach is that it can address the challenges posed by object-oriented paradigms. Moreover, test cases can be generated early in the development process and thus it helps in finding out many problems in design if any and even before the program is implemented.

UML [2] has become the de facto standard for object oriented modeling and design. It is widely accepted and used by software industry. The popularity of UML has lead to program development environments getting integrated with modeling tools.

A. Mantri et al. (Eds.): HPAGC 2011, CCIS 169, pp. 417–423, 2011.

UML models are important source of information for test case design. UML based automatic test case generation has gained attention in the recent past. Properly generated test scenarios are essential to achieve test adequacy and hence to assure software quality. This would also be useful for the testers to understand the behavior and dynamic properties of the system. The UML diagrams provide convenient basis for selecting test cases. With this motivation, this paper presents automatic test case generation based on UML Object diagrams.

## 2 Proposed Methodology for Generating Test Cases

We have proposed a methodology to automate the test case generation process from analysis models. With our methodology, errors could be detected at an early stage during software testing. Our proposed test case generation methodology is simple and can be implemented using computer programs. It is outlined in the following steps.

1. Analyze the real world system which is to be tested and accepted by user.
2. Draw the UML diagrams of the real system using a standard modeling tool which is the input for generating test cases (we have used Rational Rose [3] for modeling).
3. Store the diagrams as files for reference.
4. Parse the model files and derive the graphs corresponding to the type of UML diagram stored.
5. The nodes and edges of the graphs are mapped to the entities of the real world system.
6. By traversing the graph, valid, invalid test cases are generated.

Our methodology is illustrated with a suitable case study in the following section.

## 3 Case Study on Cell Phone System

### Step 1

We have chosen the cell phone system [4] for illustration. The UML object diagram for the above cell phone system is shown in Fig. 1 which reveals the overall functionality of the system.

### Step 2

Object diagram provides a formal graphic notation for modeling objects, classes and their relationships to one another. Objects are represented as rounded rectangles and they are connected by undirected line. Object diagram can be viewed as a weighted graph [5] in which each edge is assigned a nonnegative numerical as weight. Cell phone system object diagram is transformed into a weighted graph as shown in Fig. 2. Each object in the object diagram is represented as nodes. Edges between two nodes are constructed, if there is a message transmission between corresponding objects. Message number is assigned as the weight W of the edge. Direction is assigned to the edge based on the message flow between the nodes.

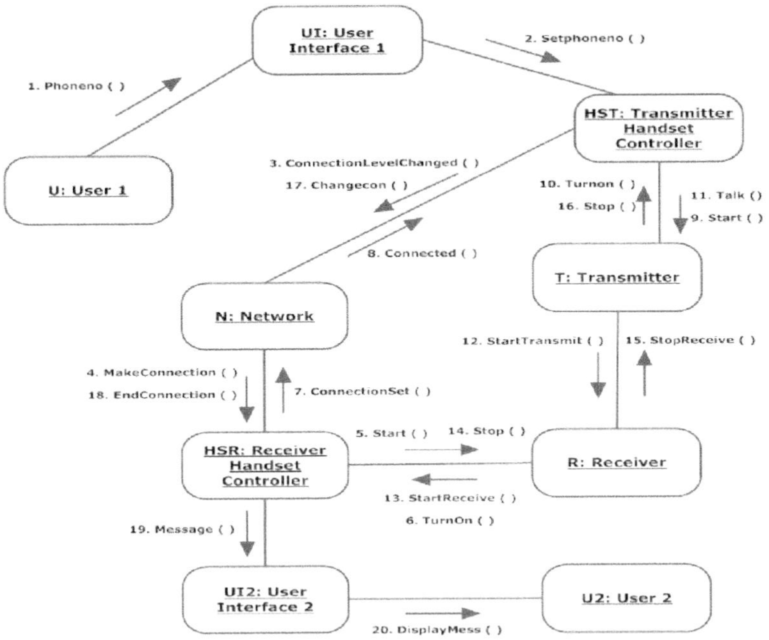


**Fig. 1.** Object diagram of cell phone system

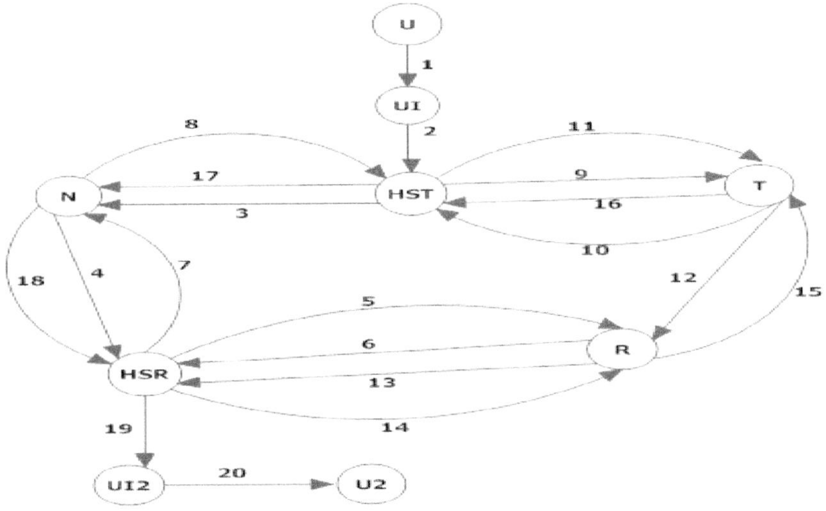


**Fig. 2.** Weighted graph for the cell phone system

**Table 1.** Node Array

| Index | Source | Destination |
|---|---|---|
| 0 | U | UI |
| 1 | UI | HST |
| 2 | HST | N |
| 3 | N | HST |
| 4 | HST | T |
| 5 | T | HST |
| 6 | N | HSR |
| 7 | HSR | N |
| 8 | UI2 | U2 |
| 9 | HSR | R |
| 10 | R | HSR |
| 11 | HSR | UI2 |
| 12 | T | R |
| 13 | R | T |

**Table 2.** Edge Array

| Index | Associated Message Numbers |
|---|---|
| 0 | 1 |
| 1 | 2 |
| 2 | 3, 17 |
| 3 | 8 |
| 4 | 9, 11 |
| 5 | 10, 16 |
| 6 | 4, 18 |
| 7 | 7 |
| 8 | 20 |
| 9 | 5, 14 |
| 10 | 6, 13 |
| 11 | 19 |
| 12 | 12 |
| 13 | 15 |

**Step 3**

**Construction of node array and Edge array from weighted graph:**

Declare Node array and Edge array as two dimensional arrays.

3a. Traverse the graph, and for every edge in the graph

I. Find the source and destination of the edge and search whether they are already present in Node array. If it is present then, find the index of the Node array and append the weight of the edge to the Edge array with the index value else continue with steps II and III.

II. Place the weight of the edge into the Edge array

III. Place the respective source and destination edges into Node array

3b. for all the index of the Edge array, arrange the weights in ascending order

The resultant node array and edge array are shown in Table 1 and Table 2 respectively.

**Step 4**

In Objects diagram, communication takes place through message passing. Hence test cases are generated based on the sequence of messages.

**Valid test case generation:**

1. From the edge array find the least weight.
2. Find the index of the weight from Edge array.
3. From the Node array, the source and destination nodes are taken using the index value.

4. Check whether the last node in the test case is same as the source node using Node array.
   i. If it is same, append only the destination node to the new test case.
   ii. Else add both source and destination nodes to the new test case.
5. Repeat the above steps for all the weights.

Sample valid test cases for Cell phone system are tabulated in Table 3.

**Invalid test case generation:**

Invalid test cases are found using Node array.

1. Repeat the steps 2 to 5 for all the index of the Node array.
2. From the source(s) and destination (s) of Node array, find whether any other index contains the destination node as their source.
3. If index value matches, then check whether any of the weight in the Edge array is in sequence with the index value.
4. If there is no sequence then that path is considered to be invalid.

Sample invalid test cases are generated and tabulated in Table 3.

**Table 3.** Sample Test cases

| Test Id | SEQUENCE | RESULT |
|---|---|---|
| T1 | U→UI | VALID |
| T2 | U →UI→ HST | VALID |
| T3 | U→UI→HST→N | VALID |
| T4 | U→UI→HST→N→HSR | VALID |
| T5 | U→UI→HST→N→HSR→R | VALID |
| T6 | U→UI→HST→N→HSR→R→ HSR | VALID |
| T7 | U→UI→HST→T | INVALID |
| T8 | U→UI→HST→T→HST | INVALID |
| T9 | U→UI→HST→T→HST→T | INVALID |
| T10 | U→UI→HST→T→HST→T→R | INVALID |

# 4 Mutation Testing

Mutation testing [6] is a technique in which multiple copies of a source code are made and each copy is altered. The altered copies are called mutants. Mutants are executed with test cases to determine whether we are able to detect the change between the original program and the mutant. A mutant that is detected by a test case is termed "killed" and the goal of mutation testing is to find a set of test cases that are able to kill groups of mutant programs. The purpose of mutation testing is to find out the effectiveness of test cases.

## 4.1 Fault Injection

We have created mutants by injecting faults in function name, guard condition, relational operator, data value, data name, parameter, omitting message function, changing the source and destination of the message in the cell phone system. One difficulty associated with whether a mutant will be killed is the problem of reaching the location where the fault is injected. Otherwise, the mutant will not be killed. The test cases derived from object diagram for the cell phone system application which is shown in Table 3 is considered for reaching the various mutants. The summary of mutation testing is shown in Table 4.

**Table 4.** Summary of Mutants

| Level of fault Injection | Faults Injected | Faults found |
|---|---|---|
| Function | **4** | **4** |
| Guard Condition | 1 | 0 |
| Relational operator | 3 | 0 |
| Data value | 3 | 3 |
| Data name | 3 | 3 |
| Parameter | 2 | 2 |
| Missing of message | 3 | 3 |
| Change in message direction | 3 | 3 |
| **Total** | 22 | 18 |

### 4.2 Mutation Testing Score

The effectiveness of the test cases is measured using mutation score, which indicates the percentage of mutants killed by a test set. Mutation score is computed using the following formula.

$$\text{Mutation score} = (\sum \text{faults found} / \sum \text{faults injected}) * 100 \qquad (1)$$

For cell phone system object diagram, we injected 22 faults and 18 were revealed from the test cases generated using our approach. Using (1), we get 81% score for cell phone system which shows efficiency level of our approach.

## 5 Conclusion

This paper suggests a model based approach to generate test cases based on the graph theory. The UML object diagram has been used as input to trace the graphical representation of the system to generate test cases. The method lists both valid and invalid test cases that are required to verify the system. This has been illustrated using cell phone system. The effectiveness of the test cases has been measured with mutation testing. It is observed that the methodology yields a mutation score of 81%. This is due to the reason that with object diagram, which is static in nature; errors introduced at guard conditions could not be identified. This is a useful method that can be employed in test case generation. The authors plan to extend this approach to other UML diagrams.

## References


1. Bertolino, A.: Software Testing: Guide to the software engineering body of knowledge. J. IEEE Trans. on Software 16, 35–44 (1999)
2. Priestley, M.: Practical Object-Oriented Design with UML, pp. 7–10. McGraw-Hill Press, New York (2006)
3. `http://www.ibm.com/software/awdtools/developer/rose`
4. Offutt, J., Abdurzik, A., Baldini, A.: A Controlled experiment evaluation of test case generated for UML diagram. Technical report, George Mason University (2004)
5. Lipschutz, S.: Theory and problems of data structures, pp. 277–278. McGraw Hill press, New York (2005)
6. Aggarwal, K.K., Singh, Y.: Software engineering, programs, documentation, operating procedures, pp. 414–415. New Age Intl. press (2005)

# Dead State Recovery Based Power Optimization Routing Protocol for MANETs (DSPO)

Tanu Preet Singh, Manmeet Kaur, and Vishal Sharma

Department of Computer Science & Engineering,
Amritsar College of Engineering & Technology
tanupreet.singh@gmail.com, mink_manu@yahoo.co.in,
vishalsharma2012@gmail.com

**Abstract.** Mobile ad hoc networks are a set of small, low cost, low power sensing devices with wireless communication capabilities. The energy concerned is the receivers processing energy, transmitter's energy requirement for transmission, loses in the form of heat from the transmitter devices. All nodes in the network are mobile and for measuring the efficiency at particular instant, the nodes are considered to be communicating in half duplex mode. In this paper, we introduce the DSPU algorithm which is an automated recovery based power awareness algorithm that deals with the self recovery of the nodes in case of recognition of dead state thus preventing network model going into state of congestion and overheads. The DSPU is an enhanced form of AODV protocol that has the ability of self recovering regarding the security issues of the network structure. The simulations are performed using the NS2 simulator [11] and the results obtained shows that the consideration of energy, bandwidth and the mobility factors enhances the performance of the network model and thus increases the throughput of the ad hoc networks by increasing the life of the nodal structure.

**Keywords:** attenuation loss, energy efficiency, mobility, automated recovery model.

## 1 Introduction

MANETS are ad hoc networks that have a route able networking environment on top of a Link Layer ad hoc network. Many academic papers evaluate protocols and the abilities assuming varying degrees of factors within a bounded space, usually with all nodes within a few hops of each other and usually with nodes sending data at a constant rate. Different protocols are then evaluated based on the packet drop rate, the overhead introduced by the routing protocol, and other measures. The concept of our model is based on CPACL-AODV protocol that has been given on basis of cross layer design [6] [7]. The DSPU algorithm given in this paper is the enhancement of the above written algorithm. In this paper, we define the efficiency of the MANETSs and include the factor of mobility and antenna range in it. The important factor of the mobility and antenna range explains the behavior of the network model when the mobility and antenna range of the node is considered, this means that instead of taking the readings by considering the nodes constant at particular instance of time, the varying

A. Mantri et al. (Eds.): HPAGC 2011, CCIS 169, pp. 424–429, 2011.

behavior is considered. The paper includes system model, and energy efficiency considering the antenna range and mobility, and numerical result along with the performance evaluation and finally the future aspects and conclusion.

## 2 System Model

The network model we considered comprises of k number of hops, hops here are the nodes, and the nodes here considered are to be single channel node. This means for k number of nodes there is k number of channels. Thus, if two nodes are communicating at a time, then we have k-1 number of relaying nodes in the network model. The distance between the source and the destination is denoted by d. the distance between the relaying nodes can be decided on basis of the dynamic routing considered or it can be given on mathematical computations, this means that the distance between the relaying nodes will be less than the actual distance between the source and destination. Thus, if we consider a constant, let this constant be $\alpha_n$ then, from the theoretical analysis [6], we obtain that this value is multiplied with the total distance to obtain the actual distance between the relaying nodes then this value should be positive and less than one. Thus, the distance between the relaying nodes will be:

$$de = \sum_{n=0}^{k} \alpha d \qquad [5].$$

Another factor considered is the , the factor included by us for mobility based analysis and § for antenna range of the MANETSs.

$$Pr = \beta \frac{Pout}{d\eta} \qquad [5]$$

The mobility introduces another simple concept. If the mobility of the structure nodes is more, the attenuation has greater effect but if the nodes are considered to be at rest then, the attenuation comes out to be so small that it can be neglected. Thus, the modified formula for the attenuation loss in a network model (Fig. 1) will be:

$$Pr = \beta \frac{Pout}{d\eta} \zeta §$$

Here, *Pr* is the attenuation loss in the MANETSs, β is the antenna constant, d is the end to end distance between source and destination, η is the path loss constant such that $2 < \eta < 4$ and ζ is the mobility factor. The mobility can be computed by analyzing the movement in terms of number of bits transferred per second per meter of the network model. Here, *Pout = fo (Pin)*, which is based on the working power amplifier present in each of the node.

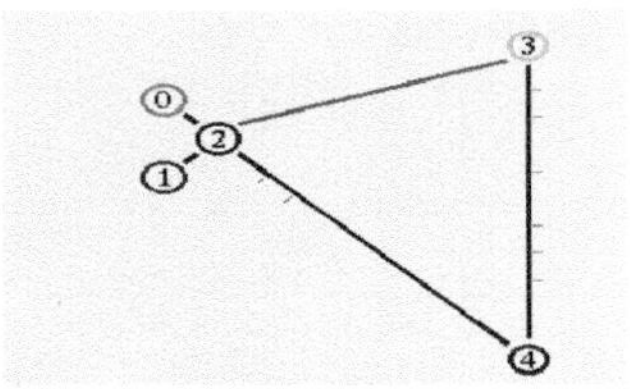


**Fig. 1.**

## 3 Dead State Recovery Based Power Optimization Protocol (DSPO)

In this section of paper, we give the automated recovery based energy efficient algorithm that has the ability of significantly solving the issue of lost data at the node and also do not allow the node to get into the dead state. In DSPU algorithm, the node on receiving the data from the previous node sends the route request and route discover signals giving the details of the energy left with it after processing the obtained data, thus, the node transferring and the node to which the data is being transferred maintains a table, that holds the dynamic values regarding the transmitters transmitting energy Tx and receivers processing energy Ep. Thus, this helps to calculate the level of correctness in transmission of data. In case, the data transmitted is incorrect or has some errors then, there is always a chance of transmission that can lead to wastage of energy of the node and the node might become dead. Thus, the node before transmitting the data checks if a node has its energy greater than the threshold energy, the threshold energy is the minimum amount of energy required by a node to process the data obtained from previous node and further transmit towards the node nearer to the destination, thus, if the node holds the above condition can participate in transmission process and in opposite case, a dynamic routing will be performed that checks for another node that holds the condition and is nearer toward the destination. This will prevent a node from getting into the dead state and thus the energy can be increased without affecting the transmission process and the performance of the MANETSs. Thus, for this the mobility of the MANETSs node has tremendous affect on the performance and efficiency evaluation of the system, if the mobility of nodes is more than the dynamic routing cannot be performed easily, the reason for this is that the routing can be performed efficiently only if mobility is less or in other words more bytes of data is transferred per second per meter of the network model. The mathematical computations for this are obtained by modifying the equations defined in paper [6], the modified form of these equations and their computation is as follows:

$(E_{tot,\,bit})_{CR}$ = energy at transmitter + energy at the receiver

$= \zeta R^{-1}_{min} [E_{tx}\ {}_{i=1}\Sigma^{k} (d_i / d_{max})^{\eta} + k (E_p + P_h Ts)]$

$= R^{-1}_{eff} [B_{k,\,CR} .\ E_{tx} / k + E_p + P_h Ts]$

$B_{k,\,CR} = {}_{i=1}\Sigma^{k} (d_i / d_{max})^{\eta \leq}\ k$

$(E_{tot,\,bit})_{CR}\, d_e^{-\eta} / N_o = \zeta R^{-1}_{eff} [B_{k,\,CR} .\ \gamma / k + \gamma_c]$

$\gamma = E_{tx}\, d_e^{-\eta} / N_o$

$\gamma_c = (E_p + P_h Ts)\, d_e^{-\eta} / N_o$

$$\mathbf{E_{eff} = R_{eff} / \zeta\ \S\ (B_{k,\,CR} .\ \gamma / k + \gamma_c)} \qquad (1)$$

$$\mathbf{Th} = ((R_{eff}/St)*8/1000) * \zeta \text{ kbps} \qquad (2)$$

$R_{eff}$ is the effective rate expresses in ratio of minimum achievable rate per channel of the network model. Here $\gamma$ is the signal to noise ratio and $\gamma c$ is the efficiency constant, k is the number of hops. $d_i/ d_{max}$ is the maximum throughput possible in the MANETSs structure, $\zeta$ is the mobility factor for the model considered. Th is the throughput of the network model and St is the simulation time.

*Example of Table maintenance by a node:

| Head Node | Linked Node | Ep (in microns) | Energy Left |
|---|---|---|---|
| A | F | 0.25 | 1.24 |
| | H | 0.25 | 1.03 |
| | K | 0.25 | 1.11 |

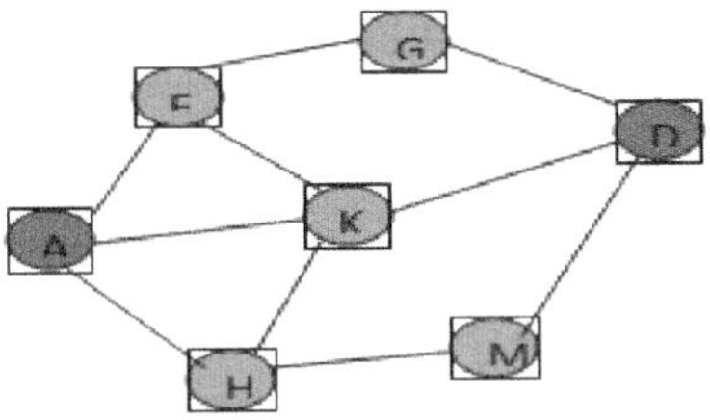


## 4 Performance Evaluation and Numerical Results

The evaluation of our algorithm is evaluated using NS2 Simulator. The number of nodes considered is 50 in 1500 * 1500 $m^2$ Network Area. The packet size was considered to be 1024. In the beginning of the simulations the battery consumptions were about 0.25 for processing and 0.34 units for transmission purpose thus giving the overall consumption of 0.59 for single node at transfer of one packet. Also, with the advent of mobility factor, it is achieved that when the mobility is more in terms of transfer of packets w.r.t. 1 $m^2$ area, the performance increases. Graphical analysis of the DSPU algorithm is done with AODV protocol as it forms the basis of our algorithm and the results shows that the performance is greatly influenced by the mobility factor and thus, gives more concerned data that can be practically adopted.

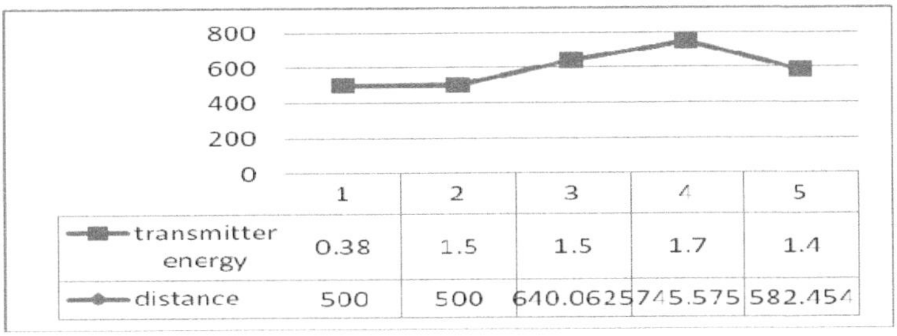


**Fig. 2.**

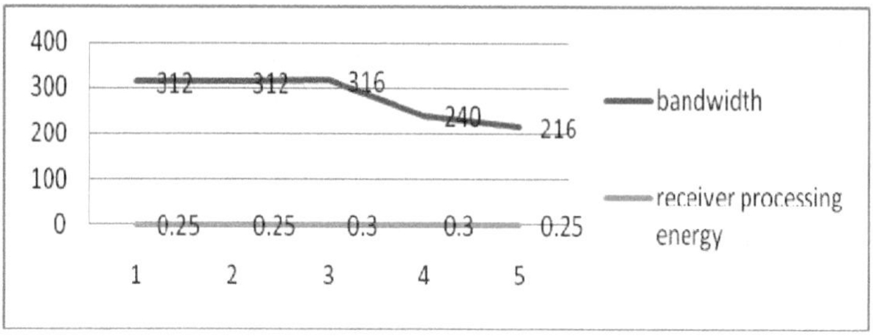


**Fig. 3.**

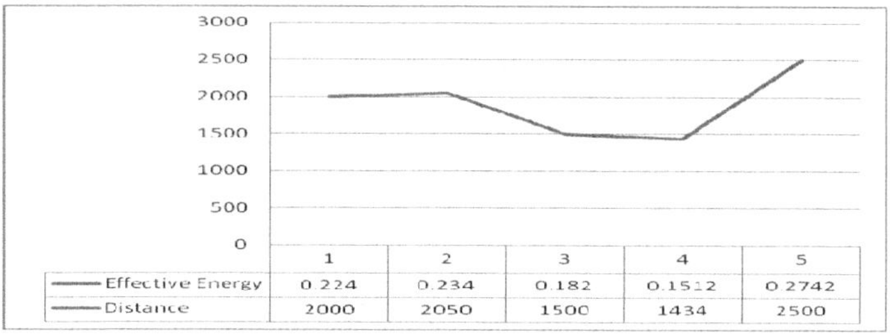


**Fig. 4.**

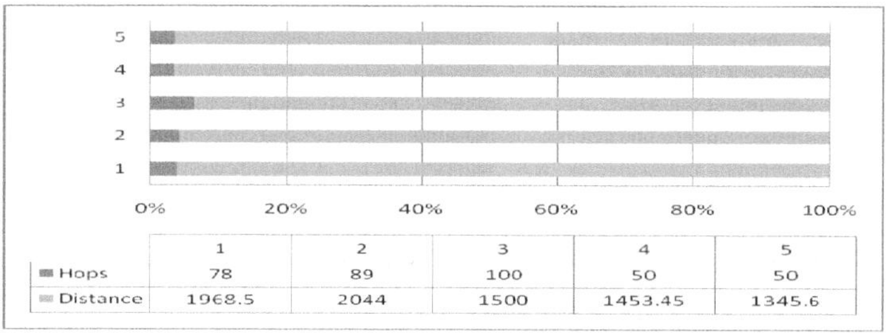


**Fig. 5.**

## 5 Conclusions

The paper proposed security based power awareness algorithm that solved the problem of security issues and the dead node with energy efficient transmission. Also, this reduced the various overheads of the network layer. The technique is practically adoptable. In this technique, it is observed that when the node gets its energy lower than the threshold energy, the alternative path is selected and thus it prevents from entering into the dead state. Thus this improves the performance of the network model. Also, as the energy packets are sent on the temporary halted route, it gets recovered and the original path is again retraced for packet transfer. In future, the work can be carried out to improve the delays that might occur due to transfer of packets through alternative path.

## References


1. Saravana, M., Murali, M., Sujatha, S.: Identifying performance metrics to maximize Manet's throughput. In: International Conference on Advances in Computer Engineering (2010)
2. Rali, M.V., Song, M., Shetty, S.: Virtual wired transmission scheme using directional antennas to improve energy efficiency in Wireless Mobile Ad Hoc Network. IEEE, Los Alamitos (2008); 978-1-4244-2677-5
3. Kim, S., Lee, J., Yeom, I.: Modeling and Performance Analysis of Address Allocation Schemes for Mobile Ad Hoc Networks. IEEE Transactions on Vehicular Technology 57(1) (January 2008)
4. Patil, R., Damodaram, A.: Cost based power aware cross layer routing protocol for Manet. IJCSNS (2008)
5. Bae, C., Stark, W.E.: A Tradeoff between Energy and Bandwidth Efficiency in Wireless Networks. IEEE, Los Alamitos (2007)
6. Rodoplu, V., Meng, T.H.: Bits-per-Joule capacity of energy-limited wireless networks. IEEE Transaction Wireless Communications 6(3), 857–865 (2007)
7. Rankov, B., Wittneben, A.: Spectral efficient protocols for half-duplex fading relay channels. IEEE Journal on Selected Areas in Communications 25, 379–389 (2007)
8. Oyman, O., Sandhu, S.: Non-ergodic power-bandwidth tradeoff in linear multihop wireless networks. In: Proc. IEEE International Symposium on Information Theory, ISIT 2006 (2006)
9. Bae, C., Stark, W.E.: Energy and bandwidth efficiency in wireless networks. In: Proc. International Conference on Communications Circuits and Systems (ICCCAS 2006), vol. 2, pp. 1297–1302 (June 2006)
10. Sikora, M., Laneman, J.N., Haenggi, M., Costello, D.J., Fuja, T.E.: Bandwidth and power efficient routing in linear wireless networks. Joint Special Issue of IEEE Transaction Information: Theory and IEEE Transaction Networking 52, 2624–2633 (2006)
11. Network simulator-2, http://www.isi.edu/nanam/ns/

# On the Potential of Ricart-Agrawala Algorithm in Mobile Computing Environments

Bharti Sharma[1], Rabinder Singh Bhatia[2], and Awadhesh Kumar Singh[2]

[1] DIMT Kurukshetra
India
bharti_kanhiya@yahoo.co.in
[2] NIT Kurukshetra
India
rsibhatia@yahoo.co.in,
aksinreck@rediffmail.co

**Abstract.** The Ricart-Agrawala protocol [1] is one of the classical solutions to mutual exclusion problem. Although, the protocol was invented, essentially, for failure free static distributed systems, it has been adapted by various researchers for almost all changing computing paradigms from classical to contemporary. The purpose of the article is to highlight the strength of the concept used in the Ricart-Agrawala protocol.

**Keywords:** Mutual exclusion, mobile computing.

## 1 Introduction

The mutual exclusion is a fundamental synchronization problem in distributed computing systems. The mutual exclusion protocols are required to ensure exclusive access to the shared resource. The processes competing for the resource cycle through *entry*, *critical section*, *exit*, and *remainder* states. Basically, designing protocol for mutual exclusion is to design *entry* and *exit* protocols. In 1981, Ricart and Agrawala (RA, for short) [1] proposed a distributed mutual exclusion (DMX) algorithm. The algorithm is based on the concept of maintaining pending request queue. Although, RA algorithm is an optimization over Lamport's mutual exclusion algorithm [2], it has introduced the novel idea of pending request queue. The pending request queue is the queue of only those outstanding requests that have priority less than the priority of the site itself, whereas the request queue used by Lamport [2] is the set of all the requesting sites. The last two decades have witnessed huge change in the computing paradigms, from distributed computing in static distributed systems to mobile computing in cellular, ad hoc, and sensor networks. In the recent past, the RA algorithm has been adapted for all the computing environments for fault free as well as fault tolerant computing. The present survey is focused on the wide applicability of the concept, introduced in RA algorithm, and to highlight its versatility and robustness.

A. Mantri et al. (Eds.): HPAGC 2011, CCIS 169, pp. 430–434, 2011.

## 2 Ricart-Agrawala Algorithm in Static Distributed Systems

The RA algorithm ensures mutual exclusion in computer networks whose nodes communicate by message passing and do not have any shared memory. The nodes are assumed to operate correctly, and no link failure occurs. The main idea behind the Ricart-Agrawala algorithm [1] is briefly explained in the following steps:

1. The requests for critical section (CS) are assigned unique priorities (determined by Lamport like timestamps [2]). Each hungry process sends a timestamped request message to every other process in the system.
2. When a site $S_i$ receives a request message, it sends a reply message in response to it, if it is not requesting or if the priority of its request is lower than that of the incoming request; otherwise, it defers the reply.
3. A site executes CS only after it has received a reply message from every other site.
4. Upon exit from CS, a process must send an acknowledgement (reply) to each of the pending requests, before making a new request or executing other actions.

Intuitively, each process, seeking entry into CS, sends (n–1) requests and receives (n–1) acknowledgements or replies to complete one trip into its CS. Therefore, in order to complete one trip into its CS, the total number of messages exchanged is only 2(n–1), unlike Lamport's algorithm [2], where it is 3(n–1).

## 3 Ricart-Agrawala Algorithm in Cellular Mobile Networks

A cellular mobile system is a distributed system consisting of a number of mobile and fixed processing units. The fixed units are called mobile support station (MSSs) and mobile units are called mobile hosts (MHs). The fixed units (MSSs) communicate with each other through a fixed wired network. The MSS is capable of directly communicating with MHs within a limited geographical area, called cell, usually via a low bandwidth wireless medium. A MH can directly communicate with a MSS (and vice versa) only if the MH is physically located within the service area of that particular MSS. A MH can move out of one cell into another cell; in such case, the MSS of the old cell has to hand over the responsibilities for the MH's communication to the MSS of the new cell. This process is called handover or handoff. The wireless channels are constrained in bandwidth than their wired counterparts. Hence, the mutual exclusion algorithms designed for static distributed systems may not work with matching performance and efficiency in the mobile computing systems. Therefore, the protocols, especially designed for mobile computing systems, always take into consideration the bandwidth limitation of wireless channel. Nevertheless, RA algorithm, with some innovations, has shown its strength in handling the conflict resolution problem, with equal ease, also in the mobile scenario.

In 1997, Singhal-Manivannan [3] proposed an algorithm for mutual exclusion in cellular networks. Although, the algorithm uses a novel 'look ahead' technique, it adapts Ricart-Agrawala protocol for mobile computing environment. The authors have partitioned the request set into 'inform set' and the 'status set'. The partitioning technique has been used to know that which sites are concurrently requesting CS.

Once it is known, the protocol uses Ricart-Agrawala method on those sites to enforce mutual exclusion. In fact, the protocol uses a heuristic approach to handle request and reply messages in Ricart-Agrawala protocol.

Each site $S_i$ maintains two arrays of processes. One is called info-set$_i$ and the other is status-set$_i$. The info-set$_i$ is set of ids of the sites to which $S_i$ informs that it is requesting. The status-set$_i$ is set of ids of the sites which inform $S_i$ that they are requesting. For any site $S_i$, the number of processes in info-set$_i$ union with the number of processes in status-set$_i$, are the processes of entire system. In order to know about concurrent CS request activity of some site $S_j$, it is required that either $S_i$ should inform the other site $S_j$ about its status or it should get informed by site $S_j$ about their status. When $S_i$ receives a request message from $S_j$, it sends a request message to $S_j$ provided $S_i$ itself is requesting CS at that time and $S_j$ belongs to Status-Set$_i$. It deletes $S_j$ from status–set$_i$ and adds $S_j$ to info-set$_i$. A site $S_i$ sends a reply message in response to a request message only if it is not requesting or if the priority of its request is lower than that of the incoming request. Afterwards, the protocol applies Ricart-Agrawala algorithm to enforce mutual exclusion. Site $S_i$ executes CS only after it has received a reply message against every request message it sent out, similar to Ricart-Agrawala algorithm. If site $S_i$ receives a reply message from $S_j$, $S_i$ should delete $S_j$ from info-set$_i$ and place in status-set$_i$. On exit CS, a site $S_i$ sends reply messages to all sites in its info-set$_i$. The authors have suggested an optimization as follows: After getting reply from those sites whose entry is in info-set, the site will delete their entries from info-set and push them in status set. Thus, on exit from CS, site has to reply only those sites whose entry is in info-set. Since info-set is now small, the site has to reply to comparatively few sites. Moreover, info-set is complete and updated. Hence, if site request those sites which were in status-set, then it delete their entry from status set and include them in info-set.

For the purpose of reducing power consumption, mobile hosts also can disconnect from network voluntarily. When a mobile host wants to disconnect, it offloads the current values of its data structure to the MSS and executes a disconnection protocol before the disconnection takes place. The MSS responds to the requests of the other mobile hosts on behalf of the disconnected mobile host. In an optimistic scenario, if a number of sites are not interested, i.e., not invoking mutual exclusion, then average message traffic reduces. It is noteworthy that the authors have succeeded in deriving advantage by exploiting the fact that a reply from a site can be assumed an indefinite reply till the site becomes hungry again. However, under heavy load condition, when all the sites are invoking mutual exclusion, the advantage of the 'look-ahead' technique is wiped out and the number of messages becomes same, as in the case of Ricart-Agrawala algorithm [1].

## 4 Ricart-Agrawala Algorithm in Mobile Ad Hoc Networks (MANETs)

The MANETs are highly constrained infrastructure-less environment for the mobile computing applications. The major constrains are low battery backup, small computation power, limited bandwidth, and highly dynamic topology. Due to these limitations, the protocols developed for cellular mobile systems do not work correctly in

MANETs. Secondly, a large number of protocols developed for cellular networks assume the communication channel to be FIFO, i.e., messages do not overtake, which is very difficult to ensure in MANETs due to the constraints, like frequent disconnection and unpredictable mobility pattern. The above mentioned Singhal-Manivannan protocol [3] also does not work correctly for MANETs if we relax the FIFO condition. Nevertheless, the underlying concept of RA algorithm has been proved advantageous because of the following reasons: (i) There is no need to maintain the logical topology, and (ii) There is no need to propagate any message if no host requests to enter CS.

These advantages make the Ricart-Agrawala approach well suited, even in MANETs. The message handling is an energy-intensive affair. Hence, the protocols, which exchange large number of messages between the MHs, are not suitable for the mobile computing applications running in MANETs. Therefore, the main challenge in designing an algorithm for MANETs is to reduce the number of messages.

In a recent paper, Wu-Ca-Yang [4] proposed a fault tolerant mutual exclusion algorithm for MANETs. They have used 'Look-ahead' technique proposed by Singhal-Manivannan [3]. However, the look-ahead technique was designed for infrastructure networks. Hence, in order to apply the technique, a numbers of issues need to be addressed. Since, there is no fixed host to support MHs in MANETs, the assumption about the FIFO channel becomes infeasible. Therefore, Singhal-Manivannan [3] algorithm faces the following challenge in MANETs.

Consider a system containing only two sites $S_i$ and $S_j$, both are hungry, and, say $S_i$ has higher priority request than $S_j$. Assume $S_i$ requested first and $S_j$ received the request. Now, $S_j$, being lower priority and hungry too, has to send reply as well as request to $S_i$. Now, there are two cases possible.

***Case* 1:** Assume the channel is FIFO. Thus, firstly, reply will be received at $S_i$. Afterwards, $S_i$ will shift $S_j$ from info-set$_i$ to status-set$_i$ and enter its CS. Now, say, request of $S_j$ is received at $S_i$. After $S_i$ exits from CS, it will send reply to $S_j$ and shift $S_j$ from status- set$_i$ to info-set$_i$.

***Case* 2:** Assume the channel is non-FIFO. Say, firstly, $S_i$ received the request from $S_j$. As $S_j$ is already in info-set$_i$, $S_i$ will not do anything. Now, say, $S_i$ received the reply from $S_j$. Hence, $S_i$ will move $S_j$ from info-set$_i$ to status-set$_i$ and subsequently enter into CS. On exit CS, $S_i$ will not send reply to $S_j$, as it is in the status-set$_i$. It will send reply only to the nodes belonging to info-set$_i$. Thus, $S_j$'s current request will remain unreplied forever. Therefore, due to violation of FIFO property, a lower priority node may be blocked in starving state.

In order to handle the challenge, Wu-Ca-Yang [4] also used the concept of RA algorithm, however, in a different way. They partitioned the request set into three components, namely, info-set, status-set and request-queue (Qreq), i.e., the set of unreplied requests. The following is advantage of the introduction of request queue. If a site Si receives a request from a site $S_j$ that is already in the infoset$_i$, site $S_i$ puts $S_j$'s request in its own request queue. With the introduction of this idea, the algorithm successfully handles non-FIFO message flows and reduces the number of messages. Assume a MENET consisting of n MHs, the communication between two MHs can be multi-hop, and both link and host failures may occur. The algorithm initializes the info-set, i.e., an array of the IDs of the hosts to which $S_i$ needs to send request messages when it wants to enter CS and status-set, i.e., an array of the IDs of the hosts which, upon requesting to access CS, would send the request messages to $S_i$. Initialization is done using $n \times n$ matrix M, where n is number of MHs. The value of each element of M, $m_{ij}$, represents the relationship between the pair of MHs $S_i$ and $S_j$. If $m_{ij} = 0$, $S_j$ is in info-set of $S_i$, if $m_{ij} = 1$, $S_j$ is in the status-set of $S_i$. All hosts satisfy these two conditions: (1) $\forall S_i$::info-set$_i \cup$ status-set$_i$ = S; $\forall S_i$::info-set$_i \cap$ status-set$_i = \phi$, (2) $\forall S_i$, $\forall S_j$:: $S_i \in$ info-set$_j \Rightarrow S_j \in$ Status-set$_i$. An arbitrary host, say $S_0$, is selected to act as the initiator. The initial value of M is determined by the initiator. When a host is hungry, it sends REQUEST message to all the hosts in its info-set. The host then waits for REPLY message corresponding to each REQUEST message. When all the REPLY messages have been received, the requesting host enters in CS. When a host $S_i$ receives a REQUEST message from another host $S_j$, it moves $S_j$ into info-set and records the request in Qreq (i.e., a queue to store pending requests). If $S_i$ is not interested in CS or it has a low priority, then it sends a REPLY message to $S_j$ and removes the record of $S_j$ from Qreq. Upon receiving of a REPLY message from host $S_j$, $S_j$ is moved to status-set$_i$. If the info-set is empty, it enters CS immediately. To tolerate link and host failure a time out is set in array of timers, called TOreq. Each site maintains TOreq associated with the REQUEST message sent to some host. Upon receiving a REPLY message from host, site removes the timeout for that host. Upon receiving a request message, requesting site is recorded in Qreq by host. When host sends a REPLY to requesting site, the record corresponding to that site is removed by the host. In addition, the protocol has capability to handle the situations when some host wants to disconnect voluntary or be in doze mode. It also guarantees that the dozing host would not receive any request message from any site unless it wakes up. The message complexity under light load conditions is $2 \times n/2$, i.e., n messages and under high load condition, it amounts to $3 \times n/2$.

## References

1. Ricart, G., Agrawala, A.: An Optimal Algorithm for Mutual Exclusion in Computer Networks. Communications of the ACM 24, 9–17 (1981)
2. Lamport, L.: Time, Clocks and Ordering of Events in Distributed Systems. Communications of the ACM 21, 558–565 (1978)
3. Singhal, M., Manivannan, D.: A Distributed Mutual Exclusion Algorithm for Mobile Computing Environments. In: International Conference on Intelligent Information Systems, pp. 557–561. IEEE Press, New York (1997)
4. Wu, W., Cao, J., Yang, J.: A Fault Tolerant Mutual Exclusion Algorithm for Mobile Ad Hoc Networks. Pervasive and Mobile Computing 4, 139–160 (2008)

# Analysis of Digital Forensic Tools and Investigation Process

Seema Yadav, Khaleel Ahmad, and Jayant Shekhar

CSE/IT Dept. S.I.T.E., SVSU,
Meerut-250002, India
Seema_yadava@yahoo.com,
khaleelamna@yahoo.co.in,
jayant_shekhar@hotmail.com

**Abstract.** Popularity of internet is not only change our life view, but change the view of crime in our society or all over the world. Increasing the number of computer crime day by day is the reason for forensic investigation. Digital forensic is used to bring justice against that person who is responsible for computer crimes or digital crimes. In this paper, we explain both type of forensic tool commercial as well as open source and comparisons between them. We also classify digital forensic and digital crimes according to their working investigation.

In this paper, we proposed a model for investigation process to any type of digital crime. This model is simple and gives efficient result to any type of digital crimes and better way to improve the time for investigation.

**Keywords:** Dead analysis; digital crime; digital evidence; digital forensic; live analysis.

## 1 Introduction

Rapid developments and lacks of proper rules and regulations to use internet becomes a crime hub. Digital forensic are the person who doing investigation on the digital type devices. This is not enough for the investigator to have only a good knowledge about computers only but must have knowledge in many other areas also. Digital forensic is a branch of forensic science that is used to recover and investigate of data in digital devices, often in relation to compute crime [1] [6]. Digital forensic is an important part of computer investigation to recovering of data [5].

Computer crime is defined as an act of sabotage ,exploitation of an individual computer system, group of interconnected system and digital technological devices such as cell phones, PDA to commit malicious and digital crime may appear novel, some features of these remains the same as those of conventional crimes [2][13][3][7][11][5]. Digital forensic differentiated in many types of forensic areas (see in figure 1).

A. Mantri et al. (Eds.): HPAGC 2011, CCIS 169, pp. 435–441, 2011.

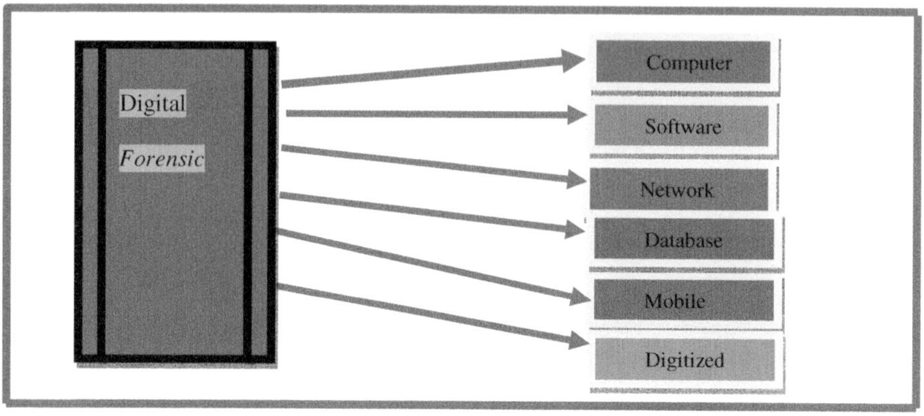


**Fig. 1.** Types of digital forensic

Digital forensic analysis of system and networks can provide digital evidences e.g., planning a murder, cyber harassment and pornography, theft of electronically stored information and data from computer system, generate fraudulent documents with the help of scanners and printers [1]. There are many types of digital crimes some of them are given in figure 2.

**Fig. 2.** Some types of digital frauds

Forensic investigation of a digital crime or digital frauds is a complicated process which is starts at the crime scene, continuous into the computer labs for investigation, and ends in the court where the final judgment is done by judge.

## 2 Literature Survey

In this section we focus on collection and recovering of digital evidence from digital devices by forensic tools. So in this section we first describe the details of digital forensic analysis, digital evidence and forensic tools.

### 2.1 Digital Forensic Analysis

Goal of forensic analysis is to find the digital evidence for any type of digital investigation. A forensic investigation uses both digital and physical evidence with scientific procedures and tools to find out the conclusions. A digital forensic investigation consist three steps [2][13].

- *Acquisition*
- *Analysis*
- *Reporting*

### 2.2. Digital Evidence

Digital evidences are probative information which is stored in digital devices in electronic form that is use as a trial in court cases. Digital evidence takes important role in a wide range of crimes such as denial of service, phishing and sniffing, hacking which is stored in digital devices such as cell phone, PDA, PC etc. Digital data can be easily modified, duplicated , restored or destroyed, so when we do investigation then take a right tool that prevent to modification of data. Goal of investigation process is to collect evidence using acceptable methods or procedures to make the evidence accepted and admitted in the court room for judgment. The final report or documentation of investigation should consists four important things

- *who did[2][11]*
- *what did[2][11]*
- *when did[11]*
- *how did[11]*

### 2.3 Forensic Tools and Their Comparison

Forensic tools are useful in our daily life to improve the security of digital devices based on stored data [4]. By using these types of tools, we can determine the security flaws in the computer system in against to the person who destroyed our computer security. There are basically two types of tools (commercial and open source) that we can use in Windows and Linux based operating system to preventing different types of attacks [9][4][2].Purpose of forensic tools is given as below [2][9].

- Ascertaining date/time stamp information
- Recovering or "un-deleting" files and directories "Carving".
- Performing keyword searches.
- Recovering Internet cookies.
- Creating forensic quality or sector-by-sector images of media
- Locating deleted/old partitions of the digital devices.

Comparisons of some commercial and open source forensic tools are given in table 1 [4][9][12][2].

**Table 1.** Comparison of digital forensic tools

| ***Tools Features*** | ***Encase*** | ***DFF*** | ***FTK*** | ***TSK*** | ***Helix*** | ***Liveview*** |
|---|---|---|---|---|---|---|
| Tools type | Commercia l | Open source | commerci al | Open source | Open source | Open source |
| Tools quality | Great quality software | Less quality software | Great quality software | Not good | Not good | Not good |
| Platform | Windows, Linux, AIX,Solara is | Cross platform | Mac OS | Windows | Windows, Linux | Windows, vista ,Linux |
| User interface | Experience type forensic uses | Easy to use for forensic | Easy to use to all | Easy to use to all | Easy to use | Easy to use |
| Language interface | Traditional Chinese | Simple English | Simple Chinese | English | English | English |
| Features support | Offers Added Features | Offer advanced features | Offers Added Features | Offers Advanced Features | Offers Advanced Features | Offers Advanced Features |
| Support to Image file creation | Yes | Yes it supports | Yes | Yes | Yes | Yes |
| Technique used for Hash value calculation | MD5 | MD5 | MD5 | MD5 and SHA-1 | MD5 and SHA-1 | MD5 |
| Tool cost in investigatio n | high | low | high | low | low | low |
| Retrieve data or file type | Graphical disk interface information | Keyword search, metadata , Retrieve file contest | Digital evidence classificat ion | Retrieve file contest, keyword, metadata | acquired Internet history, screen capture, memory | Creates a VMware machine out of a raw disk image. |

## 3 Basic Process Model for Digital Crime Investigation

After starting the investigation by forensics, we go in first phase of our proposed model.

Phase 1: in this phase forensic first see or check that the digital device is in boot states or shutdown and go to Phase 2.

Phase 2: If the digital device is in boot state then the forensic done live analysis at that device and this is simple as compare to dead analysis of any investigation, but when the digital device is not in boot state then we done dead analysis with the help of DVD/USB and RAM. After that we go to phase third.

Phase 3: Details of basic investigation process which is given in phase 3 of fig. 3 is given as below:

1. *Identification:* which recognizes an incident from digital devices and determines its type?
2. *Preparation:* which entails the preparation of forensic tools, techniques, and monitoring authorizations?
3. *Preservation:* which involves the isolation, securing and preservation of the state of physical and digital evidence?
4. *Collection:* It collects the recording of the physical scene and duplicate digital evidence using standardized and accepted procedures or methods.
5. *Examination:* which involves in depth systematic search of digital evidence relating to the suspected digital crime?
6. *Analysis:* This involves determination of the significance, reconstructing data or information and gives conclusions based on digital evidence found against the crime.
7. *Presentation:* In this we give the summary and explanation of conclusions.
8. *Reporting:* when an investigation has completed then investigator presents his data or information, usually in the form of a written report.

After the documentation or creating the report of digital evidence against the digital crime we finish the investigation process.

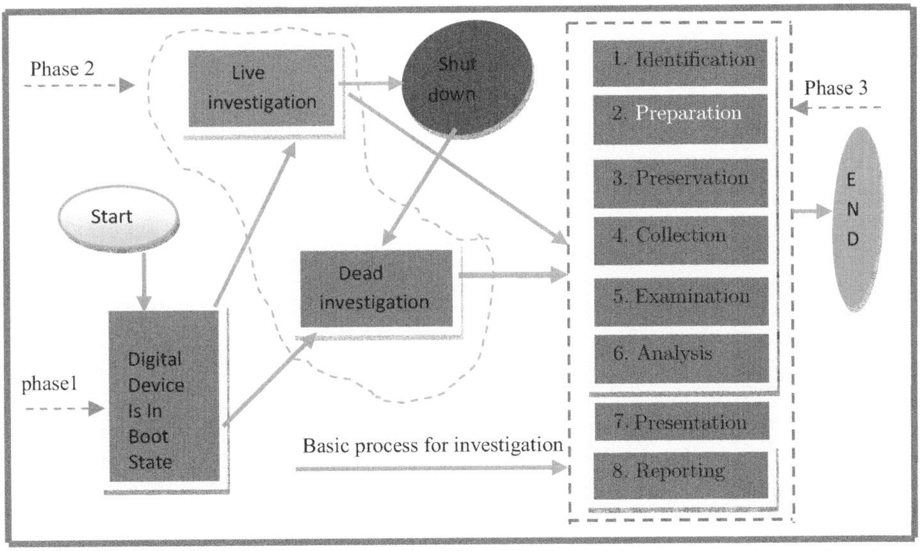


**Fig. 3.** Basic Model for investigation Process of any type of digital crime

## 4 Conclusion

This model offers a comprehensive approach to investigation of digital crimes. Current digital age gives unlimited challenges to the digital forensics. The use of computer and digital devices in the act of crime is continuously grow day by day, so this gives challenges to digital forensic that how they collect information or data from the digital devices after an incident.

This Model describes the investigation process of any type crime at that time when the initial infrastructure is put in place for investigation and going to the reporting phase of the proposed model. This model gives better and simple way for investigation to any type of digital crimes or frauds and improves the time consumption which is taken to investigation by forensics.

## References


1. Worring, M., Cucchiara, R.: Multimedia in Forensics. In: Communication of the ACM, MM 2009, October 19–24, pp. 1153–1154 (2009)
2. Yen, P.-H., Yang, C.-H., Ahn, T.-N.: Design and Implementation of a Live-analysis Digital Forensic System. In: Communication of the ACM in International Conference on Convergence and Hybrid Information Technology, pp. 239–242 (2009)
3. Simson, L., Garfinkel: Digital forensics research: The next 10 years. Digital Forensic Research Workshop in Science Direct, digital investigation 7, 64–73 (2010) `http//doi:0.1016/j.diin`
4. Olsson, J., Boldt, M.: Computer forensic timeline visualization tool. Digital Forensic Research Workshop, digital investigation 6, S78–S87 (2009); `http:// doi:10.1016/j.diin`
5. Nance, K., Hay, B., Bishop, M.: Digital Forensics: Defining a Research Agenda. In: Proceedings of the 42nd Hawaii International Conference on System Sciences, pp. 1–6. IEEE computer society, Los Alamitos (2009)
6. Balon, N., Stovall, R., Scaria, T.: Computer Intrusion Forensics research paper, pp.1-24 (2002)
7. Aggarwal, S., Duan, Z., Kermes, L., de Medeiros, B.: E-Crime Investigative Technologies. In: Proceedings of the 41st Hawaii International Conference on System Science in IEEE, pp. 1–10 (2008)
8. Cohen, M.I.: PyFlag: PyFlag-An advanced network forensic framework. In: Communication of the ACM, digital investigation, 5th edn., pp. 112–120 (2008); Digital Forensic Research Workshop, `http:// doi: 10.1016/ j.diin.2008.05.016`
9. Crim, J.: Digital forensic tools and identifications, Information Security Analyst, Virginia Commonwealth University, available at Site, `http://liveview.sourceforge.net`
10. Bradford, P.G., Perdue, M.B.J.: Towards Proactive Computer-System Forensics. In: Proceedings of the International Conference on Information Technology Coding and Computing (ITCC 2004). IEEE computer Society, Los Alamitos (2004)
11. Carrier, B.: Defining Digital Forensics Examination and Analysis Tools Using Abstraction Layers. International Journal of Digital Evidence 1(4) (2003), `http://www.ijde.org/docs/02_winter_art2.pdf`

12. Carrier, B.: Open Source Digital Forensics Tools. The Legal Argument1, as a stake search Report pp.1-11 (2002), http://www.atstake.com/researchreports/acrobat/atstake_opensource_forensics.Pdf
13. Carrier, B.: Defining Digital Forensics Examination and Analysis Tools. In: Digital Research Workshop II (2002), http://www.dfrws.org/dfrws/papers/Papers/Briancarrier.pdf

# Evaluation of Normalized Routing Load for MANET

Sunil Taneja[1] and Ashwani Kush[2]

[1] Research Scholar
Department of Computer Science & Applications,
Kurukshetra University,
Kurukshetra, India
suniltaneja.iitd@gmail.com
[2] Associate Professor
Department of Computer Science,
University College, Kurukshetra University,
Kurukshetra, India
akush20@gmail.com

**Abstract.** Mobile Adhoc Network (MANET) is a collection of wireless mobile nodes forming a temporary network without any pre-existing network infrastructure. The stable routing over such a network is a very critical task as the wireless links are highly error prone and can go down frequently due to dynamic network topology. In this paper, evaluation of prominent on-demand routing protocols i.e. DSR and AODV has been done by varying the network size. An effort has been carried out to do the performance evaluation of these protocols using random way point model. The simulator used is NS 2.34. The performance of either protocol has been studied by using a self created network scenario and by analyzing normalized routing load with respect to pause time. Based on the performance evaluation, recommendations have been made about the significance of the protocol under various circumstances.

**Keywords:** AODV, DSR, Evaluation, Mobile Adhoc Network, Normalized Routing Load.

## 1 Introduction

MANET 3 is a collection of wireless mobile nodes forming a temporary network without any fixed infrastructure where all nodes are free to move about arbitrarily and where all the nodes configure themselves. Each node acts both as a router and as a host & even the topology of network may also change rapidly. In this paper, performance evaluation of two prominent on-demand routing protocols viz. DSR and AODV has been done by carrying out simulation over network simulator and using a self created network scenario.

DSR is an adhoc routing protocol 2, 5 which is source-initiated rather than hop-by-hop and is based on the theory of source-based routing rather than table-based. DSR do not need any existing network infrastructure and this allows the network to be completely self-organizing and self-configuring. This protocol is composed of two essential parts of route discovery and route maintenance. Every node maintains a

A. Mantri et al. (Eds.): HPAGC 2011, CCIS 169, pp. 442–448, 2011.

cache to store recently discovered paths. When a node desires to send a packet to some node, it first checks its entry in the cache. If it is there, then it uses that path to transmit the packet and also attach its source address on the packet. If it is not there in the cache or the entry in cache is expired (because of long time idle), the sender broadcasts a route request packet to all of its neighbors asking for a path to the destination. The sender will be waiting till the route is discovered. During waiting time, the sender can perform other tasks such as sending/forwarding other packets. As the route request packet arrives to any of the nodes, they check from their neighbor or from their caches whether the destination asked is known or unknown. If route information is known, they send back a route reply packet to the destination otherwise they broadcast the same route request packet. When the route is discovered, the required packets will be transmitted by the sender on the discovered route. Also an entry in the cache will be inserted for the future use. The node will also maintain the age information of the entry so as to know whether the cache is fresh or not. When a data packet is received by any intermediate node, it first checks whether the packet is meant for itself or not. If it is meant for itself (i.e. the intermediate node is the destination), the packet is received otherwise the same will be forwarded using the path attached on the data packet. Since in adhoc network, any link might fail anytime. Therefore, route maintenance process will constantly monitors and will also notify the nodes if there is any failure in the path. Consequently, the nodes will change the entries of their route cache.

AODV routing protocol 2, 4 is collectively based on DSDV and DSR. It does not maintain routes from every node to every other node in the network rather they are discovered as and when needed & are maintained only as long as they are required. When a node wants to send a data packet to a destination node, the entries in route table are checked to ensure whether there is a current route to that destination node or not. If it is there, the data packet is forwarded to the appropriate next hop toward the destination. If it is not there, the route discovery process is initiated. AODV initiates a route discovery process using Route Request (RREQ) and Route Reply (RREP). The source node will create a RREQ packet containing its IP address, its current sequence number, the destination's IP address, the destination's last sequence number and broadcast ID. The broadcast ID is incremented each time the source node initiates RREQ. Basically, the sequence numbers are used to determine the timeliness of each data packet and the broadcast ID & the IP address together form a unique identifier for RREQ so as to uniquely identify each request. The requests are sent using RREQ message and the information in connection with creation of a route is sent back in RREP message. The source node broadcasts the RREQ packet to its neighbours and then sets a timer to wait for a reply. To process the RREQ, the node sets up a reverse route entry for the source node in its route table. This helps to know how to forward a RREP to the source. Basically a lifetime is associated with the reverse route entry and if this entry is not used within this lifetime, the route information is deleted. If the RREQ is lost during transmission, the source node is allowed to broadcast again using route discovery mechanism. Maintenance of routes is done using Local route repair scheme.

## 2 Related Work

Several researchers have worked on the performance evaluation of routing protocols using different performance metrics. Some of these are Georgios Kioumourtzis 7, S.Shah, A.Khandre, M.Shirole and G. Bhole 11, J. Broch, D. A. Maltz, D. B. Johnson, Y. C. Hu, and J. Jetcheva 8, D. O. Jorg 6, K. U. Khan, R. U. Zaman, A. V. Reddy 9, A. Kumar B. R., Lokanatha C. Reddy and Prakash.S.Hiremath 1, N. Vetrivelan & A. V. Reddy 10. Most of the researchers have carried out the performance comparison of routing protocols on the basis of packet delivery ratio and average end to end delay. Very few have worked on the normalized routing load. Georgios Kioumourtzis 7 and S.Shah, A.Khandre, M.Shirole and G. Bhole 11 have worked on all the key performance metrics viz. packet delivery ratio, average end to end delay and normalized routing load. The results given by them are very helpful for researchers to do further innovations in the field of mobile adhoc networks.

## 3 Performance Evaluation Metrics

RFC 2501 describes a number of quantitative metrics that can be used for evaluating the performance of a routing protocol for mobile wireless ad-hoc networks. Some of these quantitative metrics 7 are defined as follow:

### 3.1 Packet Delivery Fraction

The packet delivery fraction is defined as the ratio of number of data packets received at the destinations over the number of data packets sent by the sources.

$$\text{Packet Delivery Fraction} = \frac{\text{Total Data Packets Received}}{\text{Total Data Packets Sent}} \text{ X } 100$$

### 3.2 Average End-to-End Delay

This is the average time involved in delivery of data packets from the source node to the destination node. To compute the average end-to-end delay, add every delay for each successful data packet delivery and divide that sum by the number of successfully received data packets.

$$\text{Average End to End Delay} = \frac{\sum (\text{TimeReceived} - \text{TimeSent})}{\text{TotalDataPacketsReceived}}$$

### 3.3 Packet Loss

Packet loss occurs when one or more packets being transmitted across the network fail to arrive at the destination. It is defined as the number of packets dropped by the routers during transmission.

Packet Loss = Total Data Packets Dropped

Packet Loss = Total Data Packets Sent – Total Data Packets Received

$$\text{Packet Loss (\%age)} = \frac{\text{Total Packets Dropped}}{\text{Total Data Packets Sent}} \text{ X } 100$$

### 3.4 Normalized Routing Load (NRL)

The normalized routing load is defined as the fraction of all routing control packets sent by all nodes over the number of received data packets at the destination nodes. In other words, it is the ratio between the total numbers of routing packets sent over the network to the total number of data packets received.

$$\text{Normalized Routing Load} = \frac{\text{Total Routing Packets Sent}}{\text{Total Data Packets Received}}$$

In this paper, performance evaluation of DSR and AODV has been analyzed using normalized routing load as performance metric. This metric has been studied with respect to 20 and 50 mobile nodes by varying pause time and using UDP/TCP agents.

## 4 Performance Evaluation of DSR/AODV

### 4.1 Simulation Environment

IEEE 802.11 is used as the MAC layer protocol. The simulation experiments are carried over network simulator 2 (version 2.34) installed in Fedora Linux 12. The results have been derived by writing a *tcl* script and generating corresponding *trace* and *nam* files. Both UDP and TCP agents have been used to analyze the traffic. The mobility model used is random waypoint model in a square area. The area configurations used are 750 meter x 750 meter for 20 nodes and 1000 meter x 1000 meter for 50 nodes. The packet size is 512 bytes. The packets start their journey from a random location to a random destination with a randomly chosen speed. The simulation run time is 500 seconds during analysis of 20 nodes and 650 seconds for 50 nodes. An extensive simulation model having scenario of 20 and 50 mobile nodes is used to study inter-layer interactions and their performance implications. Same scenario has been used for performance evaluation of both DSR and AODV protocols. It has been shown that even though DSR and AODV share a similar on-demand behavior, the differences in the protocol mechanics can lead to significant performance differentials.

### 4.2 NRL for 20 Nodes with UDP Agents

In fig. 1, normalized routing load has been evaluated for DSR and AODV protocols using pause time as varying parameter with six UDP agents. Pause time has been varied from 100s to 500s. The normalized routing load values range from 0.13 to 0.50.

In this scenario, the observation is that the DSR protocol presents low normalized routing load than AODV. Hence DSR protocol outperforms than AODV in terms of efficient routing.

### 4.3 NRL for 20 Nodes with TCP Agents

In fig. 2, normalized routing load has been evaluated using pause time as a varying parameter on 20 mobile nodes having six TCP agents. Pause time varies 100s to 500s. The normalized routing load values range from 0.01 to 0.04. The observation is that on an average, the DSR protocol presents low normalized routing load than AODV in all situations and therefore DSR protocol outperforms than AODV in terms of efficient routing.

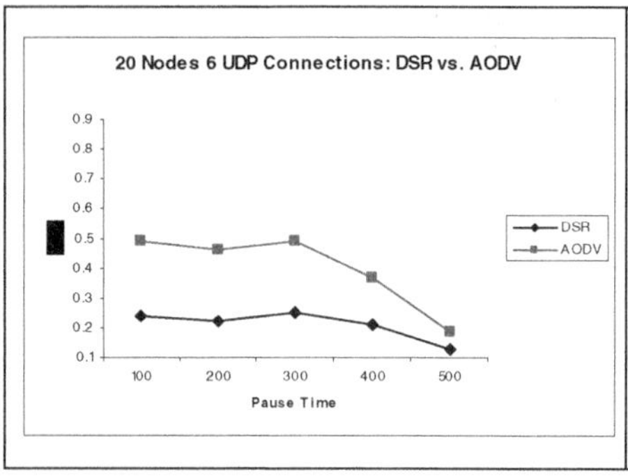


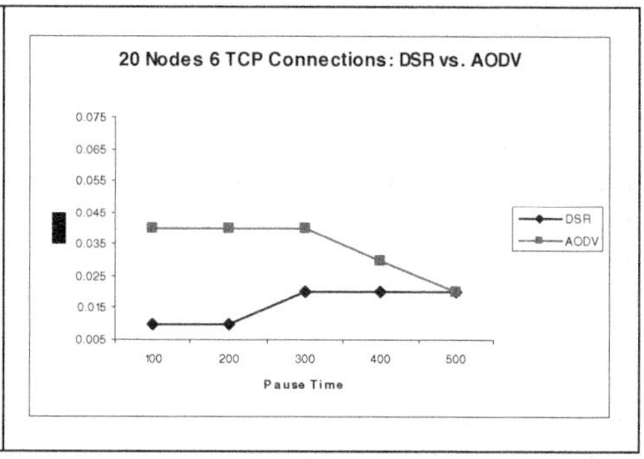


**Fig. 1.** NRL for 20 Nodes (UDP Connections) **Fig. 2.** NRL for 20 Nodes (TCP Connections)

### 4.4 NRL for 50 Nodes with UDP Agents

In fig. 3, normalized routing load has been evaluated for DSR and AODV protocols using pause time as varying parameter and ten UDP agents. Pause time has been varied from 100s to 650s. The normalized routing load values range from 0.7 to 0.25. In this scenario, again, the observation is than on an average, the DSR protocol presents low normalized routing load than AODV and therefore DSR protocol outperforms than AODV in terms of efficient routing.

### 4.5 NRL for 50 Nodes with TCP Agents

In fig. 4, normalized routing load has been evaluated for DSR and AODV protocols using pause time as a varying parameter with ten TCP agents. Pause time varies 100s to 650s. The normalized routing load values range from 0.15 to 0.75. The observation is that the results have changed now in comparison to the results obtained in sub sections 4.1, 4.2 and 4.3. It has been found that the AODV protocol presents low normalized routing load than DSR in all situations and therefore AODV protocol has started outperforming than DSR in terms of efficient routing.

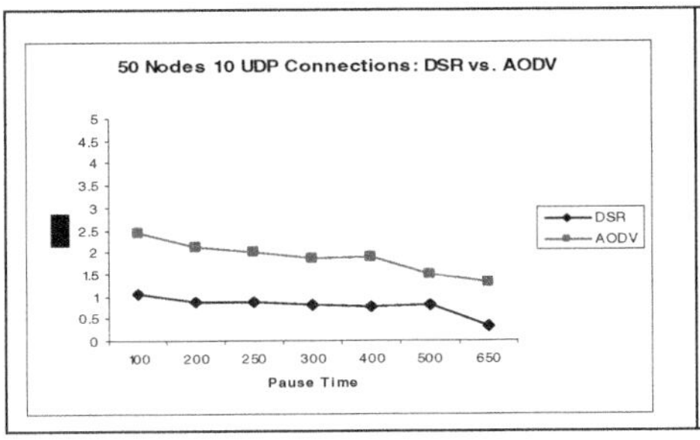


Fig. 3. NRL for 50 Nodes (UDP Connections)

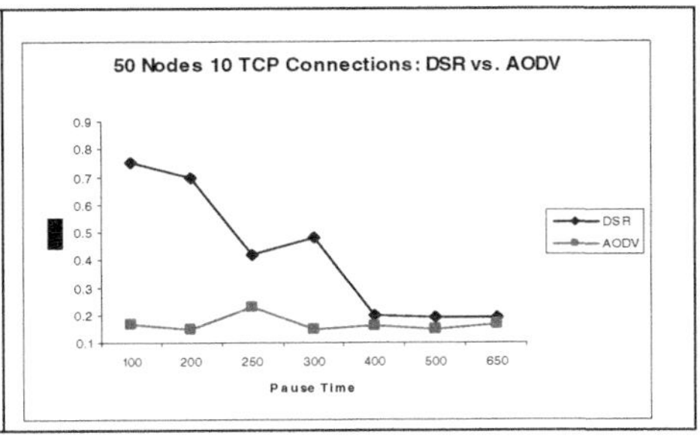


Fig. 4. NRL for 50 Nodes (TCP Connections)

### 4.6 Normalized Routing Load Comparison

It can be seen from the fig. 1, 2, 3 that DSR presents the lower normalized routing load than AODV proving that source routing proves to be an efficient routing mechanism in networks because it utilizes the wireless medium for data traffic in a better way than the other tested protocols. The key point to note from these figures is that these results are for sparse medium i.e. small number of nodes. The low normalized routing load for DSR protocol can be attributed to the caching strategy used by DSR. By virtue of aggressive caching, DSR is more likely to find a route in the cache, and hence resorts to route discovery less frequently than AODV. It can be further analyzed from fig. 4 that AODV presents low normalized routing load than DSR when TCP agents are used and number of nodes are high. If numbers of nodes are further increased to a large extent, AODV demonstrates low normalized routing load than DSR in all cases and hence it means that in denser medium, routing using AODV is more efficient than DSR. The DSR performance decreases in denser networks with higher mobility disclosing that source routing cannot efficiently adapt the network topology changes that are caused by the frequent movement of the nodes. Concluding, DSR demonstrates significantly lower routing load than AODV with the fact that it increases for DSR with growing number of nodes. When the number of nodes is less, the performance of DSR and AODV is similar regardless of mobility. With large numbers of nodes, AODV starts outperforming DSR. The major contribution to AODV's routing over-head is from route requests, while route replies constitute a large fraction of DSR's routing overhead. Furthermore, AODV has more route requests than DSR, and the converse is true for route replies.

## 5 Conclusion and Future Scope

In this paper, performance evaluation of DSR and AODV has been carried out on the basis of normalized routing load. The earlier work by researchers has been taken into consideration. The results have been analyzed using a new random way point self created network scenario. The general observation from the simulation is that the AODV protocol performs better in denser mediums and the DSR protocol performs better in sparse medium. In networks with a small number of nodes and low mobility, AODV does not suggest a good solution as a routing protocol. However, AODV has better performance in networks with higher mobility and a greater number of nodes. It is concluded that AODV is the proper protocol for any kind of application (voice, video, file transfer, etc.) in networks with high mobility that consist of large number of nodes. In most cases, DSR presented the lower normalized routing load, proving that source routing proves to be an efficient routing mechanism in networks with a small number of nodes because it utilizes the wireless medium for data traffic in a better way than the other tested protocols. However, DSR performance decreases in denser networks with higher mobility, disclosing that source routing cannot efficiently adapt the network topology changes that are caused by the frequent movement of the nodes. Since in real life scenario, the emphasis is on denser mediums, we can generalize that the use of AODV protocol is better choice over DSR protocol for efficient routing over mobile adhoc network. The performance of AODV under highly denser medium is still to be compared with TORA, STAR and ZRP. This aspect is still under our consideration. A sincere effort will also be made to evaluate the performance using throughput. Work also need to be done in the field of energy efficient and secure routing.

## References


1. Kumar, A., Lokanatha, B.R., Reddy, C., Hiremath, P.S.: Performance Comparison of Wireless Mobile Ad-Hoc Network Routing Protocols. IJCSNS International Journal of Computer Science and Network Security 8(6) (2008)
2. Kush, A., Taneja, S.: A Survey of Routing Protocols in Mobile Adhoc Networks. International Journal of Innovation, Management and Technology 1(3), 279–285 (2010)
3. Perkins, C.E.: Adhoc Networking. Addison-Wesley, Reading (2005)
4. Perkins, C., Royer, E.B., Das, S.: Adhoc On-Demand Distance Vector (AODV) Routing. IETF Internet Draft (2003)
5. Johnson, D.B., Maltz, D.A., Hu, Y.C.: The Dynamic Source Routing Protocol for Mobile Adhoc Networks (DSR). IETF Internet Draft (2003)
6. Jorg, D.O.: Performance Comparison of MANET Routing Protocols in Different Network Sizes. Computer Networks & Distributed Systems (2003)
7. Kioumourtzis, G.: Simulation and Evaluation of Routing Protocols for Mobile Adhoc Networks. Thesis, Master of Science in Systems Engineering and Master of Science in Computer Science, Naval Postgraduate School, Monterey, California (2005)
8. Broch, J., Maltz, D.A., Johnson, D.B., Hu, Y.C., Jetcheva, J.: A Performance Comparison of Multi-Hop Wireless Network Routing Protocols. In: Proceedings of the Fourth Annual ACM/IEEE International Conference on Mobile Computing and Networking (MobiCom 1998), Dallas, Texas, USA, pp. 25–30 (1998)
9. Khan, K.U., Zaman, R.U., Venugopal Reddy, A.: Performance Comparison of On-Demand and Table Driven AdHoc Routing Protocols using NCTUns. Tenth International Conference on Computer Modeling and Simulation (2008)
10. Vetrivelan, N., Reddy, A.V.: Performance Analysis of Three Routing Protocols for Varying MANET Size. In: Proceeding of the International Multi Conference of Engineers and Computer Scientists, IMECS 2008, Hong Kong, vol. II, pp. 19–21 (2008)
11. Shah, S., Khandre, A., Shirole, M., Bhole, G.: Performance Evaluation of Adhoc Routing Protocols using NS2 Simulation. In: Mobile and Pervasive Computing (CoMPC–2008), pp. 167–171 (2008)

# Reliabilty and Performance Based Resource Selection in Grid Environment

Rajesh Kumar Bawa[1] and Gaurav Sharma[2]

[1] Associate Professor,
Punjabi University Patiala
rajesh_k_bawa@yahoo.com
[2] Assistant Professor,
JMIT, Radaur
gaurav13@gmail.com

**Abstract.** Over the last few decades the development of Internet and Grid technology has rapidly increased in the number of resources to which a user, program, or community may have access. When a large number of resources are available which fulfills the minimum criteria imposed by the user then the burden goes onto the user to select the best resource. Also a wrong resource selection imposes the overhead and cost burden on the user. So, an efficient and reliable resource selection is required which overcomes the burden of the user and selects the best resource. In this paper we proposed a two phase approach for efficient resource selection. The aim of this paper is to identify the most available, reliable and fastest resources for running an application. In this context, we introduced an approach for resource selection that maximizes the quality, reliability, and efficiency of resources and minimizes the other overhead.

**Keywords:** Resource Selection, Ranking, Performance, Reliability, Grid.

## 1 Introduction

Grid computing aims to connect large numbers of geographically and organizationally distributed resources to increase computational power, resource utilization, and resource accessibility. In order to effectively utilize grids, users need to be connected to the best available resources at any given time. To support application execution in the context of the Grid, a Grid Resource Broker is desirable.Grid Resource Discovery and Selection is an essential and crucial part of grid resource brokering, which provides adequate available resources for other grid resources management [1],[5]. Grid technology allows resource sharing among several entities, but selecting the best resource to run a specific job remains one of its main problems. The challenges for the best resource selection involves analysis of several factors such as affordability of resources, success rate of resources, response time and etc. This paper presents an approach to select and compute resources by predicted resource reliability and rank them according to their performance. This work is not only focused on the reliability dimension, in order to minimize failures, but also concentrate on resource performance.

A. Mantri et al. (Eds.): HPAGC 2011, CCIS 169, pp. 449–454, 2011.

## 2 Proposed Approach

In our proposed work we present a resource selection technique that mainly aims on selecting a reliable and an efficient resource. It makes use of the various performance and trust factors of the resource to make the decision. Our selection mechanism works in two phases. The first phase is Resource Trust evaluation, in which we will calculate the trust of all potential resources based upon certain factors. The resources whose Trust value is below 0.5 are rejected and only the selected resources are eligible to participate in the next phase. The second phase is Performance evaluation phase in which we calculate the performance of the resource based upon certain parameters. It gives the list of most efficient resources. Finally we rank the resources according to their performance. Depending on this broker makes the resource selection decision.

### 2.1 Resource Trust Evaluation

Resource Trust Evaluation is the first phase of our resource selection approach. Basically this step focuses on the reliability and the quality of the resource. In order to determine the reliability of the resources we evaluate the certain trust factors which include Usage Score, Affordability, Success Rate, Bandwidth, User feedback,.The resources having the trust value less then 0.5 are rejected and not considered for the next phase of resource selection. Following are the listing of factors:

Usage score or Backlink score: A usage score is calculated on the basis of submission rate of the job. If the resource of an organization links to the resource of another organization means that it is casting a vote as an indication that the other resource is good. In other words, higher the rate of submission of jobs means higher the usage score.The equation is as follows:

$$US(A)=(1-d)+d\ (US(T_1)\ /C(T_1)\ +..+\ US(T_n)\ /C(T_n)\ ) \quad (1)$$

Where, $US(A)$ =UsageScore of resource A
$US(T_1)$ =UsageScore of resource T1 which uses the resource A
$C(T_1)$ = the number of times user in the organization T1 submits job to current resource in an organization
$d$ =damping factor which usually set to 0.85

Affordability ($A_A$): The ratio between the number of times a resource being available to the grid and the number of attempts made to access the resource. 2

It is defined as:

$$A_A= N_t\ /\ N_c \quad (2)$$

Where, $N_t$ = Number of times resource being available to grid resource A
$N_c$=Number of attempts made to access the resource

Success Rate($S_A$): The number of successful executions of jobs by a computational resource against the total number of jobs submitted to the resource that indirectly reveals the expertise of the resource provider.2

It is defined as:

$$S_A= J_S\ /\ J_T \quad (3)$$

Where, $J_S$ = Number of jobs successfully executed by the resource.
$J_T$ = Total Number of jobs executed by the resource.

User Feedback($F_A$): After each job, the requesting node returns the result of the job and the operation situation of resource to resource provider. 2.According to the feedback, the value is calculated using the following table:-

**Table 1.** Feedback Assignment

| Feedback levels | Code | Value |
|---|---|---|
| Excellent | F1 | 1.0 |
| Very High | F2 | 0.8 |
| High | F3 | 0.6 |
| Medium | F4 | 0.4 |
| Low | F5 | 0.2 |
| Very Low | F6 | 0.01 |

Based on all these factors we can calculate the Tvalue of the resource as follows:

$$\text{Tvalue}(A) = \frac{US(A) * A_A * S_A * F_A}{4} \tag{4}$$

Now, by using equation 1,2,3,4 and also the tables 1 and 2 we can calculate the trust value of the resources. As higher the Tvalue of the resource, higher the relibilty of the resource. The usual scheduler (i.e. which uses only current records) follows simple match making algorithm, it does not pay attention to the past service history of the resource and consider only the usage score as a result a resource which has higher usage score is repeatedly selected although it has the poor past records in terms of trust and user may end up with low quality or inconsistent resources leading to disappointing results. Now the resource having Tvalue lower then 0.5 are treated as unreliable and not considered for participating in the next phase. The unreliable or malicious resource is never chosen by the user in our approach. The output of this phase is a list of reliable resources.

## 2.2 Resource Performance Evaluation

In the second step (Resource performance evaluation phase), the challenge consists in selecting the eligible resources among the reliable ones, in order to obtain a high level of efficiency for the application. Therefore we need to choose the fastest and least-loaded. The performance criteria are based upon certain factors like machine performance, file transfer time, and average response time.

Average response time:-The response time is a time that is elapse from a request from a resource until it receives the resource.

$$\text{Average Response Time (RT)} = 1/N \sum_{i \in JOBS} \left(EndTime_j - SubmitTime_j\right) \tag{5}$$

where SubmitTime and EndTime are the times when job j is submitted to the queue and when it is completed. The response (or turnaround) time is probably the single most important measure for an individual submitting a job.

File Transfer time(TT):- It describes as the spent time for transferring required data for running the jth job on the resource ri.4

$$TT(tj, ri) = L(rj) + vol(tj)/B(rj) \quad (6)$$

Where, vol(tj) = presents the amount of the required data for tj .

L(rj) = latency of resource

B(rj) = bandwidth of resource

Machine Performance ($MP_f$):- $MP_f$ is used to match a job to the resource with higher system performance. $MP_f$ of machine m can be determined using CPU clock speed, no of cores.

$$MP_f = nflops_i * nc_i \quad (7)$$

Where nflopsi is the number of floating point operations that can performed on machine i and $nc_i$ is number of cores on the resource3.

Based on all these factors and by using equation 6,7 and 8 we can calculate the Pvalue of the resource as follows:

$$Pvalue(P) = \frac{RT * TT * MP_f}{3} \quad (8)$$

Once the Pvalue is estimated for each resource, we rank the resources according to their Pvalue, which indicates the efficiency or supremacy of the resource. Higher the efficiency, higher will be the rank of resource. We describe both of the phases as follows:

Step 1:- Calculate Tvalue of all the available resources according to values of usage score, affordability, success rate, user feedback .

Step 2:- Select k resources with the Tvalue greater than 0.5 and discard the other one.

Step3:- Calculate Pvalue of the selected resources according to the values of Average Response time, File Transfer Time, Machine performance.

Step4:-. Finally rank the resources according to the Pvalues.

$$Rank(r_i) < Rank(r_j) \text{ if and only if } Pvalue(r_i) < Pvalue(r_j)$$

Step 5. Select the resource with higher Rank and execute the application P.

## 3 Results and Discussion

We evaluate the impact of trust and performance factor in our approach by comparing the failure rate of jobs and performance of rank based scheduler against non-rank based scheduler. In our approach we screen the resources against two types of parameters. Only those resources whose Tvalue is more than 0.5 are eligible to participate further in resource selection phase, so the failure rate decreased dramatically in our proposed approach. From Figure 1, we conclude that the number of failed jobs in rank based scheduler is less than the non-rank based scheduler. As the failure rate of the scheduler decreases, the reliability increases.

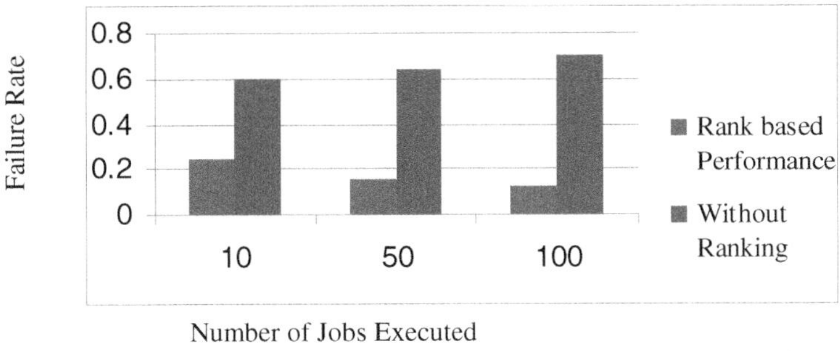


**Fig. 1.** Failure Rate of Rank based Execution versus Jobs Executed without Ranking

From Fig 2, we conclude that the performance of selected resources is much higher if we apply rank based approach on job execution while considering Average response Time, File Transfer Time and Machine Performance factors.

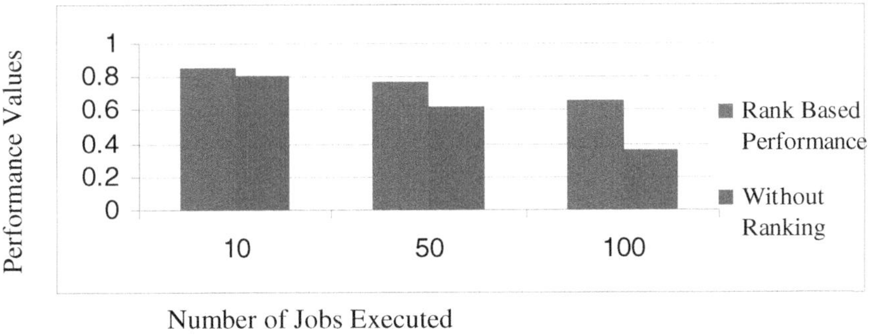


**Fig. 2.** Performance Values Analysis of the Rank Based versus non Rank based approach

## 4 Conclusions and Future Work

The effective and efficient exploitation of Grid computing facilities needs highly advanced and reliable resource management systems. This paper presents a mechanism, which allows to run the application jobs on the most reliable and most powerful resources, in respect to the application requirements. Our described approach gathers all the necessary characteristics about resources and chooses the most suitable and adequate resource for each of the jobs. This approach can get rid of the bad node with malicious behavior and provide the most efficient resources in the top order of the list. Future work include considering more dynamic factors of resource performance to make more effective selection process.

## References


1. Tung, C., Shih, Y.P.C., Li, K.C.: A High Performance Computational Resource Broker for Grid Computing. In: Proceedings of 19th International Conference on Advanced Information Networking and Applications
2. Chen, C., Li-ze, G., Xin-xin, N., Yi-xian, Y.: An Approach for Resource Selection and Allocation in Grid Based on Trust Management System. In: International Conference on Future Information Networks. IEEE, Los Alamitos (2009)
3. Addepallil, S.V., Andersen, P., Barnes, G.L.: Efficient Resource Matching in Heterogeneous Grid Using Resource Vector. International Journal of Computer Science and Information Technology 2(3) (June 2010)
4. Sedrakian, A.A., Badia, R.M., Sirvent, R., Perez, J.M.: Grid SuperScalar and Job Mapping on the Reliable Grid Resources. In: CoreGRID Workshop on Grid Programming Model Grid and P2P Systems Architecture Grid Systems, Tools and Environments (2007)
5. Kolano, P.Z.: Surfer: An Extensible Pull-Based Framework for Resource Selection and Ranking. In: Proceeding. of the 4th IEEE/ACM Intl. Symp. on Cluster Computing and the Grid, Chicago, IL, April 19-22 (2004)

# Elliptic Curve Cryptography: Current Status and Research Challenges

Sheetal Kalra[1] and Sandeep K. Sood[2]

[1] Jalandhar, India
[2] Gurdaspur, India
Department of Computer Science & Engineering
Guru Nanak Dev University, Regional Campus
sheetal.kalra@gmail.com
san1198@gmail.com

**Abstract.** Three types of standard public-key cryptographic systems that can be considered secure, efficient, and commercially practical are (i) Integer Factorization Systems (e.g. RSA) (ii) Discrete Logarithm Systems (e.g. DSA) (iii) Elliptic Curve Cryptosystems (ECC). The security of these systems is based on the relative complexity of the underlying mathematical problem. Of all these systems, for a given key size, ECC is the most secure public key cryptosystem. A survey of various protocols based on ECC has been done in the paper. The protocols have been classified according to their use in various cryptographic security mechanisms i.e. key agreement protocols, digital signature and encipherment. A comparison of ECC with conventional public key systems reveals that ECC is best suited for applications such as mobile computing, wireless sensor networks and other devices with constrained resources.

**Keywords:** Digital Signatures; Elliptic Curve Cryptography; Encryption/Decryption; Key Exchange; Smart Cards; Wireless Sensor Networks.

## 1 Introduction

With the rapid development of information technology, networks have become an important part of everyone's lives. The confidentiality and authenticity of information against illegal access, interruption and modification must be safeguarded in a network environment. Techniques like Encryption/Decryption, Digital Signature are some of the efficient solutions to safeguard the information of the user and confirm his/her authenticity. However, the overheads associated with communication and its security must be minimal. In 1985, Neil Koblitz and Victor Miller independently proposed the Elliptic Curve Cryptosystem (ECC). ECC is a public key cryptosystem based on the Elliptic Curve Discrete Logarithm Problem (ECDLP) for its security. ECC is being accepted as an alternative to conventional cryptosystems such as RSA and ElGamal as it provides the highest strength-per-bit of any other cryptosystem known today. The security of ECC depends on the difficulty of solving discrete logarithm problem over the points on an elliptic curve i.e. Elliptic Curve Discrete Logarithm Problem (ECDLP). The best known method to solve ECDLP (pollard's rho algorithm) is fully

A. Mantri et al. (Eds.): HPAGC 2011, CCIS 169, pp. 455–460, 2011.

exponential and substantially smaller key sizes as compared to other public key cryptosystems are used to obtain equivalent security. The length of cryptographic keys in ECC is comparatively much smaller than any other public key systems e.g. a 163-bit ECC cryptosystem provides as much security as a 1024-bit RSA cryptosystem would. This is an ideal feature especially for applications such as PDAs, smart cards, wireless sensor networks where resources such as memory, computing power etc are limited.

This paper is organized as follows: In Section 2, the survey of existing protocols based on ECC has been classified. In Section 3, the comparison of ECC with conventional public key systems has been done. In Section 4, we propose future directions and Section 5 concludes the paper.

## 2 Survey of Existing Protocols Based on ECC

### 2.1 Key Agreement Protocols Based on ECC

In 1976, Diffie and Hellman proposed the first public key exchange algorithm for key distribution based on discrete logarithm problem (DLP) which allows two users to exchange a key securely that can be used for subsequent encryption of messages. This algorithm itself is limited to the exchange of keys and forms the basis of many key exchange protocols. Also, one of the major roles of public key encryption has been to address the problem of key distribution. The majority of key agreement protocols based on public key cryptography use RSA. But, many efficient key agreement protocols based on Elliptic Curve Cryptosystems have been proposed recently. In 1998, the use of elliptic curve to implement public key cryptosystem was suggested as experimental studies done by Certicom showed that ECC provides greater efficiency in terms of key size and bandwidth saving than either integer factorization systems or discrete logarithm systems of relative security. At that time, it was said that "Elliptic curve cryptosystems appear promising and deserve further analysis". In 2003, Dr S A Vanstone, Founder Strategic Technology, Certicom, mentioned in one of his articles that ECC is the next generation public key system for wireless communication.

Depending on its structure, wireless networks can be categorized into two types: infrastructure and ad-hoc. The authentication and key agreement protocols play a vital role for secure communication in wireless networks. Authentication and key agreement protocols provide mutual authentication and secure means of deriving a shared secret key for communication between entities. In 2005, a key distribution protocol based on ECC for infrastructure topology wireless networks was proposed. Two different versions of this protocol provide security for basic service set (BSS) network and extended service set (ESS) network [1]. Wireless ad hoc networks allow peer to peer communication between mobile units without any central access point. The topology of such networks changes frequently because of rapid movement of the network nodes. A group key agreement protocol in ad hoc networks is used to establish a cryptographic key for secure communication between the group participants. A protocol based on ECDLP for secure group communication for wireless ad hoc networks was proposed in year 2006 [2]. Wireless sensor networks (WSN) are the latest advancement in the domain of wireless communication. Sensors are also low power devices with limited memory space and CPU power. In 2010, a protocol for key agreement and secure communication for heterogeneous sensor networks was proposed based on ECC [3].

Mobile devices such as smart cards, cellular phones, PDAs etc. are constrained devices and have limited computational power. ECC is most suitable for mobile devices with limited resources. In 2008, an efficient key agreement protocol for smart cards was proposed by Juang et al. [4]. This protocol is based on zero-knowledge proof and solves the lost-smart-card problem using ECC. In 2009[5], Yang and Chang proposed an ID-based remote mutual authentication with key agreement scheme for mobile devices on elliptic curve cryptosystem. This protocol provides mutual authentication and session key agreement between the user and the server for secure communication. In 2009, an efficient three-party authenticated key exchange protocol using ECC for mobile-commerce environments was also proposed by Yang and Chang [6].

## 2.2 Digital Signatures Schemes Based on ECC

A digital signature is an electronic signature that is used to verify the authenticity of the sender and integrity of the electronic message. At present, public key systems are almost exclusively used for key distribution protocols, integrity and authenticity purposes. Most digital signature schemes currently available are based on conventional public key systems, such as RSA and the ElGamal system. But, over the last few years, owing to its computational benefits, ECC is also being used to sign electronic documents. If a digital signature scheme provides message recovery then the original message does not have to be sent to the verifier. The verifier can then recover the message by using signer's publicly known information. Another digital signature scheme based on ECDLP with message recovery was proposed in 2004 [7]. Its efficiency was further enhanced by the use of self certified public key. But, Shao, [8] pointed out that this scheme was susceptible to insider forgery and nonrepudiation attack and improved the scheme by integrating point multiplication over elliptic curves in the original scheme.

A proxy digital signature is a scheme where the original signer can delegate his signing responsibility to another signer who acts as a proxy signer in the absence of the original signer. In 2003, a proxy signature scheme based on ECC was proposed where multiple signers can delegate their responsibility to a single proxy signer [9]. It was named as proxy multisignature digital scheme. In 2004, Hwang et al. [10], proposed a generalized proxy digital scheme where a group of original signers can delegate their signing authority to a designated proxy group. The group of proxy signers can cooperatively generate a proxy signature on behalf of the original group. Using ECDLP along with one-way hash functions, a computationally efficient proxy signature scheme was proposed in 2009 [11]. In Threshold Signature Scheme, the secret key is distributed among $n$ parties with the help of a trusted third party and at least $t$ parties are required to participate in the signing process where $t$ is a subset of $n$. Such a scheme is represented as $(t, n)$ threshold signature scheme. In 2004, a group oriented signature scheme was proposed based on ECC [12]. Blind Signature scheme is a system of digitally signing the message such that the contents of the message are hidden from the signing authority. In 2010, elliptic curves were used to implement threshold blind signature scheme [13].

### 2.3 Encryption Decryption Based on ECC

Encryption is the process of converting a plain text into cipher text using an encryption algorithm and decryption is the process of converting the cipher text back into plain text using a decryption algorithm. Encryption/ Decryption are techniques for providing data confidentiality for the information exchange over networks. Public key algorithms are mostly used for key exchange and authentication purposes due to their slow speed as compared to secret key algorithms. Some of the algorithms for encryption have been proposed using ECC also. The ElGamal algorithm which is based on DLP for its security can very well be implemented using ECC. In 1998, an efficient signcryption scheme based on elliptic curve was proposed. In this scheme ElGamal and DSS (Digital Signature Standard) were extended to elliptic curves to implement signatures and encryption in one single procedure. A similar scheme for implementing signatures and encryption in a single procedure was proposed by Chen et al. in 2004 [14]. In the proposed scheme the use of threshold signatures along with encryption/ decryption implemented over ECC considerably reduced the communication overheads.

## 3 Comparison of ECC with Conventional Public Key Systems

Benefits of ECC, compared to RSA (Rivest, Shamir, Adleman), DSA (Digital Signature Algorithm), DH (Diffie Hellman) is that it offers considerably greater security for same key size than other public key systems thereby reducing processing overhead. The benefits of this higher-strength per-bit include higher speeds, lower power consumption, bandwidth savings, less heat production and storage efficiencies. The smaller key size also makes possible much more compact hardware and software implementations for a given level of security, which makes faster cryptographic operations run on smaller chips with compact software. All these properties are of particular advantage in devices where bandwidths, processing capacity, power availability or storage are constrained. Such applications include smart cards, PDAs, cellular telephones etc. This leads to a significant improvement in efficiency in the operation of the ECC over both integer factorization and discrete logarithm systems. The difference becomes more pronounced with the increase in the key size as is indicated by the following table.

**Table 1.** Comparison of ECC and RSA based on key size for same security levels

| ECC key size (bits) | RSA key size (bits) | Key size ratio |
|---|---|---|
| 163 | 1024 | 1:16 |
| 256 | 3072 | 1:12 |
| 384 | 7680 | 1:20 |
| 512 | 15360 | 1:30 |

Supplied by NIST (National Institute of Standards and Technology)

A recommended key size of RSA for most applications is 2048 bits. For the equivalent security ECC only needs a key of 224 bits.

## 4 Future Directions

Majority of the handheld wireless devices have limited resources and operate in constrained environments. ECC is the ideal public key cryptosystem for such devices. The security mechanism in which ECC is being predominantly used is key exchange protocols and digital signatures. Use of ECC considerably reduces the cost of transmission as compared to RSA. There is a tremendous opportunity available for researchers to implement protocols for such wireless devices using ECC instead of RSA. The prime elliptic curves GF(p) are best suited for software implementations because the extended bit-fiddling operations needed by binary curves are not required whereas binary curves $GF(2^m)$ are best suited for hardware implementations as remarkably few logic gates are required to create a powerful and fast cryptosystem. Domain parameters play a vital role in secure communications using ECC. They must be selected, such that they are not susceptible to the known attacks (e.g. pollard rho attack). The selection of secure prime elliptic curves, binary elliptic curves is the focus of the research. According to NIST the elliptic curves suitable for cryptographic operations are: GF(p):P-192,P-224,P-256,P-384,P-521. $GF(2^m)$: K-163, K-233, K-283, K-409, K-571. Researchers are continuously working in this direction so that more secure curves for cryptographic purpose can be identified. Point multiplication in ECC which is done using "double and add method", is an expensive operation even today. Hence research is being done to evolve cost efficient and faster point multiplication algorithms. If such algorithms evolve in near future the entire cryptographic mechanisms including encryption/decryption of the data might be done using ECC instead of other conventional cryptosystems.

## 5 Conclusion

In the recent years, Elliptic Curve Cryptography has gained widespread exposure and acceptance. Due to its highest security-per-bit and low computation cost features as compared to other public key systems, it has already been included in many security standards such as IEEE P1363, ANSI X9.62 and ANSI X9.63. From just being a cryptosystem of theoretical importance alone, it has now emerged out as a cutting edge technology. Owing to its many computational benefits it is very well suited for wireless technology. One can go so far as to say that the RSA technology which is being used in many applications today may be replaced by ECC altogether.

## References

1. Azim, M.A., Jamalipour, A.: An Efficient Elliptic Curve Cryptography based Authenticated Key Agreement Protocol for Wireless LAN Security. In: IEEE International Conference on High Performance Switching and Routing (2005)
2. Wang, Y., Ramamurthy, B., Zou, X.: The Performance of Elliptic Curve Based Group Diffie-Hellman Protocols for Secure Group Communication over Ad Hoc Networks. In: IEEE International Conference on Communication (2006)
3. Rahman, M.M., El-Khatib, K.: Private key agreement and secure communication for heterogeneous sensor networks. J. Parallel and Distributed Computing 70, 858–870 (2010)

4. Juang, W.S., Chen, S.T., Liaw, H.T.: Robust and Efficient Password –Authenticated Key Agreement Using Smart Cards. IEEE Transactions on Industrial Electronics 55(6) (2008)
5. Yang, J.H., Chang, C.C.: An ID-based remote mutual authentication with key agreement scheme for mobile devices on elliptic curve cryptosystems. J. Computer & Security 28, 138–143 (2009)
6. Yang, J.H., Chang, C.C.: An efficient three-party authenticated key exchange protocol using elliptic curve cryptography for mobile-commerce environments. J. Systems and Software 82, 1497–1502 (2009)
7. Tzeng, S.F., Hwang, M.S.: Digital Signatures with message recovery and its variants based on elliptic curve discrete logarithm problem. J. Computer Standards & Interface 26, 61–71 (2004)
8. Zuhua, S.: Improvement of digital signatures with message recovery and its variants based on elliptic curve discrete logarithm problem. J. Computer Standards & Interface 27, 61–69 (2004)
9. Chen, T.S., Chung, Y.F., Huang, G.S.: Efficient proxy multisignature scheme based on the elliptic curve cryptosystem. Computer & Society 22(6), 527–534 (2003)
10. Hwang, M.S., Tzeng, S.F., Tsai, C.S.: Generalization of proxy signature based on elliptic curves. J. Computer Standards & Interface 26, 73–84 (2004)
11. Sun, X., Xia, M.: An improved Proxy Signature Scheme Based on Elliptic Curve Cryptography. In: International Conference on Computer and Communications Security. IEEE Computer Society, Los Alamitos (2009)
12. Chen, T.S.: A specifiable verifier group-oriented threshold signature scheme based on the elliptic curve cryptosystem. J. Computer Standards & Interface 27, 33–38 (2004)
13. Jianfen, P., Yajian, Z., Cong, W., Yixian, Y.: An application of Modified Optimal –Type Elliptic Curve Blind Signature Scheme to Threshold Signature. In: International Conference on Networking and Digital Society. IEEE, Los Alamitos (2010)
14. Chen, T.S., Huang, K.H., Chung, Y.F.: A practical authenticated encryption scheme based on the elliptic curve cryptosystems. Computer Standards & Interface 26, 461–469 (2004)

# SBFDR: Sector Based Fault Detection and Recovery in Wireless Sensor Networks

Indrajit Banerjee[1], Prasenjit Chanak[2], and Hafizur Rahaman[3]

[1] Department of Information Technology
[2] Purabi Das School of Information Technology
[3] Bengal Engineering and Science University, Shibpur, Howrah, India
ibanerjee@it.becs.ac.in
prasenjit.chanak@gmail.com
rahaman_h@it.becs.ac.in

**Abstract.** Sensor Networks are usually large collection of sensing nodes collecting data from monitoring environment and transmit to base station by multi hop wireless communication. The occurrences of faults in wireless sensor network are very high due to wireless communication and random deployment policy. Energy conservation in sensor network is another challenge to improve applicability of wireless sensor networks. In this paper, we propose sector based fault detection and recovery technique (SBFDR) which is also energy competent. In SBFDR, sensor nodes are arranged into some clusters. Cluster head and sensor nodes are jointly detect the fault of sensor nodes. Sensor node's faults are recovered by the cluster head's fault recovery policy. The simulation result depicts that the SBFDR technique more capable to detect sensor nodes faults and recover the faulty nodes in an energy efficient manner. Energy loss and fault recovery time is very low in SBFDR technique compare to other popular fault detection and recovery techniques.

**Keywords:** wireless sensor networks (WSN), base station (BS) cyclic cellular automata (CCA), fault detection, fault recovery.

## 1 Introduction

In Wireless sensor Network, a large amount of low cost sensor device is deployed in monitoring environment [1]. In a sensor networks sensor nodes may damage in monitoring area by different factors, therefore, fault detection and recovery is very important for a sensor network for efficient application. The proposed SBFDR technique detects maximum faults which are occurred due to the physical damage of sensor nodes and power fault. The SBFDR also recover these faults in energy efficient manner. Energy constrain is a major drawback for increasing applicability of sensor network. Therefore, different energy saving techniques are designed to improve battery utility of sensor nodes such as, LEACH [2], cellular automata based method for energy conservation solution in wireless sensor network [3], etc. The cluster based energy saving techniques is most popular techniques of them. In SBFDR technique clusters are generated in an energy efficient way with the help of modified cyclic cellular automata rule (mCCA) [4].

A. Mantri et al. (Eds.): HPAGC 2011, CCIS 169, pp. 461–469, 2011.

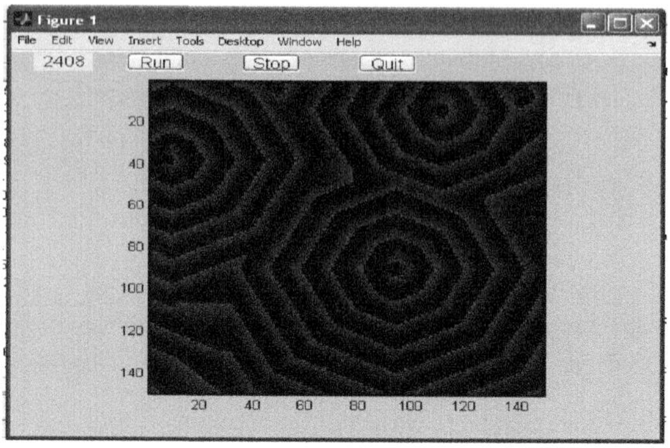


**Fig. 1.** Atomic structure of node pattern generated by mCCA

The rest of the paper is organized as follows: The section II describes modified-CCA which is used for cluster formation. The Section III briefly describes mathematical theories. The section IV represents the SBFDR algorithm. The section V represents experimental result. We have concluded our paper in section VI.

## 2 Modified Cyclic Cellular Automata

In our cyclic cellular automata based SBFDR scheme every cell changes its state according to the nine neighbour cells state condition [4]. The state change rules of the cells are define as, (a) If cells present $\delta_t(p) = n$ where $n > 0$ *and* $n < I - 1$, then next state of the cells is $\delta_{t+1}(p) = n + 1$. (b) If cells present $\delta_t(p) = n$ where $n = I - 1$, then next state of the cells is $\delta_{t+1}(p) = 0$. (c) If cells present $\delta_t(p) = 0$ state they check their neighbours cells state and if $\theta$ number (threshold value) of nodes are present in nonzero state then $\delta_{t+1}(p) = 1$ otherwise they are not changing their state i.e. $\delta_{t+1}(p) = 0$. Where $\delta_t(p)$ is the state of the cells at time t and $\delta_{t+1}(p)$ is the state of cells at t+1 time. Number of state is {0, 1...I-1}. In this scheme every node is changing their state in a time interval. The nodes pattern are controlled from the primordial soup where with respect to time every node is changing their state i.e. spirals are propagated (Fig. 1). The mCCA automata generate cell pattern like atomic structure.

## 3 Mathematical Model of Energy Consumption

Sensor nodes are losing their energy for data transmission and data receiving process. We use standard communication model as we have shown in our previous work [5]. In this model, $\rho$ size packet transmission losses energy from k distant is $P_E$. The $P_E$ is defining by following formula:

$$P_E = \rho(\beta_1 + \beta_2 k^n) \quad (1)$$

In which $\beta_1$is the energy per bit consumed by the transmitter electronics and $\beta_1$is the transmission independent variable. The $\beta_2$represent dissipated energy in the transmit op-amp. The n is a real value which is always 2$\leq n \leq$4 for the free-space and short – to-medium range communication. If sensor nodes receiving packet size $\rho$ then receiving energy loss $P_r$is represented by following formula:

$$P_r = (\rho\beta_r) \tag{2}$$

Where, $\beta_r$ denotes the receiver electronics circuit energy loss per bit. If any node forward it's data packet $\rho$ into k distance nodes at time t then total energy consumed ($P_t$) by that node, $P_t$is define in the following formula:

$$P_t = \rho((\beta_1 + \beta_r) + \beta_2 k^n \tag{3}$$

In SBFDR technique cluster are divided into some sectors. The load of the sector is calculated by the *theorem 1*. The sector head's load depends on number of message passing through sector head in a particular time. Therefore, with increased message passing in sector head, the life span of sector head is decreased. Hence, sector life span depends on the size of the sector. If size if sector is increasing then life time of sector head decreasing because in this case number of message passing is increase.

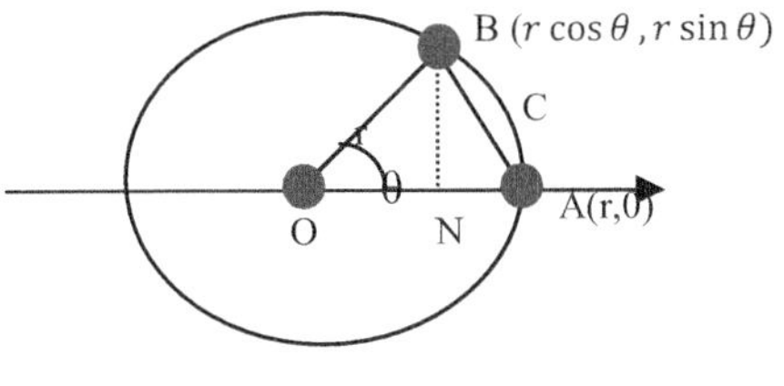


**Fig. 2.** Sector load calculation

**Theorem 1:** Load of a sector is

$$L_{sec} = \frac{r^2(\theta-\sin\theta)+\sqrt{s(s-r)^2(s-\frac{\sin\theta}{\cos\theta-1}(r\cos\theta-r)}}{2\emptyset_{area}} \qquad \text{Where } s = \frac{2r+\frac{\sin\theta}{\cos\theta-1}(r\cos\theta-r)}{2}$$

The $\theta$ is generating angle in cluster head and r is cluster radius and $\emptyset_{area}$is sensing area of a sensor node.

**Proof:** let cluster is represented by equation $x^2 + y^2 = r^2$ (Fig. 2). The cluster is divided in some sector and sector under two radiuses, one OA, x axis and OB, x make $\theta$angle with OB. Therefore, point $B = (r\ \cos\theta, r\sin\theta)$

$\therefore$ AB line equation $y = \frac{\sin\theta}{\cos\theta-1}(x-r)$

ABCA area = NBCA area – NBA area (shown in Fig. 2)

$= \int_{r\cos\theta}^{r}(y_1 - y_2)dx$, where $y_1$and $y_2$ are the value of y in $x^2 + y^2 = r^2$ and $y\frac{\sin\theta}{\cos\theta-1}(x-1)$ $= \int_{r\cos\theta}^{r}\left[\sqrt{r^2-x^2} - \frac{\sin\theta}{\cos\theta-1}(x-r)\right]dx$

$$= \left[\frac{1}{2}x\sqrt{r^2-x^2}+\frac{r^2}{2}\sin^{-1}\frac{x}{r}+\frac{\sin\theta}{1-\cos\theta}\left(\frac{x^2}{2}-rx\right)\right]_{r\cos\theta}^{r} = \left[0+\frac{r^2}{2}\sin^{-1}\frac{x}{r}+\frac{\sin\theta}{1-\cos\theta}\left(\frac{x^2}{2}-rx\right)\right] -$$

$$[\frac{1}{2}r\cos\theta\,.\,r\sin\theta+\frac{r^2}{2}\sin^{-1}\frac{x}{r}+\frac{\sin\theta}{1-\cos\theta}\left(\frac{r^2\cos\theta^2}{2}-r^2\cos\theta\right)$$

$$=\frac{r^2\pi}{4}-\frac{r^2\sin\theta}{2(1-\cos\theta)}-\frac{r^2}{2}\sin\theta\cos\theta-\frac{r^2}{2}\sin^{-1}\sin\left(\frac{\pi}{2}-\theta\right)-\frac{r^2(\sin\theta\cos\theta)}{2(1-\cos\theta)}(\cos\theta-2)$$

$$=\frac{r^2\pi}{4}-\frac{r^2}{2}\left(\frac{\pi}{2}-\theta\right)-\frac{r^2}{2}\sin\theta\cos\theta\left(\frac{\cos\theta-2}{1-\cos\theta}+1\right)-\frac{r^2\sin\theta}{2(1-\cos\theta)}$$

$$=\frac{r^2\theta}{2}-\frac{r^2}{2}\sin\theta\cos\theta\frac{-1}{1-\cos\theta}-\frac{r^2\sin\theta}{2(1-\cos\theta)} = \frac{r^2\theta}{2}+\frac{r^2\sin\theta}{2(1-\cos\theta)}(\cos\theta-1) = \frac{r^2}{2}\theta-\frac{r^2}{2}\sin\theta = \frac{r^2(\theta-\sin\theta)}{2}$$

In the above equation we add area of OAB. If every sensor node cover $\emptyset$ area then,

Load of a sector= $\frac{r^2(\theta-\sin\theta)+\sqrt{s(s-r)^2(s-\frac{\sin\theta}{\cos\theta-1}(x-r)}}{2\emptyset_{area}}$

# 4 Proposed SBFDR Algorithm

This section proposes a scheme, referred to as the SBFDR, for effective fault detection and recovery among the sensor nodes. The SBFDR targets energy efficient fault detection and recovery with minimum time elapse. A SBFDR node handles five types of messages for fault detection and recovery.

***LiMsg:*** Life messages (*LiMsg*) are used to send life information of a sensor node. The sensor node are sending their life information periodically to their supervision node like cluster head, base station.

***UpgMsg:*** The supervision node if not getting the life message from a particular node within a stipulated period of time, then the supervision node sent upgrade message *(UpgMsg)* for requesting update corresponding node's life information.

***ADV*:** When a node is detected as a faulty node then supervision node send advertisement (*ADV*) message to other nearest active nodes for distribution of responsibilities.

***POS*:** (position information) this message is used to replay *ADV* message.

***ASS*:** The assignment (*ASS*) message is used to allocate newly elected node responsibility.

In SBFDR technique, clusters are generated by modified cyclic cellular automata (mCCA) rule [4]. This section divided into two part A) fault detection B) fault recovery.

a. *Fault Detection*

In this section, we briefly describe sensor node fault detection (i.e. transmitter circuit fault, battery failure, microcontroller fault) technique in SBFDR. In SBFDR scheme every cluster head send *LiMsg* periodically. The base station checks the cluster head's working condition by reviewing the *LiMsg* message. If any cluster head is not sending *LiMsg* then the base station decide that the node is faulty because of either power fault or microcontroller circuit damage or transmission faults. Therefore, the base station is requesting to cluster head by sending *UpgMsg* to the faulty node to send the life information and wait for interval of time ($T_\alpha$). If the base station is not received the *LiMsg* because of transmission fault, then cluster head after receiving *UpgMsg* retransmits *LiMsg* to base station again. However, if microcontroller circuit of the cluster head is damaged then it is permanently destroyed. If any *LiMsg* is not receive after transmission of *UpgMsg* then base station declare this cluster head as faulty and start cluster head fault recovery process. On the other hand, power fault of the cluster head are checked by cluster head itself. If power reading ($P_w$) less than threshold value ($T_w$) then cluster head inform that to base station with *LiMsg*. On the same way cluster head detect sector head faults. Sector head detects sub-sector head's fault. Sub-sector head detect member nodes failure. Member nodes, sub-sector head nodes, sector head nodes and cluster head nodes fault are detected in top down manner.

**Algorithm 1. (Fault Detection)**

**WHILE** network is alive **DO**
Cluster and sector size are calculated by theorem 1
Clusters are generated by mCCA rule
**IF** *LiMSG* message is received by BS and CH **THEN**
    1. Base station set all cluster head condition is Good
    2. Cluster Head set all member nodes of the cluster is Good
**ELSE**
    1. Base station Send *UpgMsg* to all clusters head
       Set waiting time $T_{ch}$
    2. Cluster head send *UpgMsg* to all member nodes
       Set waiting time $T_{mn}$
        **IF** $T_{ch}$ and $T_{mn}$= $T_\alpha$
        1. Based station not receive any *LiMsg* from cluster head, set cluster head= FAULTY
           Call fault recovery process
        2. Cluster head not receive any *LiMsg* from cluster member node, set cluster member = FAULTY
    **ELSE**
        Transmission FAULT occur
        1. Base station set cluster head = GOOD
        2. Cluster head Set Cluster Member =GOOD
    **END IF**
**END IF**
**FOR** each node **DO**
    **IF** $P_w$ <=$T_w$ **THEN**
            Send power condition to supervise node.
        Supervise node CALL fault recovery process.
    **ELSE**
        Continue their work
    **END IF**
**END FOR**
**END WHILE**

A. *Fault recovery*

In this section, we have discussed the fault recovery mechanism of SBFDR. The base station when detects the cluster head fault (microcontroller, power or transmitter circuit fault) then base station send *ADV* message to faulty cluster head's neighbour active nodes. The neighbour active nodes when receive *ADV* message then they sent back a response by the *POS* message. The *POS* message contains node power level information. Therefore, after receiving all *POS* message from applicant nodes, base station decides which node is most efficient for next cluster head according to applicant response time and energy level. The base station is then calculating the average response time $AR_{time}$ shown in equation number 4 and on the basis of average power level $AP_{response-nodes}$ as shown in equation number 5.

$$AR_{time} = \frac{\sum R_T}{N_{response}} \tag{4}$$

The $AR_{time}$ is average response time, $R_T$ is response time of replying node and $N_{response}$ is total number of node reply with *POS* message.

$$AP_{response-nodes} = \frac{\sum P_W}{TP_{response-nodes}} \tag{5}$$

The $AP_{\text{response-nodes}}$ is average power level of replying nodes, $P_W$ is power level of each node and $TP_{\text{response-nodes}}$ is total power level of replying nodes. The next cluster head is selected if it has $AR_{time} \geq R_T$ and $AP_{response-nodes} \geq P_W$.

The base station declares next cluster head by *ASS* message. In the same way cluster head recovers sector head fault, sector head recovers sub-sector head fault and member fault is recovered by the sub-cluster head.

**Algorithm 2. (Fault Recovery)**

```
FOR each faulty node DO
    IF nodes= FAULTY THEN
        Supervise node send ADV message nearest active nodes
        Active nodes send POS Massage to supervise node.
        Supervise node check response time (R_T) and power condition of the nodes
        IF R_T <= AR_time THEN
            IF P_w> =AP_response-nodes THEN
                Send ASS message for next round work.
            ELSE
                Discard POS message
            END IF
        END IF
    END IF
END FOR
```

## 5 Performance Evolution

### A. Simulation Model

The SBFDR technique is divided into two parts one is cluster formation by the mCCA rule another is fault detection and recovery. In order to evaluate the performance of SBFDR four traditional metrics of WSN have considered as follows:

*Active nodes*: The active nodes are sensing data from monitoring environment and transmit to base station. The nodes are responsible for fault detection and fault recovery process in SBFDR scheme.

*Stand- by nodes*: These types of nodes are in sleeping state, therefore they are not participating in any data collection and fault detection process.

*Average time delay*: the average time delay is time recurred for fault detection and recovery by the supervised node after the fault has been occurs.

*Time elapsed*: the time elapsed is the fault detection time by any supervised node.

**Table 1.** Simulation parameters

| | |
|---|---|
| Number of node | 10000 to 22500 |
| Data Packet Size $\rho$ | 800bit |
| Initial Energy | 0.5J |
| energy consumed in the transmitter circuit $\beta_1$ | 50 nJ/bit |
| energy consumed in the amplifier circuit $\beta_2$ | 10pJ/bit |

The total number of nodes in terrain is varied from 10000 to 22500. The transmission rang of the sensor nodes are varied from 50m to 300m. The initial energy of the sensor nodes is 0.5J [5]. The different parameter values are given in Table I which are used in simulation.

*B. Simulation Results*

The two popular fault detection techniques are cash fault detection (CFI) [6] and cluster based approach [7]. The CFI technique only detect sensor node fault with loss large time and energy. This technique does not give any solution of fault recovery of sensor nodes. On the other hand cluster based approach provides sensor nodes fault detection and recovery technique but this technique take also more time and more energy to detect sensor fault.

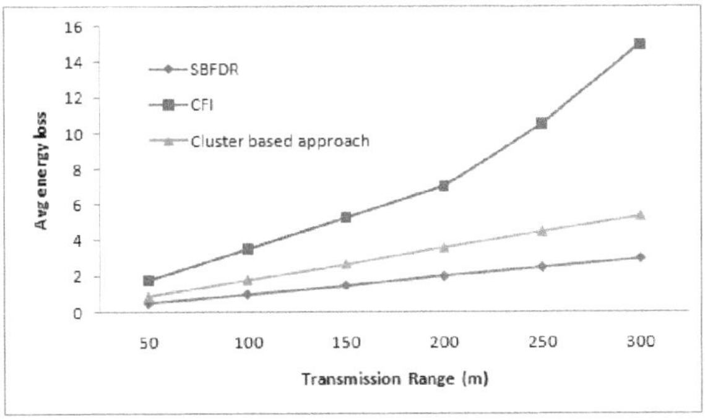


**Fig. 3.** Average energy loss for fault detection

In Fig.3 shows the comparison between our proposed methods (SBFDR), CIF [6] and cluster based approach [7] for the average energy loss of any sensor nodes with respect to transmission range. The graphs show that the energy loss for fault detection using SBFDR is less than comparison to other two popular fault detection techniques CFI and cluster base approach. In SBFDR technique fault are detected in top down approach.

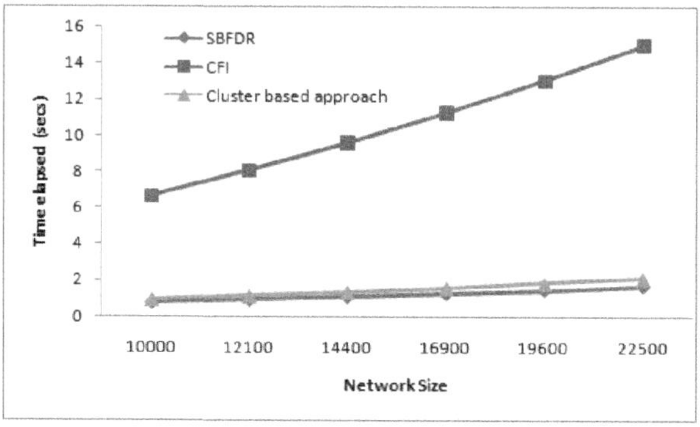


**Fig. 4.** Time Elapsed for fault Detection

The fault detection time comparison is show in Fig. 4. The graph shows that proposed SBFDR technique requires less time compression to CFI and cluster based approach. The CFI technique detects any fault locally. CFI technique takes more time to detect any sensor fault, because it take any decision on the basis of data that it collects from neighbouring node.

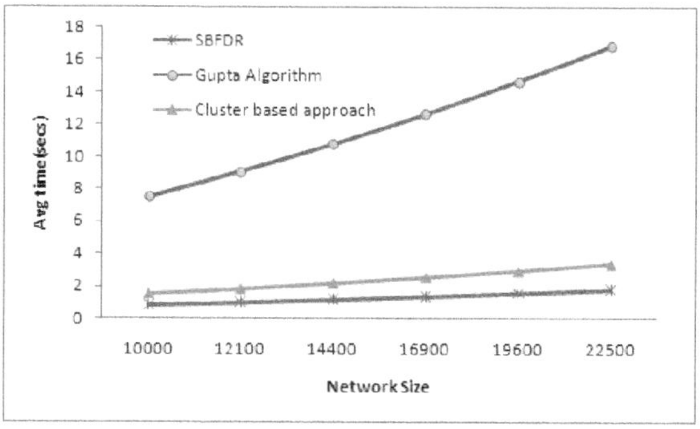


**Fig. 5.** Average time for fault detection and recovery

The Fig. 5 shows the comparison of average time taken for any faulty node detection and recovery by our proposed SBFDR and other existing algorithm. The SBFDR technique takes less time to detect any fault and recovery. The SBFDR technique recovers sensor microcontroller, transmitter, and power failure. On the other hand Gupta algorithm [8] and cluster based approach takes more time to detect any fault and recovery. The Gupta algorithm takes more time to recover any fault because in this technique when a gateway fails, the cluster is dissolved and all its nodes are re-allocated to other healthy gateways.

## 6 Conclusions

In this work, we introduce sector based fault detection and recovery (SBFDR) scheme for efficient fault detection and recovery in an energy constrained sensor network. The proposed SBFDR technique holds the promise of achieving high performance at a low cost in terms of energy, computation and communication. This approach would challenge the speed and adaptive-ness of the state-of-the-art mobile networks employed in WSN. The mobile network model works with SBFDR, may show the better performance in real application.

## References

[1] Banerjee, I., Rahaman, H., Sikdar, B.K.: UDDN: Unidirectional Data Dissemination via Negotiation. In: IEEE International Conference on Information Networking, Pusan, Korea, January 23-25 (2008)

[2] Heinzelman, W.R., Chandrakasan, A., Balakrishnan, H.: Energy-Efficient Communication Protocol for Wireless Micro sensor Networks. In: IEEE International Conference on System Sciences (2000)

[3] Adabi, S., Zadeh, A.K., Dana, A., Adabi, S.: Cellular Automata Base Method for Energy Conservation Solution in Wireless Sensor Network. IEEE, Los Alamitos (2008); 978-1-4244-2108-4/08

[4] Banerjee, I., Chanak, P., Sikdar, B.K., Rahaman, H.: EERIH: Energy Efficient Routing via Information Highway in Sensor Network. In: IEEE International Conference on Emerging Trends in Electrical and Computer Technology, kanyakumari, India, March 23 -24 (2011)

[5] Banerjee, I., Chanak, P., Sikdar, B.K., Rahaman, H.: EER: Energy Efficient Routing In Wireless Sensor Network. In: IEEE International Technology Symposium, IIT kharagpur, India, January 14 – 16 (2011)

[6] Chessa, S., Santi, P.: Crash Fault Identification in Wireless Sensor Network. Computer Comm. 25(14), 1273–1282 (2002)

[7] Venkataraman, G., Emmanuel, S., Thambipillai, S.: A Cluster-based approach to Fault Detection and Recovery in Wireless Sensor Networks. In: IEEE ISWCS (2007)

[8] Gupta, G., Younis, M.: Fault-tolerant clustering of wireless sensor networks. Wireless Communications and Networking (2003)

# Study and Analysis of Incremental Apriori Algorithm

Neeraj Kumar Sharma and N.K. Nagwani

National Institute of Technology (NIT) Raipur, CG, India
neerajksharma2007@gmail.com,
nknagwani@gmail.com

**Abstract.** Study of this paper is based on finding the threshold value of database change up to which incremental Apriori algorithm performs better. A new incremental Apriori algorithm is also proposed which performs better than the existing algorithm in terms of computation time. The performance of frequent sets generation algorithms for dynamic databases is major problem, since numbers of runs are required to accommodate the database changes. It determines the value of change percentage of original database that decides whether the user can go for re-run the actual algorithm or use the previously computed result and generate the frequent sets in incremental fashion. The purpose of this paper is two folds. First is to avoid the scans of the older database, its corresponding support count effort for newly added records by using intermediate data and results. And second is to solve the efficient updating problem of association rules after a non-trivial number of new records have been added to a database.

**Keywords:** Apriori, Incremental Apriori, Database Change, Frequent Itemsets.

## 1 Introduction

Association rule mining (ARM) is one of the most important and well researched techniques of data mining. The Apriori algorithm provides an Innovative way to find association rules on large scale.

The problem with dynamic databases is when databases grow over time, as a result there is need to undertake mining again for verifying frequent item sets obtained earlier and also for discovering new frequent itemsets. I.e. in a large dynamic database, when some new records are added to database, it may invalidate the existing frequent pattern thus maintenance of frequent itemsets for dynamic databases is an important problem due to scanning the whole database again. However, it has been realized that applying the existing algorithm frequently for updated databases may be too costly. Moreover it is very time consuming process due to its very high time complexity. So the existing algorithm is not efficient.

Incremental data mining means applying Data mining Algorithms on incremental database. Here we create one such algorithm that will use candidate and large itemset and their corresponding support count for older database during earlier mining and incremental database. And these corresponding itemset are combined together for finding candidate and large itemset for updated database. It becomes better for data change percentage ($\delta$ %) up to which the incremental technique will be least time

A. Mantri et al. (Eds.): HPAGC 2011, CCIS 169, pp. 470–472, 2011.

consuming. And it eliminates the problem of multiple scans of older database. By which we improve the efficiency of newly added record updating problem.

## 2 Literature Survey

A lot of research work has been done for developing efficient data mining algorithms. Various ARM techniques are used nowadays. There also exist Top 10 algorithms for data mining [6] including Apriori algorithm. Opting the widely used Apriori algorithm for Incremental Association rule mining (IARM) is major problem in dynamic database environment .Here we mainly focus on the IARM. Significant work is done in this field. This section represents a brief overview of some work done in ARM and its Incremental version. An influential algorithm for ARM is Apriori [1]. However it becomes cumbersome for dynamic database. They commonly require multiple scans of the given database.

Researchers are still working on IARM for finding the efficient methods. Few techniques used for increasing efficiency of IARM are described in [3].

In incremental approach, rule maintenance can be done by processing the incremental part separately, and scanning the older database only once. Limited work is done for maintenance of association rules in incremental mining [2] [4] [5].

## 3 Incremental Data Mining

The term incremental means "% of $\delta$ change in the original database" i.e. insertion of some new data items into the already existing database. Since mostly databases are dynamic databases which are used in current scenario. The percent of delta change in the database can be calculated by this given formula.

$$\% \text{ change in DB} = \frac{(\text{NEW DATA}-\text{OLD DATA})}{\text{OLD DATA}} \times 100 \,. \tag{1}$$

The following are the steps for performance comparison of the proposed algorithm with actual Apriori algorithm:

**A.** Run the Apriori algorithm over the existing data. For different values of delta changes run both the actual Apriori and incremental algorithm for comparison in terms of optimum computation time.

**B.** Again run actual Apriori algorithm on updated database and preserve candidate and large itemsets with the computation time that it takes to process it. Suppose it takes time T1.

**C.** In this step the candidate and large itemsets obtained from old and incremental database are combined to find candidate and large itemsets for updated database. Suppose it takes time T2 to process it.

**D.** Now compare these computation times T1 and T2 to find which one approach is more effective and faster.

**E.** Finally compute actual delta changes threshold value up to which incremental Apriori algorithm performs much better ( time and effort analysis) as compared to actual Apriori algorithm (T2<T1).

### 3.1 Proposed Model

**Problem:** In dynamic database environment, if new records are appended in timely Manner. Then for finding the frequent Itemset requires the whole database scan again.

**Solution:** Incremental algorithm is used to deal with this problem. The proposed algorithm identifies the value of percentage of size of original database x, which can be added to original database will be optimal. Now there might be two following cases:

**1.** Up to x% change in original database, better to use previous result
**2.** For more than x %, rerun the algorithm again

As a result, Incremented Apriori algorithm will provide faster execution than the previously used Apriori algorithm because the number of scans for the database will be decreased.

## 4 Conclusion and Future Scope

Incremental Apriori algorithm proposes faster Execution time as compared to the existing Apriori algorithm up to some certain point of change (x %) in the original database.Its future scope is in all other Association rule mining techniques in incremental approach.

## References

1. Rakesh, A., Srikant, R.: Fast algorithms for mining association rules in large databases. In: Proc. 20th Int'l Conf. Very Large Data Bases (VLDB 1994), Santiago, Chile, pp. 487–499 (1994)
2. Ratchadaporn, A., Worapoj, K.: Incremental ARM using promising Frequent Itemset:Information. In: Communicatios & Signal Processing,6th Int'l Conf. IEEE, pp. 1–5 (2007)
3. Sotiris, K., Dimitris, K.: Association Rules Mining: A Recent Overview. GESTS Int'l Transactions on Computer Science and Engg. 32(1), 71–82 (2006)
4. Ravindra, P., Swami, D.K., Pardarsani, K.R.: Lattice Based Algorithm for incremental mining of Association rules. Int'l Journal of Theoretical and Applied Computer Sciences 1(1), 119–128 (2006)
5. Sarda, N.L., Srinivas, N.V.: An Adaptive Algorithm for Incremental Mining of Association Rules. In: Proc. Ninth Int'l Workshop on Database and Expert Systems Applications, p. 240 (1998)
6. Wu, X., Kumar, V., Quinlan, J.R., Ghosh, J., Yang, Q., Motoda, H., McLachlan, G.J., Ng, A., Liu, B., Yu, P.S., Zhou, Z.-H., Steinbach, M., Hand, D.J., Steinberg, D.: Top 10 algorithms for Data Mining. In: IEEE International Conference on Data Mining (ICDM) (December 2006)

# Energy Aware and Energy Efficient Routing Protocol for Adhoc Network Using Restructured Artificial Bee Colony System

B. Chandra Mohan[1] and R. Baskaran[2]

[1] Department of CSE,
Anna University, Chennai, India
Tel.: +91-9976611188
abc.phd@hotmail.com
[2] Assistant Professor,
Department of CSE,
Anna University, Chennai, India
Tel.: +91-9094022331
baaski@cs.annauniv.edu

**Abstract.** Wireless communication is one of the fastest growing technologies all over the world. Especially, Adhoc Network is applied wide spread across the world in many different applications, which includes all major engineering systems, vehicular network etc...The optimal routing is an issue in the adhoc network and many researchers focused their attention and developed various methodologies which are feasible for certain situations. This paper proposes a honey bee mating algorithm for adhoc routing, which is a swarm intelligence technique, and this technique is already applied for data clustering; scheduling and resource allocation; optimization problems. The various benchmark proposed by the researcher for the artificial honey bee shows better result than the existing techniques. This paper has restructured the artificial bee colony algorithm from the initialization phase to the implementation phase, and shows better result than the existing methodology.

**Keywords:** Wireless Communications, Adhoc Network, Routing, Artificial Bee Colony, Swarm Intelligence.

## 1 Introduction

The wireless communication has many segments such as Infra-structure based Wireless communications, Adhoc Wireless communications, Satellite based Wireless communications, and Wireless LAN. In which, the Mobile Adhoc Network (MANET) is still an interesting research area in the field of computer network [1]. There are many issues in the Routing, Scalability, QoS, Energy Management, MAC, Transport Layer, and Security to be achieved optimal solution [2][3] in the MANET. This paper concentrates on routing issue, and is focused to provide optimal solution for routing as well as improving energy management and avoiding the congestion in the network.

A. Mantri et al. (Eds.): HPAGC 2011, CCIS 169, pp. 473–484, 2011.

The task of wired routing protocol may be exchanging route information and finding feasible path, but in the wireless routing there are some more tasks added to meet the wireless environment which includes minimum power requirement, utilizing minimal network resources like bandwidth, gathering and updating link failures. So in order to provide optimal routing, the routing protocols need to fulfil the following major challenges and requirements [4]. The challenges are mobility, bandwidth constraint, Error-prone and shared channel, Location dependent contention, and Resource Constraint like battery power, buffer size. The requirements are quick route configuration, loop free routing, minimum route acquisition delay, distributed routing approach, minimum control overhead, scalability, QoS, time sensitive traffic, and security.

The existing wireless routing protocol which is already implemented by the researchers are classified into three groups: table driven routing (such as DSDV, CGSR, STAR, GSR, FSR, HSR, WRP, etc.), source initiated on demand routing (such as AODV, DSR, TORA, ABR, SSR, FORP, PLBR etc.) and hybrid routing (CEDAR, ZRP, ZHLS). In addition to these routing protocol, Evolutionary Fuzzy multi-objective routing [5], probability routing algorithm using Ant colony optimization [6], Cross Layer Routing [7], Reverse AODV [8], Reliable adhoc routing [9], boundary mapping and boundary state routing [10] are some of the research proposals in the last few years.

The disadvantages of these existing routing protocols are, 1) All of these routing protocols need large uncontrolled overheads to solve the routing problem. The number of routing packets increases dramatically as the network size increases. This large routing overhead affects the scalability of the network and affects the network performance, since it uses a significant part of the wireless bandwidth, node's limited energy and processing power; 2) In addition, most of these algorithms are optimizing only one parameter, which in most cases is the number of hops; 3) Existing routing protocol is not useful for both wired and wireless environment.

## 2 Artificial Bee Colony Algorithm

Swarm intelligence is a new discipline of study that contains relatively optimal approach for problem solving which are the imitations inspired from the social behaviours of insects and animals, for example, Ant Colony Optimization (ACO) algorithm, Honey Bee Algorithms, Fire Fly Algorithm. The "ACO Algorithm" is a study derived from the observation of real ants' behaviour, and uses these models as a source of inspiration for the design of novel algorithms, which is the solution for optimization and distributed control problems. The Honey Bee Mating algorithm is the growing technique, which is proposed in late 2005, for many engineering applications.

Honey bees are insects that live in large colonies (around 50,000 bees as a colony) usually containing one queen and her progeny, some 20,000–40,000 female workers and 200–300 male drones. Michelle et al [11] is discussed more detailed study of honeybee in the biological aspect and about the foraging behaviour. There are many syndromes are observed like aggression syndrome, waggling dance, from the honey bee colony which is used for solving optimization problems.

Although honey bees are depicted in many cave paintings dated from 6000BC, the first recorded observations of bee behaviour were made by Aristotle. Dervis et al [12] implemented the Artificial Bee Colony (ABC) with various benchmark functions and compared with Particle Swarm Optimization (PSO), Differential Evaluation (DE) Algorithm, and Genetic Algorithm (GA). The author concluded that the performance of ABC algorithm is better than the other algorithms even though it uses less control parameters and it can be efficiently used for solving multimodal and multidimensional optimization problems. Michael Meisel et al [13] reviewed the various nature-inspired algorithms such as ACO, ABC, GA, Fireflies Algorithm and concluded that the nature-inspired algorithms could hybridize together with other algorithms to enhance it to be faster, more efficient, and more robust. Alok Singh [14] designed and implemented the ABC for leaf-constrained minimum spanning tree problem and concluded that computation time in the ABC is quite small and it completely outperforms both in terms of solution quality as well as running time. The ABC is also applied to identifying the optimal location of bio-mass power plant [15], Resource Allocation [16], Continuous Optimization Problem [17], Constraint Optimization Problem [18], Economic power dispatch [19], data Clustering in data mining [20][21][22], and Path management in the computer network [23]. In [15], the authors used a new calculation tool to determine the optimal location, biomass supply area and power plant size that offer the best profitability for investor. The detailed honey bee mating algorithm is explained in the coming section. In [16], the authors illustrate the practical utility of the theoretical results and algorithm of honey bee algorithm, and shows that how it can solve a dynamic voltage allocation problem to achieve a maximum uniformly elevated temperature in an interconnected grid of temperature zones. In [17], the authors proposed a novel hybrid Artificial Bee Colony (ABC) and Quantum Evolutionary Algorithm (QEA) for solving continuous optimization problems. ABC is adopted to increase the local search capacity as well as the randomness of the populations. In [18], For constraint handling, the proposed a modified ABC algorithm which uses Deb's rules consisting of three simple heuristic rules and a probabilistic selection scheme for feasible solutions based on their fitness values and infeasible solutions based on their violation.

## 2.1 Outlines of Honey Bee Algorithm

The honey bees are the diffuse creature which can extend itself over long distances in multiple directions in order to find the large number of food sources and at the same time to find the best food source from the collection of food sources. For ex, the flower patches with plentiful amounts of nectar or pollen that can be collected with less effort should be visited by more bees, whereas patches with less nectar or pollen should receive fewer bees.

The foraging process begins in a colony by scout bees being sent to search for promising flower patches. Scout bees search randomly from one patch to another. When they return to the hive, those scout bees that found a patch which is rated above a certain threshold (measured as a combination of some constituents, such as sugar content) deposit their nectar or pollen and go to the "dance floor" to perform a dance known as the "waggle dance". This dance is essential for colony communication, and contains three vital pieces of information regarding flower patches: the direction in which it will be found, its distance from the hive and its quality rating (or fitness). This information guides the bees to find the flower patches precisely, without using guides or maps. Each individual's knowledge of the outside environment is gleaned solely from the waggle dance. This dance enables the colony to evaluate the relative merit of different patches according to both the quality of the food they provide and the amount of energy needed to harvest it.

### 2.1.1 The Detailed Pseudo-code of Existing ABC Algorithm Is Follows

```
Initialization
Generate the initial population of the bees
Selection of the best bee as the queen
Selection of the maximum number of mating flights ('n')
Main Phase
do while i ≤ 'n'
    Initialize queen spermatheca, energy and speed.
    Select α
    do while energy > threshold and spermatheca is not full
        Select a drone
        if the drone passes the probabilistic condition then
            Add sperm of the drone in the spermatheca
        endif
        Update Speed
        Update Energy
      enddo
    do j = 1, Size of Spermatheca
        Select a sperm from the spermatheca
        Generate a brood by applying a crossover operator between the queen, the selected
              drones and the adaptive memory
        Select, randomly, a worker
        Use the selected worker to improve the brood's fitness
        if the brood's fitness is better than the queen's fitness then
              Replace the queen with the brood
        else
              if the brood's fitness is better than one of the drone's fitness then
              Replace the drone with the brood
        endif
      endif
   enddo
enddo
return The Queen (Best Solution Found)
```

### 2.1.2 The Detailed Mathematical Model of Existing ABC Algorithm Is as Follows

The algorithm requires a number of parameters to be set, namely: number of scout bees (n), number of elite bees (e), number of patches selected out of n visited points (m), number of bees recruited for patches visited by "elite bees" (nep), number of bees recruited for the other (m-e) selected patches (nsp), size of patches (ngh) and stopping criterion. The algorithm starts with the n scout bees being placed randomly in the search space.

The bees search for food sources in a way that maximizes the ratio

$$F(\theta) = \frac{E}{T} \tag{1}$$

Where, E is the energy obtained, and T is the time spent for foraging. Here 'E' is proportional to the nectar amount of food sources.

In a maximization problem, the goal is to find the maximum of the objective function F ($\theta$), $\theta \in R^P$. $R^P$ represents the region of search area.

Assume that $\theta_i$ is the position of the $i^{th}$ food source; F ($\theta_i$) represents the nectar amount of the food source located at $\theta_i$ and it is proportional to the energy E ($\theta_i$).

Let $P(C) = \{\theta_i(C) \mid i = 1, 2... S\}$ represent the population of food sources being visited by bees, in which, C is cycle, and S is number of food sources around the hive. The preference of a food source by the worker bee depends on the nectar amount F ($\theta$) of that food source. As the nectar amount of the food source increases, the probability with the preferred source by the worker bee increases proportionally. Therefore, the probability with the food source located at $\theta_i$ will be chosen by a bee can be expressed as

$$P_i = \frac{F(\theta_i)}{\sum_{k=1}^{S} F(\theta_k)} \tag{2}$$

The position of the selected neighbour food source is calculated as the following:

$$\theta_i(C+1) - \theta_i(C) \tag{3}$$

And the stop criteria of the system is

$$N_i(Q) - N_i(E) \geq H_{th} \tag{4}$$

Where,

$N_i$ (Q) represents the values of nectar of Queen,

$N_i$ (E) represents the values of nectar of Elite bee, And $H_{th}$ represents the minimum threshold value of the Hive.

At the end of iteration, the colony will have two parts to its new population - representatives from each selected patch and other scout bees assigned to conduct random searches.

## 3 Proposed Work

### 3.1 Elite Bee Formation (Process at the Food Source and Patch)

The proposed algorithm requires only two parameters to be set, namely the number of scout bees (n) and 'ngh' is the size of the patch. The 'n' is equal to number of flowers (nodes) in the garden (sub-net).

The bees search for food sources in a way that maximizes the ratio

$$\forall(E,H) \Leftrightarrow F(\theta_i) = \frac{E}{H} \tag{5}$$

Where, E is the energy obtained, H is the hop count, number of Inter Mediate Peer (IMP), between hive to food source. Here 'E' is proportional to the nectar amount of food sources discovered by bees and it works to maximize the honey being stored inside the hive. In a maximization problem, the goal is to find the maximum of the objective function, F (θ).

F (θ) is the nectar ratio, shown in equation (5), $\theta \in R^P$. $R^P$ represents the region of search area.

Assume that $\theta_i$ is the position of the $i^{th}$ food source; F ($\theta_i$) represents the nectar ratio of the food source located at $\theta_i$ and it is proportional to the energy E ($\theta_i$).

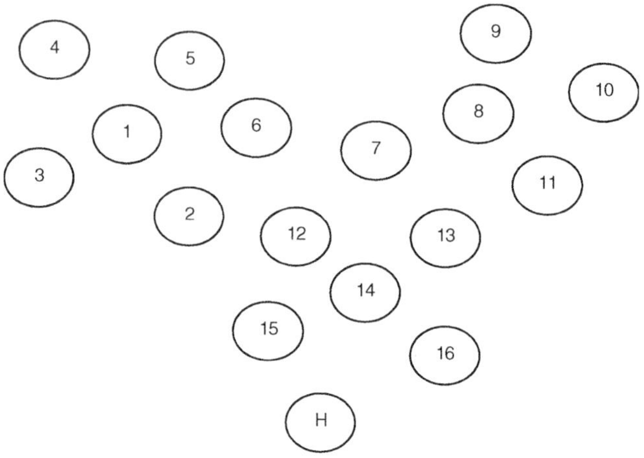


**Fig. 1.** Sample wireless adhoc network with 16 nodes (flowers) and a control centre (hive)

**Table 1.** Details of each node present in the fig 1

| Flower (Node) | Nectar (Energy) |
|---|---|
| 1 | 90 |
| 2 | 75 |
| 3 | 80 |
| 4 | 40 |
| 5 | 50 |
| 6 | 60 |
| 7 | 90 |
| 8 | 75 |
| 9 | 80 |
| 10 | 40 |
| 11 | 76 |
| 12 | 60 |
| 13 | 45 |
| 14 | 75 |
| 15 | 80 |
| 16 | 40 |

If the nectar ratio, F (θ), of the food source is higher than minimum threshold, then the scout bee initialises the waggling dance with rhythm above the food source (which is called as dance floor). This waggling dance is a visualization technique that to transfer information to the in-sight worker bees. If the worker bees are beyond in-sight, the rhythm of scout bee may reach the worker bee. Based on the visual and or audio information from the scout bee, the worker bee from one hive or more hives will reach the dancing floor (food source) for collecting the nectar.

$$T(\theta_i) = \begin{cases} \alpha \bullet F(\theta_i) & F(\theta_i) > F_{th} \\ 0 & otherwise \end{cases} \tag{6}$$

$$R(\theta_i) = \begin{cases} \beta \bullet F(\theta_i) & F(\theta_i) > F_{th} \\ 0 & otherwise \end{cases} \tag{7}$$

Where the $T(\theta_i)$ is the duration of waggling dance, $R(\theta_i)$ is the volume of rhythm, $F_{th}$is the minimum threshold of the nectar value and $\alpha$, $\beta$ are the constant which is termed as time scale factor and volume scale factor.

$$0 < \alpha < 1 \tag{8}$$

$$0 < \beta < 1 \tag{9}$$

If the value of $\alpha$ and $\beta$ are small, then convergence become fast. If the value of the same is high, more precise result will occur.

The bees search for food sources and collect the nectar (E). It initiates the waggling dance in the floor for 'T' time units (based on the equation 6) with 'R' volume of rhythm (based on the equation 7).

If the dancing time of bee is elapsed, then it will search the neighbouring dancing bee and goes to the dancing floor of neighbouring bee to watch the dance as guest bee. Suppose more than one dancing bee found near, then the bee choose the one with higher rhythm (Rhythm of bee proportional to nectar), which is illustrated in the figure 2.

When a guest bee enters into the dancing floor, the data from guest bee is stored in the nectar (routing) table of dancing bee.

The above process is explained here under with example.

A sample wireless adhoc network shown in the figure 1. And the sample nectar (energy / battery power) of each node is shown in the figure 1 is listed in the table 1.

As the first step, the hive generates 16 scout bees (hello message) and these bee flies (flood) into the region of food source (wireless network). These all scout bee access any one food source (flower) and collect the nectar (energy). After collecting the nectar, the scout bee starts waggling dance in the food source, if the nectar value is above the threshold value. The duration of the dance and the volume of rhythm are given in the equation (6 and 7). If the duration of the dance is elapsed, then the bee will search the neighbouring bee which is dancing and producing high volume of rhythm.

**Table 2.** Values of $F(\theta_i)$ for nodes 1-7 of figure 1

| Node | Energy | IMP | $F(\theta_i)$ |
|---|---|---|---|
| 1 | 90 | 4 | 23 |
| 2 | 75 | 3 | 25 |
| 3 | 80 | 4 | 20 |
| 4 | 40 | 5 | 8 |
| 5 | 50 | 4 | 13 |
| 6 | 60 | 3 | 20 |
| 7 | 90 | 3 | 30 |

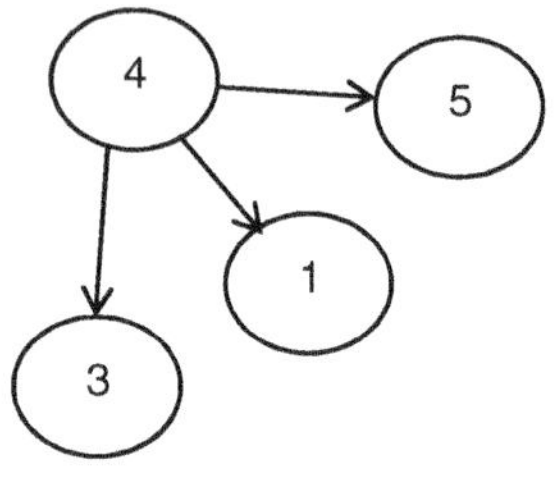


**Fig. 2.** Bee on node 4, Searching Neighbourhood bee to join as guest bee after elapsing the duration of its waggle dance

At the 8th time unit (refer the table 2), the dance of scout_bee_4 in the food_source_4 is elapsed, here for the simple calculation, the value of $\alpha$ and $\beta$ are assumed as 1. So it search the neighbouring bee, here scout_bee_1, 3, 5 in the food_source_1, 3, 5 is dancing with the rhythm values of 23, 20 and 13 respectively, so the scout_bee_4 enter into food_source_1. When the guest bee (scout_bee_4) enters, the scout_bee_1 will update its guest nectar table, which is shown in the table 3.

**Table 3.** Nectar Table of scout_bee2

| Food Source ID | Nectar Ratio $F(\theta)$ |
|---|---|
| 1 | 8 |

If no bee is found to be dancing in the neighbouring food source and at-least ‘b’ (b is given in the equation 10) number of bees entered as guest bee then the concern bee is designated itself as patch head (elite bee) and the flower is designated as elite site. Then it flies to the hive with its own nectar (routing) table.

$$b \approx ngh \pm 2 \tag{10}$$

From the given example, at 23rd time units, all the neighbouring bees appeared in the dancing floor of scout_bee2 with the value of (3,20),(5,13),(1,23),(4,8),(6,20), where the first value indicates the food source id and the second value indicates the nectar value of the concerned food source. Now there are 5 bees are appeared, so the scout_bee_2 is elected as the elite bee of the patch (consider this as patch 1). The nectar table of patch1 is shown in the table 4.

**Table 4.** Nectar Table of elite bee (scout_bee_2) in the patch1

| Food Source ID | Nectar Ratio $F(\theta)$ |
|---|---|
| 1 | 23 |
| 3 | 20 |
| 4 | 8 |
| 5 | 13 |
| 6 | 20 |

In the hive, elite bee from patch1, patch2 and patch3 is reached, and then the Elite Bee and Patch Routing table of Hive is formed, which is shown in the table 5 and table 6 respectively. In the patch routing table, '*' indicates the head of the patch and the number indicates the id of the concern elite bee.

**Table 5.** Elite bee Table in the Hive

| Food Source ID | Patch ID |
|---|---|
| 2 | 1 |
| 7 | 2 |
| 15 | 3 |

**Table 6.** Patch Routing Table in the Hive

| Food Source ID | Next Hop | Patch ID |
|---|---|---|
| 1 | 2 | 1 |
| 2 | * | 1 |
| 3 | 2 | 1 |
| 4 | 2 | 1 |
| 5 | 2 | 1 |
| 6 | 2 | 1 |
| 7 | * | 2 |
| 8 | 7 | 2 |
| 9 | 7 | 2 |
| 10 | 7 | 2 |
| 11 | 7 | 2 |
| 12 | 15 | 3 |
| 13 | 15 | 3 |
| 14 | 15 | 3 |
| 15 | * | 3 |
| 16 | 15 | 3 |

### 3.2 Formation of Communication Path (Process in the Hive)

From the nectar table (shown in table 6), the flow (routing) diagram is designed using the following formulas:

Let $N = \{1,2, \ldots., n\}$ be the set of food sources in the region and define

$C_k$ = Set of elite food sources that connected at iteration k (k=0, as the elite food sources found in the initial)

$C'_k$ = Set of food sources as yet to be connected to Hive via Elite food source at iteration k

Set $C_0 = \phi$ and $C'_0 = N$

Start with any food source 'i' in the unconnected food source set $C'_{k-1}$ that yields the shortest arc to a food source in the connected elite source / food source set $C_{k-1}$.

$$C_k = C_{K-1} + \{'i'\} \tag{11}$$

$$C'_k = C'_{K-1} - \{'i'\} \tag{12}$$

Let $u_i$ be the shortest distance from food source 1 to elite food source i, and define $d_{ij}$ ($\geq 0$) as the length of arc (i, j). Then the algorithm defines the connection for an immediately succeeding food source j as

$$[u_j, i] = [u_i + d_{ij}, i], d_{ij} \geq 0 \tag{13}$$

Compute the temporary connection as shown in equation 10 for each food source i, provided is not permanently connected with hive.

If food source j is already connected with $[u_j, k]$ through another food source k and if $u_i+d_{ij} < u_j$, replace $[u_j, k]$ with $[u_i+d_{ij}, i]$.

If all food sources have permanently connected, and verify that the set of unconnected food source set is empty, ($C'_k$ is empty), and then Stop. Otherwise, set k = k+1 and repeat the step.

## 4 Conclusion

The proposed system is implemented and tested in 7 different situations (20 Nodes of 4 Clusters, 20 Nodes of 5 Clusters, 30 Nodes of 5 Clusters, 30 Nodes of 6 Clusters, 50 Nodes of 5 Clusters, 100 Nodes of 5 Clusters, 100 Nodes of 10 Clusters, and 100 Nodes of 20 Clusters). And the three types of transmission (Two Transmissions as light load, Four Transmissions as medium load, and 5 Transmissions as heavy load) are also applied on all the seven situations. The comparison of response time of AODV, DSDV, DSR, ACO and Proposed system in the all seven situation is shown in the table 8. And the comparison of throughput of DSR, DSDV, AODV, ACO, and Proposed system is also shown in the table 9. The result shows that the proposed system provides optimal value in all the aspects like response time and throughput. Comparing all other methodology, the ABC will require less running overhead and also the size of overhead also very light. In the proposed restructured ABC, very less number of parameters is applied; it leads to less energy, less traffic in the network and consequent improvement in the life time of battery as well as the network efficiency. The proposed algorithm (in general, the ABC) is energy aware routing methodology; this behaviour is most useful and very important aspects for the current industry requirements.

## References


1. Lima, M.N., dos Santos, A.L., Pujolle, G.: A Survey of Survivability in Mobile Ad Hoc Networks. IEEE Communications Surveys & Tutorials 11(1) (2009)
2. Calafate, C.T., Malumbres, M.P., Oliver, J., Cano, J.C., Manzoni, P.: QoS Support in MANETs: A Modular Architecture Based on the IEEE 802.11e Technology. IEEE Transactions on Circuits and Systems For Video Technology 19(5) (2009)
3. Hanzo II, L., Tafazolli, R.: A Survey of QoS Routing Solutions For Mobile Ad Hoc Networks. IEEE Communication Survey 9(2) (2007)
4. Siva Ram Murthy, C., Manoj, B.S.: Adhoc Wireless Networks, 2nd edn. Pearson Education, London (2007)
5. Manvaha, S., Srinivasan, D., Tham, C.K., Vasilakos, A.: Evolutionary Fuzzy Multi-Objective Routing For Wireless Mobile Ad Hoc Networks. Congress on Evolutionary Computation 2, 1964–1971 (2004)
6. Hussein, O.H., Saadawi, T.N., Lee, M.J.: Probability Routing Algorithm for Mobile Ad Hoc Networks' Resources Management. IEEE Journal on Selected Areas in Communications 23(12) (December 2005)
7. Canales, M., Gallego, J.R., Hernandez-Solana, A., Valdovinos, A.: Performance Evaluation of Cross-Layer Routing for QoS Support in Mobile Ad Hoc Networks. In: International Federation for Information Processing, pp. 322–333 (2006)
8. Kim, C., Talipov, E., Ahn, B.: A Reverse AODV Routing Protocol in Ad Hoc Mobile Networks. In: International Federation for Information Processing, pp. 522–531 (2006)
9. Tang, J., Xue, G., Zhang, W.: Reliable ad hoc routing based on mobility prediction. Journal of Combinatorial Optimization 11, 71–85 (2006)
10. Lemmon, C.J., Musumeci, P.: Boundary Mapping and Boundary State Routing (BSR) in Ad Hoc Networks. IEEE Transactions on Mobile Computing 7(1), 127–139 (2008)
11. Elekonich, M.M., Roberts, S.P.: Honey bees as a model for understanding mechanisms of life history transitions. Comparative Biochemistry and Physiology, Part A 141, 362–371 (2005)
12. Karaboga, D., Akay, B.: A comparative study of Artificial Bee Colony algorithm. Applied Mathematics and Computation 214, 108–132 (2009)
13. Meisel, M., Pappas, V., Zhang, L.: A taxonomy of biologically inspired research in computer networking. Computer Networks 54, 901–916 (2010)
14. Singh, A.: An artificial bee colony algorithm for the leaf-constrained minimum spanning tree problem. Applied Soft Computing 9, 625–631 (2009)
15. Vera, D., Carabias, J., Jurado, F., Ruiz-Reyes, N.: A Honey Bee Foraging approach for optimal location of a biomass power plant. Applied Energy 87, 2119–2127 (2010)
16. Quijano, N., Passino, K.M.: Honey bee social foraging algorithms for resource allocation: Theory and application. Engineering Applications of Artificial Intelligence 23, 845–861 (2010)
17. Duan, H.-B., Xu, C.-F., Xing, Z.-H.: A Hybrid Artificial Bee Colony Optimization and Quantum Evolutionary Algorithm for Continuous Optimization Problems. International Journal of Neural Systems 20(1), 39–50 (2010)
18. Karaboga, D., Akay, B.: A modified Artificial Bee Colony (ABC) algorithm for constrained optimization problems. Applied Soft Computing 11, 3021–3031 (2011)
19. Kumar, R., Sharma, D., Sadu, A.: A hybrid multi-agent based particle swarm optimization algorithm for economic power dispatch. Electrical Power and Energy Systems 33, 115–123 (2011)

20. Fathian, M., Amiri, B., Maroosi, A.: Application of honey-bee mating optimization algorithm on clustering. Applied Mathematics and Computation 190, 1502–1513 (2007)
21. Karaboga, D., Ozturk, C.: A novel clustering approach: Artificial Bee Colony (ABC) algorithm. Applied Soft Computing 11, 652–657 (2011)
22. Zhang, C., Ouyang, D., Ning, J.: An artificial bee colony approach for clustering. Expert Systems with Applications 37, 4761–4767 (2010)
23. Heegaard, P.E., Wittner, O.J.: Overhead reduction in a distributed path management system. Computer Networks 54, 1019–1041 (2010)

# Implementing Key Management for Security in Ad Hoc Network

Avinash Sharma[1], Narendra Agarwal[2], Satyabrata Roy[3], Ajay Sharma[4], and Pankaj Sharma[5]

Associate Professor & Research Scholar,
SGVU, Jaipur, India
Research Scholar, Arya College of engineering & IT,
Jaipur, India
Lecturer, Jaipur Engineering College,
Jaipur, India
Research Scholar, Singhania University,
Rajasthan, India
Associate Professor, Rajasthan college of Engineering & IT,
Jaipur, India
Lecturer, Govt. Mahila Engineering College, Ajmer
sh-avinash@yahoo.com, naren.agarwal2k@gmail.com,
satya_2k6@yahoo.com, ajay@yahoo.com

**Abstract.** Key management is important to security of Mobile Ad Hoc NETwork (MANET). Based on the (t, n) threshold cryptography, this paper introduced mobile agent to exchange private key and network topological information with nodes in the network. This method can not only reduce the network overload, but also improve the service velocity and success ratio of authentication. Any t nodes in the network sized n can cooperate to perform an authentication upon a new node wanting to join the network. Carrying private key and some state variables such as survival time, mobile agent navigated in the network according to visits-balance policy, namely, node with the least visits would be first visited by mobile agent.

**Keywords:** MANET, Ad Hoc, Routing.

## 1 Introduction

Ad Hoc Network solving problems in real world increasingly attracts attention from industrial and academic fields [1]. Using a wireless radio communication channel, the main advantage of Ad Hoc Network is low cost of deployment and maintenance, since the nodes and wireless hardware are inexpensive and easily available.

### I. RELATED WORK

Many researches show that centralized approaches work well in wired network. However, it does not in Ad Hoc Network [2]. To solve this problem, many researches have

A. Mantri et al. (Eds.): HPAGC 2011, CCIS 169, pp. 485–489, 2011.

been done. L. Zhou and Z. J. Haas [3] first introduced a key management system for Ad Hoc Network based on threshold cryptography. In their solution, a group of *n* servers together with a master public-private key pair are deployed by CA.

## II. CL-PKEOVERMANET

**1) Generic Model:** We assume that at the beginning of the network there is a Key Generator Center (KGC) which generates partial secret keys for all the users. We also denote *n* to be the number of original nodes and *t* to be the pattern of security level of the threshold system. Those *n* nodes collectively form a Distributed Key Generator Center(DKGC). After the initiation, the KGC will go offline, and the network becomes self-organized. We denote those nodes that get partial secret keys from the KGC to be the original nodes, those nodes that get partial secret keys from DKGC to be the new-joint nodes and those nodes that collectively form the DKGC to be DKGC nodes.

**Setup:** This algorithm takes as input a security parameter 1*k* and returns the master private key *msk* and master public key *mpk*. This algorithm is run by the KGC, in order to setup a certificate less ad hoc system.

**Extract-partial-secret-key:** This algorithm takes as input the master public key *mpk*, the master private key *msk* and an identity ID= *i belongs to {0, 1}**. It outputs a partial private key d_*i*. This algorithm runs by KGC once at the initiation of the network.

**Extract-master-secret-key-shares:** This algorithm takes as input the master private key msk and n identity ID=i belongs to {0, 1}*.It outputs a master secret key shares msksi. This algorithm runs by KGC once at the initiation of the network.

**Extract-partial-secret-key-share-and master- secret-key- share**

This algorithm takes as input the master public key mpk, the master private key share $msk_{si}$ from a DKGC node and an identity new of a new-jointly node. It outputs a share of partial user private key $ds_{new,i}$ and a share of master secret key share $msks_{new,I}$ , I belongs to {1,2,....,n}. This algorithm runs by DKGC nodes.

### (2) Proposed Scheme

The first certi.cateless public key encryption scheme was proposed by Al-Riyami and Paterson. We incorporate their work and adopt it to MANET key management with CL-PKE. The scheme is as follows:

**Setup**

We assume *IG* is a Bilinear Di.e-Hellman parameter generator and *k* is the security parameter for the system. This algorithm has four steps.

1. Run the IG generator on an input k, it outputs (G1,G2, e) where G1 and G2 are groups of prime order q. e: G1 × G1 → G2 is a pairing.

2. Choose an arbitrary generator P belongs to G1.

3. Select a master private key msk uniformly at random from $Z_q^*$ and set P0 = msk × P.

4. Choose four cryptographic hash functions $H1 : \{0, 1\}^* \rightarrow G1$, $H2 : G2 \rightarrow \{0, 1\}$, $H3 : \{0, 1\}^m \times \{0, 1\}^m \rightarrow Z_q^*$ and $H4 : \{0, 1\}^m \rightarrow \{0, 1\}^m$, here $m$ will be the bit length of plaintexts.

The master public key mpk = (G1,G2, e,m, P, P0,H1,H2,H3,H4). The master private key is msk belongs to $Z_q^*$. The message space is $\{0, 1\}^m$ and the cipher text space is C= $\{0, 1\}^{2m}$ x G1

**Extract-partial-secret-key:** This algorithm takes as input an *ID* . *{0, 1}** and carries out the following steps.

1. Compute Q*ID* = *H*1(*ID*) . G1.

2. Output the partial private key *_ID* = *msk*× Q*ID* . G*1. Any user can verify its partial secret key by checking *e*(*_ID*, P) = *e*(Q*ID*, P0).

**Extract-partial-secret-key-share and master-secret-key- share**

Giving a master secret key share of node i *msksi* and a new-joint node's ID= *new*, this algorithm takes the following steps.

1.A partial secret key share is calculated by $ds_{new,i}$ = ˆ*L*(0, *IDi*) × *msksi* × Q*new* =ˆ*L*(0, *IDi*) × *f*(*IDi*) × Q*new belongs to* G1

2. A master secret key share is calculated by *msksnew,i* = ˆ*L*(*IDnew*, *IDi*) × *msksi belongs to Z*q*

**Set-user-keys**

This algorithm takes as select a user's secret value *xID* . *Z*q* , input the master public key *mpk* and user's partial secret key *_ID*. It outputs user's

secret key *__ID* = *xID* × *_ID* and user's public key *pkID* =< X*ID*,Y*ID* >, where X*ID* = *xID*P and Y*ID* = *xIDmsk*P.

**Encryption**

For a message *msg* . *M* and an identity ID . *{0, 1}** with its public key *pkID* =< X*ID*,Y*ID* >, the encryption algorithm takes as follows:

1. Check the public key by e(XID, P0) =e(YID, P). If the result is negative, abort the encryption and output an error symbol.

2. Compute $Q_{ID} = H_1(ID)$ belongs to $G_1^*$.

3. Choose a random number σ belongs to $\{0, 1\}^m$.

4. Set r = H3(σ,msg)

5. Compute and output ciphertext:

c= <rP, σ + $H_2(e(Q_{ID},Y_{ID})^r)$,msg + $H_4(σ)$>

**Decryption**

Suppose c =< U, V,W >belongs to C. To decrypt this cipher text with private key $sk_{ID}$:

1. Compute V + H2(e($SK_{ID}$,U)) = σ'.

2. Compute W + H4(σ') = msg'.

3. Set r' = H3(σ',msg') and test if U = r'P. If not, output an error symbol and reject the cipher text.

4. Output msg' as the decryption of c.

III. SIMULATION

**a. Simulation with C**

Setup In this simulation, we implement our scheme with C codes. The programming is based on Pairing Based Cryptography library (PBC) and GNU MP library (GMP), which define a large amount of efficient functions over pairing calculations. The programming environment is showing as follows: Result In this simulation, we assume that the network propagation delay is 0ms, which means once the partial secret key is generated, it will be sent to the correspondent node immediately.

**Table 1.** Programming Environment

| | |
|---|---|
| CPU | Intel T2250 1.73GHz |
| RAM | 1GB |
| Hard Disk | 80GB at 5400rpm |
| OS | Ubuntu 7.01 |
| GCC Version | 4.1 |
| PBC lib version | 0.4.17 |
| GMP lib version | 4.2.2 |

**Table 2.** Simulation Result

| | | | |
|---|---|---|---|
| Number of nodes | 5 | 10 | 20 |
| Keys from KGC | 142.756 | 142.756 | 142.756 |
| Key shares | 13.165 | 11.315 | 10.189 |
| Keys from DKGC | 156.739 | 224.295 | 313.790 |

## References

[1] Ramanathan, R., Jason, R.: A brief overview of mobile Ad Hoc networks: Challenges and Directions. IEEE Communications Magazine 50th anniversary commemoratives issue, 20–26 (2002-2005)
[2] Buttyan, L., Hubaux, J.P.: Report on a working session on security inwireless Ad Hoc networks. MobileComputing and Communications Review 7(1), 74–94 (2003)
[3] Zhou, L., Haas, Z.J.: Securing ad hoc networks. IEEE Networks Special Issue on Network Security 13(6), 24–30 (1999)
[4] Kong, J., Zerfos, P., et al.: Providing robust and ubiquitous security support for mobile Ad-Hoc networks. In: IEEE 9th International Conference on Network Protocols (ICNP 2001), Riverside, California (2001)
[5] Pham, V.A., Karmouch, A.: Mobile Software Agents: An Overview. IEEE Communication Magazine, 15–28 (2002-2007)
[6] Yi, S., Kravers, R.: Practical PKI for ad hoc wireless networks. Tech. rep. UIUCDCS-R-2002- 2273, UILU-ENG-2002-1717.Department of Computer Science, University of Illinois at Urbana- Champaign, Urbana, IL
[7] Di Caro, G., Dorigo, M.: Mobile agents for adaptive routing. In: Proceedings of the 31st Hawaii International Conference on Systems, pp. 88–103 (2001)
[8] Fall, K., Varadhan, K.: Ns notes and documentation. The VINT Project, UC Berkeley, LBL,USC/ISI, and Xerox PARC (1999-2007), http://www-mash.cs.berkeley.edu/ns/
[9] Yi, S., Kravets, R.: MOCA: Mobile certificate authentication for wireless Ad Hoc Networks. In: Proc of 2nd Annual PKI Research Workshop Program (PKI 2003), Gaithersburg, Maryland, pp. 65–79 (2003)

# Performance Evaluation of MAC- and PHY-Protocols in IEEE 802.11 WLAN

Vishal Sharma[1,*], Jagjit Malhotra[2], and Harsukhpreet Singh[2]

[1] Shaheed Bhagat Singh College of Engineering and Technology, Ferozepur, Punjab, India
Tel.: +91-8872544401
er_vishusharma@yahoo.com
[2] DAVIET, Jalandhar, Punjab, India

**Abstract.** This work evaluates and compares the performance of IEEE 802.11 WLAN scenario by evaluating QoS parameters such as medium access delay, end to end delay and re-transmission attempts at data rate of 2Mbps by means of different mechanisms in PHY- & MAC-layers. The Point coordination function (PCF) and Distribution Coordination Function (DCF) of MAC layer together with mechanism used in Physical layer i.e. Frequency hopping- and Direct sequence- spread spectrum (FHSS & DSSS) are reported and investigated to provide better QoS using OPNET simulator.

**Keywords:** Direct Sequence Spread Spectrum (DSSS) Scheme Distributed Coordination Function (DCF), Frequency Hopping Spread Spectrum (FHSS) Scheme, IEEE 802.11, Point Coordination Function (PCF), Quality of Service (QoS).

## 1 Introduction

The increasing demand for larger capacity and higher transmission speeds to accommodate for data-intensive multimedia in conjunction with real-time applications, the wireless networks have experienced an explosive growth in last few years. Apart from mobile telephone communications, WLANs, which came on the scene less than a decade ago (1997), have also experienced phenomenal growth. The rapid proliferation of WLAN hotspots in public places, such as airport terminals has been astounding. As a result, the number of wireless Internet subscribers has overtaken the number of wired internet users. The rapid growth of wireless communications is mainly attributed to their ease of installation in comparison to fixed networks [1]. Technological advancement & competition among mobile operators have also contributed to this growth reached to 1700 millions in 2010. It is predicted that this growth will continue to rise, and by 2015 there will be more than 2500 million mobile subscribers worldwide [2]. The IEEE 802.11 WLAN standard covers the MAC sub-layer and the physical (PHY) layer of the open system interconnection (OSI) network reference model. On the other hand, multimedia applications require some QoS support such

* Corresponding author.

A. Mantri et al. (Eds.): HPAGC 2011, CCIS 169, pp. 490–496, 2011.

as guaranteed bandwidth, end to end delay, medium delay, jitter and error rate and retransmission attempts. Provision of supporting guaranteed QoS in 802.11 WLAN is a challenging task due to uninformed functions of its MAC layer such as controlling channel access, maintaining QoS, and providing security together with noisy and variable PHY layer characteristics. Alternatively, wireless links are more prone to some problems like high loss rate, bursts of frame loss, packet re-ordering, large packet delay and jitter that are further variable in nature in terms of time and place. Once the mobile wireless subscriber moves from its current point of attachment to the new one, the subscriber expect to receive the same QoS. In this paper, we primarily focus on QoS issues such as medium access delay, end to end delay and retransmission attempts at data rate of 2Mbps by means of different mechanisms and comparing them at 802.11WLAN MAC- and PHY-layer to provide better QoS.

## 2 Simulation Setup and It's Working

Using OPNET simulator, we have designed a IEEE 802.11 WLAN based simulation scenario consist of six PCF- and two DCF- based immobile workstations in a BSS infrastructure as shown in Figure 1 to evaluate the performance of the immobile wireless system by implementing of DSSS and FHSS schemes at PHY layer level. The buffer size of data is set to 2024Kbps of each immobile workstation at data rate of 2Mbps and the access point beacon interval is set to 0.020.

The traffic flows between the different PCF-and DCF-workstations placed at different distances as shown in Fig. 1 have been configured as: PCF_wkstn 1 to PCF_wkstn 2; DCF_wkstn 3 to DCF_wkstn 4; PCF_wkstn 5 to PCF_wkstn 6;

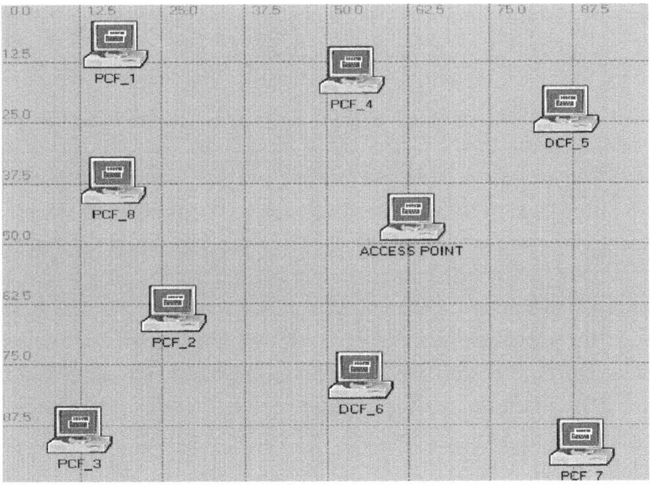


**Fig. 1.** Model of WLAN Network using OPNET Simulator

PCF_wkstn 7 to PCF_wkstn 8. The PCF, polling based mechanism, require the presence of a base station that act as access point [3]. In our hybrid WLAN network (both PCF- and DCF-supported network), the time is divided into two super frames of IEEE 802.11 WLAN as shown in Fig. 2. Each super frame consists of a contention free period (CFP) where PCF is used and is started by a special frame (a beacon) sent by the base station. The access category (AC) keeps a list of mobile stations that have requested to be polled to send the data. During CFP, it sends a poll frame to the station when they are cleared to access the medium. Upon the reception of a poll frame, the station sends a data packet if it has any queued-packet. To ensure that no DCF station are able to interrupt this mode of operation, the inter frame space (IFS) between PCF data frame is kept shorter than the usual DCF inter-frame space (DIFS) [4].

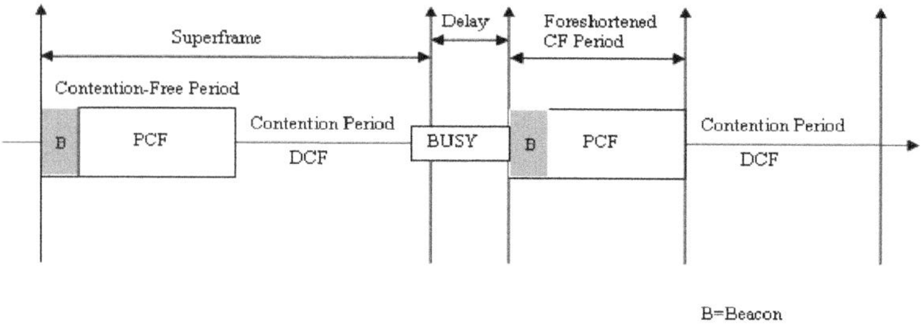


**Fig. 2.** Super-frame of IEEE 802.11

On the other hand, the CFP period is shortened in DCF access method. The DCF uses a carrier sense multiple access with collision avoidance algorithm (CSMA/CA) to immediate the access to the shared medium to process higher priority data and management packets. Because DCF was originally designed for data applications, its main weakness is the lack of QoS support [12-13] (absolute throughput, relative throughput, or delay support). Whenever a data is to be sent, the station senses the medium, if it is free at least a DIFS period of time, the frame is transmitted. Otherwise if the medium is busy, a back-off time B (measured in time slots which depend on the physical characteristics of physical layer) is chosen randomly in the interval (0, contention window) [14]. After the medium has been detected idle for at least one DIFS frame, the back-off timer is decremented by one for each time slot and the medium remains idle. If the medium become busy during the back-off process, the back-off timer is paused and is restarted when the medium has been again sensed idle for another DIFS period. When the back-off timer reached to zero, the frame is transmitted. Upon detection of collision, a new back-off timer is chosen and the back-off procedure starts over. After a successful transmission, the contention window is reset to $CW_{min}$.

## 3 Results and Discussion

To evaluate the overall performance of MAC- and PHY-layer protocols, we have determined the various parameters of IEEE 802.11 WLAN such as medium access delay, retransmission attempts and end to end delay both the protocols. The Fig. 3 shows that the end-to-end delay for the both PCF enabled stations with different spread spectrum sheme are almost same and very small increase in the end to end delay with simulation intervals. But in case of DCF enabled stations, the end-to-end delay is high and sharply increases as simulation time increases.

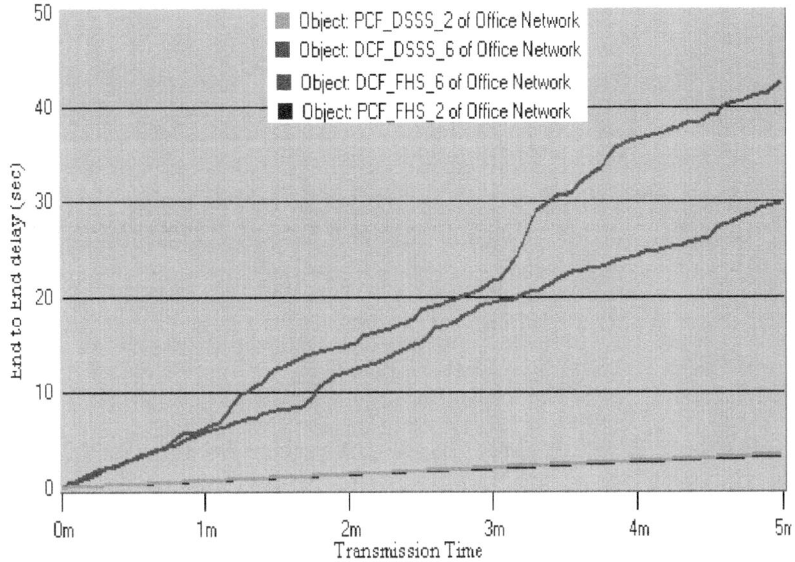


**Fig. 3.** End to End delay evaluation of PCF vs DCF-stations in WLAN (IEEE 802.11)scenario using different spread spectrum schemes

This is due to the increase in the number of nodes of DCF stations with simulation interval. It is also observed that the PCF enabled stations have much less retransmission attempt as compare to DCF stations as depicted in Fig. 4. Also, the retransmission for DCF stations is higher for first 30 seconds of simulation, but after that, it decreases with time and is stabilized. On the other hand, the PCF enabled stations are experienced with low and almost stable retransmissions attempts. Further, it is also observed that DSSS scheme shows less retransmission attempts with less end-to-end delay as compare to FHSS scheme as shown in Fig. (3-4) due to its redundant pattern which helps the station to recover demaged bits of data frames. It is also experimental that the PCF enabled stations have much less retransmission attempt as compare to DCF stations as depicted in Figure 4. On the other hand, the PCF enabled stations are experienced with low and almost stable retransmissions attempts. Further, it is observed that DSSS scheme shows less retransmission attempt as compare to FHSS stations. This is due to the fact that the redundant pattern of DSSS makes it possible to recover data without retransmitting it if one or more bits are damaged.

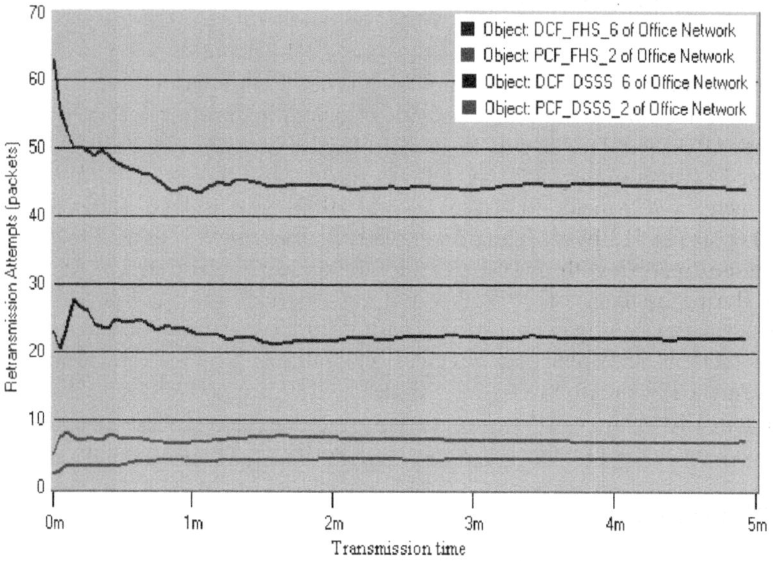


**Fig. 4.** Retransmission Attempts of PCF vs. DCF-stations in WLAN (IEEE 802.11) scenario using different spread spectrum schemes

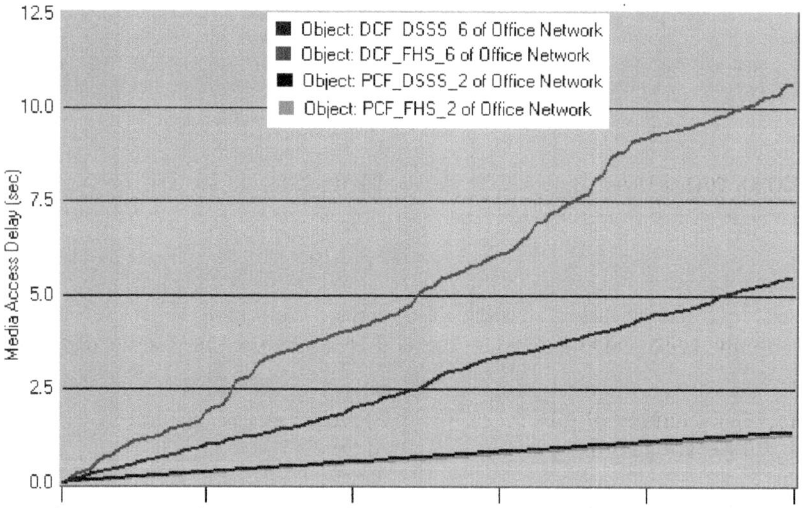


**Fig. 5**. Medium Access delay of PCF vs. DCF-stations in WLAN (IEEE 802.11) scenario using different spread spectrum schemes

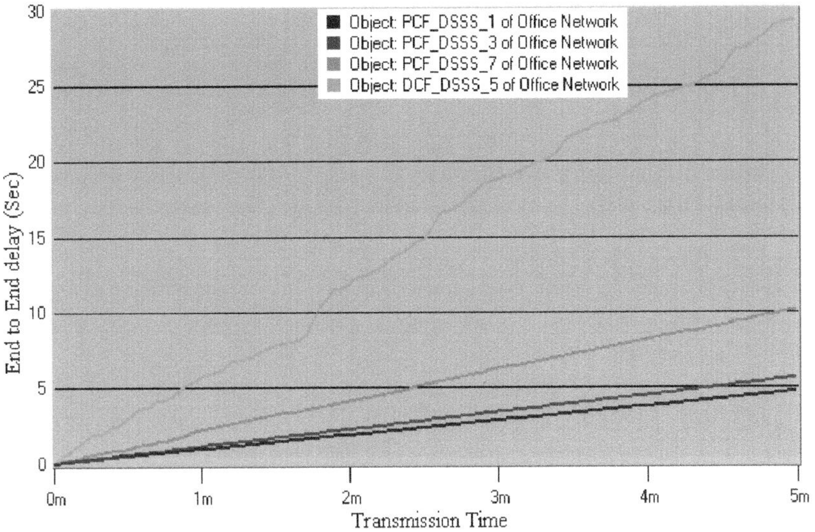


**Fig. 6.** End to End Delay of PCF vs. DCF-stations in WLAN (IEEE 802.11) scenario using DSSS scheme at different distances

In Fig. 5, the medium access delay for different PCF- and DCF-enabled stations with different medium access schemes is reported. It is analyzed that this delay is almost same for the PCF enabled stations and increases slowly with simulation period for both medium access schemes. But, this delay for DCF enabled stations increases sharply and more in case of FHSS scheme i.e. almost 5 seconds as compare to DSSS as depicted in Figure 5. Further, we have compare and calculated the end-to-end delay for different PCF- and DCF-enabled stations placed at different locations with respect to access point with DSSS access schemes. It is observed that the delay increase with the distance for both PCF and DCF schemes and maximum in case of DCF enabled station as shown in Fig. 6.

## 4 Conclusion

Our simulative work evaluated the QoS parameters of IEEE 802.11 WLAN scenario at data rate of 2Mbps by means of different mechanisms in PHY- & MAC-layers. The simulation results concluded that the PCF enabled stations have much less retransmission attempt as compare to DCF stations. Further, it is also reported that the impact of end-to-end delay is much less on the PCF enabled station as compare to DCF enabled stations for which this delay increases sharply. Also physical parameter DSSS (IEEE 802.11b) perform better than FHSS (IEEE 802.11a) for this high performance hybrid WLAN network as DSSS scheme shows less retransmission attempts with less end-to-end delay as compare to FHSS scheme due to its redundant pattern which helps the station to recover demaged bits of data frames.

## References


1. ITU, World Telecommunication Development Report 2002: Reinventing Telecoms (March 2002)
2. Kim, Y., Jeong, B.J., Chung, J., Hwang, C.-S., Ryu, J.S., Kim, K.-H., Kim, Y.K.: Beyond 3G: Vision, Requirements, and Enabling Technologies. IEEE Communications Magazine, 120–124 (March 2003)
3. Schwaderer Curtis A.: Wireless networks: Past, present and a very bright future Web site at, http://www.ivifoundation.org
4. Shaaban, S., Badawy, H.M., Hashad, A.: Performance Evaluation of the IEEE 802.11 Wireless LAN Standards. In: Proceedings of the World Congress on Engineering 2008, London, vol. 1 (July 2008)
5. Sobrinho, J.L., Krishna Kumar, A.S.: Quality-Of-Service in ad hoc carrier sense multiple access networks. IEEE Journal on Selected Areas in Communications 17, 1353–1368 (1999)
6. Draft Standard IEEE 802.11, Wireless LAN Medium Access Control (MAC) and Physical Layer (PHY) Specifications (November 1999)
7. Gu, D., Zhang, J.: QoS enhancement in IEEE 802.11 wireless local area networks. IEEE Communications Magazine 41(6), 120–124 (2003)
8. IEEE Standard 802.11, Wireless LAN Medium Access Control (MAC) and Physical Layer (PHY) specifications (1999)
9. Almquist, L.A., Schelén, O.: Evaluation of Quality of Service schemes for IEEE 802.11 wireless LANs. In: Proceedings of the 26th Annual IEEE Conference on Local Computer Networks, LCN 2001 (2001)
10. Puschita, E., Palade, T., Pitic, R.: Wireless LAN Medium Access Techniques QoS perspective. In: International Symposium Signals, Circuits and Systems, ISSCS 2005, vol. 1, pp. 267–270 (July 2005)
11. Mohammad, M., Andreas, K.: Advanced Communication Lab Opnet Tutorial (2006)
12. Siddique Mohammad, M., Andreas, K.: WLAN Lab Opnet Tutorial. University Bermen Press (2007)
13. Schiller, J.: Mobile Communication, 2nd edn. Pearson.Edu, London (2007)

# Key Authentication for MANET Security

Vijay Kumar[1], Rakesh Sharma[2], and Ashwani Kush[3]

[1] Research Scholar, Department of Computer Science,
Guru Nanak Khalsa College Karnal – Haryana, India
vk.gnkc@gmail.com
[2] Professor, Graphic Era University Deharadun India
colonelrakesh1953@gmail.com
[3] Associate Professor, Department of Computer Science,
University College, Kurukshetra University, Kurukshetra, India
akush20@gmail.com

**Abstract.** Securing a mobile ad hoc network (MANET) is a big challenge because of the nature of the MANET. A particular challenging problem is how to detect and defend possible attacks on routing protocols feasibly. Security in mobile ad-hoc networks is difficult to achieve, mainly because of the vulnerability of wireless links, the limited physical protection of nodes, the dynamically changing topology, the absence of a certification authority, and the lack of a centralized monitoring or management point. The major difficulty in ad-hoc network occurs when a new node join network but not having any trust based relation with other nodes of network. In this paper a mechanism has been proposed that when mobile node needs secure data communication, it will generate a dynamic secret session key with the desired destination mobile node directly or via proxy mobile nodes. These dynamic secret session keys are generated using Diffie-Hellman protocol

**Keywords:** AODV, DSR, Evaluation, Mobile Adhoc Network.

## 1 Introduction

MANET are made up of collaborative mobile nodes equipped with wireless network interfaces, where each node is able to communicate with other nodes within its transmission range without any fixed infrastructure, such as a name server or switches to set up connections The security services of ad-hoc networks are not together different from those of other network. The goal of these services is to protect information and resources from attacks and misbehavior. These security services such as privacy, integrity and authentication cannot be achieved without a prior solid key management. The major problem in providing security service in ad-hoc networks is how to manage the key that provide trustworthiness and privacy in data communication. In order to design practical and efficient key management system, it is necessary to understand the characteristics of ad-hoc networks and why traditional key management system is not suitable to such environments. To establish a secure communication between two mobile nodes in an ad-hoc manner, i.e. secure peer- to -peer communication, it is necessary for the two nodes to share a secret key [1]. This can be easily

A. Mantri et al. (Eds.): HPAGC 2011, CCIS 169, pp. 497–504, 2011.

achieved if there is existence of a public key infrastructure (PKI) [2, 3]. However, many mobile ad-hoc networks cannot afford to deploy public key cryptosystem due to their high computational overheads.

## 2 Related Work

### Security Evaluation Models [6, 11]

The security of cryptographic primitives and protocols can be evaluated under several different models. The most practical security metrics are computational, provable, and ad-hoc methodology, although the latter is often dangerous. The confidence level in the amount of security provided by a primitive or protocol based on computational or ad-hoc security increases with time and investigation of the scheme. However, time is not enough if few people have given the method careful analysis.

### 2.1 Unconditional Security [8 ]

The most stringent measure is an information-theoretic measure – whether or not a system has unconditional security. An adversary is assumed to have unlimited computational resources, and the question is whether or not there is enough information available to defeat the system. Unconditional security for encryption systems is called perfect secrecy. For perfect secrecy, the uncertainty in the plaintext, after observing the ciphertext, must be equal to the a priori uncertainty about the plaintext – observation of the ciphertext provides no information whatsoever to an adversary.

A necessary condition for a symmetric-key encryption scheme to be unconditionally secure is that the key be at least as long as the message. The one-time pad is an example of an unconditionally secure encryption algorithm. In general, encryption schemes do not offer perfect secrecy, and each ciphertext character observed decreases the theoretical uncertainty in the plaintext and the encryption key. Public-key encryption schemes cannot be unconditionally secure since, given a ciphertext, the plaintext can in principle be recovered by encrypting all possible plaintexts until c is obtained.

### 2.2 Complexity-Theoretic Security [9]

An appropriate model of computation is defined and adversaries are modeled as having polynomial computational power. (They mount attacks involving time and space polynomial in the size of appropriate security parameters.) A proof of security relative to the model is then constructed. An objective is to design a cryptographic method based on the weakest assumptions possible anticipating a powerful adversary. Asymptotic analysis and usually also worst-case analysis is used and so care must be exercised to determine when proofs have practical significance. In contrast, polynomial attacks which are feasible under the model might, in practice, still is computationally infeasible. Security analysis of this type, although not of practical value in all cases, may nonetheless pave the way to a better overall understanding of security. Complexity-theoretic analysis is invaluable for formulating fundamental principles and confirming intuition. This is like many other sciences, whose practical techniques are discovered early in the development, well before a theoretical basis and understanding is attained.

### 2.3 Provable Security [9]

A cryptographic method is said to be provably secure if the difficulty of defeating it can be shown to be essentially as difficult as solving a well-known and supposedly difficult (typically number-theoretic) problem, such as integer factorization or the computation of discrete logarithms. Thus, "provable" here means provable subject to assumptions. This approach is considered by some to be as good a practical analysis technique as exists. Provable security may be considered part of a special sub-class of the larger class of computational security considered next.

### 2.4 Computational Security [10]

This measure the amount of computational effort required, by the best currently-known methods, to defeat a system. It must be assumed here that the system has been well-studied to determine which attacks are relevant. A proposed technique is said to be computationally secure if the perceived level of computation required to defeat it (using the best attack known) exceeds, by a comfortable margin, the computational resources of the hypothesized adversary. Often methods in this class are related to hard problems but, unlike for provable security, no proof of equivalence is known. Most of the best known public-key and symmetric key schemes in current use are in this class. This class is sometimes also called practical security.

### 2.5 Ad-Hoc Security [12]

This approach consists of any variety of convincing arguments that every successful attack requires a resource level (e.g., time and space) greater than the fixed resources of a perceived adversary. Cryptographic primitives and protocols which survive such analysis are said to have heuristic security, with security here typically in the computational sense. Primitives and protocols are usually designed to counter standard attacks such as those given in [5]. While perhaps the most commonly used approach (especially for protocols), it is, in some ways, the least satisfying. Claims of security generally remain questionable and unforeseen attacks remain a threat.

Other Security Goals [6] are Availability, Confidentiality, Integrity, Authentication, Non-repudiation etc

## 3 Proposed Solution

Out of some key areas in Ad-hoc networks are key exchange and management, Ad-hoc routing, Power management, Intrusion detection etc. In this paper emphasis has been on key exchange using Password authenticated Diffie - Hellman key exchange for the security. Cryptographic schemes such as digital signatures are often employed to protect both routing information as well as data. Public key systems are generally espoused because of its upper hand in key distribution. In public key infrastructure each node has a public/private key pair. Public keys distributed to other nodes, while private keys are kept to nodes themselves and that too confidentially. Third party (trusted) called Certification Authority (CA) is used for key management.CA has a public/private key pair, with its public key known to every node and signs certificates binding public keys to nodes. The trusted CA has to stay online to reflect the current bindings, since the bindings could change overtime. Public key should be revoked if the owner node is no longer trusted or is out of network. A single key management service for an Ad-hoc network is probably not a good idea, since it's likely to become Achilles' heel of the network. If CA is down/unavailable nodes cannot get the current public keys of other nodes to establish secure connection. Also if a CA is compromised, the attacker can sign any erroneous certificates with the private key. Naive replication of CA can make the network more vulnerable, since compromising of a single replication can cause the system to fail. Hence it is more prudent to distribute the trust to a set of nodes by letting these nodes share the key management responsibility.

### 3.1 Proposed Password Authenticated Diffie - Hellman Key Exchange [5]

In classical cryptography you cannot avoid exchanging secrets without encrypting them. Transmission safety here must be achieved using non-cryptographic methods. A secret channel is needed for exchanging secrets. In Diffie and Hellman [14] protocol in place of secret channel, a public channel can be used. The creation of the session key at the end of the protocol is based on such a one-way-function. Even if a potential attacker has knowledge of the shared keys of both parties, it is hardly possible for him to compute the session key.

1. The key exchange protocol according to Diffie-Hellman is *NOT an encryption method*; it is merely used to agree on a secret and common key.
2. The Diffie-Hellman protocol is vulnerable against "Man-In-The-Middle"[14] attacks:

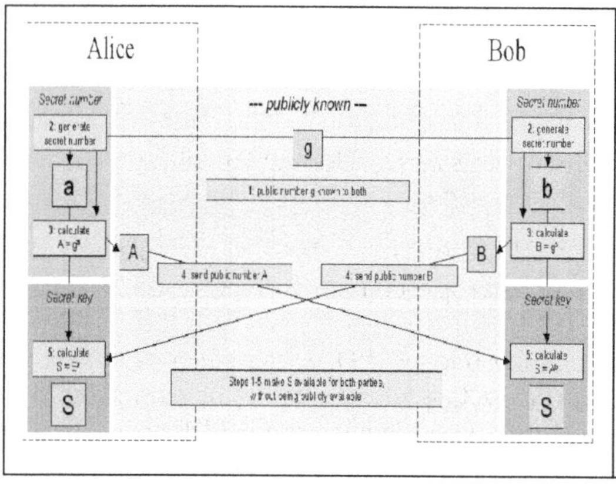


**Fig. 1.** A snapshot of CRYPTOOL

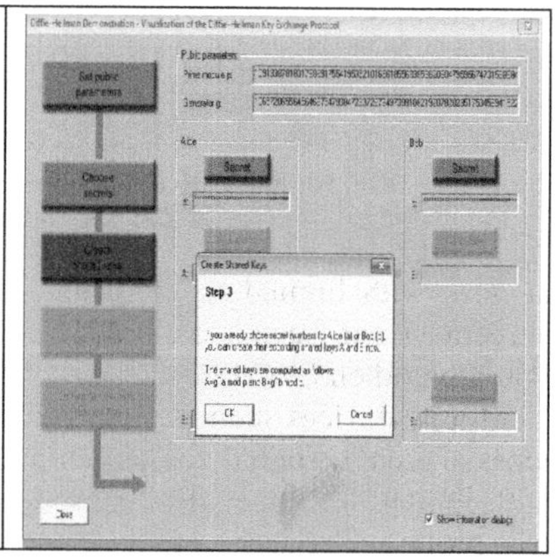

**Fig. 2.** Generation of Prime number

A cryptool[16] has been used to calculate session key. A case study has been taken with Alice as Source and Bob as destination or neighbor node to take packet to next node.

Alice and Bob use a one way function to obtain a key S, the session key, for subsequent correspondence. It becomes a secret that is only known to the two of them. Alice selects a random number and keeps it secret. She applies a one way function to calculate the number 'A = ga' and sends it to Bob. He does the same, by selecting a secret random number b, calculating 'B = gb' and sending it to Alice. The number 'g' is random and can be publicly known. Alice applies the one way function together with her secret number 'a' to B, while Bob does the same with his secret number 'b' and the received number A. The result S is the same in each case because the one way function is commutative:

$$(ga)b = (gb)a.$$

But even Bob cannot reconstruct Alice's secret number 'a' from the data available to him, while Alice cannot determine Bob's secret number 'b', and a perpetrator who knows 'g' and has intercepted both A and B cannot use this knowledge to determine 'a', 'b' or 'S'.

### 3.2 Procedure of Diffie-Hellman to Create a Session Key

Alice and Bob want to negotiate a secret session key S via a channel that may be intercepted.

1. They select a prime number 'p' and a random number 'g' and exchange this information publicly.
2. Alice now selects 'a', a random number less than 'p' and keeps it secret. Similarly, Bob selects 'b', a random number less than 'p' and keeps it secret.

Prime number p, generator g, secret key by Alice and Bob and how to create a shared key has been explained in step 3. This is shown in Figure 2.

3. Alice now calculates A = g^a (mod p). Bob calculates B = g^b (mod p)

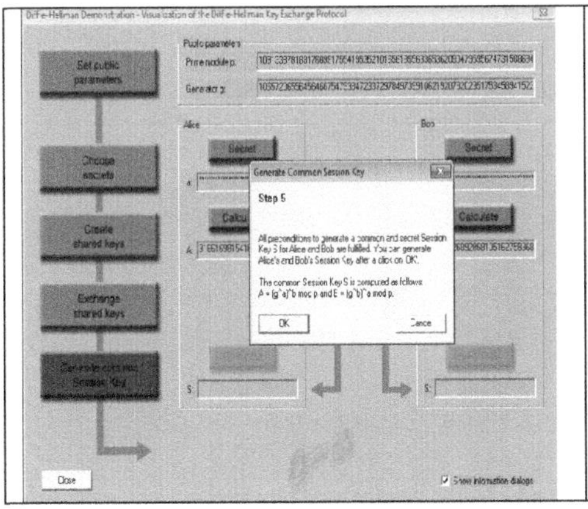

**Fig. 3.** Exchange of Keys

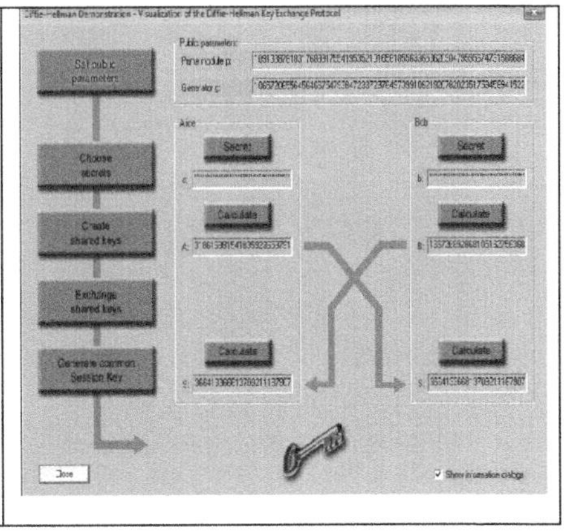

**Fig. 4.** Generation of final key session

Figure 3 explains how Alice and Bob Exchange their shared keys ($4^{th}$ step), and on the behalf, create a Session Keys which are common for the both Alice and Bob. This is the last step ($5^{th}$ step) to generate a common Session Keys.

4. Alice sends the result A to Bob. Bob sends the result B to Alice.
5. In order to now determine the session key to be used by both, they both separately raise the respective results they have received to the power of their secret random number modulo p. This means: - Alice calculates A= (g^a) ^b mod p and Bob calculates B = (g^b) ^a mod p

Figure 4 shows the ready mode of common Session keys. This can be used in data transmission.

*As a case study the protocol has been executed using Prime numbers:*

At first, Alice and Bob agreed on the public parameters. So they chose a prime p and a generator g:

*P:92038743936301164953712873660984604887845679805438497287669741648753811462
847*
*g:14454944990154641620061908682513859434056099255856747254484182599024449757 8
06*

Alice chose her secret number 'a' while Bob chose his secret number 'b'

*a:4096834980280693618399231226965737587343193497052777535921559580303389708 14*
*b:46834700599117142407861068384694339909316981533268324115960656013770147502 0
33*

If the chosen secret values a and b are greater or equal the prime module p, then they need to be reduced modulo p. The actual values are given below:

*a (reduced mod p):*
*409683498028069361839923122696573758734319349705277753592155958030338970814*
*b (reduced mod p):*
*46834700599117142407861068384694339909316981533268324115960656013770147502 03
3*

On the basis of the previously chosen secret numbers, Alice and Bob created their respective shared keys. Alice computed her shared key A, while Bob computed his shared key B:

*A:48857501745962930179317304260605660294609359951590123742325953208179925961
263*
*B:67499027538477650786003517129909707696916711140652111481252730662159667899
090*

In order to calculate their secret and common Session Key, Alice and Bob exchanged their shared keys: Alice sent her shared key A to Bob and Bob sent his shared key B to Alice.

Alice and Bob were able to calculate the secret and common Session Key now. Alice computed the Session Key SA, Bob computed the Session Key SB:

*SA:817381765167759565716939901283770534914336405961273691942904814414615410 3
0460*
*SB:817381765167759565716939901283770534914336405961273691942904814414615410 3
0460*

Theoretically it is now possible for Alice and Bob to use their Session Keys to encrypt documents they could like to exchange covertly.

## 4 Conclusion

An analytical study has been done for contemporary secured routing protocols for ad-hoc networks. Areas have been identified where further work can be done. A new solution has been proposed as hash key generation. It is clear that different protocols will have different solutions. Efforts are on to simulate the proposed scheme with different topologies and compare it with existing secured routing schemes. The proposed scheme is expected to work better in dense environments as selection of path becomes easy in case of failures. The research on MANET security is still in its early stage. The existing proposals are typically attack oriented in that they first identify several security threats and then enhance the existing protocol or propose a new protocol to thwart such threats. Because the solutions are designed explicitly with certain attack models in mind, they work well in the presence of designated attacks but may collapse under unanticipated attacks. Therefore, a more ambitious goal for ad-hoc network security is to develop a multifence security solution that is embedded possibly into every component in the network, resulting in in-depth protection that offers multiple lines of defense against many both known and unknown security threats.

## References


1. Yu, M., Zhou, M., Su, W.: A Secure Routing Protocol Against ByzantineAttacks for MANETs in Adversarial Environments. IEEE Transactions on Vehicular Technology 58(1) (January 2009)
2. Zhu, S., Xu, S., Setia, S.: Establishing Pairwise Keys for Secure Communication in Adhoc Networks:A Probabilistic Approach. In: ICNP 2003, pp. 326–331 (2003)
3. Housley,R., Ford, W., Polk, W., Solo, D.: Internet X.509 Public Key Infrastructure Certificate and CRL Profile (January 1999), http://www.faqs.org/rfcs/rfc2459.hunl
4. Yi, S., Kravets, R.: Key Management for Heterogeneous Adhoc Wireless Networks. In: IEEE ICNP 2002, pp. 12–15 (November 2002)
5. Lamport, L.: Password Authentication with Insecure Communication. Comm. of ACM 24(11), 770–772 (1981)
6. Kush, A., Hwang, C., Gupta, P.: Secured Routing Scheme for Adhoc Networks. International Journal of Computer Theory and Engineering (IJCTE) 3, 1793–1799 (2009)
7. Aberer, K., Despotovic, Z.: Managing Trust in a Peer-2-Peer Information System. In: The Xth International Conference on Information Knowledge Management (2002)
8. Gomathi, S., Duraiswamy, K.: Guaranteed Packet Transfer in MANET In and Out of Coverage Areas and Energy Saving Using Random Casting (IJCSE) International Journal on Computer Science and Engineering 02(03), 865–869 (2010)
9. Burmester, M., de Medeiros: On the Security of Route Discovery in MANETs. IEEE Transactions on Mobile Computing ISSN: 1536-1233
10. Garg, N., Mahapatra, R.P.: MANET Security Issues. IJCSNS International Journal of Computer Science and Network Security 9(8) (August 2009)

11. Haibing, M., Changlun, Z.: Key Lab. of Commun. & Inf. Syst. In: 2nd International Conference on Computer Engineering and Technology (ICCET), Beijing Jiaotong Univ., Beijing (2010); ISBN: 978-1-4244-6347-3
12. Monika, Kumar, M., Rishi, R.: Article:Security Aspects in Mobile Ad Hoc Network (MANETs): Technical Review. International Journal of Computer Applications 12(2), 37–43 (2010); Published By Foundation of Computer Science
13. Pirzada, A.A., Datta, A., McDonald, C.: Propagating Trust in Pure Ad-hoc Networks for reliable Routing. In: Proceeding of the International Workshop on Wireless Ad-hoc Networks, IWWAN (2004)
14. Cha, W., Wang, G., Cho, G.: Pair-Wise Key Agreement Scheme in Ad Hoc Networks. In: ICCS 2003. LNCS (2004)
15. Gehrmann, C., Mitchell, C.: Manual authentication for wireless devices. In: Cryptobytes 2004, vol. 7(1) (2004)
16. Esslinger, B.: CrypTool, Version 1.4.10, Deutsche Bank AG, Frankfurt/Main, University of Siegen and Darmstadt (July 2007)

# Biometric Encryption: Combining Fingerprints and Cryptography

Mini Singh Ahuja[1] and Sumit Chabbra[2]

[1] Computer Science & Engineering,
Guru Nanak Dev University, Regional Campus, Gurdaspur, India
[2] Computer Science & Application,
Khalsa College for Women, Amritsar, India
{minianhadh,Sumitchabbra_12}@yahoo.co.in

**Abstract.** These days, Biometric technologies are used to analyze human characteristics for security purposes. The most common physical biometric patterns analyzed for security purposes are the fingerprint, hand, eye, face and voice. The advantages of using biometrics to verify a person's identity over using passwords or token have been broadly presented in many research papers. However recent research has revealed that biometric technologies can be defeated with low –tech and cheap materials. This provides a new challenge when people are encouraged to use biometrics as a means to enhance network security. In this paper many approaches have been discussed to counteract security threats. We have also proposed a new method called Biometric Encryption which uses fingerprint and cryptography for enhanced security

**Keywords:** Bifurcations, Minutiae, Multimodal, PCA (principal component analysis), Ridge ending.

## 1 Introduction

Biometric refers to automatic system that uses measurable physiological characteristics or behavior traits to recognize the identity or verify/authenticate the claimed identity of an individual. The advantage to a biometric is that it doesn't change or lose. Many body parts, personal characteristics and imaging methods have been used for biometric systems such as fingers, hands, feet, eyes, ears, teeth, veins, voices, signatures, typing styles and gaits. Each biometric has its own strength and limitations and accordingly each biometric is used in identification (authentication) applications. It is not difficult to steal a biometric, create a copy and use the fake trait to attack biometric systems. This a serious issue as the people these days are using biometric as a means to enhance network security. Different technologies have been developed to defeat the spoofing attack. As biometrics is not secret they cannot be protected like passwords. People leave their biometrics everywhere without being aware that their biometric information can easily be captured, copied or forged. Another challenge to a biometric system is the speed i.e. the system must make an accurate decision in real time. This paper presents various techniques and methodologies to improve the security of a biometric system.

A. Mantri et al. (Eds.): HPAGC 2011, CCIS 169, pp. 505–514, 2011.

## 2 Attacks on Biometric Systems

Even though biometric systems offer several advantages over traditional token (e.g. key) or knowledge (e.g. password) based authentication schemes. They are still vulnerable to attacks. These attacks can be grouped into eight classes.

*Class I: Spoof attack*: In this type of attack a fake biometric e.g. (finger made from silicon, face mask, lens including iris texture) can be presented to a sensor.

*Class II:* The second class of attack is called *replay attack.* In it an interspected biometric data is submitted to the feature extractor by passing the sensor. To detect the replay attack, the authenticator as to ensure that the data is captured through the sensor and has not been injected.

*Class III: Substitution attack*: In the third type of attack the feature exactor module is replaced by a Trojan horse program that functions according to its designer specifications. Then the attacker gets an access to storage either locally or globally. He can overwrite the legitimate user's template with his /her own -in essence stealing their identity

*Class IV:* In the fourth type of attack a genuine feature values are replaced with values (synthetic or real) selected by the attacker or an imposter

*Class V:* In this type of attack the matcher is replaced with a Trojan horse program. This class of attack is called *Trojan horse Attack.*

*Class VI:* This type of attack occurs on the *template database*. The template database can be added, modified or removed. The templates can also be stolen which can be most dangerous.

*Class VII: Transmission attack:* A man in the middle attack is possible while the data is transmitted from one component to another. The attacker can manipulate the input data stream, send a fake template as an enrolled user, inject an artificial matching score or even generate a forged response.

*Class VIII:* Lastly the matured result (accept or reject) can be overridden by the attacker.

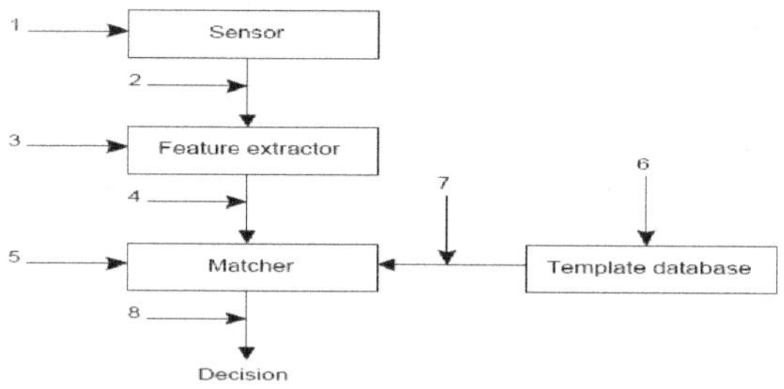


**Fig. 1.** Location of attacks in Biometric System

# 3 Multimodal Biometric Systems

Biometric systems [1], [2], [3] used in real world applications are unimodal. They rely on the evidence of a single source of information for authentication. These systems have to deal with variety of problems such as:

*Noise* in the sensed data. (e.g., due to repeated use of fingerprint sensor)
*Intra-class variation:* User who is incorrectly acting with the sensor typically causes these variations.

*Inter-class similarities:* In a Biometric System where there are large no of users, there may be inter–class overlap in the feature space of multiple users.

*Non-Universality:* The Biometric System might not be able to acquire a meaningful Biometric data from a subset of users.

*Spoof Attack:* This attack occurs when signature or voice are used in Biometric System.

Not all but some of the limitations of the unimodal can be overcome by including multiple source of information for identification. These types of system are called as *Multimodal Biometric Systems*. These systems are more reliable and have better performance, as it would be difficult for an imposter to spoof multiple biometric traits of a genuine user simultaneously.

## 3.1 Levels of Fusion

The information of the multimodal system [2], [3], [4], [5] can be fused at any of the three modules.

*Fusion at the Feature Extraction Level:* The data or the feature set originating from multiple sensors or sources are fused together.

*Fusion at the Matcher Score Level:* The scores of the each matcher are combined.

*Fusion at the Decision Level:* The final outputs of the multiple classifiers are combined. A majority vote scheme can be used to make final decision.

Biometric systems that integrate information at the early stages are more effective than those in which integration is done in later stages. So fusion at the feature level is expected to give better recognition results but it is difficult to integrate at this level because feature sets of the various systems may not be compatible. More over all commercial Biometric systems don't provide access to the feature sets, which they use in their products. Fusion at the matcher score level is usually preferred because it is relatively easy to access and combine the scores presented by different modalities.

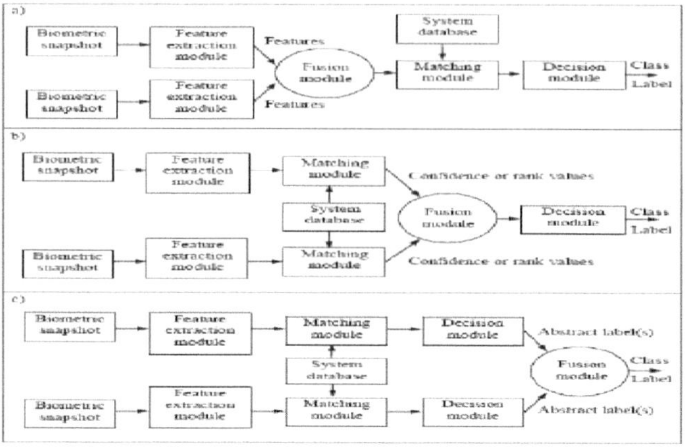


**Fig. 2.** Fusion at different levels

### 3.2 Types of Multimodal Systems

Depending on the traits, sensors and feature sets many different types of multimodal systems are there:

*Single biometric trait, multiple sensors:* Multiple sensors are used to record the same biometric characteristic. The raw data taken from different sensors can then be combined at the feature level or matcher score level to improve the performance of the system.

*Multiple biometrics:* Multiple biometric traits such as fingerprints and face can be combined. Different sensors are used for each biometric characteristic. The interdependency of the traits ensures a significant improvement in the performance of the system. A commercial product BioID uses voice, lip motion and face of a user to verify identity.

*Multiple units, single biometric traits:* Two or more fingers of a single user can be used as a biometric trait. It is inexpensive way of improving system performance, as it doesn't require multiple sensors or incorporating additional feature extraction or matching modules. Iris can also be included in this category.

*Multiple snapshots of single biometric:* In this more than one instance of the same biometric is used for the recognition. For e.g. multiple impressions of the same finger or multiple samples of the voice.

*Multiple matching algorithms for the same biometric:* In it different methods can be applied to feature extraction and matching of the biometric characteristic.

### 3.3 Modes of Operation

A multimodal biometric system can work in three modes:

*Serial mode:* In the serial mode the output of one biometric characteristic is used to reduce the no of possible identities before the next characteristic is used. So multiple source of information is not collected simultaneously.

*Parallel mode:* In it the information from multiple characteristics is taken together to perform recognition.

*Hierarchical mode:* In it individual classifiers are combined in a tree like structure. This mode is well suited where we have large no of classifiers.

## 4 Biometric Encryption

In order to improve the security and privacy of the biometric, we can combine cryptography and biometrics. The conventional cryptography uses encryption keys, which are usually very long. A person cannot remember long random keys so the key is generated from the PIN or password after various transformations. This means that password is the weakest point in the cryptography. Biometrics on the other hand provides person with unique characteristics, which are with him always. But these cannot be used as cryptographic keys as every new biometric sample is always different. Moreover the biometric system always gives output in the form of yes/no, which is one bit of information. So biometric image or template cannot be used as cryptographic key. A solution to this is that we can combine biometric images or templates to a 128 bit key i.e. *Biometric Encryption*. Biometric Encryption is a process that securely binds a PIN or cryptographic key to a biometric so that neither the key nor the biometric can be retrieved from the stored template. The key is recreated only if the correct biometric sample is presented on verification. The PIN can be of any length, as we don't need to remember it. The key is independent of the biometrics so it can be changed or updated easily. Here we have combined fingerprints with the cryptography.

## 5 Fingerprints

A fingerprint [9],[10] is an impression of the friction ridges on all parts of the finger. A friction ridge is a raised portion of the epidermis on the palmar (palm) or digits (fingers and toes) or plantar (sole) skin, consisting of one or more connected ridge units of friction ridge skin. These are sometimes known as "epidermal ridges" which are caused by the underlying interface between the dermal papillae of the dermis and the interpapillary pegs of the epidermis. These epidermal ridges serve to amplify vibrations triggered when fingertips brush across an uneven surface, better transmitting the signals to sensory nerves involved in fine texture perception. Fingerprints may be deposited in natural secretions from the eccrine glands present in friction ridge skin (secretions consisting primarily of water) or they may be made by ink or other contaminants transferred from the peaks of friction skin ridges to a relatively smooth surface such as a fingerprint card. The term fingerprint normally refers to impressions transferred from the pad on the last joint of fingers and thumbs, though fingerprint cards also typically record portions of lower joint areas of the fingers (which are also used to make identifications).

# 6 Fingerprint Features and Classification

Before a print can be classified the relevant features must be extracted, either manually or automatically. These features are of three kinds:

*Level 1 features:* Large scale patterns of the ridge flow.

*Level 2 features:* Finer scale features of the placement and behavior of the ridges themselves, especially instances where ridges meet or terminate. These features are referred to as minutiae.

*Level 3 features:* Details of the individual ridges, e.g. the location of the sweat pores on the ridges, or irregularities in the ridge edges. These entities are much more difficult than the level 1 and 2 features to capture in a fingerprint image.

## 6.1 Level 1 Feature (Ridge Flow Patterns)

Penrose [12] describes how a system of lines (such as the epidermal ridges on the fingertips) that are locally parallel but can curve gradually has certain topological constraints. The lines may fan out and form cusps without disturbing the continuity of the field. However there are two kinds of discontinuity that can arise:

1. A loop is formed when the parallel field turns through 180 and meets itself.
2. A triradius or delta is formed when three fields surround a point, leaving a gap in the centre. The presence of a loop in a pattern requires that a triradius also be present in order to maintain broad scale continuity of the ridge line flow.

These patterns can be divided into 4 types: arch, tented arch, loop and whorl.

An arch fingerprint has ridges that enter from one side, rise to a hump, and leave from the opposite side.

A tented arch is similar to the plain arch, but contains one ridge with a high curvature and contains one loop and one delta.

A loop has one or more ridges that enter from one side, curve back, and leave on the same side they entered. A loop can be further classified as a right loop or a left loop according to whether the ridges enter and leave from the right side or the left side.

A whorl contains at least one ridge that makes a complete 360 path around the centre of the feature. The whorl class can be further divided into two categories: twin loop or plain whorl; the latter may be thought of as two loops that coincide. In either case, a whorl is always associated with exactly two deltas.

## 6.2 Level 2 Features (Ridge Placement and Minutiae)

Several classes of feature may be identified, but, there are seven basic types of fine scale ridge features. However these can all be described as combinations of the basic two: ridge endings and ridge bifurcations.

### 6.3 Level 3 Features (Fine Ridge Details)

Level 3 details include details such as the individual sweat pores are frequently visible in good quality database prints and improved latent fingerprint extraction techniques can often reveal these features in a latent print.

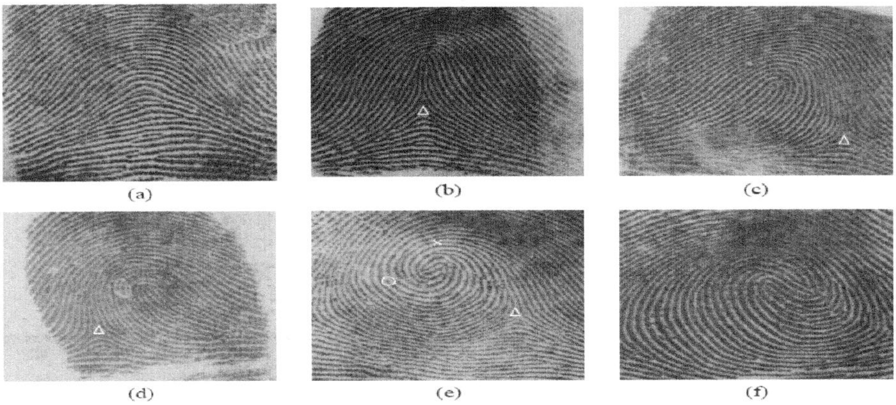

**Fig. 3.** Fingerprint a) arch b) tented arch c) right loop d) left loop e) whorl f) twin loop

## 7 Fingerprint Processing and Feature Extraction

Feature extraction algorithm is:

1. From the input image, estimate the ridge orientation at each point.
2. Generate a ridge representation:
   - Identify the locations of the ridges.
   - Refine the ridges to eliminate imperfections such as spurious gaps.
   - Use a thinning algorithm to reduce the ridges to one pixel width, creating a \skeleton image".
3. Analyze the ridges to identify the location and orientation of the minutiae.

Most existing classification approaches make use of the orientation image. The orientation field may be computed by the application of appropriate image filters sensitive to particular texture orientations For example a 2-D Gabor filter may be applied to detect periodicity in a particular direction. Another way to compute orientation is to employ Principal Component Analysis. This proceeds by taking the intensity gradients in the x and y directions at each point, and constructing a correlation matrix. The orientation derived in this way is effectively the direction in which the intensity variation is a minimum. [11] Describe a method for deriving a high-resolution orientation field via Principal Component Analysis, extracting the singular points, identifying each as a core or delta, and finding their orientations. Minutiae extraction can be performed by using the Crossing Number or Condition Number. The image is first preprocessed, in the course of which it is converted to a binary image. The crossing number is then calculated at each pixel by traversing the eight neighboring pixels and counting the number of transitions between the two possible binary values:

$$CN = \sum_{k=1}^{8} |(P_{k+1} - P_k)|$$

Where Pk has the value 0 or 1, and P9 = P1.

A CN of 2 identifies a ridge ending, while a CN of 6 identifies a bifurcation. The spacing between the ridges, or equivalently the ridge frequency, may be a useful quantity for classification purposes, and can be computed in a number of ways. Most researchers have proceeded by first producing an estimate of the orientation field, then using one or more projections perpendicular to the orientation to infer the ridge frequency.

## 8 Working of a Biometric Encryption

The features extracted from the biometric sample (finger print) is taken which is provided to the BE algorithm to securely bind the key to it to create a protected BE template. This BE template is called a private template. The private template can now be stored either in a database or locally on a smart card, laptop, cell phone etc. At the end of the enrollment both the key and fingerprints are discarded.

On verification, the user gives his/her fresh biometric sample (finger print) which is then provided to the BE template which will then be used by BE algorithm to retrieve the same key/password. So biometric acts as a decryption key. At the end of the verification fingerprint sample is again discarded. Attacker cannot get access, as his fingerprint sample will be different, so he will not be able to retrieve the password.

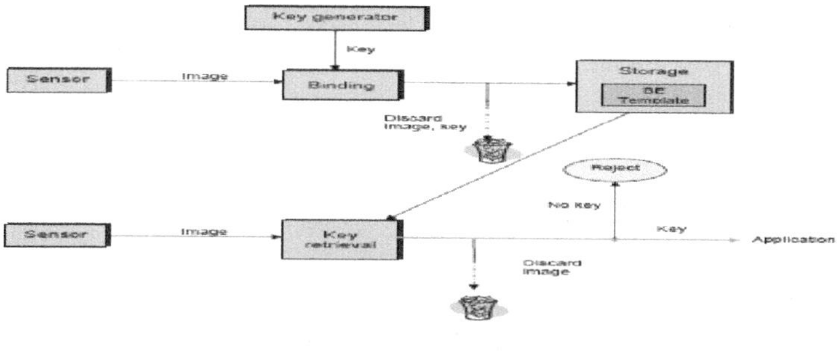


**Fig. 4.** Biometric Encryption Process

## 9 Advantages of Biometric Encryption

There are many advantages of Biometric Encryption over conventional Biometric system.

1. No need to keep the image or template in the memory. The template/image and the key can be discarded at the end of the enrollment process.
2. BE can enable the creation of a private and highly secure anonymous database structure for personal information in large private or public database.
3. BE allows individuals to use single biometric for multiple accounts and purposes. There is no fear that these separate identifiers or users will be linked together by a single biometric image or template.
4. Improved security of personal data and communications.
5. No large databases of biometrics are created. Only biometrically encrypted keys.
6. Suitable for large-scale applications.

## 10 Applications of a Biometric Encryption

Biometric Encryption can be employed anywhere where traditional biometric can be used i.e. for authentication, access control and to encrypt data. BE is well suited for use with smart cards and mobile devices. It can also be used to create biometric tickets such as personalized boarding passes. With the help of BE, employees can easily and quickly encrypt, digitally sign electronic documents and messages without the fear of unnecessarily revealing personal data. BE generates alphanumeric text string so its output can be easily reproduced anywhere on disposable media such as bar codes, on boarding passes, on pill bottles or on any sensitive document or product over which ownership or authenticity is needed. As BE techniques are configurable, new identifiers can be generated from the same biometric to replace existing ones if they are lost, stolen or comprised by the organization. Identity frauds and thefts arising from stolen identifiers can be reduced by the BE technology.

## 11 Conclusion

Biometric technology adds a new layer of security by ensuring secure identification and authentication. But biometric authentication systems like any other technology are also vulnerable to attacks such as transmission, replay and spoofing. There are many proposed methodologies that are used to defeat them. Multimodal biometric system is a major approach to defeat spoofing attacks. Biometric Encryption is a positive sum technology that achieves strong authentication, security and privacy. It uses both the traditional cryptography and fingerprints (biometric) features.

## References

1. Jain, A.K., Ross, A.: Multibiometric systems. Communications of the ACM, Special Issue on Multimodal Interfaces 47(1), 34–40 (2004)
2. Ross, A., Jain, A.K.: Multimodal biometrics: An overview. In: Proc. of 12th European Signal Processing Conference, pp. 1221–1224 (September 2004)

3. Indovina, M., Uludag, U., Snelick, R., Mink, A., Jain, A.K.: Multimodal biometric authentication methods: A COTS approach. In: Proc. of Workshop on Multimodal User Authentication, Santa Barbara, CA, pp. 99–106 (December 2003)
4. Jain, A.K., Uldang, U.: Attacks on biometric system: A case study in fingerprints. In: Proc.SPIE-EI (January 2004)
5. Qiughan Security Issues in Biometric Authentication. Proc. of the IEEE (2005)
6. Cavoukian, A., Stoianov, A.: Biometric Encryption: A Positive-Sum Technology that Achieves Strong Authentication, Security and Privacy (March 2007), http://www.ipca.on.ca/images/resources/up-1bio-encypt
7. Biometric Encryption FAQ, http://www.ipca.on.ca
8. Webb, J.: The myth of fingerprints. New Scientist 2517(3) (September 2005)
9. Maltoni, D., Jain, A.K.: Handbook of Fingerprint Recognition. Springer, Heidelberg (2003)
10. Bazen, M., Gerez, S.H.: Methods for the computation of the directional fields and singular points of fingerprints. IEEE Transactions on Pattern Analysis (July 2002)
11. Bolle, R.M., Connell, J.A., Pankanti, S., Ratha, N.K., Senior, A.W.: Guide to Biometrics. Springer, Heidelberg (2004)
12. Coetzee, L., Botha, E.C.: Fingerprint recognition in low quality images. Pattern Recognition 26(10), 1441–1460 (1993)

# Node Architectures and Its Deployment in Wireless Sensor Networks: A Survey

Sumit Kushwaha[1], Vinay Kumar[2], and Sanjeev Jain[2]

[1] ECE Department,
S.H.I.A. Technology and Sciences,
Allahabad, (U.P.)
India
sumit.kushwaha1@gmail.com
[2] ECE Department,
Motilal Nehru National Institute of Technology,
Allahabad, (U.P.)
India
{Vinay2008el03mnnit,snjece}@gmail.com

**Abstract.** In conventional wireless sensor networks (WSNs), adding a few mobile nodes can greatly improve the control and sensing capabilities of the networks and can help researchers solve many challenges such as network deployment and scalability etc. The video capture, processing, and communication in wireless video sensor networks depend on the resources of the nodes forming the sensor networks. The major challenge in designing wireless sensor networks (WSNs) is the support of the functional, such as data latency, and the non-functional, such as data integrity etc. Careful sensor node placement can be a very effective optimization, means for achieving the desired design goals. In this paper we thoroughly survey and contrast the existing sensor node architectures. We also survey the research on optimized node placement in WSNs. We have provided the placement strategies into static and dynamic depending on whether the optimization is performed at the time of deployment or while the network is operational.

**Keywords:** Sensor Node Architectures, Node placement, Positioning, Node relocation, Wireless sensor networks.

## 1 Introduction

Recent years have witnessed an increased interest in the use of WSNs in numerous applications such as disaster management, multimedia surveillance sensor networks, traffic congestion avoidance systems, advanced health care delivery, industrial process control, storage of potentially relevant activities etc [1], [2]. In this survey we have focused on two important things first one is sensor node architectures and second one is the deployment of the sensor nodes in WSNs. In sensor node architectures firstly we have introduced the classical node architecture after that we have introduces

A. Mantri et al. (Eds.): HPAGC 2011, CCIS 169, pp. 515–526, 2011.

the enhanced version of the node architectures like wireless video sensor nodes, multimedia sensor nodes etc. We comprehensively survey the existing wireless video sensor node platforms (WVSNPs) considering the hardware and software components required for implementing the wireless video sensor node functionalities. All functional aspects of a wireless video sensor network ranging from the video capture and compression to the wireless. Mostly sensor nodes are deployed to measure scalar phenomena such as temperature, pressure, humidity etc. The integration of low power wireless networking technologies with inexpensive hardware such as Complementary Metal-Oxide Semiconductor (CMOS), cameras and microphones is now enabling the development of distributed networked systems that we refer to as Wireless Multimedia Sensor Networks (WMSNs). WMSNs have ability the retrieval of multimedia data streams, inbuilt signal processing and transmit it [3]. Different design schemes for WSNs pursue optimization at the various layers of the communication protocol stack. One of the design optimization strategies is to deterministically place the sensor nodes in order to meet the desired performance goals. In such case, the coverage of the monitored region can be ensured through careful planning of node densities and fields of view and thus the network topology can be established at setup time. However, in many WSNs applications, sensors deployment is random and little control can be exerted in order to ensure coverage and yield uniform node density while achieving strongly connected network topology. Therefore, controlled placement is often pursued for only a selected subset of the employed nodes with the goal of structuring the network topology in a way that achieves the desired application requirements. In addition to coverage, the nodes positions affect numerous network performance metrics such as energy consumption, delay and throughput. Given the importance of sensor node architecture and its deployment in WSNs, rest of paper is organized in following structure Section 2 presents the sensor node architectures in WSN. Section 3 presents deployment of the sensor nodes in WSNs. Section 4 presents the conclusion of the paper.

## 2 Sensor Node Architectures

### 2.1 Classical Architectures

Fig. 1 shows the architecture of a wireless sensor node, it consists of four main subsystems (1) a sensing subsystem that has one or more sensors for data acquisition (2) a radio communication subsystem for wireless data communication; (3) a processing subsystem that has a micro-controller and memory for data processing and (4) a power supply unit. Obviously, the power breakdown depends heavily on the specific node type. It is shown that the power characteristics of stargate sensor node are different from Mote-class sensor node [36]. There are number of sources for power consumption in Sensor including signal sampling, conversion of physical signals to electrical and signal conditioning Analog to digital conversion.

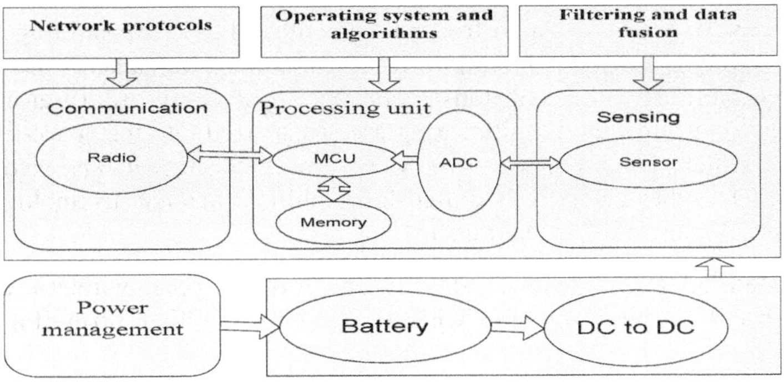


**Fig. 1.** Sensor Node System Architecture

## 2.2 Wireless Multimedia Sensor Networks

The group of wireless multimedia sensor nodes is called a wireless multimedia sensor network (WMSN). The multimedia sensor node consists of sensing, processing, communication, and power subsystems. The processor subsystem is the central unit of the multimedia sensor node and a processor unit must be considered in terms of both energy-efficiency and performance basis. The sensing unit is detecting the sound, still or moving pictures including the relevant scalar parameters such as temperature, humidity, pressure etc. After detecting the real-world signals, this fed in to the processing unit in the digital form with the help of Analog to digital converter (ADC). The processed signal is interfaced with the memory, communication sub-systems and coordination sub-systems. The memory has some data mining and feature extraction algorithms. These algorithms play a role to identify the important characteristics of the event. The mobility and actuation unit enables the movement of motor according to the instruction or manipulation of receiving data. The whole architecture (system) is powered by a power unit with energy conservation technique and also having an energy scavenging unit, such as solar cells, as shown in Fig. 2.

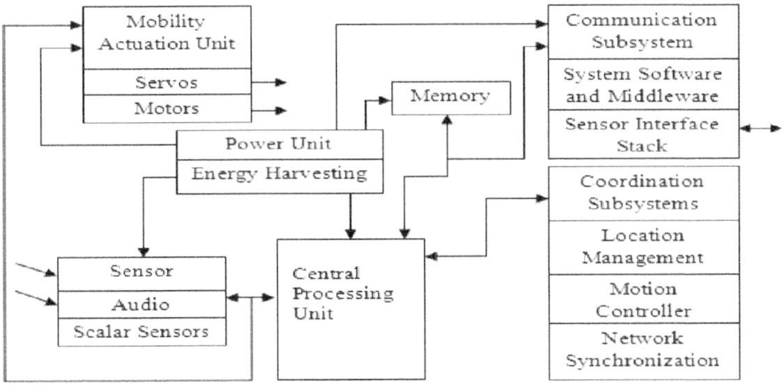


**Fig. 2.** Internal Architecture of wireless multimedia sensor network

### 2.3 Wireless Video Sensor Node Platforms (WVSNPs)

Wireless Video Sensor Node (WVSN) is very good medium for the analysis of different types of phenomena occur in the real world. For example, in the coastal setting cameras can be used to monitor sea state including wave height and period, evaluate land erosion, and observe a variety of animal species. A group of many WVSNs creates Wireless Video Sensor Networks Platforms (WVSNPs). Here, platforms provide a method for the transmission of analyzed data. There are several limitations in the deployment of WVSNs in the designed area because of energy, data rate, bandwidth, lifetime, coverage, high cost etc. Conversely, WVSNs design plays a major role in the avoidance of all the limitations and improves performance. WVSNs are performing some tasks such as video capturing and compression and transmitting this data to the sink node in the WVSNPs. After detailed study of the WVSNPs [4], [5], [6], we conclude three basic requirements for WVSNPs which are power consumption, throughput, and cost. First one, The power requirements are influenced by a wide range of design choices, including power source type, component selection, power management hardware and software, and importantly sensor node and network management algorithms, such as implemented by a real time operating system (RTOS) [7] or sensor network duty cycling schedules [8]. To make the WVSNs energy-efficient, WVSNs has to provide following power modes: on, Ready, Doze, Sleep, Idle, Hibernate etc  for general background on microprocessor design and their power-efficient design and operation [9], [10], [11], [12], [13].

### 2.4 XYZ Node Architecture Design

WVSN is designed around the 57.6 MHz 32-bit OKI Semiconductors ML67Q500x ARM THUMB. The MCU has an internal 256 KB of Flash, 32 KB of RAM, and 4 KB of boot ROM as well as external SRAM, as shown in Fig. 3. The Omni vision off-the-shelf OV7649 camera module and the 32×32 pixel event-based ALOHA CMOS imager have been connected to the XYZ node in separate research efforts [15], [16]. The OV7649 can capture VGA (640×480) and quarter VGA (QVGA, 320×240) images. The image data is transferred from the camera to the on-board SRAM with an 8-bit parallel port using direct memory access (DMA), which does not involve the MCU.

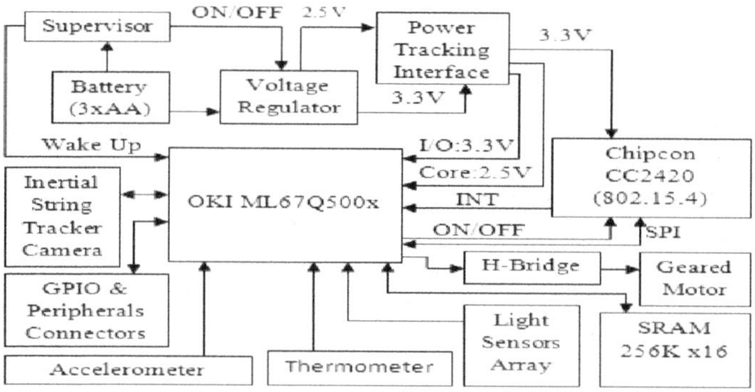


**Fig. 3.** XYZ Sensor Node Architecture

## 2.5 Efficient-XYZ Sensor Node Architecture

The rapid progress in WVSNPs is constantly revealing new unexplored problems and applications that create the requirement for a diverse set of sensing platforms. After the detailed study of XYZ: WVSN architecture, we suggested a new enhanced Efficient-XYZ: WVSN architecture. In this architecture, we replace Inertial String Tracker (IST) camera by CMOS Imagers, and add two more blocks in its architecture, named as Power Booster and CPLD, to make this WVSN energy-efficient, increase performance and sensibility. Here, we provide an overview about the Enhanced Efficient-XYZ WVSN architecture. **Long Term Deep Sleep (LDS) Mode** [14], to support long term deep sleep we have implemented a supervisor circuit outside the processor operated by a real-time clock (RTC) with two interrupts [17]. With this circuit, the node has software control to transition into a deep-sleep mode by disabling its main power supply regulator. **Mobility Support Mechanism** [14], the mobility subsystem consists of an H-bridge and a miniature geared motor from a pager device. The mobility subsystem is implemented on an additional accessory board that transforms XYZ to a 2-D motion enabled sensor node. Two output pins of the processor control the motor direction and braking. The motor is controlled via an H-bridge which converts the 3.3V logic of the CPU to a 5V DC supply necessary for supporting speed and motor power requirements. A LED/Photo-transistor pair focused on a 4 segment black and white pattern that is pasted directly onto a wheel acts as an optical encoder and is used as odometer. A transition from one segment to another, detected by the sensor, occurs when the sensor node is moving. The transition is fed directly to a counter on the sensor node, and by accumulating the counter values, a position value is determined. Advantage of this architecture is the MCU provides numerous peripherals which can be turned on and off as required by the application. The on and off switching is accomplished through software enabling/disabling of clock lines to MCU peripherals. The node is therefore capable of a myriad of power management algorithms.

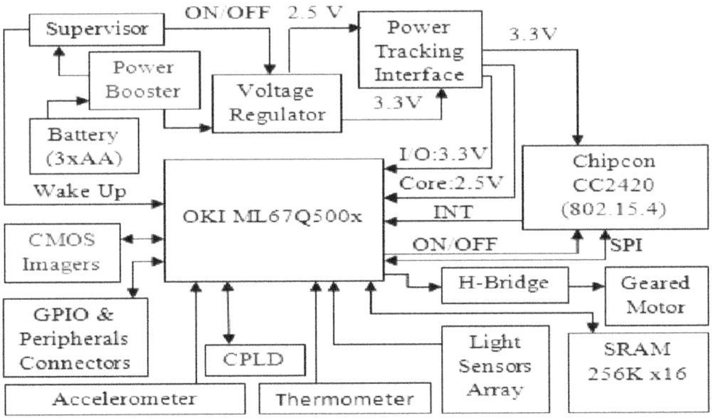


**Fig. 4.** Enhanced Efficient XYZ Sensor Node Architecture

The XYZ is a motion-enabled and power aware sensor platform targeting distributed sensor network applications. As illustrated in Figure, the platform consists of several subsystems, including subsystems for sensing (light, temperature, and accelerometer), communication (TI CC2420 Zigbee radio), mobility (geared motor), power (voltage regulator, power tracker, supervisor, and three AA 1.2 V Ni-MH rechargeable batteries pack) and a camera. The capacities of the batteries range from 1200 to 2000 mAh.

### 2.6 Enhanced Efficient-XYZ

**WVSN Design Concept and Architecture,** we design Efficient-XYZ node as a video sensor node capable of wireless video streaming via Zigbee. Such efficient-xyz node architecture can be made by modifying the old xyz node architecture in term of adding three blocks in the design architecture such as (1) power booster, (2) CMOS imager, and (3) CPLD, and these three enhanced blocks are showing in red color in figure (4) which functioning are describe ahead. **Power Booster,** a boost converter (step-up converter) is a power converter with an output DC voltage greater than its input DC voltage. It is a class of switching-mode power supply (SMPS) containing at least two semiconductor switches (a diode and a transistor) and at least one energy storage element. Filters made of capacitors (sometimes in combination with inductors) are normally added to the output of the converter to reduce output voltage ripple. Power can also come from DC sources such as batteries. A process that changes one DC voltage to a different DC voltage is called DC to DC conversion. A boost converter is a DC to DC converter with an output voltage greater than the source voltage [18]. **CMOS Imager,** CMOS Imager is a system of reducing power consumption in and active pixels sensor. The sensor is broken into different blocks, and each of the blocks is individually optimized. The optimization may include minimizing the parasitic capacitance on the readout bus, turning off biases when not in use and operating in a way that minimizes static power consumption of different elements such as A/D converters [37]. **CPLD,** [38] any engineer involved with portable or handheld products knows that minimizing power consumption is an absolute requirement for today's designs. But only the veterans understand the subtle yet important details that can stretch a systems' battery life to the maximum. In this white paper we will focus on how those seasoned experts use ultra-low-power complex programmable logic devices (CPLDs) to wring out every last microwatt from the I/O subsystems of their embedded designs.

## 3 Network Deployment in WSNs

### 3.1 Static Sensor Node Deployment Scheme

In WSNs, the sensor nodes are deployed in such a manner, in real-world, to make the network and operation energy-efficient in all respect. Before the sensor node deployment, consider about the strategies how we deployed the nodes to make the network efficient. Suppose the location of the sensor nodes are fixed in such a way, to get the maximum throughput and life time of the sensor nodes. The static metrics are applied for area coverage, inter-node connectivity, and among others. The routing path for transmission of information in static network scheme in almost fixed. In this section, as Fig. 5 shows, we explore the static sensor node deployment scheme including the sensor node deployment methodology, optimization objectives of deployment and role of sensor node in WSNs. **Deployment Methodology**, sensor nodes are deployed either in sequence manner or random manner. The deployments of sensor nodes are depend upon the various conditions. The condition is nothing but the situation of the real-world where sensor nodes to be deployed and work for efficient operation. The nodes are deployed in controlled manner when it is applied for the viable purpose and on the other side; the nodes are deployed in random manner when it is applied for the harsh environments and others. This methodology is of two types: **Controlled Node Deployment**, in this sensor nodes are deployed in controlled manner for the indoor applications such as in surveillance, medical field, etc. The Active Sensor Network (ASN) project at the University of Sydney in Australia [19], the Multiple Sensor Indoor Surveillance (MSIS) project at Accenture Technology Labs in Chicago [20], and the Sensor Network Technology (SNT) projects at Intel [21] are the examples of indoor applications. The ASN and MSIS projects applied for border security and surveillance applications which are safe and secure in installation and all the available property of person or company. **Random Node Deployment**, the random deployment of the sensor nodes can be done only when such types of conditions required. The conditions are as there is no possibility to deploy the sensor nodes in deterministic manner such as in harsh environments, forest fire detection, animal life cycle, border security in sensitive area etc.

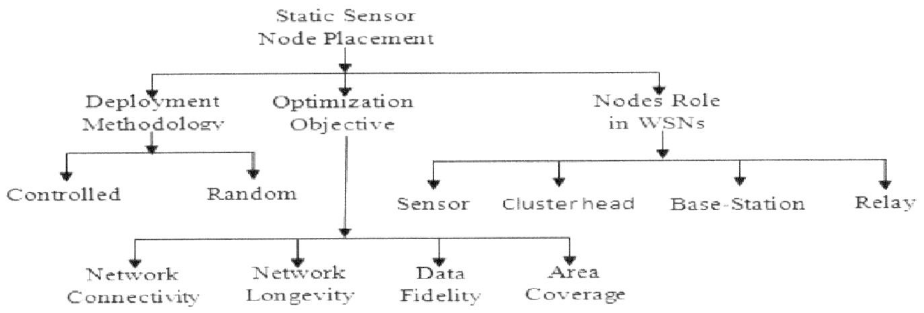


**Fig. 5.** Classification of Static Sensor Node Placement

**Optimization Objectives,** researchers always think about the deployment of sensor node in such a way that it covers whole design objectives. So, they suggested various sensor node deployment techniques in the literature that have concentrated on increasing the area coverage, data fidelity, network longevity, and network connectivity. Another considerable objective is to traffic load balancing and tolerance of node failure. Here, we are facing the problem in design objective through the random sensor node distribution. In WSNs, to keep the minimum utilization of required network resources are very difficult. **Area Coverage:** To fix the data rate in the coverage of sensor node's field are used to measure the collective coverage of deployed sensors. Some of the published papers, especially early ones, use the ratio of the covered area to the size of the overall deployment region as a metric for the quality of coverage [22]. The advantage of exposure-based area coverage with the constant data rate is the inclusion of a real object detection probability i.e. it is based on signal processing formulations, e.g., signal distortion, as applicable to specific sensor types [23]. **Network Connectivity:** Network connectivity is another problem in WSNs design, to maintain it up to the sensor nodes life-time. There are optimal number of sensor nodes are deployed and they are connected to each other with the transmission of data. Sensor network connectivity is necessary to secure that data messages are propagated to the appropriate base station and the loss of network connectivity if often treated as the end of network life. **Network Longevity:** Challenges of network longevity in the design technique of WSNs because of the all required energy resources are fixed. The recharging or replacing of the sensor node's battery in the sensor network may be quite difficult. It is facing severe problems in the data communication and signal processing time between all sensor nodes in the sensor network. **Data Fidelity:** It is important for considering the design objective of WSNs probably to ensure the credibility of the received data. A sensor network basically provides a collective assessment of the detected phenomena by fusing the readings of multiple independent sensors [24]. **Nodes Role in WSNs:** The sensor nodes position can affected both area coverage as well as the properties of the network topology. Mostly researches are based on the architecture of the sensor network to optimize the performance of the sensor nodes such as considering the longevity, latency, data rate etc. These architectures are assigning for the active sensor nodes and applied a node specific positioning strategy on the sensor node depend upon the role play in the network. Since the previous section covered the sensor nodes deployment scheme **Relay Nodes:** a wireless sensor network consists of many low-cost, low-power sensor nodes. These sensor nodes can perform detecting, analyzing, and transmission of detected data. **Sensor Nodes:** Sensor node is a node in WSNs that is capable of performing some processing, gathering, sensory information and communicating with other connected nodes in the network [25]. **Cluster-Head:** Cluster is defined as the group of similar objects or the process of finding a natural association among some specific objects or data. It finds applications in many fields. Clustering, specifically in sensor networks, could be used to solve a verity of problems. It uses clusters to transmit processed data to base stations, hence minimizing the number of nodes that take part in long distance communication. This directly affects the overall system energy dissipation. A wireless sensor network is to use the sensor nodes limited energy efficiently by using the clustering technique and also can control the energy consumption of WSNs. **Base- station:** In WSNs, a base station is a radio transceiver that operates as a hub of the local wireless sensor network. It may also be serves as a gateway between a wired sensor network and the wireless sensor network. It is a low-power transceiver (i.e. energy-efficient) and also acts as a wireless router [26]. The base stations are also called a control or fixed stations.

### 3.2 Dynamic Repositioning of Nodes

Mostly techniques discussed above, analyzed the optimal position for the sensor nodes and don't consider moving them, once they have been fixed. The deployment design scheme has made at the time of sensor network establishment and it doesn't

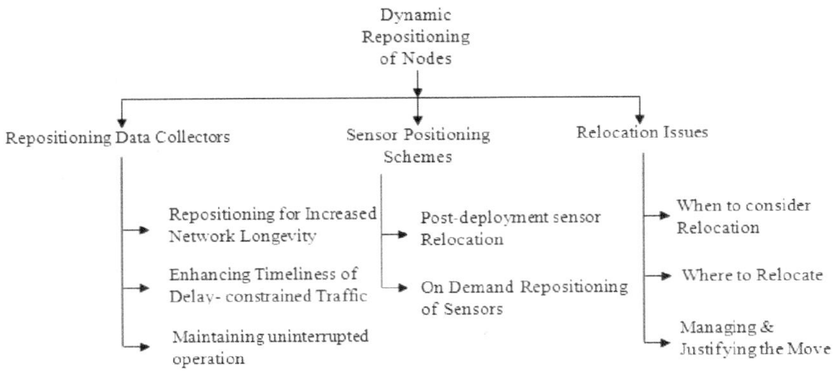


**Fig. 6.** Classification of Dynamic Repositioning of Nodes

allow for dynamic changes at the time of sensor network operation, as shown in Fig. 6. For example, the traffic pattern can be dynamically changed but the traffic load can not be distributed equally among all the sensor nodes, because of bottlenecks.

**Relocation Issues,** the need of relocation of sensor nodes can be processed very carefully since it may create problem in data delivery. There are some points to be considering at the time of relocate the sensor nodes. They are: **When to Consider Relocation:** Sensor nodes are needs to relocation in the sensor network because of its unwanted performance measure or a wish to boost such measures at that point where acceptable nodes performance measure. Example are the analysis of bottlenecks in the data relaying, decrease performance in coverage of the sensor nodes in the sensor networks, increase in data latency or consumes more energy on the per deliver data. So, "when" and "where" relocation issue of nodes deployment needs to qualify the impact of repositioning at new location on the network performance. **Where to Relocate:** To find out an optimal relocation for the sensor nodes in a multi-hop sensor network is a very complex problem. A mathematical formulation of the sensor nodes repositioning problem can be involve large number of parameters including the positions of all deployed nodes (including energy reserve, transmission range, etc.) and the data sources in the network. The dynamic nature of the network makes the sensor nodes state and sources of data may change rapidly; thus the optimization process may have to be repeated frequently [28]. **Managing and Justifying the Move:** After the new relocation decided for the sensor nodes, the nodes should identity a travel path to the new relocation. The minimization of the travel distance for the sensor nodes are decisive, since the mechanical parts consumes more energy than the communication and computation energy. Since, a WSN is usually deployed in harsh environments to detect and track dangerous targets or events, the node should avoid exposure to harm or getting trapped. For example, the node should not go through a fire to reach the new location [27]. **Sensor Repositioning Schemes:** Mobile sensor nodes have been exploited to increase the performance of WSNs. For example, mobile sensor nodes can be repositioned in designed area to get uniform coverage, move closer to loaded nodes in order to prevent bottlenecks, and increase bandwidth of the channel for transferring the data to the base station [29]. **Post-deployment Sensor Relocation:** This type of repositioning schemes is continued at the end of the sensor node deployment phase when the sensor nodes are already deployed in the designed area. The main aim is to maximize the designed area covered within the shortest time duration and with minimum over-head in terms of travel distances and inter-sensor message traffic. There are three methods have been proposed to decide where relocation of sensor nodes can be takes place: vector-based (VEC), Voronoi-based (VOR), and minimax [30]. **Repositioning Data Collectors:** The data collect by the sensor nodes is gathered at base station or channels for collection and some additional signal processing consistent manner with the computational capabilities of such data collectors. In dynamic relocation, data collectors have been continued as a means for boosting the network performance, dealing traffic bottlenecks and preventing interruption in network operation. **Repositioning for increased network longevity**, dynamically changes in the energy-efficient routing technique and traffic pattern can be accepted by the sensor nodes in designed network area, sensors near data collectors die quickly as they are the most utilized nodes in the network [31], [32], because they are continuously receive data from neighbor sensors and transmit it to the data collectors. So, it can consume more energy due to the heavy traffic load and for the result, quickly die. The sensor nodes can be spread in a spiral manner and loosing the nodes energy. Hence the longevity of the sensor network goes to quiet decrease [33]. To over come out of this problem, the data collectors can be repositioned continuously in the sensor network and increase the longevity of the network, before repositioning of the data collectors. In **Enhancing Timeliness of Delay**- constrained Traffic, in few cases, repositioning of the data collectors is very useful to increase the life-time of the sensor networks. But in real-time, the route of data collectors is busy due to the increasing of traffic load. Cause of this, the data packets are not receive in timely manner and due to delay in reception, possibility to lose of the data packets [34],[27]. In **Maintain Interrupted Operation**, the repositioning of data collectors have been continued to keep WSNs operational without interruption.

## 4 Conclusion

In this paper, we have survey the different architectures of sensor nodes in WSNs. These architectures are classical, multimedia and XYZ node architectures. General purpose platforms in Video Sensor Node Platforms (WVSNPs) are classified into, heavily coupled platforms, and externally dependent platforms. We have found that the existing platforms do not meet practical WVSNP power consumption limits, cost constraints etc. In this survey paper we have also discussed the effect of sensor node placement strategies on the operation and performance of WSNs. We have divided the various approaches for node positioning into static and dynamic. On other hand static approaches, optimized sensor node placement is pursued in order to get some required properties for the sensor network topology and its coverage.

## References


1. Akyildiz, I.F., Su, W., Sankarasubramaniam, W., Cayirci, E.: Wireless sensor networks: a survey. Computer Networks 38, 393–422 (2002)
2. Chong, C.Y., Kumar, S.P.: Sensor networks: Evolution, opportunities, and challenges. Proceedings of the IEEE 91(8), 1247–1256 (2003)

3. Akyildiz, I.F., Melodia, T., Chowdhury, K.R.: A survey on wireless multimedia sensor networks (December 2007)
4. Akyildiz, I.F., Melodia, T., Chowdhury, K.R.: A survey on wireless multimedia sensor networks. Computer Networks 51(4), 921–960 (2007)
5. Hill, J.L.: System architecture for wireless sensor networks. Ph.D. dissertation, University of California at Berkeley (2003)
6. Seema, A.: Blueprint for a low-cost piggyback wireless video sensor node/platform architecture. M.S. thesis, Arizona State University, Electrical Engineering, pp. 1–201 (Apr 2009)
7. Dong, W., Chen, C., Liu, X., Bu, J.: Providing OS support for wireless sensor networks: Challenges and approaches. IEEE Commun.Surveys & Tutorials, in print (2010), doi:10.1109/SURV.2010.032610.00045
8. Anastasi, G., Conti, M., Francesco, M.D., Passarella, A.: Energy conservation in wireless sensor networks: A survey. Ad Hoc Networks 7(3), 537–568 (2009)
9. Chandrakasan, A., Bowhill, W., Fox, F.: Design of High- Performance Microprocessor Circuits. Wiley, Chichester (2000)
10. Parhi, K.: VlSI Digital Signal Processing Systems: Design And Implementation. Wiley, Chichester (2007)
11. Raghunathan, A., Jha, N.K., Dey, S.: High-Level Power Analysis and Optimization. Kluwer Academic Publishers, Dordrecht (1998)
12. Agarwal, K., Singh, H., Sylvester, D., Nowka, K.: Enhanced leakage reduction techniques using intermediate strength power gating. IEEE Trans. Very Large Scale Integr (VLSI) Syst. 15(11), 1215–1224 (2007)
13. Yeap, G.: Practical low power digital VLSI design. Kluwer Academic Publishers, Dordrecht (1998)
14. Lymberopoulos, D., Savvides, A.: XYZ: a motion-enabled, power aware sensor node platform for distributed sensor network applications. In: Proc. Int. Symposium on Information Processing in Sensor Networks (IPSN), pp. 449–454 (2005)
15. Teixeira, T., Andreou, A.G., Culurciello, E.: Event-based imaging with active illumination in sensor networks. In: Proc. IEEE International Symposium on Circuits and Systems (ISCAS), pp. 644–647 (2005)
16. Culurciello, E., Andree, A.G.: ALOHA CMOS imager. In: Proc. IEEE Int. Symposium on Circuits and Systems (ISCAS), pp. 956–959 (2004)
17. Akyildiz, I.F., Melodia, T., Chowdhury, K.R.: A survey on wireless multimedia sensor networks. Computer Networks 51(4), 921–960 (2007)
18. Boost converter - Wikipedia, the free encyclopedia, `http://en.wikipedia.org/wiki/Boost_converter`
19. Brooks, A., Makarenko, A., Kaupp, T., Williams, S., Durrant-Whyte, H.: Implementation of an indoor active sensor network. In: the Proceedings of the 9th International Symposium on Experimental Robotics, Singapore (June 2004)
20. Petrushin, V.A., Wei, G., Shakil, O., Roqueiro, D., Gershman, V.: Multiple-Sensor Indoor Surveillance System. In: The Proceedings of the 3rd Canadian Conference on Computer and Robot Vision (CRV 2006), Québec city (June 2006)
21. Krishnamurthy, L.: Design and Deployment of Industrial Sensor Networks: Experiences from a Semiconductor Plant and the North Sea. In: The Proceedings of the 3rd ACM Conference on Embedded Networked Sensor Systems (SenSys 2005), San Diego, CA (November 2005)

22. Huang, C.F., Tseng, Y.C.: The Coverage Problem in a Wireless Sensor Network. In: The Proceedings of the ACM 9th Annual International Conference on Mobile Computing and Networking (MobiCom 2003), San Diego, CA (September 2003)
23. Meguerdichian, S., Koushanfar, F., Potkonjak, M., Srivastava, M.B.: Coverage problems in wireless ad-hoc sensor networks. In: The Proceedings of the 20th International Annual Joint Conference of the IEEE Computer and Communications Societies (INFOCOM 2001), Anchorage, Alaska (April 2001)
24. Ganesan, D., Cristescu, R., Beferull-Lozano, B.: Power-efficient sensor placement and transmission structure for data gathering under distortion constraints. In: The Proceedings of the 3rd International Symposium on Information Processing in Sensor Networks (IPSN 2004), Berkeley, CA (April 2004)
25. Wireless sensor nodes - Wikipedia, the free encyclopedia, `http://en.wikipedia.org/wiki/wireless_sensor_nodes`
26. Base stations - Wikipedia, the free encyclopedia, `http://en.wikipedia.org/wiki/base_stations`
27. Younis, M., Akkaya, K.: Strategies and Techniques for Node Placement in Wireless Sensor Networks: A Survey
28. Efrat, A., Har-Peled, S., Mitchell, J.S.B.: Approximation Algorithms for Two Optimal Location Problems in Sensor Networks. In: The Proceedings of the 3rd International Conference on Broadband Communications, Networks and Systems (Broadnets 2005), Boston, Massachusetts (October 2005)
29. Wu, J., Yang, S.: SMART: A Scan-based Movement Assisted Sensor Deployment Method in Wireless Sensor Networks. In: The Proceedings of the 24th International Annual Joint Conference of the IEEE Computer and Communications Societies (INFOCOM 2005), Miami, FL (March 2005)
30. `http://www.symbio.jst.go.jp/symbio/morph.html`
31. Younis, M., Youssef, M., Arisha, K.: Energy-Aware management in Cluster-Based Sensor Networks. Computer Networks 43(5), 649–668 (2003)
32. Heo, N., Varshney, P.K.: Energy-Efficient Deployment of Intelligent Mobile Sensor Networks. IEEE Transactions on Systems, Man, Cybernetics, Part A 35(1), 78–92 (2005)
33. Heinzelman, W., Chandrakasan, A., Balakrishnan, H.: Energy-efficient communication protocol for wireless sensor networks. In: The Proceeding of the Hawaii International Conference System Sciences (HICSS 2000), Hawaii (January 2000)
34. Younis, M., Bangad, M., Akkaya, K.: Base-Station Repositioning For Optimized Performance of Sensor Networks. In: The Proceedings of the IEEE Vehicular Technology Conference (VTC-Fall 2003), Orlando, Florida (October 2003)
35. Akkaya, K., Younis, M.: Relocation of Gateway for Enhanced Timeliness in Wireless Sensor Networks. In: The Proceedings of the IEEE Workshop on Energy-Efficient Wireless Communications and Networks (EWCN 2004), in conjunction with the 23rd IEEE International Performance Computing and Communications Conference (IPCCC 2004), Phoenix, Arizona (April 2004)
36. Raghunathan, V., Schurghers, C., Park, S., Srivastava, M.: Energy-aware Wireless Microsensor Networks. IEEE Signal Processing Magazine, 40–50 (March 2002)
37. CMOS image sensor with a low-power architecture, `http://www.freepatentsonline.com/7286180.html`
38. Practical Low Power CPLD Design. A Lattice Semiconductor White Paper, Lattice Semiconductor, 5555 Northeast Moore Ct., Hillsboro, Oregon 97124 USA (August 2009), `http://www.latticesemi.com`

# New Innovations in Cryptography and Its Applications

Saurabh Sharma[1] and Neeraj Kumar Mishra[2]

[1] Dept. of Computer Science & Engineering,
Gyan Ganga Institute of Technology & Sciences, Jabalpur (M.P.)
saurabh.sharma44@gmail.com
[2] Dept. of Computer Science & Engineering,
Shri Ram Institute of Technology,
Jabalpur (M.P.)
neerajmishra127@gmail.com

**Abstract.** The invention of public-key cryptography was of central importance to the field of cryptography and provided answers to many key management problems for large scale networks. For all its benefits, however, public-key cryptography did not provide a comprehensive solution to the key management problem. Indeed, the possibilities brought forth by public-key cryptography heightened the need for sophisticated key management systems to answer questions such as the following:

"How can I easily encrypt a file once for a number of different people using public-key cryptography?"

"If I lose my keys, how can I decrypt all of my files that were encrypted with those keys?"

"How do I know that I really have Alice's public key and not the public key of someone pretending to be Alice?"

"How can I know that a public key is still trustworthy?"

The paper discusses public key cryptography and its use in applications such as Key Agreement, Data Encryption and Digital Signature. The paper discusses some public key algorithms such as DH, RSA, and DSA and also gives working explanations of these algorithms.

## 1 Introduction

The data transferred from one system to another over public network can be protected by the method of encryption. On encryption the data is encrypted/scrambled by any encryption algorithm using the 'key'. Only the user having the access to the same 'key' can decrypt/de-scramble the encrypted data. This method is known as private key or symmetric key cryptography.

The concept of securing messages through cryptography has a long history. Indeed, Julius Caesar is credited with creating one of the earliest cryptographic systems to send military messages to his generals.

A. Mantri et al. (Eds.): HPAGC 2011, CCIS 169, pp. 527–538, 2011.

Throughout history, however, there has been one central problem limiting widespread use of cryptography. That problem is *key management*. In cryptographic systems, the term *key* refers to a numerical value used by an algorithm to alter information, making that information secure and visible only to individuals who have the corresponding key to recover the information. Consequently, the term key management refers to the secure administration of keys to provide them to users where and when they are required.

Historically, encryption systems used what is known as symmetric cryptography. Symmetric cryptography uses the same key for both encryption and decryption. Using symmetric cryptography, it is safe to send encrypted messages without fear of interception (because an interceptor is unlikely to be able to decipher the message); however, there always remains the difficult problem of how to securely transfer the key to the recipients of a message so that they can decrypt the message.

A major advance in cryptography occurred with the invention of public-key cryptography. The primary feature of public-key cryptography is that it removes the need to use the same key for encryption and decryption. With public-key cryptography, keys come in pairs of matched "public" and "private" keys. The public portion of the key pair can be distributed in a public manner without compromising the private portion, which must be kept secret by its owner. An operation (for example, encryption) done with the public key can only be undone with the corresponding private key.

A shared secret can be established between two communicating parties online by exchanging only public keys and public constants if any. Any third party, who has access only to the exchanged public information, will not be able to calculate the shared secret unless it has access to the private key of any of the communicating parties.

Apart from Key Agreement the other important applications of public key cryptography are Data Encryption and Digital Signature.

## 1.1 One-Way Function

In public key cryptography, keys and messages are expressed numerically and the operations are expressed mathematically. The private and public key of a device is related by the mathematical function called the one-way function. One-way functions are mathematical functions in which the forward operation can be done easily but the reverse operation is so difficult that it is practically impossible. In public key cryptography the public key is calculated using private key on the forward operation of the one-way function. Obtaining of private key from the public key is a reverse operation. If the reverse operation can be done easily, that is if the private key is obtained from the public key and other public data, then the public key algorithm for the particular key is cracked. The reverse operation gets difficult as the key size increases. The public key algorithms operate on sufficiently large numbers to make the reverse operation practically impossible and thus make the system secure. For e.g. RSA algorithm operates on large numbers of thousands of bits long.

## 1.2 Key Agreement

Key agreement is a method in which the device communicating in the network establishes a shared secret between them without exchanging any secret data. In this method the devices that need to establish shared secret between them exchange their public keys. Both the devices on receiving the other device's public key performs key generation operation using its private key to obtain the shared secret.

The public keys are generated using private key and other shared constants. Let P be the private key of a device and U(P, C) be the public key. Since public key is generated using private key, the representation U(P, C) shows that the public key contain the components of private key P and some constants C where C is known by all the device taking part in the communication. Consider two devices A and B. Let PA and UA(PA, C) be the private key and public key of device A, and PB and UB(PB, C) be the private key and public key of device B respectively. Both device exchanges their public keys.

Device A, having got the public key of B, uses its private key to calculate shared secret

KA=Generate_Key(PA, UB(PB, C))

Device B, having got the public key of A, uses its private key to calculate the shared secret KB=Generate_Key(PB, UA(PA, C))

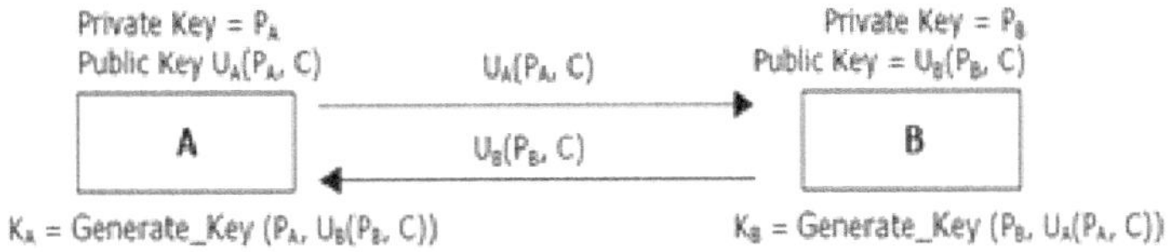


The key generation algorithm 'Generate_Key' will be such that the generated keys at the device A and B will be the same, that is shared secret KA=KB=K(PA, PB, C).

Since it is practically impossible to obtain private key from the public key any middleman, having access only to the public keys UA(PA, C) and UB(PB, C), will never be able to obtain the shared secret K. Examples of key agreement algorithms are DH and RSA.

During the key exchange process the public keys may pass through different intermediate points. Any middleman can thus tamper or change the public keys to its public key. Therefore for establishing shared secret it is important that device A receives the correct public key from device B and vice versa. Digital Certificate helps to deliver the public key in authenticated method.

### 1.3 Confidentiality and Encryption

Once the electronic check is digitally signed, it can be encrypted using a high-speed mathematical transformation with a key that will be used later to decrypt the document. This is often referred to as a *symmetric key* system because the same key is used at both ends of the process. As the check is sent over the network, it is unreadable without the key. The next challenge is to securely deliver the symmetric key to the bank.

Encryption is a process in which the sender encrypts/scrambles the message in such a way that only the recipient will be able to decrypt/ descramble the message. Consider a device B whose private key and public key are PB and UB respectively. Since UB is public key all devices will be able to get it. For any device that needs to send the message 'Msg' in a secured way to device B, it will encrypt the data using

B's public key to obtain the cipher text 'Ctx'. The encrypted message, cipher text, can only be decrypted using B's private key. On receiving the message the B decrypts it using its private key PB.

Since only B knows its private key PB none other including A can decrypt the message.

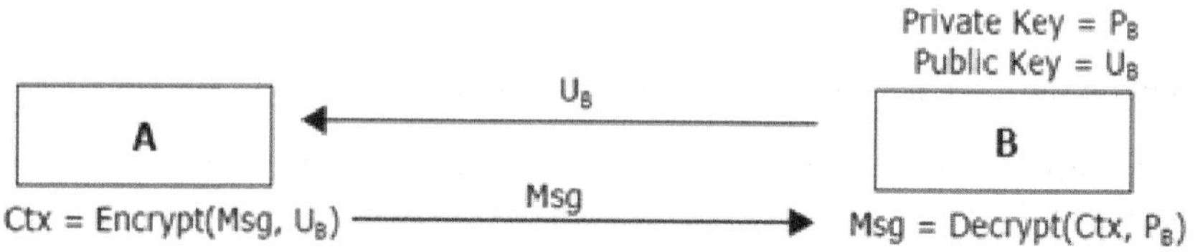


It is important that device A receives the correct public key from device B, i.e. no middleman must tamper or change the public key to its public key. Digital Certificate helps to deliver the public key in authenticated method.

## 2 Digital Signature and Certificate

The process of digitally signing starts by taking a mathematical summary (called a *hash code*) of the check. This hash code is a uniquely-identifying digital fingerprint of the check. If even a single bit of the check changes, the hash code will dramatically change. The next step in creating a digital signature is to sign the hash code with your private key. This signed hash code is then appended to the check.

How is this a signature?

Well, the recipient of your check can verify the hash code sent by you, using your public key. At the same time, a new hash code can be created from the received check and compared with the original signed hash code. If the hash codes match, then the recipient has verified that the check has not been altered. The recipient also knows that only you could have sent the check because only you have the private key that signed the original hash code.

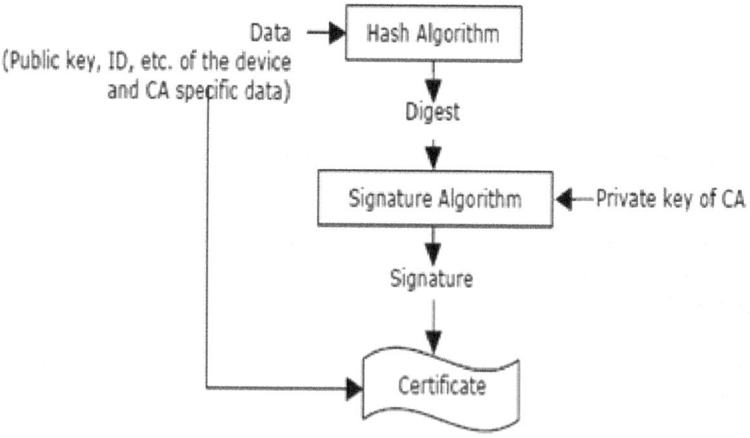

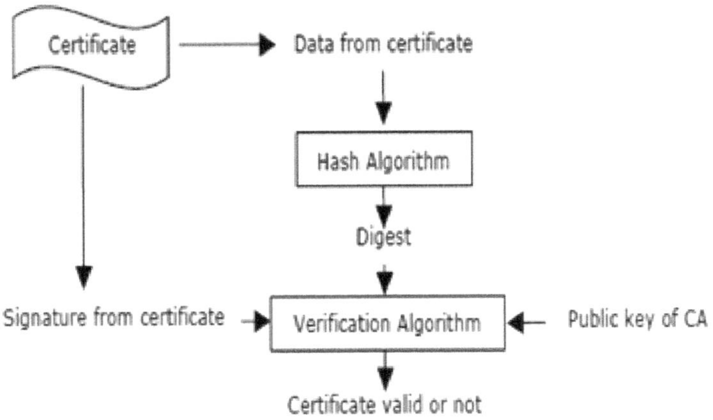


We can easily understand this process by the following example-

Bob (Bob's private key)

Bob has been given two keys. One of Bob's keys is called a Public Key, the other is called a Private Key.

Bob's Co-workers:

Anyone can get Bob's Public Key, but Bob keeps his Private Key to himself

Pat Doug Susan

Bob's Public key is available to anyone who needs it, but he keeps his Private Key to himself. Keys are used to encrypt information. Encrypting information means "scrambling it up", so that only a person with the appropriate key can make it readable again. Either one of Bob's two keys can encrypt data and the other key can decrypt that data.

Susan (shown below) can encrypt a message using Bob's Public Key. Bob uses his Private Key to decrypt the message. Any of Bob's coworkers might have access to the message Susan encrypted, but without Bob's Private Key, the data is worthless.

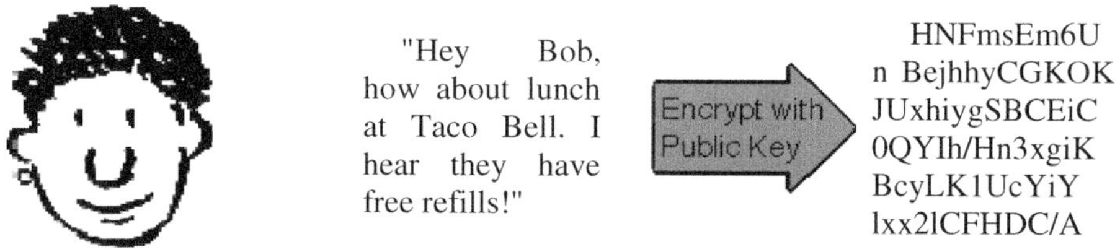

With his private key and the right software, Bob can put digital signatures on documents and other data. A digital signature is a "stamp" Bob places on the data which is unique to Bob, and is very difficult to forge. In addition, the signature assures that any changes made to the data that has been signed can not go undetected.

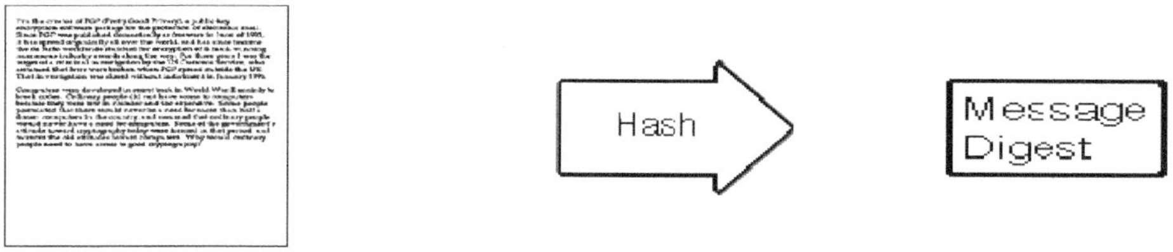

To sign a document, Bob's software will crunch down the data into just a few lines by a process called "hashing". These few lines are called a message digest. (It is not possible to change a message digest back into the original data from which it was created.)

Bob's software then encrypts the message digest with his private key. The result is the digital signature.

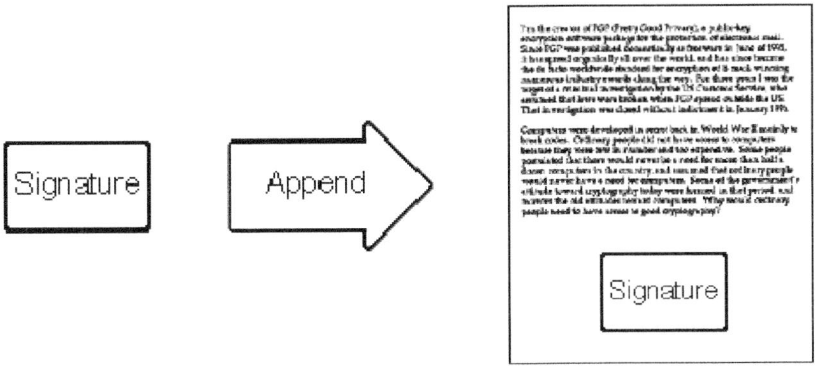


Finally, Bob's software appends the digital signature to document. All of the data that was hashed has been signed.

Bob now passes the document on to Pat.

First, Pat's software decrypts the signature (using Bob's public key) changing it back into a message digest. If this worked, then it proves that Bob signed the document, because only Bob has his private key. Pat's software then hashes the document data into a message digest. If the message digest is the same as the message digest created when the signature was decrypted, then Pat knows that the signed data has not been changed.

Doug (our disgruntled employee) wishes to deceive Pat. Doug makes sure that Pat receives a signed message and a public key that appears to belong to Bob. Unbeknownst to Pat, Doug deceitfully sent a key pair he created using Bob's name. Short of receiving Bob's public key from him in person, how can Pat be sure that Bob's public key is authentic?

It just so happens that Susan works at the company's certificate authority center. Susan can create a digital certificate for Bob simply by signing Bob's public key as well as some information about Bob.

Now Bob's co-workers can check Bob's trusted certificate to make sure that his public key truly belongs to him. In fact, no one at Bob's company accepts a signature for which there does not exist a certificate generated by Susan. This gives Susan the power to revoke signatures if private keys are compromised, or no longer needed. There are even more widely accepted certificate authorities that certify Susan.

Let's say that Bob sends a signed document to Pat. To verify the signature on the document, Pat's software first uses Susan's (the certificate authority's) public key to check the signature on Bob's certificate. Successful de-encryption of the certificate proves that Susan created it. After the certificate is de-encrypted, Pat's software can check if Bob is in good standing with the certificate authority and that all of the certificate information concerning Bob's identity has not been altered.

Pat's software then takes Bob's public key from the certificate and uses it to check Bob's signature. If Bob's public key de-encrypts the signature successfully, then Pat is assured that the signature was created using Bob's private key, for Susan has certified the matching public key. And of course, if the signature is valid, then we know that Doug didn't try to change the signed content.

## 3 Modular Arithmetic

Modular arithmetic is the commonly used arithmetic in public key cryptography. Modular arithmetic deals only with integers. Since it involves no floating-point operations, the mathematical calculations are more accurate and efficient than the real number arithmetic. Modular arithmetic over a number n involves arithmetic operations on integers between 0 and n – 1, where n is called the modulus. If the number happens to be out of this range in any of the operation the result, r, is wrapped around in to the range 0 and n – 1 by repeated subtraction of the modulus n from the result r. This is equivalent in taking the remainder of division operation r/n.

For e.g. for modulo 23 arithmetic n=23, Let a=15, b=20 (a+b) mod n = (15+20) mod 23 = 35 mod 23 = 12.

Since the result of a+b=35 which is out of the range [0,22], the result is wrapped around in to the range [0 22] by subtracting 35 with 23 till the result is in range [0,22] a mod b is thus explained as remainder of division a/b Subtraction and multiplication can also be explained similarly.

A negative number is added repeatedly with n till it can be represented in the range [0,n-1] The modular division a/b mod p is defined as a*b-1 mod p. b-1 is the multiplicative invers of b.

Multiplicative inverse of number b with respect to mod p is defined as a number b-1 such that b*b-1 mod p = 1.

### 3.1 Congruent Relation

Modular arithmetic is a congruent relation. Congruence is shown by the symbol '≡'. For a modulus n two numbers a and b are said to be congruent if a mod n = b mod n. i.e. a ≡ b (mod n) if, a mod n = b mod n.

For example consider the modulus 7 i.e. n = 7

Then the numbers 2, 9, 16, 23 etc are congruent to each other since (2 mod 7) = (9 mod 7) = (16 mod 7) = (23 mod 7) etc.

### 3.2 Properties of Modular Arithmetic

P1- a≡b mod n implies a–b=k*n, where k is an integer

P2- a mod n + b mod n ≡ a+b (mod n), also true for other operators '-', '*'and '/'

P3- a+b≡b+a (mod n) , also true for other operators '-', '*'and '/'

P4- a≡a mod n

P5- a≡b mod n implies b≡a mod n

P6- Fermat's little theorem, if M and p are coprime then Mp-1≡1(mod p)

P7- if p and q are coprime and also if a≡b(mod p) and a≡b(mod q) then a≡b(mod pq).

## 4 RSA Algorithm

RSA is a public key algorithm that is used for Encryption, Signature and Key Agreement. RSA typically uses keys of size 1024 to 2048. The RSA standard is specified RFC 3447, RSA Cryptography Specifications Version 2.1 [3] . Overviews of RSA algorithms are given below.

## 4.1 RSA Encryption

### 4.1.1 Parameter Generation

R1- Select two prime numbers p and q.

R2- Find n=p*q, Where n is the modulus that is made public. The length of n is considered as the RSA key length.

R3- Choose a random number 'e' as a public key in the range $0<e<(p-1)(q-1)$ such that gcd(e,(p-1)(q-1))=1.

R4- Find private key d such that ed≡1(mod (p-1)(q-1)).

### 4.1.2 Encryption

Consider the device A that needs to send a message to B securely.

R5- Let e be B's public key. Since e is public, A has access to e.

R6- To encrypt the message M, represent the message as an integer in the range $0<M<n$.

R7- Cipher text C = Memod n, where n is the modulus.

## 4.2 Decryption

R8- Let C be the cipher text received from A.

R9- Calculate Message M = Cd mod n, where d is B's private key and n is the modulus.

## 4.3 RSA Key Agreement

Since public key cryptography involves mathematical operation on large numbers, these algorithms are considerably slow compared to the symmetric key algorithm. They are so slow that it is infeasible to encrypt large amount of data. Public key encryption algorithm such as RSA can be used to encrypt small data such as 'keys' used in private key algorithm. RSA is thus used as key agreement algorithm.

### 4.3.1 Key Agreement Algorithm

For establishing shared secret between two device A and B

R10- Generate a random number, key, at device A.

R11- Encrypt key by RSA encryption algorithm using B's public key and pass the cipher text to B

R12- At B decrypt the cipher text using B's private key to obtain the key.

## 4.4 RSA Signature

RSA Signature is similar to RSA encryption except that the private key is used for signing and public key is used for verification.

### 4.4.1 Parameter Generation

The parameter generation process is same as that in RSA Encryption.

#### 4.4.2 Signing

Consider the device A that needs to sign the data that it sends to B.

R13- Let d be A's private key

R14- To sign a data M, represent the data as an integer in the range $0<M<n$

R15- Signature C = Mdmod n

#### 4.4.3 Verification

R16- Let M be the message and C be the signature received from A

R17- Calculate M'=Cemod n, where e is A's public key. Since e is public, B has access to e

R18- If M'=M, the signature is verified, else failed.

## 5 DSA – Digital Signature Algorithm

DSA is a public key algorithm that is used for Digital Signature. The DSA standard is specified FIPS 186-2, Digital Signature Standard [2] An overview of the algorithm is given below.

### 5.1 Parameter Generation

S1- Choose a 160-bit prime q.

S2- For an integer z, choose an L-bit prime p, such that p=qz+1, $512 \leq L \leq 1024$, and L is divisible by 64.

S3- Choose h, where $1<h<p-1$ such that $g=hZ \bmod p>1$.

S4- Choose a random number x, where $0<x<q$.

S5- Calculate y=gx mod p.

S6- Public key is (p, q, g, y). Private key is x.

### 5.2 Signing

Consider the device A that sign the data M that it sends to B.

S7- Let x be A's private key and (p, q, g, y) be A's public key.

S8- Generate a random per-message value k, where $0<k<q$.

S9- Calculate r = (gk mod p) mod q.

S10- Calculate s = (k-1(M+x*r))mod q, where M is the hash SHA1 of the message

S11- The signature is (r, s).

### 5.3 Verification

S12- Let M be the message and (r, s) be the signature received from A

S13- Let (p, q, g, y) be A's public key. Since (p, q, g, y) is public, B has access to it.

S14- Calculate w = s-1 mod q.

S15- Calculate u1 = (M*w) mod q, where M is the hash SHA1 of the message.

S16- Calculate u2 = (r*w) mod q.

S17- Calculate v = ((gu1*yu2) mod p) mod q.

S18- The signature is valid if v=r, invalid otherwise.

## 6 Conclusion

Public key cryptography is an innovation and is an unavoidable part of almost all security protocol and application. Being able to negotiate a shared secret between two devices online without the need of any exchange of secret data created a breakthrough in secure network/internet communication. Though theoretically it is possible to find the shared secret from the available public information, it will take exponentially longer time making it practically impossible. It is the belief in age-old mathematics, that finding an easy method for reverse process of one-way function is unlikely, keeps the public key cryptography going.

## References

[1] Menezes, A.J., van Oorschot, P.C., Vanstone, S.A.: Handbook of Applied Cryptography. CRC Press, Boca Raton (1996)

[2] FIPS PUB 186-2, Digital Signature Standard (DSS) (January 2000), http://csrc.nist.gov/publications/fips/fips186-2/fips186-2-change1.pdf

[3] RSA Laboratories, PKCS 1 v2.1: RSA Cryptography Standard (June 2002), http://www.rsa.com/rsalabs/node.asp?id=2125

[4] RFC 2631, Diffie-Hellman Key Agreement Method (June 1999), http://tools.ietf.org/html/rfc2631

[5] Certicom, Standards for Efficient Cryptography, SEC 1: Elliptic Curve Cryptography,Version 1.0 (September 2000), http://www.secg.org/download/aid-385/sec1_final.pdf

[6] Certicom, Standards for Efficient Cryptography, SEC 2: Recommended Elliptic Curve Domain Parameters, Version 1.0 (September 2000), http://www.secg.org/download/aid-386/sec2_final.pdf

[7] ITU, Recommendation X.509, http://www.itu.int/rec/T-REC-X.509-200508-I

[8] Anoop MS, Elliptic Curve Cryptography - An Implementation Guide (January 2007), http://hosteddocs.ittoolbox.com/AN1.5.07.pdf

[9] Openssl, http://www.openssl.org

[10] Certicom, http://www.certicom.com/index.php?action=ecc_tutorial,home

[11] RSA Laboratories, http://www.rsa.com/rsalabs/node.asp?id=2193

# Competitive Equilibrium Theory and Its Applications in Computer Science

J. Ujwala Rekha[1], K. Shahu Chatrapati[2], and A. Vinaya Babu[1]

[1] Dept of Computer Science and engineering,
JNTUH College of Engineering,
Hyderabad
ujwala_rekha@yahoo.com, dravinayababu@gmail.com
[2] Dept of Computer science and Engineering,
JNTUH College of Engineering,
Jagitial
shahujntu@gmail.com

**Abstract.** In the capitalist market, vital regulatory functions such as ensuring, stability, competency, and fairness are relegated to pricing mechanisms. Thus, competitive equilibrium theory of equilibrium prices acquired a prominent place in mathematical economics. With the advent of internet, there has been an extensive research done at the boundary across computer science and economic theory, over the past few years. We discuss in this paper about the competitive equilibrium theory and its applications to computer science.

**Keywords:** Competitive Equilibrium Theory, Walrasian Market, Fisher Market.

## 1 Introduction

Competitive equilibrium theory can be opined as a specialized branch of game thoery that deals with decision making in large markets. It is extensively used in the analysis of economic activities dealing with fiscal or tax policy, in finance for analysis of stock markets and commodity markets, to study interest, and exchange rates, and other prices. It serves as a yardstick for efficiency in economic analysis. It relies on the assumption of competitive market, where each trader decides upon a quantity that is so small compared to the total quantity traded in the market, such that their individual transactions have no influence on the prices. Competitive markets are an ideal, and a standard that other market structures are evaluated by [5].

Competitive equilibrium is a state of market, characterized by a set of prices and an allocation of commodities such that at equilibrium prices, each agent maximizes his objective function subject to his technological limitations and resource constraints, and the market clears i.e., the aggregate supply and demand for the commodities traded are equal.

At competitive equilibrium, the allocation is pareto-optimal which stipulates an important social justification for this theory i.e., there is no other feasible allocation that can increase the benefit of at least one agent in the economy without reducing the benefit of some other agent.

A. Mantri et al. (Eds.): HPAGC 2011, CCIS 169, pp. 539–547, 2011.

In section 2 we discuss about the history of competitive equilibrium theory. In sections 3 and 4, we describe two market models- Walrasian market model and Fisher's market model respectively. Section 5 presents some important defintions and theorems related to competitive equilibrium theory. Section 6 presents applications of competitive equilibrium in computer science and we conclude with section 7.

## 2 History of Competitive Equilibrium Theory

In the capitalist market, vital regulatory functions such as ensuring stability, competency, and fairness are relegated to pricing mechanisms. The chief principal in pricing mechanisms is that prices be such that demand equals supply; that is the economy should function at equilibrium. Since the nineteenth century, economists have presented models that represent the concept of market equilibrium. In 1874, Walras published the "Elements of Pure Economics", in which he introduced a model that indicates the state of an economic system in terms of demand and supply, and expresses the supply equal demand equilibrium conditions [9]. He also introduced a price-adjustment process called tâtonnment process performed by a fictitious auctioneer. The prices are announced by auctioneer, and agents report their demands at these prices. No transactions and no production can happen at disequilibrium prices. Instead, prices are lowered for goods with non negative prices and excess supply and prices are raised for goods with excess demand. The question to be answered is, under what conditions such a process will terminate in equilibrium in which demand equals supply for goods with non negative prices, and demand does not exceed supply for goods with a zero price. Walras was not able to give a conclusive answer to this question.

The first proof for the existence of an equilibrium was given by Wald in 1936, although under severe restrictions[4]. Later in 1954, Arrow and Debreu jointly gave the proof for the existence of an equilibrium under much milder assumptions [7]. Also, they have validated using The First Fundamental Welfare Theorem that such equilibrium is pareto optimal [7].

In 1891 Irving Fisher [22] independently modeled a market. Fisher and Walrasian market models have been studied extensively in mathematical economics. Two techniques have been mostly used for computing equilibria for these models- the primal-dual scheme [13] and an auction-based approach [17].

## 3 Walrasian Market Model

The Walrasian model is also known as the Arrow and Debreu model or the exchange model. In Walrasian model the market consists of **m** agents and **n** divisible goods. Each agent $i$ has an initial endowment of goods $\mathbf{w_i}=(\mathbf{w_{i1}},\ldots,\mathbf{w_{in}})\in R^n_+$, and preferences for consuming goods described by utility function $u_i:R^n_+\rightarrow R_+$. At given prices, each agent $i$ sells their initial endowment and then uses the income to buy a bundle of goods $\mathbf{x_i}=(\mathbf{x_{i1}},\ldots,\mathbf{x_{in}})\in R^n_+$, such that their utility $u_i(\mathbf{x})$ is maximized.

The problem is to find prices $\mathbf{p}=(\mathbf{p_1},...,\mathbf{p_n})\in R^n_+$, for the goods, such that if each agent sells her initial endowment at these prices and buys her optimal bundle, the market clears; i.e., there is no shortage or surplus of any good. In other words, an equilibrium is a set of prices **p**, such that

i. For each agent *i*, the vector $\mathbf{x_i}$ maximizes $u_i(\mathbf{x})$ subject to the constraints $\sum_{i=1}^{n} \mathbf{p_i} * \mathbf{x_{ji}} \leq \sum_{i=1}^{n} \mathbf{p_i} * \mathbf{w_{ji}}$ for all $j$=1,...

ii. For each good $j$, $\sum_{i=1}^{n} \mathbf{x_{ji}} = \sum_{i=1}^{n} \mathbf{w_{ji}}$

# 4 Fisher's Market Model

Fisher's model is a market of **n** goods and **m** utility maximizing buyers of the goods. Each buyer $i$, is endowed with money $\mathbf{e_i}$ and a concave utility function $u_i : R^n_+ \rightarrow R_+$ for specifying the preference for consuming various goods. Each good $j$ has an initial endowment $\mathbf{b_j}$ of the good. An equilibrium is then a non negative vector of prices $\mathbf{p}=(\mathbf{p_1},...,\mathbf{p_n})\in R^n_+$ at which there is a bundle $\mathbf{x_i}=(\mathbf{x_{i1}},...,\mathbf{x_{in}})\in R^n_+$ of goods for each trader $i$ such that the following two conditions hold:

i. For each buyer $i$, the vector $\mathbf{x_i}$ maximizes $u_i(\mathbf{x_i})$ subject to the constraint $\sum_{k=1}^{n} \mathbf{p_k} * \mathbf{x_{ik}} \leq \mathbf{e_i}$

ii. For each good $j$, $\sum_{k=1}^{m} \mathbf{x_{kj}} = \mathbf{b_j}$

It can easily be observed that the Fisher's model is a special case of Walras model, when money is considered a good.

# 5 Properties and Characterization of General Equilibrium

The most important issues in competitive equilibrium analysis are concerned with the conditions under which an equilibrium will be efficient, which efficient equilibrium can be achieved, when an equilibrium is guaranteed to exist.

Let **S** be a commodity space, where each vector represents a basket of commodities. Let there be **I** consumers, 1,2,..,I. Each consumer $i$ chooses a basket of commodities in the set $\mathbf{X_i} \subseteq \mathbf{S}$, giving each commodity point (that is, basket of commodities) a value called the utility function $u_i : \mathbf{X}_i \to R$. Let there be **J** firms, 1,2,…,J. Each firm *j* produces a basket of commodities in the set $\mathbf{Y}_j \subseteq \mathbf{S}$, subject to its technological limitations.

The firms and consumers, interact with each other. Firms demand factors of production and supply final goods, whereas consumers supply factors of production (for example, in the form of labour) and demand final goods.

The selections made by firms and consumers are constrained such that the market clears i.e., $\sum_i \mathbf{x}_i - \sum_j \mathbf{y}_j = 0$. This means that all produced goods are consumed, and there is no unemployment. The pair $((\mathbf{x}_i),(\mathbf{y}_j))$ called the allocation describes the consumption $\mathbf{x}_i$ of each consumer and the production $\mathbf{y}_j$ of each firm. The allocation is feasible if $\mathbf{x}_i \in \mathbf{X}_i$, for all i and $\mathbf{y}_j \in \mathbf{Y}_j$, for all j, and $\sum_i \mathbf{x}_i - \sum_j \mathbf{y}_j = 0$. Finally, the price system is a continuous linear function $\Phi : \mathbf{S} \to \mathbf{R}$. This means that a vector in **S** representing a basket of commodities has an associated price $\mathbf{P}_i$ for each commodity and can be assigned a single number which is the expenditure of purchasing that basket of commodities for consumers.

**Definition 1.** An allocation $((\mathbf{x}_i),(\mathbf{y}_j))$ is Pareto Optimal if it is feasible and if there is no other feasible allocation $((\mathbf{x}_i'),(\mathbf{y}_j'))$ such that $u_i(\mathbf{x}_i') \geq u_i(\mathbf{x}_i)$ for all *i* and $u_i(\mathbf{x}_i') > u_i(\mathbf{x}_i)$ for some *i*. That is, an allocation is Pareto Optimal if no individual can improve his utility without decreasing the utility of others.

**Definition 2.** An allocation $((\mathbf{x}_i^0),(\mathbf{y}_j^0))$ together with a price system $\Phi : \mathbf{S} \to \mathbf{R}$ is a competitive equilibrium if the following three conditions are satisfied.

(C1) The allocation $((\mathbf{x}_i^0),(\mathbf{y}_j^0))$ is feasible

(C2) For each *i*, $\mathbf{x} \in \mathbf{X}_i$ and $\Phi(\mathbf{x}) \leq \Phi(\mathbf{x}_i^0)$ implies $u_i(\mathbf{x}_i) \leq u_i(\mathbf{x}_i^0)$.

Equivalently, $u_i(\mathbf{x}_i) > u_i(\mathbf{x}_i^0)$ implies $\Phi(\mathbf{x}) > \Phi(\mathbf{x}_i^0)$.

(C3) For each *j*, $\mathbf{y} \in \mathbf{y}_j$ implies $\Phi(\mathbf{y}) \leq \Phi(\mathbf{y}_j^0)$

We note from conditions (C3) and (C2) that the allocation is profit maximizing for each firm and, the allocation is utility maximizing for each consumer respectively at the given price system.

**Definition 3.** The local nonsatiation condition for consumers is satisfied if for each consumer *i* and each $\mathbf{x} \in \mathbf{X}_i$, and for every ε > 0, there exists $\mathbf{x}' \in \mathbf{X}_i$ such that

$\|x' - x\| \leq \varepsilon$ and $u_i(x') > u_i(x)$. That is, for any basket of commodities there is another basket of commodities arbitrarily nearby that is strictly preferred to it.

**Lemma 1. (Existence)** If in a economy $\mathbf{X_i}$ is a closed covex set bounded from below and utility function $u_i : \mathbf{X_i} \rightarrow R$ is continous and concave, then there exists an equilibrium.

Generally the proofs of the existence theorem depend upon Brouwer's fixed point theorem for functions, and Kakutani's fixed point theorem for set-valued functions. There are other methods of proof that use Sard's lemma and Baire category.

**Theorem 1. (First Welfare Theorem).** If the local nonsatiation condition for consumers is satisfied and $((\mathbf{x}_i^0),(\mathbf{y}_j^0),(\Phi))$ is a competitive equilibrium, then the allocation $((\mathbf{x}_i^0),(\mathbf{y}_j^0))$ is pareto optimal.

*Proof. For the sake of contradiction, let the initial allocation* $((\mathbf{x}_i^0),(\mathbf{y}_j^0),(\Phi))$ *be not pareto optimal. Then there is another feasible allocation* $((\mathbf{x}_i'),(\mathbf{y}_j'))$ *such that* $u_i(\mathbf{x}_i') \geq u_i(\mathbf{x}_i^0)$ *for all i, with strict inequality for some i. Employing condition (C2), this strict inequality* $u_i(\mathbf{x}_i') > u_i(\mathbf{x}_i^0)$ *implies* $\Phi(\mathbf{x}_i') > \Phi(\mathbf{x}_i^0)$.

*Since Φ is linear, summing across all i gives* $\sum_i \Phi(\mathbf{x}_i') > \sum_i \Phi(\mathbf{x}_i^0)$ *Since* $\sum_i \mathbf{x}_i - \sum_j \mathbf{y}_j = 0$ *we have* $\sum_j \Phi(\mathbf{y}_j') = \sum_i \Phi(\mathbf{x}_i') > \sum_j \Phi(\mathbf{y}_j^0) = \sum_i \Phi(\mathbf{x}_i^0)$ *and this contradicts condition (C3)*

**Theorem 2. (Second Welfare Theorem).** Assume the economy satisfies the following conditions

1. For each consumer i=1,…,m
   a. $\mathbf{x}_i$ is non empty and convex
   b. $u_i(\mathbf{x}_i)$ is continuous, locally nonsatiated, and convex
2. For each producer j=1,…,n
   a. $\mathbf{y}_j$ is non empty and convex

Let $((\mathbf{x}_i^0),(\mathbf{y}_j^0))$ be an efficient allocation. Then there is non-zero price system $\Phi : S \rightarrow R$ such that

(S1) For each consumer i=1,..,m $\mathbf{x} \in \mathbf{X_i}$ and $u_i(\mathbf{x}) \geq u_i(\mathbf{x_i^0})$ implies $\Phi(\mathbf{x}) > \Phi(\mathbf{x_i^0})$

(S2) For each j, $\mathbf{y} \in \mathbf{Y_j}$, implies $\Phi(\mathbf{y}) \leq \Phi(\mathbf{y_j^0})$

According to definition of competitive equilibrium conditions (S2) and (C3) are identical since both satisfy the profit maximizing condition for firms. Condition (S1) is slightly weaker one because it only minimizes the expenditure for consumers, rather than maximizing utility as in condition (C2). However, it is possible to improve (S1) if we impose conditions that the consumption set be convex and, the utility function $u_i : \mathbf{X_i} \rightarrow R$ be continuous, and locally non-satiated.

*Proof: Since* $\left(\left(\mathbf{x_i^0}\right),\left(\mathbf{y_j^0}\right)\right)$ *is efficient, it is impossible to make every one better off. Let us define the set V as*

$$V = \sum_i \mathbf{X_i}$$

*and set A by*

$$A = \sum_j \mathbf{Y_j}$$

*By pareto optimality* $A \cap V = \phi$. *Suppose* $\mathbf{x} \in A \cap V$, *since* $\mathbf{x} \in V$, *we can write* $\mathbf{x} = \sum_i \mathbf{x_i}$, *where each* $\mathbf{x_i} \in \mathbf{X_i}$. *Since* $\mathbf{y} \in A$, *we can write* $\mathbf{y} = \sum_j \mathbf{y_j}$. *But then* $\left(\left(\mathbf{x_i}\right),\left(\mathbf{y_j}\right)\right)$ *is an allocation, and* $u_i(\mathbf{x_i}) > u_i(\mathbf{x_i^0})$ *for each i, contradicting the efficiency of* $\left(\left(\mathbf{x_i^0}\right),\left(\mathbf{y_j^0}\right)\right)$

*V is open and convex since each summand is, and is nonempty by local nonsatiation. Similarly A is convex. Thus by the separating hyperplane theorem, there is a non-zero price system* $\Phi : S \rightarrow R$ *satisfying*

$$\Phi(\mathbf{x}) > \Phi(\mathbf{y}) \textit{ for each } x \in V \textit{ and } \mathbf{y} \in A$$

*Each* $\mathbf{x_i^0}$ *belongs to the closure of* $\mathbf{X_i}$, *so* $\sum_i \mathbf{x_i^0}$ *belongs to the closure of V. Now* $\sum_i \mathbf{x_i} = \sum_j \mathbf{y_j}$ *so it belongs to A. It follows that*

$$\Phi(\mathbf{x}) \geq \Phi(\sum_i \mathbf{x_i}) = \Phi(\sum_j \mathbf{y_j}) \geq \Phi(\mathbf{y}) \textit{ for each } \mathbf{x} \in V, \ \mathbf{y} \in A$$

*from the summation principle we have*

$$\Phi(\mathbf{x_i^0}) \leq \Phi(\mathbf{x}) \textit{ for all } \mathbf{x} \in \mathbf{X_i}$$

*and*

$$\Phi(\mathbf{y_j^0}) \geq \Phi(\mathbf{y}) \textit{ for all } \mathbf{y} \in \mathbf{Y_j}$$

## 6 Applications of Competitive Equilibrium

Over the past few years, there has been an extensive research done in applying economic solutions to the problems of computer science. We list below some of the domains in computer science for which competitive equilibrium solutions are proposed.

### 6.1 Congestion Control and Avoidance

Over the years computer networks have seen an explosive growth. While available bandwidth is increasing, the demand for bandwidth is growing even more. Congestion control is therefore an important engineering topic. A good congestion control algorithm should consider the utilities of users while being fair. Competitive equilibrium approach for congestion control and avoidance is studied in [8],[3],[18],[2].

### 6.2 Information Security and Network Security

With the advent of internet, an increasing amount of information is available on-line and cyber crime is becoming one of the most profitable criminal activities. Technology based solutions for security are available, but they fail because incentives are wrong. For instance, the people who should guard the system are not the one who suffer the full costs of failure. Consequently, they make less effort than would be socially optimal. The concepts of competitive equilibrium theory are becoming just as important as the mathematics of cryptography to the security problems ([6],[14],[1],[15]).

### 6.3 Distributed Computing

Distributed applications generate a large number of tasks/jobs that need resources scattered on various sites that are connected to each other by a network. Since resources are distributed and used by many users having different requirements, users are likely to behave in a selfish manner. Load balancing is a technique, that distributes workload evenly across the resources, such that resource utilization and throughput are maximized, and response time and overload are minimized. Competitive equilibrium approach for load balancing is considered in ([19],[20],[21]), which simultaneously minimizes the mean response time of all jobs, and the response time of each job individually.

### 6.4 Video Streaming

In direct broadcast satellite, cable TV, video-on-demand service, and video surveillance, multiple video steams are transmitted simultaneously through a shared channel. Generally, while multiplexing several video streams on a single channel, the quality of high motion videos is improved at the expense of reduction in the quality of low motion videos. Competitive equilibrium approach for simultaneously improving the quality of all video streams is studied in [11], [12].

### 6.5 Spectrum Allocation

With the explosion of various radio devices and services, multiple wireless systems sharing a communication spectrum must co-exist. In such a multi-user system, each user's performance, measured by shannon utility function, depends on not only the power allocation of its own, but also of other users in the system. A decentralized method for spectrum allocation management and optimization using competitive equilibrium approach is discussed in [10], [16] which maximize each user's utility simultaneously.

## 7 Conclusion

Competitive equilibrium theory of a market is a powerful tool for modeling, analyzing, and understanding the situations where there are multiple self-interested agents who interact with each other while making their decisions. Each agent strives to maximize its utility by choosing their strategies depending on the strategies chosen by other agents and its limitations. Most problems in computer science can be modeled as a market. Therefore, competitive equilibrium theory concepts can be applied for many problems in computer science and provide a useful insight. In this paper an introductory overview of basic concepts of competitive equilibrium theory and their applications in networking, security, and many more are discussed.

## References

[1] Bisin, A., Gottardi, P.: Competitive Equilibria with Asymmetric Information. Journal of economic Theory 87(1), 1–48 (1999)
[2] Tang, A., Wei, D., Low, S.H.: Heterogeneous Congestion Control: Efficiency, Fairness, and Design. In: The Proceedings of IEEE ICNP, pp. 127–136 (2006)
[3] Tang, A., Wei, D., Low, S.H., Chiang, M.: Equilibrium of Heterogeneous Congestion Control: Optimality and Stability. IEEE/ACM Transactions on Networking 10(2), 272–286 (2002)
[4] Wald, A.: On Some System of Equations of Mathematical Economics. Econometrica 19(4), 368–403 (1954)
[5] http://en.wikipedia.org/wiki/Competitive_equilibrium
[6] Huang, P., Wu, H.M.: Competitive Equilibrium of Incomplete Markets for Securities with Smooth Payoffs. Journal of Mathematical Economics 23, 219–234 (1994)
[7] Arrow, K.J., Debreu, G.: Existence of an Equilibrium for Competitive Economy. Econometrica 22(3), 265–290 (1954)
[8] La, R.J., Anantharam, V.: Utility Based Rate Control in the Internet for Elastic Traffic. IEEE/ACM Transactions on Networking 10(2), 272–286 (2002)
[9] Walras, L.: Elements of Pure Economics, or the Theory of Social Wealth. Richard Irwin (1954)
[10] Lin, M.-H., Tsa, J., Ye, Y.: Budget Allocation in a Competitive Communication Spectrum Economy. EURASIP Journal on Advances in Signal Processing (2009)
[11] Tiwari, M., Grover, T., Cosman, P.: Bitrate Allocation for Multiple Videostreams at Competitive Equilibria. The Proceedings of IEEE Conference on Signals, System, and Computers (2008)
[12] Tiwar, M., Grover, T., Cosman, P.: Competitive Equilibrium Bitrate Allocation for Multiple Videostreams. IEEE Transactions on Image Processing 19(4), 1009–1021 (2010)
[13] Devanur, N., Papadimitriou, C.H., Saberi, A., Vazirani, V.V.: Market Equilibrium via a Primal-dual-type algorithm. In: The Proceedings of 43rd Annual IEEE Symposium on Foundations of Computer Science (2002)

[14] Shetty, N., Schwartz, G., Felegyhazi, M., Walrand, J.: Competitive Cyber-Insurance and Internet Security. In: The Proceedings of the Workshop on Economics of Information Security (2009)
[15] Shetty, N., Schwartz, G., Walrand, J.: Can Competitive Insurers Improve Network Security. In: The Proceedings of the 3rd International Conference on Trust and Trustworthy Computing (2010)
[16] Mochaourab, R., Jorswieck, E.: Walrasian Equilibrium in Two-User Multiple-Input Single-Output Interference Channels. In: The proceedings of IEEE International Conference on Communications (2011)
[17] Garg, R., Kapoor, S.: Auction Algorithms for Market Equilibrium. In: The Proceedings of 36th Symposium on Theory of Computing (2004)
[18] Gibbens, R.J.: Control and Pricing for Communication Networks. Research Report, Statistical Laboratory, University of Cambridge (1999)
[19] Shahu Chatrapati, K., Ujwala Rekha, J., Vinaya Babu, A.: Competitive Equilibrium Approach for Load Balancing in Computational Grids. In: The Proceedings of the International Conference on Advances and Emerging Trends in Computing Technologies, pp. 63–68 (2010)
[20] Shahu Chatrapati, K., Ujwala Rekha, J., Vinaya Babu, A.: Competitive Equilibrium Approach for Load Balancing a Computational Grid with Communication Delays. Journal of Theoretical and Applied Information Technology 19(2), 126–133 (2010)
[21] Shahu Chatrapati, K., Rekha, J.U., Babu, A.V.: Recursive Competitive Equilibrium Approach for Dynamic Load Balancing a Distributed System. In: Natarajan, R., Ojo, A. (eds.) ICDCIT 2011. LNCS, vol. 6536, pp. 162–174. Springer, Heidelberg (2011)
[22] Brainard, W.C., Scarf, H.E.: How to compute Equilibrium Prices in 1891, Cowles Foundation Discussion Paper 1270 (2000)

# A Novel Approach for Information Dissemination in Vehicular Networks

Rakesh Kumar[1] and Mayank Dave[2]

[1] Department of IT, M. M. University, Mullana, Haryana, India
[2] Department of Computer Engineering, N. I. T. Kurukshetra, Haryana, India
{raakeshdhiman,mdave67}@gmail.com

**Abstract.** Vehicular Networks or simply VANETs are important component for the development of Intelligent Transportation System. Due to the features of VANET, data dissemination is an important issue that has to be addressed. In this paper we discuss the types of information involved in the dissemination process. We discuss the existing approaches of data dissemination. A new approach is proposed for the dissemination of data using network coding. In the proposed solution the packet is forwarded to vehicles after coding only. The simulations show that the proposed method reduces the number of broadcast packets being sent on the network. Also increases the channel throughput.

**Keywords:** Coding, Information dissemination, throughput, VANET.

## 1 Introduction

Over the last few years, interest in vehicular networks has become resurgent. This is mainly due to the allocation of 75 MHz of *licensed* spectrum by the Federal Communications Commission (FCC) in the 5.850-5.925 GHz range [1] for so-called *dedicated short range communications* (DSRC) for intelligent transportation system applications. This spectrum opened up new opportunities for apprehending a variety of communications between vehicles – both vehicle-to-vehicle communications (V2V) and vehicle to roadside infrastructure (V2I). Vehicular networks are thus expected to become an important part of community networks of the future [1,2]. This paper surveys the different types of information exchanged in the vehicular networks with common practices and methodologies that have been considered in research literature (e.g.: opportunistic exchange of resources between vehicles, vehicle assisted data delivery, cooperating downloading of information, etc.) with a special emphasis to the network coding technique. In this paper a new approach for data dissemination is proposed with which network throughput can be increased.

## 2 Network Architecture

In vehicular networks, generally there is limited access to an infrastructure network that is to be supported by roadside base stations. Such access is limited in its nature for two reasons. First, the deployment of the infrastructure is expected to be slow and incremental leading to wide areas where, there is no access to the infrastructure.

A. Mantri et al. (Eds.): HPAGC 2011, CCIS 169, pp. 548–556, 2011.

Second, a complete deployment is probably being sparse because of cost. The coverage provided by a roadside base station may be of the order of 200-300 m while roadside base stations may be placed after every km and so on.[3]

Consequently, not all vehicles are connected to the infrastructure at all times. To obtain access to safety or other types of information, it becomes necessary to rely on vehicle-to-vehicle communications.

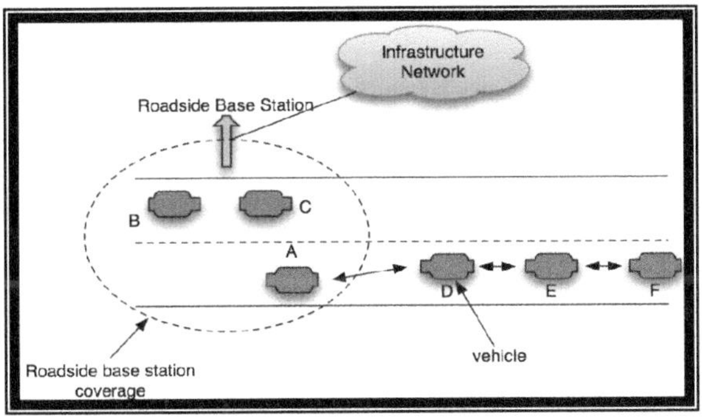


**Fig. 1.** Vehicular network architecture

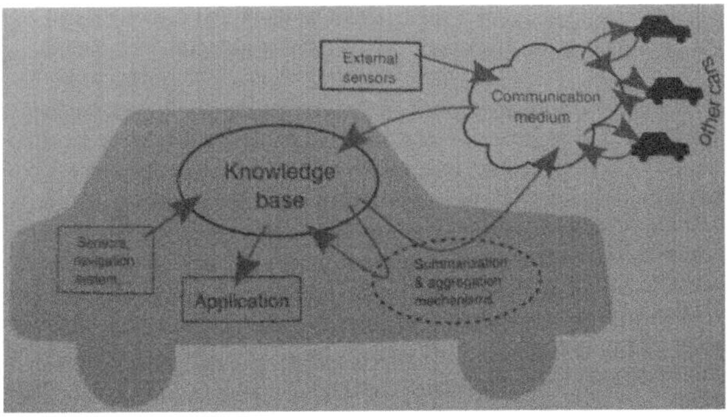


**Fig. 2.** A dissemination-based VANET application

As shown in Figure 1, vehicles A, B, and C have access to a roadside infrastructure, which has limited coverage. These vehicles can obtain information from the roadside base station. However, vehicles D, E, and F have no communications with the fixed infrastructure. For instance, Vehicle F will have to rely upon information from vehicle E, which in turn has obtained information that has passed through vehicles A and D. Vehicles that are in the range of a roadside infrastructure may be connected to the infrastructure for extremely small durations of time because of small coverage and high vehicular speeds. So the total information that can be pulled from the infrastructure is inevitably limited. It is also possible that vehicles move into the range of the roadside infrastructure with some information obtained from cooperating vehicles they have encountered. The issue then becomes one of updating the information, increasing the reliability or relevance of information or acquiring information that complements that already available to the vehicle.

# 3 Information Dissemination

Information dissemination using dedicated short range communication (DSRC) is quite smart due to the large bandwidth and the possibility of using multiple channels. The IEEE standards propose employing multiple 10 MHz channels, each capable of carrying 27 Mbps of data for vehicular communications. Up to seven channels are available in the 5.9 GHz bands and one channel is supposed to be dedicated for safety applications [7]. The remaining channels could potentially be used for content distribution and delivery. In this section, we describe the different types of information that need to be disseminated in a vehicular network and the methodologies that have been considered in the research literature.

Figure 2 shows a basic sketch of a dissemination based system. Information is collected from external and internal sources (i.e., from sensors in the car and potentially also from roadside infrastructure) and is stored in a knowledge base. The contents of this knowledge base are made available to the application, and are further shared and exchanged with other cars by means of wireless communication. The amount of data exchanged between cars can be reduced by applying summarization and aggregation techniques to the data in the knowledge bank.[5]

## 3.1 Types of Information Involved in the Dissemination Process

The information types that need to be communicated and then shared in a vehicular network can be roughly categorized into four types:

- Safety information: Safety information is perhaps the key information type that is communicated in a vehicular network. US Department of Transportation and the Vehicle Safety Communications Consortium (VSCC) have identified different applications that rely on security information.
- Traffic information: Congestion of roads is also very important information to share on vehicular networks in order to let personal electronic navigators to choose the best route and avoid traffic. This can be extremely useful in case of car crash or even under emergency circumstances because mobile nodes can be informed in advance about the area where there is congestion, so that service vehicles can quickly reach the destination.
- Infotainment information: With infotainment information we intend all that kind of messages that provide entertainment to the passenger and, more in general, useful messages to the driver.
- Content: This kind of information differs from the others discussed earlier especially in the quantity of data exchanged between nodes. Vehicles are interested in obtaining exceptionally large files (e.g. videos and movie clips) that cannot be shared through a limited transaction with either the infrastructure or the nearby mobile nodes. The finding and distribution of large quantities of information is still a challenging problem especially in a dynamic environment such as a vehicular network.[8]

### 3.2 Data Dissemination Techniques

The process of information dissemination determines a cluster of problems because of several reasons. At a first glance, we have to consider the network size but also the vehicles speed and the patchy and intermittent connectivity between mobile nodes; moreover, there is another problem which can badly affect the entire process: latency requirements. As a result, content information has to be quickly discovered and distributed among nodes. In the literature, several methods of information delivery are given and, generally speaking, we can distinguish the following data dissemination approaches:

- Opportunistic: information is pulled from other vehicles / infrastructure as a target vehicle encounters them.
- Vehicle-Assisted: a vehicle carries information along with it and delivers it either to the infrastructure (roadside base station) or to other vehicles when it encounters them. This process involves mobility in addition to wireless transmissions in order to disseminate the information.
- Cooperative: vehicles can download partial units of some content and then share them afterwards to obtain the complete content. This process is particularly suitable for content dissemination (where the amount of information is larger and rather important in terms of file size).

## 4 Network Coding

Network coding is a novel mechanism that promises optimal utilization of the resources of a network topology. Or in other words network coding is a particular in-network data processing technique that exploits the characteristics of the broadcast communication channel in order to increase the capacity or the throughput of the network. With network coding, every transmitted packet is a linear combination of all or a subset of the packets available at the sender (similar to XORing multiple packets). Observe that encoded packets can be further recombined to generate new linear combinations, enabling nodes to generate encoded packets without having the full file. The original information can be reconstructed after receiving enough linearly independent packets [4,6]. In this paper we focus on the use of Network Coding for vehicular networks. In particular, network coding improves performance when the number of nodes or vehicles in the system increases while the information that each node has about others remains constant.[5]

### 4.1 Superposition Coding

In this paper we discuss a more general multi-source network coding in which more than one mutually independent information sources are generated at possibly different nodes, where each information source is transmitted to a certain set of nodes in the network. We continue to assume that the communication channels in the network are free of error.

The achievable information rate region for a multi-source network coding problem refers to the set of all possible rate tuples at which multiple information sources can be multicast simultaneously on a network. In a single source network coding problem, a primary goal is to characterize the maximum rate at which information can be multicast from the source node to all the sink nodes. In a multi-source network coding problem, we are interested in characterizing the achievable information rate region. Multi-source network coding turns out not to be a simple extension of single-source network coding. In this section the unit of information is assumed to be bit.[9]

Figure 3 shows a network for which superposition coding may be suboptimal. Here, we assume that each channel has a unit capacity. For $i = 1, 2$, the source node $i$ generates a bit $b_i$ which is sent to the node $t_i$.

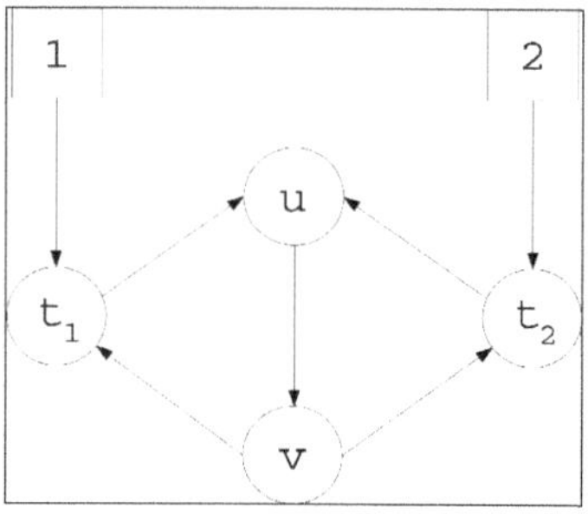


**Fig. 3.** A simple network for which superposition coding is suboptimal

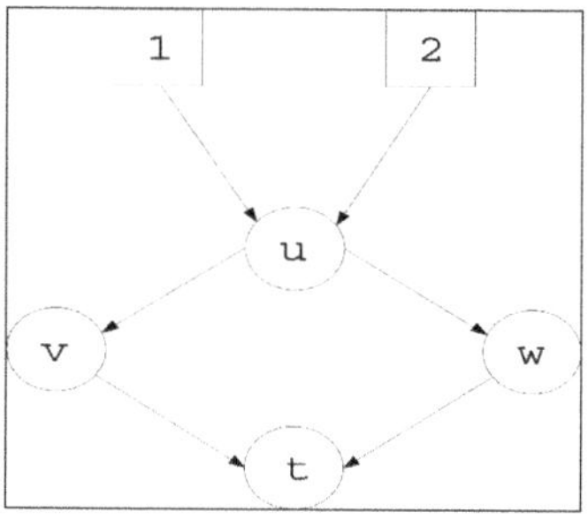


**Fig. 4.** A network for which superposition coding is optimal

In order to exchange the two bits $b_1$ and $b_2$ for the nodes $t_1$ and $t_2$, network coding must be performed at the node u. This example in fact has a very interesting implication. Imagine that on the Internet a message in English and a message in Chinese are generated at two different locations. These two messages are to be transmitted from one point to another point within the network, and we can assume that there is no correlation between the two messages. Then according to the above example we may have to perform joint coding of the two messages in the network in order to achieve bandwidth optimality. A method of coding individual information sources separately is known as superposition coding. The above example simply shows that superposition coding can be suboptimal. Figure 4 shows an example for which superposition coding does achieve optimality. We set the capacities of the channels *1u* and *2u* to infinity so that the information generated at both source nodes is directly available to the node *u*. For all the other channels, we set the capacity to 1. If the information generated at source node 1 is multicast to the nodes *v*, *w* and *t*, and information generated at the source node 2 is transmitted to the node t.

Let $X_1$ and $X_2$ be independent random variables representing the information generated respectively at the source nodes 1 and 2 for one unit time. The rate of the information generated at the source node s is given by $\omega_s = H(X_s)$ for s = 1, 2. Let $U_{ij}$ be the random variable sent on the channel *ij*, where $H(U_{ij}) \leq 1$ due to the bit rate constraint for the channel. Then for any coding scheme achieving the prescribed communication goals, we have,

$2\omega_1 + \omega_2 = 2H(X_1) + H(X_2)$
$= 2H(X_1) + H(X_2/X_1)$
$a)$
$\leq 2H(X_1) + H(U_{vt},U_{wt}/X_1)$
$b)$
$\leq 2H(X_1) + H(U_{uv},U_{uw}/X_1)$
$\leq 2H(X_1) + H(U_{uv}/X_1) + H(U_{uw}/X_1)$
$= H(U_{uv},X_1) + H(U_{uw},X_1)$
$c)$
$=H(U_{uv}) + H(U_{uw}) \leq 2,$

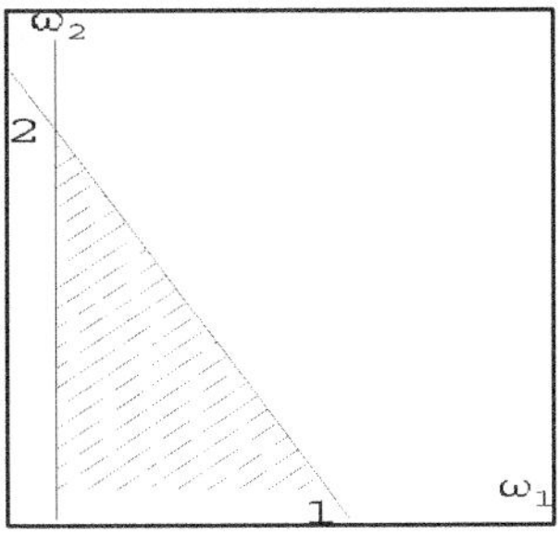


**Fig. 5.** Information rate region for the network in figure 4

where a) follows because $X_2$ is a function of $U_{vt}$ and $U_{wt}$, b) follows because $U_{vt}$ is a function of $U_{uv}$ and $U_{wt}$ is a function of $U_{uw}$, and c) follows because $X_1$ is a function of $U_{uv}$ and a function of $U_{uw}$. This region is shown in figure 5. To see that the whole region is achievable by superposition coding, let $r^{(s)}_{ij}$ be the bit rate on the channel $ij$ for transmitting the information generated at the source node $s$. Due to the bit rate constraint for each channel $ij$, the following must be satisfied:

$$r^{(1)}_{ij} + r^{(2)}_{ij} \leq 1$$

Then the rate pair $(\omega 1, \omega 2) = (1,0)$ is achieved by taking

$$r^{(1)}_{uv} = r^{(1)}_{uw} = r^{(1)}_{vt} = 1$$

and

$$r^{(1)}_{wt} = r^{(2)}_{uv} = r^{(2)}_{uw} = r^{(2)}_{vt} = r^{(2)}_{wt} = 0,$$

while the rate pair (0,2) is achieved by taking

$$r^{(1)}_{uv} = r^{(1)}_{uw} = r^{(1)}_{vt} = r^{(1)}_{wt} = 0$$

and

$$r^{(2)}_{uv} = r^{(2)}_{uw} = r^{(2)}_{vt} = r^{(2)}_{wt} = 1$$

Then the whole information rate region depicted in figure 5 is seen to be achievable via a time-sharing argument.

## 4.2 Modified Superposition Coding (MSC)

In this paper, the superposition coding is not applied rather it is embedded into the following algorithm.

In this algorithm first each node builds a coding list from its 1-hop neighbors according to the distance. (The distance between nodes is estimated by the number of shared neighbors, and the nodes are inserted into the coding list one by one to maximize the distance from the source node and existing nodes in the list.) After that during the first round, each node broadcasts its own information unit to other nodes. After during the *j-th* round ($j \leq 2$), each node sequentially picks up *j* nodes from its coding list and sends out the linear combination of their information units. If the end of the list is reached, it should wrap to the head of the list. (The coding vectors are linearly independent and stored at each node beforehand). After each round, each node recursively decodes all the information it receives from previous rounds until no more new neighbors can be decoded. The algorithm terminates after the round when j equals the number of the node's 1-hop neighbors or termination conditions are met.

In the above algorithm coding vectors are stored at each node beforehand; the only overhead added to transmission is the identification of neighbor nodes associated with linear combinations. In practice, we can add one round before all transmissions for each node to broadcast its coding list, so that no overhead is needed for later transmissions. Regarding the decoding complexity, the simulation shows that the proposed algorithm achieves very high 2-hop coverage in less than 10 rounds, which indicates that the rank of the equations to be solved is also very limited compared to the node density.

## 5 Simulation

To evaluate the efficiency of the integration of the MSC algorithm with the superposition coding for broadcasting in vehicular networks, we implemented MSC in a custom network simulator ns2. In the beginning of the simulation, nodes are placed randomly in the simulation area and don't move in between the simulation process. Transmissions are received by all nodes within transmission range. We have assumed that a packet transmission takes exactly one time unit. A node can either send or receive only one packet at a time unit. For network traffic, we assume that each node has one packet to broadcast to all the nodes in the network. All of these packets are generated at the beginning of the simulation, and after that the simulation continues to run without inserting further packets until all the packets are delivered to the entire network. We assume that only one node can send at a given time unit and in this way we avoid packet collisions. We have compared the performance for different density (average number of neighbors) values.

Figure 6 shows the number of transmissions required to broadcast all the packets into the entire network and average delay for a packet to be delivered to all the nodes. Figure shows the efficiency of our approach MSC in term of the number of transmissions required for flooding all the packets and the packet delay. Clear gain can be seen in comparison to simple store and forward approach, and is especially notable for low densities.

Figure 7 shows that thee optimization accomplished in the encoding function increases the number of packets delivered in single transmission. Also distribution function decreases the average number of coded packets and also it speeds up the decoding process.

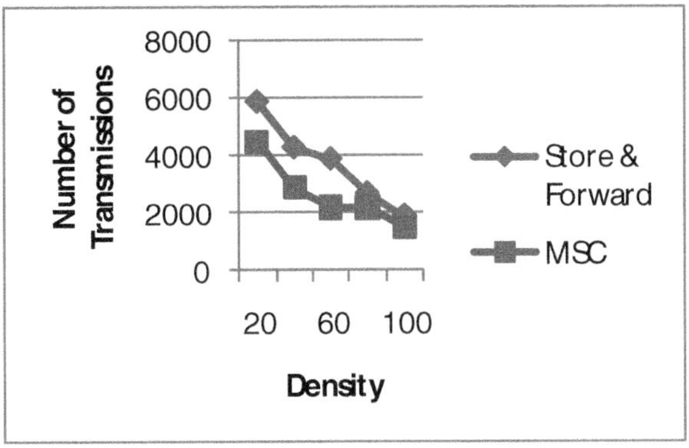


**Fig. 6(a).** Number of Transmission

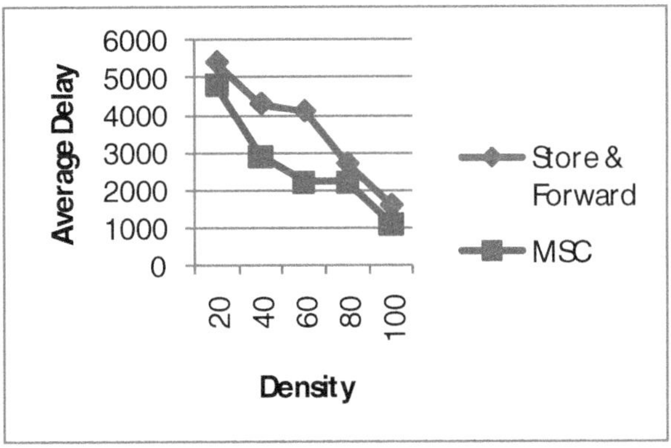


**Fig. 6(b).** Average Packet Delay

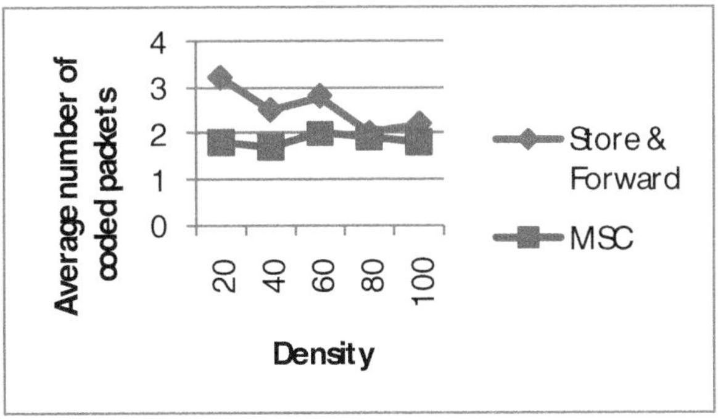


**Fig. 7(a).** Average packet degree

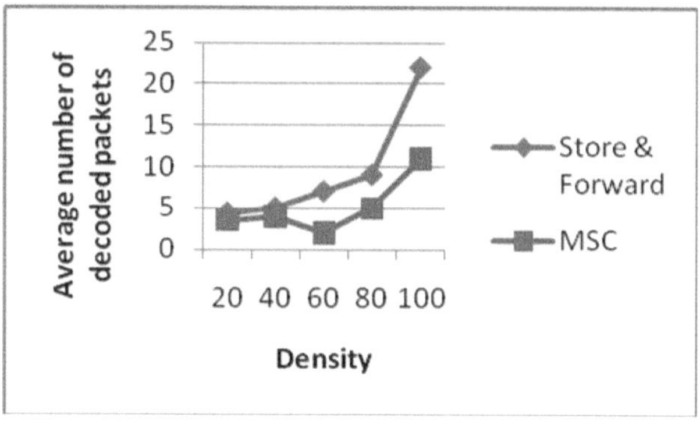


**Fig. 7(b).** Packet delivery

## 6 Conclusion

In this paper we have discussed the architecture of vehicular networks. We studied the information dissemination in Vanet. Various types of messages used in information dissemination are mentioned. We have discussed the superposition algorithm and how it can be used in vanet to disseminate the information effectively with its optimality. A network coding based algorithm MSC is proposed. The core idea of MSC is from the superposition coding algorithm. The simulation results show that the MSC algorithm outperforms the simple superposition algorithm or just store and forward methodology. Our future work will involve network coding utilization in improving the scalability of the information dissemination.

## References

[1] Hartenstein, H., Laberteaux, K.P.: VANET: Vaehicular Applications and Inter-Networking Technologies, 1st edn. John Wiley and sons, Chichester (2010); ISBN:9780470740569

[2] Huang, C.M.: Telematics Communication Technologies and Vehicular Networks: Wireless Architectures and Applications, 1st edn. IGI Global (2010); ISBN:9781605668406

[3] Dimakis, Prabhakaran, V., Ramchandran, K.: Decentralized Erasure Codes for Distributed Networked Storage. IEEE Trans. Inform. Theory (June 2006)

[4] Wang, D., Zhang, Q., Liu, J.: Partial Network Coding: Theory and Application for Continuous Sensor Data Collection. In: Proc. 14th IEEE International Workshop on Quality of Service, IWQoS (2006)

[5] Fragouli, J.W., Boudec, J.: A Network Coding Approach to Energy Efficient Broadcasting, Technical Report LCA-REPORT-2005-009 (2005)

[6] Widmer, J., Boudec, J.-Y.L.: Network coding for efficient communication in extreme networks. In: ACM SIGCOMM Workshop on DTNs (2005)

[7] Torrent-Moreno, M., Festag, A., Hartenstein, H.: System Design for Information Dissemination in VANETs. In: The Proceedings of 3rd International Workshop on Intelligent Transportation (WIT), Hamburg, Germany, pp. 27–33 (March 2006)

[8] Durresi, M., Durresi, A., Barolli, L.: Emergency Broadcast Protocol for Inter-Vehicle Communications. In: 11th International Conference on Parallel and Distributed Systems (ICPADS 2005), pp. 402–406 (2005)

[9] Kadi, N., Agha, K.A.: Network coding based flooding using fountain codes. Technical Report 1500 LRI, Univ. Paris-Sud XI, Tech. Rep. 1500 (2008)

# Understanding the Generation of Cellular Technologies

Manjit Sandhu[1], Tajinder Kaur[2],
Mahesh Chander[1], and Anju Bala[3]

[1] SBBSIET Jalandhar
manjit_sandhu67@yahoo.com
[2] MMU Mullana
kaurtajinder07@gmail.com,
maheshchander14@gmail.com
[3] LPU Jalandhar
anjubala28@yahoo.com

**Abstract.** Due to the increase in demand for speed, multimedia support and other resources, the wireless world is looking forward for a new generation technology to replace the third generation. This is where the fourth generation wireless communication comes into play. 4G wireless communication is expected to provide better speed, high capacity, lower cost and IP based services. The main aim of 4G wireless is to replace the current core technology with a single universal technology based on IP. This paper deals with understanding the features and challenges, the proposed architectural, multimedia support, applications and multiple access schemes for 4G.

**Keywords:** Architecture, features, multiple access schemes, challenges, applications.

## 1 Introduction

4G refers to the fourth generation of cellular wireless standards. It is a successor to 3G and 2G families of standards. These are fourth generation mobile communications systems that are characterized by high-speed data rates at 20 to 100 Mbps, suitable for high-resolution movies and television and efficient multimedia services 3G is necessary but not sufficient for the demands today. So the world is taking its leap towards the fourth generation wireless communication that promises to bring an end to most of the problems faced. 4G wireless is expected to be launched by 2010, but there are numerous challenges faced by researchers in achieving the desired features. Most of the ongoing researches are in the area of distributed computing, mobile agents, multimedia support etc

## 2 History of Wireless Technology

- 1864: James Clark Maxwell, Predicts existence of radio waves.
- 1886: Heinrich Rudolph Hertz , Demonstrates radio waves.

A. Mantri et al. (Eds.): HPAGC 2011, CCIS 169, pp. 557–567, 2011.

- 1895-1901:Guglielmo Marconi, Demonstrates wireless communications over increasing Distances.
- Also in the 1890s: Nikola Tesla, Alexander Stepanovich Popov, Jagdish Chandra Bose and others, demonstrate forms of wireless communications.
- 1924: first mobile radio telephone came.
- 1957 to 1960: cellular mobile radio technology was developed by Bell labs.
- Since 1980 various generations of wireless technology came which are as follows:

- 1G: 1G was based on analog technology and basically intended for analog phones. It was launched in the early 1980s. It introduced the first basic framework for mobile communications like the basic architecture, frequency multiplexing, roaming concept etc. Access technology used was AMPS (Advances Mobile Phone Service).
- 2G: 2G was a revolution that marked the switching of mobile communication technology from analog to digital. It was introduced in the late 1980s and it adopted digital signal processing techniques. GSM was one of the main attractive sides of 2G and it introduced the concept of SIM (Subscriber Identity Module) cards. Main access technologies were CDMA (Code Division Multiple Access) and GSM (Global System for Mobile Communication).
- 2.5G: 2.5G was basically an extension of 2G with packet switching incorporated to 2G. It implemented hybrid communication which connected the internet to mobile communications.
- 3G: The basic idea of 3G is to deploy new systems with new services instead of just provide higher bandwidth and data rate. Support for multimedia transmission is another striking feature of 3G. It employs both circuit switching and packet switching strategies. The main access technologies are CDMA (Code Division Multiple Access), WCDMA (Wideband CDMA), and TSSDMA (Time division Synchronous CDMA).
- 4G: 4G is the latest technology that is being launched. We will discuss in detail about this technology in this paper.

## 3 Need of 4G over Other Technologies

There are various drawbacks of other technologies which needs to overcome in 4G. need of 4G is due to :

- Difficulty of CDMA to provide higher data rates
- Need for continuously increasing data rate and bandwidth to meet the multimedia requirements
- Limitation of spectrum and it's allocation
- Inability to roam between different services
- To provide a seamless transport end-to-end mechanism
- To introduce a better system with reduces cost.

## 4 Objectives of 4G

Main objectives of 4G are:

- Flexible channel bandwidth of 5MHz to 20MHz and upto 40MHz.
- Data rates of 100 mbps when user is mobile and 1gbps when user is stationary.
- Link spectral efficiency of 15 bits/sec/Hz for uplink and 6.75 bits/sec/Hz for downlink.
- System spectral efficiency of 3 bits/sec/Hz/cell for uplink and 2.25 bits/sec/Hz/cell for indoor usage.
- Smooth handoff across heterogeneous networks.
- Seamless connectivity and global roaming across heterogeneous networks.
- High quality of service for next generation multimedia support.
- Interoperability with existing wireless standards.
- IP packet switched network
- Femtocells: these are home nodes connected to fixed internet broadband infrastructure.

## 5 Main Features of 4G

- High usability and global roaming: The end user terminals should be compatible with any technology, at any time, anywhere in the world. The basic idea is that the user should be able to take his mobile to any place, for example, from a place that uses CDMA to another place that employs GSM.
- Multimedia support: The user should be able to receive high data rate multimedia services. This demands higher bandwidth and higher data rate.
- Personalization: This means that any type of person should be able to access the service. The service providers should be able to provide customized services to different type of users.

## 6 Components of 4G

**Access scheme.** New access schemes like orthogonal FDMA (OFDMA), single carrier. FDMA (SCFDMA), interleaved FDMA (IFDMA), multicarrier CDMA (MCDMA) are being used. Among all these techniques OFDMA is more important as it is used in 4G. OFDMA is based on orthogonal frequency division multiplexing (OFDM). In OFDM, usable bandwidth is divided into a large number of smaller bandwidths that are mathematically orthogonal using fast Fourier transforms (FFTs). Reconstruction of the band is performed by the inverse fast Fourier transform (IFFT). One challenge in today's wireless systems is an effect called 'multipath.' Multipath results from reflections between a transmitter and receiver whereby the reflections arrive at the receiver at different times. The time span separating the reflection is referred to as delay spread. This type of interference tends to be problematic when the delay spread is on the order of the transmitted symbol time. Typical delay spreads are microseconds in length, which are close to CDMA symbol times. OFDMA symbol

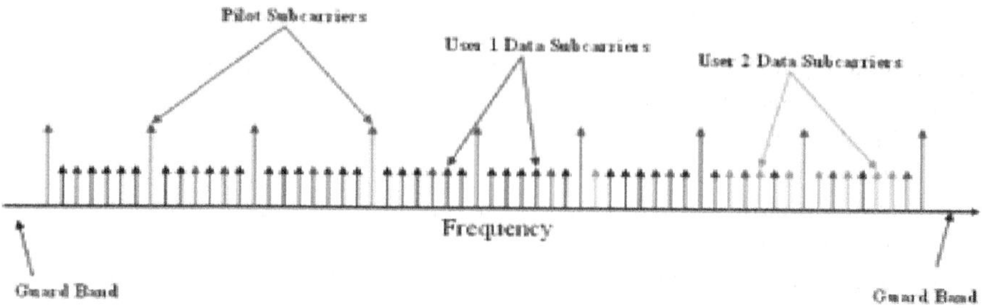


**Fig. 1.** OFDM

times tend to be on the order of 100 microseconds, making multipath less of a problem. In order to mitigate the effect of multipath, a guard band of about 10 microseconds, called the cyclic prefix, is inserted after each symbol.

OFDMA was developed to move OFDM technology from a fixed-access wireless system to a true cellular system with mobility. In OFDMA, subcarriers are grouped into larger units, referred to as subchannels, and these subchannels are further grouped into bursts which can be allocated to wireless users. Each burst allocation can be changed from frame to frame as well as within the modulation order. This allows the basestation to dynamically adjust the bandwidth usage according to the current system requirements.

**IPv6 support.** The process of IPv4 address exhaustion is expected to be in its final stages. So in 4G technology, IPv6 addressing is required to support a large number of users. It removes the need of NAT (Network Address Translation) i.e. a method of sharing a limited number of addresses among a various devices.

**Advanced Antenna Systems.** multiple antenna technologies are emerging to achieve the goal of 4G systems such as high rate, high reliability, and long range communications. In the early 90s, to cater the growing data rate needs of data communication, many transmission schemes were proposed. One technology, spatial multiplexing, gained importance for its bandwidth conservation and power efficiency. Spatial multiplexing involves deploying multiple antennas at the transmitter and at the receiver. Independent streams can then be transmitted simultaneously from all the antennas. This technology, called MIMO (as a branch of intelligent antenna), multiplies the base data rate by (the smaller of) the number of transmit antennas or the number of receive antennas. Apart from this, the reliability in transmitting high speed data in the fading channel can be improved by using more antennas at the transmitter or at the receiver. This is called transmit or receive diversity. Both transmit/receive diversity and transmit spatial multiplexing are categorized into the space-time coding techniques, which does not necessarily require the channel knowledge at the transmitter. The other category is closed-loop multiple antenna technologies, which require channel knowledge at the transmitter.

**Software-Defined Radio (SDR).** SDR is one form of open wireless architecture (OWA). Since 4G is a collection of wireless standards, the final form of a 4G device will constitute various standards. This can be efficiently realized using SDR technology, which is categorized to the area of the radio convergence.

## 7 4G Standards

In September 2009 the technology proposals have been submitted to ITU-R as 4G candidates. Basically all proposals are based on two technologies:

- LTE Advanced standardized by the 3GPP
- 802.16m standardized by the IEEE -Wimax

**LTE Advanced**

- 3GPP Long Term Evolution is the name given to the new standard developed by 3GPP to cope with the increasing throughput requirements of the market. 3GPP RAN working groups started LTE/EPC standardization in December 2004. The process of defining the future IMT-Advanced family was started with a Circular Letter issued by ITU-R calling for submission of candidate Radio Interface Technologies (RITs). Just after receiving the Circular Letter, the 3GPP organized a workshop on IMT-Advanced where the following decisions were made.

(i) LTE-Advanced will be an evolution of LTE. Therefore LTE-Advanced must be backward compatible with LTE Release 8.
(ii) LTE-Advanced requirements will meet or even exceed IMT-Advanced requirements following the ITU-R agenda.
(iii) LTE-Advanced should support significantly increased instantaneous peak data rates in order to reach ITU requirements. Primary focus should be on low mobility users. Moreover, it is required a further improvement of cell edge data rates. These are some of the current agreements on the requirements for LTE-Advanced.

(i) Peak data rate of 1 Gbps for downlink (DL) and 500 Mbps for uplink (UL).
(ii) Regarding latency, in the C-plane the transition time from Idle to Connected should be lower than 50ms. In the active state, a dormant user should take less than 10ms to get synchronized and the scheduler should reduce the U-plane latency at maximum.
(iii)The system should support downlink peak spectral efficiency up to 30 bps/Hz and uplink peak spectral efficiency of 15 bps/Hz with an antenna configuration of $8 \times 8$ or less in DL and $4 \times 4$ or less in UL.
(iv)The 3GPP defined a base coverage urban scenario with intersite distance of 500m and pedestrian users. Assuming this scenario, average user spectral efficiency in DL must be 2.4 bps/Hz/cell with MIMO $2 \times 2$, 2.6 bps/Hz/cell with MIMO $4 \times 2$ and 3.7 bps/Hz/cell with MIMO $4 \times 4$, whereas in UL the target average spectral efficiency is 1.2 bps/Hz/cell and 2.0 bps/Hz/cell with SIMO $1\times2$ andMIMO$2\times4$, respectively.
(v) In the same scenario with 10 users, cell edge user spectral efficiency will be 0.07 bps/Hz/cell/user in DL $2 \times 2$, 0.09 in DL $4 \times 2$ and 0.12 in DL $4 \times 4$. In the UL,

this cell edge user spectral efficiency must be 0.04 bps/Hz/cell/user with SIMO 1 $\times$ 2 and 0.07 with MIMO 2 $\times$ 4.

(vi)The mobility and coverage requirements are identical to LTE Release 8. There are only differences with indoor deployments that need additional care in LTE Advanced.

**802.16m standardized by the IEEE –Wimax.** IEEE 802.16m incorporates some advanced functions relative to the legacy system including:

New subframe-based frame structure that allows faster air-link transmissions/ retransmissions, resulting in significantly shorter user-plane and control plane latencies. New subchannelization schemes and more efficient pilot structures in the downlink and uplink to reduce L1 overhead and to increase spectral efficiency.

New and improved control channel structures in the downlink and uplink to increase efficiency and reduce latency of resource allocation and transmission as well as system entry/re-entry.

### Requirements

The following are backward compatibility requirements:

1. An IEEE 802.16m MS shall be able to operate with a legacy BS, at a level of performance equivalent to that of a legacy MS.
2. Systems based on IEEE 802.16m and the Wireless-OFDMA Reference System shall be able to operate on the same RF carrier, with the same channel bandwidth; and should be able to operate on the same RF carrier with different channel bandwidths.
3. An IEEE 802.16m BS shall support a mix of IEEE 802.16m and legacy MSs when both are operating on the same RF carrier. The performance of such a mix should improve with the fraction of IEEE 802.16m MSs attached to the BS.
4. An IEEE 802.16m BS shall support seamless handover of a legacy MS to and from legacy BS
5. An IEEE 802.16m BS shall be able to support a legacy MS while also supporting IEEE 802.16m MSs, at a level of performance equivalent of what a legacy BS provides to a legacy MS when operating on the same RF carrier.

### Complexity

The IEEE 802.16m PHY/MAC should enable a variety of hardware platforms with differing performance and complexity requirements. IEEE 802.16m shall minimize complexity of the architecture and protocols and avoid excessive system complexity. It should enable interoperability of access networks, support low cost devices and minimize total cost of ownership. IEEE 802.16m should only provide enhancements in areas where the Wireless MAN-OFDMA Reference System does not meet the requirements. In addition, the complexity of MSs and BSs shall be minimized by adhering to the following:

a) The performance requirements shall be met with mandatory features only.

b) Optional features shall be considered only if they provide significant functional and performance improvements over mandatory features.

c) Support of multiple mandatory features which are functionally similar and/or have similar impact on performance shall be avoided.

d) The number of states of protocols and procedures should be minimized

**Services.** IEEE 802.16m should support existing services more efficiently as well as facilitate the introduction of new/emerging types of services. IEEE 802.16m and its services architecture shall be flexible in order to support services required for next generation mobile networks, such as those identified by Report ITU-R M.2072 and IMT-Advanced (IMT.SERV). IEEE 802.16m shall support different quality of service (QoS) levels for different services. IMT21 Advanced QoS requirements shall be supported including end-to-end latency, throughput, and error performance.

**Operating frequencies.** IEEE 802.16m systems shall operate in RF frequencies less than 6 GHz and be deployable in licensed spectrum allocated to the mobile and fixed broadband services and shall be able to operate in frequencies identified for IMT-Advanced. IEEE 802.16m shall be capable of coexisting with other IMT-Advanced technologies. IEEE 802.16m shall support scalable bandwidths from 5 to 20 MHz. Other bandwidths shall be considered as necessary to meet operator and ITU-R requirements.

**Duplex schemes.** IEEE 802.16m shall be designed to support both Time Division Duplex (TDD) and Frequency Division Duplex (FDD) operational modes. The FDD mode shall support both full-duplex and half-duplex MS operation. Specifically, a half-duplex FDD MS is defined as a MS that is not required to transmit and receive simultaneously. IEEE 802.16m shall support both unpaired and paired frequency allocations, with fixed duplexing frequency separations when operating in full-duplex FDD mode. System performance in the desired bandwidths specified in Section 5.5 should be optimized for both TDD and FDD independently while retaining as much commonality as possible. The UL/DL ratio should be configurable. In TDD mode, the DL/UL ratio should be adjustable. In FDD mode, the UL and DL channel bandwidths may be different and should be configurable (e.g. 10MHz downlink, 5MHz uplink). In the extreme, the IEEE 802.16m system should be capable of supporting downlink-only configurations on a given carrier. Asymmetrical operation should be supported in addition to symmetrical operation.

**Support of advanced antenna techniques.** IEEE 802.16m shall support MIMO and beam forming operation. The IEEE 802.16m standard shall define minimum antenna requirements for the BS and MS. For the BS, a minimum of two transmit and two receive antennas shall be supported. For the MS, a minimum of one transmit and two received antennas shall be supported. This minimum is consistent with a 2x2 downlink configuration and a 1x2 uplink configuration.

## 8 Architecture

There are three proposed architectures in 4G technology that are as follows:

**Multimode Devices:** In this configuration, a single terminal employs multiple interfaces to access different wireless system. Figure 2 shows the framework of this architecture. The requirement for this scheme is that the device should incorporate the required hardware necessary to access the different technologies. The flaw with this is that it increases the complexity of the user device which might make it more expensive to the common user. One advantage of this architecture is that it does not require any network modification or internetworking devices. The QoS handling for this type of architecture still remains an open issue.

**Overlay network:** In this type of architecture, a user can access an overlay network which consists of several UAP (universal access points). The architectural model for this is shown in figure 3. The UAP selects an access point depending on user choice, availability and desired QoS. In this case the overlay network performs the major operations such as hand off, frequency translation, content adaptation etc, instead of the terminal performing it. So the overlay network will suffer an increase in complexity.

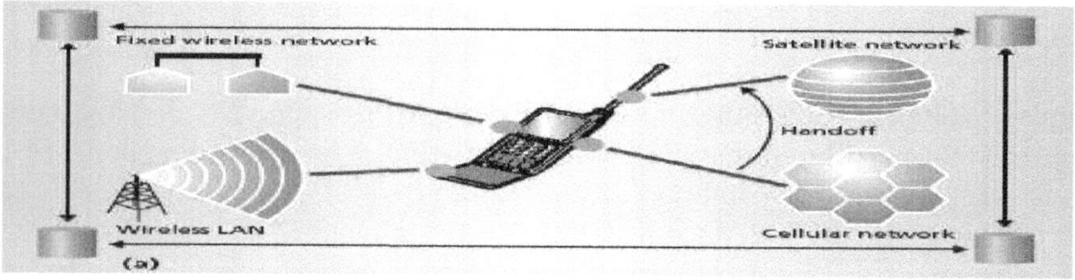


**Fig. 2.** Multimode Device Architecture

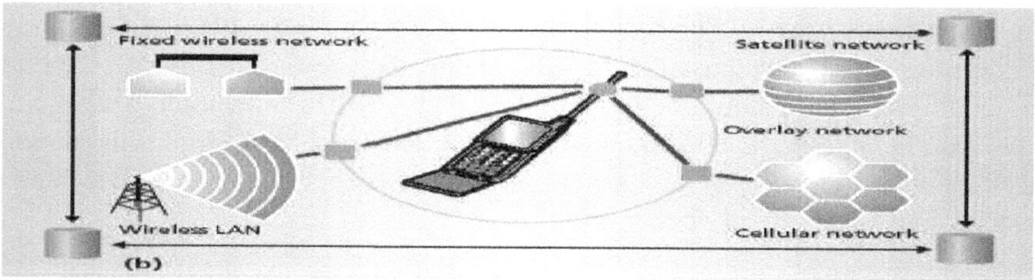


**Fig. 3.** Overlay Device Architecture

**Common Access Protocol:** This architecture can be used if a wireless network can support one or two different protocols. Figure 4 shows the framework for this architectural model. One solution to this is to use wireless ATM (Asynchronous Transfer Mode), which might need internetworking between different networks. To implement

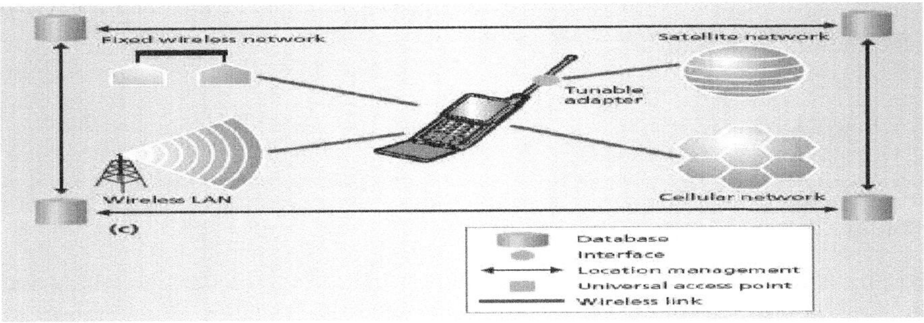


**Fig. 4.** Common Acess Protocol Architecture

this all wireless networks must be capable of transmitting ATM cells with additional headers. This allows the user to communicate with different wireless networks using the same protocol.

## 9 Handoff Mechanism

In the traditional mobile systems only horizontal hand off has to be performed where as in 4G systems both horizontal and vertical hand off should be performed. As shown in figure 5, horizontal hand off is performed when a mobile movies from on cell to another and vertical handoff is performed when a mobile moves between two wireless systems.

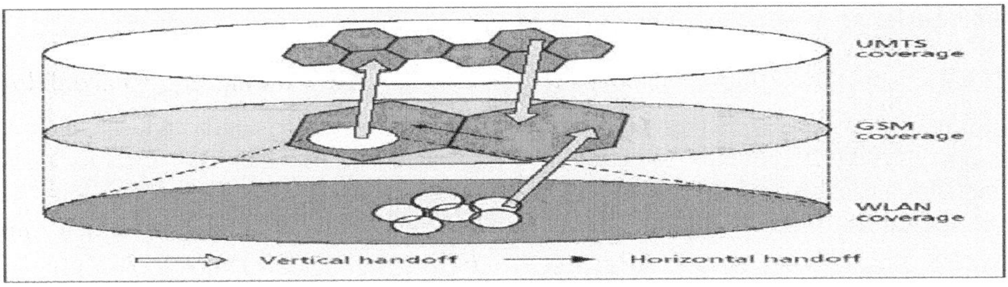


**Fig. 5.** Handoff in 4G

When a mobile moves from its home to another location, it obtains a new care-of address by contacting the subnet of that particular network. When a mobile node moves from one network to another, it detects the new subnet by analyzing the router advertisement that is periodically sent out by the access router. By using the subnet of the new network, which the mobile gets from the router advertisement, the mobile node configures itself a new care-of address. The mobile then needs to verify if there is any duplicate address in the same radio range. So it performs the DAD (Duplicate Address Detection) process and it scans the neighbourhood for any duplicate addresses. If the mobile duplicate address, it has to reconfigure itself a new care-of address. Once a new care-of address is obtained, the mobile updates the home agent with its new care-of address

## 10 Applications

**Virtual Presence:** This means that 4G provides user services at all times, even if the user is off-site.

**Virtual navigation:** 4G provides users with virtual navigation through which a user can access a database of the streets, buildings etc of large cities. This requires high speed data transmission

**Tele-Medicine:** 4G will support remote health monitoring of patients. A user need not go to the hospital and can get videoconference assistance for a doctor at anytime and anywhere.

**Tele-geo-processing applications:** This is a combination of GIS (Geographical Information System) and GPS (Global Positioning System) in which a user can get the location by querying.

**Crisis management:** Natural disasters can cause breakdown in communication systems. In today's world it might take days or weeks to restore the system. But in 4G it is expected to restore such crisis issues in a few hours.

**Education:** For people who are interested in lifelong education, 4G provides a good opportunity. People anywhere in the world can continue their education online in a cost effective manner.

## 11 Future

In future wireless technologies. Swarm intelligence technique will be used to find the destination of packets and also to find shortest path between source and destination nodes. In this technique, behavior of ants is studied i.e. how they find their food and then shortest path of food from their nest. After that all ants in a colony follow that shortest path to collect food to their colony. It is very efficient technique and will lead to good results in future.

## 12 Conclusion

4G technology will change the people's life in the wireless world. 4G has many attractive features which ensures a very high data rate, global roaming etc. New ideas are being introduced by researchers throughout the world, but new ideas introduce new challenges. There are several issues yet to be solved like incorporating the mobile world to the IP based core network, efficient billing system, smooth hand off mechanisms etc.

## References


[1] Sun, J.Z., Sauvola, J., Howie, D.: Features in future: 4G visions from a technical perspective. In: Global Telecommunications Conference GLOBECOM 2001, November 25-29, vol. 6, pp. 3533–3537. IEEE, Los Alamitos (2001)

[2] Bria, A., Gessler, F., Queseth, O., Stridh, R., Unbehaun, M., Wu, J., Zander, J.: 4th-generation wireless infrastructures: scenarios and research challenges. Personal Communications 8(6), 25–31 (2001); see also IEEE WirelessCommunications

[3] Hu, J., Lu, W.W.: Open wireless architecture - the core to 4G mobile communications. In: International Conference on Communication Technology Proceedings, ICCT 2003, April 9-11, vol. 2, pp. 1337–1342 (2003)

[4] Montavont, N., Noel, T.: Handover management for mobile nodes in IPv6 networks. Communications Magazine 40(8) (August 2002)

[5] Zhen, L., Wenan, Z., Junde, S., Chunping, H.: Consideration and research issues for the future generation of mobile communication. In: Canadian Conference on Electrical and Computer Engineering, IEEE CCECE 2002, May 12-15, vol. 3, pp. 1276–1281 (2002)

[6] The Future of Wireless Technologies By Olu Lafe, PhD Presented at John Carroll University Student. Free Enterprise Event (March 7, 2005)

[7] Understanding OFDMA- the interface of 4g technology by By Arnon Friedmann, Software Manager,Texas Instruments Inc.

[8] Issues in Emerging 4G Wireless Networks Upkar Varshney and Radhika Jain, Georgia State University

[9] Wireless Tutorial by Brough Turner NMS Communications

[10] Beyond 3G / 4G Radio Access Technologies (RATs) and Standards Roadmaps- eMobility Technology Platform Whitepaper, Editors: Didier Bourse (Motorola Labs), Didier.Bourse@motorola.com, Rahim Tafazolli (University of Surrey, CCSR), R.Tafazolli@surrey.ac.uk

[11] WIRELESS NETWORK (3G & 4G) Arti Rana1, Kamaljeet Kaur2, Anuj Aggarwal3, Department of Computer Science, AIMT, Ambala,(Kurukshetra University)

# Evaluation of Routing Schemes for MANET

Sima Singh[1] and Ashwani Kush[2]

[1] Department of Computer Science & Engg.
Karnal Institute of Technology & Management Kunjpura Karnal
(Haryana, India)
simasingh.2009@gmail.com
[2] Department of Computer Science,
University College, Kurukshetra University,
Kurukshetra, India
akush20@gmail.com

**Abstract.** The recent advancements in wireless technology have opened new vistas in the development of a new wireless system. A Mobile Adhoc Network is a self configuring network of wireless devices connected by wireless links. Reactive routing protocols have been found to be user friendly and efficient when compared to other routing protocols. In this study a comparison and performance evaluation of two reactive routing protocols AODV and DSR is done using NS-2 Simulator to identify the protocol that is best suited for MANET's. An effort has been carried out to do the performance evaluation of these protocols using random way point model. The simulator used is NS 2.34.

**Keywords:** AODV, DSR, Mobile Adhoc Network.

## 1 Introduction

Electronic technology has advanced a lot in the last few years. This is evident from the recent developments in various fields such as Medicine, Computer science and Information technology. In no other field has these developments been more evident than in field of wireless technology. Though wireless systems have existed since the 1980's it is only in recent times that wireless systems have started to make inroads into all aspects of human life. Mobile Ad Hoc Networks (MANETs) are advanced wireless communication networks. Mobile Ad hoc Network is an autonomous system of mobile nodes connected by wireless links. Each node operates as an end system and a router for all other nodes in the network. A mobile Ad hoc Network is a self configuring network of mobile routers connected by wireless links –the union of which forms an arbitrary topology. An Ad hoc network is often defined as an "infrastructure less" network means that a network without the usual routing infrastructure, link fixed routers and routing backbones. Performance comparison has been done for critical evaluation of MANET main on demand routing protocols. Rest of the paper is organized as : Section 2 describes issues in MANET, Routing have been discussed in section 3, Simulation setup has been described in Section 4 and conclusion is in Section 5.

A. Mantri et al. (Eds.): HPAGC 2011, CCIS 169, pp. 568–573, 2011.

## 2 Critical Issues in MANET

An ad hoc network is a dynamic type of network with similarities and great differences to its parent fixed communication network. The properties of an ad hoc network with its shortcomings and highlight security challenges has been explained by many researchers vide [7,8,9,10,11and 12]. A fixed entity structure such as a base station or central administration is crucial for security mechanisms. The trusted third party member who is expected in traditional networks often defines security services; the absence of such a control entity introduces new opportunities for security attacks on the network. The network instead of relying on a central administrator for network and security service, the network relies upon the nodes for these duties in a self-organized manner. Connectivity is a problem in ad hoc networks as networks are created spontaneously and nodes are mobile. Therefore connectivity between the nodes is sporadic. In ad hoc networks nodes may have no prior relationships with other nodes within the network. Prior acquaintance between nodes can be seen as pre-trust relationships between nodes. It cannot be assumed that pair wise secrets exist between nodes. Physical vulnerability is a major problem in Mobile adhoc network. Mobile node capture or compromised nodes are of higher probability in ad hoc networks than in traditional wired networks with stationary hosts. Lack of Resources is a problem with a network which has no central administrator to perform network and security tasks, and rather relies upon nodes to accomplish such services. This creates a heavy burden upon nodes to perform their own tasks as well as network tasks therefore nodes will have limited resources compared to fixed wired nodes.
The main objective of the study is to compare and analyze the performance of reactive routing protocols in mobile ad hoc network.

- Study focuses on identification of reactive routing protocols which emphasizes on quality of service in mobile ad hoc networks.
- To compare the performance of two reactive routing protocols which focuses on quality of service namely AODV and DSR.
- To deduct the reactive routing protocol which is most efficient in enhancing quality of service and which may lead to optimal increase in performance

## 3 Routing Protocols in MANET

Routing is the Exchange of information from one station of the network to the other. The major goals of routing are to find and maintain routes between nodes in a dynamic topology with possibly uni-directional links, using minimum resources. There are several unicast protocols such as proactive, reactive and hybrid routing protocols. In MANET , routing is achieved via o demand routing protocols.

**Reactive Protocols: It** acquires routing information only when it is actually needed. The Advantage is that due to the high uncertainty in the position of the nodes, however, the reactive protocols are much suited and perform better for ad-hoc networks. The Disadvantages of reactive protocols include High latency time in route finding and excessive flooding leading to network clogging. Some of the Reactive Routing Protocols are[14-18]: Admission Control Enabled On Demand Routing (ACOR), Associativity Based Routing (ABR), AODV (Adhoc on-demand Distance Vector),

DSR (Dynamic Source Routing), CHAMP (CacHing and MultiPath Routing), LAR (Location Aided Routing ).

# 4 Performance Evaluation Metrics

RFC 2501 describes a number of quantitative metrics that can be used for evaluating the performance of a routing protocol for mobile wireless ad-hoc networks. Some of these quantitative metrics [19] are defined as follow:

## 4.1 Packet Delivery Fraction

The packet delivery fraction is defined as the ratio of number of data packets received at the destinations over the number of data packets sent by the sources.

$$\text{Packet Delivery Fraction} = \frac{\text{Total Data Packets Received}}{\text{Total Data Packets Sent}} \text{ X } 100$$

## 4.2 Average End-to-End Delay

This is the average time involved in delivery of data packets from the source node to the destination node. To compute the average end-to-end delay, add every delay for each successful data packet delivery and divide that sum by the number of successfully received data packets.

$$\text{Average End to End Delay} = \frac{\sum (\text{TimeReceived} - \text{TimeSent})}{\text{TotalDataPacketsReceived}}$$

# 5 Performance Evaluation of DSR/AODV

IEEE 802.11 is used as the MAC layer protocol. The simulation experiments are carried over network simulator 2 (version 2.34) installed in Fedora Linux 12. The results have been derived by writing a *tcl* script and generating corresponding *trace* and *nam* files. The mobility model used is random waypoint model in a square area. The area configurations used are 750 meter x 750 meter for 20 nodes and 1000 meter x 1000 meter for 30 and 50 nodes. The packet size is 512 bytes. Same scenario has been used for performance evaluation of both DSR and AODV protocols. It has been shown that even though DSR and AODV share a similar on-demand behavior, the differences in the protocol mechanics can lead to significant performance differentials.

Figure 1 to 3 denotes graphical representation for PDR, End to end delay and throughput for 30 nodes using UDP packets. Packet delivery ratio is stable in DSR than in AOV for 30 nodes. Both reaches the target of approximately 99%. Pause time has been varied from 0 to 500. Pause time of 0 denotes moving nodes instantly and 500 denotes that nodes start moving after 500s. End to end delay is also stable for AODV. DSR has less delay at start and it remains very stable till pause time is 300. It is due to the reason that all initial calculations are carried out by DSR before start of routing. Throughput denotes AODV performing better than DSR.

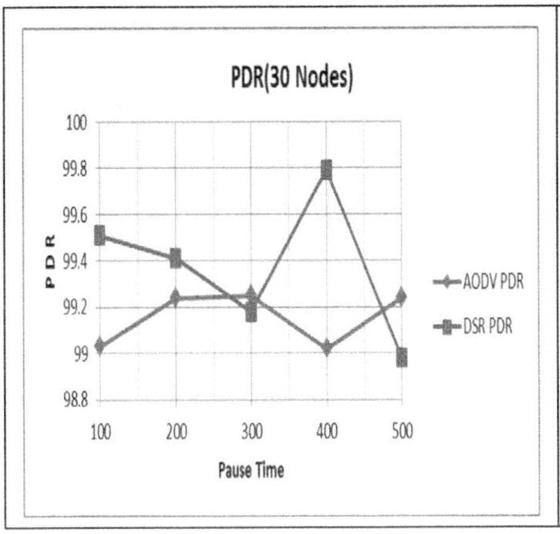


**Fig. 1.** PDR 30 nodes

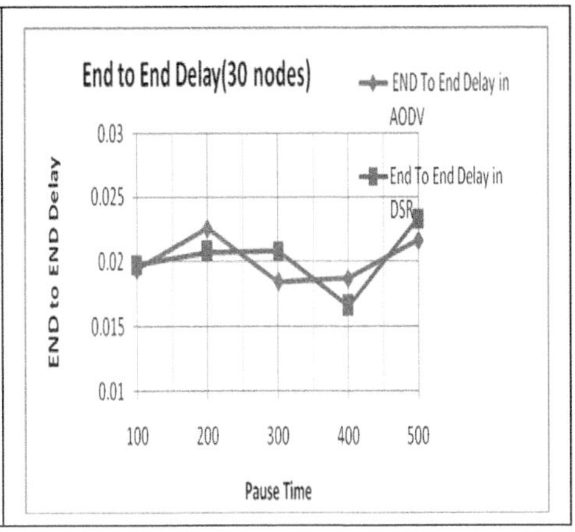


**Fig. 2.** End to end delay 30 nodes

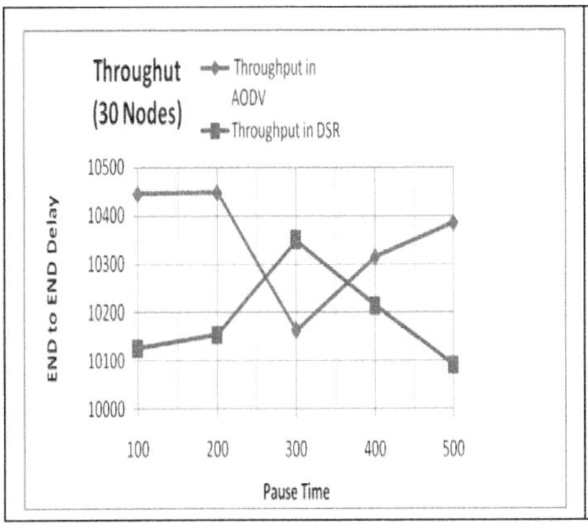


**Fig. 3.** Throughput 30 nodes

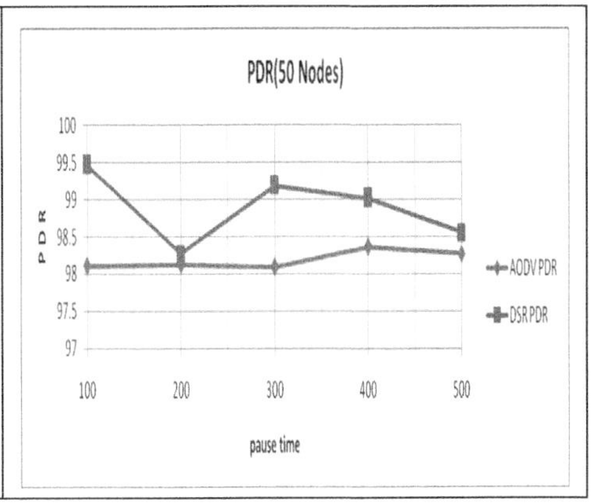


**Fig. 4.** PDR 50 nodes

Packet delivery ratio is stable in DSR than in AOV for 50 nodes. It has been shown in figures 4,5and 6. AODV starts approximately 99% and reaches at appx 98%. Pause time has been varied from 0 to 500. End to end delay is also stable for AODV. The DSR has less delay at start and it remains very stable till pause time is 300. It is due to the reason that all initial calculations are carried out by DSR before start of routing. Throughput denotes AODV performing better than DSR.

The DSR performance decreases in denser networks with higher mobility disclosing that source routing cannot efficiently adapt the network topology changes that are caused by the frequent movement of the nodes. Concluding, DSR demonstrates significantly lower routing load than AODV with the fact that it increases for DSR with

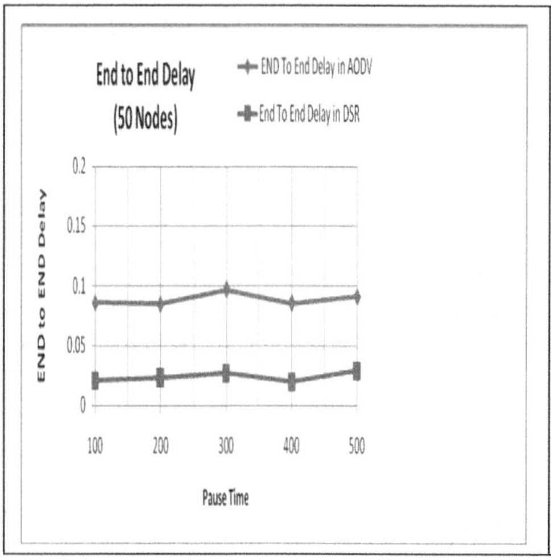


Fig. 5. Delay 50 nodes

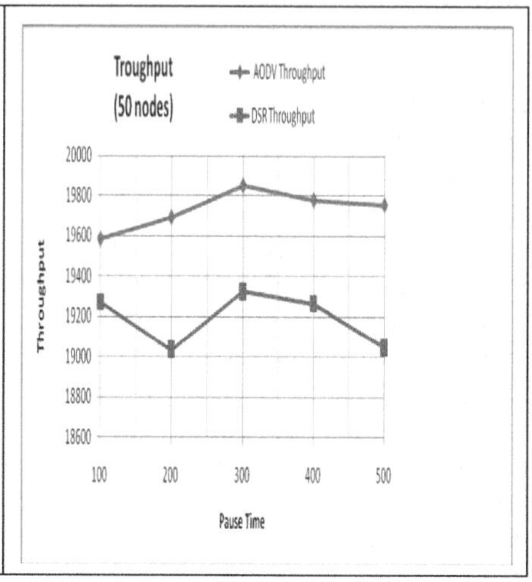


**Fig. 6.** Throughput 50 nodes

growing number of nodes. When the number of nodes is less, the performance of DSR and AODV is similar regardless of mobility. With large numbers of nodes, AODV starts outperforming DSR. The major contribution to AODV's routing overhead is from route requests, while route replies constitute a large fraction of DSR's routing overhead. Furthermore, AODV has more route requests than DSR, and the converse is true for route replies.

## 6 Conclusion

In this paper, performance evaluation of DSR and AODV has been carried out on the basis of Packet delivery ratio, end to end delay and Throughput. The results have been analyzed using a new random way point self created network scenario. The general observation from the simulation is that the AODV protocol performs better in denser mediums and the DSR protocol performs better in sparse medium. In networks with a small number of nodes and low mobility, AODV does not suggest a good solution as a routing protocol. However, AODV has better performance in networks with higher mobility and a greater number of nodes. It is concluded that AODV is the proper protocol for any kind of application (voice, video, file transfer, etc.) in networks with high mobility that consist of large number of nodes.

Since in real life scenario, the emphasis is on denser mediums, we can generalize that the use of AODV protocol is better choice over DSR protocol for efficient routing over mobile adhoc network. The performance of AODV under highly denser medium is still to be compared with TORA, STAR and ZRP. This aspect is still under our consideration. A sincere effort will also be made to evaluate the performance using throughput. Work also need to be done in the field of energy efficient and secure routing.

## References


[1] Johnson, D., Maltz, D.: Dynamic source routing in adhoc wireless networks. In: Imielinksi, T., Korth, H. (eds.) Mobile Computing, vol. 353, pp. 151–181. Kluwer Academic Publishers, Dordrecht (1996)

[2] Perkins, C.E., Royer, E.M.: Adhoc on-demand distance vector routing. In: Proceedings of the 2nd IEEE Workshop on Mobile Computing Systems and Applications (WMCSA), pp. 90–100 (1991)

[3] Moghim, N., Hendessi, F., Movehhedinia, N.: An improvement on adhoc wireless network routing based on aodv. In: The 18th International Conference on Communication Systems ICCS, vol. 2, pp. 1068–1070. IEEE, Los Alamitos (2002)

[4] zeinalipour-yazeti demetrios A glance at quality of service in Mobile Ad-hoc networks final research report for CS 120- Seminar in Mobile Ad-hoc networks, Fall (2001)

[5] Reddy, P.C., Chandrasekar Reddy, P.: Performance analysis of Ad hoc network Routing protocols (2006)

[6] Acs, G., Buttyan, L., Vajda, I.: Provably Secure on-demand Source routing in Mobile Ad-Hoc Networks. IEEE Transactions on Mobile Computing (November 28, 2005)

[7] Capkun, S., Buttyan, L., Hubaux, J.-P.: Self-organised Public- Key Management for Mobile Ad-Hoc network. IEEE Transcations on Mobile Computing 2, 52–64 (2003)

[8] Haas, Z.J., Deng, J., Liang, B., Papadimitratos, P., Sajama, S.: Wireless Ad-Hoc Networks. In: Encyclopedia of Telecommunications. John Willey, New York (2002)

[9] Lidong, Z., Zygmunt, H.: Secuting Ad- Hoc Networks. Cornell University, Ithica (1999)

[10] William, S.: Cryptography and network security principles and practice, 2nd edn. Prentice-Hall,Inc, Englewood Cliffs (1999)

[11] Salem, N.B., Buttyan, L., Hubaux, J.-P., Jakobsson, M.: Node Cooperation in HybridAdhoc Networks. IEEE Transactions on Mobile Computing (2005)

[12] Eschenauer, L., Gligor, V.D.: A KeyManagement Scheme for Distributed Sensor Networks. In: Proc. 9th ACM Conf. on Computer and Communication Security ( ACM CCS 2002) (2002)

[13] Trung, H.D., Benjapolakula, W., Duca, P.M.: Performance evaluation and comparison of different ad hoc routing protocols. Computer Communication: Special issue on security on wireless ad hoc and sensor networks 30(11-12), 2478–2496 (2007)

[14] Sarma Dhulipala, V.R., Chandrasekaran, R., Prabakaran, R.: Timing analysis and repeatability issues of mobile adhoc networking application traffics in large scale scenarios. International Journal of Recent trends in Engineering 1(1) (May 2009)

[15] Aaron, A., Weng, J.: Performance Comparison of Ad-hoc Routing Protocols for Networks with Node Energy Constraints. Stanford university, Stanford

[16] Abolhasam, M., Wysocki, T., Dutkiewicz, E.: A review of routing protocols for mobile adhoc networks. AdHoc Networks (2004)

[17] Kumar, A., B.R., Reddy, L.C., Hiremath, P.S.: Performance Comparison of Wireless Mobile Ad-Hoc Network Routing Protocols. IJCSNS International Journal of Computer Science and Network Security 8(6) (2008)

[18] Kush, A., Taneja, S.: A Survey of Routing Protocols in Mobile Adhoc Networks. International Journal of Innovation, Management and Technology 1(3), 279–285 (2010)

[19] Kioumourtzis, G.: Simulation and Evaluation of Routing Protocols for Mobile Adhoc Networks. Thesis, Master of Science in Systems Engineering and Master of Science in Computer Science, Naval Postgraduate School, Monterey, California (2005)

# Fuzzy Logic Based Routing Algorithm for Mobile Ad Hoc Networks

Sonia Gupta , P.K. Bharti, and Vishal Choudhary

AIMT, Knowledge, Park-3 ,Greater Noida -201308
{soniagee,pa_dutt}@gmail.com,
vishalhim@yahoo.com

**Abstract.** Mobile ad hoc networks consist of mobile nodes that communicate without an infrastructure. It is a self configuring network connected by wireless links. All the nodes move around randomly, thus changing the network topology dynamically. The primary challenge in building a MANET is equipping each device to continuously maintain the information required to properly route the traffic. In this paper a routing algorithm based on Fuzzy Logic is proposed which is having low communication overhead and storage requirements. The proposed algorithm takes three input variables: signal power, mobility and delay. The absolute value of each parameter can take a large range at different points on the network.

**Keywords:** Mobile ad hoc Networks, Fuzzy Logic, Fuzzification, Defuzzification, wireless routing.

## 1 Introduction

A mobile ad hoc network (MANET) also called a mesh network, is a self configuring network of mobile devices connected by wireless links [1]. It consists of mobile nodes that autonomously establish connectivity via multi-hop wireless communications. There is no use of static network infrastructure such as base station or any centralized administration in MANET. They can set up anywhere. These types of networks have many advantages, such as self configuring and adaptability to highly variable mobile characteristics like transmission, conditions, propagation channel, distribution characteristics and power level. They are useful in many situations such as military applications, conferences, emergency search, and rescue operations. However, such benefits come with new challenges which mainly reside in the unpredictability of the network topology due to mobility of nodes and the limited available bandwidth due to the wireless channel. These characteristics demand a new way of designing and operating these types of networks. For such networks, an effective routing protocol is critical for adapting node mobility as well as a feasible path for data transmission. [2, 3, 4].

## 2 Related Work

There are two classes of protocols: one is based on the preparation of information tables i.e. Pro-active or Table Driven routing protocols and the other is without them called Reactive or On Demand routing protocols.

A. Mantri et al. (Eds.): HPAGC 2011, CCIS 169, pp. 574–579, 2011.

## 2.1 Proactive or Table Driven Routing Protocol

These protocols attempt to maintain consistent, up-to- date routing information from each node to every other node in the network. Each node maintains one or more table to store routing information and respond to changes in the network topology. Routing is continuously updated regardless of the network traffic.

### 2.1.1 Destination Sequenced Distance Vector Routing Protocol (DSDV)

DSDV protocol [5, 6] has been specifically targeted for mobile networks. Every mobile station needs to maintain a routing table, the no. of hops to reach the destination and the sequence no. assigned to the destination node. The main drawback of DSDV is that it requires a regular update of its routing tables, thus reducing the bandwidth efficiency. It is not suitable for very large network i.e. less scalability and it is not also suitable for highly dynamic network.

### 2.1.2 Wireless Routing Protocol (WRP)

WRP [7] belongs to the class of path finding algorithms. It avoids count to infinity problem. Each node in the network maintains four tables: distance table, routing table, link cost table and message retransmission list.

### 2.1.3 Clusterhead Gateway Switch Routing Protocol (CGSR)

CGSR elects a node as a cluster head using a distributed algorithm within the cluster.

## 2.2 Reactive or on Demand Routing Protocol

This type of protocol creates rote only when desired by the source node. When a node requires a route to a destination, it initiates a route discovery process within the network

### 2.2.1 Dynamic Source Routing Protocol

In DSR [8] the packet contains the full route to destination and the intermediate nodes do not have to make any routing decisions. It is not scalable to large networks and even requires significantly more processing resources than other protocols. Basically, In order to obtain the routing information, each node must spend lot of time to process any control data it receives, even if it is not the intended recipient.

### 2.2.2 Ad Hoc on Demand Vector Routing Protocol (AODV)

AODV [9] is based on DSDV but it minimizes the no of required broadcasts by creating routes on the demand basis. It broadcasts a route request packet to its neighbors, and so on, until the destination is located. In AODV as the network size increases the performance metrics begin to decrease. It is vulnerable to various kinds of attacks as it requires the cooperation of other nodes.

### 2.2.3 Temporarily Ordered Routing Algorithm (TORA)

TORA is highly adaptive loop-free distributed routing algorithm based on the concept of link reversal [10]. It is designed to minimize reaction to topological changes. It guarantees that all routes are loop free and typically provides multiple routes for any source/destination pair. It depends on synchronized clocks among nodes in the ad hoc network.

#### 2.2.4 Associativity Based Routing Protocol (ABR)

ABR [11] is totally different routing protocol for ad hoc wireless network which is free from loops, deadlocks and packet duplication and defines a new routing metric for the same.

## 3 Proposed Scheme

### 3.1 The Proposed Fuzzy Logic Based Routing Scheme

The proposed routing algorithm takes three input variables – signal power, mobility and delay. We take normalized values of each parameter. Crisp normalized values are converted into fuzzy values. For this, three fuzzy sets are defined for each variable:

**Signal power:**

**(1)** low (from 0 to 0.4)
**(2)** medium (from0.2 to 0.8)
**(3)** high (from 0.6 to 1.0)

**Delay:**

**(1)** Poor (from 0 to 0.4)
**(2)** Average (from0.2 to 0.8)
**(3)** Excellent (from 0.6 to 1.0)

**Mobility:**

**(1)** low (from 0 to 0.4)
**(2)** medium (from0.2 to 0.8)
**(3)** high (from 0.6 to 1.0)

The normalized value of each parameter is mapped into fine sets. Each value will have some grade of membership function for each set. The memberships that have been defined for each of the fuzzy set for any particular input variable are triangular in shape. Next the inference rules are written. The crisp value of input variable is given and a defuzzified crisp value for selected variable is calculated from the derived algorithm. An output linguistic variable is used to represent the route.

Proposed optimal routes are based upon the fuzzy rules for different ranges of the metric availability. The routes are defined as

- below optimal (from 0 to 0.4)
- suboptimal (from 0.2 to 0.8) and
- optimal (from 0.6 to 1.0)
  between two mobile hosts.

The below optimal indicates not optimal path, the sub optimal indicates good path and the optimal path indicates the best path. The proposed routing algorithm can apply to different routing metrics. These routes have to satisfy the mobility, signal power and delay requirements of the network. The grade of membership function can be any where between 0 and 1 for each fuzzy set. The defuzzified crisp value for selected variable is calculated from the derived algorithm.

### 3.2 Fuzzy Inference Rules

**1.** If (signal power is low) and (mobility is low) and (delay is poor) then the route will be below optimal.

**2.** If (signal power is low) and (mobility is low) and (delay is average) then the route will be below optimal.

**3.** If (signal power is low) and (mobility is low) and (delay is excellent) then the route will be sub optimal.

**4.** If (signal power is low) and (mobility is medium) and (delay is poor) then the route will be below optimal.

**5.** If (signal power is low) and (mobility is medium) and (delay is average) then the route will be sub-optimal.

**6.** If (signal power is low) and (mobility is medium) and (delay is excellent) then the route will be sub optimal.

**7.** If (signal power is low) and (mobility is high) and (delay is poor) then the route will be below optimal.

**8.** If (signal power is low) and (mobility is high) and (delay is average) then the route will be below optimal.

**9.** If (signal power is low) and (mobility is high) and (delay is excellent) then the route will be sub-optimal.

**10.** If (signal power is medium) and (mobility is low) and (delay is poor) then the route will be sub- optimal.

**11.** If (signal power is medium) and (mobility is low) and (delay is average) then the route will be optimal.

**12.** If (signal power is medium) and (mobility is low) and (delay is excellent) then the route will be optimal.

**13.** If (signal power is medium) and (mobility is medium) and (delay is poor) then the route will be below optimal.

**14.** If (signal power is medium) and (mobility is medium) and (delay is average) then the route will be sub optimal.

**15.** If (signal power is medium) and (mobility is medium) and (delay is excellent) then the route will be optimal.

**16.** If (signal power is medium) and (mobility is high) and (delay is poor) then the route will be below optimal.

**17.** If (signal power is medium) and (mobility is high) and (delay is average) then the route will be sub optimal.

**18.** If (signal power is medium) and (mobility is high) and (delay is excellent) then the route will be optimal.

**19.** If (signal power is high) and (mobility is low) and (delay is poor) then the route will be sub-optimal.

**20.** If (signal power is high) and (mobility is low) and (delay is average) then the route will be optimal.

**21.** If (signal power is high) and (mobility is low) and (delay is excellent) then the route will be optimal.

**22.** If (signal power is high) and (mobility is medium) and (delay is poor) then the route will be sub-optimal.

**23.** If (signal power is high) and (mobility is medium) and (delay is average) then the route will be sub-optimal.

**24.** If (signal power is high) and (mobility is medium) and (delay is excellent) then the route will be optimal.

**25.** If (signal power is high) and (mobility is high) and (delay is poor) then the route will be below - optimal.

**26.** If (signal power is high) and (mobility is high) and (delay is average) then the route will be sub-optimal.

**27.** If (signal power is high) and (mobility is high) and (delay is excellent) then the route will be optimal.

## 4 Conclusion

There are number of qualitative and quantitative metrics that can be used to compare reactive routing protocols. Most of the existing routing protocols ensure the qualitative metrics. Therefore, the following different quantitative metrics have been considered to make the comparative study of these routing protocols through simulation.

1. **Routing Overhead:** This metric describes how many routing packets for route discovery and route maintenance need to be sent so as to propagate the data packets.
2. **Average Delay:** This metric represents average end-to-end delay and indicates how long it took for a packet to travel from source to the application layer of the destination. It is measured in seconds.
3. **Throughput:** This metric represents the total number of bits forwarded to higher layers per second. It is measured in bps. It can also be defined as the total amount of data a receiver actually receives from sender divided by the time taken by the receiver to obtain the last packet.
4. **Media Access Delay:** The time a node takes to access media for starting the packet transmission is called as media access delay. The delay is recorded for each packet when it is sent to the physical layer for the first time.
5. **Packet Delivery Ratio:** The ratio between the amount of incoming data packets and actually received data packets.
6. **Path Optimality:** this metric can be defined as the difference between the path actually taken and the best possible path for a packet to reach its destination.

## 5 Future Work

In this paper, an effort has been made to concentrate on the comparative study and performance analysis of various reactive/proactive routing protocols on the basis of the performance metrics. AODV is still better in Route updating and maintenance process. It has been further concluded that due to the dynamically changing topology and infrastructure less, decentralized characteristics, security and power awareness is hard to achieve in mobile ad hoc networks. Hence, security and power awareness mechanisms should be built-in features for all sorts of applications based on ad hoc network. The focus of the study is on these issues in our future research work and effort will be made to propose a solution for routing in Ad Hoc networks by tackling these core issues of secure and power aware/energy efficient routing.

## References

1. Krag, T., Büettrich, S.: Wireless Mesh Networking. O'Reilly Wireless Dev Center, Sebastopol (2004)
2. Chang, C.Y., Chang, C.T., Chen, T.S., Chang, H.R.: Hierarchical management protocol for constructing a QoS communication path in wireless Ad hoc networks. Information Sciences 177(13), 2621–2641 (2007)
3. Perkins, C., Bel ding-Royer, E., Das, S.: Ad Hoc On-Demand Distance Vector (AODV) Routing
4. Sun, B.L., Li, L.Y.: A QoS Multicast Routing Optimization Algorithms Based on Genetic Algorithm. Journal of Communications and Networks 8(1), 116–122 (2006)
5. Perkins, C.E., Bhagwat, P.: Highly dynamic destination sequenced distance vector routing (DSDV) for mobile computers. In: ACM SIGCOMM, vol. 24(4), pp. 234–244 (October 1994)
6. Charles Perkins, E., Bhagwat, P.: Highly Dynamic Destination-Sequenced Distance- Vector Routing (DSDV) for Mobile Computers. In: The Proceedings of the Conference on Communication Architecture, Protocols and Applications, pp. 234–244. ACM Press, New York (1994)
7. Murthy, S., Garcia-Luna-Aceves, J.J.: An efficient routing protocol for wireless networks. ACM Mobile Networks and App. J., Special issue on Routing in Mobile communication Networks, 183–187 (October 1996)
8. Broch, J., Maltz, D.A.: The dynamic source routing protocol for mobile ad hoc networks. Internet Draft, draft-ietf-manet-dsr-00.txt (March 1998)
9. Perkins, C.E., Royer, E.M.: Ad hoc on-demand distance vector routing. In: Proc. 2nd IEEE Wksp. Mobile Comp. Sys. and Apps., pp. 99–100 (February 1999)
10. Park, V.D., Corson, M.S.: A highly adaptive distributed routing algorithm for mobile wireless networks. In: Proc. INFOCOM, p. 9 (April 1997)
11. Toh, C.K.: A novel distributed routing protocol to support ad hoc mobile computing. In: Proc. IEEE 15th Annual Int'I. Phoenix Conf. Comp. and Commun. (March 1996)
12. Stojmenovic, I.: Position based routing in ad hoc networks. IEEE Communications Magazine, 128–134 (July 2002)

# Analysis of Security and Key Management Schemes for Authenticated Broadcast in Heterogeneous Wireless Sensor Networks

P. Kalyani[1] and C. Chellappan[2]

[1] Prof., Department of CSE, I.R.T.T., Erode, India
kalyani_irtt@yahoo.co.in
[2] Prof., Department of CSE, Anna University, Chennai, India

**Abstract.** The security in wireless sensor networks (WSN) is a critical issue due to the inherent limitations of computational capacity, storage capacity and power usage. Packets are dropped or discarded completely, or selectively forwarded by an anonymous party. Also the network is flooded with global suspicious broadcasts. These kinds of attacks may be avoided when using multi path and authenticated broadcasts, which has to be facilitated by the underlying key management architecture. Key management only makes sure the communicating nodes possess the necessary keys, at the same time providing the confidentiality, integrity and authenticity of the communicated data. The proposed work, made an effort to survey the well-known security issues in WSNs and study the various asymmetric key or public key algorithms which are used for key distribution as well as encryption/decryption in sensor network for authenticated message broad cast. Based on the analysis, proposed a new method which will improves the performance of an existing RSA algorithm by using Chinese Remainder Theorem (CRT) algorithm for decryption phase. Also it offers countermeasures for the attacks in the network layer of WSN.

**Keywords:** Wireless Sensor Network, Security, key management, Cryptography.

## 1 Introduction

Wireless sensor networks (WSN's) are quite useful in many applications since they provide a cost effective solution to many real life problems. But it appears that they are more prone to attacks than wired networks. They are susceptible to a variety of attacks, including node capture, physical tampering, and denial of service, prompting a range of fundamental research challenges [1], an attacker can easily eavesdrop on, inject or alter the data transmitted between sensor nodes. Security allows WSNs to be used with confidence and maintains integrity of data. Providing security in wireless sensor networks is pivotal due to the fact that sensor nodes are inherently limited by resources such as power, bandwidth, computation, and storage. Key management only makes sure the communicating nodes possess the necessary keys, at the same time protecting the confidentiality, integrity and authenticity of the communicated data.

A. Mantri et al. (Eds.): HPAGC 2011, CCIS 169, pp. 580–587, 2011.

We proposed an enhanced RSA algorithm for providing security for a wireless sensor network and it is also compared with traditional public cryptography techniques. Also RSA-CRT being an asymmetric key with simple algorithm, it consumes less energy and thus increases the life time of the network.

## 2 Issues in WSN Security

Security mechanisms in WSN are developed in view of certain constraints and are classified into two types. One is security needed for operations and another is security for information. The objectives of these securities are, first the network should continue its function even when some of its components attacked and CIA of information should never be disclosed respectively. In the second objective the information security can be achieved with cryptography. The security in WSNs is critical issue due to the inherent hardware limitations and constraints: 1) Energy efficiency, 2) No public-key cryptography, 3) Physically tamperable, 4) Multiple layers of defense.

In this paper an effort made to analysis various asymmetric keys algorithms ELGAMAL, RSA (Rivest Shamir Adelman), Public Key Encryption (PKE )which are used for key distribution as well as encryption/decryption in sensor network for authenticated message broad cast. The analysis shows that RSA is better than ELGAMAL and PKE. We proposed a method to enhance and improve the performance of RSA by applying Chinese Remainder Theorem (CRT) in decryption phase of RSA. This concept is utilized in Hardware fault attacks and has been tested in hardware like CPU, RAM, EPROM, Smart card processors fault attacks and shows improvement in speed and reduced computation time and space for the RSA cryptography algorithm.[11,12,13]. The same concept of applying CRT in the decryption phase of RSA algorithm used for message authentication in WSN will give advantages over the methods studied (ELGAMAL,RSA,PKE) with respect to energy, computation time, storage space, speed of processing in turn reduces the communication over heads.

In the literature of WSN, Priority and Random Selection for Dynamic Window Secured Implicit techniques [14], Optimal Power Multicast [15], Low Power Phase Locked Loop Frequency Synthesizer [16], Transportation Infrastructural Health [17], and Ethical Issues in E-Commerce [18] are available resources for further studies.

Against wormhole attacks and hence sinkhole attacks, so far there is no resource-lean and energy-efficient countermeasure, i.e. with or without key management, wormhole and sinkhole attacks are still an open issue. Furthermore, the corresponding key management architecture has to support node-specific key pre-distribution (i.e. every node has to share one key with every other node in the network), in addition to authenticate broadcasts. These two issues multipath routing support and node specific key pre-distribution support are well analyzed and a new efficient key management scheme has been proposed.

## 3 Proposed Method

RSA operations are modular exponentiations of large integers with a typical size of 512 to 2048 bits. RSA encryption generates a cipher text $C$ from a message $M$ based on a modular exponentiation $C = M^e \bmod n$. Decryption regenerates the message by computing $M = C^d \bmod n^1$. Among the several techniques that can be used to accelerate RSA, we specifically focused on those applicable under the constraints of sensor nodes and considered the CRT technique.

RSA private-key operations, namely decryption and signature generation, can be accelerated using the Chinese Remainder Theorem (CRT). RSA chooses the modulus $n$ as the product of two primes $p$ *and* $q$, where $p$ and $q$ are on the order of $\sqrt{n}$ (e.g. for a 1024-bit $n$, $p$ and $q$ are on average 512 bits long). Using the CRT, a modular exponentiation for decryption $M = C_d \bmod n$ can be decomposed into two modular exponentiations $M_1 = C_1^{d1} \bmod p$ and $M_2 = C_2^{d2} \bmod q$, where $C_1$, $d_1$, $C_2$ and $d_2$ are roughly half the size of $n$. Assuming schoolbook multiplication with operands of size $m/2 = [log_2(n)]/2$, modular multiplications can be computed in roughly 1/4 of the time as $m$-bit modular multiplications. Thus the CRT reduces computation time through Montgomery multiplication by nearly 3/4 resulting in up to a 4x speedup.

RSA is more efficient in Chinese Remainder Theorem mode than in straightforward mode. As it handles data with half the RSA modulus size, RSA with CRT is theoretically about four times faster and is therefore better suited to wireless sensor networks. The proposed implementation only requires the usual parameters needed for the computation, the input message m and the classical RSA-CRT key set (p, q, dp, dq, iq). The RSA signature in CRT mode is described below:

---

Input: message m, key (p, q, dp, dq, iq)
Output: signature $m^d \in ZN$

---

$Sp = m^{dp} \bmod p$
$Sq = m^{dq} \bmod q$
$S = Sq + q \cdot (iq \cdot (Sp - Sq) \bmod p)$
Return (S)

---

CRT implementation of RSA

## 4 Result and Performance Analysis

The methodology is implemented in NS2. The performance analyses of the proposed methodology are given in the table 1, 2, 3 and in the figures 1, 2, 3, & 4. The requirement of Storage Space in Mega-byte is shown in table 1 and figure 1. Energy Consumption of each node in mili-Watt is shown in table 2 and figure 2. Time Consumption for Key Exchange in client and server side are shown in table 3 and figure 3, 4.

**Table 1.** Storage Space Required (in MB)

| No of Nodes | ELGAMAL | RSA | PKC | Proposed |
|---|---|---|---|---|
| 20 | 0.30 | 0.20 | 0.10 | 0.15 |
| 40 | 0.39 | 0.26 | 0.13 | 0.20 |
| 60 | 0.51 | 0.34 | 0.17 | 0.25 |
| 80 | 0.63 | 0.42 | 0.21 | 0.32 |
| 100 | 0.79 | 0.53 | 0.26 | 0.40 |
| 120 | 0.99 | 0.66 | 0.33 | 0.50 |
| 140 | 1.19 | 0.79 | 0.40 | 0.59 |
| 160 | 1.43 | 0.95 | 0.48 | 0.71 |
| 180 | 1.71 | 1.14 | 0.57 | 0.86 |
| 200 | 1.88 | 1.25 | 0.63 | 0.94 |

**Table 2.** Energy Consumption (in mW)

| No of Nodes | ELGAMAL | RSA | PKC | Proposed |
|---|---|---|---|---|
| 200 | 1120 | 273.8 | 412.3 | 102.7 |
| 400 | 1321.60 | 350.46 | 527.74 | 126.32 |
| 600 | 1559.49 | 448.59 | 675.51 | 155.37 |
| 800 | 1840.20 | 574.20 | 864.66 | 191.11 |
| 1000 | 2355.45 | 734.98 | 1106.76 | 235.07 |
| 1200 | 3014.98 | 940.77 | 1416.65 | 289.13 |
| 1400 | 3859.17 | 1204.19 | 1813.31 | 355.63 |
| 1600 | 5132.70 | 1541.36 | 2321.04 | 437.43 |
| 1800 | 6826.49 | 1972.94 | 2970.93 | 538.04 |
| 2000 | 9079.23 | 2525.36 | 3802.80 | 661.78 |

**Table 3.** Time Consumption for Key Exchange in client and server sid

| Key Size (in bits) | Time Consumption for Key Exchange in Server | | | | Time Consumption for Key Exchange in Client | | | |
|---|---|---|---|---|---|---|---|---|
| | ELGAMAL | RSA | PKC | Proposed | ELGAMAL | RSA | PKC | Proposed |
| 160 | 3 | 3 | 4 | 3 | 1.32 | 1.23 | 1.78 | 1.11 |
| 256 | 9 | 7 | 11 | 8 | 4.80 | 4.30 | 5.20 | 4.06 |
| 512 | 17 | 16 | 18 | 15 | 8.28 | 7.37 | 8.62 | 7.01 |
| 1024 | 26 | 22 | 28 | 19 | 11.76 | 10.44 | 12.04 | 9.96 |
| 2048 | 161 | 166 | 178 | 122 | 15.24 | 13.51 | 15.46 | 12.91 |

network layer of WSNs and a new efficient key management scheme proposed. In this view we proposed a new efficient key management scheme RSA-CRT algorithm to support both multi path and node specific key pre-distribution for authentication of message broadcast in Wireless Sensor Networks(WSNs). The proposed method takes the advantages of the self enforcing scheme ie public key algorithm and key pre distribution scheme and are combined together to achieve efficient key management scheme which will reduce the energy consumption and communication overheads even with limited resources. The proposed new algorithm implemented and simulated in NS2 simulator and the results shows that improvement in performance and reduced energy consumption and time delay thus increases the network life time and reduced communication over heads.

## References

[1] Perrig, A., Stankovic, J., Wagner, D.: Security in wireless sensor networks. Communication ACM 47(6), 53–57 (2004)

[2] Djenouri, D., Khelladi, L.: A Survey of Security Issues in Mobile AdHoc and Sensor Networks. IEEE Communication Surveys and Tutorials 7(4), 2–28 (2005)

[3] Gaubatz, G., Kaps, J.-P., Sunar, B.: Public Key Cryptography in Sensor Networks—Revisited. In: Castelluccia, C., Hartenstein, H., Paar, C., Westhoff, D. (eds.) ESAS 2004. LNCS, vol. 3313, pp. 2–18. Springer, Heidelberg (2005)

[4] Perrig, A., Szewczyk, R., Wen, V., Culler, D., Tygar, J.D.: SPINS: Security Protocols for Sensor Networks. Wireless Networks 8(5), 521–534 (2002)

[5] Baggio, A.: Wireless sensor networks in precision agriculture. In: ACM Workshop on Real-World Wireless Sensor Networks (REALWSN 2005), Stockholm, Sweden (June 2005)

[6] Mainwaring, A., Polastre, J., Szewczyk, R., Culler, D., Anderson, J.: Wireless SensorNetworks for Habitat Monitoring. In: First ACM Workshop on Wireless Sensor Networks and Applications, Atlanta, GA, USA (September 2002)

[7] Fuchs, G., Truchat, S., Dressler, F.: Distributed Software Management in Sensor Networks using Profiling Techniques. In: 1st IEEE/ACM International Conference on Communication System Software and Middleware (IEEE COMSWARE 2006): 1st International Workshop on Software for Sensor Networks (SensorWare 2006), New Dehli, India (January 2006)

[8] Zhang, W., Cao, G.: Group Rekeying for Filtering False Data in Sensor Networks: A Predistribution and Local Collaboration-Based Approach. In: 24th IEEE Annual Joint Conference of the IEEE

[9] Yang, H., Luo, H., Ye, F., Lu, S., Zhang, L.: Security in mobile ad hoc networks: challenges and solutions. IEEE Wireless Communications 11(1), 38–47 (2004)

[10] Delgosha, F.: Senier member of IEEE A multivibrate key establishment scheme for wireless sensor networks. IEEE Transaction on wireless communication 18 (April 2009)

[11] Xiao, Y., Rayi, V., Sun, B., Du, X., Hue, F.: A survey of key management schemes in wireless sensor networks. Computer Communications 30, 2314–2341 (2007)

[12] Gura, N., Patel, A., Wander, A., Eberle, H., Shantz, S.C.: Comparing Elliptic Curve Cryptography and RSA on 8-bit CPUs

[13] Vigilant, D.: RSA with CRT: A new cost-effective solution to thwart fault attacks. In: Oswald, E., Rohatgi, P. (eds.) CHES 2008. LNCS, vol. 5154, pp. 130–145. Springer, Heidelberg (2008)

[14] Hanapi, Z.M., Ismail, M., Jumari, K.: Priority and Random Selection for Dynamic Window Secured Implicit Geographic Routing in Wireless Sensor Network. American Journal of Engineering and Applied Sciences 2, 494–500 (2009), doi:10.3844/ajeassp.2009.494.500
[15] Maalla, A., Wei, C., Taha, H.J.: Optimal Power Multicast Problem in Wireless Mesh Networks by Using a Hybrid Particle Swarm Optimization. American Journal of Applied Sciences 6, 1758–1762 (2009), doi:10.3844/ajassp.2009.1758.1762
[16] Ismail, N.M.H., Othman, M.: Low Power Phase Locked Loop Frequency Synthesizer for 2.4 GHz Band Zigbee. American Journal of Engineering and Applied Sciences 2, 337–343 (2009), doi:10.3844/ajeassp.2009.337.343
[17] Chang, C., Mehta, R.: Fiber Optic Sensors for Transportation Infrastructural Health Monitoring. American Journal of Engineering and Applied Sciences 3, 214–221 (2010), doi:10.3844/ajeassp.2010.214.221
[18] Delgosha, F.: Senier member of IEEE A multivibrate key establishment scheme for wireless sensor networks. IEEE Transaction on Wireless Communication 18 (April 2009)

# Simulative Analysis of Bidirectional WDM/TDM-PON Using NRZ and RZ Downstream Signals and Narrowband AWG

Rajniti[1], Suman Anita[1], Sheetal Anu[2], and Kumar Parveen[1]

[1] BCET, Gurdaspur, Punjab, India
suman_a1@yahoo.com
[2] GNDU, Regional Campus,
Gurdaspur, Punjab, India
nripanu@ yahoo.co.in

**Abstract.** In this paper, we compare the performance of WDM-PON system for NRZ and RZ data formats operating at bit rates up to 20Gb/s by varying fiber length. It is observed that the multicast signals can be delivered to the designated subscribers with acceptable performance almost upto 45km. At Optical Network Unit (ONU), upstream data can be remodulated on the downstream wavelength by using Array Wave Guide (AWG). It is found that the performance of the NRZ data format is better at low bit rates, Further it is reported that system gives optimum performance for the middle channel due to lower interchannel interference. The upstream signals can be effectively transmitted up to 20Gb/s, however downstream signals can be transmitted only at 2.5Gb/s. The faithful transmission of signal can be carried up to 50km for upstream and 60km for downstream, as the Bit Error Rate (BER) rises above $10^{-9}$ beyond this distance.

**Keywords:** WDM-PON, NRZ, RZ, OLT, ONU.

## 1 Introduction

The bandwidth demand in the access network has been increasing rapidly over the past several years. The predominant broadband access solutions deployed today are the digital subscriber line (DSL) and community antenna television (CATV) / (cable TV) based networks. Passive optical networks (PONs) have evolved to provide much higher bandwidth in the access network. A PON is a point-to-multipoint optical network, where an optical line terminal (OLT) at the Central Office (CO) is connected to many optical network units (ONUs) at remote nodes through one or multiple 1:N optical splitters. [1-2].

Various blends of the PON have emerged in recent years: the Ethernet PON (EPON), broadband PON (BPON), generic framing procedure based PON (GFP PON). Although the PON provides higher bandwidth than traditional copper-based access networks, there exists the need for further increasing the bandwidth of the

A. Mantri et al. (Eds.): HPAGC 2011, CCIS 169, pp. 588–593, 2011.

PON by employing wavelength-division multiplexing (WDM) so that multiple wavelengths may be supported in either or both upstream and downstream directions [3-4]. WDM-PONs are the next generation in development of access networks. Ultimately, they can offer the largest bandwidth at the lowest cost. In principle, the architecture of WDM-PON is similar to the architecture of the PON. A WDM-PON solution provides scalability because it can support multiple wavelengths over the same fiber infrastructure, is inherently transparent to the channel bit rate, and it does not suffer power-splitting losses [5-7]. Different wavelengths may be operated at different bit rates, if necessary; hence, different varieties of services may be supported over the same network. In the downstream direction of the WDM-PON, the wavelength channels are routed from the OLT to the ONUs by a passive arrayed waveguide grating (AWG) router, which is deployed at a "remote node" (RN), which is where the passive splitter used to be in a TDM-PON [8-10].

Bock et al. [1] described WDM/TDM-PON architecture by using free spectral range FSR periodicity and arrayed waveguide grating AWG. Transmission test shows correct operation at 2.5Gb/s to a 30km. by mean of optical transmission test. Calabretta et al.[2] presented an innovative architecture to realize a single feeder bidirectional WDM/TDM-PON on modified NRZ Differential Phase Shift Keying (NRZ-DPSK) downstream signals at 20kb/s and narrowband AWG. In this remodulated upstream signals are obtained at 1Gb/s. the experimental result shows error free transmission for both upstream and downstream signals. Feny et al. [3] discussed a scheme in which modified NRZ format is used to realize multicast WDM-PON by adjusting downstream extinction ratio and achieved good BER rate performance for upstream signals. Han et al. [4] proposed a WDM-PON model with multicast capability like high scalability multiwavelength converter and single copy broadcast capability by employing multistage AWGs at remote node.

However, the feasibility of the WDM passive optical network for higher bit rate (>1Gb/s) and transmission distance greater than 30km is not available as such in the literature, and thus is explored here to investigate the performance of WDM-PON system for varying data rates and length. The system performance has also been investigated for various data formats. In section 2, the system description and simulation are described. In section 3, comparison results of the simulated system have been reported for various data formats and bit rates. And finally in section 4, conclusions are made.

## 2 System Description and Simulation

The schematic of optical communication system simulation setup is shown in Fig.1. An externally modulated CW DFB Laser source having line width =10MHz with variable input power is used. Pseudo-Random Bit Sequence (PRBS) generator with variable bit rate 2.5, 5, 10, 20 Gb/s has been used.

The logical sequence generated by PRBS is converted into electrical signal using NRZ electrical pulse generator. The rise time and fall time of NRZ is taken to be 0.05 bit with amplitude of 1a.u. the modulated signal is generated by CW DFB Laser optical source operating at frequency of 1550 nm. WDM Multiplexer is also operating at frequency of 1550nm and bandwidth of 10 GHz. The extinction ratio of the Mach-Zehnder Amplitude Modulator is 30 dB which modulates the electrical signal. An EDFA Ideal amplifier with gain=17dB and Noise Figure=6dB is used in the transmitter to boost the optical signal to the desired power level. EDFA is followed by power splitter and upstream circulator with return loss and isolation of 60dB. AWG 8x8 with frequency = 193.4THz and bandwidth =10GHz is used for upstream and AWG 1x8 with frequency = 1550 nm and bandwidth=10GHz is used for downstream. The SMF fiber parameters include attenuation=0.24db/km, dispersion slope=0.075 ps/km-nm$^2$ and dispersion at 1550nm is 16.75ps/km-nm.

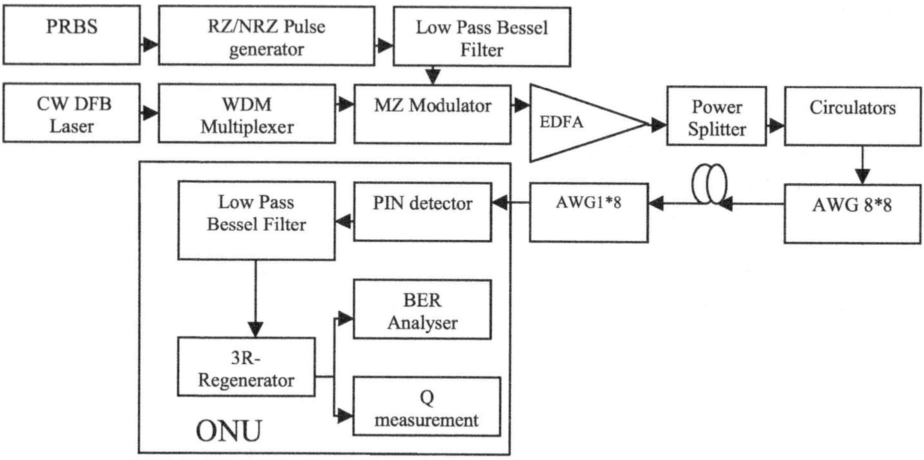


**Fig. 1.** Schematic of Simulation Setup

In the receiver the signal is detected by a PIN photodiode (PD). It has responsivity [A/W]=0.7 and dark current=[10nA] having thermal noise 79.99w/hz. It is then pass through the low pass Bessel filter with 3db cut-off frequency=0.75*bit rate, order of the filter=4, depth 100db. Thereafter, 3R Regenerator is used to regenerate an electrical signal that can be connected directly to the BER analyzer, which is used as a visualizer to generate graphs and results such as eye diagram, BER, Q value, eye opening etc. we have also considered the ASE noise, shot noise thermal noise estimated receiver noise and ASE-ASE noise effects in the optical receiver.

## 3 Results and Discussion

To estimate the performance, the BER and Q value [dB] from the eye diagrams of electrical scope have been considered. Fig. 2(a) & (b) shows BER, for upstream and downstream data respectively.

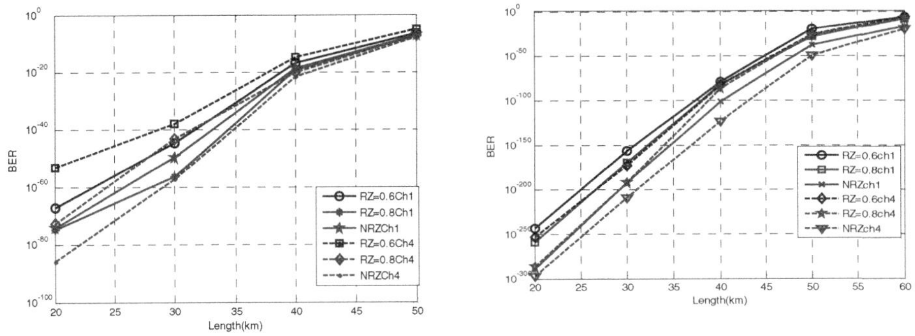


(a) (b)

**Fig. 2.** BER Vs Length upstream for various data formats

The simulation results show that the BER increases with increase in the transmission distance and at longer distance greater than 45km. WDM-PON system becomes almost independent of the data format. for distance more than 45km, NRZ at ch4 gives better performance followed by RZ(0.8) and RZ(0.6), As the faithful transmission of signal can be carried upto 50km for upstream and 60km for downstream as bit rate rises above $10^{-9}$.

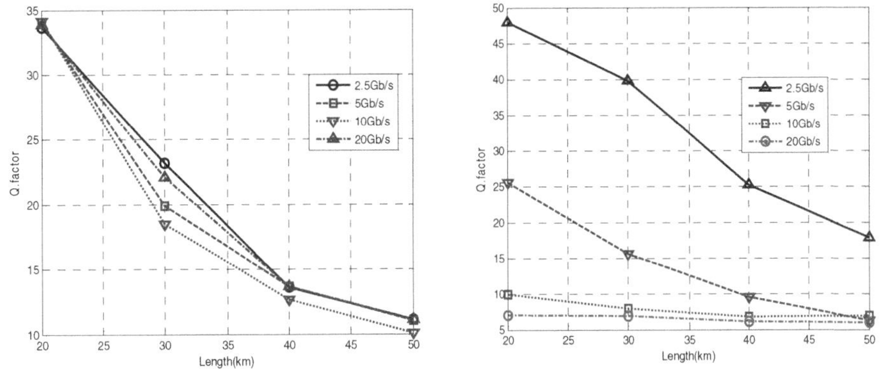


(a) (b)

**Fig. 3.** Q.factor Vs Length upstream CH4 for various data rate(NRZ)

Fig.3(a) & (b) shows the Q value of data formats for various data rates i.e at 2.5, 5, 10 and 20Gb/s and NRZ data formats for upstream and downstream data respectively. It is clear from Fig. 3(a) that simulation system gives almost same Q value for different data rates, it shows the feasibility of the system at higher bit rate, Fig. 3(b) shows that system gives excellent Q value for data rate 2.5Gb/s only. The upstream data is independent of data rates & modulations formats.

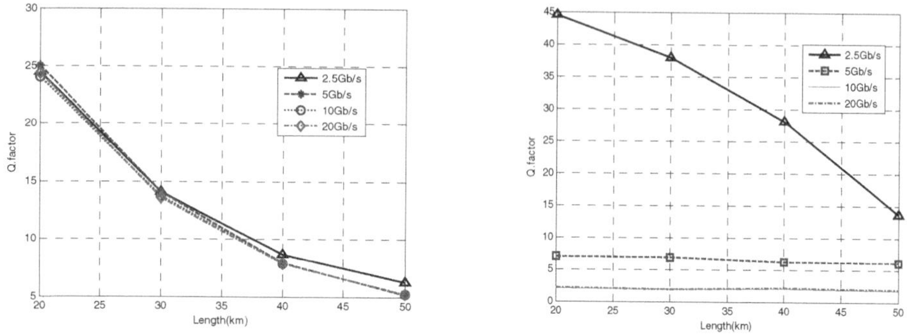


(a) (b)

**Fig. 4.** Q Factor Vs Length upstream CH4 for various data rate(RZ=0.8)

Fig.4(a) & (b) shows the Q value for various data rate that is at 2.5, 5, 10 and 20Gb/s and RZ=0.8 data formats for both upstream and downstream data respectively. It is clear from Fig. 4(a) that simulation system gives almost same Q value for different data rates, it shows the feasibility of the system at higher bit rate, Fig. 4(b) shows that system gives excellent Q value for data rate 2.5Gb/s only.

## 4 Conclusions

We have simulated WDM-PON system over a transmission distance of 50km and bit rate upto 20Gb/s. It is concluded that BER increases with increase in the transmission distance, NRZ data formats shows improved performance as compared to RZ(0.8), Also ch4 gives good performance, the faithful transmission of the signal can be carried up to 50km for upstream and 60 km for downstream beyond which the bit rate rises above $10^{-9}$. The Q value decreases with increase in the length of the fiber also it show that downstream data gives better Q value than upstream data. It is quite evident from the results that the Q value decreases and BER increases with increase in the fiber length. Further it is seen that the system performance degrades due to interchannel interference and non linearity at higher bitrates.

## References

[1] Bock, C., Prat, J., Walker, S.D.: Hybrid WDM/TDM-PON using AWG FSR and featuring centralized light generation and dynamic bandwidth allocation, Spanish ministerio technology project TIC2002-00053 (2005)

[2] Calabretta, N., Presi, M., Proietti, R., Contestabile, G., Ciaramella, E.: A bidirectional WDM/TDM-PON using DPSK downstream signals and a narrowband AWG. IEEE Photonics Technology Letters 19(16) (2007)

[3] Feny, H., Liu, F.: A novel scheme of multicast WDM-PON using modified NRZ signal format. In: Proceedings of the IEEE INFOCOM (2009)
[4] Han, K.-E., Yoo, K.-M., Yang, W.H., Kim, Y.-C.: Design of AWG based WDM-PON architecture with multicast capability. In: Proceedings of the IEEE INFOCOM (2008)
[5] Xu, Z., Wen, Y.J., Zhong, W.-D., Attygalle, M., Cheng, X., Wang, Y., Cheng, T.H., Lu, C.: WDM-PON architecture with a single shared interferomatric filter for carrier reuse upstream. Journal of Lightwave Technology 25 (2007)
[6] Chac, C.-J., Oh, N.-H.: WDM.TDM-PON system employing a wavelength- selective filter and a continuous- wave shared light source. Optical Communications Research Department, korea Telecom Access Networks Laboratories (1998)
[7] Choudhary, M., Kumar, B.: Analysis of next generation PON architecture for optical broadcast access networks. IEEE Comm. Mag. (2006)
[8] Maier, M.: WDM passive optical networks. J. Opt. Comm. Network 1(4) (2009)
[9] Kani, J.-I.: Enabling technologies for future scalble and flexible WDM-PON system. IEEE Journal of Quantum Electronics 16(5) (September 2010)
[10] Zhang, J., Ansari, N.: Design of WDM-PON with tunable laser upstream scenario. Journal of Lightwave Technology 28(2) (January 2010)

# Data Mining Techniques for Prefetching in Mobile Ad Hoc Networks

Naveen Chauhan, L.K. Awasthi, and Narottam Chand

Department of Computer Sc. & Engineering
National Institute of Technology, Hamirpur (H.P.)-177005
{naveen,lalit,nar}@nitham.ac.in

**Abstract.** Caching is a means of providing faster access of data to the requester. The data which is frequently accessed/ required is kept in cache (fast memory) so as to improve the query latency. This technique has proved its worth in various environments vis-à-vis, web environment, mobile communication, mobile computing as well as in processor architecture. The data items in the cache are kept based on the experience or the past record of data items which are accessed frequently. In this paper we propose a data caching and prefetching scheme for mobile ad hoc networks (MANETs). Through prefetching system will try to sense the future needs of mobile nodes (MNs). Based on the assessment, the requisite data is prefetched from server and query latency is reduced and data availability is improved further. We have applied the data mining techniques to prefetch the data.

**Keywords:** MANETs, Intelligent Prefetching, Wireless Communication, Caching Rules.

## 1 Introduction

With the rapid development in the field of wireless technology & mobile devices, the devices comes with more capabilities like more processing power, more storage capacity fast computation, more battery power etc. As such there is no dearth of resources in present scenario, but at the same time it is also true that as system is growing in terms of maturity everything comes with "more" i.e. more energy, more storage capacity, more computational power, even then there is ever increasing demand for higher bandwidth, higher processing. In the state of sufficient cache size in mobile nodes, it is possible to hold all /majority of data items in the cache. Neither it is possible to hold all accessed data items in the cache of each mobile node nor it is advisable to keep the data at each and every location in the network, this will create a unnecessary mess in the entire network. Moreover the information which should be kept handy might not get a place to reside.

In this paper we proposed a novel algorithm that prefetches the data based on association among data items. Our scheme prefetches highly related data items and considers confidence of association rules. To find the relationship among data items, association rules base data mining technique is used [4]. Since caching and prefetching are both well recognised for improving client perceived response time, the

A. Mantri et al. (Eds.): HPAGC 2011, CCIS 169, pp. 594–602, 2011.

integration of both strategies may be exploited to improve the system performance. In the prefetching access to remote data is anticipated and data is fetched before it is required. If the requested data item is not prefetched earlier, the client has to send an uplink request to ask for data item when the query comes. This not only increases the query latency but also the uplink bandwidth requirement.

The remainder of this paper is organised as follows. In Section 2 we briefly review the related studies on cache replacement and prefetching in mobile ad-hoc networks and mobile environment. Section 3 gives description of system model. Section 4 describes the proposed algorithm and Finally, Section 5 concludes the paper and Section 6 describes the scope of future work.

## 2 Related Work

Caching frequently accessed data on client side is an effective technique to improve performance in mobile ad-hoc networks [1]. A lot of research has been done on cache invalidation in past few years [3, 11], with relatively little work being done on cache prefetching methods. In the following, we briefly review related studies on prefetching in mobile environments.

Prefetching has been widely employed to reduce the response time in the web environment [12, 13]. Most of these techniques concentrate on estimating the probability of each file being accessed in near future. Since these techniques are designed for point-to-point communication environment, they are not suitable for mobile ad-hoc networks. Also they do not consider the constraints of mobile ad hoc networks. In [14], Cao proposed an adaptive prefetch scheme, where clients record the number of times a cached item being accessed and prefetched respectively. The client calculates the Prefetch access ratio (PAR), which is the no. of prefetches divided by the no. of accesses for each item. If PAR is less than one, prefetching item is useful since prefetched data may be accessed multiple times. Gitzenis and Bambos [15] proposed a prefetched scheme considering the quality of wireless channel. Clients prefetch aggressively when channel quality is good but reduce the prefetch rate when the channel quality becomes poor. Yin and Cao [10] proposed a power aware prefetch scheme called value-based adaptive prefetch (VAP) scheme, which can dynamically adjust the no. of prefetches based on the current energy level to prolong the system running time. All the previous schemes have ignored the relationship between the data items. Song and Cao [9] realized that cache misses are not isolated events, and a cache miss is often followed by a series of cache misses. They addressed the prefetching issue among related data items by using a cache-miss-initiated prefetching scheme, which is based on association rule mining technique, but the disadvantage of this scheme is that the item appearing in the consequent of mined prefetched rules can not be prioritized. M. Denko et. al [5] proposed a two level prefetching strategy based on prefetching at the level of caching agent and mobile host and other one between caching agent and a cluster head. A popularity based prediction algorithm for prefetching the data item was proposed. This calculates the popularity of a data item based on past access history.

In this section we present our improved distributed weighted clustering algorithm. This algorithm can be divided into three phases. These phases are described as:

## 3 System Model

This paper studies the prefetching of requisite data in mobile ad hoc networks. Figure depicts a typical system model used for study. As illustrated, there is a cache management mechanism in a client. The client employs thea rule generating engine to derive caching rules from the client's access log. The derived caching rules are stored in the caching rule depository of the client.

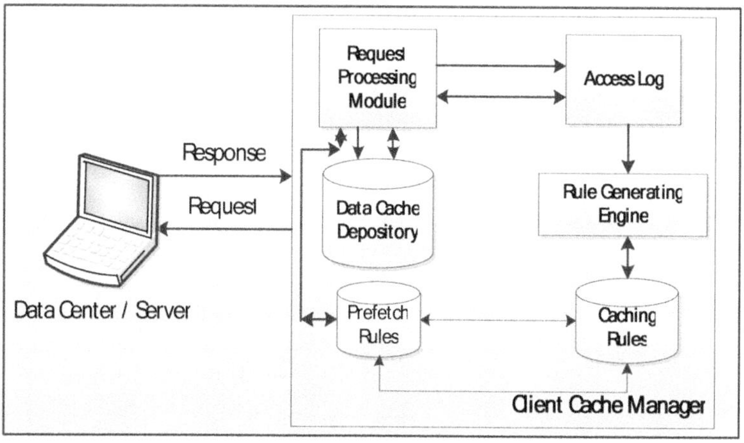


**Fig. 1.** System Model

### *A. Intelligent Prefetching*

Whenever a mobile node issues a request, the cache request processing module first logs this request into record and checks whether the desired data item is available in local cache of MN or in any of the MN in the cluster. If it is a cache hit, the cache manager still needs to validate the consistency of the cached item with the copy at the original server. To validate the cached item, the cache manager checks the validation of data item from its TTL value. If the data item is verified as being upto date, it is returned to the MN immediately. If it is a cache hit, but the value is obsolete, the cache manager sends an uplink request to the server and waits for the data broadcast. When the requested data item appears, the cache manager returns it to the requester and retains a copy in the cache. In the case that a cache miss occurs, the client cache manager checks the caching rule repository to derive the prefetching rules corresponding to the requested item. If this request triggers some prefetching rules, the *ids* of the item implied by these prefetching rules will also be piggybacked to the server alongwith *id* of missed cache item.

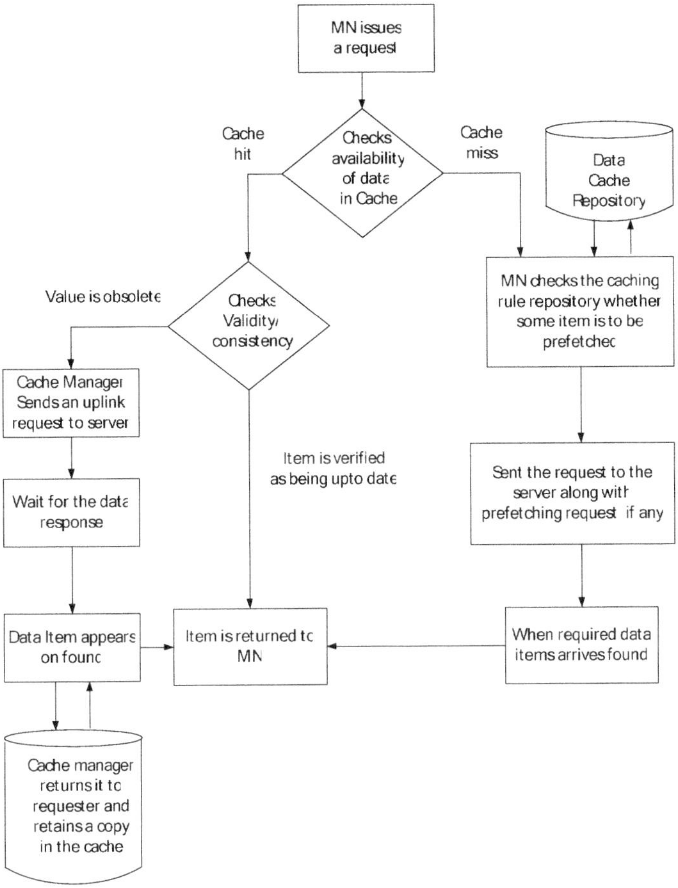


**Fig. 2.** Flowchart representing intelligent prefetching

### *B. Generating Caching Rules*

At a MN, data items queried during a period of time are related to each other. Observation of the history of data items queried by the client may lead to find relationship among the items. These relationship can take the form of patterns of behavior that can tell us that if the client has accessed certain items during a period of time then it is likely that one particular item is accessed  will be accessed in near future e.g. if a client access $d_1$ and $d_2$ then it accesses $d_3$ 80% of the times. The if part of the rule is called the antecedent while the then part is called consequent. These rules are known as association rules in the data mining literature [4]. We propose to use data mining techniques to discover the association rules in the access history and apply the rules to make caching decisions ( prefetching, replacement etc.).

Let $D = \{d_1, d_2, ..., d_N\}$ be the set of data items at the server. Suppose a client's access trace S consists of a set of consecutive parts: $\{part_1, part_2, ..., part_i, ..., part_n\}$. Let $A = \{d_1, d_2, ..., d_m\}$ denotes the set of data items accessed by the client. Let $S_i$ denotes the data items contained in part parti. $S_i$ is called a session and $S_i \subset A$. We say a session $S_i$ contains x if $S_i \supseteq x$, where $x \subset A$. A caching rule $r_{x,y}$ is an expression of the form $x \rightarrow d_y$, where $x \subset A$, $d_y \in A$, and $x \cap \{d_y\} = \phi$. x is called antecedent of the rule and $d_y$ is called consequent.

In general, a set of data items is called an itemset. The number of data items in an itemset is called the size of the itemset and an itemset of k size is called k-itemset. The support of an itemset x, support(x), is defined as the percentage of sessions that contains x in the client's access trace S [9, 16]. The support of a caching rule $r_{x,y}$, support($r_{x,y}$), is defined as the support of the itemset that consists data items in both the antecedent and the consequent of the rule, i.e., support($r_{x,y}$) = support($x \cup \{d_y\}$).

The confidence of a caching rule $r_{x,y}$, confidence($r_{x,y}$) is defined as the support of the rule divided by the support of the antecedent [9, 16].

$$Confidence\ (r_{x,y}) = \frac{support\ (x \cup \{d_y\})}{support\ (x)} \times 100\%$$

In general, $c_{x,y}$ denotes confidence($r_{x,y}$) and $r_{x,y}$ is expressed as $X \xrightarrow{c_{x,y}} d_y$.

The confidence of a caching rule is the conditional probability that a session in S contains the consequent given that it contains the antecedent. Given an access trace S, the problem of mining caching rules is to find all the association rules that have support and confidence greater than the user defined minimum support (*minsup*) and minimum confidence (*minconf*), respectively.

The problem of mining caching rules can be decomposed into the following subproblems [4]:

1. Find all the itemsets x such that support(x) ≥ minsup. An itemset x that satisfies this condition is called frequent itemset.
2. Use the frequent itemsets to generate association rules with minimum confidence.

## 4 Proposed Algorithm

*A. Algorithm to Generate Frequent Itemsets*

In this section, we present an algorithm to generate frequent itemsets from the client's access trace. Table 1 show the notations used in the algorithm. Figure 3 shows main steps of the algorithm. It accepts an access trace S, a minimum support (minsup), and the maximum number of items NR to be used in a rule as parameters. In line 1, S is analyzed to generate the frequent 1-itemset. This is done by calculating the support of each data item and comparing it to the minimum support. Every data item that has minimum support forms one frequent 1-itemset.

Loop from line 3 to line 20 is used to generate all the frequent 2-, 3-, ..., k-itemsets. Each iteration of the loop, say iteration k, generates frequent k-itemsets based on the (k-1)-itemsets generated in the previous iteration. This loop continues until it is not

**Table 1.** Notations

| | |
|---|---|
| *NR* | Maximum number of items in a rule |
| *k*-itemset | An itemset with k items |
| $F_k$ | The set of frequent k-itemsets (those with minimum support) |
| $f_i, f_j$ | Any of the frequent (k-1)-itemsets within $F_{k-1}$ |
| $f_i[m]$ | m-th item in itemset $f_i$ |
| $f$ | A new frequent k-itemset obtained by combining a frequent (k-1)-itemset with one item |
| $Z$ | The set of caching rules |

possible to generate new itemsets or the number of items in an itemset exceeds the predefined maximum NR. Lines 3-14 generate all the new candidate frequent k-itemsets out of the frequent (k-1)-itemsets. Lines 15-18 remove those candidate frequent k-itemsets that do not fulfill the minimum support requirement. In line 21 the algorithm returns all the frequent itemsets generated.

```
1)   F_1 = {frequent 1-itemsets}
2)   k = 2
3)   while F_{k-1} ≠ φ ∧ k ≤ NR do
4)      F_k = φ
5)    for each itemset f_i ∈ F_{k-1} do
6)       for each itemset f_j ∈ F_{k-1}
7)          if f_i[1] = f_j[1] ∧ ...∧ f_i[k-2] = f_j[k-2] ∧ f_i[k-1] < f_j[k-1]
8)                then f = {f_i} ∪ {f_j[k-1]}; F_k = F_k ∪ {f}
9)                for each (k-1)-subsets s ∈ f do
10)                   if s ∉ F_{k-1}
11)                   then F_k = F_k −{f}; break
12)               end
13)         end
14)      end
15)      for each itemset f_i ∈ F_k do
16)         if support(f_i) < minsup
17)         then F_k = F_k −{f_i}
18)      end
19)  k = k+1
20)  end
21)  return ∪_k F_k
```

**Fig. 3.** Algorithm to generate frequent itemsets

For example, Table 2a shows a sample trace for a client divided into five sessions. Table 2b shows the frequent itemsets generated from the sample trace by applying frequent itemsets generating algorithm (shown in Figure 2). Here, minsup = 60%, and NR = 3.

*B. Algorithm to Generate Caching Rules*

We are interested in generating, from a frequent *k*-itemset $f_i$, rules of the form $x \rightarrow d_y$, where $x$ is a $(k\text{-}1)$-itemset, $d_y$ is a 1-itemset and $f_i = x \cup \{d_y\}$. Table 1 shows the notations used in the algorithm.

**Table 2a.** The sample trace in sessions

| Session # | Data item requests |
|---|---|
| 1 | $d_1\, d_2\, d_3\, d_4\, d_5$ |
| 2 | $d_1\, d_2\, d_3\, d_4$ |
| 3 | $d_1\, d_2\, d_3$ |
| 4 | $d_1\, d_2\, d_5$ |
| 5 | $d_1\, d_2\, d_3\, d_5$ |

**Table 2b.** The frequent itemsets

| | |
|---|---|
| 1-itemset | $\{d_1\}, \{d_2\}, \{d_3\}, \{d_5\}$ |
| 2-itemset | $\{d_1, d_2\}, \{d_1, d_3\}, \{d_1, d_5\}, \{d_2, d_3\}, \{d_2, d_5\}$ |
| 3-itemset | $\{d_1, d_2, d_3\}, \{d_1, d_2, d_5\}$ |

```
k = 2
while F_k ≠ φ do
   for each itemset f_i ∈ F_k do
      for each item fi[j] ∈ f_i
         if support(f_i) / support(f_i − f_i[j]) ≥ minconf
         then Z = Z ∪ {{f_i-f_i[j]}→f_i[j]}
      end
   end
   k = k+1
end
return Z
```

**Fig. 4.** Algorithm to generate the caching rules

Figure 4 illustrates the main idea of the algorithm. The algorithm accepts the frequent itemsets and a minimum confidence (minconf) as parameters. For each frequent itemset, the rules are generated as follows. Of all the data items within frequent itemset, one item becomes the consequent of the rule, and all other items become the antecedent. Thus, a frequent k-itemset can generate at most k rules. For example, suppose $\{d_1, d_2, d_3\}$ is a frequent 3-itemset. It can generate at most three rules: $\{d_1, d_2\}\rightarrow d_3$, $\{d_1, d_3\}\rightarrow d_2$, and $\{d_2, d_3\}\rightarrow d_1$. After the rules have been generated, their confidences are calculated to determine if they have the minimum confidence. Only the rules with at least the minimum confidence are kept in the rule set Z. For example, for the rule $\{d_1, d_2\}\rightarrow d_3$, confidence $conf = support(\{d_1, d_2, d_3\})/\text{support}(\{d_1, d_2\})$. If $conf \geq minconf$, the rule holds and it will be added to the rule set Z.

## 5 Conclusion

In this paper we have proposed an algorithm which maximizes the performance improvement due to caching in mobile ad hoc networks. We devise an algorithm which not only considers the caching parameters of a data item, but also the relationship of this item with the cache set. Association rule based data mining is applied to find the relationship among data items. To enhance the caching performance, the generated association rules are used to prefetch the data item(s).

## 6 Future Work

For the next step in this algorithm, simulation will be performed to evaluate the performance of our algorithm under several circumstances.

## References

1. Acharya, S., Alonso, R., Franklin, M., Zdonik, S.: Broadcast Disk: Data Management for Asymmetric Communication Environments. In: ACM SIGMOD Conference on Management of Data, pp. 199–210 (1995)
2. Cao, G.: A Scalable low latency cache invalidation strategy for mobile environments. IEEE Transactions on Knowledge and Data Engineering 15(5), 1251–1265 (2006)
3. Chand, N., Joshi, R.C., Misra, M.: Energy Efficient Cache Invalidation in a Disconnected Wireless Mobile Environment. International Journal of Ad Hoc and Ubiquitous Computing (IJAHUC) 1(3) (2006)
4. Agarwal, R., Imielinski, T., Swami, A.: Mining Association Rules Between Sets of Items in Large Database. In: ACM SIGMOD Conference on Management of Data, pp. 207–216 (May 1993)
5. Denko, M., Tian, J.: Cooperative Caching with Adaptive Prefetching in Mobile Ad Hoc Networks. In: IEEE International Conference on Wireless and Mobile Computing and Communication, pp. 38–44 (2007)
6. Cheluvaraju, B., Srinivas Ramachandra Kousik, A., Rao, S.: Anticipatory Retrieval and Caching of Data for Mobile Devices in Variable-Bandwidth Environments. In: 5th IEEE, International Systems Conference (IEEE Sys Con) (2011)
7. Han Bae, I., Olariu, S.: Design and Evaluation of Fuzzy Cooperative Caching Scheme for MANETs. In: 6th International Conference on Wireless Communication Networking & Mobile Computing (WiCom), pp. 1–5 (2010)

8. Wu, S.-Y., Hsu, J., Chen, C.-M.: Headlight Prefetching for Mobile Media Streaming. In: 6th International Workshop on Data Engineering for Wireless and Mobile Access, pp. 67–74 (2007)
9. Song, H., Cao, G.: Cache-Miss-Initiated Prefetch in Mobile Environments. Computer Communication Journal 28, 741–753 (2005)
10. Yin, L., Cao, G.: Adaptive Power Aware Prefetch in Wireless Networks. IEEE Transactions on Wireless Communication 3(5), 1648–1658 (2004)
11. Shen, H., Kumar, M., Das, S., Wang, Z.: Energy-Efficient Data Caching and Prefetching of Mobile Devices Based on Utility. ACM/Kluwer Journal of Mobile Networks and Applications (MONET)
12. Padmanabhan, V., Mogul, J.: Using Predictive Prefetching to Improve World Wide Web Latency. Computer Communication Review, 22–26 (1996)
13. Jiang, Z., Kleinrock, L.: An Adaptive Network Prefetch Scheme. IEEE Journal on Selected Areas in Communications 16(3), 1–11 (1998)
14. Cao, G.: Proactive Power -Aware Cache Management for Mobile Computing Systems. IEEE Transactions on Computers 51(6) (2002)
15. Gitzenis, S., Bambos, N.: Power-Controlled Data Prefetching/Caching in Wireless Packet Networks. In: INFOCOM (2002)
16. Saygin, Y., Ulusoy, O.: Exploiting Data Mining Techniques for Broadcasting Data in Mobile Computing Environments. IEEE Transactions on Knowledge and Data Engineering 14(6), 1387–1399 (2002)

# An Image Steganography Approach Based upon Matching

Sukhpreet Kaur[1,*] and Sumeet Kaur[2]

[1] Department of Computer Science & Engineering, Baba Farid College of Engg. &Technology, Bathinda, India
preetsukhpreet@gmail.com
[2] Department of Computer Engineering, Yadavindra College of Engineering, Punjabi University Guru Kashi Campus, Talwandi Sabo, India

**Abstract.** Use of internet has made the data transfer very easy but at the same time very risky also. The major concern of these days is the security of the data being transferred over the Internet. Steganography is a technique that proves to be very good to achieve this. Steganography can be defined as a technique of embedding data inside some other object by altering its properties. This paper includes discussion of a technique that has been designed to hide more number of bits per pixel. The technique maps the secret data to one of the channels of the cover object and use LSBs of other channels to mark the presence of data in that channel. The results show that the technique achieves high security and provides more data hiding capacity.

**Keywords:** Attacks, image steganography, PSNR, security, steganography.

## 1 Introduction

Internet has become an important part of the today's life. People use internet for data transfer but the major concern of these days is, the security of data being communicated over internet. Cryptography was developed as a technique for making the data communication secure. Cryptography makes the data secure by making it difficult to understand but still the attacker knows about the presence of the data. So, there is need of a technique that can even hide the presence of data from attacker. The technique that can achieve this objective is called steganography. Steganography is the art and science of invisible communication. This is accomplished through hiding information in other information, thus hiding the existence of the communicated information. The word steganography is derived from the Greek words "*stegos*" meaning "cover" and "*grafia*" meaning "writing" defining it as "covered writing" [1]. Steganography can be defined as the technique used to embed data or other secret information inside some other object commonly referred to as cover, by changing its properties. Classification of Steganography techniques depend upon the type of digital media used as cover object. Image steganography is most widely used. The reason behind the popularity of image steganography is the large amount of redundant

* Corresponding author.

A. Mantri et al. (Eds.): HPAGC 2011, CCIS 169, pp. 603–608, 2011.

information present in the images that can be easily altered to hide secret messages inside them.

This paper presents an image steganography technique that attempts to achieve more capacity by using more number of bits to hide secret data. At the same time technique is very much secure from statistical and visual attacks. The paper is organized as follows. In the next section we introduce the related work. Section three describes the proposed technique. Section four contains the results after implementing the technique. Finally conclusion and future work is included in section five.

## 2 Related Work

A very interesting survey of different information hiding techniques has been presented in [2]. The survey contained details of information hiding techniques along with their applications and limitations. A technique of hiding a large amount of data with high security using steganographic algorithm has been proposed in [3]. In this technique, an attempt to improve the efficiency of the payload has been done. In the technique adaptive image filtering and adaptive image segmentation with bits replacement on the appropriate pixels has been used. A new method of hiding data in $3^{rd}$ LSB and still achieving higher PSNR has been introduced in [4]. The method is to replace the $3^{rd}$ LSB of cover with the message bits and compensate the difference change the $1^{st}$ and $2^{nd}$ LSB accordingly. An approach has been presented in [5], which use 7 bits of every pixel for hiding data. The technique actually does not hide data; it maps the character values to the pixel values by matching them to each other. The technique increases the data hiding capacity but there is change in 5 bits to store 7 bits in a pixel. The concept of storing variable number of bits in each channel (R, G or B) of pixel based on the actual color values of that pixel has been implemented in [6]. The lower color component stores higher number of bits. The technique uses actual color of the channel to decide number of data bits to store. But the maximum number of bits that can be stored per pixel is 5. That too can result in changed statistical properties of the cover image. Pixel Indicator technique introduced in [7] uses LSB of one channel to decide the number of bits that can be hidden inside other two channels. The technique offers good results when 1 or 2 bits are hidden in the data channels. So, it is clear from the discussion that techniques that already exist they have achieved a good capacity of hiding 4 to 5 bits per pixel but the security from statistical attacks and visual attack remains the major concern. Some techniques like [8] and [9] provide great security against these attacks but at the cost of capacity.

## 3 Proposed Technique

The technique starts with conversion of secret message into corresponding ASCII code values. As we know that each pixel of the BMP image is made up of three bytes, one each for Red, Green and Blue components of the pixel. Each character of thesecret message is converted into ASCII code, which is 7-bit code. Therefore, to match a character to a pixel component we need only 7 bits of the pixel. The technique has been designed to match the data with the pixel values of the cover image and use LSB of pixel value as marker to mark presence of data inside a pixel.

The LSB of the pixel has been used as marker because the data matching will be done only with the 7 MSBs of selected the data channel of every pixel. So the LSBs of all the three channels are free to be used as marker. In our proposed technique, different combinations of Least Significant Bits are used to show that component of the pixel which contains secret data. We use LSB of two channels to mark presence of data in any of the three channels. To decide which channels will act as marker channels, we use a pseudo random number which is generated for every character of secret message. Then, depending upon the value of pseudo random number it is decided that which two channels will act as marker channels. After generating the number, we convert it into binary bit sequence and count the number of 1s & 0s present in sequence. Depending upon binary bit sequence of pseudo random number, following three cases are generated and one is used to select a set of marker channels. The selection procedure is shown in table 1. We work on all the three channels turn wise. First we select red channel as data channel. In case the matching of all the characters of secret message does not occur in red channel then we select green channel as data channel. If again there is no match in green channel then we move to blue channel. After selecting set of marker channels we start from the first row of cover image. We

**Table 1.** Selection of marker channel

| Case | Marker channel set | Order1(if parity is even) | Order2(if parity is odd) |
|---|---|---|---|
| If no. of 1s are more than number of. Zeros | RG | RG | GR |
| If no. of 0s are more than number of 1s | GB | GB | BG |
| If no. of 0s are equal to number of 1s | RB | RB | BR |

**Table 2.** Criteria to set value of marker channel

| Data channel | LSB of Marker 1 | LSB of Marker 2 |
|---|---|---|
| RED Channel | 0 | 0 |
| GREEN Channel | 0 | 1 |
| BLUE Channel | 1 | 0 |
| No match | 1 | 1 |

hide length of secret message in first row using LSB method. Then start tracing from the second row to match first character of secret message with 7 MSBs of red component of first pixel. If there is a match, then value of marker channels is set according to the criteria given in Table 2. As clear from the Table 2, we set the values of marker channels as 00 if data matches in red channel.

### 3.1 Algorithm

The encoding part of the algorithm is as follows.

1. Read cover image, convert it into three channels R, G, B.
2. Read secret message, convert it into ASCII code, store in variable C.
3. Take number of pixels in cover in variable max.
4. Extract message length and store it into the variable L, the message length is terminated by a "/" as it will act as a termination symbol.
5. Hide message length into the first row of cover image, inside blue channel only, by using LSB method.
6. Select Red channel as data channel.
7. Start from second row of cover image. Take first pixel pi of second row with i=0.
8. Take next character from C, store it in a temporary variable B.
9. Select Marker channel pair, depending upon the pseudo random number.
10. Match the character with 7 MSBs of red channel of the pixel pi=pi+1.
11. If there is a match between character bit sequence and 7 bits of red channel, then set LSB of marker channel1 and marker channel2 equal to 0.
12. Set L = L-1.
13. Check if L>0 and i<max. If yes then go to step no 8.
14. If not matched, then set LSB of marker channel1 =1 and marker channel2 = 1. Goto step 9.
15. Stop when all characters are consumed and L is equal to zero.
16. If L>0 and i>max, then choose green channel as data channel and repeat steps 7 to 15 with green channel.
17. If L>0 and i>max, then select blue channel as data channel and repeat steps 7 to 15 for blue channel as data channel.

The decoding part is just reverse of the encoding part.

## 4 Results

A number of experiments were performed to show the effectiveness of the proposed technique. The histograms of cover and stego images have been used to show that the techniques discussed are very much statistically strong. The histograms are the commonly used tools to find out the effect of hiding data inside the cover image. PSNR computes Peak Signal-to-noise Ratio, in decibels between two images. PSNR (Peak Signal-to-noise Ratio) is again used to evaluate the visual degradation of the stego image due to secret data hidden inside it. The higher value of PSNR means the stego image has a good quality. To evaluate the results, we have used a number of color images, out of which the results of the image baboon.bmp having size 2621444, has been shown in this paper. Fig 1. Shows cover image and Fig 2. Shows the stego image after hiding data with proposed technique. It is clear from figure 1 and 2 that there is no visual difference between the cover and the stego image. It is not possible to identify the stego image based upon the visual analysis.

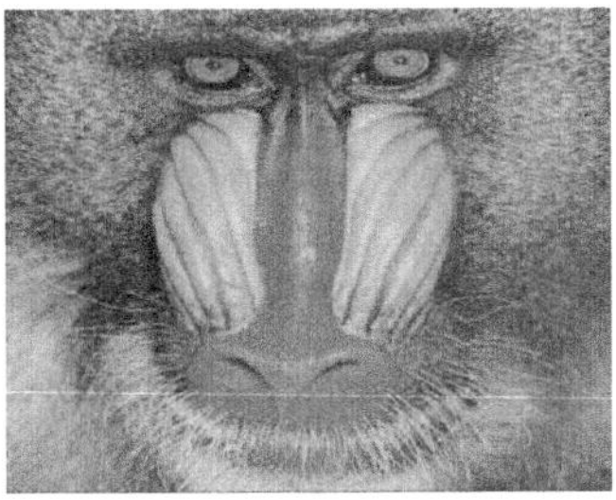

**Fig. 1.** Cover image baboon.bmp

**Fig. 2.** Stego image baboon.bmp

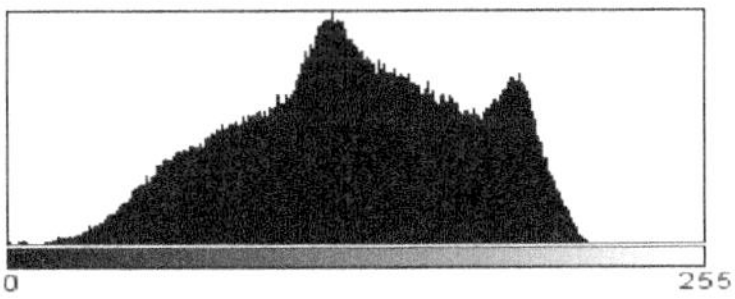


**Fig. 3.** Histogram of stego Baboon.bmp

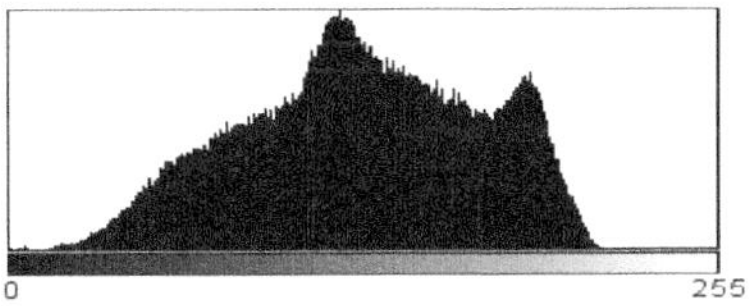


**Fig. 4.** Histogram of stego Baboon.bmp

Fig 3. Shows the histogram of cover image and Fig 4. Shows the histogram of stego image. It is clear from the histograms that there is negligible change in the histogram of stego image. So, proposed technique is secure fom statistical attacks. The Table 3 shows the results of hiding different size messages inside the cover using proposed technique.

**Table 3.** Effect of increasing message length on PSNR

| Image | Message Length | PSNR |
|---|---|---|
| Baboon.bmp | 100 | 60.037 |
| Baboon.bmp | 50 | 63.023 |
| Baboon.bmp | 25 | 64.70 |

## 5 Conclusion and Discussion

In this paper, we have introduced a new technique of image steganography. The main objective was to achieve high data hiding capacity and more security against statistical and visual attacks. The results show that we have been successful in achieving the same. The technique provides more security against visual attacks as the cover and stego images does not show the visible differences. The technique introduced in this paper is also secure against the statistical attacks as the cover and stego images do not show much difference in their histograms. More data hiding capacity has been achieved by using the 7 bits per pixel to hide data. Results show a very good value of PSNR that means technique shows better imperceptibility. The main strength of the technique lies in the fact that actually there is no alteration in the cover image to hide secret data inside it. There is only mapping of data and alteration is only there to mark presence of data inside a pixel. The future scope lies in the modification to increase the chances of matching of secret data with the pixel patterns.

## References

1. Morkel, T., Eloff, J.H.P., Olivier, M.S.: An Overview of Image Steganography. In: Proceedings of the Fifth Annual Information Security South Africa Conference (ISSA 2005), Sandton, South Africa (2005)
2. Petitcolas, F., Anderson, R.J., Kuhn, M.G.: Information hiding: A survey. Proceeding of the IEEE 87(7), 1062–1078 (1999)
3. Nameer, N., EL-Emam.: Hiding a Large Amount of Data with High Security Using Steganography Algorithm. Journal of Computer Science 3(4), 223–232 (2007)
4. Ibraheem, A., Sada, A.: Hiding Data Using LSB-3. J. Basrah Researches (Sciences) 33(4), 81–88 (2007)
5. Al-Husainy, Mohammed, A.F.: Image Steganography by Mapping Pixels to Letters. Journal of Computer Science 5(1), 33–38 (2009)
6. Parvez, M.T., Gutub, A.: RGB Intensity Based Variable-Bits Image Steganography. In: Proceedings of 3rd IEEE Asia-Pacific Services Computing Conference, Yilan, Taiwan, pp. 653–658 (2008)
7. Gutub, A.: Pixel Indicator Technique for RGB Image Steganography. Journal of Emerging Technologies In Web Intelligence 2(1), 56–64 (2010)
8. Vijayalakshmi, V., Palanivelu, T.G., Nagaraj, V.: A Modulo Based Image Steganography Algorithm Against Statistical Analysis. IEEE Transactions on Signal Processing 57(6), 1053–1056 (2009)
9. Zhang, H., Tang, H.: A Novel image steganography algorithm against statistical analysis. Proceeding of the IEEE 19(22), 3884–3888 (2007)

# From Calculus to Number Theory, Paves Way to Break OSS Scheme

G. Geetha and Saruchi

Department of CSE/IT, Lovely Professional University, Phagwara, Punjab
gitaskumar@yahoo.com, ganpati.saruchi@gmail.com

**Abstract.** An authentication technique that also includes measures to counter repudiation by the source is termed Digital Signature. Most cryptographic schemes rely on some hard mathematical problem. Ong-Schnorr-Shamir is a digital signature scheme with the advantage of ease of implementation. It was believed that this was of similar difficulty to that of factoring the modulus n. In this paper, we break the OSS scheme by appealing to a famous calculus problem.

**Keywords:** Calculus, Cryptanalysis, Digital signature, Ong Schnorr Shamir Scheme, Number theory.

## 1 Introduction

The concept of digital signature was proposed by Diffie and Hellman[1] together with that of public key cryptosystem. The RSA[2] method was considered the best for both the uses. Ong, Schnorr and Shamir[3] led to a new method for signatures which seemed to be much easier to implement. They used polynomials modulo n. In this paper we propose a solution to break OSS. Section 2 deals with the OSS scheme, Section 3 deals with solution from Pollard and Schnorr, Section 4 describes our proposed scheme to cryptanalyze the OSS scheme and Section 5 presents the conclusion.

## 2 OSS Scheme

In this system, the public key consists of two integers n and k. The modulus n is a large odd composite number whose factorization is kept secret. A valid signature of the message m, where $0<m<n$, is any pair of integers x,y with

$$x^2 + ky^2 \equiv m \pmod n \tag{1}$$

It was believed that this was of similar difficulty to that of factoring the modulus n.

## 3 Solution by Pollard and Schnorr

The solution of the congruence was proposed by J.M. Pollard and C.P. Schnorr [4]. Their solution was based on the assumption of generalized Riemann Hypothesis. A probabilistic algorithm was extended to solve the equation.

A. Mantri et al. (Eds.): HPAGC 2011, CCIS 169, pp. 609–611, 2011.

Algorithm:

Step 1: If n is pure prime power, then solve equation (1) by computing square roots in Zn*

Step 2: Replace m by an equivalent m' such that $0<m'<=\sqrt{4k/3}$ in case k>0, and $0<|m'|<=\sqrt{|k|}$ in case k<0

Step 3: If m' is a perfect square, or m'=k , solve $x^2 + ky^2 = m'(\mathrm{mod}\ n)$ with y =0 or x=0, and goto step 5

Step 4: Apply the algorithm recursively to solve $x'^2 - m'y'^2 = -k(\mathrm{mod}\ n)$ such that (y',n) =1. Solve $x^2 + ky^2 = m'(\mathrm{mod}\ n)$ with x:=x'/y', y:=1/y'(mod n)

Step 5: Work back to a solution of the original equation.

## 4 The Proposed Method–Cryptanalysis of OSS Scheme

The diophantine equation

$$a^2 + b^2 - ab = c^2$$

obtained through a familiar calculus problem [5] studied by James Duemmel can be transformed to

$$\alpha^2 + 3\beta^2 = c^2 \qquad (2)$$

by substituting $a = \alpha - \beta$, $b = \alpha + \beta$.

The general form of solutions are [91]

$$\beta = 2rs$$
$$\alpha = 3r^2 - s^2$$
$$c = 3r^2 + s^2, \quad r>=s$$

where r and s are arbitrary parameters

Proposed solution to OSS scheme:

Equation (1) from OSS scheme is compared with the diophantine equation (2)

$$x^2 + ky^2 \equiv m(\mathrm{mod}\ n)$$
$$\alpha^2 + 3\beta^2 = c^2$$

The general forms of solutions are

$$\beta = 2\,rs$$
$$\alpha = 3r^2 - s^2$$
$$c = 3r^2 + s^2$$

where $r \geq s$ and r and s are arbitrary parameters.

Analyzing the problem numerically, r and s has the following representations for the x-value to be an integer where in n, k, s, $l \in Z^+$

**Table 1.** Representations for s and r

| Representations | s | r |
|---|---|---|
| (i) | $n^2 + n - 1 + 6k$ | $s + 2l$ |
| (ii) | $n^2 - n + 5 + 6k$ | $s + 2l$ |
| (iii) | $6n - 5 + 2k$ | $s(2l + 1)$ |
| (iv) | $2n$ | $s(2l + 1)$ |

**Table 2.** The primitive solutions of $\alpha^2 + 3\beta^2 = c^2$

| Representations | $\beta$ | $\alpha$ | c |
|---|---|---|---|
| (i) & (ii) | $s(s + 2l)$ | $s^2 + 6l^2 + 6ls$ | $2s^2 + 6l^2 + 6ls$ |
| (iii) & (iv) | $(2l + 1)$ | $1 + 6l + 6l^2$ | $2 + 6l + 6l^2$ |

Here k=3

$$m(\text{mod } n) = c^2 \tag{3}$$

Now the generalized diophantine equation is

$$\alpha^2 + k\beta^2 = c^2$$

for which $c = kr^2 + s^2$ (4)

Comparing (3) and (4)

$$(kr^2 + s^2)^2 = m \bmod n$$

$$(kr^2 + s^2)^2 - m = \text{multiple of } n$$

$$(kr^2 + s^2)^2 - m = np$$

$$(kr^2 + s^2)^2 - np = m$$

## 5 Result

We conclude that the OSS can be broken with the above choice of k and n which is the public key. We shall try to generalize the choice for k.

## References


[1] Diffie, W., Hellman, M.E.: New directions in cryptography. IEEE Trans. Infom. Theory IT-22, 644–654 (1976)

[2] Rivest, R.L., Shamir, A., Adleman, L.M.: A method for obtaining digital signatures and public-key cryptosystems. Communications of the ACM 21(2), 120–126 (1978)

[3] Ong, H., Schnorr, C.P., Shamir, A.: An efficient signature scheme based on quadratic equations. In: Proc. 16th Symp. on the Theory of Computing, Washington, pp. 208–216 (1984)

[4] Pollard, J.M., Schnorr, C.P.: An efficient solution of the congruence x 2+ky 2 ° m(mod n). IEEE Log number 8613566

[5] Duemmel, J.: From Calculus to Number theory. Amer. Math. Monthly 96, 140–143 (1989)

# Digital Image Watermarking Technique Based on Dense Descriptor

Ekta Walia[1] and Anu Suneja[2]

[1] Professor and Head, Department of IT
Maharishi Markandeshwar University
Mullana, Haryana, India
wekta@yahoo.com
[2] Research Scholar, MMICT & BM
Maharishi Markandeshwar University
Mullana, Haryana, India
anusuneja3@gmail.com

**Abstract.** Enthused by the robustness and simplicity of WLD(Weber's Local Descriptor) descriptor, a new digital image watermarking is proposed. The proposed watermarking technique is based on WLD Descriptor. WLD descriptor is a histogram representation of an image that consists of two components of a pixel: its differential excitation and orientation. Differential excitation is the difference of center pixel with its surrounding neighbors and orientation is the gradient orientation of center pixel. WLD descriptor of an image is robust against various geometric and photometric attacks. This feature of WLD descriptor engrossed it to be used in digital image watermarking.

## 1 Introduction

Digital Image Watermarking is the technique used to protect the ownership of an image by inserting hidden information in the host image. The objectives of watermarking techniques are to add a transparent pattern in an original image so that, it may not be copied by an unauthorized person.

A number of watermarking techniques have been developed. The watermarking techniques are broadly categorized into two types: Watermarking in spatial domain of an image and watermarking in frequency domain of an image.

Watermarks are inserted by modifying the original image to carry hidden information. As images are represented in either spatial or frequency domain, watermark can be inserted within images by modifying either pixel values or the transformed domain coefficient values of the image.

A number of watermarking techniques based on feature points have developed. *Hae-Yeoun Lee et. al* [1] have proposed a watermarking method that is robust to geometric distortions. They used a local invariant feature of image called Scale Invariant Feature Transform (SIFT), that is invariant to translation and scaling distortion. The problem with this technique is that due to the computation time for the SIFT descriptor, this scheme can not be used for real time applications.

*Jin-guang Sun* [8] presented an effective watermarking algorithm that is robust against geometric attack based on improved SIFT. In this technique some suitable

A. Mantri et al. (Eds.): HPAGC 2011, CCIS 169, pp. 612–616, 2011.

feature points have been selected to form circular patches of the carrier image. The circular patches are then divided into sectors, and watermark is embedded into the sectors using spatial odds and even quantification. Watermark is extracted from an odd-even detector before detecting the rotated angle and the scale. The drawback of this method is the high time complexity of algorithm. *Lei-Da et al.* [9] presented a robust image watermarking scheme for resisting both traditional and geometric attacks. In this method, Harris interest points are first extracted from the scale normalized image and the locally most stable points are employed to generate some non-overlapped circular regions. The watermark synchronization is achieved using these local circular regions. Thereafter, the rotation invariant pseudo-Zernike moments (PZMs) are computed over each circular region and the watermark signal is composed of the PZM vectors extracted from all the local regions. Then the PZM vectors are modified and reconstructed, producing some error images. The watermark is embedded by adding the error images into the corresponding local circular regions in spatial domain directly. During watermark detection, local circular regions are first extracted from the distorted image and PZMs are computed over each local region. A minimum distance decoder is used to detect the watermark blindly. In this paper, a new digital image watermarking technique has been proposed using dense descriptor that is based on Weber's law.

## 2 WLD Descriptor

To find the similarity among two images, descriptor of both the images can be compared. A number of descriptors have been proposed for the images that contain robust information about the image. Descriptors have been categorized into two types: Sparse Descriptor and Dense Descriptor.

In sparse Descriptors, Feature points are first find out using some feature point detection technique and then descriptor information is computed and stored only for local region around the detected feature points. Feature points can be find out using any of existing feature points detection technique like: SIFT(Scale Invariant Feature Transform), Harris Corner Detection method etc. Whereas, in Dense Descriptors, descriptor information is find out for each pixel depending upon the two components of a pixel: Differential Excitation( $e_i$) and Orientation($t_i$ ).

*Jie et. al* [10] have extended the Weber Local Descriptor(WLD) that is a kind of dense descriptor and used WLD descriptor in texture classification and face detection. In [10] following steps have been followed to compute WLD histogram:

1. Divide the image into 3 X 3 sized blocks.
2. Compute the Differential Excitation($e_i$) for the center pixel $x_c$ of each block using the equation:

$$e_i = \arctan\left(\sum_{i=0}^{p-1} \frac{x_i - x_c}{x_c}\right) \quad (1)$$

where $x_i$ 's are the neighbors of center pixel $x_c$ and p is the number of neighbors.

3. Compute the orientation($t_i$) of $x_c$ using the following equation:

$$t_i = \arctan\left(\frac{x_7 - x_1}{x_7 - x_3}\right) \tag{2}$$

4. Quantize $t_i$ into T number of dominant orientations using Eq.(3 ) and define 2D
histogram $W(e_i,q_i)$

$$q_i = \frac{2i}{T}\pi \tag{3}$$

where i=0,1,……..$T$-$1$ .

5. Rearrange the excitations into sub histograms corresponding to each column of dominant orientation.
6. Divide each sub histogram into the *'M'* number of segments.
7. Divide each segment into *'s'* number of bins.

## 3 Proposed Technique

In this paper, a watermarking technique has been proposed based on WLD Descriptor. WLD descriptor consists of two components of a pixel: its differential excitation and orientation. Here, differential excitation is computed as difference of center pixel with its surrounding neighbors and orientation is computed as the gradient orientation of center pixel. In WLD descriptor image is represented in the form of a histogram.

In watermarking technique based on WLD Descriptor, descriptor is computed for watermarked image at sending end and this descriptor information is stored in a register file. At the receiving end, descriptor is recomputed for watermarked image that was transformed due to some geometric transformation or other signal processing attack. Now this descriptor is compared with descriptor information stored in register file and decided whether received image is original watermarked image or not.

The steps that will be followed in the proposed watermarking technique are:

1. Load the host image $I$ and watermark to be inserted $W$.
2. Embed the watermark W into the host image I using some signal embedding method to get watermarked image $I_w$.
3. Compute the WLD descriptor or the watermarked image $I_w$ and store in some register file .
4. At the receiving end, let the received watermarked image $I_w$ is transformed into $I_w'$ due to some geometric or signal processing attacks.
5. Compute the WLD descriptor $D'$ for the transformed watermarked image $I_w'$ and compare it with the descriptor stored in register file.
6. Apply some matching technique to find similarity among descriptors $D$ and $D'$.
7. If similarity measure is above some threshold value $T$, received image is authenticated as original image.

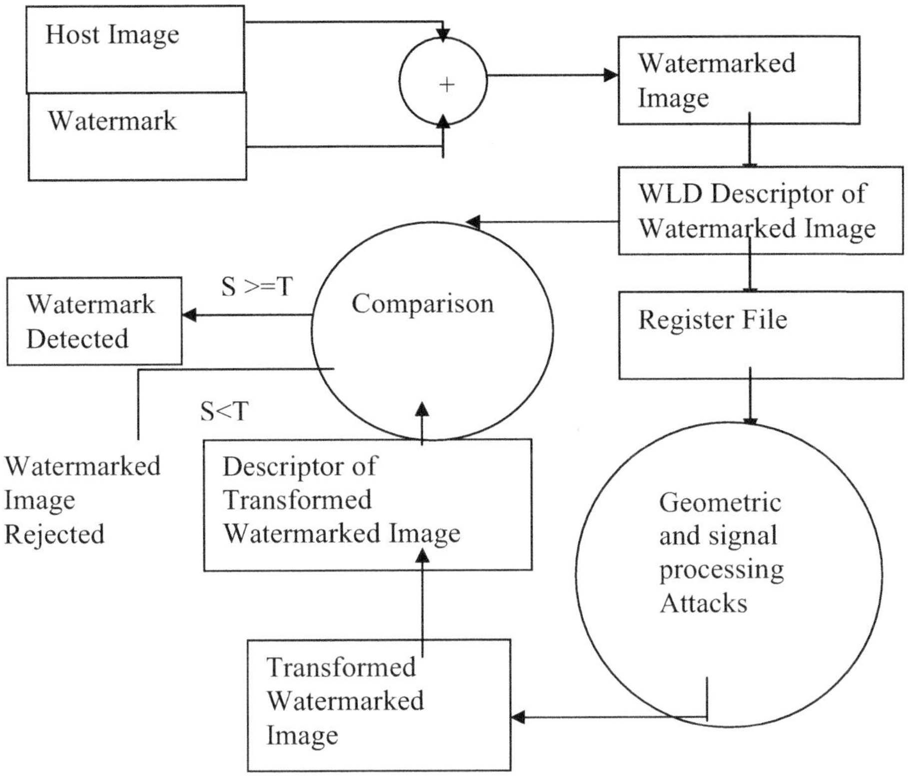


**Fig. 1.** Proposed watermarking technique using WLD descriptor

## 4 Conclusion

It has been concluded that WLD descriptor based watermarking techniques are more robust to geometric and signal processing attacks and have following advantages over other descriptor based watermarking techniques:

1. Size of WLD Descriptor is very less as compared to existing SIFT descriptor.
2. Computational complexity of WLD descriptor algorithm is also very less as compared to other descriptor algorithms as it is involves computation of Differential Excitation and Orientation of each pixel and then divide this information into segments and then further inn histogram bins.
3. Due to easy implementation of WLD Descriptor , watermarking technique based on WLD descriptor will also be least complex as compared to existing descriptor based watermarking techniques.

## References


[1] Lee, H.-Y., Kim, H., Lee"Robust, H.-K.: image watermarking using local invariant features. Optical Engineering 45(3) (2006)

[2] Bas, P., Chassery, J.-M., Macq, B.: Geometrically invariant watermarking using feature points. IEEE Trans. Image Processing 11(9), 1014–1028 (2002)

[3] Tang, C.W., Hang, H.M.: A feature based robust digital image watermark scheme. IEEE Trans. Signal Processing 51(4), 950–995 (2003)

[4] Pham, V.Q., Miyaki, T., Yamasaki, T., Aizawa, K.: Geometrically invariant object-based watermarking using SIFT feature. In: IEEE International Conference on Image Processing, vol. 6 (2007)

[5] Dajun, H., Sun, Q., Tian, Q.: A secure and robust object based authentication system. Eirasip J. on Applied Signal Processing 14, 1–14 (2004)

[6] Sun, J.-g., He, W.: RST invariant watermarking scheme based on SIFT feature and pseudo-Zernike moment. In: Second International Symposium on Computational Intelligence and Design (2009)

[7] Singhal, N., Lee, Y.-Y., Kim, C.-S., Lee, S.-U.: Robust image watermarking based on local Zernike moments. In: IEEE 9th Workshop on Multimedia Signal Processing, pp. 401–404 (2007)

[8] Sun, J., Lan, S.: Geometrical attack robust spatial digital watermarking based on improved SIFT. In: International Conference on Innovative Computing and Communication and 2010 Asia-Pacific Conference on Information Technology and Ocean Engineering, pp. 98–101 (2010)

[9] Li, L.-D., Guo, B.-L., Guo, L.: Combining interest point and invariant moment for geometrically robust image watermarking. Journal of Information Science and Engineering 25, 499–515 (2009)

[10] Chen, J., Shan, S., He, C., Zhao, G.: WLD: A Robust Local Image Descriptor. IEEE Transactions on Pattern Analysis band Machine Intelligence 32(9), 1705–1719 (2010)

# Novel Face Detection Using Gabor Filter Bank with Variable Threshold

P.K. Suri[1], Walia Ekta[2], and Verma Amit[3]

[1] Dean,Chairman,Professor, KUK, Kurukshetra, India
pksurif25@yahoo.com
[2] Professor & Head, MMU, Mullana, India
wekta@yahoo.com
[3] Assistant Professor, RBIEBT, India
ervermaamit@gmail.com

**Abstract.** Face detection is the method of locating human faces in a given image under all lighting conditions, scales and orientations. Face is a unique feature of every person and the same is applicable to pupil, iris and fingerprints which are also unique as well. With the improvement of technology, neural network and processors' high capacity have resulted in induction of informatics in this area. Automatic face detection and recognition has been drawing the main attention in the recent years. We have proposed here an accurate face detection system that can detect faces under different contrast with hurdles like faces with spectacles, heavy beard and even closed eyes. We use Gabor filter bank with varying threshold for feature extraction and face detection.

**Keywords:** Contrast, Detection, Gabor Technique, Illumination, Neural Network, Threshold.

## 1 Introduction

The paper is presented in seven sections. Section 2 includes the general introduction about face detection. In Section 3, certain significant methods for face detection have been elaborated. Section 4 is enlightenment about the system overview as how the algorithm [14] [32] actually works. Section 5 represents Gabor filter bank and its mathematical representation with algorithm for training. Section 6 gives details regarding neural [1] [2] network and Section 7 finally gives output and conclusion of the work.

## 2 Introduction to Face Detection

Automatic face processing includes numerous issues for example like face detection in a picture. There are a lot of challenging issues in accurate detection of faces. The most straightforward variety of this problem is the detection of a single face at a known scale and orientation. It is used as initial step of an automated face recognizer.

A. Mantri et al. (Eds.): HPAGC 2011, CCIS 169, pp. 617–625, 2011.

Thought of in this sense, face detection [13][33] can be applied to systems for things such as automated surveillance and human traffic census. It is an analogue to face tracking that requires no knowledge of previous frames. In some cases or at certain conditions in which a picture is taken can be fully controlled. For example picture taken for standard data set like from CMU and MIT[28][30] or specifically the face position can be easily specified in passport sized photographs. Yet in many situations; position, size and orientation of a face is not known in advance. (as shown in figure 1)

**Fig. 1.** Inaccurate Detection

The face of the persons in the figure 1(2nd from left and first from right-Picture taken from MIT database-as output of previous detectors) is missed and incorrectly detected due to orientation and localization. Another source of complexity for faces is the complex 3-dimensional shape of one's face, and the resulting difference in the appearance of a given face under different lighting conditions, while even in same environment. A face is naturally recognizable to a human being despite its many points of variation (e.g. skin tone, hairstyle, facial hair, glasses, etc.

## 3 Basic Method and Strategy for Face Detection

A human being is able to detect a face in the context of an entire person, but we want a simple, context-free approach to detection [22-25] that can be used by an automated system. The methods proposed to far generally fall into two categories one is feature based. The first technique has grouped different facial features according to geometrical relationship. The second approach is classification based, which can further be divided into statistical or learning based approach. We emphasize on learning based methods specifically neural network(feed forward) [1][16][19].This method aims to find a decision line between faces and non faces. Due to discrimination ability of learning based methods, these are more robust against pose variations and light illuminations. However, classification method gives better accuracy; it is associated with large computation complexity since the performance of classification method depends on features used. Normally gray value of the input image is taken as feature for the classifier [8] [17].

## 4 System Overview

To detect face [27] [31] of variable size under different illumination with varying pose [3] [10][12], the detector needs to be accurate and efficient. In our system, input image is convolved with Gabor filter bank after reshaping or rescaling the output of the convolution. This input is given to neural network and training accomplished at this phase. We train the network for number of epochs till the error become very less. Bigger is the data base more the number of epochs required to train the network to achieve lowest error rate. If the training goal is not achieved, we reduce the threshold limits and train the network again (as clear from the algorithm given in 5.2).Then we check the various images as input for the detector, to test the output/performance of the detector. At the end of the paper, we have achieved highest detection rate which is remarkable. No earlier method (even those that use Gabor technique before) have achieved detection of faces with as much accuracy that we achieved. For example, our algorithm detects faces even with spectacles ,heavy beard and large varying illumination. Moreover, with this approach, we detect all the faces in one go. Our algorithm is much more accurate though face at angle more than 45 degree is hard to detect.

## 5 Gabor Filter

The Gabor function has been recognized as a very powerful tool in image detection and recognition process, due to its optimal localization properties in the both spatial and frequency domain. There are lot many papers published on its application since Gabor (1946-1947) proposed one dimensional functions. The family has then been extended to two dimensions by Daugman(1980-1981)[5][6] as a framework for understanding the orientation selective and spatial selective receptive field properties of neurons in the brains visual cortex and mathematically implemented the.

The 2-D Gabor function is a harmonic oscillator composed of sine wave of a particular frequency and orientation. A complex 2-D Gabor function is described as below, for any image *f(x,y)*:

$$G(x,y) = \frac{\exp(-\{x-x_o)^2}{2\sigma x^2 - (y-y_o)^2/2\sigma_y^{\ 2}} \times \exp(-2\pi\, i\, (u_o(x-x_o)+y_o(y-y_o)) \tag{1}$$

The 2-D Gabor filter can also be represented for even real symmetric configuration as, $G(x,y;T,\Phi)$ which can be decomposed into two parts: one parallel and the other perpendicular to orientation $\Phi$. The parallel part behaves like BPF (band pass filter [9][29])where as perpendicular part behaves like LPF (low pass filter[9][29]).It clearly means the two scales are orthogonal to each other.

Here $(x,y)$ are the location in the image $(u,v)$ particular modulation or or instantaneous modulation, $\sigma_x$ and $\sigma_y$ are standard deviations along horizontal and vertical axis.

Many people define Gabor faces with different pattern [21] though all derivations are same as one more implementation of Gabor is as follows.

$$\Psi_{u,v}(z)=\frac{\left|k_{u,v}\right|^2}{\sigma^2}\times e^{\frac{\left|k_{u,v}\right|^2\left|z\right|^2}{2\sigma 2}}\times(e^{(ik_{u,v})}-e^{(\frac{\sigma^2}{2})}) \tag{2}$$

where

$$z=(x,y),\quad k_{u,v}=k_v\, e^{\frac{iu\pi}{8}}$$

$$e^{i(k_{u,v})}=\cos(k_{u,v})+i\sin(k_{u,v})$$

is an oscillatory wave function. Here $v$ controls the scale and $u$ controls the orientations.

### 5.1 Gabor Features Extraction

Since facial features[11][20][26] are mainly in horizontal or vertical direction, Gabor filter[4][7] with four orientation is considered here, so $\Phi=\pi(L-1)/4$, where L=1,2,3,4. In order to get accurate results four filters(in all directions)should not overlap. This means aliasing effect shouldn't be there.
The Gabor representation of the image is convolution of the input image with Gabor filter bank (5*8=40 features).

**Fig. 2.** Gabor repersentation

$$C_{m,n}(x,y)=f(x,y)*GaborFeature \tag{3}$$

For detection process we derive feature vector which is image to vector transformation. Before adding or checking the effect of all features sampling rate for convolved[18] image that is $C_{m,n}(x,y)$ (the rate at which we can sample the input data at the rate of uniform interval of $2\pi$*(frequency of the interest) apart), is down sampled that is below the max limit. We apply masking for removing corners (as they contain no such information regarding faces [15-16]).

### 5.2 Algorithm for Learning and Training

Give labeled example set $T$ and threshold $б_1$
Gabor Detection

- *I=1;* The first stage error rate $F_1$,

Train the first n-stage classifier based on gabor features
While the first I stage error rate $F_i$> *б*

- Train the network based on gabor features
- Calculate $(i+1)^{th}$ stage error $e_{i+1}$calculate the error rate

Train the network till $e_{i+1}$>predefined error rate.
If last stage error $e_l$> predefined error rate,

- Training not accomplished perfectly
  Else
- Target achieved.

Test the algorithm,
If Detection=Accurate
Stop,
Else take *б* $_2$
*such that.* *б* $_2$< *б*$_1$
Goto Gabor Detection.

# 6 Neural Network

Feed forward neural networks with back propagation are the first and arguably the simplest type of artificial neural networks devised. Two types of feed forward neural networks are commonly used in modelling; either normal or intrusive patterns [21]. We use Feed forward net with 100 hidden layers (as shown in figure 3),

$$\text{Here } x = (x_1, x_2, x_3 \ldots\ldots\ldots \quad \ldots x_n)$$

is input pattern., where $x_n$ is the last pixel value of image vector (reshaped vector from figure 2).The output can be computed from algorithm [14][32] given below.

## 6.1 Algorithm for Feed Forward Neural Network

*Initialize the weight*
While stoping condition is false
For each training pair do step 1 to 3

1. Each input unit receives the input signal xi and transmit to hidden layer
2. Hidden unit sums its weighted input signal $H = v_{o_i} + \Sigma x_i v_{ij}$ apply

activation function $H_j = f(H)$.

3. Output $Y = w_{ok} + \Sigma H_j w_{jk}$ *and* $Y_k = f(y)$

*Error calculation*

1. $E=(t_k- Y_k) f(H)$ and $E_o= \Sigma E(hidden\ unit)\ w_{jk}$
2. Error is calculated as $E(hidden\ unit)=E_o{*}f(H)$

Test for stopping condition

1. Minimization of error
2. Number of epochs.

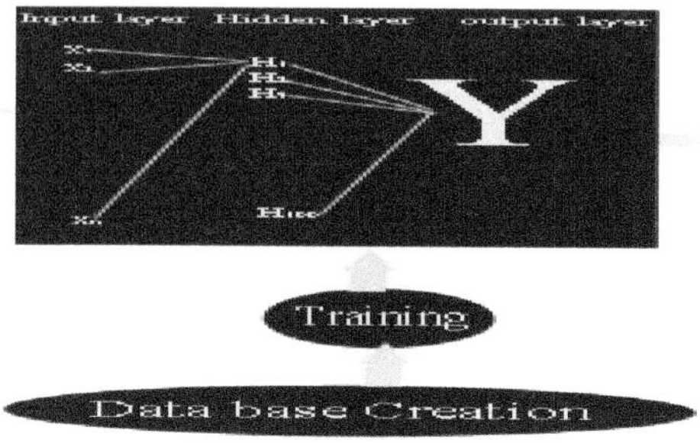


**Fig. 3.** Feed Forward Net

## 7 Conclusion and Tabular Comparison

After training the network, we can check the performance of the algorithm by giving image under test as input. The image will be detected and rectangle will be drawn around the faces in picture under test (as shown in figure 4 and figure 5). We check our algorithm on the image given under and hence the output.

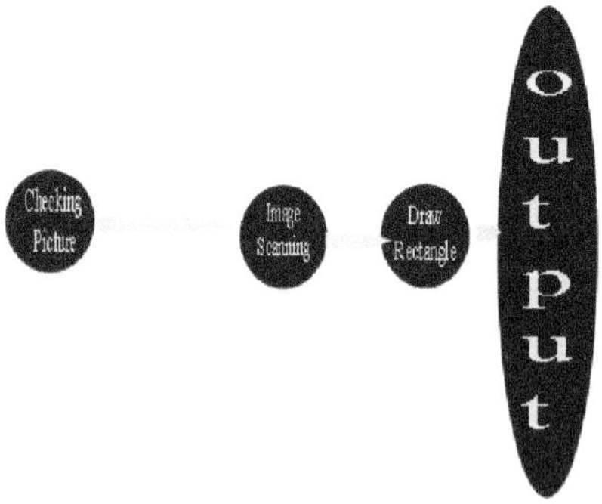


**Fig. 4.** Output Phase

**Fig. 5.** Detector output for face

Fig. 6. Elaboration of above figure

Fig. 7. Face with angle more than $40^0$

Fig. 8. Faces with different angles and threshold

Fig. 9. Comparative better performance

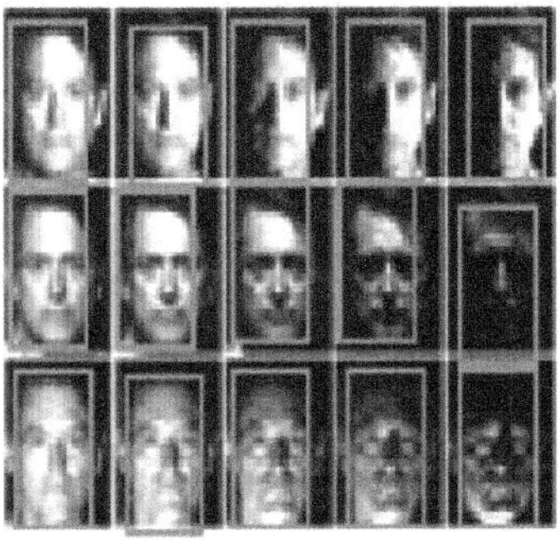

Fig. 10. Detection with Varying Illumination

Fig. 11. With varying size

Though detection time may vary depending on the machine being used, we have used core to duo machine with 4GB RAM for our algorithmic check purpose.

**Table 1.** Detector Performance

| S. No | Test Image | False Positive | False Negative | Detection Rate |
|---|---|---|---|---|
| 1 | Figure-5 Φ=upto 40° | - | - | 100% |
| 2 | Figure-9 Φ=upto 40° | - | - | 100% (against 90% of previous[data from [28][30] and MIT) |
| 3 | Figure-7 Φ> 40° | 11% | 5% | 70% |
| 4 | Figure- 8 | - | 3% | |
| 5 | Figure10 | - | - | 100% |
| 6 | Figure 11 | - | - | 60% |
| Overall Efficiency | 86% | | | |

## References


[1] Amit, Y., Geman, D.: A computational model for visualselection. Neural Computation 11, 1691–1715 (1999)

[2] Crow, F.: Summed-area tables for texture mapping. Proceedings of SIGGRAPH 18(3), 207–212 (1984)

[3] Fleuret, F., Geman, D.: Coarse-to-fine face detection. Int. J. Computer Vision 41, 85–107 (2001)

[4] Freeman, W.T., Adelson, E.H.: The design and use of steerable filters. IEEE Transactions on Pattern Analysis and Machine Intelligence 13(9), 891–906 (1991)

[5] Daugman, J.G.: Two-dimensional spectral analysis of cortical receptive field profiles. Vision Research 20, 847–856 (1980)

[6] Daugman, J.G.: Uncertainty relation for resolution in space, spatial frequency, and orientation optimized by two dimensional visual cortical filters. J. Optical Soc. Amer. 2(7), 1160–1169 (1985)

[7] Gabor, D.: Theory of communication. J. IEE 93, 429–457 (1946)

[8] Freund, Y., Schapire, R.E.: A decision-theoretic generalization of on-line learning and an application to boosting. In: Vitányi, P.M.B. (ed.) EuroCOLT 1995. LNCS, vol. 904, pp. 23–37. Springer, Heidelberg (1995)

[9] Greenspan, H., Belongie, S., Gooodman, R., Perona, P., Rakshit, S., Anderson, C.: Over complete steerable pyramid filters and rotation invariance. In: Proceedings of the IEEE Conference on Computer Vision and Pattern Recognition (1994)

[10] Itti, L., Koch, C., Niebur, E.: A model of saliency-based visual attention for rapid scene analysis. IEEE Patt. Anal. Mach. Intell. 20(11), 1254–1259 (1998)

[11] John, G., Kohavi, R., Feger, P.: Irrelevant features and the subset selection problem. In: Machine Learning Conference Proceedings (1994)
[12] Osuna, E., Freund, R., Girosi, F.: Training support vector machines: An application to face detection. In: Proceedings of the IEEE Conference on Computer Vision and Pattern Recognition (1997a)
[13] Papageorgiou, C., Oren, M., Poggio, T.: A general framework for object detection. In: International Conference of Computer Vision (1998)
[14] Quinlan, J.: Induction of decision trees. Machine Learning 1, 81–106 (1986)
[15] Roth, D., Yang, M., Ahuja, N.: A snow based face detector. Neural Information Processing 12 (2000)
[16] Rowley, H., Baluja, S., Kanade, T.: Neural network-based face detection. IEEE Patt. Anal. Mach. Intell. 20, 22–38 (1998)
[17] Schapire, R.E., Freund, Y., Bartlett, P., Lee, W.S.: Boosting the margin: A new explanation for the effectiveness of voting methods. Ann. Stat. 26(5), 1651–1686 (1998)
[18] Simard, P.Y., Bottou, L., Haffner, P., LeCun, Y.: Boxlets: A fast convolution algorithm for signal processing and neural networks. In: Kearns, M., Solla, S., Cohn, D. (eds.) Advances in Neural Information Processing Systems, vol. 11, pp. 571–577 (1999)
[19] Sung, K., Poggio, T.: Example-based learning for view based face detection. IEEE Patt. Anal. Mach. Intell. 20, 39–51 (1998)
[20] Tsotsos, J., Culhane, S., Wai, W., Lai, Y., Davis, N., Nuflo, F.: Modeling visual-attention via selective tuning. Artificial Intelligence Journal 78(1/2), 507–545 (1995)
[21] Webb, A.: Statistical Pattern Recognition. Oxford University Press, New York (1999)
[22] Zhao, W., Chellappa, R., Phillips, P.J., Rosenfeld, A.: Face recognition: A literature survey. ACM Computing Surverys (CSUR) 35(4), 399–458 (2003)
[23] Li, S.Z., Lu, J.: Face recognition using the nearest feature line method. IEEE Trans. Neural Netw. 10, 439–443 (1999)
[24] Turk, M.A., Pentland, A.P.: Eigenfaces for recognition. Cognitive Neurosci. 3(1), 71–86 (1991)
[25] Nefian, A.V., et al.: Hidden markov models for face recognition. In: Proceedings International Conference on Acoustics, Speech and Signal Proceeding, pp. 2721–2724 (1998)
[26] Huang, J., Heisele, B., Blanz, V.: Component based face recognition with 3D morphable models. In: Proceedings of International Conference on Audio and Video-based Person Authentication, vol. 5, pp. 27–34 (2003)
[27] Belhumeur, V., Hespanda, J., Kiregeman, D.: Eigenfaces vs. fisherfaces: recognition using class specific liear projection. IEEE Trans. on PAMI 19, 711–720 (1997)
[28] Osuna, E., Girosi, F.: Reducing the run-time complexity in support vector machines. In: Schölkopf, B., Burges, C.J.C., Smola, A.J. (eds.) Advances in Kernel Methods —Support Vector Learning, pp. 271–284. MIT Press, Cambridge (1999)
[29] Burges, C.J.C.: Simplified support vector decision rules. In: International Conference on Machine Learning, pp. 71–77 (1996)
[30] Burges, C.J.C., Schölkopf, B.: Improving the accuracy and speed of support vector machines. In: Mozer, M.C., Jordan, M.I., Petsche, T. (eds.) Advances in Neural Information Processing Systems, vol. 9, p. 375. MIT Press, Cambridge (1997)
[31] Viola, P., Jones, M.: Rapid object detection using a boosted cascade of simple features. In: Proceedings IEEE Conference on Computer Vision and Pattern Recognition (2001)
[32] Blanchard, G., Geman, D.: Hierarchical testing designs for pattern recognition. Technical Report 2003-07, Universit Paris-Sud (2003)
[33] Rowley, H.A., Baluja, S., Kanade, T.: Neural network-based face detection. IEEE Transactions on Pattern Analysis and Machine Intelligence 20(1), 23–38 (1998)

# When to Stop Testing

Ajay Jangra[1], Gurbaj Singh[1], Chander Kant[2], and Priyanka[3]

[1] CSE Department U.I.E.T. Kurukshetra University, Kurukshetra, India
[2] DCSA Kurukshetra University, Kurukshetra, India
[3] ECE Department U.I.E.T. Kurukshetra University, Kurukshetra, India
er_jangra@yahoo.co.in, gurbaj1986@yahoo.com,
ckverma@rediffmail.com, priyanka.jangra@gmail.com

**Abstract.** Testing process is to detect variance between actual and expected results, to make good quality software. Testing is necessary but to test each and every part of software is not feasible. There should be threshold point to stop testing, in this paper we proposed an analytical scheme which provides threshold values to stop testing without compromising the quality of software. By using proposed scheme, developer can easily find out sufficient and enough level of testing.

**Keywords:** Software testing, Quality, Test cases, testing metrics.

## 1 Introduction

Testing refers to the process of evaluating attributes like correctness, completeness, security, consistency, unambiguousness, quality etc. of a software and determine whether it meets its required functionality or not. Software testing doesn't only for fixing the bug (in the code), but also to check whether the program behaves according to its given specifications and requirements.[14,1]

Scope of testing directly dependent upon the following factors:

- The size and complexity of the software.
- How much new software integration work is required?
- Whether any new or cutting edge technology is involved.
- The performance of the system integrator to date.
- The knowledge and experience of customer's project team.

Software testing is an essential activity in software engineering, it is a discipline as well as an iterative process, which consists of Tests Designing, Tests Execution, Problems Identifying and Problem Fixing, for validating functionality and as well as for attempting the software break. Testing built the confedence of user/ developer about the software but it can't guarantee the production of high quality software. We do Software testing to find problems and to fix them for improving software quality. Software testing may consume 35%-40% of a software development budget. Manual and Automated Testing methods seems complementary to each other. Manual software refers to the process to test the software statically, may be carried out by an individual or individuals, where automated software testing involves creation of test scripts, which can be run then automatically, repetitively, and through a number of

A. Mantri et al. (Eds.): HPAGC 2011, CCIS 169, pp. 626–630, 2011.

iterations. Automated software testing helps to minimiz the variability in the results, speed up the testing process, increase test coverage, and ultimately provide greater confidence in the quality of the software being tested.[2,13].

## 2 Test Cases

Software tester must negate the hypothesis that the software satisfies its required specification and assume the software contains at least one fault. They must then build a test set such that if the software contains a fault, then at least one test will fail. This method defines the ability to detect faults as a set adequacy criterion. The satisfaction of this method can be evaluated before the software is used. The degree of satisfaction can be estimated by:

1. Determining if the test set satisfies necessary conditions related with particular fault types.
2. Determining if the test set satisfies sufficient conditions associated with particular fault types.
3. Inserting different types of faults in the software and Determining whether the test set can detect them.

Designing test cases is a time consuming process, but they are worth giving time because they can really avoid unnecessary retesting or debugging or at least lower it. Designing proper test cases is important for software testing plans as a lot of bugs, ambiguities, inconsistencies and slip ups can be recovered in time as also it helps in saving your time on continuous debugging and re-testing test cases.[5,6,12]

Before designing of effective test cases, a software tester must understand the basic principles of software testing:

- **All tests should be traceable to the requirements:** the objective of software testing is to uncover errors. It follows that the most severe defects are those that cause the software to fail to meet its requirements.
- **Tests should be planned long before testing begins:** Detailed designing of test cases can begin as soon as the design model has been solidified. Therefore, all tests should be planned and designed before any code has been generated.
- **The Pareto principle applies to software testing:** Pareto principle states that 80 percent of all errors uncovered during testing will likely be traceable to 20 percent of all software components. The problem, of course, is to separate these suspect components and to thoroughly test them.
- **Testing should start in the small and ends in the large:** The first tests planned and executed generally focus on single components. As testing progresses, focus shifts to find errors in integrated group of components and ultimately in the entire software.
- **Complete testing is not possible:** The number of path permutations for even moderately sized software is exceptionally. Therefore, it is impossible to execute all combination of paths during testing. It is possible, to adequately cover program logic and to ensure that all conditions in the component-level design have been exercised.[3,4]

## 3 Proposed Scheme

Testing is endless process. Tester can not stop testing until all the defects are removed, it is simply impossible. At some point, tester has to stop testing and ship the software. After observing all the details of Software testing we come to a conclusion that "testing can never be considered complete". Tester can never be proved theoretically or scientifically that our software is free from errors now.[7]
Basically testers stop testing when:

- The planned testing deadlines are about to expire.
- Not able to detect any errors even after execution of all the planned test Cases.

These two statements do not have any meaning and are contradictory since we can satisfy the first statement even by doing nothing while the second statement is equally meaningless since it can not ensure the quality of our test cases. Pin pointing the time when to stop testing is difficult. Most of the today's softwares are so complex and run in such an Interdependent environment, that complete testing can never be done. Other important factors which are helpful in deciding when to stop the testing are:[8,9]

- When deadlines like release or testing deadlines have reached.
- When the test cases have been finished with some pass percentage.
- When the testing budget comes to its end.
- When Functional coverage, code coverage, the client requirements meets to certain point.
- When bug rate drops below a prescribed level.
- When the period of beta testing / alpha testing gets over.
- Beta or Alpha testing period finished.
- When Resources are availability finished.

Some Scientific Methods to decide when to stop testing are:

***1) Decision based upon Number of Pass / Fail test Cases:***

- Prepare defined number of test cases before test execution cycle.
- Execution of all test cases in every testing cycle.
- Stop testing process when all the test cases get Passed
- Stop testing when percentage of failure in the last testing cycle is observed to be extremely low.

***2) Decision based upon Metrics:***

- *Mean Time between Failures*: - recording the average operational time before the software failure.
- *Coverage metrics*: - recording the percentage of number of executions during tests.
- *Defect density*: - recording the defects related to size of software like the number of open bugs and their severity levels.

Testing metrics helps the testers to take better and accurate decisions; like when to stop testing or when the application is ready for release, how to track testing progress & how to measure the quality of a product at a certain point in the testing cycle. The best way for tester is to have a fixed number of test cases ready well before the beginning of test execution cycle. Finally measure the testing progress by recording the total number of test cases executed using the following metrics which are quite helpful in measuring the quality of the software product.[10,11]

1) ***Percentage Completion***: - (Total executed test cases) / (Total test cases).
2) ***Percentage Test cases Passed*** :- (Total passed test cases) / (Total executed test cases)
3) ***Percentage Test cases Failed*** :- (Total failed test cases) / (Total executed test cases).

A test case is declared Failed when just one bug is found while executing it, otherwise it is considered as Passed.

Practically tester's feels that the decision of stopping testing is based on the level of the risk acceptable to the management. As testing is a never ending process tester can never assume that 100 % testing has been done, we can only minimize the risk of shipping the software to end user with some type of testing done. The risk can be measured for the small duration, low budget, low resources project; risk can be reduced simply by Risk analysis.

Other important methods from which a tester/ testers can decide to stop testing are:

a) If the coverage of the code is good.
b) Mean time between failures is quite large.
c) Defect density is very low.
d) Number of high severity Bugs is very low.

## 4 Conclusion

Tester can defines the time to stop testing is when, all the test cases, derived from white-box testing methods like equivalent partitioning, cause-effect analysis & boundary-value analysis are executed without detecting errors. This paper elaborates of when to stop software testing, and the methods discussed should be executed sequentially so that the tester can assure the customer that the software quality is on the higher side and all types of requirements are fulfilled and the software can have negligible number of errors at the end of testing phase.

## References


1. Nita, M., Notkin, D.: White-Box Approaches for Improved Testing and Analysis of Configurable Software Systems. IEEE, Los Alamitos (2009)
2. Saha, G.K.: Understanding Software Testing Concepts. Published in ACM Ubiquity 9(6) (2008)
3. Do, H., Mirarab, S., Tahvildari, L.: An Empirical Study of the Effect of Time Constraints on the Cost-Benefits of Regression Testing. ACM, New York (2008)

4. Kapfhammer, G.M.: Software Testing. ACM, New York (2008)
5. Nilsson, R., Offutt, J.: Automated Testing of Timeliness: A Case Study. IEEE, Los Alamitos (2007)
6. Seo, K.I., Choi, E.M.: Comparison of Five Black-box Testing Methods for Object-Oriented Software. IEEE, Los Alamitos (2006)
7. Bentlet, J.E., Bank, W., C.NC. : Software Testing Fundamentals-Concepts, Roles, and Terminology, pp. 1–12.
8. Cem Kaner, J.D.: Ph.D.The Ongoing Revolution in Software Testing, in Software Test & Performance Conference, December 8 (2004)
9. Kaner, C.: Inefficiency and Ineffectiveness of Software Testing: A Key Problem in Software Engineering (2004)
10. Kusinitz, A.: Software Testing—How Much Is Enough? (June 2003)
11. Tse, T.H., Chen, T.Y., Feng, X.: On the completeness of Test Cases for Atomic Arithmetic Expressions. IEEE, Los Alamitos (2000)
12. Martin, E.: Testing and analysis of access control policies. In: International conference on software engineering (ICSE 2007 companion) (2007)
13. Chen, T., Sahinoglu, M., von Mayrhauser, A., Hajjar, A.: How Much Testing is Enough? Applying Stopping Rules to Behavioral Model Testing (2001)
14. Jangra, A., Singh, G., Singh, J., Verma, R.: Exploring Testing Strategies. International Journal of Information Technology & Knowledge Management IV-I (December 2010); ISSN 0973-4414

# An Efficient Power Saving Adaptive Routing (EPSAR) Protocol for Mobile Ad Hoc Networks (MANETs)

Ajay Jangra[1], Nitin Goel[1], Chander Kant[2], and Priyanka[3]

[1] CSE Department UIET, Kurukshetra University, Kurukshetra- India
[2] DCSA, Kurukshetra University, Kurukshetra- India
[3] ECE Deptt. Kurukshetra Institute of Technology and Management, Kurukshetra, India
er_jangra@yahoo.co.in, goelnitin0887@gmail.com,
ckverma@rediffmail.com, priyanka.jangra@gmail.com

**Abstract.** Mobile ad hoc networks (MANETs) are networks that have no centralized body and in order to communicate with the nodes it has no fixed topology. Also it is difficult to find the route from source to destination in MANETs, because of its arbitrary mobility of nodes and in general MANETs works on multihop environment to select a route from source to destination. For this a new scheme has been proposed that results in an efficient and reliable path named as Efficient Power Saving Adaptive Routing (EPSAR). In this paper we would apply some amendments to FRENSA and make a new algorithm that could make it more efficient and reliable enough and also apply it to three different scenarios like Best, Average and Worst case for selecting a path and can check the performance of the network and the complexity of selected route. In all the cases we would take some parameters that have been used in EPSAR and analyze the performance of this algorithm in all the cases.

**Keywords:** Mobile Ad-hoc Networks (MANETs), multi-hoping, complexity, multicasting, node selection algorithm, FRENSA, EPSAR.

## 1 Introduction to MANETs

Mobile ad hoc network (MANET) is an appealing technology that has attracted lots of research efforts. Ad hoc networks are temporary networks with a dynamic topology which doesn't have any established infrastructure or centralized administration. Mobile ad hoc networks (MANETs) are dynamically configure, multi-hop wireless networks with varying topology. Mobile nodes in such networks are continuously associated or disassociated with each other, according to their topological arrangements. Thus, the network topology varies with time due to the ataxic locomotion of the participating nodes. In recent years the concern over the security of computer networks has been widely discussed and popularized. The discussion has, however, typically involved only static and wired networking while the mobile or ad-hoc networking issues have not been handled extensively. The emergence of such new networking approaches sets new challenges even for the fundamentals of routing since the mobile ad-hoc networks (MANET) are significantly different from the wired networks.[4,6]

A. Mantri et al. (Eds.): HPAGC 2011, CCIS 169, pp. 631–639, 2011.

## 2 Previous Work

1. Multicasting used for one to many and many to many node communications. Performance of MAODV (multicast ad hoc on demand distance vector routing protocol) and ODMRP (on demand multicast routing protocol) are evaluated in terms of PDR (packet delivery ratio and latency). It observed that MAODV performs well in case of large area and ODMRP performs better in high speed. [15]
2. A certificate based protocol named ARAN is proposed to reduce security threats to AODV and DSR and avoid all identified attacks.[16]
3. A new protocol MOCA (mobile certificate authority) based on PKI (public key infrastructure) and CA (certificate authority) proposed for efficient communication.[17]
4. For multi-hop network a new mobile intrusion detection system (MIDS) is proposed, which efficiently detect the misbehavior, packet drop and delay by appointing a special node (monitor) in the network.[18]
5. FRENSA (Farthest Reliable Efficient Node Selection Algorithm), MANETs works on multi-hopping mechanism in order to get efficient communication we need to get a secure reliable and efficient too route from source to destination. So in this connection a new algorithm is designed for next node selection to enhance the quality of next node selection method for MANETs. For packet forwarding the proposed algorithm select the next node with respect to its distance fron sender node, power backup and reliability, which reduces the overall communication head and improves the reliability of communication in a multi-hop environment. We could call that algorithm farthest reliable efficient node selection algorithm (FRENSA). A new approach is proposed which selects the farthest reliable efficient node for packet forwarding firstly a table of neighbor nodes is arranged with respect to their distance fro the sender node. The nodes having less power backup then Minimum Qualifying power value (MQp) are removed from the table NAT. then select the highest node and search it from RnR table to check its reliability if the node is in R column of RnR table then its our next node for forwarding the packet ( Farthest Reliable Efficient Node is Selected) and if the node is listed with nR column that means its an unreliable node we cannot trust on this node so leave that node and select the second highest node from NAT table and check out its reliability from RnR table. If the node does found in RnR table then off line certification is demanded to its parent node/multicast group by sending CVP (certificate verification packet). [1,2,3]

## 3 Parameters Used

In order to work with EPSAR, some parameters we need to follow i.e Beacon request message that is used to send the request to start the communication (format as shown in fig 1(a)) and Beacon Reply message used for acknowledgement means a particular node is ready for communication (format as shown in fig 1 (b)). The next parameter we would use is Network awareness Table *(timestamp)* NATts (format as shown in fig 1(c)) that tells sender node maintains a network awareness table with current timestamps; this table contains the information about the neighbor nodes collected from beacon_reply packet. The nodes are arranged in the decreasing order of their distance from the sender node (distance is estimated by the signal strength of beacon_reply packet). Generally weakest the signal means farthest the distance. For every communication the sender node defines Minimum Qualifying power value (TMQp) depends on the length of the transmitting data and communication duration required. Sender node defines the value of the TMQp, means the next node (selection) bettry power should be higher then defined Minimum Qualifying power value (TMQp). The NATts table enrolls only that nodes having reliable battery life (RBL) > TMQp. The next parameter to work with this algorithm is a reliable –unreliable table (RnR) (format as shown in fig 1(d)) is maintained, which contains the list of reliable and unreliable nodes with respect to their IP address and port number. Whenever a node experience multicast, multi-hop communication the behavior of node travels (during packet transmission) is reported in RnR table.

The reliable node information is stored in R column and information about unreliable, malicious etc. is recorded in nR column which is further communicated to the nodes of its multicast group. The next parameter is CVP (certificate verification packet) for a new node (not found in RnR table) a special packet is transmitted to the node for providing the off line verification certificate and the new node supplies the verification certificate generated by its parent node (format as shown in fig 1(e,f)). This certificate gives the guaranty of geniuses and reliability of the bearer node and RnR table updated with the new node. Otherwise two cases are formed in case-I when *node produce a unreliable proof*, the corresponding node is enrolled in nR column of RnR table and in case-II when node fails to provide certificate/ timeout, for both the cases (case-I & case-II) the current node is left and the next highest node from NATts table is selected to check the reliability and one parameter is Data seen table (DST) Every node maintains a DST (format as shown in fig 1(g))which contains the log of packet forwarded. It reduces the duplicate packet forwarding because Whenever a node receives a packet it check the information to its DST if information is found then it discard the packet otherwise information is inserted to DST and packet is forwarded. At last a data packet sends by the source node to the selected node to begin the communication. (Format as shown in fig 1(h)) [3,11,14].

## 4 Proposed Routing Algorithm

As from the EPSAR approach we can clearly see that how we can select a route to destination but if we try to change some parameters then we can easily enhance the performance of previous algorithm and we are taking these scenarios in term of multi-hopping by using clusters in MANETs i.e.

**Beacon Packet Call**

| 2 | 4 | 4 | 2 |
|---|---|---|---|
| Packet Type | Source Address | Destination Address | TMqp |

(a)

**Beacon Packet reply**

| 2 | 4 | 4 | 2 |
|---|---|---|---|
| Packet Type | Source Address | Destination Address | RBL |

(b)

**Network awareness Table *(timestamp)* NATts**

| Neighbor Node ID | Port No. | RBL (in Sec) |
|---|---|---|

(c)

**RnR Table**

| R (Reliable Node) | | nR (Unreliable node) | |
|---|---|---|---|
| Id | Port No. | Id | Port No. |

(d)

**Certificate Verification packet**

| 2 | 4 | 4 | 4 |
|---|---|---|---|
| Packet Type | Source Address | Destination Address | Certification Required |

(e)

**Certificate Verifier Packet**

| 2 | 4 | 4 | 4 |
|---|---|---|---|
| Packet Type | Source Address | Destination Address | Verifier Data |

(f)

**Data Seen Table**

| Packet Id | Multicast Id | Source Id |
|---|---|---|

(g)

**Data Packet Header:**

| Packet Type | Packet Id | Source Address | Destination Address | Time-To-Live | Hop Count |
|---|---|---|---|---|---|

(h)

**Fig. 1.**

**EPSAR (Efficient Power Saving Adaptive Routing Protocol)**

```
        Input    // Nodes receive Beacon Messages from other node with in its territory//
        Output   // Efficient, Farthest, Reliable node is selected having less computational
                 Power//

 1.  Maintain NATts table.
 2.  Synchronization between the nodes with in its clusters.
 3.  Beacon request is send and beacon Reply is received
 4.  if
           (RBL>=TMqp)              // comparing efficiency with Threshold value//
 5.    then
           Nodes reply who' having more and equal power and stores all the nodes
           into NATts with priotity
 6. else
                  Node checks
                         if (node id==destination id)
                         then                              go to 11
                         else                              go to 2
 7. if
           (Node is at NATts ==1)   //select the highest and farthest node//
 8. Then
           Traverse the RnR table   // for reliability and certification//
 9.  else
           Select he highest node from the NATts and goto then of 8.
 10. if
           Node is in R column
 11. Then
           Elect the EPSAR node     // efficient, farthest and reliable node//
 12. if
           Node is in nR column
           Release the CVP packet   // call for offline certificate//
           go to 9
```

***Efficient (E)*** means when we start communication with the help of Beacon messages including TMqp, only those nodes reply who fulfill the said power criteria. So that our overall network communication will be efficient because only efficient/qualified would participate in communication.

***Power Saving (PS)*** it drastically reduces the number of communication attempts (save bandwidth) in terms of Beacon reply (only qualified nodes may reply not all). It leads to the less computation requires for marinating NATts, RnR table and obviously the Next Node selection algorithm.

***Adaptive (A)*** refers to the dynamic packet routing which depends upon the status of current node. Although following shortest path remains prime motto of any routing protocol, in the proposed routing scheme data packet is forwarded to the next node after evaluating its reliability and efficiency.

Flowcharts (fig 2) shows the functioning of proposed routing protocol

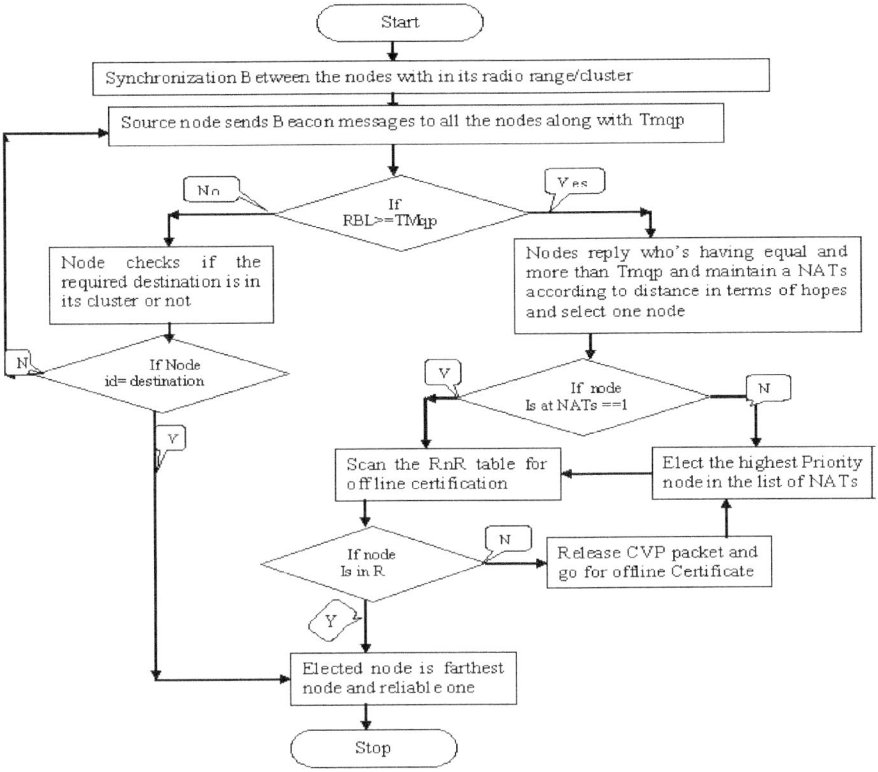


## 5 Analysis on Different Problem Scenarios

In order to find the route from source to destination in terms of hops we can have three cases i.e. Best case, Average case and Worst case. Best case results in minimum hop and intermediate nodes (if any) should be reliable and efficient too. Average case results in optimal number of hopes (i.e depends upon the distance and the geographical region) from source to destination. Worst case results in huge number of hops that may not be the optimum path.[9,10]

We are taking an example of few nodes with their geographic areas and generating a scenario that gives us a clean view for both the cases while communication. Before that let us take an example of few nodes with source node and destination node on which we are going to implement all the cases.[11]

**Fig. 3.** General Scenario of MANETs

1. Now take an example of Best case in fig. 4, where when a source node wants to communicate with destination node then in this case source node can easily get the path to the destination only with in a single node and that destination node should be in the geographical region of the source node and vice-versa.

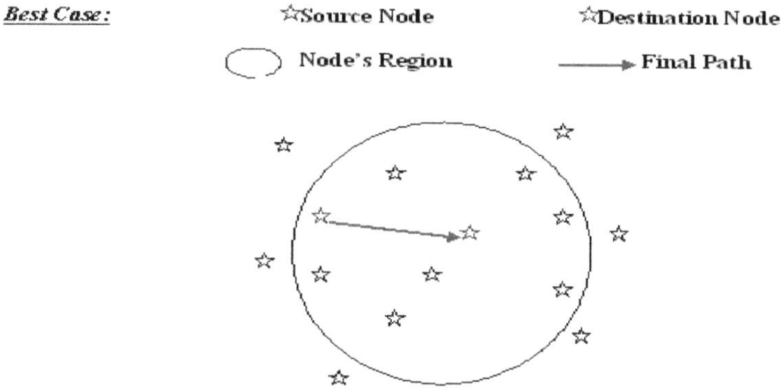


**Fig. 4.** Example of Best Case

In figure 4 clearly we can see that how source node can able to communicate with the destination node only with in single hop.

2. Fig. 5 shows an example of few nodes to represent Average Case. In this case when a source node wants to communicate with the destination node and this destination node is not in the range of source node. So in this case source node sends beacon messages to all the other nodes with in their range and elects a node that can act as a gateway node in between the two adjacent regions. Similarly the same phenomenon is applied to all the respective elected gateway nodes' region (i.e. in terms of iteration). And at last we can able to get the destination node and path in term of hopes can easily be determined. In the first iteration source node sends the beacon message to all the nodes within its region. After those nodes who wish to communicate with the source node they send reply message to source node with their trusted ids. Now in the first iteration finally source node can easily get the first hop towards the destination and in the same way next final hop node starts the same iteration process to get the next hop towards the destination. In the same way again

nodes who wish to communicate they send reply packet to first hop node same as that of first iteration. From this way the source node can easily get the second hop node toward the destination. At last in this multi-hopping mechanism how a source node can easily get the destination in terms of hop and on the bases of some parameters hopes can be chosen (i.e network requirements). As in fig 5.

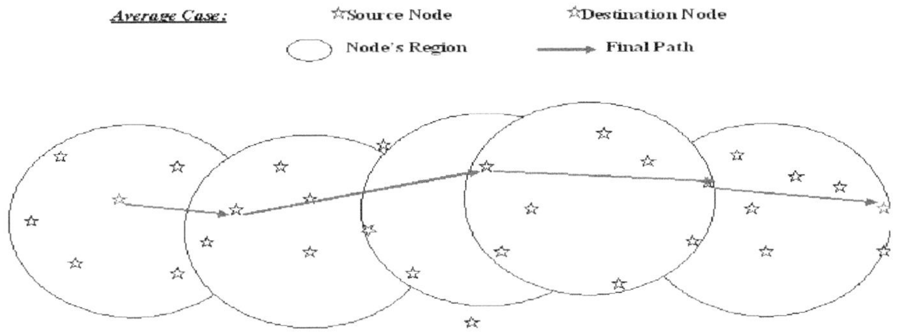


**Fig. 5.** Example of Average Case

3. Fig. 6 represents an example for Worst Case. The working phenomenon of worst case is same as that of the Average Case. (i.e when a source node wants to communicate with the destination node and this destination node is not in the range of source node. So in this case source node sends beacon messages to all the other nodes with in their range and elects a node that can act as a gateway node in between the two adjacent regions. Similarly the same phenomenon is applied to all the respective elected gateway nodes' region. But at last when we are able to get the actual path from source to destination, we have a path that is different from Average case in terms of hope. In worst case we have path that is high in terms of hopes as compare to average case.

In this worst case same process is followed by network but we can clearly see the number of hopes increases for the same network as that of in average case. As in fig 6.

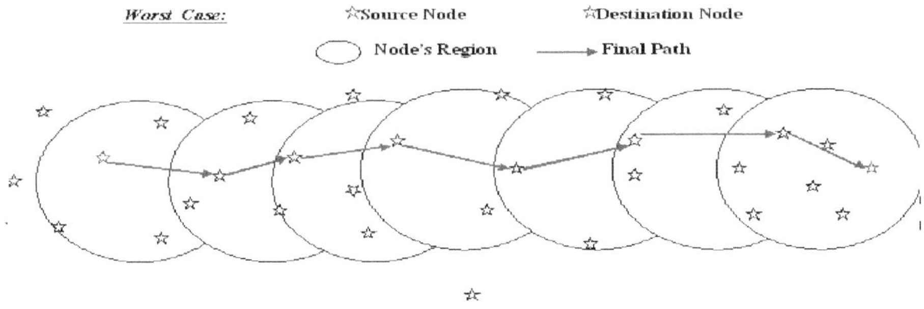


**Fig. 6.** Example of Worst Case

## 6 Important Findings

Now we can represents the three cases into tabular form and can compare them easily. By this we can check the performance of each case and can easily determine the case is better for which kind of network.

**Table 1.** Comparison of Performances

| Attributes / Cases | *Hopes Counts* | *Malicious nodes* | *unefficient node Encountered* |
|---|---|---|---|
| *Best Case* | 1 | 1 | 1 |
| *Average Case* | 4 | 6 | 7 |
| *Worst Case* | 7 | 9 | 4 |

On the bases on above tabular form for each cases by taking some parameter we plot a graph that can easily look how these differ from each other in terms of different attributes and we can easily choose our path that must be reliable and efficient too on the bases of our type of network (i.e on the bases of number of nodes and the scenario we choose).

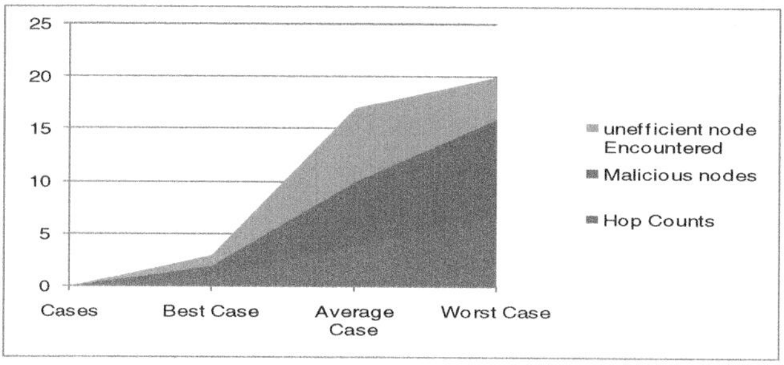


## 7 Conclusion

Mobile ad hoc network (MANETs) is a kind of network that works on multi-hopping mechanism and in order to get a reliable and efficient path in multi-hopping we have followed EPSAR and analyzed it for three different cases (i.e Best case, Average case and Worst case). In this paper by taking an example of few nodes we have examined the actual working of EPSAR with different parameters like malicious nodes encountered, unreliable battery power/signal strength for each node that are competing for communication and finally the path that must be examined on the bases of hopes in distance and also have done little amendments to FRESNA that makes our algorithm far better then the existing one. The number of hopes in distance for a reliable and efficient path may wary according to the cases on which we are going to analyze EPSAR. On the bases of these scenarios we can say that this algorithm works better to find path in multi-hopping environment of mobile ad hoc networks (MANETs).

## References


1. Sharma, A.K., Goel, A.: Best Neighbor Node Selection Algorithm for MANET. Journals of Institution of Engineers (January 2005)
2. Sharma, A.K., Goel, A.: Moment to Momant Node Transition Awareness Protocol (MOMENTAP). International Journal of Computer Application (IJCA) special issue. IASTED 27(1) (January 2005)
3. Priyanka, Bhatia, K.K., Jangra, A.: FRENSA: Farthest, Reliable and Efficient Node Selection Algorithm for Mobile Ad-hoc Networks (MANETs). IJCST International Journal of computer Science and Technology 1(2) (December 2010)
4. Komninos, N., Vergados, D., Douligeris, C.: Detecting unauthorized and compromised nodes in mobile ad hoc networks" see front matter _ 2005 Elsevier B.V. All rights reserved doi:10.1016/j.adhoc2005.11.005, http://www.elsevier.com/locate/adhoc
5. Jangra, A., Goel, N., Priyanka, Bhatia, K.: Security Aspects in Mobile Ad Hoc Networks (MANETs). A Big Picture International Journal of Electronics Engineering 2(1), 189–196 (2010)
6. Chakrabarti, G., Kulkarni, S.: Load balancing and resource reservation in mobile ad hoc networks. Ad Hoc Networks 4, 186–203 (2006), doi:10.1016/j.adhoc.2004.04.012
7. Razak, S.A., Furnell, S.M., Clarke, N.L., Brooke, P.J.: Friend-assisted intrusion detection and response mechanisms for mobile ad hoc networks. Ad Hoc Networks 6, 1151–1167 (2008), doi:10.1016/j.adhoc.2007.11.004
8. Arboit, G., Peau, C.C., Davis, C.R., Maheswaran, M.: A localized certificate revocation scheme for mobile ad hoc networks. Ad Hoc Networks 6, 17–31 (2008), doi:10.1016/j.adhoc.2006.07.003
9. Komninos, N., Vergados, D., Douligeris, C.: Detecting unauthorized and compromised nodes in mobile ad hoc networks. Ad Hoc Networks 5, 289–298 (2007), doi:10.1016/j.adhoc.2005.11.005
10. Philip, S.J., Anand, V.: Mobility Aware Path Maintenance in Ad hoc Networks. In: SAC 2009, Honolulu, Hawaii, U.S.A, March 8-12 (2009) Copyright 2009 ACM 978-1-60558-166-8/09/03
11. Mei, A., Stefa, J.: Routing in Outer Space: Fair Traffic Load in Multi-Hop Wireless Networks. In: MobiHoc 2008, Hong Kong SAR, China, May 26–30 (2008) Copyright 2008 ACM 978-1-60558-073-9/08/05
12. Vaidya, B., Yeo, S.-S., Choi, D.-Y., Han, S.J.: Robust and secure routing scheme for wireless multihop network. Springer, London (2009)
13. Vasiliou, A., Economides, A.A.: Evalution of multicasting algorithm in MANETs. In: Proceedings of World Acadmy of Science, Engineering and Technology, vol. 5 (April 2005); ISSN 1307-6884
14. Sanzgiri, K., Dahill, B., Levine, B.N., Shields, C., Belding-Royer, E.M.: A secure routing protocol for AdHoc Networks. In: Proceedings of 10th IEEE International Conference on Network Protocols, ICNP 2002 (2002)
15. Yi,S., Kravets, R.: MOCA: Mobile Certificate Authority for Wireless AdHoc Networks. University of Illinois at Urbana-Champaign Urbana, IL61801, {seungyi,rhk}@cs.uiuc .edu
16. Madhavi, S., Kim, T.H.: An intrusion detection system in mobile adhoc networks. International Journal of Security and its Applications 2(3) (July 2008)

# Agile Software: Ensuring Quality Assurance and Processes

Narinder Pal Singh[1] and Rachna Soni[2]

[1] Dept. Of Computer Science, Gobindgarh Public College, Alour, Khanna
narindersonu@gmail.com
[2] Dept. of Computer Science, D.A.V. College for Girls, Yamuna Nagar
rsoni63@gmail.com

**Abstract.** In the present scenario when the software systems are getting increasingly complexed. The time lines and schedule are getting tightened day by day. The processes need to be expected as adaptable rather than rigid. The process of the development also needs to be redesigned. The old concept of sequential phase must be updated with the iteration. The processes must ensure the user acceptance with the accepted level of quality of the software. The need of time is that process of software development to be reviewed. The concept of Agility can be used to provide good quality solutions for the upcoming software system. The sustainable solution will be considered now onward that will be capable of maintaining quality, acceptance of changes at any time with a minimum cost and rescheduling of each phase of development. Agile Software Development can be a winner of coming future. But on the downside for which many times the agile processes are criticized are inability to work with the CMMI environment, this paper proposes a new model for the agile software development, that includes Customer feedback and project documentation as its major element that makes the agile development more auditable, accountable and process centric.

**Keywords:** Agile software, Agile processes, Software Quality, Agile solution. Planning.

## 1 Introduction

The software in itself is a very large entity which contains a large number of processes to be carried out and a number of conditions to be satisfied before it is to be delivered as a golden release to the client. One of the very hard question to answer for the industry yet is, what will be criteria of acceptance, what will be the processes to be followed which can meet with the need of the time and ongoing project requirements. As each project has its distinct requirements and need. The question becomes even severe because the software is not an physical entity/commodity for which levels of quality can be framed exactly. In order to design a quality solution a number of proposal are made with a different theme and architecture. But one question which none of them was capable to answer is change management. All the solution somewhere takes the software requirement as a sequential/static entity, but actually this is the most volatile entity in the whole project development life cycle. This one single assumption makes whole of the development at a risk. There are

A. Mantri et al. (Eds.): HPAGC 2011, CCIS 169, pp. 640–648, 2011.

available "a number of solution for the project management, but still the solutions for the change management are still lacking". The unmanaged changes start hampering whole of the project life cycle and all the bad devils of Schedule Slippage, Cost slippage, Product adherence to the Standards, Quality Assurance become the biggest myths of the project documentation. As the product and requirement fails to agree with each other. In order to deal with the problem of change management with quality assurance, agile solutions can be adapted which are yet not accepted at a large. The paper will try to highlight some of the open problems that are limitation of the practiced approaches of software development. This paper is also an effort towards how agile development can be helpful in quality assurance.

## 2 Literature Review

Agile manifesto is a step towards the new practice of software development, In which more emphasis is given on the working people i.e. development team, communication and Customer. Agile implementation considers software as a continuously evolving dynamic entity while accepting the changes. In agile methodology emphasis is given more on the developing working software [Mike Cohn][11] by stepping back from heavy documentations. Agile methods have introduced Customer as an integral part of Project Development Life Cycle. The project is started as small daily meetings and works in the months long sprints [Hesam Chiniforooshan][9].
During these meetings each development progress is considered with consultation of customer and planned for the next iteration. The benefit of agile methods are remarkable as feedback on the development is readily available to keep the project in direction. Agile process having many good things are still sometimes negated for valuing active working code development only and ignoring the Planning and process driven approach. The methods of communication is also direct from one member to other. No process and plan centric approach is followed by the agile. In other words each person is managing his processes at its own, and there is no otherwise control over the autonomous development teams [Minna Pikkarainen ] [10]. There are some serious queries in this regards.

- The Process is going to be distributive rather then centric, what will be the Planning to control the overall project.
- Agile process assumes the self disciplined [Minna Pikkarainen] [10] independent team members as its core. which is not possible all the time.
- In Agile processes that lack central management, how team competency, team player and role selection for current assignment, communications [Hesam Chiniforooshan] [9] and reporting channel will be established. What will be flow of information?
- Management opened queries like what will the project schedule, progress reports, cost analysis for features updates done time to time for customers.

Agile processes steps back from documenting process which yields uncertainty for the project management team, because the project completed and ongoing project does not produces any matrices, creating an illusion between change, maintenance and rework. agile methodology says embrace the changes, but what will be the final cost estimations [Minna Pikkarainen] [10], schedule, final delivery dates, quality checks which is no where written or documented. Which makes a good methodology week enough to be accepted for large projects over the years. The project which will last over a couple of years, it is likely that the team member continues to join and leave the team. How such project can be looked after and managed for quality which lack the documentation and does not have any control over the autonomous teams.

A new model in this direction has been proposed which consider Documentation, Phased Customer feedback and Iterative development of software as its core components. The model and introduces documentation at its core will be able to answer all the opened queries of management, quality assurance ensuring correct project direction. It has been seen that agile concept mixed with waterfall methodology leads to better far more stable results.

## 3 Pros and Cons of Waterfall Model

Till the time, the most used and accepted model has remained Waterfall Model, with its pros and cons. The model presents a very rigid structure to the development community. As the model ignores and does not welcome the changes that are likely to be made to the software time to time [Victor Szalvay][5]. There are distinct phases in the model which are almost isolated from each other. Because the *"flow of information is not by communication between the phases, but it is by transition of one phase from other"*. So the feedback concept is totally missing in this case. The other horrible story is that there is nothing available in between to review, until one has crossed all the stages of development. "*There is no concept of testing or reviewing what is going on. Because there is nothing as output till the design phase is over*". The Testing and End User also comes in the role when the deliverable is prepared

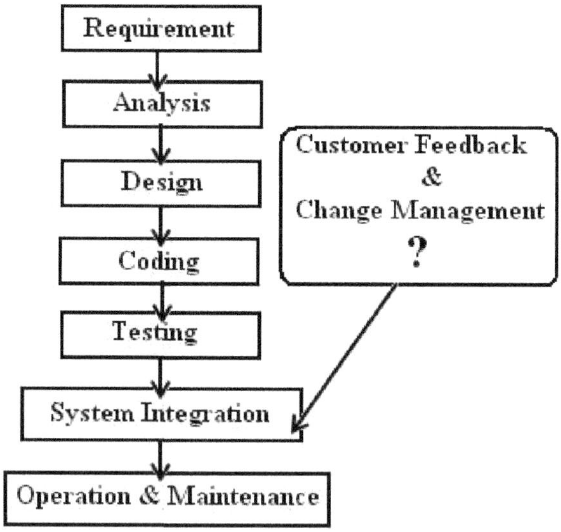


**Fig. 1.**

after coding. After the first release is made, it is quite practiced that changes are sidelined and this may be due to defective processes or the lack of synchronization of project documentation with the changes suggested.

There is no question to get the UAT (User Acceptance Testing) get performed and knowing about the limitations of the product or change management. The assumption followed in the model is very rigid and opposes the changes during the development. Working with this approach leads to serious drawbacks in the system. This has been proved by a number of surveys conducted all over the world.

- The Chaos Report (1995), [1] The Standish Group research survey shows a staggering 31.1% of projects will be canceled before they ever get completed. Further results indicate 52.7% of projects will cost over 189% of their original estimates.
- The OASIG Study (1995) [1] survey says, The IT project success rate quoted revolves around 20-30% based on its most optimistic interviews. Bottom line, at best, 7 out of 10 IT projects “fail” in some respect.
- The KPMG Canada Survey (1997) [1] findings are, Over 61 % of the projects that were analyzed were deemed to have failed by the respondents. More than three quarters blew their schedules by 30% or more; more than half exceeded their budgets by a substantial margin.
- The Bull Survey (1998), [1] the French computer manufacturer and systems integrator, a survey in the UK was conducted to identify the major causes of IT project failure in the finance sector. The key findings of survey was missed deadlines (75%), exceeded budget (55%) poor communications (40%) inability to meet project requirements (37%).
- The Robbins-Gioia Survey (2001) [1] - 51 % viewed their ERP implementation as unsuccessful 46 % of the participants noted that while their organization had an ERP system in place, or was implementing a system, they did not feel their organization understood how to use the system to improve the way they conduct business.
- IT Cortex Conclusion [1] The statistics presented here all converge to establish that: an IT project is more likely to be unsuccessful than successful about 1 out of 5 IT projects is likely to bring full satisfaction the larger the project the more likely the failure
- Larman, in the UK [8] shows that of 1,027 projects, only 13% did not fail, and waterfall-style scope management was the "single largest contributing factor for failure, being cited in 82% of the projects as the number one problem.
- Larman 1995, in a study of over $37 billion USD [8] worth of US Defense Department projects concluded that "46% of the systems so egregiously did not meet the real needs (although they met the specifications) that they were never successfully used, and another 20% required extensive rework" to be usable.
- Larman , In a study of 6,700 projects [8], it was found that four out of the five key factors contributing to project failure were associated with and aggravated by the waterfall model, including inability to deal with changing requirements, and problems with late integration.

- Larman, over 400 waterfall projects[8] reported that only 10% of the developed code was actually deployed, and of that, only 20% was used.

There is a long list of stories like these showing too high failure reports. *"The only factor responsible was non responsive waterfall modeling for the change management, and ignorance of the customer from beginning to end. There was no role and importance given to customer feedback given anywhere from the customer side, who has to ultimately approve the product. Which leads to such devastating results?"*

The Quality of such project is always under the question. *"In case of the Waterfall model it is no where guided or there is no such phase in which one can assure that whether the project is aligned in the right direction with the project specifications after consideration of all the changes. Another major issue with the model that Quality Assurance has not been given place during the phase transition to feedback over the work done"*. It is the major drawback of the model that Quality Assurance comes into the action when the all phases of the project are over. It is then tested whether the project has gone in right direction. *"Till the first release of the project, the end user or customer is not able to express his feedback over the development that has been over. Which leads to serious loop wholes and gaps between specifications and implementation. The project developed in such a way becomes very vulnerable to requirement and specification gaps, design issues and integration issues and ultimately collapses at the UAT (User Acceptance Testing).* Agile lifecycle can be an answer to all these problems of changes management and Quality Assurance.

## 4 Barrier to Agile

- Existing process of software industry that does not allows the development teams to communicate with the customer directly.
- Project Mangers does not want to engage the customers because it is taken as hindrance to pace of development process.
- Development team is use to develop the software as a sequential entity.
- Customer Feedback is taken as disturbance to the development, although this is the only communication that can bridge all the gaps.
- Not Suitable for CMMI Processes.

## 5 Agile Solution

Software is an intangible entity without any physical attributes so it is very hard to lay down the acceptance and rejection criteria's and any scaling parameters for ensuring the quality. In order to ensure the quality of the software, there must be direction alignment if the Customer Specification and its Implementation. The Traceability matrix must be able to clear the gaps of Specifications and Implementation. *In order to bridge the gap of Requirement, Change Management, Actual Specifications and Implemented Specifications and to improve the quality of the software the "Customer" must be brought in the development cycle. The Process of Software Development must be redesigned to 'Release Development'.* Each release will cover a part of the overall functionality. The process of system integration must be changed to the cumulative release integration. Each new release must cover the existing functionality with the new feature and must be thoroughly tested and sent for the customer feedback. *The end of the project will always be in a high quality software already gone through the UAT(User Acceptance Testing).* We proposes the following model of Release Development in accordance to the Agile Software.

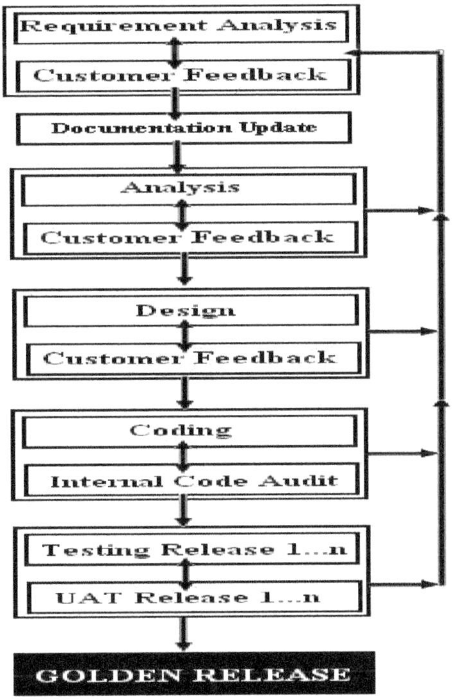


**Fig. 2.**

In the model proposed, all the stages have been designed ensuring customer satisfaction and quality of the software. In our terms *"Software Quality can be defined as developing a solution to perform in accordance to the customer Specifications, adaptable to changes for which it is developed and rejects the domain gracefully for which it is not developed"*.

The solution proposed meet the every need of the dynamic environment of software development and ensuring the quality. There is no phase left in the model for which the customer feedback or audit is not performed. The model is based on *"Release Development"* Concept. In which each release has gone for the approval of customer satisfaction. The software developed on this approach ensures a high degree of acceptance rate. As each step has undergone through the UAT(*User Acceptance Testing).*

## 6 Discussion

The Agile process for which they are criticized and does not give acceptance for the CMMI processes is lack of documentation. As "Agile manifesto [2][3], "Prepare the document when it is necessary". On the other hand the CMMI is largely process driven and is based on continuous improvement and maturity model [Martin Fritzsche] [6]. But this is very hard to accept that Agile methods can not be compatible with CMMI processes. The Agile Architecture has changed one thing that is bringing the customer in the developmental model [Martin Fritzsche] [6]. *"This only one change has a major positive change on the Cost, Schedule, Requirement Analysis, Change Management as all that will be finalized will be going to prove itself in the direction of customer satisfaction only, which can be named as Quality"*

- The Agile process for which are criticized is lack of proper and validated documentations. *The proposed model has given due advantage to this to make it CMMI and Quality compatible.* There will not be a single change that will not be recorded and communicated to whole of the developmental team, which is a definite step in the direction of process improvement and synchronization. For the CMMI process the User Stories [Martin Fritzsche] [6] can be taken as base of requirement analysis. The record keeping for whole of the discussion with the customer can be done to present as proof on some later stage to resolve an issue.
- On the basis of discussion with the customer, and suggested changes and there impact, More Effective schedule and Cost Estimates can be made that is again a requirement of CMMI processes [6].
- This is the only feature of Agile Processes that customer is approving each functionality of the software as it is made available. so the Cost to implement the changes are minimum in the Agile Processes because the changes are acknowledged at the same stage in which they are to be made. This always results in a positive impact on the cost and schedule of the project, which is also one of the objective in CMMI.
- The major role of the project management is to give the right direction to the project, and avoid any kind of schedule slippages. Agile process are an

answer to this question, as the direction of the project is always aligned with customer, which goes in parallel in the customer feedback. Each and every thing of project is in consultation with customer which gives a great quality improvement, effective management.

- CMMI all objectives can be met with agile solution with a little changes in the system. The process will give there maximum in the right direction reducing the failure rate of the project almost to negligible.

## 7 Conclusion

*Software development is not a Sequential process. So it is not possible for a Sequential Model to meet the needs of development process. It is a dynamic and repetitive Process, where each revision is equally important and impart a major impact on the overall development of the software. The documentation is the integral part of any project. This is the repository from where all the functionality is clear and visible. But the problems with which process lack is not the heavy and large documentation, but lack of synchronization ,change management and regular customer feedbacks. There are processes to update the documentation in every company, whatever they are so, but each one is lacking the process of managing the changes in a positive way and to accept them as a part of regular development. Agile process can be an answer to the solution. If we incorporate one change in the development process that is currently going, the success rate can be far better then at present. The need of change to be incorporated in* ***Every Phase of Waterfall Model Must be Converted to a Release of that phase and must be sent for the customer feedback and the changes must be reviewed for their impact and seriousness on the rest of module and priority of the change must be accessed properly. The product made by following this approach will be already free from the flaws of changes, and will clear always its final test from the user side that us User Acceptance Testing.*** The developed software using this method will be maturing a prototype towards the final direction and will always be in the shape and type it was assumed to be.

## References


1. The IT-Cortex website, Failure rate, `http://www.it-cortex.com/Stat_Failure_Rate.htm`
2. The Agile System Development Life Cycle website (2005-2010), `http://www.ambysoft.com/essays/agileLifecycle.html`
3. The Manifesto for Agile Software Development (2001) website, `http://agilemanifesto.org`
4. Vijayasarathy, L.R., Turk, D.: Aagile software development: a survey of early adopters. Journal of Information Technology Management XIX(2) (2008)
5. Szalvay, V.: An Introduction to Agile Software Development (November 2004), `http://www.danube.com`

6. Fritzsche, M., Keil, P.: Agile Methods and CMMI:Compatibility or Conflict? e-Informatica Software Engineering Journal 1(1) (2007)
7. The Benefits of Agile Development webiste, `http://www.versionone.com/Agile101/Agile_Benefits.asp`
8. The Waterfall Approach: a Critique website (November/18/2003), `http://www.adaptionsoft.com/adios_waterfall.html`
9. Esfahani, H.C., Et. Al.: Adopting Agile Methods: Can Goal-Oriented Social Modeling Help? In: 4th International Conference on Research Challenges in Information Science (RCIS). IEEE, France (2010)
10. Pikkarainen, M., Wang, X.: An Investigation of Agility Issues in Scrum Teams Using Agility Indicators. In: The Irish Software Engineering Research Centre (2008), `http://www.lero.ie`
11. Cohn, M., Ford, D.: Introducing an Agile Process to an Organization, pp. 74–78. IEEE Computer Society, Los Alamitos (2003) 0018-9162/03/$17.00

# Measure Complexity in Heterogeneous System

Kuldeep Sharma

Assistant Professor, Department of Computer Science and Engineering
Chitkara University (Himachal Pradesh) India
kuldeep.sharma@chitkarauniversity.edu.in

**Abstract.** Heterogeneous systems are becoming advanced and more complex day by day. To acknowledge the complexity of such system is a significant challenge. Distributed systems reminiscent of grid systems, internet systems, ever-present computing environments, storage systems, sensor networks and online enterprise systems often contain massive numbers of heterogeneous and mobile nodes. These systems are highly dynamic and fault-prone as well. As an end result, it's difficult to debug and developers find it easier said than done to program new applications and services for these systems; administrators find it complicated to manage and configure these complex, device-rich systems and end-users find it difficult to use these systems to perform tasks. In this context the aspects of complexity have been studied extensively. The concepts of time and space complexity of different kinds of algorithms are well understood. The complexity of heterogeneous systems for people has been broadly acknowledged to be a key problem.

**Keywords:** Heterogeneous Complexity, Complexity in Online banking system, Complexity Analysis in Heterogeneous system.

## 1 Introduction

Heterogeneous systems consist of vast verities of hardware and software working together in accommodating fashion to solve problems. There may be dissimilar representations of data in the system. This might include different representations for integers, byte streams, floating point numbers, and character sets. Most of the data can be transport from one system to another without losing importance. Attempts to make available a worldwide canonical form of information are lagging. There may be many different instructions sets. An application compiled for one instruction set cannot be easily run on a computer with another instruction set unless an instruction set interpreter is provided.

Many enterprises have large comprehensive computing systems which are intricate to manage, keep up and change. Developers constantly face step learning curves before they can start programming large systems. Organizations also tend to spend huge amounts in administration and maintenance costs, which often exceed the cost of buying the system in the first place. End-users too are commonly weighed down with the complexity of a single computer. In spite of the fact that system complexity has been broadly talked about, the term 'complexity' is often used with a loose knot in

A. Mantri et al. (Eds.): HPAGC 2011, CCIS 169, pp. 649–656, 2011.

connection with computer systems. There are no standard definitions of complexity or ways of measuring the complexity of Heterogeneous systems. As with so many intricate things, computer system complexity means different stuff to different people. This paper presents how different algorithms that take part in the complexity of heterogeneous system? Like fault tolerance algorithm, routing algorithms, backup algorithm, error detection and correction algorithms etc. In this the underlying principle behind these different aspects and how these aspects manifest themselves in heterogeneous systems has been described. How these aspects of complexity impact the overall system? The computer systems research community has been enthusiastically looking at various approaches to reduce the complexity of systems. These approaches often take the form of middleware or programming frameworks to make things easier. The task of developers is various system organization tools for administrators and intuitive user interfaces for end-users. The main contribution of this paper is calculating the complexity of Online banking system in a more formal and scientific way.

## 2 Different Aspects of System Complexity

Different people use the term complexity in different contexts. In general system complexity can be described how difficult it is to perform tasks in the system and a measure of how understandable a system is? A system with high complexity requires great intellectual or cognitive effort to comprehend and use, while a system with low complexity is easily valued and brought into play.

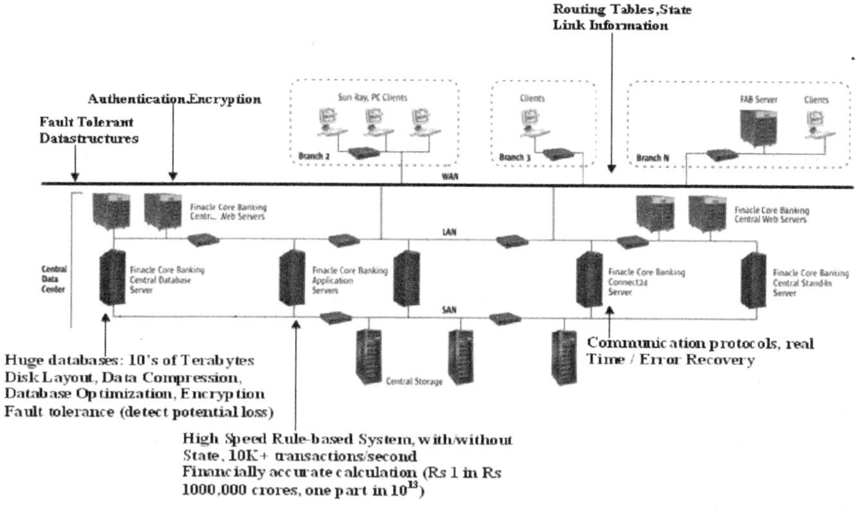


**Fig. 1.** Heterogeneous System

## 3 Complexity Measure

For instance online banking system: It has both different hardware as well as software in it. Above stated illustrates the concept to measure the complexity of a heterogeneous system. This banking application utilizes all kinds of algorithms from symbolic through real time through fault tolerant. Figure 1 demonstrates a simplified form the architectural building blocks .In this section, we attempt to capture some of the aspects of systems that make them difficult to understand. We will find the complexity of Internet banking system. Internet Banking has built up due to the increasing demand of online banking transactions. The biggest advantages of Internet Banking consist of complex banking solutions, 24 hours availability, quick and secure access to the back-end application through Internet. Customer requests for quick access and no matter the location at their bank accounts or at financial/banking transactions, all of these have determined the banking institutions worldwide to adopt the internet as the optimal solution for the presented demands. In the cause of the internet network banks are able to connect front-end applications with back-end. Based on such advantages, internet banking applications were created. The evolution of bank presence on web from simple, static applications to complex, dynamic application, to complex, dynamic application with numerous transactions, is presented in figure 2.

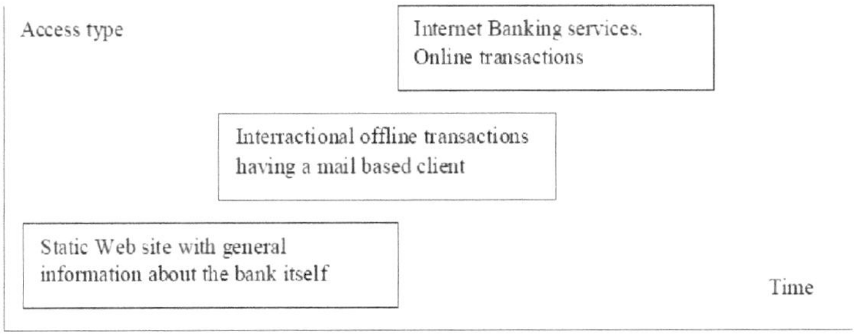


**Fig. 2.**

The advantages of internet banking applications comprises quickness, secured access to sensitive data as accounts, personal data of customers, transactions, account management; operating sale-purchase transactions in real-time and at long distance; suppressing the stress of staying in bank for a transaction; low costs for the maintenance of this kind of applications.

The example given in figure 1 terminals at a branch connected to a set of clustered web servers for authentication. The web servers are in turn connected to a set of application servers for implementing banking rules and policies. The application server's access mirrored and/or replicated data storage. Redundancy is present in the network also. Telephone banking using an Interactive Voice Response System is used as a backup if the branch terminals break down. The design of the authentication hardware and software requires fault tolerance – the users should not have to re-login if one or more servers fail – some state should be stored in the form of cookies in non-volatile storage somewhere. The banking calculations require very high accuracy (30+ digit accuracy). Various kinds of fault tolerance schemes are used for storage. For example, two mirrored disks always keep identical data. A write to one disk is not considered complete till the other is written also. The servers have to respond within seconds to each user level request (deposit, withdrawal, etc). The real time response of the system has to be evaluated using queuing theory and similar techniques. For TTS, the response output speech samples have to be guaranteed to be delivered at periodic time intervals, assume every 125*ms* (micro-seconds).In simple words in this system there are number of algorithms (like authentication, encryption, fault tolerant, routing algorithms, ACID (atomicity, consistency, isolation, durability) properties for transaction and, communication protocol etc) . A proportion is running parallel and another are in sequence. We have to calculate the complexity of overall system. In a communication network routing is one of the latency sources. Routing algorithm has task to determine the suitable port for delivering message address to one network node. Correct delivery of message from source node to destination node is considered as optimal routing. The aim is to have the fastest and optimal routing available in the network. The principle of routing with routing tables is straight forward. Ever node keeps a table with entries for each node. Using these entries, it could be determined via which outgoing port a message had to be sent. Interval routing is a space efficient routing strategy used in computer networks. The basis of interval routing idea is to label all of the nodes with integer from one set (for example $\{1, 2, 3...n\}$) and labelling of all arcs with interval (in range of number of nodes). These intervals mean that a message addressed to a node labelled with $u$, via a port labelled with interval includes.

# 4 Module-Wise Complexity of Online Banking System

In online banking system we have following modules:-

1) Fault tolerance and Recovery
2) Authentication and encryption
3) Routing
4) Backup
5) Communication.

## 4.1 Complexity of Fault Tolerance and Recovery Algorithms

Fault-tolerance or graceful degradation is the property that enables a system to continue operating properly in the event of the failure of some of its components. If its operating quality diminishes to the least, the decrease is proportional to the severity of the failure, as compared to a naively-designed system in which even a small failure can cause total breakdown. Fault-tolerance is particularly sought-after in high-availability or systems. Fault-tolerance is not just a property of individual machines; it may also characterize the rules by which they interact. For example, the Transmission Control Protocol (TCP) is designed to allow reliable two-way communication in a packet-switched network, even in the presence of communications links which are imperfect or overloaded. It does this by requiring the endpoints of the communication to expect packet loss, duplication, reordering and corruption, so that these conditions do not damage data integrity, and only reduce throughput by a proportional amount. Recovery from errors in fault-tolerant systems can be characterized as either roll-forward or roll-back. When the system detects that it has made an error, roll-forward recovery takes the system state at that time and corrects it, to make it able to move forward Check pointing with rollback recovery is a well-known method for achieving fault-tolerance in distributed systems. The check pointing algorithms can handle multiple concurrent initiations by different processes. While taking checkpoints, processes do not have to take into consideration any application message dependency. The synchronization is achieved by passing control messages among the processes. Application messages are acknowledged. Each process maintains a list of unacknowledged messages. Here a logical checkpoint is being used, which is a standard checkpoint (i.e., snapshot of the process) plus a list of messages that have been sent by this process but are unacknowledged at the time of taking the checkpoint. The worst case message complexity of the check pointing algorithm is $O(kn)$ when $k$ initiators initiate concurrently. The time complexity is $O(n)$. For the recovery algorithm, time and message complexities are both $O(n)$[1].

### 4.2 Authentication Algorithm Complexity

Encryption is the process of converting a plaintext message into cipher text which can be decoded back into the original message. An encryption algorithm along with a key is used in the encryption and decryption of data. There are several types of data encryptions which form the basis of network security. Encryption schemes are based on block or stream ciphers. The type and length of the keys utilized depend upon the encryption algorithm and the amount of security needed. Authentication means the ability to determine the identity of a party for an interaction and to ensure that a message came from, who it claims to have come from. Authentication is seldom used in isolation. Authentication is used as the root for authorization. In Alice and Bob Protocol Cost of each number is $O(\log n)$ bits, n numbers are sent so, Total cost is $O(n \log n)$ bits[2].

### 4.3 Complexity of Routing Algorithms

Routing is the process of moving information packets and messages across a network from a source host to a destination host. In a communication network routing is one of the latency sources. Routing algorithm has task to determine the suitable port for delivering message address to one network node. Correct delivery of message from source node to destination node is considered as optimal routing. The aim is to have the fastest and optimal routing available in the network. The principle of routing with routing tables is straightforward. Each node keeps a table with entries for each node. Using these entries, it could be determined via which outgoing port a message had to be sent. Interval routing is a space efficient routing strategy used in computer networks. The basis of interval routing idea is to label all of the nodes with integer from one set (for example {1, 2, 3...$n$}) and labelling of all arcs with interval (in range of number of nodes). These intervals means that a message addressed to a node labelled with $u$, via a port labelled with interval which includes a node $u$ is forwarded [3].

**Table 1.** Complexity of message passing

| | *Routing Tables* | *Interval routing* |
|---|---|---|
| *Space complexity* | $O(n \log \Delta)$ | $O(k \Delta \log n)$ |
| *Time complexity* | $O(1)$ | $O(\log(k \Delta))$ |

*Note: - Δ is maximum degree of the graph, n is number of nodes in the graph, k is interval.*

### 4.4 Backup Algorithm Complexity

The backup tree algorithm to compute a set of n-1 backup multicast delivery trees from the default multicast tree for application level multicast. For each backup multicast tree exactly one link of the default multicast tree is replaced by a backup link from the set of available links. The backup tree algorithm calculates the n-1 trees with a complexity of $O\ (m \log n)$ [4].

*Note: - m is number of messages and n are number of nodes (systems)*

### 4.5 Communication Complexity

It depends upon the algorithm used in system for Distance Vector routing scheme complexity will be $O(m)$.In case of Link state Algorithm complexity will be $O(n)$.

**Table 2.** Complexity of all Modules

| | |
|---|---|
| *Communication* | $O(m)$ |
| *Fault Tolerance* | $O(kn)$ |
| *Recovery* | $O(n)$ |
| *Authentication* | $O(n*\log n)$ |
| *Routing* | $O(1)$ |
| *Backup* | $O(m \log n)$ |

The above shown results are just an estimate; complexity may vary according to the algorithm. There are number of things that can affect the system, front end and network speed plays a vital role in this. In this type of system any algorithm may change the complexity of the system. There are number of algorithms available for particular task. In above case if we choose Open Ear Decompositions for fault tolerance routing scheme then complexity will be $O(m)$ for that particular module [5]. If we change routing scheme in the system then also the complexity will be change like we may switch to distance vector routing then the communication complexity will be $O(n)$ instead of $O(m)$.

**Table 3.** System Complexity

| | |
|---|---|
| *Communication* | $O(n)$ |
| *Fault Tolerance* | $O(m)$ |
| *Recovery* | $O(n)$ |
| *Authentication* | $O(n*log\ n)$ |
| *Routing* | $O(1)$ |
| *Backup* | $O(m\ log\ n)$ |

As the user login first the authentication is being done then routing protocol comes into existence. Then depends upon the mode of transaction, the user may only view the account statement in this case the complexity will be $\{O(n*log\ n) + O(1)+O(m)\&\&\ O(kn)\}$ ,authentication , routing will run in sequence viewing account statement fault tolerance will run parallel. If user wants to transfer the money from one account to another then complexity will be different approximate$\{O(n*log\ n)+O(n)+O(1)+O(n)+[O(m)\ \&\&\ O(m\ log\ n)]\}$.

# 5 Conclusion

The size and complexity of heterogeneous systems have been increasing inexorably in the recent past. We are now trying to evaluate online banking system using the complexity metrics. In particular, we are trying to evaluate large, complex diverse system like ubiquitous computing systems, enterprise systems and internet-based systems. This would help establish the generality of the metrics. Some preliminary evaluation in this regard as well as a description of some of the strategies for alleviating complexity in a ubiquitous computing system is discussed [6]. Another area of future work is to try to broaden the set of metrics to include a measure for debugging and failure-recovery. Heterogeneous systems may fail due to a quantity of reasons - OS faults, software bugs, hardware errors, human error, etc. While self-repair and high-level programming can trim down the episode of failures and make available automated, graceful failure-recovery, systems. Above and beyond, self-configuring and self-repairing systems may misbehave or take wrong actions, for this reason, heterogeneous systems should provide mechanisms for uncovering the different layers of abstraction and deducing the cause of faults. The complexity of debugging a system is a significant measure of the ease of using and maintaining the system. System complexity forms just one of the factors that manipulate the productivity of administrators, developers and end-users. Other factors that affect complexity include communication changing requirements, technology churn, etc. While we are still far-off from being able to quantify these other factors, they would all form part of any evaluation of the productivity of different kinds of users.

Calculating the complexity of this type of system is just a ballpark figure because complexity may vary on the mode of transaction. One algorithm may affect the system. In this system there are various factors that will engage in recreation the overall complexity like one minor change may increase or decrease the complexity of the system e.g. if we change authentication algorithm then the overall system complexity will affected.

## References


[1] Mandal, P.S., Mukhopadyaya, K.: Concurrent checkpoint initiation and recovery algorithms on asynchronous ring networks. Journal of Parallel and Distributed Computing 64(5) (May 2004); ISSN:0743-7315.

[2] Based on lecture by Dr. Ely Porat, Complexity course, Computer Science Department Bar-Ilan University (January 2008)

[3] Complexity Analysis of Routing Algorithms in computer network, M. Bieliková (Ed.), IIT.SRC, pp. 69-76 (April 27, 2005)

[4] Braun, T., Arya, V., Turletti, T.: A Backup Tree Algorithm for Multicast Overlay Networks. In: Boutaba, R., Almeroth, K.C., Puigjaner, R., Shen, S., Black, J.P. (eds.) NETWORKING 2005. LNCS, vol. 3462, pp. 1430–1434. Springer, Heidelberg (2005); ISBN 978-3-540-25809-4

[5] Berman, K.A., Paul, J.L.: Algorithms, by Course Technology (2002); ISBN-13:978-81-315-0512-2

[6] Ranganathan, A.: An Task-Execution Framework for Autonomic Ubiquitous Computing. Phd Thesis, University of Illinois at Urbana-Champaign (2005)

[7] Knuth, D.: The Art of Computer Programming, Volume 3: Sorting and Searching, vol. 3. Addison-Wesley, Reading (1997); ISBN 0-201-89685-0

[8] Cormen, T.H., Leiserson, C.E., Rivest, R.L., Stein, C.: Introduction to Algorithms, 2nd edn. MIT Press and McGraw-Hill (2001); ISBN 0-262-03293-7

[9] Atkinson, Kendall, A.: An Introduction to Numerical Analysis. John Wiley & Sons, New York ISBN 978-0-471-50023-0

# Author Index

GPSR Compliance

*The European Union's (EU) General Product Safety Regulation (GPSR) is a set of rules that requires consumer products to be safe and our obligations to ensure this.*

*If you have any concerns about our products, you can contact us on ProductSafety@springernature.com*

In case Publisher is established outside the EU, the EU authorized representative is:

Springer Nature Customer Service Center GmbH
Europaplatz 3
69115 Heidelberg, Germany

Batch number: 10838023

Printed by Printforce, the Netherlands